The 226 City Maps of Jacob van Deventer in Figures

Mapping the Past

VOLUME 8

The titles published in this series are listed at *brill.com/mtp*

The 226 City Maps of Jacob van Deventer in Figures

Street Lengths, Accuracy in Relation to Modern Maps, Cross-Comparisons

By

Peter van Druenen

BRILL

ISSN 2589-9945
ISBN 978-90-04-75502-4
ISBN 978-90-04-75503-1 (PDF)
DOI https://doi.org/10.1163/9789004755031

Library of Congress Control Number: 2026939938

Brill and Koninklijke Brill BV are part of De Gruyter Brill.
www.degruyterbrill.com

Questions about General Product Safety Regulation: productsafety@degruyterbrill.com

Cover illustrations: The map of Brugge. ca. 1566. Respectively: unprocessed, measured, analyzed. Jacob van Deventer, *Planos de ciudades de los Países Bajos. Parte II* [Manuscrito], Res. 207, no. 68. © Biblioteca Nacional de España, Madrid. In the three smaller circles: details from the map of Vlissingen. 1548. Jacob van Deventer, *Planos de ciudades de los Países Bajos. Parte III* [Manuscrito], Res. 200, no. 1. © Biblioteca Nacional de España, Madrid. From top to bottom: the bulwark, the gallows field, and the Onze-Lieve-Vrouwekerk (Church of Our Lady).

For Hanna, Charlotte, Maartje, and Linne

Contents

Figures and Graphs

Figures

Graphs

Tables

CHAPTER 1

Prologue

The 226 maps drawn by Jacob van Deventer between 1542 and 1575 of cities in present-day Netherlands, Belgium, Luxembourg, and parts of France and Germany are simultaneously an open book and a mystery. They are an open book because of the very precise view of the structure, streets and fortifications in these cities halfway through the sixteenth century. Yet, they are also a mystery because, at first glance, they leave much untold. For the researcher willing to invest time and effort into delving into the deeper layers of these maps, a new world unfolds: the world of the sixteenth century in which Van Deventer lived and worked. It's a world in which, using the limited means available at the time, he managed to create highly reliable cartographic representations of reality. These maps also reveal the different underlying layers: the processes and motivations that can be explained from military, political, judicial, socio-economic, ecclesiastical-cultural, demographic and urban morphological history.

1.1 Measure and Count

The 226 city maps of Jacob van Deventer: dimensions, accuracies, comparisons (hereinafter referred to as: 'The present study') will primarily focus on the latter layer: the demographic one. Regarding the possibility of using Van Deventer's city maps for historical-demographic research, in 2022 the study *Meten en Tellen* (Measure and Count) was published.[1] It presents a method – the Passus method – that can be used to calculate urban population figures in the Netherlands, Belgium, Luxembourg, and northern France during the sixteenth and seventeenth centuries, based on the street lengths in old city maps. This method does not rely on common sources such as baptism, marriage, burial, tax, or citizen registers because historiographical research indicates that they are unreliable for application across time and space due to their incompleteness and inaccuracy.[2] The literature review in the field of historical demography in the first part of *Meten en Tellen* revealed that this discipline has undergone enormous development over the past half-century in terms of the number of research fields included, the range of disciplines involved, and the methods and techniques used. However, the oldest aspect – calculating urban population figures over multiple centuries on a broader geographic scale than just local and regional – has lagged behind. As a result, the development of reliable quantitative datasets for studying urbanization and migration patterns has also been limited.

The development of the Passus method aims to fill this gap and more specifically to systematically, consistently, and extensively map urban population numbers in the sixteenth, seventeenth, and eighteenth centuries. The method is based on a type of source that has rarely been used for this purpose: historical city maps. The result is an alternative method of calculation, founded on the combination of: (a) measuring the built-up street sides on the city maps of Jacob van Deventer from the mid-sixteenth century, (b) counting houses on the city maps of Joan Blaeu from the mid-seventeenth century, and (c) interpreting the first national census of 1795–1796. The first two steps were supported by incidental data from tax registers and by alternative city maps. The method was developed and tested using data from seven cities in Zeeland and six in Holland and Vlaanderen.

The final quantitative research result is a linear time series of population figures for the test cities between 1548, 1649, and 1796. Additionally, the study provides the theoretical insight that, through the applied multidisciplinary approach – where city history, historical demography, and historical cartography are complementary – methods can be developed that allow for the calculation of urban population figures over a period of multiple centuries. Such methods enable researchers to objectively compare city maps that – like those of Jacob van Deventer – were drawn by a single cartographer or a single team of cartographers within a defined period, using the same techniques and cartographic principles. In line with this, the number of houses and inhabitants per city can also be derived and compared. The primary uncertainty lies in the applicability of the two reduction factors used in the method: the number of houses per measured street side and the number of inhabitants per house. It is therefore crucial to always take into account local deviations, depending on the specific location, period, and urban scale, as well as local developments.

The part of the *Meten en Tellen* method specifically focused on the work of Van Deventer forms the basis

1 Van Druenen, 2022.
2 Ibid., pp. 69–79.

 | HTTPS://DOI.ORG/10.1163/9789004755031-002

for the present study, in which, for the first time, the 16th-century population figures of all 226 cities depicted by Jacob van Deventer are systematically and consistently calculated. Additionally, the study follows up on one of the key recommendations from *Meten en Tellen*: analyzing all 226 city maps by Jacob van Deventer.[3] The questions related to this are addressed in the first four chapters.

- Chapter 1. Why is an alternative method for calculating urban population figures from the early modern period necessary, which scientific disciplines would benefit from such a method and in which research questions and debates can it play a role (1.2)? What are the developments in the use of Jacob van Deventer's city map by historians? (1.3)? How does the demographic layer relate to the military, political, judicial, socio-economic, church-cultural and urban morphological layers (1.4)? How does the research design of the study compare with that of similar previous studies (1.5)? What are the objections to using old city maps for demographic research, instead of using more common sources, such as baptism, marriage, burial, tax, or citizen registers, for this purpose (1.6)? This primarily historiographical chapter ends with the conclusions and the follow-up (1.7).
- Chapter 2. What do Jacob van Deventer's city maps reveal about the numbers of houses and inhabitants during the period these maps were developed and is it possible to develop a method for consistent research and analysis based on all maps of a coherent group of cities – in this case, the seven test cities in Zeeland, Holland and Vlaanderen mapped by Jacob van Deventer – regarding the size of the built environment and the population? The presented procedure and tests in this chapter are an update from those in *Meten en Tellen*.[4]
- Chapter 3. Is it possible to apply this method to all of the 226 citie maps of Jacob van Deventer, so that the research findings of each separate city can be accurately compared with those of the other 225?
- Chapter 4. Is such a method applicable to other city maps or groups of city maps from different countries/regions or time periods?
- Chapter 5. Contains the summary, the sub-conclusions, the final conclusion and the recommendations for follow-up research.

3 Van Druenen, 2022, p. 239.

4 Ibid., pp.104–147.

1.2 Demography and Cartography

As already noted, the field of historical demography has undergone tremendous development over the past three-quarters of a century. It has grown into a fully-fledged discipline with many practitioners around the world and a continuously increasing flow of research and publications, offering ever more precise insights into population trends in past centuries. This development can be described and summarized by focusing on its four most striking characteristics: (a) the importance of life-course and family history, (b) the diversity within the field, (c) its multi- and interdisciplinary nature, and (d) the systematization and standardization of research methods.

1.2.1 Life-Course and Family History

As early as 1903, Henri Pirenne, professor of history at the University of Ghent, advocated for detaching historical demography from the mere recording of population figures and instead gaining more knowledge about individuals and their social environments, something for which a wealth of source material was available. His remarks were recorded in the Démographie report he prepared for the 11th Congrès International d'Hygiène et de Démographie, held in Brussels that year. In addition to broadening the scope of the field, Pirenne also called for a systematic processing of source material. He mentioned (1) genealogies, which allow researchers to study the development and extinction of families, (2) registrations of new citizens, which – even if incomplete – provide the means to map urban immigration, (3) trade registers, which offer high-quality information about the petty bourgeoisie, and (4) numerous other documents such as death notices, house books, and records of real estate transactions.

Almost 90 years later, Flemish historians Michel Cloet and Chris Vandenbroeke concluded, in the anthology *Ten Contributions to Local and Regional Demography in Flanders*, that little had been done with Pirenne's recommendations up to the mid-20th century. Until then, the field had largely focused on 'an approximate representation of population figures.' This view, however, needs to be nuanced, in the sense that in the half-century following – and certainly in response to Pirenne's statement – there were indeed developments that served as the foundation for the way in which the field was able to expand and deepen after 1960.[5] One of the later historians who was

5 Cloet and Vandenbroeke, 1989, p. 6. The introduction to this 1989 collection was already written in 1983. See also Engelen, 1992, p. 318.

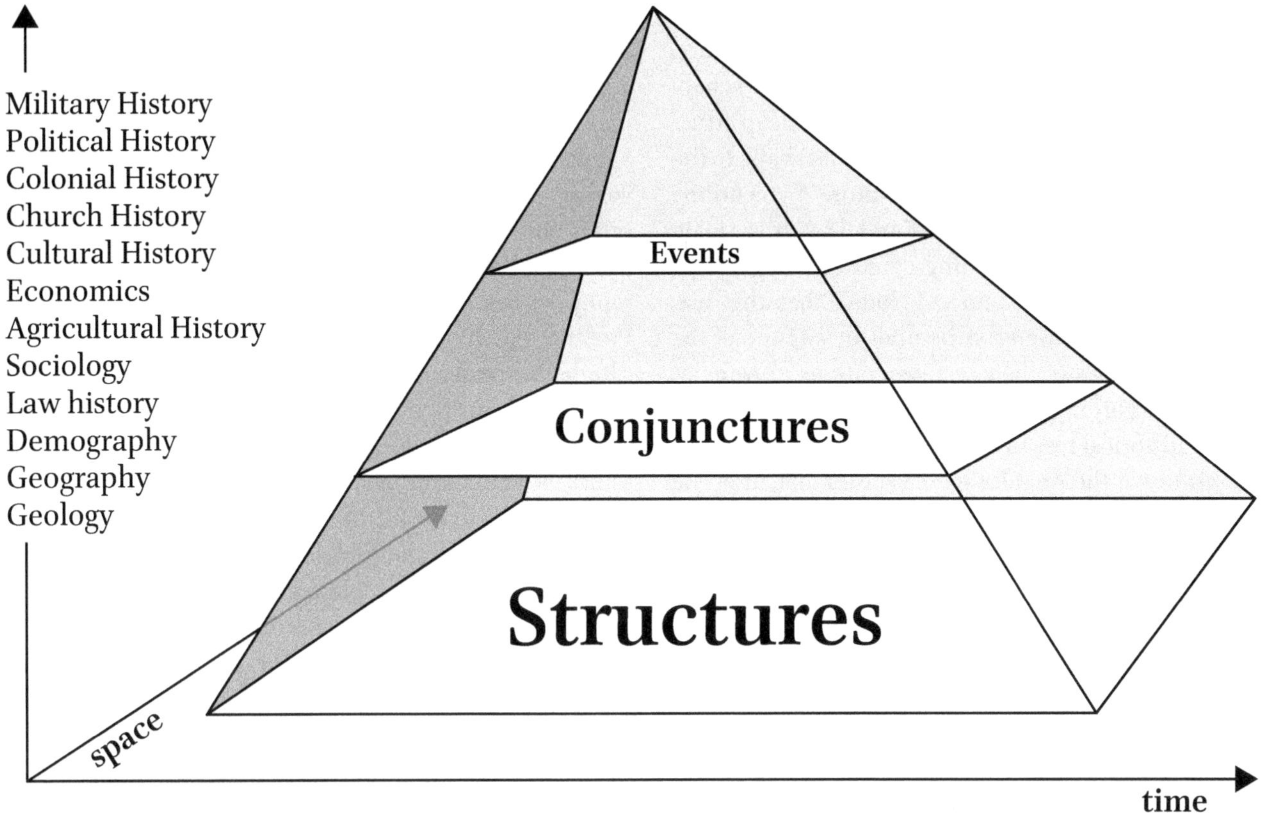

FIGURE 1.1 The structuralist model of Fernand Braudel, who was part of the second generation of Annales historians. History could be told and analyzed based on three layers: the structures (longue durée), the conjunctures (moyenne durée) and the events (courte durée). These layers represent a hierarchy of interconnected driving forces through which the course of history could be understood. The model can be represented as a three-dimensional matrix, where history, along with all related disciplines, space, and time, is projected onto the height, depth, and width axes, respectively.
Note: Burke, 2015, pp. 36–43; Van der Woude, 1980, pp. 12–17; Breure et al., 2008, p. 3; Gencúrová, 2016, p. 14.

inspired by him was Fernand Braudel, who, in the 1950s and 1960s, was the leading figure of the group of historians associated with the French journal *Les Annales* and the author of the three-volume *La Méditerranée et le Monde Méditerranéen à l'époque de Philippe II* (1947).[6] The predecessors of Braudel, Lucien Febvre and Marc Bloch, who founded the journal in 1929, considered the Ghent professor as one of their major sources of inspiration.[7] They aimed for a change in the way history was written: from a primarily political and military approach to a historiography that gave a prominent place to the social sciences. In the early years, these were primarily economics and sociology, which also provided the knowledge and experience regarding the use of quantitative methods and techniques.[8] In the period after World War II, alongside economics and sociology, interest grew in anthropology, geography, cultural history, political science, theology, history of mentalities, and sciences with a beta character, such as ecology and demography. This approach not only provided new perspectives on the course of history but also introduced previously unused methodological frameworks within which historical research could be structured. The storytelling about great men and their great deeds, with narrative history as the primary technique, was relegated to the background.

In 1979, the Dutch historian Willem Frijhoff referred to the Annales movement, to which he also considered himself a part, as not a true school working from the same theoretical background. Rather, it was a gathering of scholars with diverse backgrounds, working in different fields, but united in their agreement on the scientific foundations of their collaboration.[9] According to Frijhoff, these principles were already established at the founding of the

6 Braudel, 1949; Burke, 2015, p. 43.
7 Burke, 2015, pp. 14, 23.
8 The full name of the magazine was: *Annales d'histoire économique et sociale*. This name changed in 1939, 1944, 1946 and 1994 respectively to *Annales d'histoire sociale*; *Mélanges d'histoire sociale*; *Annales. Economies. Sociétés. Civilisations*; *Annales. Histoire Sciences Sociales*.
9 Frijhoff, 1981, p. 115.

movement. The focus of historical research should no longer be on isolated events but on their mass and structure. By the latter term, the founders of the Annales school were referring not to surface structure, but to deep structure: 'the interconnections that give form and shape to the well-being and woes of society and culture.'[10] According to the socio-economic historian Karel Davids, who in 2011 wondered whether anything could still be learned from the Annales school and concluded that this was indeed the case, this layered structuralism was one of the main characteristics of the latest generations of Annales historians.[11] According to him, it had become a solid foundation for historical research in and outside France.

The history of the Annales school shows that, after the 1940s, the field of historical demography gained an independent place within the broader spectrum of disciplines through which history was researched, described, and analyzed. It was no longer just about 'establishing population figures,' but much more – and even primarily – about gaining knowledge 'about the human being and his social environment.' Family reconstructions and the linking of life events such as birth, marriage, and death to economic variables such as the development of grain prices became important research themes. In Braudel's three-dimensional model, demography can be placed at the bottom of the layer of conjunctures, assessed as dynamic, with structural elements.

A portion of the generation of researchers after Braudel even broke away from the cliometric approach to historical-demographic research in order to focus entirely on qualitative aspects. However, a common denominator remained the complex of methodological guidelines that were largely indebted to the social sciences, in which the objectivity of the research method, clearly formulated problem statements, and a systematic application of proven methods and techniques were central. As such, the development of historical demography within the Annales movement can be seen as a missing link between the publication of Pirenne's work in 1903 and the direction in which the field would develop in the decades after 1960.

1.2.2 Diversity

Although life-course and family history attracted a large number of practitioners, the number of other areas of interest where research was conducted and about which publications appeared increased rapidly. One example of this diversity is the 1977 edition *Demografische evoluties en gedragspatronen van de 9de tot de 20ste eeuw in de Nederlanden*, compiled by the Ghent professor and pioneer of Flemish historical demography, Christiaan Vandenbroeke, and his colleague, Walter Prevenier.[12] The volume was number 200 in the *Studia Historica Gandensia* series and the first of a new tradition to include unpublished studies. The theme chosen was historical demography, with a reference to Ghent's greats such as Henri Pirenne and his early students Hans van Werveke and Charles Verlinden. The authors of the volume, including several students of these 'masters,' as Prevenier referred to them in his preface, filled the volume with a wide range of topics, both in terms of space and time. The common factor, according to the introduction, was that all the articles addressed difficult methodological problems, resulting in 'a broad spectrum of the many approaches to one of the most challenging core sectors in our field.'[13]

In *The Economic History Review*, Herman van der Wee, professor of social and economic history at the Catholic University of Leuven, reported a year later that the articles were highly significant and further emphasized the vitality and reputation of the Ghent historical school.[14] This praise undoubtedly stemmed from the immense diversity of the edition: all the major topics in historical demography were represented – population development, fertility and birth, marriage, migration, and mortality. In addition, there was room for multidisciplinary approaches, with sociology and economics playing a role. Those interested in methodology could turn to an article on a reconstruction of the rural population of the Brugse Vrije around 1520 using regression comparisons, as well as to an analysis of the use of a ninth-century polyptych (a document recording land ownership) from the Saint-Bertin Abbey in Northern France for calculating population figures at both local and national levels. Eight of the nine articles dealt with a local or regional case study. All eight contained numerical series with the corresponding calculations and analyses. The numbers were never an end in themselves; rather, the story behind them was what mattered. One contribution did not present a case study but instead proposed a uniform instrument designed to provide insight into the distribution of statistical data, enabling comparisons between different studies.

Due to the enormous diversity of subjects, geographical spread, multidisciplinarity, applied methods and techniques, and the interrelation of various source types, the

10 Ibid., pp. 117–118.
11 Davids, 2011, p. 3.
12 Prevenier and Vandenbroeke, 1977.
13 Ibid., preface.
14 Van der Wee, 1978, pp. 697–698.

volume serves as a good example of how the field had developed both in breadth and in depth. Broadly speaking, the kind of micro-level research presented in this volume – with its many variations in time, space, subject, and method – would continue to play an important role in the hundreds of historical-demographic publications that appeared in the Netherlands and Belgium alone.

At the same time, there was a growing number of synthetic works, conducted on a larger geographical scale, multidisciplinary or interdisciplinary in nature, or primarily focused on research methodologies. A major landmark for the latter was the research platform of the Scientific Research Community for Historical Demography established in the 2000s. The main objectives of this platform were the exchange and comparison of research data and methodologies, the promotion of multidisciplinary research, the implementation of international comparative and longitudinal studies, and the provision of infrastructure for database management. Researchers from the universities of Leuven, Hasselt, Ghent, Brussels, Amsterdam, Nijmegen, Utrecht, Tilburg, Wageningen, and Geneva participated in this platform, alongside institutions such as the International Institute of Social History (IISH) and the Netherlands Interdisciplinary Demographic Institute (NIDI).[15]

Between 2006 and 2015, the platform saw the publication of nearly 200 articles, books, book chapters, and working papers. Roughly one-fifth of these can – with some caution and necessary caveats – still be classified as small-scale case studies. While there can never be too many of such studies to eventually piece together the bigger picture, the overall view emerging from this survey is indicative of the development of historical demography since the 1980s: it has become a mature and wide-ranging discipline within which a number of full-fledged specializations have emerged. A current example of such specialization is the rapid growth of migration history over the past 30 to 40 years. The subject plays an important role not only in calculating population totals but also in reconstructing family and life-course histories. Moreover, the recent international public debates on migrant and refugee flows into Europe have undoubtedly contributed to its prominence.[16]

1.2.3 Multi- and Interdisciplinarity

An indispensable characteristic of modern historical-demographic research is its connection with other sciences. This was already one of the methodological principles behind the founding of *Les Annales* by Marc Bloch and Lucien Febvre in 1929: collaboration with economics and sociology. Later, as the movement developed around the journal, many other disciplines were added. In the early years, this collaboration was primarily multidisciplinary; methods and techniques from the social sciences were applied, but without true integration. Today, the field can, in part, be described as interdisciplinary; the various disciplines depend on one another to achieve research outcomes. For example, the fields of urban history and the historical demography of cities have become virtually inseparable. Explaining fluctuations in birth and migration balances is only possible with the help of disciplines such as economics and social history. Economic cycles, in turn, can only be fully understood through data on population developments – numerical, social, medical, and economic. And what would social geography – where the relationship between people and space is central – be without knowledge of demography, and vice versa? The integration of these two sciences has even given rise to a new discipline: population geography, in which birth, death, and migration are studied from a spatial perspective.

One example of a field in which economic and demographic history have been almost fully integrated is the study of the secular trend. The statistical term *secular* indicates that a development – charted, for instance, through a time series – extends over a long period and includes both upward and downward movements. Secular trends reveal the long waves that have occurred in economic life over a given, often centuries-long, period. One of the first to use the concept was the German economist and economic historian Wilhelm Abel, in the first edition of his seminal work *Agrarkrisen und Agrarkonjunktur* (*Agrarian Crises and Agrarian Conjuncture*), published in 1935. While medium- and long-term trends, such as those seen since the beginning of the Industrial Revolution, had already been the subject of study for some time, Abel argued that much longer trends existed in history, trends

15 The research community was continued in 2021 through a new project until the end of 2025. See: https://research.ugent.be/ (accessed on 19 March 2025). HiDo is organized around five working groups (WG), each focused on a specific theme:
WG I: international comparison of causes of death;
WG II: long-term trends in partner choice, love, and marriage;
WG III: historical demography of colonial societies;
WG IV: citizen science in historical demography;
WG V: genetics and historical demography.

16 See among others: https://iisg.amsterdam/nl/onderzoek/projecten/migratiegeschiedenis and https://vijfeeuwenmigratie.nl/ (accessed on March 20, 2025).

that could be traced back as far as the Middle Ages.[17] He traced the course and interrelation of prices, wages, rents, and population, and catalogued the similarities in economic trends across Western European countries between the thirteenth and nineteenth centuries. In 1966, he extended his survey to include the twentieth century. This was also the period in which the Dutch historian Slicher van Bath, in his handbook *De agrarische geschiedenis van West-Europa (500–1850)* (*The Agrarian History of Western Europe, 500–1850*), charted the major overarching patterns in agricultural history over a very long timespan: first the gradual transition from a closed to a more market-oriented economy, and from the second half of the Middle Ages onwards, to a system characterized by secular trends.[18]

His student Ad van der Woude defined the 'secular trend' shortly thereafter as the growth period between roughly 1500 and 1650, followed by contraction with fluctuations until the early nineteenth century. Van der Woude elaborated on this in his 1972 dissertation, focusing on the *Noorderkwartier* in North Holland. He presented it as an attempt to combine the historical-sociographic method of his doctoral advisor, Slicher van Bath – developed in *Samenleving onder spanning* (*Society Under Strain*) – with the *histoire sérielle* approach that was particularly influential in France.[19] His aim was to gain insight into the secular development of the population in terms of place of residence, home, family composition, social background, status, religion, migration, nuptiality, gender and means of subsistence in the period between the end of the Middle Ages and the beginning of the nineteenth century.

In the 2012 volume *Rural Economy and Society in Northwestern Europe, 500–2000*, an international group of authors tried to find an answer to the question of what happened to family structures in the rural societies surrounding the North Sea over the past 1,500 years.[20] Although it could be demonstrated that this long period saw a gradual process of deruralization – and thus urbanization – the conclusion is that the processes involved, as well as the ways in which families adapted to them, varied according to time and place, and can only be understood within their specific regional and social contexts. The overarching theme in the various contributions highlights two major turning points: (1) towards the end of the Middle Ages, feudalism and manorialism gave way to greater independence, a more market-oriented approach among farming households, and the rise of cottage industries; and (2) in the last quarter of the nineteenth century, agricultural trade became globalized, prices declined, a process of scaling-up began, and cottage industry gradually lost its importance.[21]

In the seventh and eighth chapters, Isabelle Devos, Erwin Karel, Thijs Lambrecht, Richard Paping, and Eric Vanhaute explore the integrated history of the Low Countries in relation to the volume's central theme. Market orientation, employment outside of agriculture, and deruralization had already begun to take shape here as early as the fourteenth century. However, significant regional differences emerged within this area. In the coastal regions, patterns of commercialization and large-scale farming were particularly evident. In contrast, inland areas such as Vlaanderen, Brabant, and parts of Twente were characterized by high population density, proto-industrialization, and fragmented land ownership. Drenthe presented an opposite picture: low population density, large areas of uncultivated land, little cottage industry, and scarcely any land fragmentation. In the Gelderland river region and the Ardennes, large-scale landownership prevailed. Despite these differences, the regions were interdependent due to the movement of seasonal laborers and reciprocal trade.[22]

1.2.4 Systematization and Standardization

The fourth and final important feature of the development that historical demography has undergone concerns the systematization and standardization of the research methodologies used. Pirenne already pointed this out in 1903 when he advocated for a systematic processing of source material. After him, Rogier Mols argued in 1956 for comparative analyses and the necessary uniformity in the methods used, and in 1967 Slicher van Bath advocated for the functional application of quantitative data.[23] The uncovering of as much numerical data as possible – with the help of increasingly sophisticated digital techniques – followed by systematic organization, analysis, interpretation, and comparison undoubtably plays a prominent role in this process. That the compilation of data series on higher levels of time and space must always take nuances and details into account is a necessary condition. And here the subject of the present study returns: the calculation

17 Abel, 1980.
18 Slicher van Bath, 1960.
19 Slicher van Bath, 1957; Van der Woude, 1972, p. 15.
20 Vanhaute, Devos, Lambrecht, 2012.
21 Ibid., pp. 323–346.
22 Ibid., pp. 178–179.
23 Mols, 1954–1956, Deel I, pp. XVIII–XXV; Slicher van Bath, 1967, p. 157.

of urban population figures in the early modern period in the Low Countries.

1.2.5 An Article of Crucial Influence

It is remarkable that in the rapid and extensive development of the discipline, the original core of historical demography – the determination of population figures in cities and rural areas – has demonstrably lagged behind. This is not so much the case in terms of generating the figures themselves, but rather in the methods and techniques used in the research. Most population estimates are based on the subdisciplines of life course and family history, which now encompass many thousands of regional and especially local studies focused on hamlets, villages, small towns, and neighborhoods of larger cities. Source materials for these studies include population records such as baptism, marriage, burial, tax, and citizenship registers. These studies have yielded – and continue to yield – a wealth of data, primarily serving the subdiscipline itself, but also used for calculating the population figures of larger areas. Apart from the criticism of such sources discussed in paragraph 1.6, questions can be raised regarding the use of the research data that these sources yield, particularly the extrapolations of this data across space and time. A telling example of this is an article from 1965, the results of which are still used today in research on population figures: *Population changes and economic developments in the Netherlands* by Bernhard Slicher van Bath and his doctoral students Ad van de Woude, Johannes Faber, Henk Roessingh, and Harry van Xanten.[24] This research group was the driving force behind the rise of historical demography as an independent discipline in the Netherlands and was inspired by the French Annales school. The article was also remarkable because all the authors – who would go on to earn their doctorates between 1972 and 1976 based on regional studies with a demographic orientation – were involved in its creation: Faber with *Drie eeuwen Friesland* in 1972, Van der Woude with *Het Noorderkwartier* in the same year, and Roessingh with *Inlandse tabak* in 1976. The exception was Van Xanten, who conducted research in 1965 on demographic and economic developments in the Meierij of Den Bosch. His dissertation on this topic was announced as early as 1965, but it ultimately did not result in a PhD.[25] In their joint article from 1965, the authors combined demographic and socio-economic research data from five regions in the Netherlands: Friesland, Overijssel, the Veluwe, Holland, and Brabant. From the combined data, total population figures were extrapolated for the entirety of the Netherlands for the period 1500 to 1950. In the article, the authors presented their research methodologies, source materials, and source criticism in detail. The French adaptation and translation was published in *Annales de Démographie Historique*.[26]

To offer a critical commentary in 2024 on an article from 1965 may seem an anachronism. It certainly would be if, for instance, the statistical techniques used in the 1960s were to be compared with the (primarily digital) possibilities available today or if the research results from that time were to be critiqued on the basis of new studies or newly discovered sources. Fortunately, that is not necessary, as the five regional case studies in the article appear exemplary even from a contemporary perspective. This also applies to the incorporation of insights from the Annales school.

And yet, something doesn't sit quite right. On a map of the Netherlands, the authors had marked the five research areas, including the 26 cities studied. Altogether, these areas covered 62 percent of the country's surface. The figures derived from them were, as the authors themselves put it, 'interpolated' to apply to the entire Netherlands, after which the total population size of the country could be determined. The question now is: in what way were these 'interpolations' actually carried out?[27] A first important note is that the term *interpolation* is not appropriate in this context. This technique allows a series of numbers to be extended with values that fall *within* that series. A better term would have been *extrapolation*: the extension with values *outside* the series. In the remainder of this chapter, the term *extrapolation* will therefore be used, as that is what the authors actually meant.

However small the Netherlands may have been, the regional differences were significant, and a village on the Veluwe could hardly be compared to a village in Zeeland. The same held true for urban comparisons. The authors carried out comparisons based on assumptions, for example, that developments in Holland and in large parts of Friesland and northwestern Overijssel were similar to those in Utrecht, the Gelderland river area, and Zeeland. These assumptions may have been correct, but they were

24 Faber et al., 1965, pp. 47–114.

25 Van Xanten and Van der Woude, 1965, p. 30, footnote 59: "In the soon-to-be-published dissertation by H.J. Van Xanten, these and other issues concerning the economic and social history of the Meierij will be addressed." Van Xanten, however, did not earn his doctorate until 2007, and then as a theologian, with the dissertation: *Katholieken, wat zijt gij toch een wonderlijk volk.*

26 Dupâquier, 1965, pp. 199–226.

27 Faber et al., 1965, p. 110.

TABLE 1.1 Calculation of the Likely Population Figures of the Netherlands in the Period 1500–1800, as presented by Faber et al. in 1965 and Van der Woude in 1980[a]

Year	Faber et al. (1965)	Van der Woude (1980)
1500	0.9 to 1.0 million	0.9 to 1 million
1550	1.2 to 1.3 million	1.2 to 1.3 million
1600	1.4 to 1.6 million	1.4 to 1.6 million
1650	1.85 to 1.9 million	1.85 to 1.9 million
1700	1.85 to 1.95 million	1.85 to 1.95 million
1750	1.9 to 1.95 million	1.9 to 1.95 million
1795	2.078 million	
1800		2.1 million

a Faber, 1965, p. 110; Van der Woude, 1980, p. 134.

never substantiated, which means that the estimates of Dutch population development between 1500 and 1795 could not be definitively verified.

The question, then, is: why was it necessary to do this? The five studies were already valuable in their own right, and bringing them together in a single article was valuable as well. The unfortunate consequence, however, was that – because the article was highly regarded – many researchers in the following years adopted its figures without source criticism and based new research on them. The article received no critical reviews in 1965 or in the years that followed, not even regarding the extrapolation of regional data into national series. This observation is of great importance for this justification section of the present study. Bibliometric research has shown that in the more than half a century since 1965, the publication has been used frequently in articles and books about urban population figures in the Netherlands.

The major driving force behind this was the demographic chapter that one of the article's authors, Ad van der Woude, published in 1980 in Volume V of the *Nieuwe Algemene Geschiedenis der Nederlanden* (NAGN), a 15-volume work partly based on the ideas of the Annales school. The quantitative section of that chapter was entirely based on the 1965 article, as shown in Table 1.1, which compares the estimates from 1965 with those from 1980. The question now is whether the demographic data in the publications by Van der Woude from 1965, 1980, and 1995 are reflected in the historiography of the period after 1980. To get an impression of this, a selection of 40 titles was made, 18 of which were published after 2000.[28] It should be explicitly stated that the selection was not made based on whether or not Van der Woude is cited. The main criterion, however, was that the publications had to address historical demographic topics to a greater or lesser extent. The publications are divided into five categories:

1. Historical demographic publications by Van der Woude himself after 1980 (6);
2. Publications in the field of general history, which gave substantial attention to demographic processes (12);
3. Publications in the field of demographic history (8);
4. Publications in the field of migration history (10);
5. Publications in the field of methods and techniques of historical demographic research (4).

The selection is illustrated in Graphic 1, showing the influence of the three key publications from 1965, 1980, and 1995 on the 40 selected publications. The period after 1980 is divided into four time periods, each spanning ten years. At first glance, the graphic presents the expected pattern: the 1965 publication was cited in the 1980s, 15 years after its release, and throughout the first decade after the publication of the *NAGN*, by 8 publications. After that, the number of citations decreased: from 8 to 5, then to 3, and finally to 1.

A similar pattern emerged with the demographic chapter from 1980: in the 1980s, it was used by 12 publications. In the 30 years that followed, this number declined, but the chapter remained influential, even though 1995 marked the publication year of the third key work: *The First Modern Economy: Success, Failure, and Perseverance of the Dutch Economy, 1500–1815*, which Ad van der Woude co-authored with Jan de Vries in 1997.[29] The book contained 20 tables and 3 graphics dedicated to demographic indicators. Fifteen of these tables and two of the graphics referred directly, or through an earlier or later publication by Van der Woude, to the *NAGN*.[30] Thus, the Wageningen

28 The 40 titles are: Blondé, 2015; Briels, 1985; Brusse, 2006; Brusse, 2011; Brusse, 2013; De Vries, 1995; Van Druenen, 2015; Flinn, 1981; Foldvari, 2012; Frijhoff, 1998; Groenveld, 1980; Hélin, 1980; Kossmann, 1986; Leeuwen, 1993; Lourens, 1997; Lucassen, 1984; Lucassen, 2002; Lucassen, 2010; Lucassen, 2011; Noordhoff, 2011; Nusteling, 1985; Paping, 2014; Prak, 2013; Puschmann, 2018; Schöffer, 1985; Schrover, 2014; Schuurman, 1991; Van Cruyningen, 2006; Van Cruyningen, 2012; Van der Schoor, 1999; Van der Woude, 1979; Van der Woude, 1981; Van der Woude, 1982; Van der Woude, 1983; Van der Woude, 1988; Van der Woude, 1990; Van der Woude, 1996; Van der Woude, 2000; Van Lottum, 2007; Van Vliet, 2003.

29 Vries, J. de A.M. van der Woude, *The First Modern Economy: Success, Failure, and Perseverance of the Dutch Economy, 1500–1815*. Cambridge 1997.

30 Ibid., p. 71 (table 3.1, from: *NAGN*, Volume 5, p. 134), p. 72 (figure 3.1, from: *NAGN*, Volume 5, p. 135), p. 74 (table 3.2, from: *NAGN*, Volume 5, p. 131), p. 75 (table 3.3, from: *NAGN*, Volume 5, p. 131), p. 77 (table 3.4, from: *NAGN*, Volume 5, p. 131), p. 77 (table 3.5,

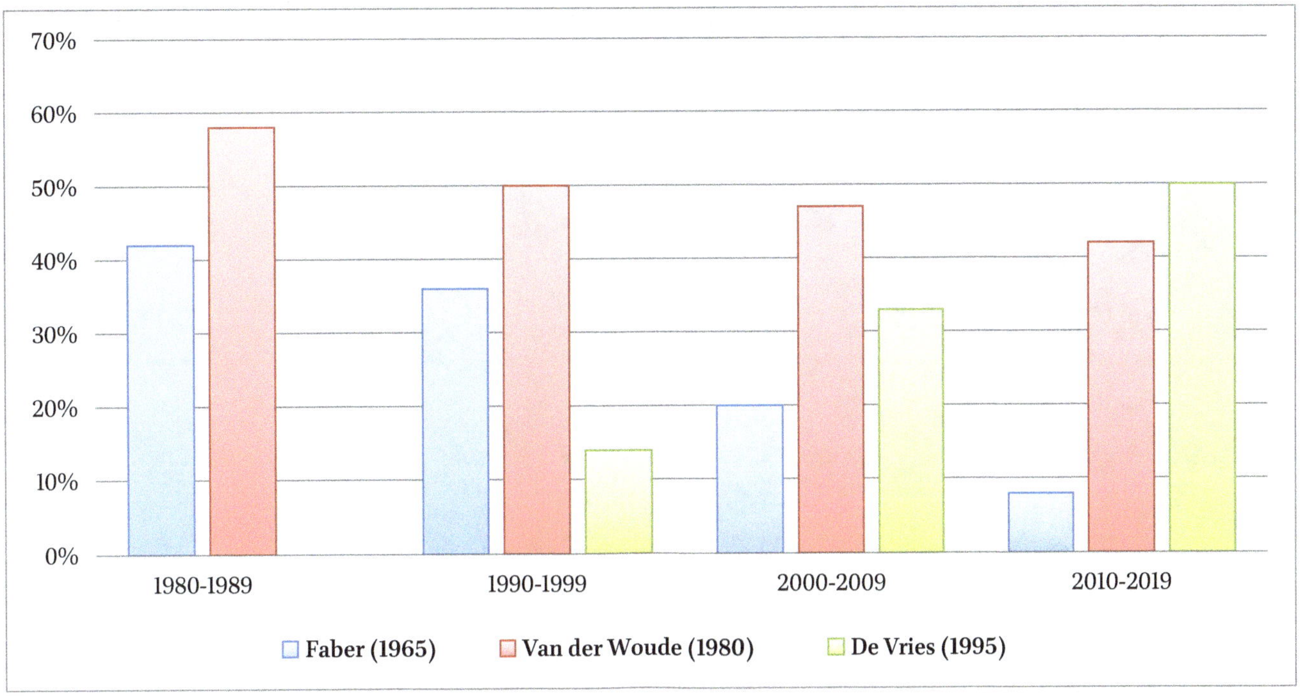

GRAPH 1.1 The trend between 1980 and 2020 of the number of publications, in percentages of the total (N = 40) per decade, referring to one of the three key publications from 1965, 1980, and 1995

article from 1965 was referenced three times, and *Het Noorderkwartier* once. The most frequently cited – five times – was a French-language article from 1983: *Les villes néerlandaises* from 1983.[31] In all cases, the data presented in the NAGN were exact copies of those in the original article from 1965. Since the empirical sections in the demographic chapter of the NAGN were largely copies of the 1965 article, it can be concluded that the demographic data for the *Netherlands, 1500–1815*, was mostly compiled in 1965, partly from research in the primary sources of a limited number of cities and regions, and for a substantial part through extrapolations that, with the insights of 2025, no longer withstand critical scrutiny.

In the first decade after publication, there were two referencing publications, in the second five, and in the last six. Between 2010 and 2020, this book took the lead. Faber had one reference, and Van der Woude had five. Since it was earlier established that the demographic pages in the book by De Vries and Van der Woude were nearly identical to those in Volume 5 of the NAGN, it can be concluded from this statistic that the influence of the NAGN remained almost constant over the 40 years following its publication. When the references to the two publications are added together, a steadily increasing line emerges: 1980–1989: 58%, 1990–1999: 64%, 2000–2009: 80%, and 2010–2020: 92%.

The most comprehensive and detailed historical demographic publication regarding the cities in the Northern Netherlands is undoubtedly *Inwonertallen van Nederlandse steden ca. 1300–1800* by Lourens and Lucassen from 1997. The authors also used, in addition to recent literature, including the standard publications from 1965, 1980, and 1995 discussed here, published sources from before 1965 and primary sources. Graph 1.2 shows the ratio of 1007 source references for 38 cities across the Netherlands.

The selection of these cities was based on the inventory made by Van der Woude in the *NAGN* of cities for which population figures could be determined or estimated in 1675.[32] In the section 'Publications Wageningen School',

from: *NAGN*, Volume 5, p. 131), p. 77 (table 3.6, from: other), p. 80 (table 3.7, from: *NAGN*, Volume 5, p. 139), p. 82 (table 3.8, from: *NAGN*, Volume 5, p. 125), p. 83 (table 3.9, from: *NAGN*, Volume 5, p. 83), p. 84 (table 3.10, from: *NAGN*, Volume 5, p. 135), p. 88 (table 3.11, from: *NAGN*, Volume 5, p. 139), p. 91 (table 3.12, from: *NAGN*, Volume 5, p. 131), p. 92 (table 3.13, from: *NAGN*, Volume 5, p. 131), p. 93 (table 3.14, from: *NAGN*, Volume 5, p. 131), p. 101 (table 3.15, from: *NAGN*, Volume 5, p. 151), p. 102 (figure 3.2, from: other), p. 211 (figure 5.1, from: *NAGN*, Volume 7, p. 261), p. 276 (table 6.5, from: other), p. 281 (table 6.8, from: *NAGN*, Volume 5, p. 83), p. 334 (table 8.1, from: other), p. 335 (table 8.2, from: other), p. 336 (table 8.3, from: other).

31 Van der Woude, 1983, pp. 309–385.

32 Van der Woude, 1980, p. 137. In addition to the data for the year 1675, he also provided figures for the years 1525, 1750, and 1795. The choice of 1675 can be justified as follows: the 1525 inventory

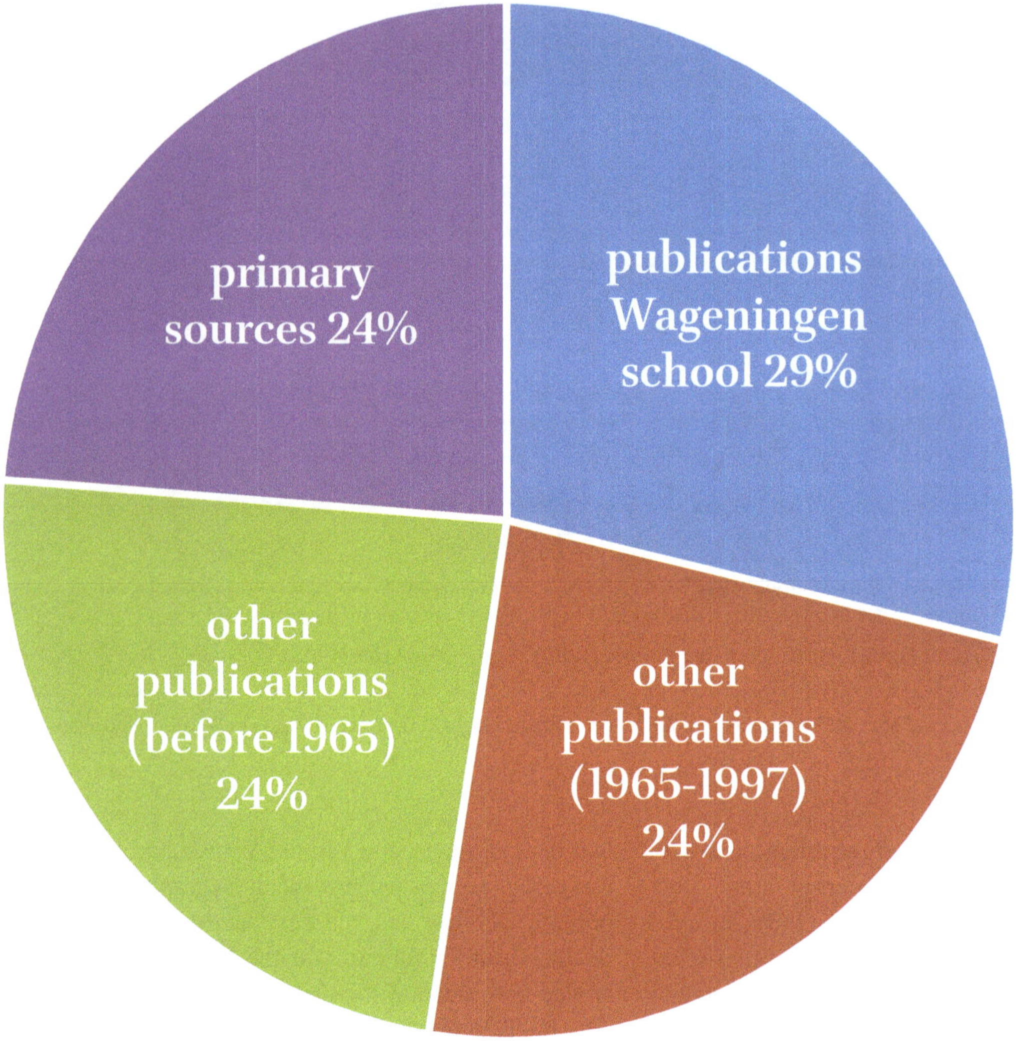

GRAPH 1.2 The percentages of cities in a selection of 38 cities (in Lourens and Lucassen, Inwoneraantallen, 1997) in which references (number of references = 1007) are made to a publication or a primary source. Note: Lourens and Lucassen, 1997, pp. 11, 11, 12, 12, 15, 17, 22, 24, 28, 32, 34, 36, 37, 39, 54, 55, 59, 61, 63, 64, 69, 71, 76, 83, 85, 87, 92, 93, 100, 101, 103, 107, 108, 112, 116, 117, 119, 120. The cities are: Alkmaar, Amersfoort, Amsterdam, Arnhem, Bergen op Zoom, Bolsward, Delft, Brielle, Den Haag, Deventer, Dokkum, Dordrecht, Enkhuizen, Franeker, Goes, Gouda, Haarlem, Harderwijk, Harlingen, Kampen, Leiden, Maastricht, Medemblik, Middelburg, Monninckendam, Nijkerk, Roermond, Rotterdam, Schiedam, Schoonhoven, Sneek, Utrecht, Venlo, Vlaardingen, Weert, Weesp, Zutphen en Zwolle. The cities that were included by Van der Woude but that he marked with a question mark have been omitted from the inventory. Maassluis, Tilburg, and Roosendaal – although mentioned by Van der Woude – are also absent, as they are not listed in Lourens's overviews.

all the publications cited by Lourens and Lucassen from the authors responsible for the standard publication from 1965 are included.

was too limited to produce a representative picture, the 1750 data contained too many gaps, and those of 1795 concerned the population census from that year.

1.2.6 Discussion: Migration and Urbanization

The above bibliometric analysis demonstrates that the 1965 publication *Population Changes and Economic Developments in the Netherlands: A Historical Survey* has had a significant impact on historical demographic research in the decades that followed – initially directly, and after 1980 through an ever-evolving lineage of epigones. As a result, the key figures concerning population development in the Netherlands between 1500 and 1800 have become standardized; they consistently reappear in new publications and have lost none of their influence to this day.[33]

The figures have also permeated the relatively new discipline of migration history. Their role is prominent: the population size of cities at the beginning and end of a given period reveals a great deal about net migration and urbanization balances. This means that current estimates of migration during the Early Modern Period are still based on the same, predominantly secondary sources. The same applies to the various datasets used in research on urbanization.

The basis of all calculations – the knowledge of how many people lived in a given city at specific points in time – is, in all its simplicity, unreliable. In this context, reference can certainly be made to the more than century-old debate about the correct number of Flemish immigrants who migrated to the Northern Netherlands between 1550 and 1650 and who played a foundational role in the Dutch 'Golden' Seventeenth Century. Estimates here range from 60,000 to 600,000 immigrants. The Ghent-based jurist Johan Eggen estimated the total number of southern immigrants who settled in the Northern Netherlands in the last quarter of the sixteenth and the first quarter of the seventeenth century at 604,000.[34] Eleven years later, in 1919, the Northern Dutch response came from professor and minister Aart Arnout van Schelven. He drew on newly uncovered research material and pointed to errors he had found in Eggen's work. In the end, his revised calculations led to an estimate of 60,000 immigrants, just one-tenth of his Flemish colleague's figure.[35] New estimates since then have never fallen below or risen above these two extremes. In 1978, the Utrecht art historian Jan Briels arrived at a figure between 100,000 and 150,000.[36] In 1980, the Wageningen professor of Agrarian History, Ad van der Woude, counted 200,000 immigrants.[37] This figure is based on the influential article discussed earlier, *Population changes and economic developments in the Netherlands: a historical survey* from 1965.[38]

In the closely related debates on urbanization and de-urbanization, the situation is not much better. For example, regarding the city of Middelburg around 1650, population estimates range from 18,500 to 30,000. The legal scholar, economist, and representative Gerrit Adriaan Fokker calculated in 1878 that Middelburg, at the height of its prosperity in the mid-seventeenth century, had 30,000 inhabitants.[39] In 1998, the agrarian historian Peter Priester revised this number to 18,500 inhabitants.[40] A year later, the city archivist Peter Sijnke argued that Fokker was closer to the actual numbers than Priester: the population was therefore more likely to have been 30,000 rather than 18,500. In 2006, historian Paul Brusse attempted to uncover the causes of these differences. These could have been related, among other factors, to whether or not the population in the areas around Middelburg, where hundreds of people lived, had been included in the count.[41] Brusse ultimately arrived at a total population of 17,500 around 1700, which could suggest a trend of de-urbanization. However, calculations in *Meten en Tellen* from 2022 show that de-urbanization was not yet occurring: Middelburg had 11,750–14,100 inhabitants in 1650, and by 1739, the population had risen to 25,000.[42] The decline did occur in the second half of the eighteenth century: the population census of 1795 recorded 17,687 inhabitants for that year.[43]

In the period after 2002, a stream of scientific and popular-scientific publications emerged about earlier migration movements, with the implicit, but often also explicit, message that most receiving countries benefited from these migrations in the past.[44] The present study aims to play a role in providing a method for conducting discussions – such as those about urbanization and de-urbanization processes – based on correct quantitative data. It provides a method that uses a uniform source of information, shaped according to the same standards and within the same defined time period. In the specific case of the present study, this source is city maps drawn by a single cartographer or a single team of cartographers

33 See, for example, the *Bosatlas van de geschiedenis van Nederland* (*The Historical Atlas of the Netherlands*) from 2011, in which the demographic pages (pp. 155, 312) draw on Van der Woude (1980) for data prior to 1500 and on Lourens and Lucassen (1997) for the figures thereafter.

34 Eggen, 1908, p. 211.

35 Van Schelven, 1919, p. 16. The two works by Eggen and Van Schelven are extensively described and analyzed in: Van Druenen, 2024, pp. 49–61.

36 Briels, 1978, p. 220.

37 Van der Woude, 1980, p. 141.

38 Faber et al., 1965.

39 Fokker, 1875, p. 95.

40 Priester, 1998, pp. 52, 481, 485.

41 Brusse, 2006, pp. 58–65.

42 Van Druenen, 2022, pp. 183, 218.

43 The Commission for the Design of a Constitution for the People of the Netherlands, *Volks-tellinge in de Nederandsche republiek 1796*. Den Haag, 1796, pp. 40–43.

44 See, among others: Bade, 2003; Foldvari et al., 2012; Lucassen, 2002; Lucassen et al., 2011; Lucassen et al., 2016; Obdeijn et al., 2008; Schrover, 2014; Haas, 2024 (and a critical review on this book: Hofhuis, 2024).

within a defined period, using the same techniques and cartographic insights. To calculate migration figures for a specific period, two sets of city maps are required: one from the beginning and one from the end of the chosen period. For each city, population figures can then be calculated. The difference between the two figures, after subtracting the birth balance, represents the positive or negative migration balance, including data on urbanization and de-urbanization. Primary demographic research, therefore, is just as closely aligned with the exact sciences as it is with the humanities. As the French historian Ernst Labrousse remarked in 1970: '*Pour être historien, il faut savoir compter*' (to be a historian, one must know how to count).[45] The analysis, interpretation, and assessment of the final results, which form the basis for opinions and debates, can only take place after the population has been calculated using with methods that are valid and have sufficient scale in space and time.

1.3 The City Maps of Jacob van Deventer

Jacob van Deventer was born between 1500 and 1505, probably in the Hanseatic city of Kampen, in the region of Oversticht, later known as Overijssel. His mother was unmarried and married a man from Deventer after his birth, where the family also moved to live. Jacob attended the Latin school there and later moved to Leuven, where he first studied medicine and mathematics, and later surveying and cartography. Jacob van Deventer created maps of Brabant, Holland, Gelderland, Friesland, and Zeeland between 1536 and 1549. These maps were primarily of an administrative nature.[46] The maps were etched onto copper plates, printed, and thus also published and put into circulation.[47] The basis for Van Deventer's maps included, among other things, drawings made in the context of water management control. It is not unthinkable that Italian maps from the early sixteenth century also served as examples for Van Deventer. The similarity between his city plans and, for example, the map of the city of Imola by Leonardo da Vinci from 1502 is striking (see Figure 1.2): both used the orthographic projection method, where the projection lines are parallel to each other and perpendicular to the projection plane. However, Van Deventer added the skewed parallel projections of important buildings.[48]

Another new technique that Van Deventer is believed to have used is triangulation, which was developed by his contemporary Gemma Frisius from Friesland.[49] Using this technique, distances in an imaginary triangle can be calculated by measuring at least three known features, such as two angles and another side of the triangle. For mapping geographic units like cities and regions, multiple triangles were necessary, systematically projected across the landscape. Recent literature suggests that Van Deventer likely used, or even exclusively used, polygonal surveying. In this method, the surveyor draws a closed polygon along the perimeter of the area, which can be divided into smaller sections by intermediate polygons. Triangulation is less suitable for measuring small areas on large-scale maps.[50]

Around 1540, Van Deventer was appointed Imperial Geographer. After the abdication of the Holy Roman Emperor Charles V in 1555 and the coronation of Philip II as King of Spain and Lord of the Netherlands, this title was changed to Royal Geographer. Van Deventer had already begun his impressive series of city maps in the 1540s. It is said that in March of 1540, he visited Charles V to show him his map of Brabant and a map of the Northern Netherlands.[51] At the time, Charles was in Ghent with his advisors to explore the possibilities of constructing a citadel. One would indeed be built shortly thereafter, designed by the Italian architect Donato de Bono. It became the first bastioned fortress in the Netherlands, a model that would be replicated in the decades that followed at strategically located sites along borders and coastlines. It is not unlikely that Van Deventer had already begun making plans to produce reliable city maps at that time.[52]

There is some disagreement about the dating of this work. Estimates range from 1542 to 1569. On the first Zeeland map – of Vlissingen – the year 1545 is written in Roman numerals at the bottom, within the brown drawn border surrounding the map. The prevailing opinion in the literature is that this annotation was added later. At the top of the folio, which features both the town of Vlissingen and, within the same frame, the village of 'Vlißing', it states that the map was created by order of Emperor Charles V in 1545.[53] Based on the handwriting used, Rutte

45 Darnton, 2004, p. 725.

46 Koeman, 2007, p. 1257. Koeman mentions 1547 as the final year. This has been corrected to 1549 based on the reference in the *Cronijcke van Zeelandt* by Jan Jansz Reygersberch from 1551.

47 Ibid., p. 1258.

48 Ibid., p. 1251.

49 Koeman, 2007, p. 1257.

50 Martens, 2022, p. 63. Rutte and Vannieuwehuyze 2018, p. 32.

51 Rutte, 2018, p. 17; Dupont, 2019, pp. 51, 66.

52 Martens, 2022, pp. 65–66.

53 Biblioteca Nacional de España. Res/200, nr 1 (Vlißinge): 'Estas Plantas le ize Dibujar Carlos V = Imperador en Flandres 1545'.

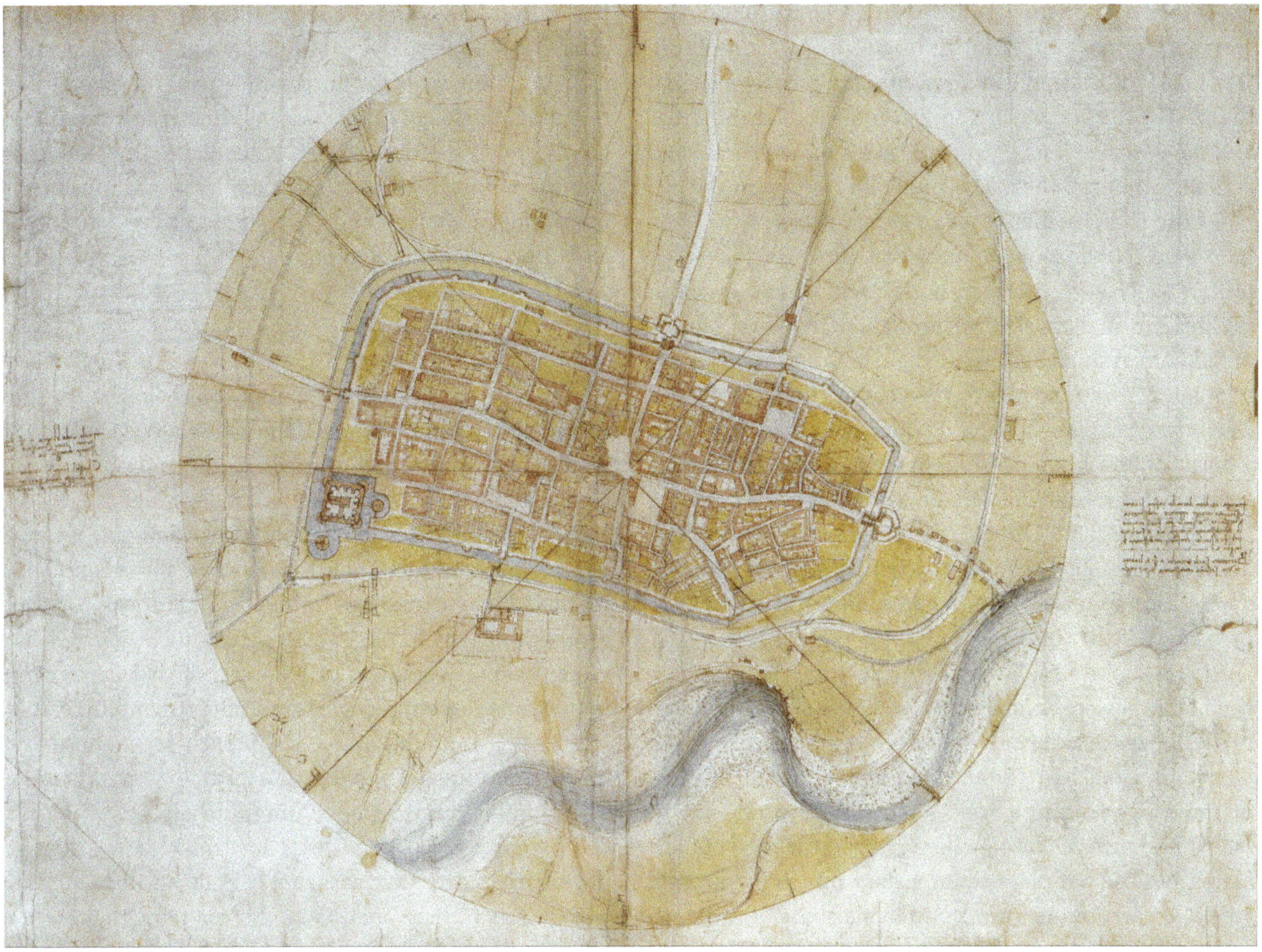

FIGURE 1.2 Plan of the town of Imola, Italy, by Leonardo Da Vinci, 1502

and Vannieuwenhuyze suggested that this annotation was only added in the eighteenth century.[54] Ramaer wrote that it was not until 1558 that Philip II commissioned the mapping of Dutch cities, including the rivers, villages, access roads, and surrounding fortifications.[55] Because this work was completed in the early 1570s, he dated the situation depicted on the maps to around 1564. According to the physicist, apothecary, and historical cartography enthusiast Henk Deys, Philip II issued the commission in 1559. In the royal decree dated May 29 of that year, it was stipulated that Van Deventer would receive an annual salary retroactive to April 1, 1558. He was also granted a safe-conduct to travel throughout the Netherlands to visit and describe all cities, towns, and rivers. This was to result in the city maps as we know them today. Deys also dated the Vlissingen map to after 1559. He questioned the appearance of the year 1545 and suspected that it was caused by 'tampering' with the sequence of maps by Van Deventer and a misinterpretation by a later bookbinder.[56]

Rutte and Vannieuwenhuyze point to the royal decree and safe-conduct issued in 1559, but also to the correspondence between Joachim Hopper, an advisor to Philip II, and the president of the Council of State in the Netherlands, Viglius. The latter reported in 1570 that Van Deventer had already been working on the project for twelve years.[57] The authors did point out that a number of city maps were created before 1555. According to the overview map in their city atlas showing the 'hypothetical dating of the fieldwork,' the maps of Zeeland should belong to this category, at least as far as the fieldwork is concerned.[58] The dating of the map of Dordrecht is considered as certain: 1545.[59] There are also indications that Van Deventer

54 Rutte and Vannieuwenhuyze, 2018, p. 38.

55 Ramaer, Middelpunten, p. 30; see also Koeman, p. 1272.

56 Deys, 1989, p. 39. See also: Van 't Hoff, 1953, p. 13; Visser, 1992–2001, p. 5; Van der Krogt, 2006, pp. 116–120.

57 Rutte and Vannieuwenhuyze, 2018, pp. 35–41.

58 Ibid., p. 40.

59 Koeman, 2007, p. 1272; Martens, 2022, p. 55.

was already conducting fieldwork in Zeeland around 1543. In 1548, he mapped Vlissingen, and in 1549, the entire province.[60] Rutte and Vannieuwenhuyze pointed out that the period during which Van Deventer visited the cities and the years in which he processed his research data and actually produced the maps could differ significantly.[61]

Of the 260 city maps by Van Deventer, 226 have been preserved, either as a *'minuutkaart'* (first version), *'nettekaart'* (revised version), or both.[62] Of the first versions, 135 are known, most of which are kept in the Royal Library of Belgium. A small number are held in Dutch archives. 179 atlas maps are bound in two atlas volumes in the Biblioteca Nacional de España in Madrid. There must also have been a third volume, but it has disappeared. Volume II contains Hainaut, Artois, and Vlaanderen, and Volume III Zeeland, Holland, Friesland, Utrecht, Groningen, and Gelderland. Volume I most likely contained Brabant, Limburg, Namur, and Luxembourg. Of this volume, only the individual maps have been preserved.[63]

Some maps – usually of the smaller cities located near each other – were combined on a single folio. Most of the revised maps also include an inset map showing only the city walls, canals, important buildings, and streets. Most of the maps feature a scale bar with the units: *passus* (step) and *pedum* (foot), where one step represents five feet. The scale bar plays an important role in the present study. To carry out fieldwork and to make use of the assistance of local citizens, Van Deventer had access to official letters of recommendation that he could present to the local authorities. He undertook his journeys to the cities until around 1570 from his place of residence in Mechelen. When the revolt against the harsh rule of the governor on behalf of Philip II, the Duke of Alba, began to take the shape of a bloody war, Van Deventer moved in 1572 to Cologne, the city where he died in 1575.[64]

Jacob van Deventer may be regarded as one of the founding figures of cartographic production in the sixteenth, seventeenth, and eighteenth centuries. In addition, his work is of great importance to the historiography of the sixteenth century because the maps provide an accurate depiction of the layout and structure of the cities, including the surrounding areas, waterways, and access roads. In many cases, the maps offer the earliest known image of the respective cities. Remarkably, Van Deventer used the same scale (approximately 1:8000), graphic cartographic language, projection method (projection lines parallel to each other and perpendicular to the projection plane, combined with oblique parallel projections of the important buildings), and orientation (always with north at the top) for all his city maps.

1.3.1 Layering

The number of studies and publications on the work of Jacob van Deventer is very large. A relatively new field of research focuses on uncovering the information not immediately visible on his maps: the layering.

In 2008, the Ghent historian Bram Vannieuwenhuyze developed a new technique for analyzing old maps: digital thematic deconstruction. In his doctoral dissertation, he investigated which cores the medieval city of Brussels had grown from. His aim was to demonstrate that the city had not originated from a single point, but from multiple ones. He chose a multidisciplinary approach for his research, drawing on institutional history, toponymy, historical geography, and cartography, among others.

In applying the latter discipline, his idea was that an old city map can serve as a source in its entirety, but that within a map there are objects and patterns that are not visible to the naked eye – usually due to the overwhelming number of elements or because the eye is drawn to other features. Vannieuwenhuyze therefore considers a map to be a palimpsest: a document containing multiple layers. When maps are used in their entirety, these underlying layers remain invisible and thus unused. Through the use of digital thematic deconstruction, these layers can be made visible. The technique involves digitizing a map and then segmenting it into small pieces that can be analyzed in different layers using digital methods. Vannieuwenhuyze emphasizes that the researcher must base their work solely on the data present in the map itself, and not on external knowledge or assumptions. The latter can lead to expectations and, consequently, to interpretive errors: 'Someone who assumes in advance that there are seven city gates, will absolutely want to find seven on the map. No more, no less'.[65]

This technique makes it possible to reveal the cartographic elements and structures, with the ultimate aim of studying them more thoroughly. Not only does cartography stand to gain a great deal from this approach, but the

60 For the evidence of the year 1548 for the map of Vlissingen, see paragraph 1.4, first example. For the year 1549 (map of Zeeland), refer to Reygersberch van Cortgene, 1551 (the 1634 reissue is used, p. 42).

61 Rutte and Vannieuwenhuyze, 2018, pp. 38–39.

62 Koeman, 2007, p. 1275.

63 Martens, 2022, p. 55; Rutte and Vannieuwenhuyze, 2018, p. 23; Koeman, 2007, pp. 172–1274.

64 Ibid., pp. 1257–1275.

65 Vannieuwenhuyze, 2008, p. 21.

in-depth study of the history of the mapped region or city also acquires a new dimension. Vannieuwenhuyze introduced his method in his dissertation *Brussel, de ontwikkeling van een middeleeuwse stedelijke ruimte* (*Brussels, the Development of a Medieval Urban Space*), which spans more than 1,400 pages.[66] He chose two maps for his research: that of Jacob van Deventer from 1550–1554 and that of Braun and Hogenberg from the period 1572–1617.[67] In the years following his doctoral defense, Vannieuwenhuyze succeeded in motivating a group of students to apply the same approach to other maps by Van Deventer. This resulted in several master's theses, all structured according to the same methodology. Together with Vannieuwenhuyze's own work, they contributed significantly to the understanding of Van Deventer's maps and to the historiography of the cities in question.[68]

In line with Vannieuwenhuyze's work, the commercial edition of Colin Dupont's dissertation was published in 2019 under the title *Cartographie et pouvoir au XVIe siècle. L'atlas de Jacques de Deventer* (*Cartography and Power in the 16th Century: The Atlas of Jacob van Deventer*).[69] Dupont developed a method based on 'digital thematic deconstruction,' in which the maps of Jacob van Deventer are digitized and implemented in a Geographic Information System (GIS). This allows the objects on the maps (such as roads, rivers, fortifications, and official buildings) to be documented and organized in a database, and also enables a deeper understanding of Van Deventer's 'cartographic language'. Additionally, new insights may emerge regarding what the maps depict – or intend to depict between the cartographic lines. His findings lead to several striking conclusions that are important to some long-standing debates about Van Deventer's city maps. One such debate concerns the purpose for which the maps were made. Since the Habsburg Netherlands – the region in which most of the cities mapped by Van Deventer were located – was in a constant state of revolt and war between 1540 and 1570, and given the existence of concrete sources from around 1559 indicating that Van Deventer worked on behalf of Philip II, it has long been assumed that the maps were primarily intended to serve a military purpose.[70]

Based on his digital reconstructions, however, Dupont concludes that the maps hold a broader political and administrative significance. The object counts on the maps support this: fortifications make up, on average, only 6.8 percent of all depicted objects. In contrast, the road network and the surrounding landscape account for 40 and 24 percent, respectively. In line with this, Dupont notes that Van Deventer mapped the cities that were represented in the provincial States assemblies, suggesting that his selection was likely based on the administrative structure of the Netherlands.[71] The maps are therefore also an instrument of state formation. They show what belongs to the king: the provinces, the cities, and the boundaries between them.[72] Based on his analyses, Dupont also argues that Van Deventer did not merely map the topography, but also the political and religious urban community. Had the purpose of his city plans been purely military, the former would have sufficed.[73]

1.4 The Seven Layers

The seven layers (military, political, judicial, socio-economic, ecclesiastical-cultural, demographic and urban morphological) are not the only fields of study for which city maps can be useful. For example, the geological and geographical layers are equally important and, by comparing earlier and later maps, we can see the spatial development over long periods of time. However, these two layers have been omitted here because the aim is to specifically explore the period in which the draftsman and his commissioners lived. In Braudelian terminology, we

66 Ibid. The method is described in Chapter 2 of Vannieuwenhuyze's dissertation, pp. 16–54.

67 Braun, 1572. The map of Brussels was depicted in Part 1, published in 1572.

68 Amongst others: D'haene, 2010; Moerman, 2010 and Carton, 2010. Marieke Moerman applied the method in her thesis, but not to a city map by Jacob van Deventer. However, there is an analysis of Van Deventer's city maps of Leuven, Zoutleeuw, Gistel, Tienen, Landen, and Dendermonde: (1) Jelle Lisson, *Een papieren stad. Het ontstaan en de ontwikkeling van Zoutleeuw aan de hand van een morfologische analyse van Jacob van Deventers stadsplattegrond (10e–16e eeuw)* (unpublished master's thesis). Leuven, 2012; (2) Pieter-Jan Gorissen, *Een studie van de middeleeuwse ontwikkeling van Tienen en Landen gebaseerd op de stadsplannen van Jacob van Deventer* (unpublished master's thesis). Leuven, 2013; (3) Jan Bogaert, *Jacob van Deventers Gistel. Een onderzoek naar de betrouwbaarheid van een zestiende-eeuwse stadsplattegrond* (unpublished master's thesis). Gent, 2011; (4) Céline Neervoort, *De vorming van een stad: Dendermonde (14de–16de eeuw). Een cartografische studie op basis van de kaarten van Jacob van Deventer, Sanderus en Alphons De Vlaminck* (unpublished master's thesis). Gent, 2011; (5) Benjamin Symons, *Leuven door de ogen van Jacob van Deventer. Een grondige analyse en bronnenkritiek van zijn Leuvense stadsplattegrond* (unpublished master's thesis). Leuven, 2012.

69 Dupont, 2019.

70 Vannieuwenhuyze, 2018, p. 35; among others Koeman, 2007, p. 1272; Martens, 2022, p. 60.

71 Dupont, 2019, pp. 265–285, 299–300.

72 Ibid.

73 Ibid.; Martens, 2022, p. 62.

FIGURE 1.3 Map of Vlissingen by Jacob van Deventer, 1548

aim to explore not the structures but the conjunctures and events of military, political, judicial, socio-economic, church-cultural, and demographic history. It is also important to answer the question of how the demographic layer – in terms of weight – relates to the six other layers.

As this particular study is primarily of a quantitative nature, and a qualitative and multidisciplinary history of all 226 cities would require many times more than the over 600 pages found in the present study , this section suffices with one representative example city: Vlissingen, situated at the mouth of the Western Scheldt.[74] The city was among the first ones Van Deventer drew. In Part I of the atlas, which is actually Part III and which includes the Northern Dutch cities, Vlissingen's map is the first one displayed.[75] Preliminary research in *Meten en Tellen* indicates that it is the most accurate of the eleven city maps examined.[76] Additionally, Vlissingen has the advantage of having a total of four city maps and -views preserved from between 1548 and 1588: Cornelis Poel (1570: a possible update of this surveyer of an unknown 1491 map by an unknown draftsman), Jacob van Deventer (1548), Antoon

74 Biblioteca Nacional de España, Planos de ciudades de los Países Bajos. Parte iii. Res/200, nr 1 (Vlißinge). Opinions are divided regarding the dating of the city map of Vlissingen. The calculations range from 1545 to 1569. The former date can be found on Jacob van Deventer's map of Vlissingen, noted in Roman numerals at the bottom of the brown drawn frame surrounding the map. The prevailing assessment in the literature is that this annotation was added later. At the top of the folio, where, alongside the city of Vlissingen, the village 'Vlißing' is also depicted in the same frame, it is stated that the map was made by order of Emperor Charles V in 1545 ('Estas Plantas le ize Dibujar Carlos V = Imperador en Flandres 1545'). Rutte and Vannieuwenhuyze, 2018, p. 38, suggested based on the handwriting used that this text was added only in the eighteenth century.

75 The books in Madrid have an original numbering on the front: II and III. Two I's have been scratched off from part III so that it is generally considered as part I. See also Rutte and Vannieuwenhuyze, 2018, p. 23.

76 Van Druenen, 2022, pp. 104–147.

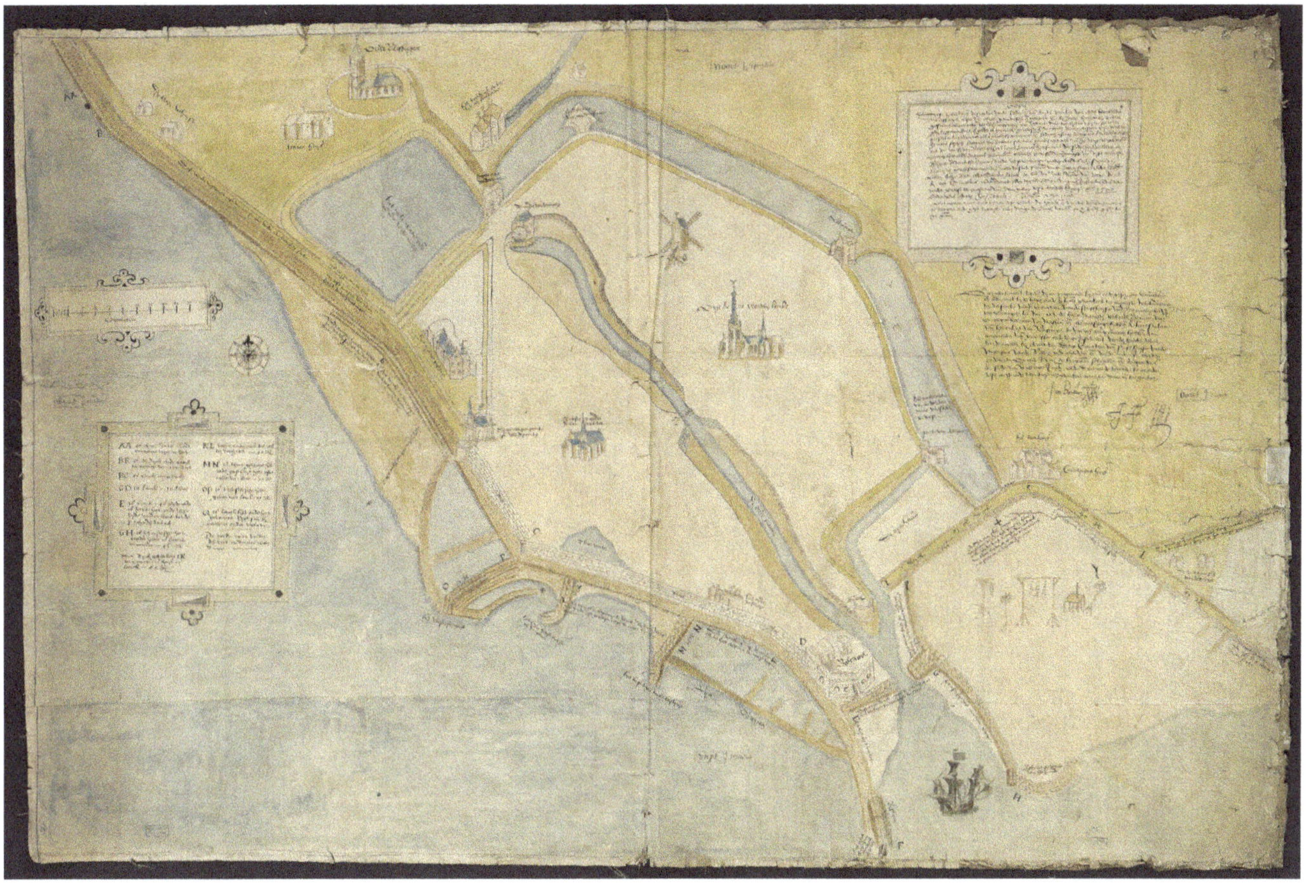

FIGURE 1.4 Map of Vlissingen by Cornelis Poel, 1491 and 1570

FIGURE 1.5 Panorama of Vlissingen by Antoon van den Wijgaerde, 1550

FIGURE 1.6 Map of Vlissingen by John Adams, 1588

van den Wijngaerde (part of a panorama of the island of Walcheren for which research was conducted between 1549 and 1551), and Robert Adams (a detailed military map made in 1588 on behalf of Elizabeth I of England during the period when Vlissingen was pledged to England). Besides that, in 1567 a house count was conducted by order of Philip II.[77] This allows for comparisons that are not feasible with most other maps due to either the absence or scarcity of comparative and reference material.

1.4.1 First Example: Military History (the bulwark)

On the map of Vlissingen, there is a structure depicted by Van Deventer in a way that gives the impression the mortar was still wet and the construction was in full swing: the bulwark at the Westerhavenhoofd, later known as the Keizersbolwerk, named after Charles V, who commissioned the construction of this defensive work. Vlissingen held significant strategic importance for him due to its location directly along the deep waters of the Scheldt and the North Sea. Throughout most of his reign, Charles had to continually defend the vast Habsburg empire and his position therein against external and internal attacks. Much of his time was spent in conflict with the French

77 Groenveld and Vermaere, Zeeland en Holland, pp. 103–174.

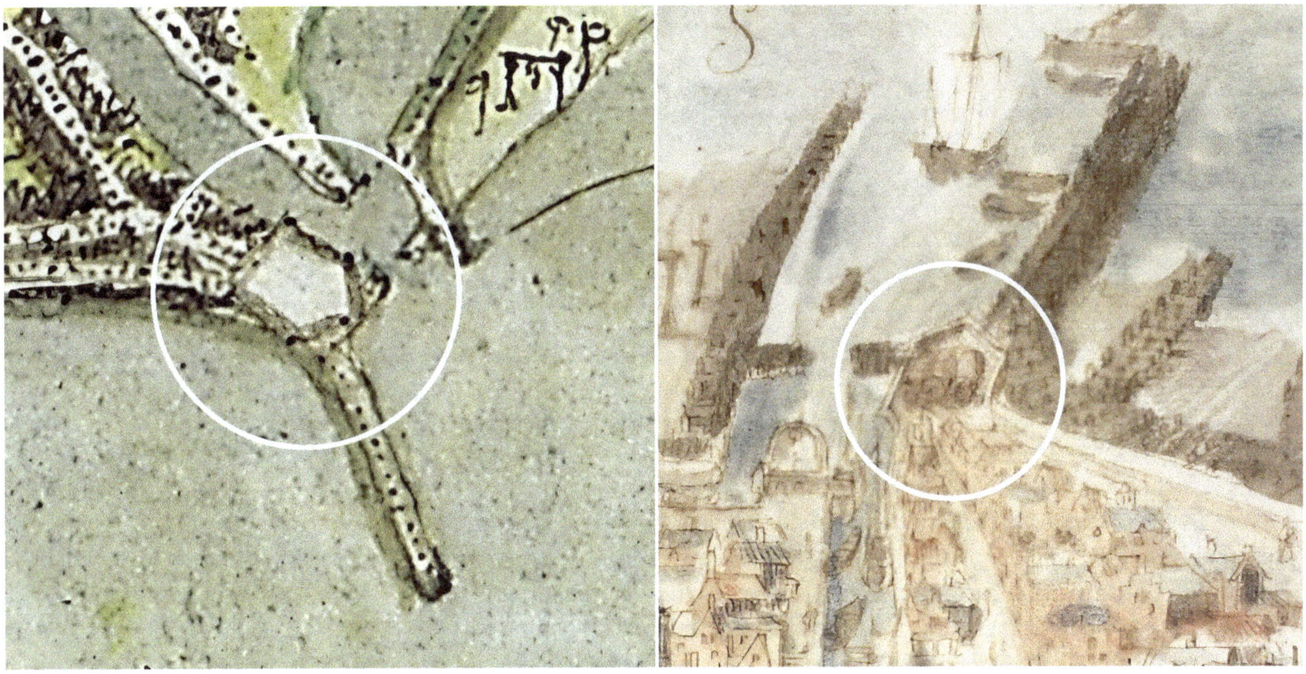

FIGURE 1.7 The bulwark on Jacob van Deventer's map of Vlissingen, 1548 and on the Panorama of Vlissingen by Antoon van den Wijngaerde, 1550

king, primarily contesting power in Europe, particularly in Italy, where the individual states were not definitively divided between the two major powers. On the eastern front, it was the Ottomans who repeatedly breached the empire's borders, reaching the gates of Vienna in 1529 after the conquest of Hungary in 1526. The fact that neither France nor Turkey managed to create a permanent breach in Charles's defense was due to ample financial resources derived from taxes imposed in the many territories and cities belonging to the Habsburg realm.[78] With these resources, Charles could build and maintain his armies but also fortify defenses around cities and important ports, such as those of Vlissingen.

The bulwark under construction was designed by Donato de Boni, the architect who had been responsible for the construction of Fort Rammekens a year earlier, located a few kilometers east of Vlissingen. An archival document from 1560 reveals that Maximilian of Burgundy, the lord of Vlissingen, had already commenced the construction of the bulwark in 1544.[79] In 1548, to be precise on April 29 of that year, the cornerstone was laid.[80] The buildings erected on the bulwark can be observed in the city panorama of Vlissingen created by Antoon van den Wijngaerde in 1550.[81] However, they were not present on Van Deventer's map.[82] This indicates that Van Deventer had gathered the data for his map well before 1550.[83] The following scenario now seems likely: in 1544, Maximilian of Burgundy initiated the construction of the bulwark, which was unfinished in 1545. It is plausible that it was

78 About Charles V: Blockmans, 2000, De Iong, 1981, and more specifically in relation to the history of Zeeland and Vlissingen: Van Druenen, 2015, pp. 140–192 and 194–292.

79 This information originates from a book written by the then Mayor of Vlissingen, H.P. Winkelman, in 1873. He used an archive document from 1560, which no longer exists today. He also wrote that the lord of Vlissingen, Maximilian of Burgundy, had already commenced fortifications in 1544. His appendix indicated that this concerned the bulwark, later known as the Keizersbolwerk. This served as a source for the later city archivist P.K. Dommisse for his extensive article from 1910 on the early history of the city. See: Dommisse, 1910, p. 161, and Winkelman, 1873, pp. 12, 273, Appendix 1. Vlissingen, Old archive (no longer available) no. 1: Vertoogh 't welke Baillu, Burgemrs. enden Schepenen der stede Vlissynghe doende syn aen den Raedt ordin. van Syne Co Maj.

80 The 17th-century chronicler Marcus Zuerius Boxhorn, who possibly used the same archive document, wrote that in 1548, specifically on April 29, the cornerstone was laid for the new bulwark. Whether he referred to the bulwark itself or the buildings is not clear (Boxhorn, 1644, p. 190).

81 Van den Wijngaerde, 1550.

82 The possibility that Van Deventer did not draw the buildings or possibly even forgot about them is not likely: other maps of similar defenses also show the buildings. See, for example, the maps of Gorinchem, Harlingen, Hesdin, Mariembourg and Vianen.

83 The map of Flushing has the year MDXLV noted in the lower brown box: 1545. The prevailing assessment of this in the literature is that the annotation was made later. At the top of the folio is written that the map was commissioned by Emperor Charles V in 1545 ('Estas Plantas le ize Dibujar Carlos V = Imperador en Flandres 1545'). See also: Rutte and Vannieuwenhuyze, 2018, p. 38.

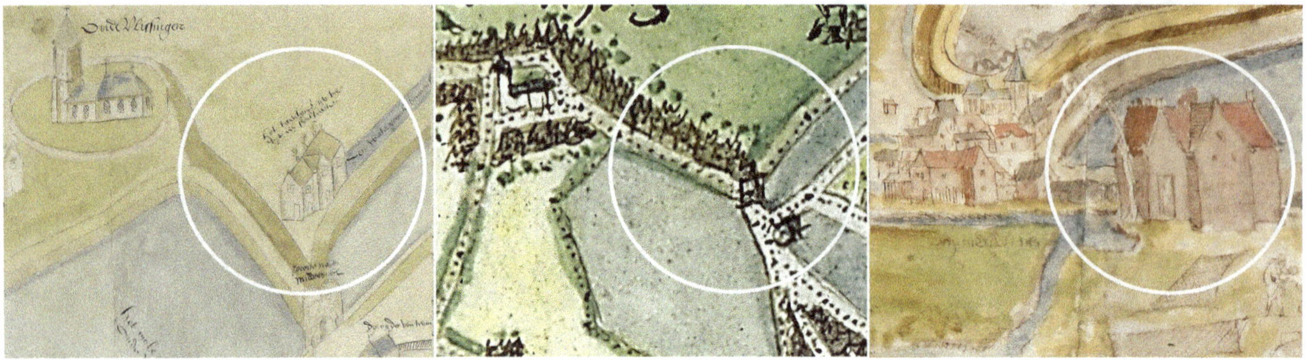

FIGURE 1.8 Left: the 'Taenhuys' on Cornelis Poel's maps of Vlissingen, 1491 and 1571. In the middle: the map by Jacob van Deventer, 1548, without the 'Taenhuys'. Right: the 'Taenhuys' on the Panorama of Vlissingen by Antoon van den Wijngaerde, 1550

finished by 1548, as another cornerstone was laid that year, presumably for the buildings on the bulwark. Where Jacob van Deventer observed the bare structure just before or in 1548, Antoon van den Wijngaerde, two years later, could see and sketch the final outcome. This suggests that Van Deventer conducted his observations for the map of Vlissingen in early 1548 or slightly earlier: right between the completion of the bulwark and the commencement of building the structures.[84] In conjunction with the panorama from 1550 and other source materials, the few square millimeters of 'still wet mortar' on the map of Vlissingen provide insight into the construction history of the defensive work, which would continue to play a crucial role in the many international conflicts involving Vlissingen as a frontline city over the next four centuries.

1.4.2 Second Example: Socio-Economic History (the 'Taenhuys')

The map also tells a story about a building that is not depicted but that the cartographic researcher will seek when examining the other two available Vlissingen maps from the mid-sixteenth century: the 'Taenhuys', located just outside the northernmost entrance to the city, the Middelburgse Poort. The building dates back to the time when the inhabitants of the village of Vlissingen, located northwest of the later city, were engaged in salt extraction, partially serving the fishing industry, alongside the ferry to Vlaanderen. This activity occurred mainly during the 13th and early 14th centuries, primarily from the creek connecting the village to the Scheldt. Along this creek lay the 'Taenhuys', where sails, ropes, and fishing nets were treated with 'taen', a substance made from tree bark that prevents mold and rot. Working in this industry was highly unhealthy due to the tar vapors released. As was the case with salt mining, 'taening' near densely populated areas was eventually prohibited.

On a map from 1491, the 'Taenhuys' is clearly visible. This map, of which no copies exist, was updated in 1571 by Pieter Cornelis Poel, the land surveyor of Zeeland.[85] He added new elements and for elements that no longer existed, he wrote a brief text. Next to the 'Taenhuys', he wrote: 'het taenhuys uu ter tyt een soutkeete,' indicating that the salt industry no longer existed in 1571.[86] The same applied to 'taening'. In accordance with his mandate, Van Deventer always drew according to reality, as will be demonstrated in the following chapter. Concerning Vlissingen, this was the situation in 1548 when the 'Taenhuys' had already disappeared and was replaced by a row of residential houses. Interestingly, in the later panorama by Van den Wijngaerde from 1550, the 'Taenhuys' is still visible without any text indicating its non-existence. The panorama includes at least one more element that did not exist during its creation. In the upper left, a huge whale is depicted being slaughtered on the beach. This was an event not based on current facts but added by the draughtsman to liven up the image of Vlissingen. Incidentally, the carcass did not resemble a whale, but was

84 This calculation aligns with the chronology in Rutte and Vannieuwenhuyze's City Atlas, 2018, p. 40. They date the creation of the city maps of the Zeeland cities Arnemuiden (p. 272), Middelburg (p. 272), Vlissingen (p. 267), Veere (p. 269), and Zierikzee (p. 275) around 1550, and of Goes (p. 279) between 1550 and 1554. No information is available regarding Brouwershaven (p. 277) and Tholen (p. 282), but it is plausible that Van Deventer also drew these cities during his work in Zeeland between 1550 and 1554.

85 Map of Vlissingen by Cornelis Poel from ca. 1570 (Nationaal Archief, Collection Hingman no. 3492).

86 The disappearance of the salt industry from Vlissingen had been a long process, beginning in the first half of the fifteenth century and ending in the second half of the sixteenth century. Van Druenen, 2015, pp. 137–138, 153 and 297.

a greatly enlarged herring. Van Wijngaerde was aware of the story of a stranded whale in 1517, but he had probably never seen one himself. The story was probably known by oral tradition to the first generations after 1517 and was recorded around 1550 by the Antwerp chronicler Jan Janz. Reygersbergh, a contemporary of Van den Wijngaerde.[87] This illustrates that for Van den Wijngaerde, who mainly drew visually appealing images, reality was less important than it was for Van Deventer.

The comparison between different maps from the same period of the same city or region raises many questions that can only be answered after thorough analysis and consideration of the historical backgrounds. For Van den Wijngaerde, who was primarily an artist, the aesthetic aspect was especially significant; for Poel, it was the documentation of the history of buildings, and for Van Deventer, it was the exact reality.[88]

1.4.3 Third Example: Political History (the Breached Sea Dike)

The above characterization of the three cartographers is also reflected in the attempts of Vlissingen to dig a new harbor just east of the city, just outside the city gate, wich ultimately did not proceed. On September 29, 1545, the then lord of Vlissingen, Maximilian of Burgundy, also lord of Beveren, ordered the sea dike there to be breached.[89] Maximilian had conceived a plan upon his appointment in 1541 to establish a new harbor at that location, officially to accommodate the significantly increased fleet of Vlissingen and Westkapelle fishing vessels, and unofficially for trade in other goods. Maximilian pursued an active policy to foster the city's economic growth. This required a substantial expansion of infrastructure, notably the available harbor capacity in the case of Vlissingen. The most recent new harbor dated back to 1443, and in the interim the city had grown significantly: the population had increased eightfold, the number of fishing vessels exceeded the capacity of the harbors, and the activity due to the thriving fishing industry was larger than ever before.

For such significant infrastructure projects, permission was required from the emperor, Charles V. Because Middelburg also opposed the plans due to the looming competition, Maximilian's request lay dormant for four years. Anticipating a favorable decision from the emperor for Vlissingen, Maximilian thought it wise to commence breaching the dike. Yet, this action drew criticism from all sides for Maximilian. Even though the dike fell under his administration, the area of the future harbor belonged to West-Souburg. Additionally, the Estates of Walcheren – the island on which Middelburg, Vlissingen and West Souburg were located – protested the breach due to the significant risks it posed to the island's water management, ordering immediate repair.[90] Nevertheless, in 1546, Vlissingen received permission from Charles V to construct the harbor, on the condition that it would only accommodate fishing vessels. This satisfied Middelburg's wishes as well.

However, the new harbor ultimately did not materialize because the city lacked sufficient financial resources to implement the plans. Moreover, the citizens of Middelburg remained active in presenting recurrent objections, which naturally had to be addressed by the court, leading to repeated delays in the project.

Thus far the history. The question now is to what extent the source of the events – the dike breach – can be observed on the two city maps of Vlissingen from this period. It is clearly visible on Van den Wijngaerde's map from 1550, as well as the area that consequently became submerged and fell victim to the influence of ebb and flow.[91] However, on the map by Van Deventer from 1548, there is nothing to indicate this event. The only original source, a municipal account from Middelburg from 1552, confirms the story but does not specify a year. The impression is given that the incident took place not long before that time. Van Deventer would surely have depicted the flooded area if it still existed in 1548. We have previously noticed Van den Wijngaerde drawing elements that no longer existed in 1550. Therefore, it is plausible that the dike was indeed breached in 1545 but repaired by 1548. In that case, Van den Wijngaerde portrayed the breach, similar to the washing ashore of the sperm whale, as a historical event. Alternatively, the breach might have occurred around 1550, as indicated by the Middelburg municipal account. If so, Van den Wijngaerde did depict the actual situation.

87 Reygersberch, 1551, p. 369.

88 Rutte, 2020, pp. 1–24.

89 Referring to the Altenapoort. Dommisse, 1910, pp. 156, 157.

90 The fear of the Estates of Walcheren was not unfounded. Later, in 1552, when the dikes collapsed at various locations during the great storm surges of January 13 and February 15 of that year, the citizens of Middelburg did not hesitate to blame the Vlissingers, referring to Maximilian of Burgundy's breaching of the dike. See: Gottschalk, 1971–1977, pp. 542, 547.

91 From Wijngaerde, 1550, part of Vlissingen.

FIGURE 1.9 Left: the place of the breached dike on Jacob van Deventer's map of Vlissingen, 1548. Right: the breached dike in the Panorama of Vlissingen by Antoon van den Wijngaerde, 1550. Please note that on this map, south is at the top and north is at the bottom.

1.4.4 Fourth Example: Judicial History (the Gallows Field)

Another example of how city maps convey more about a city than just geographical elements is how Jacob van Deventer depicted the gallows fields. On his map of Vlissingen, it is drawn on the beach south of the city. Executions – usually by hanging – took place during that time in the market area of Vlissingen, near the town hall. The lifeless bodies were then transferred to the gallows field, where they were hung with iron chains on a beam laid across two stone walls. Sometimes they were also fastened to a stake, a tall pole with a horizontally placed wheel at the top. These structures are clearly visible on the map. The gallows field primarily served the purpose of deterring the population and warning them not to deviate from the straight and narrow. In this manner cities could also display that they had received the so-called 'right of execution' from the higher administrative authority. In the case of Vlissingen, this was granted by Charles V in 1548.[92] With this right, the city held a higher legal status than cities that did not possess it. It is plausible that Van Deventer depicted the gallows field in consultation with the city council. Some of his city maps, such as that of Zierikzee, lack a gallows field, even though many of the cities depicted surely possessed the right of execution. The Vlissingen field was drawn by the first three cartographers: Van Deventer, Van den Wijngaerde, and Poel. The latter even added the lifeless bodies. On Adams' map from 1588, the gallows field has been replaced by a part of the new district.[93]

1.4.5 Fifth Example: Ecclesiastical-Cultural History (from One to Six Churches)

Church and culture were inseparably linked during the Middle Ages and the early modern period. Anyone who doesn't consider this close relationship and searches for cultural elements on Van Deventer's city maps will achieve little result. Buildings specifically oriented towards theater, music, and visual arts, which especially from the seventeenth century onwards became part of the cityscape, are hardly to be found. This, of course, does not imply the absence of culture. In this period they were much more integrated into daily life: markets were the backdrop for theater and singing, art adorned the houses of the affluent, often evolving into private museums, and the art of writing and poetry was strongly intertwined with faith

92 Grol, 1930, p. I.

93 Plan of the town of Flushing, drawn by Robert Adams, 1588. Location: British Library, Shelfmark: Cotton MS Augustus I ii 105.

FIGURE 1.10 Top left: The gallows field on the map of Vlissingen by Van Deventer, 1548. Top right: the gallows field on the map of Vlissingen by Poel, 1491 and 1571. Bottom left: the gallows field on the Panorama of Vlissingen by Van den Wijngaerde, 1550. Bottom right: the place of the gallows field on the map of Adams, 1588

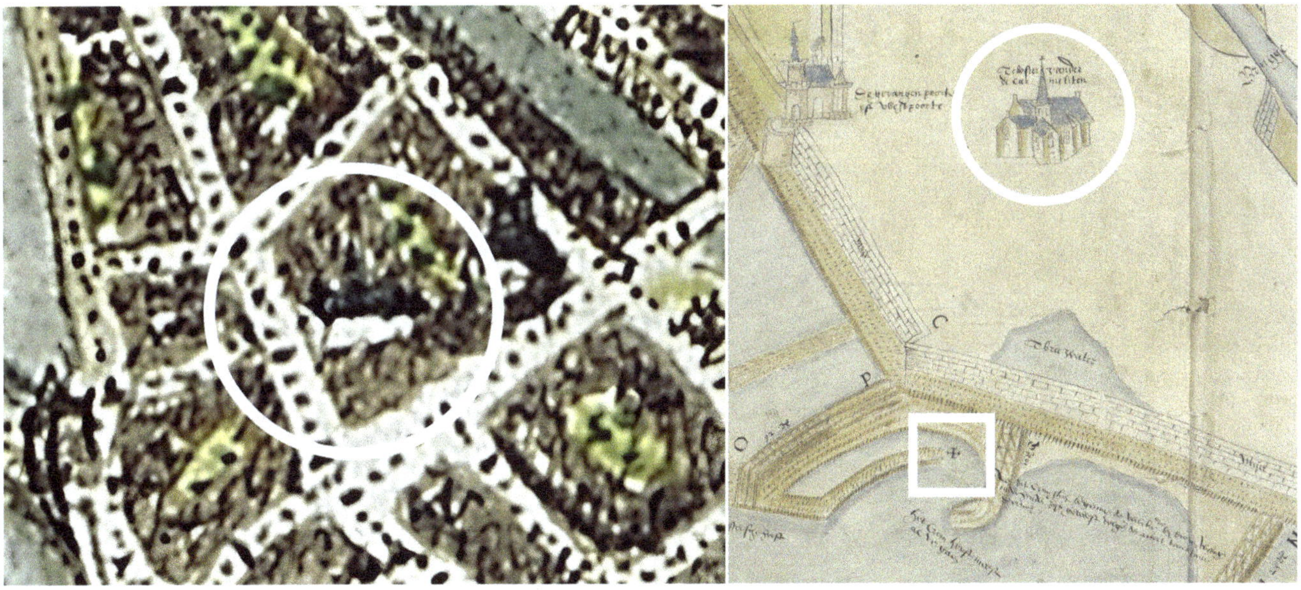

FIGURE 1.11 Left: the Carmelite monastery on Jacob van Deventer's map of Vlissingen, 1548. Right: the Carmelite monastery (circle) and the the pilgrimage statue (square) on Cornelis Poel's map, 1491 and 1570

FIGURE 1.12 Top: The Onze-Lieve-Vrouwekerk (Church of Our Lady) in Vlissingen on Van Deventer's map, 1548. Bottom: arrival and reception of the Count Palatine Frederick and his wife Elisabeth in Vlissingen in 1613 with the six church towers (painting by Hendrick Corneliszoon Vroom)

and the church.[94] This was true in the fifteenth century for the monks of the Carmelite monastery in Vlissingen.

On the maps of Van Deventer and Poel, this monastery is clearly visible, especially because Vlissingen grew into a place of pilgrimage due to the monastery and the Marian statue established there, leading to several monks gaining national and international recognition through their published writings. The Marian statue was moved to a location on the beach by the end of the fifteenth century, becoming known as 'Onze Lieve vrouw op 't stokske', and was still attracting pilgrims to the city. Poel kept it as a historical element on his map and provided commentary.[95] Van Deventer's map lacks the statue.

One of the first priors of the monastery was Goswinus Hexius, who was born in Loenhout and around 1455 received a theological doctorate in Paris. In 1460, he settled in Vlissingen, where he led the monastery and began writing books. His work on religious matters was among the most internationally influential church literature of that period. The monastery was closed and demolished in 1582 following the Calvinist upheaval of 1572. Hence, it is not visible on Adams' map.[96] After that year, Vlissingen had numerous religious opinion leaders as residents, particularly those rooted in strict Calvinist ideology.[97] From the 1550s, the city became a significant arrival point, especially for Flemish and Walloon Protestants. The new faith initially found more support there than in the Northern territories. The Iconoclastic Fury started in 1566 in Vlaanderen, leaving a trail of destruction. The arrival of these groups greatly influenced Vlissingen. The Protestant Church dominated societal life and even shaped legislation. Over the course of the seventeenth century, the number of churches in the city increased to six.[98] The preachers, most of whom had Flemish or Walloon origins, had significant influence and formed the largest group of prominent figures.

Their sermons played a crucial role in the spiritual life of all layers of the population and were often written down and printed for publication. Some reached audiences beyond Zeeland. Many writers published their manuscripts with Dutch publishers. There was also a dense network of publishers and bookshops within Vlissingen itself. Between 1573 and 1795, there were 42 active businesses in the publishing sector. Additionally, in the seventeenth and first half of the eighteenth century, Vlissingen housed 61 writers, artists, and scientists.[99]

On the map by Van Deventer from 1548, there is one church: the Roman Catholic Our Lady's Church. The cityscape from 1613 by Hendrick Cornelis Vroom shows six. Vlissingen, with fewer than six thousand inhabitants, was thus very well provided with places of worship, all Protestant, in the seventeenth century.

1.4.6 Sixth Example: Demographic History (Migration and Urbanization)

Reliable population figures – established through systematic research – are of great importance for mapping demographic processes such as migration and urbanization. In these two complex research areas, data on population size can quickly provide an initial idea of the quantity and nature of arrival and departure to and from cities. When an indication of the birth rate is also known, it can be calculated whether there is growth or decline and where it originates from: due to arrivals or departures, due to a surplus of births or deaths, or a combination of both. Subsequently, the breakdown of incoming and outgoing individuals can be the subject of further study: what percentage can be classified as belonging to immigration or emigration from or to more distant provinces or countries, and what percentage to urbanization or de-urbanization from or to nearby rural areas?

The new district mentioned in the fourth example, emerged after the Vlissingen uprising in April 1572 against the quartering and oppression by the Spanish and Walloon troops of the then governor Alva. The uprising led to two important processes. Firstly, Vlissingen became the primary destination for refugees from Vlaanderen, a region where the armies of Alva and his successors had not been defeated. The motives of the predominantly Protestant immigrants for crossing the Scheldt were partly economic and partly religious, and these were often intertwined. Indeed, since the events in Vlissingen were the prelude to a much broader uprising that in 1584 would lead to the secession of the Northern Netherlands from the Habsburg Empire of then-King Philip II, control over passage via water to the main southern port city of Antwerp could be

94 See among others: Pleij, 1976, pp. 53–64 en Pleij, 1977, pp. 34–41.

95 Van Druenen, 2015, pp. 134–136, 1047–1048.

96 Ibid., pp. 66–67, 127–129, 130–136, 262–263, 198, 324, 518, 1036–1038, 1047, 1050–1051.

97 Van Druenen, 2019, p. 73.

98 Ibid. The churches were: the Grote Kerk (Great Church), the Waalse Kerk (Walloon Church), the English Church, the Doopsgezinde Kerk (Mennonite Church), the Middelkerk, and the Oostkerk (East Church). Additionally, there was a small temporary synagogue.

99 Ibid., pp. 73–75.

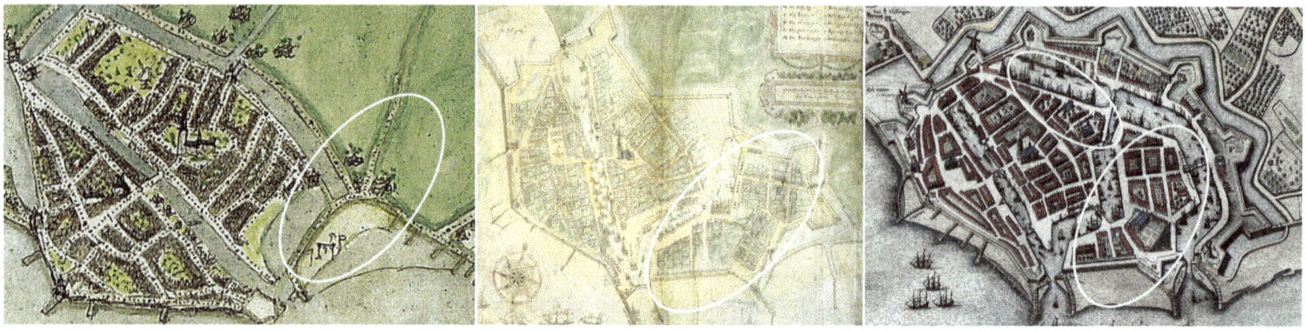

FIGURE 1.13 City expansions between 1548, 1588 and 1649 (maps by Van Deventer, Adams, and Blaeu)

established, seemingly signaling a bleak future for trade along this route. Therefore, many traders and manufacturers chose the northern route, arriving via the most western route in Vlissingen. Many stayed there, while others continued traveling to other Zeeland or Holland cities.[100]

The second process that enabled the city to grow involved rewards from the leader of the Dutch revolt, Willem van Oranje. Because Vlissingen was the first town that freed itself from the Spanish, he rewarded the city generously with, among other things, a new harbor, the establishment of – alongside that of Amsterdam – a naval headquarters, and numerous trade privileges.[101] These measures attracted traders, sailors, laborers, and fortune seekers. Immigration to Vlissingen was strongest for these reasons in the last quarter of the sixteenth century. In the first half of the seventeenth century, the growth continued, albeit at a slower pace.[102] Figure 1.22, which compares the maps from 1548 (Van Deventer), 1588 (Adams), and 1648 (Blaeu), illustrates this growth.[103]

1.4.7 Seventh Example: Urban Morphological History

In general terms, urban morphological history can be defined as the study of the development of urban form. As it is practiced today, the field is multidisciplinary, since the physical structure of cities has, over the centuries, been shaped by a wide variety of factors spanning multiple disciplines: socio-economic, military, political, cultural, climatic, demographic, geological, and ecclesiastical. Changing elements in this structure include the settlement itself, the road network, the parcel structure, and built or unbuilt spaces. In 1991, it was the architectural historian Spiro Kostof who, in his seminal work *The City Shaped*, elevated historical morphological research beyond merely tracing spatial changes and explored how cities have been shaped by a range of cultural, social, political, and economic forces.[104]

Architectural historian Elwin Koster demonstrates in his 2001 doctoral dissertation *Stadsmorfologie* (Urban Morphology) – through a thorough historiographical analysis – that the development of the field into a multidisciplinary discipline was, up to that point, far from a settled matter. He cites a 1990 work by historical geographer Michael Conzen, in which the author notes that 'during the course of history the townscape is created and transformed by its local society.'[105] Koster adds that this element is often neglected in morphological analysis: 'The problem is that the disciplinary foundations of the former still need to be more clearly defined before reliable and precise historical research can be conducted into the ever-changing relationship between spatial and social morphology.'[106] In that same year, Koster developed a method to study urban development through map comparison. Automated techniques proved to be the most suitable for this purpose. He designed a computer program that allowed historical maps to be overlaid and compared. He applied this method to a district in the historic city center of Groningen, which had been developed starting in 1620. This approach provided insight into the historical transformation processes the city had

100 Ibid., pp. 349–355.

101 There is sufficient evidence to suggest that it was not Den Briel on April 1, 1572, but rather Vlissingen on April 6, 1572, that was the first city to liberate itself from Spanish occupation. The main source for this is the letter of congratulations that Willem van Oranje wrote to the people of Vlissingen on May 7, 1572: 'You will reap praise and honor from other lands because you were the first to render such a good and loyal service to the Fatherland in these troubled times. In doing so, you have set an example for all others: they too, like you, can cast off the yoke of tyranny and slavery.' Van Druenen, 2022, pp. 4–7.

102 Ibid., pp. 256–307.

103 Blaeu, J., *Toonneel der steden van de Vereenighde Nederlanden*. Amsterdam, 1652.

104 Kostov, 1991, 2005.

105 Koster, 2001, pp. 44–116.

106 Ibid.

undergone.[107] Koster points to the shifting relationship between spatial and social morphology, as well as the connection between morphological and functional approaches within urban morphological history.[108]

The Dutch medievalist Catrien Santink made this approach to historical morphological research the core of her critique of the 2010 work by her Belgian colleague Wim Blockmans, *Metropolen aan de Noordzee*. According to her, the book presents a 'consistent image of the origins of the Flemish-Dutch entrepreneurial spirit, but unfortunately, it prevents the non-economic and non-political aspects of both the southern and northern Netherlands and their broader European context from receiving the attention they deserve. Even within Blockmans' chosen concept of core regions and urban networks, more attention should have been given to other economic and cultural dynamics.'[109]

That attention was indeed present in the *Atlas van de verstedelijking in Nederland* (*Atlas of Urbanization in the Netherlands*) published in 2014. The authors, urban and architectural historian Reinout Rutte and his colleague Jaap Evert Abrahamse, investigate how Dutch cities have acquired their present form.[110] The authors take physical space as their starting point and distinguish four periods: (1) the emergence of cities along navigable and continuous waterways or major trade routes in the second half of the Middle Ages, (2) expansion and contraction up until the mid-nineteenth century, (3) growth and new urban forms due to industrialization up to shortly after World War II, and (4) explosive growth due to the rapid increase in wealth and built-up areas. Most of the older cities have undergone all these phases, while a twentieth-century city like Almere has only experienced the last.

The origin of the city of Vlissingen can be classified within the first period of Abrahamse and Rutte's model: founded in the Middle Ages along navigable and continuous waterways and later subject to a pattern of alternating growth and decline. In the period when industrialization gripped the Netherlands after the first half of the nineteenth century, the city underwent developments in the third and fourth phases, though on a much smaller scale. Still, Vlissingen holds an exceptional position regarding the first phase compared to most other cities, which typically have a history of gradual growth from an early settlement. The city of Vlissingen was deliberately planned and built as a city in the first decade of the fourteenth century. In historiography, this is often referred to as a 'planned city' (as opposed to a 'grown city'), though this view can be nuanced; usually, there were older settlement cores before the city was founded, around which the new city was built.[111]

In the case of Vlissingen, however, this was not the case. At the location where the city emerged around 1300, peat was being extracted, sheep grazed, and the salt industry of the nearby village of *Vlissinghe* was established. The first inhabitants were only a few dozen in number, and they settled on the creek ridge that had formed after the storm surge of 1134.[112] In the adjacent creek, they moored their fishing boats. They also built an inn where travelers could eat, drink, and sleep. In the area where the city of Vlissingen now stands, they constructed salt kilns where peat was burned into salty ash, as they did not want this hazardous activity to take place in the village. Thus, the city of Vlissingen effectively originated as the industrial area of the village of *Vlissinghe*. Gradually, more buildings were erected, and during the 13th century, the fishermen began planning a new harbor entrance, aimed more toward the southeast. The creek's entrance faced directly west, making it far from ideal.

This plan gained momentum around 1290 when Floris V conceived the idea of building a full-fledged city at the southwestern tip of Walcheren. The Count of Holland wanted a strategic base along the increasingly important Western Scheldt, to defend against possible attacks from the Flemish Count. The two counties had been in conflict for centuries over the possession of Zeeland. Floris V, who had spent part of his upbringing at the castle of *Vlissinghe*, which belonged to his guardian Aleid of Avesnes, commissioned the design of a so-called bastide city with a long, straight harbor in the center and a street plan that connected perpendicularly. In 1996, Reinout Rutte defined a bastide city as a new town founded on 'ideological motives to establish a political and economic power base'. 'Often on virgin land, a city was deliberately planned by a single person, a prince, a lord, or a vassal. Prestige, the symbolic sealing of power relations, and territorial boundaries played a significant role in this process.'[113] The regular street plan was considered 'aesthetically pleasing and could be seen as a representation of order and good governance'. However, its realization was delayed because the count was murdered in 1296 by a group of nobles. These

107 Vergelijk: Koster, 2001, p. 13.
108 Ibid., p. 47.
109 Santing, 2012, pp. 75–80 (about: Blockmans, 2010).
110 Rutte, 2014.
111 Rutte, 2005, p. 134.
112 These paragraphs about Vlissingen are based on Van Druenen, 2015, pp. 10–140.
113 Rutte, 1996, pp. 189, 190, 196 and 198. See also: Blockmans, 2010, pp. 90–98.

nobles were supported by a large portion of the Zeeland nobility, who had more affinity with the Flemish than with the Dutch counts. In 1304, the plans were revived by his third successor, Count Willem III. In addition to military interests, he had commercial ambitions for Vlissingen and wanted to turn it into a true trading port. The fishing industry barely figured into his plans. In the city charter of in 1315, fishermen are scarcely mentioned. The rights are primarily focused on attracting as many merchants as possible.

Less than half a century later, there is no trace of Vlissingen's merchants. The city had become a fishing town, and the trade that existed was conducted by the fishermen. They caught herring in the North Sea, brought some of it ashore in England, took goods with them, and sold them in Vlaanderen, Zeeland, and Brabant. They were also of great importance to the herring trade in the rising city of Antwerp. Herring was a staple food, and the Brabant city sourced most of its fish from the fishing towns along the Scheldt. In addition to Vlissingen, Veere and Westkapelle were also major suppliers. The fishing town reached its peak in 1443 when Count Philip the Good granted Vlissingen an important herring privilege. All the herring caught by fishermen from the southern part of Walcheren had to be brought ashore and sold in Vlissingen. The city flourished as a fishing town, and somewhat as a trading city, though this was primarily an extension of the herring trade.

In the figure 1.23, the 1548 map of Vlissingen is compared with the schematic representation in three maps depicting the city's development. The village originated in the 12th century on a sand ridge along the creek that formed after the storm surge of 1235. The salt industry, deemed dangerous, and the inn for travelers using the ferry service were located outside the village. By 1315, the small creek on the southeast side was excavated and extended as a harbor, and around it, between 1300 and 1315, a new city was built. The salt industry had disappeared. In 1548, the year Jacob van Deventer drew the map of Vlissingen, the coastline had receded several hundred meters inland due to the storm surges of the 14th century coming up against the city. The old creek had been widened and served as a sluice basin for the harbors.

The map by Van Deventer shows the characteristic pattern of a bastide city, with streets projected perpendicular to a long, open space, usually a market square in most cities, but in Vlissingen this was the harbor. However, Vlissingen never became a true bastide city; such a city was primarily intended for military purposes, in line with the ideas of Floris V. When Willem III came to power and began to involve himself with Vlissingen, the stakes had shifted. Of course, Holland and Zeeland were still at war with Vlaanderen, but it became increasingly clear that Willem would not quickly relinquish control of the contested islands of Zeeland west of the Scheldt. His power grew due to a savvy marriage policy and his ability to choose the right strategic partners at the right moments. He married the sister of the French king in 1305 and married off his four daughters to high nobility, including the German Emperor and the English King. He sided with the French when they were at war with the Flemish but also supported the English in their conflict with France without significantly harming his own position. Willem succeeded in gaining control over Utrecht and also brought large parts of Friesland under his rule.[114] With the Treaty of Paris in 1323 between his county and Vlaanderen, he was able to definitively secure control over Zeeland west of the Scheldt.[115]

His plans for Vlissingen were therefore primarily of a commercial nature and much less military. This is why the city also deviated significantly from other bastide cities in one important way: there were hardly any fortifications. Of course, the new city benefited from its location on the water to the south, with the natural barrier of the Scheldt at its disposal. In the northwest, there was the sluice basin (Spuikom), but in between, there was nothing, at most, some earthen embankments and a fence made of palisades.[116] The same was true for the western side, which did not yet border the sea at the time, as the coastline was nearly a kilometer further west. Blockmans points out that many of the cities founded in this manner remained limited to regional centers or even no more than local fortifications. He cites examples such as Binche, Chièvres, Lessen, Zinnig, and Le Roeulx. Only a few managed to develop into fully-fledged cities, such as Dunkirk, Nieuwpoort, Damme, 's-Hertogenbosch, Harderwijk, and Amersfoort.[117] Vlissingen can also be included in this group, where fortifications only began to take shape after 1485, and even then they were only on the seaward side. Fortifications on the landward side were not constructed

114 Blok, 1939, Volume X, 1212–1213.

115 Van Druenen, 2015, p. 53.

116 This assumption is based on the absence of archaeological findings that could indicate the presence of stone fortifications, as well as on the history of the Fortress of Vlissingen, which only began in 1485 after the then 'Kabeljauwse' Vlissingen was attacked by the 'Hoekse' Sluizenaren. The local authorities in Vlissingen at the time, Anna van Borsele and her husband Philip of Burgundy, decided to strengthen the city with stone city walls and the associated gates.

117 Blockmans, 2010, pp. 97–98.

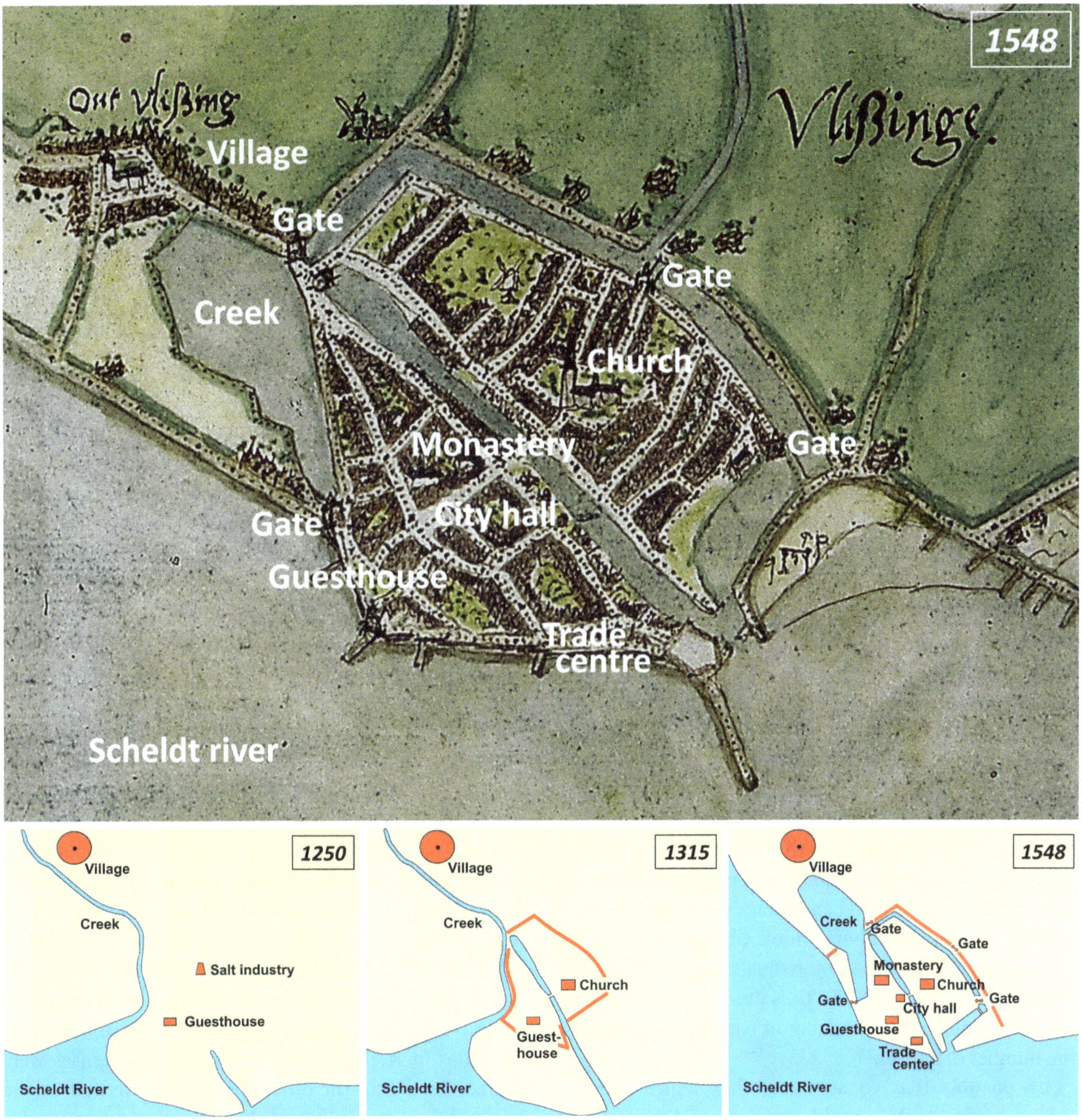

FIGURE 1.14 Excerpt from Jacob van Deventer's map of Vlissingen, 1548 and a schematic representation of the morphological prehistory (13th century), construction (1300–1315), and growth of the city (up to 1550)

until much later: during the French occupation between 1795 and 1810 to protect the city from English attacks.[118]

1.4.8 Synthesis: Locating the Demographic Layer

The seven examples demonstrating how city maps can unveil elements of history that are not immediately visible, show that the search for these elements should never rely solely on the maps. The picture becomes clearer as more diverse sources are used. In the case of Vlissingen, they initially involved a comparison with the maps of Poel, Van den Wijngaerde, and Adams. However, these comparisons usually raised more questions than answers. Other written primary and secondary sources had to be consulted to reconstruct the history.

This requires an important nuance. The demographic layer of history can indeed be explored without additional

118 Van Druenen, 2015, pp. 612–614.

source material. It can even be argued that demographic research based on city maps in general, and those created by Jacob van Deventer specifically, should initially proceed without this material. The objective is to impartially and objectively determine the population figures of a city in a specific year. These figures can then form the quantitative basis for further demographic research, such as investigating migration and urbanization processes.

1.5 Previous Comparable Research

1.5.1 Vossius and Graunt

In 1684, the philologist Isaac Vossius established a correlation between the number of houses and the population of a village or city in his book 'Variarum observationum liber,' published in London.[119] He estimated the number of houses in Paris and London at 25,000 and 50,000, respectively. For Paris, he assumed 12 people per house and for London, 6. Therefore, the population of both cities was determined to be 300,000. Vossius did not mention his sources but would have certainly been aware of the work of John Graunt, who had previously attempted to calculate the population of London in 1662. Graunt used a city map from 1658 by Richard Newcourt for his calculations. This map meticulously detailed the houses down to the smallest features. In his calculations, besides the number of houses, he also considered the dimensions, estimating the average street frontage to be 24 feet, approx 6.5 yards. On every 100 square yards, an average of 54 families lived, totaling 47,520 throughout London. Graunt estimated the average family size to be 8 individuals, resulting in a total population of 384,000.[120] He thus used the reduction factor of 8 to calculate the number of parishioners based on the number of houses.[121]

It's possible that Vossius had other sources for the house counts. It's also possible that he found the factor of 8 to be too high. If he had used it, he would have arrived at the same population figures as Graunt. In 1695, Gregory King used reduction factors of 5 and 7. Based on 105,000 houses during that period, he estimated population totals of 479,000 and 527,560.[122] In 1739, William Maitland calculated a total of 95,968 houses housing 725,703 Londoners. He used a reduction factor of 7.6.[123]

119 Vossius, 1685, p. 65.
120 Graunt, 1676, 1759, p. 31.
121 Ibid., p. 35.
122 Glass, 1946, pp. 170–183.
123 Maitland, 1756, pp. 531, 541 (page reference: Struyck, pp. 333–334.).

1.5.2 Ramaer

In the Netherlands, in 1921, the hydraulic engineer Johan Ramaer developed a method to calculate population figures between 1500 and 1800 using both censuses and city maps.[124] The method consisted of four components: (1) a coefficient to calculate the population halfway through a period, of which the start and end dates were known, (2) a ratio used to compare city maps from different periods, (3) accounting for specific circumstances that played a role in the history of a village or city, and (4) a coefficient, the reduction factor, indicating the average number of people living in a house. All of these components were used to calculate house and population numbers during periods when one or both were missing.

Using the first part of his method, Ramaer calculated the population figures of 29 Dutch cities and 'villages'. The latter refers to larger villages with urban characteristics that did not have city rights, such as The Hague. Out of these 29, the population figures for 1514 and 1632 were known from censuses. In Amsterdam, the totals were 13,500 and 120,000, respectively.[125] The growth between these two years amounted to 106,500. Ramaer assumed that this growth had not occurred gradually. The trend was that cities were becoming increasingly densely populated, and the occupancy per house was increasing. His assumption was that the growth between 1564 and 1632 had occurred slightly over four times faster than between 1514 and 1564.[126] The growth of 106,500 was then divided into 5.6 equal parts.[127] One part was added to the period between 1514 and 1578 (19,000), while the other four parts were added to the period between 1578 and 1632 (totaling 87,500).[128] Therefore, the population of Amsterdam in the period 1564–1578 could be determined as 32,500 (13,500 + 19,000). Ramaer's calculations were limited to the cities in Holland because systematic censuses were not conducted for the other cities in the Netherlands in 1514 and 1632, as was the case in Holland.

For the second part of his method, Ramaer utilized the city maps created by Jacob van Deventer from around

124 Ramaer, 1921, p. 36.
125 Ramaer, 1921, p. 33. Ter Gouw reported 115,000 inhabitants here, as indicated earlier: Ter Gouw, 1894, pp. 281–282.
126 Ramaer alternated between the years 1564 and 1578: the former because Jacob van Deventer's map was supposed to have been drawn then, and the latter because Amsterdam began to grow from that year onwards.
127 The factor of 5.6 is calculated based on the statement that the growth between 1578 and 1632 was 'a little over four times faster' than between 1514 and 1578.
128 Note: the numbers are rounded to the nearest hundred.

1564, the Municipal Atlas by J. Kuyper from 1871, and the census of 1869.[129] This enabled him to compare the surface area of houses around 1870 with those on the maps from Van Deventer and establish a formula with three known factors: the two surface areas (a and b) and the population figures in 1870 (d). The population in 1564 (c) was the only unknown factor, but could be calculated using the equation a/b = c/d.

From this calculation, an estimated population of 48,100 emerged for Amsterdam in 1564.[130] This was significantly higher than the result obtained from the first part of the method. Hence, Ramaer applied a correction – the third part of the method – based on the demographic, social, and economic history of the city under examination.[131] Here, he also used the fact that the city began growing only after 1578, indicating that the population in 1564 must have been considerably lower. According to Ramaer, this correction led to a most probable calculation of 30,900 inhabitants in 1564 for Amsterdam.[132] This method was also applied to other cities in the Netherlands.

The foundation of the fourth part of Ramaer's method consisted of house counts from 1514, 1632, 1732, and 1840. When these counts coincided with censuses, the reduction factor could be calculated, indicating the average number of people living in a house. These factors, in turn, could be used for periods where the house counts were available, but no census had taken place.[133] The underlying assumption when using these reduction factors for calculations of both population figures and house counts was that they were variable concerning time and space, thus varying per period and per city or village.[134] Ramaer assumed that, from 1514 to 1840, there had been a slow and gradual shift in the relationship between the numbers of houses and the numbers of inhabitants. The number of inhabitants per house was found to have increased by an average of 22 percent during the period 1514–1840. With this data, he could subsequently create a table showing the total population per Dutch city and 'village' between 1514 and 1840, with the intervening years being 1632 and 1732.[135]

Criticism of Ramaer's methods also exist. In 1985, urban planner Ko Visser, a professor at Delft University of Technology, reported that Ramaer's method raised so many questions that he chose not to use his series of population figures from 1921. Visser, who worked with surface area measurements on Van Deventer's maps in combination with population figures from the Wageningen Professor of Agricultural History, specializing in historical demography, Ad van der Woude from 1525, estimated a population for Vlissingen of 1000 to a maximum of 2500 inhabitants in 1560, averaging less than half of Ramaer's calculation.[136] It is noteworthy that Visser used Van der Woude's figures for the sixteenth century, who, as mentioned in the bibliography of his calculations, regarded Ramaer's work as the only available starting point for the urban situation in the sixteenth century.[137] The difference calculated by Utrecht researchers Bram van den Hoven van Genderen and Ronald Rommes for their city based on a hearth tax count from 1577 was somewhat smaller: 27,500 inhabitants compared to Ramaer's 35,000. Yet they also thought that Ramaer was generally too high in his calculations.[138]

In the decades that followed, Ramaer's work did not receive much attention. Historical demography was not yet an independent discipline and was scarcely practiced. However, this began to change in the 1930s, especially in the period following the Second World War, influenced by the ideas of the previously mentioned French Annales movement.

1.5.3 Mols

In 1956, the historical demographer Roger Mols (1909–1986) advocated for the use of old city atlases in his field of research. He did so in his monumental three-volume standard work *Introduction à la démographie historique des villes d'Europe du XIVe au XVIIIe siècle* (Introduction to the historical demography of European cities from the 14th to the 18th century). In this work, Mols focused on cities due to the generally abundant availability of source material compared to rural areas. He defined a city as an

129 Ramaer, 1921, p. 31; Biblioteca Nacional de España. Res/200; Kuyper, 1871; CBS, 1869. The year 1564 is Ramaer's assumption.

130 Ramaer, 1921, pp. 31 and 33. The surface area of houses in 1564 related to those in 1870 as 2/11 (a and b). The population figures in both years related to each other as c/d, where 'c' was the unknown. The equation then became a/b = c/d. In concrete numbers: 2/11 = c/264,700 = 48,100/264,700.

131 See, for instance, Ramaer, 1921, p. 180, where he referred to the work of Brugmans, 1911.

132 Ibid., pp. 32–33.

133 Ibid., pp. 175–178. In discussing the house counts, Ramaer did not address the difference between hearths (fireplaces) and dwellings. Since he used the maps of Jacob van Deventer for the second part of his method, it is plausible that he referred to the structures themselves and not the potential subdivision into hearths.

134 Ibid.

135 Ibid.

136 Ibid., p. 136; Visser, 1985, p. 18.

137 Van der Woude, 1980, p. 529.

138 Van den Hoven van Genderen, 1995, pp. 67, 71.

agglomeration distinguished by a sufficiently developed central function. The period over which Mols made statements ranged from the end of the 14th century, the earliest period for which population data were 'more than sporadically' preserved, to the end of the 18th century when large censuses began with a nationally standardized counting method.[139] Mols' merit lay not only in compiling and categorizing an enormous amount of source material but also in formulating guidelines based on the data from these sources for the construction of new data series, which he actively implemented.[140]

Mols explicitly highlighted the work of cartographers such as Van Deventer, Braun and Hogenberg, Janssonius, Blaeu, and De Wit. Their atlases of the Netherlands are widely available and contain maps of nearly all the cities in the region. By using these atlases for historical-demographic research, cities could be compared much more easily than with most other sources available. Because of their extensive geographical coverage, these atlases were therefore indispensable for the field.[141] One of the research possibilities Mols mentioned was the use of these maps to calculate urban population figures. Mols praised the Dutch researcher Johan Ramaer, who had already done so in 1921. He described Ramaer's work as a 'method systematically applied to all cities of the United Provinces'. According to Mols, this work deserved to be followed and emulated.[142]

1.5.4 Chandler, Bairoch, Pasciuti and others

The emulation that Mols referred to hardly materialized in the Low Countries. It wasn't until 2022 that the aforementioned study, *Meten en Tellen*, was published.[143] However, outside the Low Countries, historical-demographic research did indeed take place from the 1970s onwards, involving the use of old city maps and views. Some of these studies directly referenced the work of Mols.[144]

In 1974, American historians Tertius Chandler and Gerald Fox attempted to calculate the population figures of all major cities in world history in their work *3000 Years of Urban Growth*.[145] They utilized a wide variety of sources: censuses, travel reports, individual texts, as well as city views and maps. The combination of multiple types of sources was essential for their research because the sources were very scarce and, in terms of cities as a whole, lacked mutual coherence. Thus, achieving unity in the research methodology was impossible. The type of source that came closest to this desired unity was the latter category: maps. Chandler and Fox considered maps as a very useful tool for determining population figures. Sometimes, the area or perimeter of a city was known. In most cases, researchers had to measure it themselves, with walled cities posing the least problematic scenarios. Of these, the cities with a square perimeter were easier to handle than cities with a more rounded or intricate shape. Once the perimeter and thus the area were established, population estimates were made based on assumptions regarding population density. This involved considering similar cities for which such data were already known. An important variable was the age of the city walls. Chandler used what's known as Fox's Law, named after his co-author Gerald Fox. The law suggests that population density – the number of inhabitants per square meter – increases within a city as the city walls age. This was the result of the fact that walls weren't easily movable as a city's population grew; they had to make do with the available space within the city walls. Only when new walls with new city surfaces were built could the density decrease again. A natural outcome of this law was seen in Genoa in the late Middle Ages: surrounded by mountains, the city was the most densely populated in Europe during this period. Old buildings there sometimes reached 14 stories high.[146]

The study by Chandler and Fox underwent a significant revision in 1987 based on new data and insights. This edition sparked considerable discussion and criticism. Among other things, the definition of 'city', which included suburban areas, faced heavy scrutiny, particularly because Chandler argued that there had been hardly any rapid growth of suburbs until 1850. This was challenged since the presence of suburbs is generally well-documented, including the construction of new city walls. Consequently, Chandler's figures were questioned.[147] The calculations of the earliest cities' populations before the start of the common era based on the spatial extent of the built-up area and the average population density of the region were also subject to debate. Despite the inclusion of Fox's Law, Chandler had not adequately accounted for potential changes in population density, even in the earliest periods.

Two years earlier, Swiss researcher Paul Bairoch, based on *3000 Years of Urban Growth* and recent sources not

139 Mols, 1954–1956, Part I, pp. XVIII–XXV.
140 Van Werveke, 1957, p. 154.
141 Mols, 1954–1956, Part II, pp. 72, 83 note 3.
142 Ramaer, 1921; Mols, 1954–1956, Part II, p. 273.
143 Van Druenen, 2022.
144 Bairoch, 1988 (among others, p. 24) and Chandler, 1974 (among others: pp. 108–131).
145 Chandler, 1974.
146 Chandler, 1974, pp. 4, 5.
147 Pasciuti, 2002, p. 2; Reba, 2016, p. 9.

used by Chandler, presented new calculations in his work *De Jéricho à Mexico*, of which an English version titled *Cities and Economic Development from the Dawn of History to the Present* was released in 1988.[148] In this latter edition, despite the short period between the two editions, the author had already updated a portion of his data.[149] This was possible because, in the meantime, he and his Swiss colleagues Jean Batou and Pierre Chèvre had completed an inventory of 2,206 European cities whose populations were calculated at the beginning of each century over a period of 1,000 years, from 800 to 1850. The result was a database and a book: *La population des villes européennes, 800–1850*.[150] About forty institutions from across Europe contributed to this project, which is an indication that the realization of such large-scale projects – considering the increasing number of available research studies – requires multiple parties to collaborate and agree on the data to be provided and how they should be organized and interpreted. Bairoch also partly used perimeter and surface area measurements, focusing particularly on the correct utilization of such data. He identified three factors that must always be taken into account. Firstly, the density of the buildings within a city: the presence of many or few gardens and other open spaces is an important factor to consider. The same applies to vertical construction: an average difference of one floor can potentially double the number of inhabitants. Lastly, explicit consideration should also be given to the number of individuals per dwelling unit. Bairoch calculated this number per room and warned against too many generalizations, as the outcomes show large differences in time and space.[151]

The works of Chandler and Bairoch were not the last in which urban population figures were calculated at an international level and over centuries-long periods.[152] The continuously growing stream of historical-demographic studies ensures that comprehensive calculations and their application in multi- and/or interdisciplinary studies, at any level, can be further refined. For instance, Daniel Pasciuti and Christian Chase-Dunn expanded Chandler and Bairoch's datasets with major cities in Asia in 2002, based on recent research: *Estimating The Population Sizes of Cities. Urbanization and Empire Formation Project*.[153] The two researchers employed a so-called 'measurement error model' to determine urban population figures using estimates of (1) the built-up area of the entire city, (2) the residential area, (3) the uninhabited area, (4) the entire area within the city wall, (5) the population density of the entire city, (6) the population density of the residential area, (7) the total number of dwellings, (8) the average number of inhabitants per dwelling, and (9) events that could increase or decrease the population size of each city, such as disease epidemics, looting, becoming a capital city, the addition of new monumental buildings, construction of a new wall, and so forth.[154] In more recent works, researchers continue to build upon the idea that, based on surface area, density of construction, population density, and societal events, the population figures of cities in both the distant and less distant past can be accurately estimated. For instance, George Modelski's *World Cities: -3000 to 2000* from 2003, a study on urbanization spanning five millennia, and Jan de Vries's *European Urbanization 1500–1800* from 2013, based on the population history of 379 European cities with ten thousand or more inhabitants.[155] The databases from Chandler in 1987 and Modelski in 2003 formed the basis for an update in 2016 by Meredith Reba, Femke Reitsma, and Karen Seto: *Spatializing 6,000 years of global urbanization from 3700 BC to AD 2000*. They used software to digitize, transcribe, and geocode the tables of their two predecessors. This new dataset placed the size of urban populations over the past 6,000 years into a geographical perspective.[156]

1.5.5 Positioning

The present study relates to earlier similar research in an unsurprising way: there are hardly any similar studies. The closest comparison would be John Graunt's research over 350 years ago, where he counted the number of houses based on a city map of London from 1658 and, using a reduction factor, estimated the number of inhabitants. In the twentieth century, the previously mentioned Ramaer in 1921 and Mols in 1955 followed the example of Graunt. In the major international studies from the 1970s onwards concerning the demographic history of world cities over millennia, the use of city maps and views has occupied a crucial place, but in these cases it also revolve around surface area measurements. The method of the present study, based on length measurements, therefore largely stands on its own and has no precedent.

148 Respectively: Bairoch, 1985; Bairoch 1988.
149 Bairoch, 1988, p. xxii.
150 Bairoch, 1988. The database (copyrights Bairoch, Batou, and Chevre): https://github.com/JakeRuss/bairoch-1988 (this download: August 19, 2021).
151 Bairoch, 1988, p. 22.
152 See among others the works of De Vries, 1984, 2013, Terlouw, 1996, 2009, 2011, Modelski, 2003, Pasciuti, 2002 and Reba, 2016.
153 Pasciuti and Chase-Dunn, 2002.
154 Ibid., p. 8.
155 Modelski, 2003; De Vries, 1984.
156 Reba, 2016.

1.6 Source Criticism

The question now arises as to why systematic and sufficiently large-scale inventories of houses and population figures based on old city maps have not been used earlier. In contrast, more common sources such as baptism, marriage, burial, taxation, and citizen registers, have been frequently utilized. It is remarkable that, although source criticism plays a role in all methods used to determine house and population numbers, it has never been an obstacle to initiating the process. This further underscores the need to delineate the objections to the two research approaches: the cartographic and the traditional.

1.6.1 The Use of Old City Maps: Source Criticism

New York-based geography professor Mark Monmonier dedicated a book in 1996 to the unreliability of maps, with the telling title: *How to Lie with Maps*.[157] In it, he argued that for cartographers, it's not just easy but also necessary to lie. It is crucial that the map user realize this and take into account the possible deviations occurring in determining and presenting (a) the scale, (b) the projection, and (c) the map symbols. Furthermore, Monmonier stated that a map seeking to provide accurate information cannot avoid these deviations. In his terms, these are considered 'little white lies.' Concerning (a) the scale used, he believed that only a graphic representation was reasonably reliable. For instance, most maps are printed and resized to fit the book's format, so a centimeter indicated on the map as representing one kilometer might not remain an actual centimeter depending on the book's size. Conversely, a graphic scale where kilometers are marked on a scale bar automatically adjusts in size along with the map. Most city maps by Jacob van Deventer include such a scale bar.[158] Regarding (b), in drawing maps, there will always be a conversion from three-dimensional reality to a two-dimensional presentation, inevitably causing distortion. This distortion is visible on a world map where polar regions have the same width as equatorial areas. It is the cartographer's task to compensate for these differences. Lastly, map symbols (c), such as cities, roads, or rivers, pose a challenge themselves because they need to be clearly visible to provide useful information yet must be represented larger than reality for clarity. Monmonier calls this, in loose translation, a 'little white lie for a good cause'.

157 Monmonier, 1996.

158 The scales of several maps have been calculated. See for example J.C. Visser, 1965, pp. 116–123.

At the time Monmonier wrote this book, a new direction was emerging in the field: that of critical cartography, a response to the prevailing tradition in which map-making was considered a value-neutral representation of reality. In this view, cartographers' maps cannot be separated from the culture and power dynamics in which they were created: why was there a need for a map of a country, a region, or a city, and why did the final product look the way it did? Moreover, what was the influence of commissioners, who often represented the ruling classes?[159] The pioneers of this new way of analyzing maps were the cartographers Michael Blakemore and John Brian Harley in 1980. In their seminal work, *Concepts in the History of Cartography*, they argued that whether an old map or plan is accurate should mainly be concern the accuracy of the different facets of that map and much less the whole.[160] According to them, the analysis should always take into account the purpose the cartographers had in mind while creating the map or plan. Blakemore and Harley distinguished three main types of accuracy and, consequently, of inaccuracy: (a) chronometric, (b) planimetric, and (c) topographic.

The first type of accuracy involves determining the difference in dating between the cartography and the document. A map created in 1570 but with a map image dating back to 1520, therefore, lacks high chronological accuracy. This can be determined through research into the composition of the base materials such as paper and ink, with evidence found on the map itself, and by examining the techniques used in measurement, design, and drawing. Blakemore and Harley pointed out that there are also maps containing elements, such as waterways and railway lines, that were planned but may have never materialized.[161]

The second type is planimetric accuracy, which concerns both length measurements and surface measurements.[162] A commonly used technique in this regard involves comparing the old map with a modern map of the same area or city. However, there can be at least three problems with this approach. Firstly, it remains uncertain whether the map was intended to be precise. Its creation may have been influenced by other objectives that conflict with the accurate depiction of lengths and surfaces; for instance, the recognizable portrayal of significant buildings and (certain) houses, which might appear

159 See among others: the overview articles of Wood, 2009, pp. 340–344; Perkins, 2018, pp. 80–89.

160 Blakemore, 1980, p. 54.

161 Ibid., p. 55.

162 Ibid., p. 60.

larger than they actually were. Secondly, it is never certain whether the medium – usually a type of paper – on which the map was drawn or printed still maintains the same dimensions as at the time of production. Paper can undergo various changes due to moisture, dust, and dryness, altering its shape: shrinking, wrinkling, stretching, curling, etcetera.[163] Finally, errors could also arise because many old maps themselves were copies of existing copies, which may or may not have been updated.

The third type of accuracy focuses on the quality of information on the old map. Concerning city maps, this involves the depiction of, for instance, houses, buildings, gates, and defensive structures like walls, watchtowers, and ramparts. The question always remains whether the cartographer accurately represented these objects (such as drawing a castle as a castle and not as a monastery) and whether aesthetic elements were incorporated that did not align with the actual situation.[164]

1.6.2 The Use of 'the Common' Sources: Source Criticism

While there is criticism regarding the use of old city maps, as outlined above, there are also many reasons to criticize the use of more common sources such as baptism, marriage, burial, taxation, and citizen registers. This criticism is generally widely recognized but often overlooked as is made evident by the enormous amount of quantitative research in this field.[165] The shortcomings of these sources, whether in their original form or digitized, can be summarized and grouped as follows: (1) the accuracy regarding the correct spelling of names and places and the accurate notation of figures during data input by contemporaries cannot be estimated, (2) the extent and nature of the discrepancy between actual demographic events (marriages, births, deaths, and migrations) and recorded demographic events are unknown, (3) the level of completeness of the preserved archival documents is unknown, (4) many concepts in the source material, such as 'citizen', 'hearth', 'window', 'communicant', or 'able-bodied man', and currency types (important for comparing property values in areas using different currencies) can vary by period and by city or village, (5) significant risk of double counting due to overlapping registers, confusion between patronymics and surnames, and second, third, and subsequent marriages, (6) missing births due to delayed baptisms, (7) the transcription, carried out in the archives by non-professional volunteers, of mainly handwritten texts remains error-prone, (8) the quality of previous transcriptions of lost sources is incalculable, (9) the carelessness in digitizing sources and the aforementioned transcriptions is considerable, and (10) the digital techniques and database software used by different archival services are not uniform, making the data difficult to compare with each other.

Researchers who additionally use registers of new citizens to find data on immigration encounter, in addition to the aforementioned ten objections, an additional difficulty: (11) the limited scope of these registers. Newcomers in the cities could only become citizens after payment. They were noted in the registers and thereafter enjoyed all the rights associated with citizenship. Individuals who were already citizens and living in the city at the time they acquired this status were not listed in these registers. Registers of new citizens were, therefore, primarily immigrant registers. Newcomers without money were not recorded and consequently lacked the associated benefits. Moreover, it is uncertain whether all newcomers wanted to become citizens. Therefore, their numbers are unknown and can only be found in the baptism, marriage, burial, taxation, and citizen registers, as well as in records of real estate transactions and lease agreements. This also applies in most cases to the family members of the paying and registered citizens. There has been little research done on the relationship between the registered and unregistered individuals. Therefore, there is also a lack of ratios to extrapolate the numbers of entrants in the citizen registers to immigration totals.[166]

1.7 Conclusions and Follow-up

A cautious comparison between the two types of sources and the criticism of both yields the following: the collection of city maps by Jacob van Deventer, compared to the more common sources, in its entirety offers a consistent and unambiguous quantity of 226 primary sources of as many cities in the then Habsburg Netherlands around the middle of the sixteenth century. Despite

163 Ibid., p. 61: 'Flanning, Drawn in, Back-edge wrinkle, Center wrinkle, Around-cylinder stretch, Across-cylinder stretch, Between-printings stretch, Back-end register, Distortion, Curl, Tail hook, Blister, Pick, Dust, Blanket scum, Offsetting (coated), Offsetting (uncoated), Skitic, Hickies, Emulsification, Slow drying'.

164 Ibid., p. 68.

165 See Van Druenen, 2022, pp. 69–79.

166 This overview is compiled based on the source criticisms of Slicher van Bath, 1964 and Van der Woude, 1980 and supplemented with the research experiences of the author of the present study during the period 1984–2024.

the vast quantity of common sources from comparable areas, a comparable collection does not exist. Therefore, historical-demographic research regarding this period and area must almost always be limited to local or at most regional collections of sources, raising doubts about whether the various types of sources in the archive of one city or region are comparable or have equal coverage in terms of time periods.

Moreover, within historical cartography, the Van Deventer collection appears to be the only coherent collection of maps possessing all the following desired qualities: created within a relatively short period under the direction of a single cartographer who almost always used the same scale, design, and structure. In the century following Van Deventer, various cartographers dedicated themselves to producing similar collections of city maps, including Braun and Hogenberg, Janssonius, Blaeu, and De Wit. However, most of these collections were created based on copies of existing maps or the contributions of multiple draftsmen. As a result, each collection of maps varies significantly in terms of accuracy and consistency.[167] Where Van Deventer's work was primarily aimed at representing reality as meticulously as possible, his successors had a different primary motivation: selling as many maps and atlases as possible. The scientific objectives of these projects are generally considered as subordinate to the commercial ones. For example, Van Deventer's successors needed to consider a certain speed of documentation, graphic production, and printing. Within that limited time frame, relatively much attention was given to the graphic quality of the work, which is crucial for potential buyers, and the delineation of houses, public buildings, and churches in the respective city. These were depicted in the so-called oblique parallel projection; streets were often widened in the line of sight to enhance visibility of the facades. The important buildings such as churches, towers, town halls, and castles were disproportionately magnified as is the case on Van Deventer.[168]

In summary, Jacob van Deventer's city maps can serve as primary sources for large-scale quantitative historical-demographic research, as they meet the requirements of this kind of research. They are unique for the early modern period, even when compared to other types of sources. Comparable material only became available from the second half of the eighteenth century when the first large national censuses were organized in many countries.

From this answer, the following question can now be directly posed: is it possible to derive reliable demographic data from Jacob van Deventer's city maps? A positive evaluation of the maps as primary sources does not exempt the researcher from the responsibility to remain critical while working with this material. Many of the concerns raised by Blakemore and Harley with regard to assessing old maps apply to Van Deventer as well. Specifically, the researcher must assess the maps' (a) chronological accuracy based on evidence on the maps themselves, (b) planimetric accuracy through comparisons with new maps in geographic software packages like Open street maps and MapAnalyst, and (c) specific planimetric accuracy concerning major architectural and infrastructural works used as reference points to establish the scale ratios of the maps. Chapter 2 will attempt to determine what Jacob van Deventer's city maps can tell us about the numbers of houses and inhabitants during the period of their inventory and creation.

The results from Chapter 2 will be used in Chapter 3 to set up a consistent study to the extent of the build environment and the population in the 226 cities of Jacob Van Deventer and their mutual comparability. Finally, in Chapter 4, it will be examined whether the method can be applied to city maps from other areas and earlier or later periods.

167 Van 't Hoff, 1966, p. 31; Alkhoven, 1992, pp. 87–93.

168 Koeman, 2007, pp. 1363, 1375.

CHAPTER 2

Method

2.1 Measurements

Counting houses on the city maps by Jacob van Deventer and calculating population numbers based on these counts is not possible as it is with maps drawn in oblique parallel projection, such as those by Braun and Hoogenberg in the sixteenth century and Blaeu in the seventeenth century. Although the houses are visible on Van Deventer's maps, they are represented with seemingly randomly inserted ^-symbols. In many places where houses must have stood, the maps show blotches, unclear signs, or nothing at all. Yet, they must have been there. Furthermore, the red coloring is not very consistent. For these reasons, instead of counting houses, it has been decided to digitally measure the built-up street sides in pixels. Based on these measurements and using a coefficient, the number of houses can be determined. To establish this coefficient, a city must be selected for which the measurement results on Van Deventer's map can be compared with an actual count on an alternative city map where houses are visible and from a more traditional source. Both should be from the 40-year period when Van Deventer was making these maps. Moreover, the buildings in the city part to be measured and counted must have remained relatively stable during this period.

In the Zeeland city of Vlissingen, which is used in Chapter 1 to visualize the presence of the six layers, this combination of sources can be found. In addition to (a) Van Deventer's city map, in Chapter 1 dated 1548, and a (b) very detailed military map made by Robert Adams in 1588 on behalf of Elizabeth I of England, there is (c) a house count from 1569 available from a 1569 report prepared by the Italian military officer Chiappino Vitelli on behalf of the then governor, Fernando Alvarez de Toledo, the third Duke of Alva.[1] The report was meant to investigate the extent to which the coasts of Zeeland and Holland were resistant to external attacks. The king of the Habsburg Netherlands Philip II and Alva considered the island of Walcheren to be the most obvious location for the enemy to land. Therefore, defending the coasts became a priority. The commission began its work on March 15, 1569, in Vlissingen. The assignment was to study the defense works and access roads, locate the best places for the enemy to invade, gather information about available building materials, food supplies, and ship necessities, and finally, to count the fishing ships, houses, and able-bodied men.[2] Adams' map and Vitelli's house count will be discussed further under steps 2 and 5/6, respectively, of the eight-step procedure described below.

Using these three sources, the number of houses can be calculated and compared with the outcomes of other calculations. There is a eight-step procedure which yields a coefficient with which the population numbers of other cities mapped by Van Deventer can be calculated. It can thereby be investigated whether the new method of counting – referred to as the 'Passus method', after the units of measure on Van Deventer's scale stick – is feasible and can be validated. The procedure consists of the following eight steps:

1. Formulate the standards to be used for the counts on Van Deventer's maps:
   - the type of city map;
   - the concepts of house, hearth, hearthstone, building, and street side;
   - the buildings to be counted (houses, warehouses, public buildings, churches);
   - the reduction factor houses/inhabitants to be used;
   - the spatial delimitation of the urban area (within the city walls, walls, canals, or also outside);
   - the theoretical concepts to be used for statements about the accuracy of the maps.
2. Estimate the creation date of the maps of Vlissingen by Van Deventer and by Adams.
3. Select a city part and:
   - argue that the number of houses in this part of the city has not changed in the period between the drawing of the two maps;
   - argue that the lengths of the built-up street sides and the street pattern in this part of the city are still the same today (reference year 2024) as in the year Van Deventer drew the map;
   - measure the lengths of these built-up street sides on the 2024 map of Vlissingen in Google Maps in meters;

1 Groenveld, 1980, pp. 103–174.

2 Ibid., pp. 133–134.

 | HTTPS://DOI.ORG/10.1163/9789004755031-003

- measure the lengths of these built-up street sides on Van Deventer's map in pixels, convert the result to *passus* and meters, and check whether the average scale of the street lengths corresponds to the scale drawn by Van Deventer at the bottom right of the map;
- count on Adams' map how many houses are depicted on these built-up street sides,
- determine the quotient of (a) the number of houses and (b) the lengths of the built-up street sides: the Passus coefficient.

4. Determine the planimetric accuracy of Van Deventer's and Adams' maps using MapAnalyst, a software application for the visualization and study of the accuracy of old maps discussed in paragraph 2.2.
5. Measure all the built-up street sides of Vlissingen within the city walls on Van Deventer's map, calculate the total number of houses in the city in 1548 using the Passus coefficient, and:
   - compare the Passus method with the numbers of Ramaer (inhabitants) and Vitelli (two counts: houses and able-bodied men) and inventory the similarities and differences;
   - formulate a preliminary statement about the usability of the Passus method.
6. Check whether the scale and resolution on Van Deventer's maps in a first test group of four cities mapped in the same early period and whose houses were counted by Vitelli in 1569 correspond to those on the map of Vlissingen, and:
   - if not already the case, adjust the resolution (necessary for pixel measurement) and scale (necessary for converting the number of pixels into the number of paces and then meters);
   - measure all the built-up street sides of these cities and use the Passus coefficient of Vlissingen to get the number of houses and inhabitants per Zeeland city;
   - compare the outcomes of the Passus method on the Zeeland maps with the house count of Vitelli in Goes, Middelburg, Veere, and Zierikzee;
   - nuance the preliminary statement (procedure step 5) about the usability of the Passus method.
7. In order to investigate whether the Passus coefficient can be used to count houses in other cities, put together a second test group with a city in Vlaanderen and a city in Holland with more than 5000 inhabitants that meet at least two of the three conditions used for Vlissingen: (a) the presence of a city map made by Van Deventer, (b) an alternative city map and/or (c) an alternative source such as tax registers or house counts, and:
   - repeat the procedure for these cities as previously performed for the first group of five test cities;
   - nuance the preliminary statements (procedure steps 5 and 6) about the usability of the Passus method.
8. Compare the deviations of all existing counts and the Passus method and formulate a definitive statement about the validity of the Passus method.[3]

2.1.1 (ad 1) Standards

2.1.1.1 The Type of City Map

Of the 226 maps, 179 are bound in atlases: two dark brown books that can be found in the Biblioteca Nacional de España in Madrid.[4] The use of these atlas maps has the disadvantage over the separate copies (kept separately in various archives and libraries in the Netherlands and Belgium) that the distortion is greater, among other things due to the fold. When both types of maps are used in one study, such as is the case in the present study, this difference in quality must be taken into account. For example, when determining the margins of error to be used. There is no other choice because there are no sixteenth-century copies of the atlas maps and no atlas has been made of the separate maps that are in Brussels.[5]

A second important difference for this research is that on the atlas maps almost everywhere a scale stick is drawn, while it appears on only one of the separate maps: Mechelen. If one of the two types of maps had been chosen, an even stricter standardization could have been achieved. However, because the goal of the present study is to measure and analyze all 226 city maps drawn by Van Deventer, it was nevertheless chosen to use all of the maps, regardless of the type of map. The unifying factor is the maker: Jacob van Deventer. The second unifying

3 The procedure and tests in this section are an update (based on the recent research findings as presented in Chapter 3 of *The 226 city maps of Jacob van Deventer: dimensions, accuracies, comparisons*), of the pages 104–147 in: Van Druenen, 2022.

4 Biblioteca Nacional de España, Biblioteca Digital Hispánica, City plans of the Netherlands, Deventer, Jacob van (ca. 1505–1575). Res/207 (ii) and Res/200 (iii).

5 There is actually another choice, as in the early twentieth century lithographic reproductions were made of the separate maps of the Zeeland (and all other Dutch) cities. In theory, these could also be used, but in the assessment made, contemporary maps were preferred because it cannot be excluded that errors were made in the lithographic reproductions or graphic embellishments were applied.

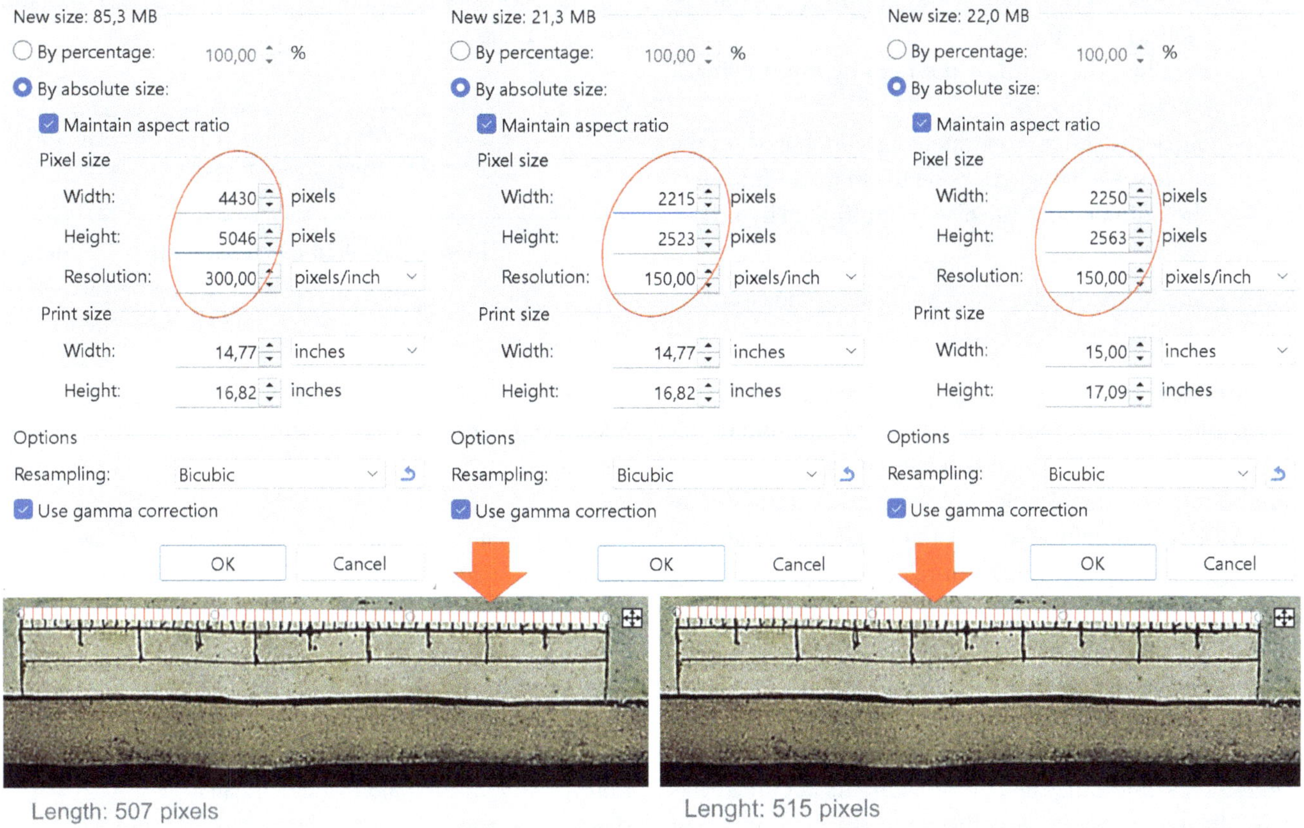

FIGURE 2.1 Adjusting resolution of atlas maps. Example: the map of Rotterdam by Jacob van Deventer, 1557–1563

factor is an artificial one: the alignment of the resolutions and scales. The resolutions must be equal for the pixel measurement. The scales must be equal for converting the number of pixels into paces and meters. After all, with only an equal resolution, the distances on the different maps cannot yet be compared with each other.

For the purpose of standardizing the measurement and analysis tools, two different sub-procedures were used, one for the atlas maps and one for the separate maps:

- *Atlas maps (189 cities)*:
  - Open the downloaded digital map in a graphics program;
  - Change the resolution from 300 pixels per inch to 150;
  - Halve the width of the map (keeping the ratio between 300 and 150);[6]
  - Measure the scale bar in pixels;
  - If this value is not 515 (as in the example in Figure 2.1: 507 pixels), change the dimensions of the entire map so that the scale bar is again 515 pixels;
  - Now, the dimensions of the map are multiplied by the quotient of two fractions: (a) the length of the scale bar on the map of Vlissingen (515) and (b) the length of the scale bar on the map to be measured (in the example: 507).
- *Individual maps (37 cities)*:

  For measuring these maps, Mechelen is used as the standard, as it is the only map with a scale bar. The assumption here is that the 37 individual maps show minor differences in terms of the relationship between dimensions and scale. The digital map of Mechelen is 9776 pixels wide and 8912 pixels high, with a resolution of 400 pixels. The scale bar measures 1053 pixels, but it represents 4 instead of 5 passus. Sub-procedure:
  - Open the downloaded digital map in a graphics program;
  - Change the resolution from 400 pixels per inch to 150;
  - Multiply the width of the map by 0.375 (the ratio between 400 and 150 pixels): 9776 * 0.375 = 3666 pixels, with a new scale bar length of 396 pixels;[7]

6 Because the 'Maintain aspect ratio' option is checked, the height of the map changes automatically.

7 Ibid.

- Increase the adjusted scale bar from 4 to 5 passus: (396/4) * 5 = 495;
- Readjust the width of the map by multiplying it by (515/495 = 1.0404);
- Since the other individual maps are based on the same scale bar length as on the map of Mechelen, their dimensions can also be multiplied by 1.0404.

2.1.1.2 The Reduction Factor to be Used for the Number of Inhabitants Per House

The existing historical-demographic literature offers a broad and disjointed spectrum of reduction factors representing the average relationship between the number of inhabitants and the number of houses. Not only do they vary widely, but they also constantly change regarding whether there are on average more or fewer people living in cities than in villages and rural areas.[8] The conclusion is therefore that no clear guidelines can be derived from the many studies conducted over the last three centuries for determining the correct reduction factor for each combination of period, region, and size of the buildings. Therefore, a 'safe' range of 4 to 7 inhabitants per house has been chosen for this study, with a derived average of 5.5. In the case of Vlissingen, there are two inventories of houses and inhabitants and the resulting reduction factors that can support this choice: the counts of Chiapinno Vitelli in 12 Dutch cities from 1569 and the censuses of 1830 and 1839.

In Table 2.1, the numbers of houses and able-bodied men are shown for four cities in Zeeland and eight in Holland, as mentioned in Vitelli's report.[9] In the third column, the number of able-bodied men is multiplied by the factor 3.85.[10] This yields the population numbers for the respective cities. The average reduction factor (inhabitants per house) for these 12 cities is 6.5. The underlying factors per city range from 3.1 to 11.6. The second support – albeit from a much later period – is provided by the reports of the censuses of 1830 and 1839, in which both houses and inhabitants of the Netherlands were counted. The average reduction factors were respectively 6.3 and 6.4. The range was much less than in 1569: 5.7–6.8 and 5.9–7.0 (Table 2.2).[11]

TABLE 2.1 Number of houses and able-bodied men in 12 Dutch cities according to Vitelli's report from 1569, converted to numbers of inhabitants and the reduction factors

	Houses	Able-bodied men	Inhabitants based on able bodied men (*3,85)	Reduction factor houses/ residents
Middelburg	800	1400	5390	6.7
Vlissingen	800	1000	3850	4.8
Veere	500	500	1925	3.9
Zierikzee	1000	1800	6930	6.9
Goeree	200	480	1848	9.2
Brielle	600	750	2888	4.8
Delft	2000	5000	19,250	9.6
Alkmaar	500	750	2888	5.8
Medemblik	300	500	1925	6.4
Enkhuizen	1000	3000	11,550	11.6
Hoorn	700	900	3465	5.0
Haarlem	5000	4000	15,400	3.1
Average				6.5

2.1.1.3 The Concepts of House, Hearth, Hearthstead, Building, and Street Side

The literature provides different definitions for these concepts. Hearthsteads and houses are seen as the same unit of calculation. The concept of hearth will not be used, and with the concept of building, it can refer to one house or several. The concept of 'street side' has been chosen for its measurable and comparable nature, as the concept of 'street' is much less so. A street can consist of several street sides. This also applies that when a street side is interrupted by a side street, a new street side begins.

2.1.1.4 The Buildings to be Counted (Houses, Warehouses, Public Buildings, Churches)

On Van Deventer's maps, where houses cannot be counted but can be measured in street lengths in groups, it is difficult to distinguish houses from warehouses and public buildings, except when the latter had such a monumental

8 Van Druenen, 2022, p. 2–7.

9 Groenveld, 1980: Alkmaar (p. 159), Brielle (p. 157), Delft (p. 158), Enkhuizen (p. 163), Goeree (p. 155), Haarlem (p. 165), Hoorn (p. 164), Medemblik (p. 162), Middelburg (p. 150), Veere (p. 150), Vlissingen (p. 150), Zierikzee (p. 154).

10 The factor 3.85 is the reversal of the factor as calculated by Van de Woude in 1962: 26 able-bodied men per 100 inhabitants (Van der Woude, 1962, pp. 35–76).

11 CBS, Volkstelling.nl, 1830 and 1839.

TABLE 2.2 Houses and population numbers in the Netherlands per province according to the censuses of 1830 and 1839, converted to the reduction factors houses/inhabitants

Census							
1830				1839			
Provinces	Houses	Inhabitants	RF	Provinces	Houses	Inhabitants	RF
Drenthe	9,957	63,868	6.4	Drenthe	11,391	72,484	6.4
Friesland	35,956	204,909	5.7	Friesland	38,660	227,859	5.9
Gelderland	47,840	309,793	6.5	Gelderland	51,691	345,762	6.7
Groningen	27,482	157,504	5.7	Groningen	28,479	175,651	6.2
Limburg	32,279	186,281	5.8	Limburg	33,402	196,719	5.9
Noord-Brabant	56,501	348,891	6.2	Noord-Brabant	61,253	378,437	6.2
Noord-Holland	61,630	413,799	6.7	Noord-Holland	63,398	443,334	7.0
Overijssel	30,637	178,895	5.8	Overijssel	33,465	197,694	5.9
Utrecht	21,167	132,359	6.3	Utrecht	22,643	145,132	6.4
Zeeland	23,353	137,262	5.9	Zeeland	24,412	151,358	6.2
Zuid-Holland	71,057	479,737	6.8	Zuid-Holland	75,984	526,020	6.9
Total	**417,859**	**2,613,298**	**6.3**	**Total**	**444,778**	**2,860,450**	**6.4**

character that they were drawn separately by the cartographer. This certainly applies to churches, which are almost always recognizable. Because these buildings were proportionally larger, this certainly affected the length of the surrounding buildings to be measured. Another element to consider in determining the error margins is the way in which detached and connected houses and buildings are depicted. The first group takes up proportionally more space than the second. For the current study it is decided not to measure recognizable street parts without buildings, clearly visible churches, or public buildings. Furthermore:

- If there are no dots in a (usually very short) street, no houses are counted;[12]
- The sides of blocks of houses are only counted when there is a street with dots.

Finally, it should be noted that grouping houses and warehouses and other buildings that cannot be clearly identified as such is acceptable because the totals include both the houses with multiple dwelling units and the buildings or houses where few or no people lived. The assumption here is that the positive and negative 'errors' will largely cancel each other out. In terms of probability calculation: the chance that the number of positive measurement errors made by the cartographer equals the number of negative errors becomes greater or smaller as the city is larger or smaller.

2.1.1.5 The Spatial Demarcation of the Urban Area (within the City Walls, Walls, Canals, or Also Outside)

As a geographical demarcation of the space in which measurements and counts are taken, the area up to and including the city walls, ramparts, or canals has been chosen. In the procedure, this means that when a city is not surrounded by water, walls, or ramparts, the street sides are counted up to and including the rows of adjoining houses. Every house that falls outside of this, no matter how close to the city, is not included. This is the only possible hard criterion because concepts like 'close to', 'in the immediate vicinity of', or 'within sight of' the city are hardly uniquely definable, even when a demarcation in a distance measure is agreed upon, for example, 'within 500 meters of the city'. Is this measured as the crow flies, or over one or more streets? What to do when the houses belong to another city or village? Do nearby villages automatically belong to the city in question? And what to do when the buildings are bisected by the 500-meter boundary?

12 According to Molders, 2020, p. 19–23, there is no fixed relationship on the city maps of Jacob van Deventer between the number of dots and the length of the street section in which they are drawn. 'The thin dots cannot represent distances measured by the cartographer, nor chains of measurement or numbers of steps. They merely indicate traces.'

2.1.1.6 The Theoretical Concepts to be Used for Statements about the Accuracy of the Maps

In the research on the city maps of Van Deventer, focusing mainly on the lengths of the inhabited street sides, the following theoretical concepts discussed in paragraph 1.6 will be addressed: (a) the chronometric accuracy based on evidence on the maps themselves, (b) the planimetric accuracy based on comparisons using MapAnalyst, and (c) the specific planimetric accuracy regarding a number of major architectural and infrastructural works, which are addressed in the elaboration of the fourth procedural step *ad 4* and as a tool for determining the dating of the maps under *ad 2*.

2.1.2 (ad 2) Dating of the Maps of Vlissingen by Van Deventer and Adams

As demonstrated in paragraph 1.4, the reconnaissance work for the map of Vlissingen by Jacob van Deventer took place around 1548. The dating of the map by Adams is absolutely clear: the spring of 1588. This is mentioned by the British Library in London, and it is also a logical year, considering the political-military developments in Vlissingen during this period.[13] When Willem van Oranje was murdered in July 1584, the independent regions of the Netherlands – mainly the northern provinces, because the south except for Antwerp was already in the hands of the Spaniards – were in a miserable state. The prince had searched in vain for support in France and England, and the already sixteen-year-old Revolt was threatening to end in failure. Maurits, Willem's son and successor, who was sixteen at the time, could not change much in the first years of his reign. Normally, Maurits' older brother Philip Willem would have succeeded his father as leader, but he had been kidnapped to Spain shortly after the Revolt broke out in 1566 to be raised further in the Catholic tradition. The power of the Oranges was limited and not comparable to that of the kings and emperors in most other European countries. In the Netherlands, initially the Council of State and later the States General held the highest power.[14]

Within this latter body, Johan van Oldenbarnevelt, the powerful 'raadspensionaris' – the principal public official – was most influential. This confidant of Willem van Oranje was a proponent of Maurits becoming the new stadhouder of Holland and Zeeland, which happened in 1585. That year, the position of the rebellious Netherlands was further undermined because on August 27, the southern stronghold, Antwerp, was captured by the Spaniards. Nothing seemed to stand in the way of further fragmentation and the eventual downfall of the Revolt, not even the efforts of the Oranges, the States General, and the national lawyer. That it would not come to this was due to a combination of events and developments that mostly took place outside the Netherlands but ensured that the Revolt ultimately became a success. In 1588, the Republic of the Seven United Netherlands could be proclaimed, and in the more than half a century that followed unprecedented prosperity was built up.[15]

One of the key elements in this turnaround in the years 1585–1588 was the role of English Queen Elizabeth I. Her attitude towards the Spaniards had always been ambivalent: she had little desire for war with them, but at the same time saw their power increasing, so she had to be constantly on her guard. Thus, in 1585 she refused to become the sovereign ruler of the Netherlands as she did not want to further harm the relationship with Spain. However, shortly thereafter she did send an army of 6000 men to stop the advancing Spaniards. In the negotiations with Prince Maurits about this aid, Elizabeth demanded temporary ownership of the two port cities of Brielle and Vlissingen, including Fort Rammekens. A Zeeland representative, Jacob Valcke, who initially refused to hand over Vlissingen as collateral, participated in these negotiations. As a result the arrival of the English troops was delayed.[16]

The Zeelanders eventually agreed because England convinced them that the Netherlands would inevitably be overrun by the Spaniards without outside help. Since support from France was no longer an option after the failed French attempt in 1583 to subdue Antwerp, only the English remained. However, they had their own agenda, in which Vlissingen played a significant role due to its strategic location and easily accessible ports. The city map from 1588 was drawn based on the work of the military engineer and cartographer Robert Adams. The client was the Queen's Secretary of State, Sir Francis Walsingham. In 1588, Adams was also involved in mapping the course of the Spanish Armada, the fleet with which the Spanish king Philip II – as it turned out, in vain – tried to invade England in the summer of that year.

13 Plan of the town of Flushing, drawn by Robert Adams, 1588. Location: British Library, Shelfmark: Cotton MS Augustus I ii 105.

14 This paragraph on political and military events is based on Tjaden, 1979, Vol. VI, pp. 244–258 and Groen, 2013, p. 79–86, 89–101.

15 Ibid.

16 Van Druenen, 2015, p. 338–349.

2.1.3 (ad 3) Selection of Built Street Sides

For the purpose of this study, a search was conducted in the center of Vlissingen for a coherent group of built-up street sides that had not changed between 1548 (map by Van Deventer) and 1588 (map by Adams) and still exist in the street pattern of 2024, so that the current length in meters can be determined based on modern city maps. The selected built street sides were located in the old center of Vlissingen, around the then Onze-Lieve-Vrouwekerk, the present-day Sint-Jacobskerk. This part of Vlissingen also had a considerable variety of houses: both the parts along or near the harbors and the parts in the narrower built street sides behind them. The spatial structure dates from the end of the thirteenth century, when the then Count of Holland and Zeeland Floris V conceived the plan to build a well-defensible city directly on the open sea outside the village of Vlissinghe according to the principle of the bastide city. This type of city was built according to a fixed plan: straight streets, running perpendicular to a market square and surrounded by defensive walls. In the case of Vlissingen, the streets were not situated around the market, but around the harbor, which ran more than half a kilometer into the city. This street plan has largely remained the same.[17] The city had grown considerably in the third quarter of the sixteenth century as a result of the uprising of April 1572 and the rewards for it from Willem van Oranje in the form of privileges and a new harbor: the Pottekaai.

However, the expansion of the city had mainly taken place in the eastern direction, beyond the old city walls, on previously unbuilt lands. This part, where the stadhouder also had his Vlissingen palace built, was called the New City.[18] In these years, there was certainly construction in the old center, but it mainly involved replacement construction of existing buildings. It is possible that some densification took place through the building of yards and the splitting of houses. However, it is questionable whether there was a substantial need for this, as a completely new neighborhood was being built just a stone's throw away.

The 22 selected built street sides are included in Figure 2.2, the counts in Table 2.3.[19] On the map of Vlissingen by Van Deventer, a scale bar is depicted with the unit Passus quinque pedum: One pass is five feet; 100 passes are then equal to 500 feet, or 157 meters. The scale bar of Van Deventer is equal to 515 pixels, 500 passes, 2500 feet, and 775 meters on the used image of the map of Vlissingen. It turns out that the foot used by Van Deventer is the Rijnland foot: 0.31 meters.[20] The ratio – the Passus

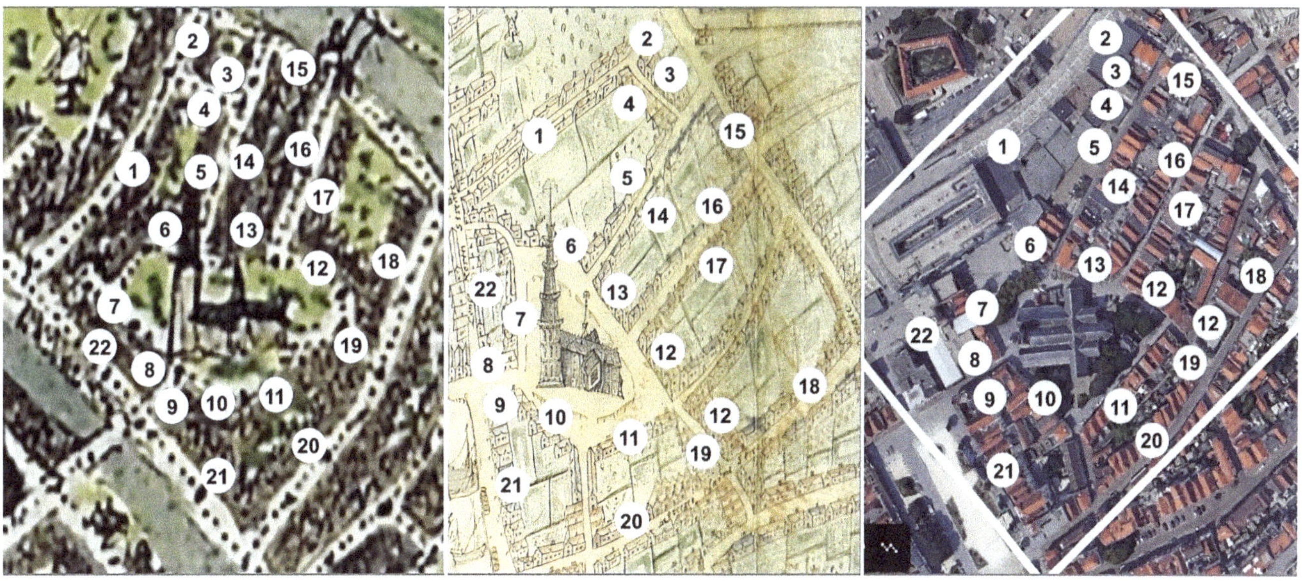

FIGURE 2.2 The three map sections of Vlissingen by Jacob van Deventer, 1548, Robert Adams, 1588, and Google Maps, 2020 with the 22 selected built street sides

17 Van Druenen, 2015, p. 43–55. See also the map of Adams on which the length of the newly dug, northwest-facing Harbor is mentioned: 'The New Towne'.

18 Ibid., pp. 195–200, 302–307, 310–326.

19 Measurements were taken on March 22, 2024, using the map of Vlissingen in Google Maps, 'Measure distance and area' function.

20 Although the Rijnland foot was in principle a fixed unit of measurement, its application in the period that Van Deventer was

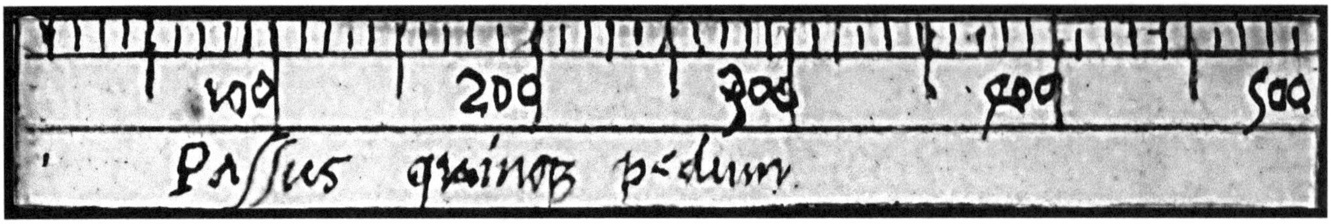

FIGURE 2.3 The scale bar used by Van Deventer on almost all his city maps with the unit 'Passus quinque pedum': One pass is five feet. This figure is from the map of Vlissingen, 1648

TABLE 2.3 Comparison between the maps of Vlissingen by Jacob van Deventer (1548), Robert Adams (1588), and Google Maps (2020). The marked number 0.1756 is the coefficient used to calculate the number of houses on the other city maps of Van Deventer. The marked number 8698 is the average and weighted scale of the total of the 22 built street sides. The marked number 0.31 is the calculated ratio between meters and feet[a]

		1548				1588	2020	
		Van Deventer, map Vlissingen				Adams	Google Maps	Scale: cm (1548) / meter (2020)
		Pixels	Centimeters	Houses / pixels	Feet / meter	Houses	Meters	
Scale stick Van Deventer >		515	8.69		0.31		775	8918
1	Vrouwestraat	94	1.6	0.09	0.33	8	149	
2	Vrouwestraat	23	0.39	0.04	0.28	1	31	
3	Vrouwestraat	27	0.46	0.22	0.26	6	34	
4	Vrouwestraat	27	0.46	0.3	0.25	8	33	
5	Sint Jacobsstraat	53	0.9	0.19	0.5	10	129	
6	Oude Mark	16	0.27	0.5	0.39	8	30	
7	Oude Markt	45	0.77	0.2	0.24	9	53	
8	Kerkstraat	19	0.32	0.16	0.39	3	36	
9	Kerkstraat	21	0.36	0.1	0.37	2	37	
10	Achter de Kerk	31	0.53	0.19	0.36	6	54	
11	Branderijstraat	52	0.88	0.12	0.3	6	75	
12	Lepelstraat	40	0.68	0.13	0.18	5	34	
13	Oude Markt	27	0.46	0.19	0.27	5	35	
14	Sint Jacobsstraat	80	1.36	0.16	0.32	13	126	
15	Walstraat	28	0.48	0.29	0.28	8	37	
16	Weteringstraat	64	1.09	0.17	0.36	11	113	
17	Weteringstraat	57	0.97	0.25	0.39	14	109	
18	Groenewoud	63	1.07	0.17	0.27	11	82	
19	Lepelstraat	21	0.36	0.33	0.67	7	68	
20	Groenewoud	84	1.43	0.19	0.32	16	132	
21	Bellamypark	63	1.07	0.16	0.31	10	95	
22	Bellamypark	67	1.14	0.13	0.29	9	95	
Totals		1002	17.03	0.1756	0.31	176	1588	8698
		(total)		(average)		(total)		(weighted average)

a Procedure and the underlying tables and calculations: Van Druenen, 2022, Appendix 2.1.
NB1: Only the street sides that show buildings on Adams' map have been measured on Van Deventer's map.
NB2: The table only shows the continuous built-up street sides; therefore, a street may be mentioned several times.

FIGURE 2.4 Comparison in MapAnalyst of the center of Vlissingen on the city map by Jacob van Deventer, 1548 and on the modern map of the city

FIGURE 2.5 Comparison in MapAnalyst of the center of Vlissingen on the city map by John Adams, 1588 and on the modern map of the city

coefficient – between the number of houses and the number of pixels has also been determined, which allows the total number of meters to be converted to the total number of houses: 0.1756. The number of houses is that on Adams' map, projected on Van Deventer's map: all strips are built-up street sides that already existed in 1548 and 1588, often under different names, and still exist today. In the counts presented later in this chapter, the factor 0.1756 is used with all four decimals. Rounding and calculating the margin of error only take place when presenting the results.

Finally, based on the totals, the scale ratio of this part of the map of Vlissingen was also calculated: 1 centimeter on the map is 8698 centimeters in reality.

MapAnalyst was used to determine the planimetric accuracy of the area with the 22 selected built street sides on the city map of Van Deventer. The area on this map is delimited with the same 10 points as those on the modern city map.

Figures 2.4 and 2.5 show the deformation grid on both maps. On the modern map, there are no deformations, while on the map from 1548, there are slight deformations. The average deviation of all ten measuring points in 1548 compared to 2024 is 11 meters, averaging 2.7 percent. This is on an area of 370 x 448 meters. Given the time margin of 470 years, this is a very small percentage. The conclusion is that the planimetric accuracy of Van Deventer's city map, particularly in the city center, is considerable. This conclusion is supported by the scale ratio calculated

active may not have been the same everywhere and among all professional groups (cartographers, architects, land surveyors). See also: Meertens Institute, *The old Dutch measures and weights* (https://www.meertens.knaw.nl/mgw/maat/10, accessed on August 20, 2023). Rutte and Vannieuwenhuyze (City Atlas, 27) estimate Van Deventer's foot at 30 centimeters.

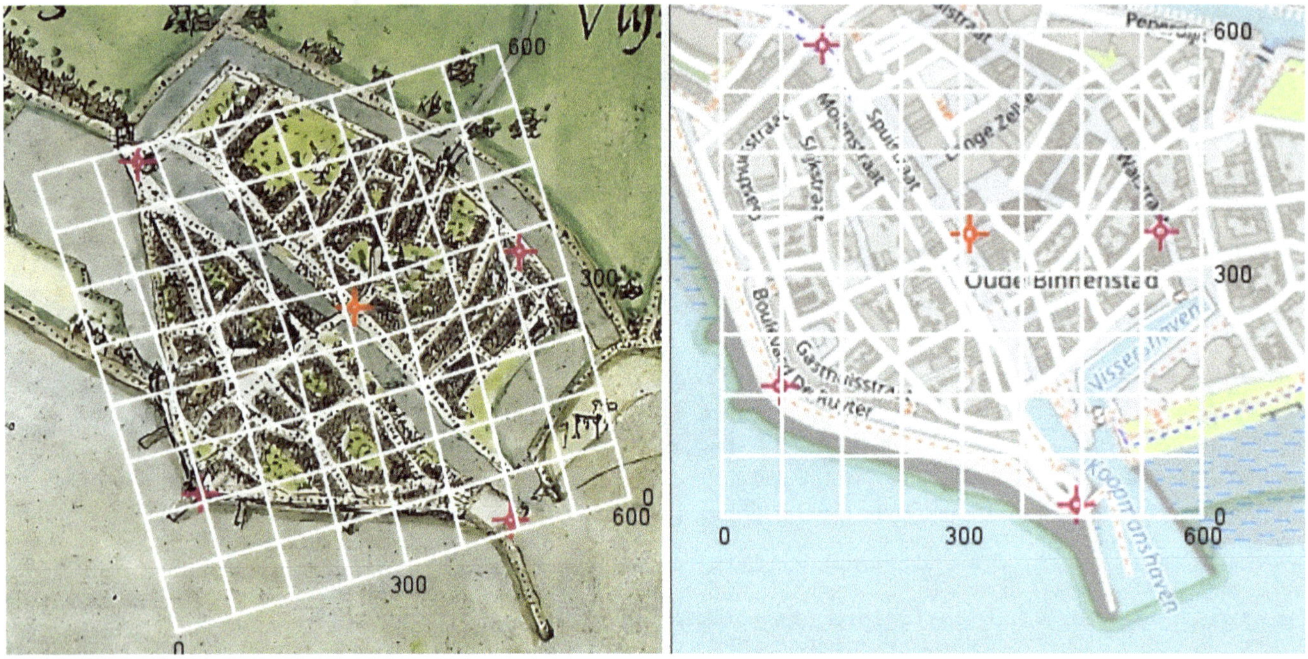

FIGURE 2.6 Comparison in MapAnalyst of reference points of the bird's-eye view measurement in Vlissingen on the city map by Jacob van Deventer, 1548 and on the modern map of the city

with MapAnalyst in the selected city section. The deviation is 4 percent: 1:8334 against 1:8698 as the scale result of measuring the 22 built street sides or parts thereof. This indicates that the Passus coefficient of 0.1756 is sufficient accurate and therefore useful. The comparison between Adams's city map from 1588 and the modern map gives slightly broader margins: The average deviation of all ten measuring points in 1548 compared to 2024 is 17 meters, averaging 4.1 percent.

2.1.4 (ad 4) Planimetric and Topographic Accuracy

The previous exercise shows that the distances on Van Deventer's map of Vlissingen are almost the same as on the modern MapAnalyst map. This indicates the high degree of accuracy of the 1548 work, which is all the more remarkable given the – by current standards – 'simple' tools Van Deventer had to work with. According to Rutte and Vannieuwenhuyze, the maps were far ahead of their time and excelled in map image, mapping style, orientation, planimetric accuracy, and scale.[21] The criticism formulated in recent years mainly concerns inaccuracies in the representation of backyards, streets, and water and land roads. No contemporary documentation of Van Deventer's measurement and mapping method has survived.

As mentioned in section 1.3, he probably used triangulation and/or polygonal measurements to measure the urban area and its immediate surroundings. The distances in the sixteenth century could be calculated by simply walking them and then counting the steps, the 'passus'. For this, pedometers and measuring chains were already in use. On Van Deventer's maps, this was reflected in the black-dotted lines that ran through every street and over every road.[22] The distances that could be walked were drawn on the maps with a high degree of accuracy. This is not surprising because, if done correctly, there was almost nothing that needed to be estimated. When measuring distances between two points where there was no street in between, but for example a block of houses, the flat triangulation could provide a solution when two sides were walkable. However, this had to be done in a straight line because the slightest curvature could already have a substantial influence on the number of degrees of the angle.

Whether the bird's-eye distances between certain points that Van Deventer could not walk because he would have had to make his way through or over the houses and buildings were also in accordance with reality can be easily determined by measuring them with the help of MapAnalyst. This time the possibilities are somewhat broader because a number of reference points outside the old center also existed in 1548: the Gevangentoren

21 Rutte and Vannieuwenhuyze, 2018, p. 31.

22 Ibid., p. 32. See also footnote 86.

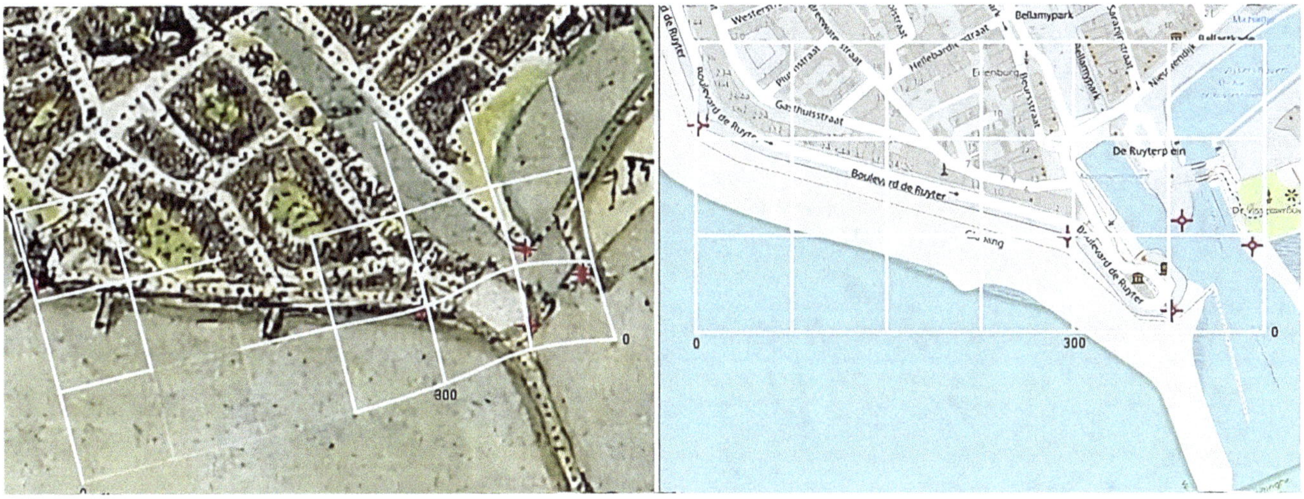

FIGURE 2.7 Comparison in MapAnalyst of the still existing defensive wall along the southern coastline of Vlissingen combined with the harbor heads on the city map by Jacob van Deventer, 1548 and on the modern map of the city

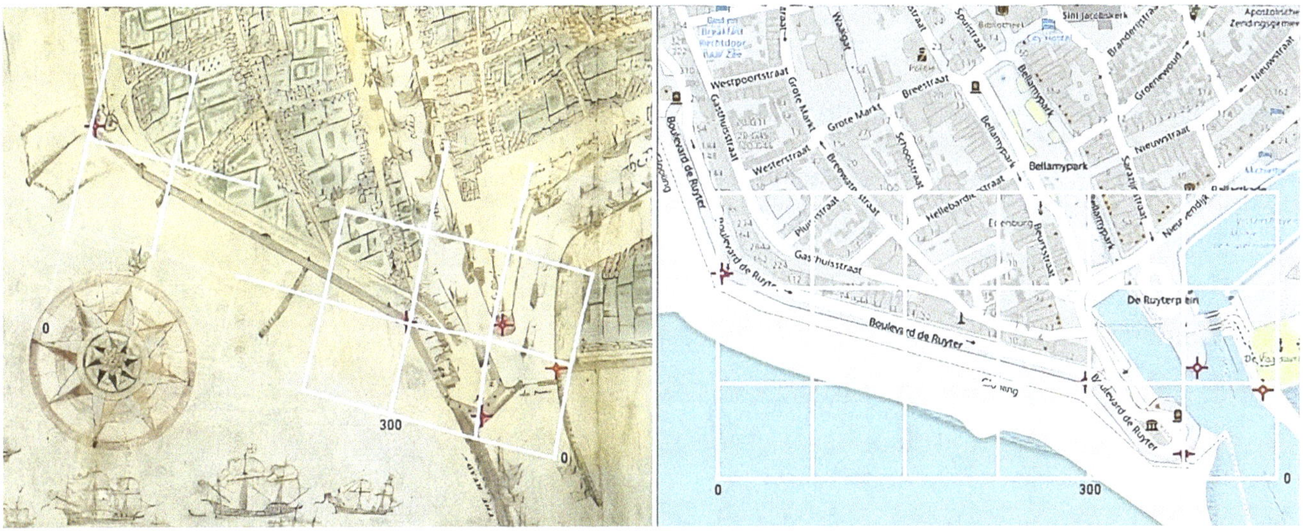

FIGURE 2.8 Comparison in MapAnalyst of the still existing defensive wall along the southern coastline of Vlissingen combined with the harbor heads on the city map by John Adams, 1588 and on the modern map of the city

from 1489 and the previously discussed Keizersbolwerk from 1548.[23] The control with the help of MapAnalyst yields an average deviation of 14 meters for seven reference points in 1548 compared to 2024. This result can be seen as a nuance of the consensus about the high quality of Van Deventer's maps.

Determining the topographic accuracy of Van Deventer's maps is hardly possible for the aforementioned reasons: the houses and buildings are not or only partially recognizable. However, as an alternative, the planimetric accuracy of a number of large architectural and infrastructural works, such as the city walls and the harbors that still exist in their original form today and have a considerable length and/or area, can be determined.[24] For comparing this with the current situation, the quality of the information on the map can also – albeit to a limited extent – be analyzed using MapAnalyst.

In the research city of Vlissingen, this applies to (1) the still existing defense wall along the southern coastline of Vlissingen in combination with the harbor heads with a total length of 799 meters and (2) the structure of the still existing inner city harbors with a total length of 863 meters. The construction of these infrastructural

23 The Gevangentoren received that name only in 1610. Before that time, the building, along with another tower, was part of the West Gate of the city. See among others: Van Druenen, 2015, p. 163.

24 For the concept of 'topographic accuracy': Blakemore, 1980, p. 68.

FIGURE 2.9 Comparison in MapAnalyst of the structure of the still existing inner city harbors of Vlissingen on the city map by Jacob van Deventer, 1548 and on the modern map of the city

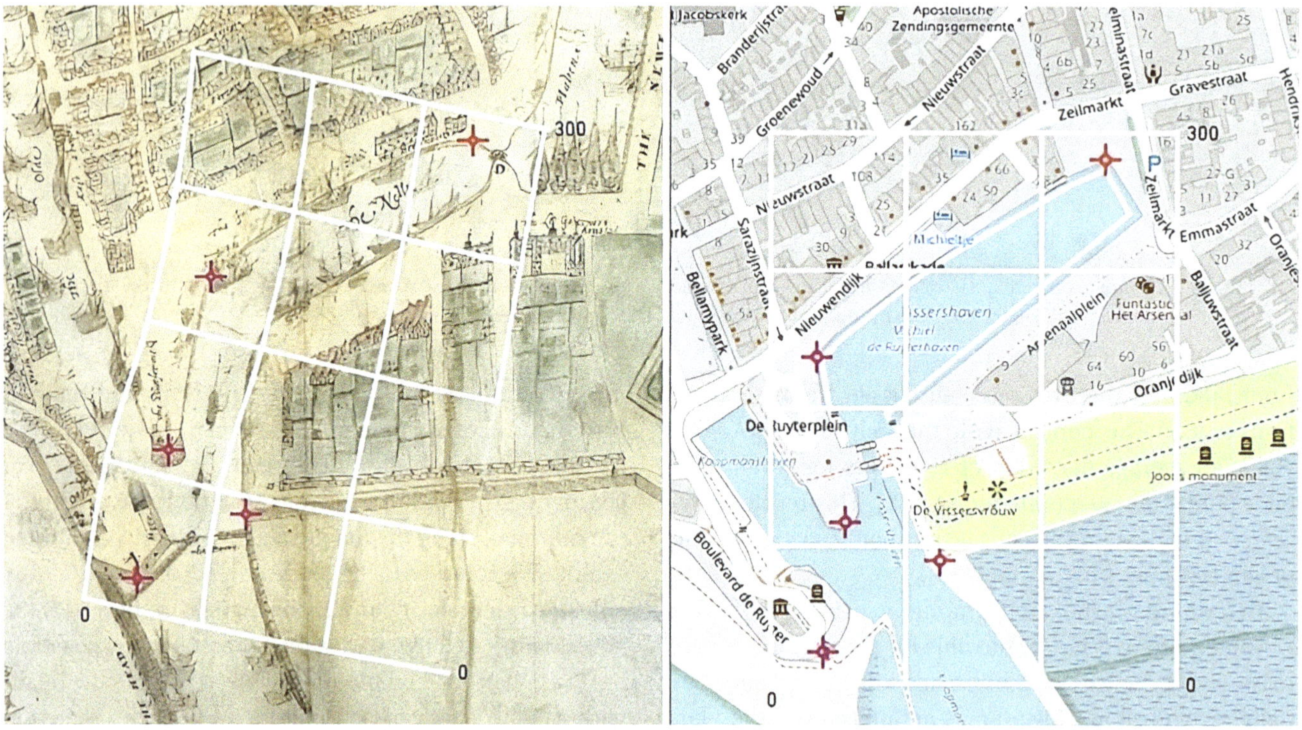

FIGURE 2.10 Comparison in MapAnalyst of the structure of the still existing inner city harbors of Vlissingen on the city map by John Adams, 1588 and on the modern map of the city

TABLE 2.4 Comparison of the number of houses in Vlissingen according to the Passus method, Vitelli, Ramaer, and the house count on Adams' map, including the conversion to population numbers, calculated with a reduction factor of 4 and of 7

Built-up streetsides		**Houses**				**Inhabitants** (reduction factors 4 / 7)				**Deviations houses and inhabitants** (reduction factor 5,5)		
pixels	meter	Passus method 1548	Vitelli 1569	Ramaer 1548	Adams 1588	Passus method 1548	Vitelli 1569	Ramaer 1548	Adams 1588	Passus method > Vitelli	Passus method > Ramaer	Passus method > Adams
4245	6495	745	800	730	830	2980 / 5215	3200 / 5600	2920 / 5110	3320 / 5810			
								Percentages		–7%	2%	–11%

elements that were primarily used for military purposes took place in the period 1300–1584.

Since then, only cosmetic adjustments have been made, while the structure has remained intact. Thus, the dimensions remain largely unchanged. The Figures 2.7, 2.8, 2.9, and 2.10 show on the left the maps of Van Deventer and Adams and on the right the current situation. The accuracy analysis with the help of MapAnalyst shows that the defensive wall and the harbor heads have an average deviation of 13 meters with Van Deventer and 7 meters with Adams. For the harbor structure, this is 9 and 14 meters, respectively. In percentages, these deviations from the total lengths are respectively 1.63 and 1.04 percent and 0.88 and 1.62 percent, which establishes that the accuracy of the measured and analyzed objects on the city maps of Vlissingen is very high.

2.1.5 (Section 5) Passus Method Vlissingen

Table 2.4 shows the total number of built street sides, houses, and inhabitants in Vlissingen in 1548 according to the Passus method, using the coefficient of 0.1756 presented in Table 2.3 for counting the houses in the whole city. These totals are compared with those of the Vitelli, Ramaer, and Adams commission. It appears that the Passus method yields 7 percent fewer inhabitants than the Vitelli counting. In comparison with Ramaer, it is 2 percent more. The house count on Adams's map yields 11 percent less. An important condition for drawing conclusions from this comparison is that the methods must differ in several aspects. When the outcomes, as is the case here, largely correspond, this says something about the usability of the procedure. The more different methods lead to the same outcomes, the more plausible it becomes that the outcomes conform to reality.

Ramaer used the areas of the built-up city parts as a starting point, where the new method is based on the lengths of the built street sides, which can partly explain the difference in outcomes between Ramaer and the Passus method. Secondly, Ramaer arrived at his coefficient by comparing a city map of Amsterdam from 1871 with the map of Van Deventer from 1560. His 'ordinary' source was the census of 1869. In the new method, two maps and an alternative source – the report of Vitelli – from the same half century were chosen: specifically from 1548, 1588, and 1569, respectively.

The difference between the Passus method and that of Vitelli can partly be explained by the fact that Vitelli also counted houses that were outside the city walls, an area that also included the village of Old Vlissingen.[25] He likely did so not for military-strategic reasons, but with an eye to the possibilities for billeting. Every little bit helped, and the distances of the houses to the city center were minimal. A count in the village and the outside area on the 1548 map itself – unlike the city, the houses outside are drawn more clearly and distinguishable from each other – yields an additional number of 80 houses. If Vitelli had counted these, the difference with the Passus method, in which only the houses within the city walls are included, would be 14 percent. Ramaer also gave no indication of the reach of his calculations. Given the large difference

25 The designation 'Out Vlissinge' comes from the map of Van Deventer of Vlissingen from 1548 (Vlissingen, Biblioteca Nacional de España. Res/200, no. 1). The village, which had already arisen in the twelfth century, was separate from the city of Vlissingen, which was built in a short time around 1300 by the Counts of Holland and received city rights in 1315 from Count Willem III. From that time, the original village was referred to as 'Old Vlissingen' with various spellings. See also: Van Druenen, 2015, pp. 12–191.

that already exists between the outcomes of Ramaer and Vitelli, the former probably only counted the area within the city walls. In the comparison between the Passus method and the count on the map of Adams, this problem does not play a role: only the areas within the city walls were counted. Additionally, the street sides built in 1588 but not in 1548 were left out of the comparison.

The deviation percentages of −7, 2, and −11 percent can be classified as low. This is a first confirmation of the usability of the Passus method.

2.1.6 (ad 6) Passus Method: First Test Group of Four Cities

These results also support the preliminary conclusion that the city maps of Jacob van Deventer are highly accurate, especially when it comes to measuring street lengths. Furthermore, the results of the Vlissingen case study indicate that the new method of counting houses also provides a picture that approximates the reality of the years 1548–1554. The sixth step in the procedure is to calculate, based on the key figures of Vlissingen, the number of houses and inhabitants of the first four test cities around 1548, whose houses – in addition to those of Vlissingen – were counted by Chiappino Vitelli in 1569.

This is done by measuring all built-up street sides per city and calculating, using the Passus coefficient and a reduction factor bandwidth of 4–7, the respective number of houses and inhabitants.

The outcomes can be compared with those of the Vitelli commission for, in addition to Vlissingen, the four test cities: Goes, Middelburg, Veere, and Zierikzee. Before conducting the second calculation (measuring all built-up street sides), it is necessary to compare the scales and resolutions used by Van Deventer on the five city maps and adjust them accordingly, if necessary, as previously described in the sub-procedures under ad.1. The scales and their deviations are listed in Table 2.5. It should be emphasized that this is not about the conventional scale that indicates the ratio between the number of centimeters on the map and the number of meters in reality. Previously, it was found that the map of Vlissingen had a scale of 1:8698. However, the scale used in Table 2.5 indicates the ratio between the number of pixels on the digital representation of the map and the number of meters in reality in 2024. It turns out that the differences here are extremely small: between 1 and 3 percent. The exception is Veere, which was depicted on one folio with Westkapelle and was given a 29 percent smaller scale.[26] The combination map of Veere and Westkapelle is drawn comparatively smaller than most other maps. This can be clearly seen in chapter 3.2, Table 3.230: the original downloaded map has the shortest scale stick of all 226 maps: 725 pixels.

The scale adjustments are accounted for in Table 2.6. It appears that the resolutions of all used city maps, as they are digitally available from the Biblioteca Digital Hispánica, are exactly the same and therefore do not need to be converted. However, the length and width of the maps do need to be adjusted, albeit slightly.

The results of the house counts in the five cities are presented in Table 2.7. Table 2.8 compares these results with the other available sources: Ramaer (inhabitants) and Vitelli (houses).[27] Alternative sources have been added for Middelburg, Veere, and Zierikzee. For these three cities, the number of houses has been estimated based on tax records from the period 1543–1576.[28] Notably, the various counts for the four test cities show significant discrepancies. This applies to the comparison between Ramaer and Vitelli, but even more so to the comparison with the Passus method. Vlissingen is the major exception in both cases. The Passus method there results in 4,100, differing by only 2 percent and 7 percent from the two earlier counts, respectively. The difference between Vitelli and Ramaer is 10 percent. Another striking observation is that the Passus method shows only minor deviations from the alternative sources for Middelburg, Veere, and Zierikzee, on average 12 percent. In contrast, the Passus method deviates by 32 percent and 27 percent from the figures of Ramaer and Vitelli for all cities except Vlissingen. When comparing the Ramaer and Vitelli figures for Middelburg, Veere, and Zierikzee with the alternative sources, the averaging percentages are 47 and 24.

If one were to look solely at the numbers from Ramaer and Vitelli in comparison to the Passus method, the initial conclusion might be that this sixth step in the procedure does not support the reliability of the Passus method. However, there is also no weakening of the argument since the lack of consensus was already known (see columns I, J, and K) and it was not expected that the Passus method would validate or reinforce the others. When the alternative sources for Middelburg, Veere, and Zierikzee are also considered in the research findings, this initial conclusion can be significantly refined: the differences between these sources and the Passus method averaging 12 percent: Middelburg 18, Veere 9 en Zierikzee 7. The

26 Biblioteca Nacional de España. Res/200. Map numbers of Vlissingen and the four test cities: Goes, Res/200, no. 7, Middelburg, Res/200, no. 4, Veere, Res/200, no. 3, Vlissingen, Res/200, no. 1, Zierikzee, Res/200, no. 5. The measurements were carried out on the digital copies of these folios.

27 Ramaer, 1921, p. 36.

28 Unger, 1956, p. 12; Unger,1966, p. 45; Blom, 1993, p. 8.

TABLE 2.5 Scale differences between the five test cities on the city maps of Jacob van Deventer (resolution and format adjusted to 50% of the original download)

	Scale indication on the map in pixels	Passus	Feet	Factor pixel / feet	Factor feet / meter	Factor pixel / meter	Deviation vs. Vlissingen
Goes	499	500	2500	0.20	0.314	0.6357	−2.70%
Middelburg	504	500	2500	0.20	0.314	0.6420	−1.80%
Veere	363	500	2500	0.15	0.314	0.4624	−29.20%
Vlissingen	513	500	2500	0.21	0.314	0.6535	0.00%
Zierikzee	503	500	2500	0.20	0.314	0.6408	−1.90%

TABLE 2.6 Overview of the scale adjustments of Jacob van Deventer's maps of the five test cities

	Deviation vs. Vlissingen	Length in pixels	Length in pixels modified	Resolution (pixels per inch) original and modified situation
Goes	−2.70%	3206	3293	150 and 150
Middelburg	−1.80%	1690	1720	150 and 150
Veere	−29.20%	2317	3274	150 and 150
Vlissingen	0.00%	3329	3329	150 and 150
Zierikzee	−1.90%	3238	3301	150 and 150

TABLE 2.7 House and inhabitant counts in the five test cities based on the maps of Jacob van Deventer from 1548 and the reduction factors 4 and 7.

	Passus method 1548		
Stad	Built-up street-sides in pixels	Houses	Inhabitants (reduction factors 4 and 7)
Goes	3188	560	2239/3919
Middelburg	7328	1287	5147/9008
Veere	4229	743	2972/5201
Vlissingen	4245	745	2982/5218
Zierikzee	7962	1398	5593/9787

findings in this sixth step of the procedure thus still support the usefulness of the method, which will be further tested in Step 7 using the Van Deventer maps of two major cities: Ghent and Leiden.

2.1.7 (ad 7) Ghent and Leiden: Control

Unfortunately, very few cities meet the three conditions based on which the Passus coefficient could be determined in Vlissingen: (1) the presence of a city map by Jacob van Deventer, (2) an alternative source that can provide insight into the number of houses and/or inhabitants of the respective city, and (3) a second city map that meets three conditions: (3a) the houses and buildings are recognizably indicated, (3b) the dating falls within the same period as the map of Van Deventer, with a margin of a maximum of 40 years, and (3c) the urban areas to be examined have hardly changed in terms of buildings during this period. Ghent meets the first two conditions, Leiden all three.

The map of Ghent, like that of Leiden, is dated between 1558/1560 and 1564.[29] It was the largest city of the more than 220 that Jacob van Deventer drew. The city map did not fit on the page size used by Van Deventer, which is why fold-out sheets had to be used to draw the entire city. Only the atlas map, which due to its large size was folded up in the book, has been preserved. The result was that the map has quite a few distortions, caused by folds and creases, which will have had some impact on measuring

29 Biblioteca Nacional de España. Res/207, nr 40. Rutte and Vannieuwenhuyze, 2018, p. 200–203.

TABLE 2.8 Three calculations of population numbers, compared with the outcomes of the Passus method, 1543–1569

	Inhabitants 1548–1569									
A	B	C	D	E	F	G	H	I	K	J
	Ramaer 1548	Vitelli 1569	Passus method (RF 5,5) 1548	Alternative figures	Deviation Passus method – Ramaer	Deviation Passus method – Vitelli	Deviation Passus method – Alternative figures	Deviation Ramaer – Vitelli	Deviation Ramaer – Alternative figures	Deviation Vitelli – Alternative figures
Goes	2600	1925	3079	–	16%	37%	–	26%	–	–
Middelburg	5500	4400	7077	5781	22%	38%	18%	20%	5%	31%
Veere	2100	2750	4084	3702	49%	33%	9%	31%	76%	35%
Vlissingen	4000	4400	4100	–	2%	7%	–	10%	–	–
Zierikzee	4500	7700	7690	7156	41%	0%	7%	71%	59%	7%
					32%	27%	12%	32%	47%	24%

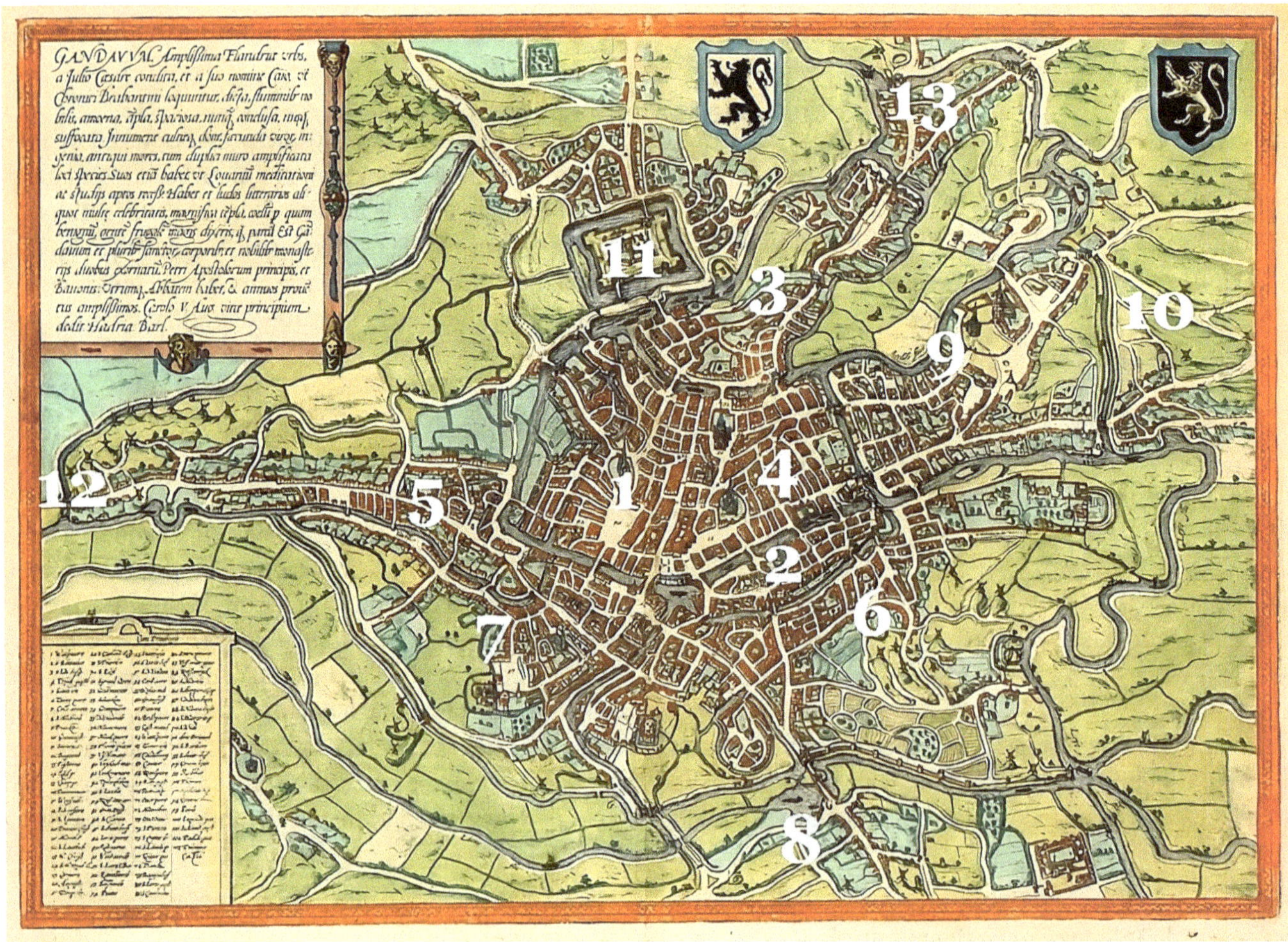

FIGURE 2.11 The thirteen tax quarters in Ghent, 1571–72, filled in on the map of George Braun and Frans Hogenberg, 1572
Note: Dambruyne, 2001, pp. 32–37, 369–376.

FIGURE 2.12 The street side measurement on the map of Jacob van Deventer, 1558–1564, in the first tax quarter (no. 1 in Figure 2.11) of Ghent, including the Minnemeersbrug, the Vleeshuisbrug, and the Sint-Jacobskerk

the streets in pixels.[30] An alternative source is also available for Ghent: the ledgers of the twentieth penny, a tax levied on the annual income from real estate, for the years 1571 and 1572. Ghent historian Johan Dambruyne analyzed the material in 2001 and concluded that it is a very accurate and complete representation of the taxpayers and houses per tax district in Ghent.[31] He found no indications of fraud and could even determine that almost all social classes, except for the homeless, are reflected in the data. The few gaps, which were there in three of the thirteen tax quarters, could be filled with the help of the ledgers of 1569 and 1577.

In Leiden, Van Deventer conducted his research for the city map before 1564.[32] Three alternative sources are available for this city: (a) the population count of 1574, (b) the

30 Rutte and Vannieuwenhuyze, 2018, p. 198, as well as a written communication about this from Vannieuwenhuyze dated December 22, 2020.

31 Dambruyne, 2001, pp. 32–37, 369–376.

32 Biblioteca Nacional de España. Res/200, nr 30; Rutte and Vannieuwenhuyze, 2018, p. 327.

TABLE 2.9 Comparison between the map of Ghent by Jacob van Deventer from ca. 1560 (pixels and meters) and the counts in the ledgers of the twentieth penny from 1571–1572 (houses and inhabitants) for Ghent (Biblioteca Nacional de España. Res/207, nr 42. Measurements were only made within the city walls and in the counted quarters). NB: in the totals in the last column, the negative numbers are made positive

Gent	Street		Houses		Inhabitants	
	Pixels	Meters	Passus method, factor 0,1756	Tax ledgers 1571–1572	Passus-method ca. 1560 (reduction factor 4 en 7)	Deviations taks ledgers > Passus method (reduction factor 5,5)
First quarter	7919	12,750	1391	1331	5564/9737	4%
Second quarter	2649	4265	465	431	1860/3255	7%
Third quarter	5007	8061	879	939	3516/6153	7%
Fourth quarter	6364	10,246	1118	1027	4472/7826	8%
Fifth quarter	4014	6463	705	666	2820/4935	6%
Sixth quarter	9221	14,846	1619	1434	6476/11,333	11%
Seventh quarter	4454	7171	782	831	3128/5474	6%
Ninth quarter	6479	10,431	1138	1052	4552/7966	8%
Thirteenth quarter	1247	2008	219	233	876/1533	6%
Gent (counted quarters)	47,354	76,240	8315	7944	33,260/58,205	7%

population count from 1581, and (c) the Streets and Canals Books of Leiden from the period 1583–1587.

a. The Leiden census of 1574 was one of the first in the Netherlands and was conducted in August 1574, during the siege of the city by the troops of Alva. The municipal government wanted to gain insight into the size of the population in order to oversee the food shortage during the siege of that year.[33] The scope was manageable because it only inventoried within the city walls, allowing for the same range to be maintained in the Passus method. It should be noted, however, that there are no data on the number of people from outside who stayed in the city, had fled Leiden, or had died as a result of the war situation that year.

   It is quite possible that these two unknown numbers cancel each other out. Unger, in a 1915 article, arrived at a total of 12,660 inhabitants after a 'precise recalculation.' This included the entire 'ortsanwesende' population: including the country people who had fled to the city and a few soldiers. The original number of inhabitants recorded in the census was 12,391.[34] In 2018, an attempt was made to derive the numbers from digital scans of the original sources. The scans were converted to a word processing program using optical character recognition (OCR) technology, and eventually to Excel and a database. The end result of this exercise was 12,453 inhabitants.[35] Because the techniques used, especially OCR, were and still are far from reliable, Table 2.10 with all Leiden results uses Unger's directly noted and 'recalculated' number of 12,660.[36]

b. The 1581 census in Leiden was organized by the city secretary Jan van Hout.[37] It is uncertain why a census was held. No reason is given in the sources. A research group from KNAW (Koninklijke Nederlandse Akademie van Wetenschappen), NWO (Nederlandse Organisatie voor Wetenschappelijk Onderzoek) and (DANS) (Data Archiving and Networked Services), which transcribed the sources again in 2018 and was able to digitally map the number of houses and inhabitants, concluded that the reason can be traced back to the person of Jan van Hout – a 'civil servant par excellence' – and that the census was part of the pursuit of a new city administration. The 2018 researchers also formulated a source criticism that was mainly focused on

33 Van Nierop, 1919, p. 193.

34 Unger, 1915, pp. 86–92.

35 DANS Dataset Census Leiden, 1574, Martin Hooymans, version 1.0, August 2018.

36 See among others, the very extensive literature study by Memon, 2020.

37 Hooymans, 2018, p. 1.

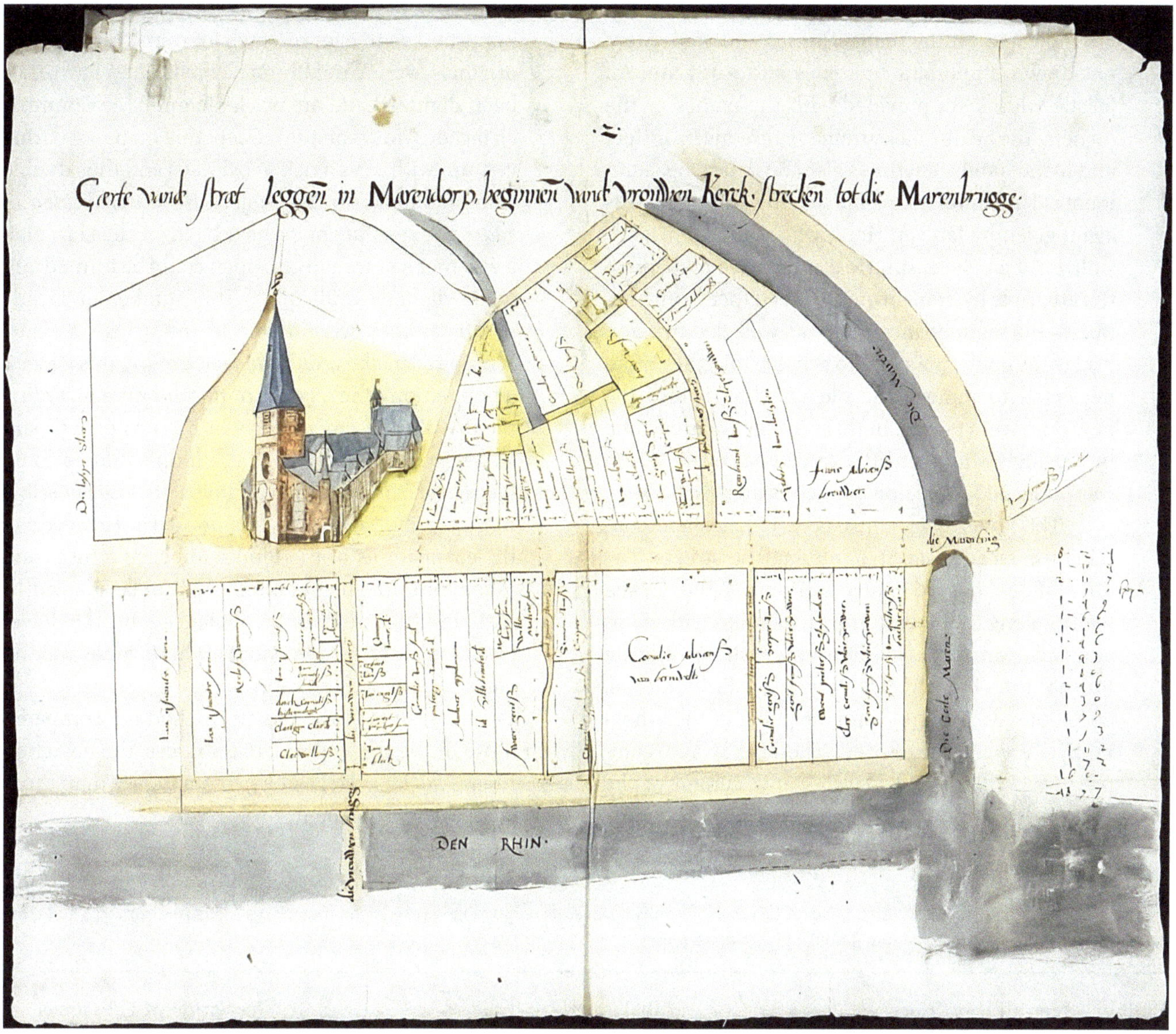

FIGURE 2.13 One of the maps from the street book of Leiden, 1583–1587: 'Map of the street layout in Marendorp starting from the Church, stretching to the Maren Bridge'
Note: Heritage Leiden and surroundings, SA II inv.nr. 5153, Straetbouc, map 28.

counting the house addresses.[38] Criteria were formulated regarding which addresses had to be excluded from the count, but at the same time, there were also indications that there were uncounted houses where families lived.

The census was thoroughly tackled. Leiden was divided into 17 administrative units, called 'bonnen,' each assigned two members of the forty-member council, totaling 33 because the Gansoord 'bon' only got one. This council consisted of the 40 richest citizens of the city who, among other things, were allowed to make nominations for the positions of mayor and aldermen.[39] This last group, including the magistrate, was also part of the census team. The final mapped census resulted in 12,324 inhabitants living at 2339 addresses. The calculated reduction factor is 5.23.[40]

c. The Street and Canal Books of Leiden were compiled in the last quarter of the sixteenth century by the land surveyors Simon van Merwen and Salomon van Dulmanhorst and contain 128 separate maps.[41]

38 Ibid., p. 11.

39 Van der Vlist, 2001, p. 88.

40 Hooymans, 2018, p. 9.

41 Heritage Leiden and surroundings (formerly: GA Leiden), SA II inv.nr. 5153, Straetbouc. Maps of all streets, with mention of the street and parcel widths and the owners of the parcels.

The special thing is that all facade widths in Leiden are indicated on the maps. Houses and yards were not drawn, important buildings and gates are. All facade widths are provided with the names of the owners, the width measurements are given in feet and inches, and sometimes also the depth measurements. The commission for making the maps was again given by Jan van Hout, who was also responsible for the 1581 census. He wanted to put an end to the attempts by homeowners to keep their contribution to the maintenance of the seawall cladding and bridges as low as possible. To achieve this, they often neglected to remove the old cladding before the new ones were placed. In this way, the walls became increasingly wider and the canals narrower, which was not good for the flow of canal water.[42]

The Street Book consists of 32 maps and contains the facade widths bordering the streets. The Canal Book has 96 maps and shows the facade widths along the canals. In 1874, all folios with maps and texts were collected in a very exclusive edition for that time by the Leiden publisher Brill. In total, 200 copies were printed and the plates of the lithography were then destroyed. The author and compiler was Willem Pleyte (1836–1903), curator of the Museum of Antiquities.[43]

This edition has since served as the most important source for research for over a century. The originals were carefully preserved. They have now been digitized and are available online.[44] Counting all facade widths on the 128 separate maps is a daring venture with a high risk of overlaps and thus double counts. Thanks to the project *Historisch Leiden in kaart*, the separate maps have been arranged in such a way that a new map of Leiden could be formed and made digitally available.[45] The number of facade widths as they were drawn in the 1580s was 2219. Multiplied by the reduction factor of 5.5, this yielded a population of 12,205. An important note regarding the use of the term 'facade width' – owned by a single registered owner – is that multiple houses could be present on the associated plot. It is not possible to determine how often this occurred. However, in the calculations of the total number of houses and inhabitants, the higher numbers from facade widths with multiple houses are balanced by the lower numbers from facade widths where there was no construction or habitation.

When the results of the Passus method are compared with those of the three alternative sources, the deviation percentages are an average of 3 percent higher than those in the tax registers of Ghent.

TABLE 2.10 Comparison between the map of Leiden by Jacob van Deventer, approx. 1560 (pixels and meters), the censuses from 1574 and 1581, and the Street and Canal Book of Leiden, 1583–1587

Built-up streetsides on the map of Jacob van Deventer	**Houses**			**Inhabitants**				**Deviation vs. The Passus method**		
Pixel/Meter	**Passus method**	**Census 1581**	**Street book 1583–87**	**Passus method ca. 1560**	**Census 1574**	**Census 1581**	**Street book 1583–87**	**Census 1574**	**Census 1581**	**Street book 1583–87**
14,313/23,044	2513	2224	2219	13,823	12,660	12,234	12,205	8%	11%	12%

1588–1597; SA II 5113, Grachtenboek. Maps of the canals and moats with details of the minimum required widths, the widths of the adjacent parcels and their owners. 1583–1589.

42 De Baar, 1985, p. 2.

43 Pleyte, 1874. See also: De Baar, 1985, pp. 5–18.

44 See www.erfgoedleiden.nl.

45 www.leidseregioinkaart.nl/kaarten/1583/.

The presence of two of the three alternative sources in Leiden that contain a count of the number of inhabitants (in 1581 and 1583–1587) and the number of houses, and moreover are close to each other in time, allows for the first time in this study a real existing reduction factor to be calculated: 5.71 and 5.51. This result is more than a 'bycatch' of the above comparisons: it shows that the reduction factor 4–7, with an average of 5.5, used in this study is very realistic. The comparison in Table 2.11 between the coefficient of 0.1550 for converting the lengths of the built-up street sides (Census 1581) in pixels to the number of houses (Street Book 1583–87) with the previously established Passus coefficient of 0.1756, results in a deviation of 11.73 percent.

TABLE 2.11 The reduction factor and the Passus coefficient in Leiden 1574–1587, calculated with the data from the two alternative sources from that city: the 1574 census and the Street and Canal Book of Leiden from 1583–1587

	Leiden				
Deviation	**Reduction factors**			**Passus coëfficiënt**	
	Chosen standard	Census 1574	Census 1581	Chosen standard	Street book 1583–87
	5.5	5.71	5.51	0.1756	0.1550
		4.00%	0.20%		11.73%

TABLE 2.12 The numbers of immigrants in Leiden and the numbers of inhabitants based on censuses in 1560, 1574, 1581, 1584–1587, and 1622[a]

Period	Immigrants	Houses	Inhabitants
1575–1579	193	Passus method 1560	13,823
1580–1584	328	Census 1574	12,660
1585–1589	400	Census 1581	12,234
1590–1594	580	Street book 1584–87	12,205
1595–1599	353	Census 1622	44,745
1600–1604	314		
1605–1609	424		
1610–1614	485		
1615–1619	577		

a The data of the immigrants are derived from Posthumus, 1912, p. 70.

Finally, it should be noted that, although the alternative counts in Ghent and Leiden differ by 10 to 25 years from the Passus method of around 1550, there was only limited substantial growth of the cities during that period. Thus, the censuses, including the error margins, can be compared with each other. It is likely that the influx of immigrants was largely offset by the consequences of the 1574 occupation: war deaths and victims of hunger and disease. Table 2.12 shows that the influx between 1575 and 1580 was still relatively low and only really got underway after 1585, along with the total population increase.

2.1.8 (ad 8) Validation

Five arguments, incorporating the seven preceding steps (not in the exact same order), support the validation of the Passus method.

a. Comparisons of the Passus method based on the Van Deventer map of Vlissingen from 1548 with the map of Adams from 1588 and the house count of Vitelli from 1569 show deviation percentages of respectively 7 and 11 percent, an average of 9 percent. This can be qualified as acceptable.
b. The representativeness of the Vlissingen case, which serves as the starting point for all measurements and counts in the present study, also substantiates the validity of the Passus method. The results of the four different counts are very close to each other, indicating that all 'registrars' – Van Deventer, Vitelli (two counts) and Adams – independently mapped the city in great detail and with great care. For Van Deventer, Vlissingen was one of the first cities visited on his mission, for Vitelli it was the first, and for Adams it was the only one in the Low Countries. Their direct commissioners all acted in line with the policy of the sovereign power and saw Vlissingen as one of the most strategic deep-water port cities in the defense of the various regions of which the Dutch territories were a part: Charles V and Philip II for the Habsburg Netherlands and Elizabeth I for England when Vlissingen was pawned to England.[46] It can be assumed that the researchers and mapmakers put a lot of effort into their work in this city. The strong similarities with regard to the quality of their work provide evidence for this.

46 Van Druenen, 2015, pp. 200–205 (Charles V), pp. 247–254 (Philip II), pp. 346–349 (English Pawnship).

c. The controls in Ghent and Leiden, which resulted in deviations of 7 percent (Ghent), 8 percent (Leiden, census 1574), 11 percent (Leiden, census 1581) and 12 percent (Leiden Streets and Canals Book 1583–1587), demonstrate that the Passus method can also be used for cities outside Zeeland. The average deviation of these percentages is 9 percent. However, a caveat should be noted here. Earlier in this study, it was stated that fiscal sources have a significant disadvantage compared to the use of city maps for measuring streets and counting houses: the possibility for the homeowner or occupant to declare less property than actually existed. Therefore, logically, fewer houses would have always been counted. This fact is in line with the differences that have occurred between the Passus method and the fiscal counts: the former always yields more houses than the latter: in Ghent 7 percent and Leiden 11 percent.
d. The tests of the results for four other Zeeland cities (Goes, Middelburg, Veere, and Zierikzee) based on the calculations of Ramaer (1921) and Vitelli (1569) show significantly higher deviation percentages than the checks using alternative local sources for Ghent and Leiden. However, tests for three Zeeland cities (Middelburg, Veere, and Zierikzee), for which alternative sources are available in addition to Vitelli and Ramaer, do yield positive results: 18, 9, and 7 percent.
e. The research in Leiden also shows that both the reduction factor of 4–7 (with an average of 5.5) chosen earlier in this chapter and the Passus coefficient of 0.1756 correspond to reality.

These five arguments provide sufficient support for the assumption that the Passus method on the city maps of Jacob van Deventer is a valid instrument for calculating urban population numbers in the Southern and Northern Netherlands in the mid-sixteenth century.

2.2 Accuracy

In the procedure development described in section 2.1, MapAnalyst, an online program for analyzing and assessing the accuracy of old maps, was used. This program, developed in the object-oriented programming language Java, will also be used to analyze all 226 city maps of Jacob van Deventer and calculate the degree of distortion, visualizing it through distortion grids. The old map needs to be entered by the researchers themselves. The new map is opened in MapAnalyst itself, derived from OpenStreetMap (OSM), which started development in 2004, a year before Google Maps began its operations. Both MapAnalyst and OpenStreetMap are developed based on the Open Source principle: anyone can participate in their (further) development. While Google Maps worked with 7000 developers in 2012, OpenStreetMap had over a million in the same year.

When using MapAnalyst, at least five reference points must be entered on both the old and new maps. The most obvious choices are buildings, street patterns, blocks of houses, or intersections that existed at the time of the map and still exist or can be determined through archaeological research. Natural reference points such as river bends and hilltops can also be used, but there must be some certainty about whether they have changed over the centuries due to natural circumstances and/or human intervention. Because of the lack of adequate source material, this is often difficult to estimate, so in the research chapter mainly buildings, street patterns, blocks of houses, and intersections will be used.

After entering the reference points on the old and new maps, analyses can be generated, both visually and numerically: the former in the form of distortion grids on both maps and the latter in the form of tables with the five reference points, the scale, the rotation angle and the weighted mean position error of the reference points. The outcomes of MapAnalyst will be supplemented in the present study with three new analyses:

a. The height of the scale determined by MapAnalyst depends on the quality of the reference points, and it is thus more or less independent of the actual scale on the Van Deventer maps. This varies by city. In subsequent research into the demographic history of individual cities, all key figures in the present study must therefore be assessed based on the reliability of the reference points. Operationalized, this means that the overview table with the 226 cities will be expanded with a reliability factor. This has six values (0–5). For each reference point, it is indicated how certain it is that the coordinates are indeed the same on both maps. In case of considerable doubt, the respective point 'scores' a '0'; with considerable certainty, a '1'. A map can achieve a maximum score of '5' (for 5 points) when there is little doubt about all chosen points, and a minimum of '0' when all points are doubted. However, a lower value does not say anything about the chosen uncertain reference points: they may have been correctly chosen 'by accident'.
b. Since all measurements and counts in the present study were done using digitized maps, it should not

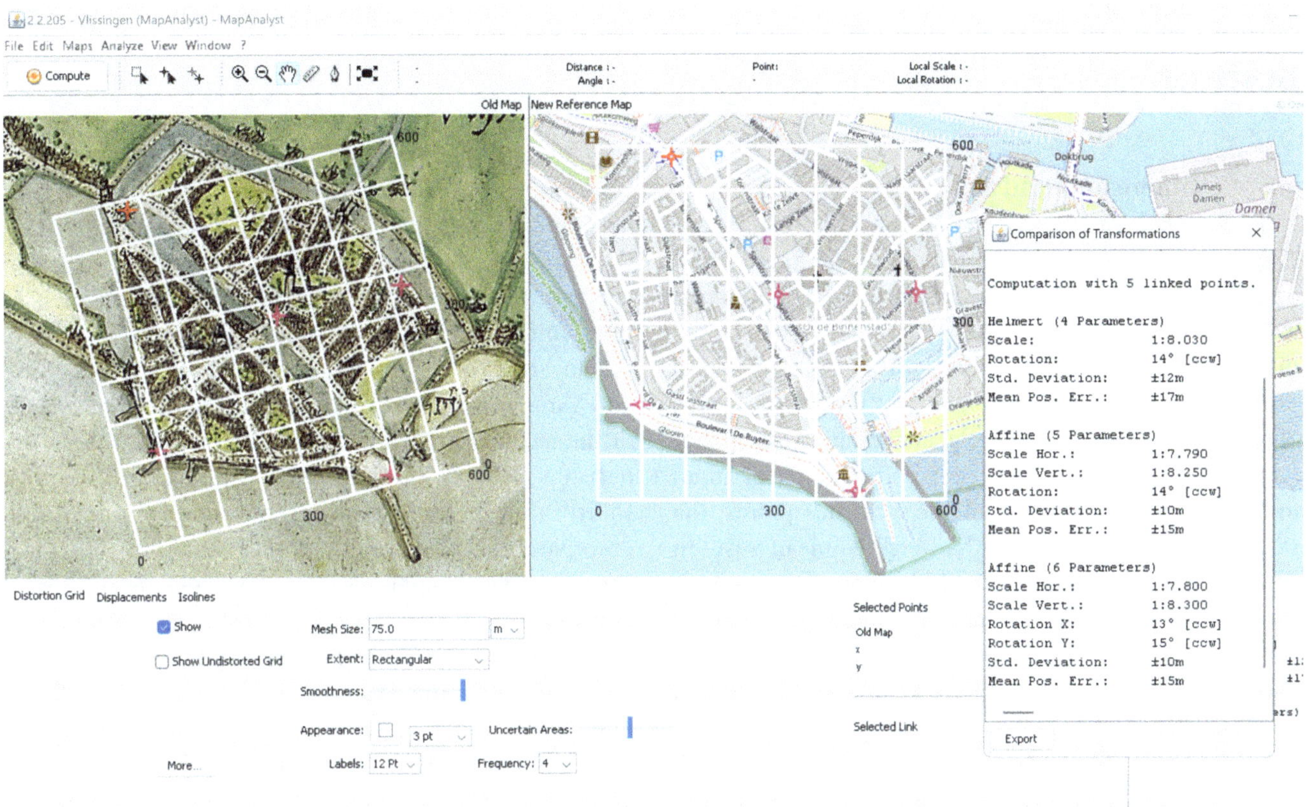

FIGURE 2.14 Screenshot of the MapAnalyst analysis of the Van Deventer map of Vlissingen, 1548

be assumed that the outcomes (in pixels) are also all in accordance with the original atlases.[47] Differences therein can therefore also be responsible for the differences in the calculations and analyses. This also applies to the averages of the other two analyses: rotation and weighted mean position scores per city. As a result, scale ratios may differ among the maps; certain parts of the maps may have a different scale than other parts. Generally, the center is close to the average scale of 1:8148, and the areas around it differ. The same applies to the combination of points lying 'close to' or 'far from' each other. These will therefore be located as much as possible within the urban built-up area and, at the same time, not too close to each other. Occasionally this is deviated from, for example, when (a) there are no or very few reliable reference points to be found within the built-up area, while they are outside the city, or (b) the only reliable points in the city are very close to each other or far apart.

c. The individual measurements and analyses of all 226 city maps will be presented in Chapter 3 and compared with each other. This makes it possible to perform calculations that can say something about the whole of the maps: the total and average number of houses and inhabitants per city and the average of the three analyses: scale, rotation, standard deviation, weighted mean position error, and reliability of the reference points. The set of analyses will be expanded with calculations that can reveal the interrelationships. This will be done using the correlation coefficient, which calculates the the relationship between the values of the series.[48] The addition of this is of considerable importance for determining the quality of the research in the present study and – more specifically – the addition of the reliability factor to the MapAnalyst analyses.

47 For the atlas maps, the most common resolution is 300 pixels per inch; for the individual maps it is 400 pixels per inch. All maps (including those with a different resolution) have been adjusted to a resolution of 300 before measuring and analyzing.

48 The range of the correlation: +1 (perfect); greater than 0.5 (strong); between 0.3 and 0.5 (moderate); between 0 and 0.3 (weak); 0 (no correlation); between 0 and –0.3 (weak); between –0.3 and –0.5 (moderate); greater than –0.5 (strong), –1 (perfect).

CHAPTER 3

Research

3.1 Dimensions and Accuracies

This section presents the results of the research conducted on all 226 city maps by Jacob van Deventer. The research was based on two operations, as discussed and assessed in chapter 2: (a) dimensions: measuring the street sides of the cities and deriving population numbers from this and (b) accuracies: analyzing the maps using MapAnalyst (version 1.3.35) and five reference points, resulting in an indication of the reliability of the reference points, the scale, the rotation angle and the weighted mean position error of the reference points. Both operations are presented in a tabular form for each city, including the three corresponding maps: the unprocessed version, the version with the street sides marked in red, and the version with the MapAnalyst grid.

No urban historical narratives were included for the 226 cities. Only measurements and analyses were conducted. The aim was to examine all city maps using the same set of protocol steps as described in sections 2.1 and 2.2, and to formulate conclusions about the whole and make comparisons between the individual maps based on the research results. If variations per city due to specific urban historical developments before Van Deventer recorded the data for the respective map had been taken into account, comparisons between the maps would no longer have been possible. It should be emphasized that this restriction only applies to the research on the maps themselves, that is to say, the way in which Van Deventer mapped the cities, not to the peculiarities of the cities themselves.

This means that researchers who want to work with the population numbers of a particular city based on the data in the present study must pay attention to the specific city history and assess whether the data can be used integrally or need to be refined. An example of this is the Flemish city of Poperinge, which Van Deventer only partially drew. Rutte and Vannieuwenhuyze state in their standard work *Stedenatlas Jacob van Deventer* from 2015 that there is nothing to be seen of the built-up outskirts to the left and right of the city on the map.[1] In current Poperinge, there are several buildings that existed in the sixteenth century but were not drawn by Van Deventer such as the Saint Bertinus Church and the Our Lady Church. The locations are only marked with white squares. Rutte and Vannieuwenhuyze suggest that the city fire of 1563 or the iconoclasm of 1566 may have played a role in the decision made by Van Deventer – who was in Poperinge shortly after these events for his inventory work – not to draw parts of the buildings. Another possibility is that Van Deventer was in Poperinge when the Dutch Revolt started and that the cartographer had to flee the region. This could also explain why he did not map Ypres.[2]

A more extreme example is Thérouanne. Until the mid-sixteenth century the town was a French enclave in the Netherlands. In 1553, the city was destroyed by order of Charles V. When Jacob Van Deventer drew the nearby town of Aire-sur-la-Lys, he drew a side map of Thérouanne. Only the ramparts can be seen on that map, so street side measurements are therefore not possible. However, the contours of the old town can be seen on the contemporary map.

For determining the number of houses and thus the number of inhabitants, not measuring the missing street sides that did exist makes a big difference. Local investigations must therefore supplement the Passus method. For comparing the map with the other 225 maps, it makes no difference.

In the tables on the following city pages, the following key figures are presented:

1 Rutte and Vannieuwehuyze, 2015, p. 229 (already stated in Vanden Bussche, 'Poperinghe'. In: *Ruelens, Atlas des villes de la Belgique*, livraison 5, 1886.).

2 Dupont, 2019, pp. 257–258.

 | HTTPS://DOI.ORG/10.1163/9789004755031-004

Dimensions

- Pixels (the length of the measured street sides in pixels).
- Scale stick on the downloaded and unprocessed digital map (in pixels).
- Houses (the total number of pixels multiplied by the Passus coefficient of 0.1756).
- Street sides (total number of street sides measured).
- Houses per street side (the number of houses divided by the number of street sides).
- Inhabitants RF 4.0 (the number of houses multiplied by the reduction factor 4).
- Inhabitants RF 5.5 (the number of houses multiplied by the reduction factor 5.5, the average of the reduction factors 4.0 and 7.0).
- Inhabitants RF 7.0 (the number of houses multiplied by the reduction factor 7).

Accuracies

- Research date (the day the analysis was performed in MapAnalyst).
- Computation with five linked points (the chosen reference points, with the points assessed as less reliable shown in italics).
- Reference points (the reliability of the total of the five reference points on a scale of 0–5).
- Scale (the scale of the Van Deventer map, with the current map of the city taken from the online geographic program OpenStreetMap as a reference point).
- Rotation (clockwise or counterclockwise).
- Mean position error ($\sqrt{2}$ times the standard deviation, a global measure of the accuracy of all reference points).

1. Aalst

Tables 3.1 (measured and analysed).
Figures 3.1 (unmeasured, measured and analysed).

Measured							
Pixels	Scale stick original map (pixels)	Houses	Street sides	Houses per street side	Inhabitants (RF 4.0)	Inhabitants (RF 5.5)	Inhabitants (RF 7.0)
6812	1030	1196	133	8.99	4785	6579	8373

Analysed	
February 9, 2023 8:08:55 PM	
Computation with 5 linked points	
(*italics* points are less reliable):	1. Sint-Martinuskerk
	2. Belfort
	3. Werfkapel
	4. Korte Zoutstraat
	5. Begijnhof
Reference points (Reliability 0–5)	4
Transformation with 3 parameters	
Scale	1:8090
Rotation	6 °
Mean position error	17m

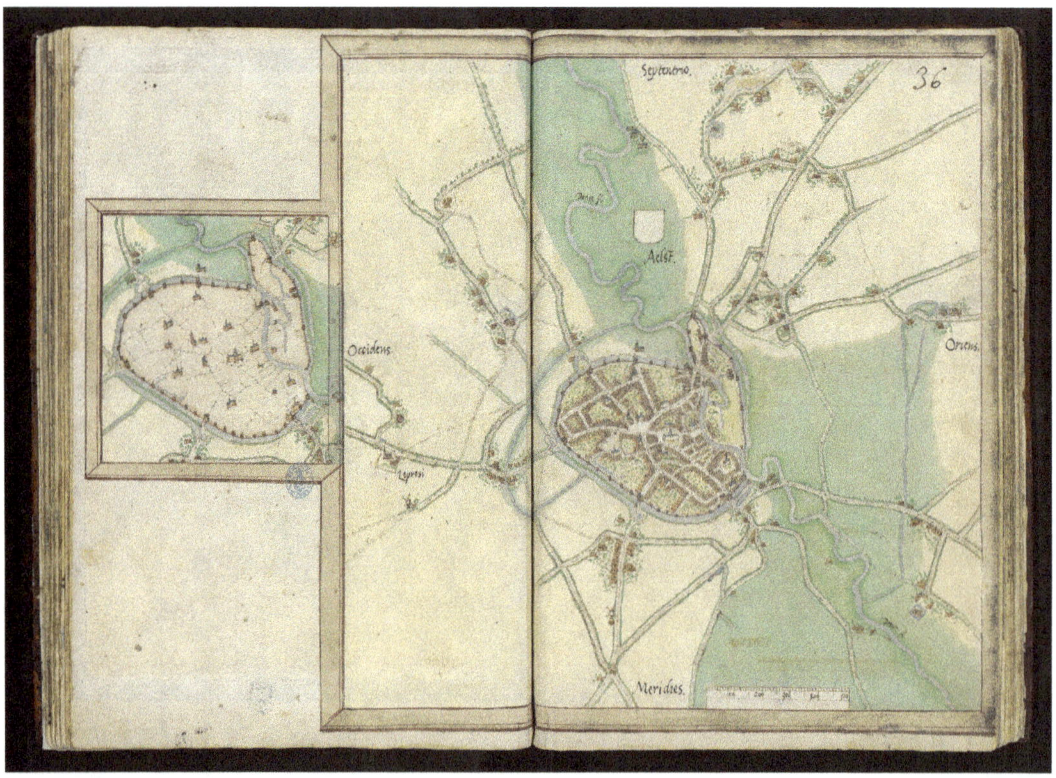

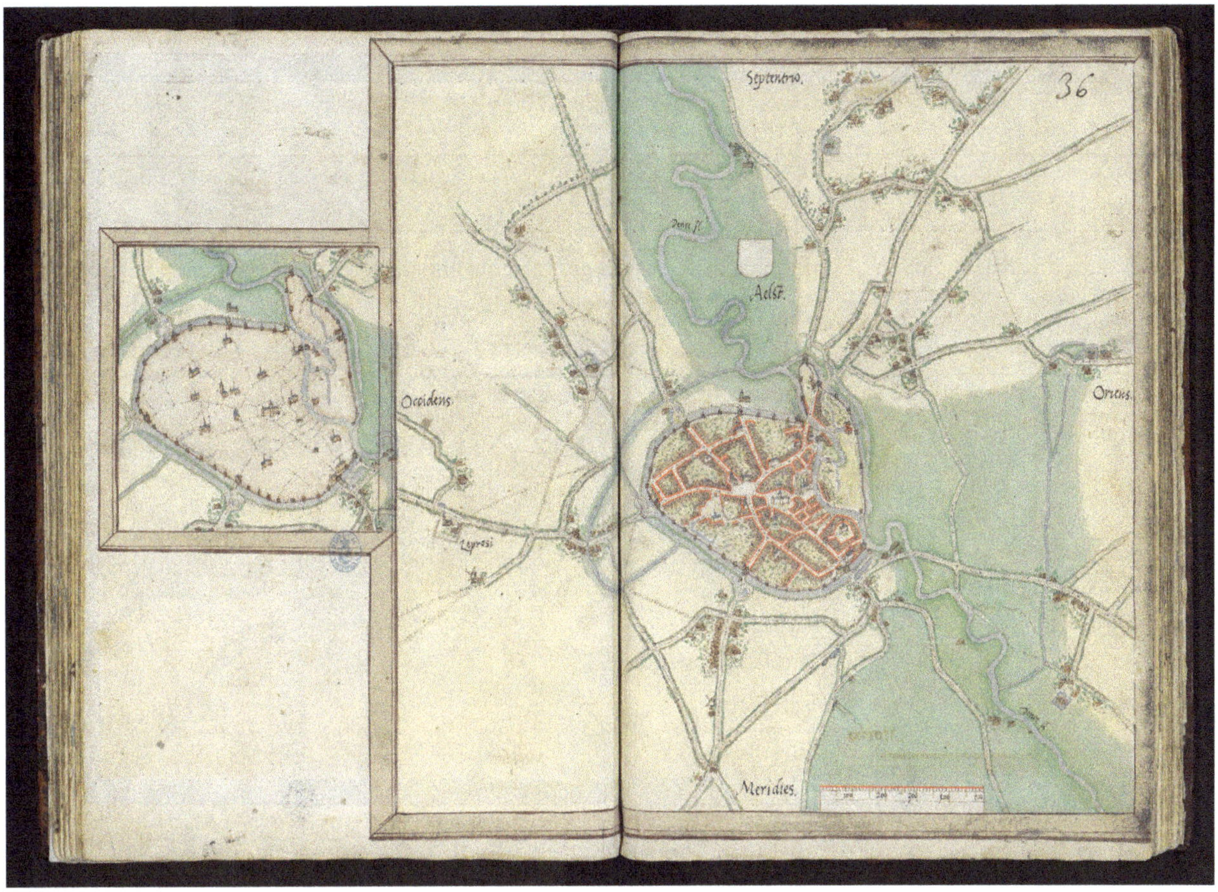
36
Septentrio.
Aelst.
Occidens.
Oriens.
Meridies.

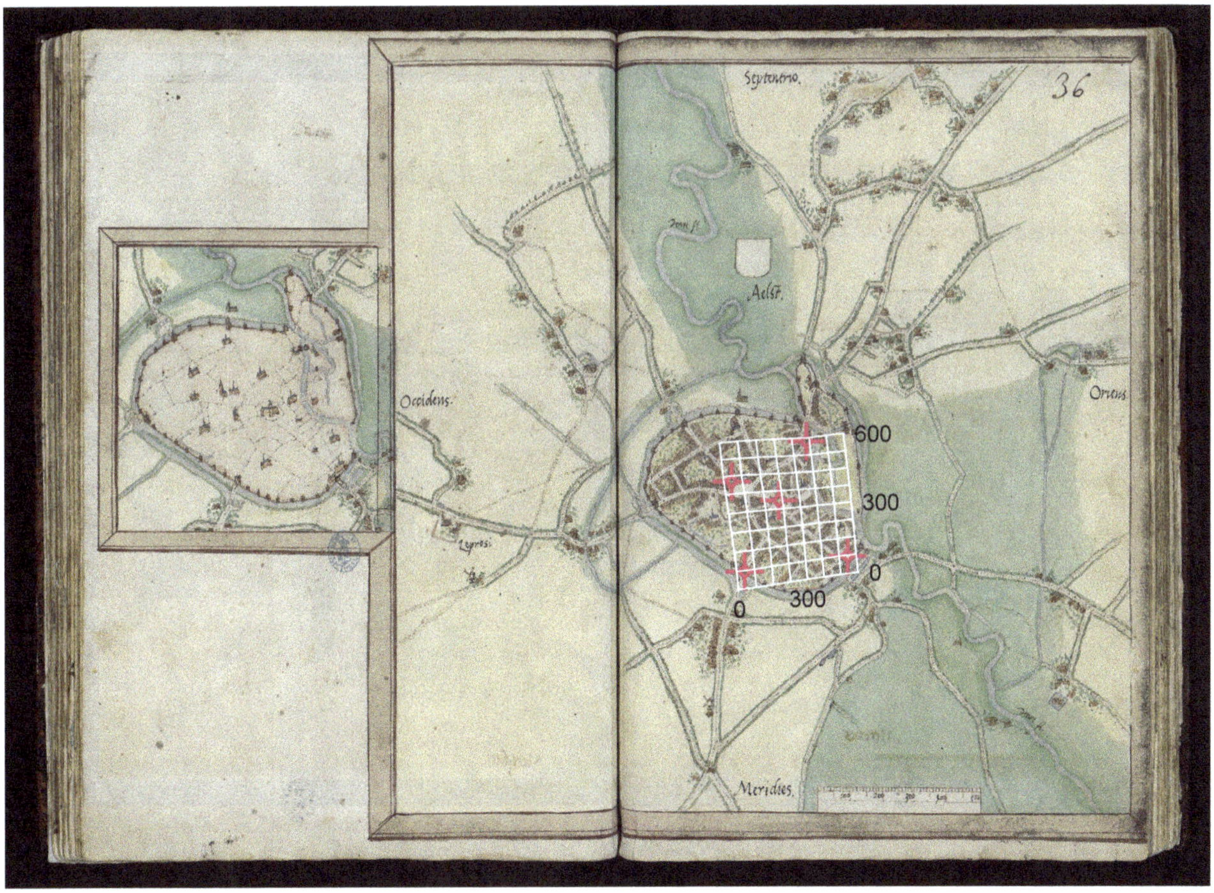
36
Septentrio.
Aelst.
Occidens.
Oriens.
600
300
0
0
300
Meridies.

2. Aardenburg

Tables 3.2 (measured and analysed).
Figures 3.2 (unmeasured, measured and analysed).

Measured

Pixels	Scale stick original map (pixels)	Houses	Street sides	Houses per street side	Inhabitants (RF 4.0)	Inhabitants (RF 5.5)	Inhabitants (RF 7.0)
3264	1030	573	84	7	2293	3152	4012

Analysed

February 9, 2023 5:45:01 PM	
Computation with 5 linked points	
(*italics* points are less reliable):	1. Sint-Bavokerk westzijde
	2. Sint-Bavokerk oostzijde
	3. Weststraat
	4. Kalverstraat
	5. Peursenstraat
Reference points (Reliability 0–5)	2
Transformation with 3 parameters	
Scale	1:7300
Rotation	1°
Mean position error	27m

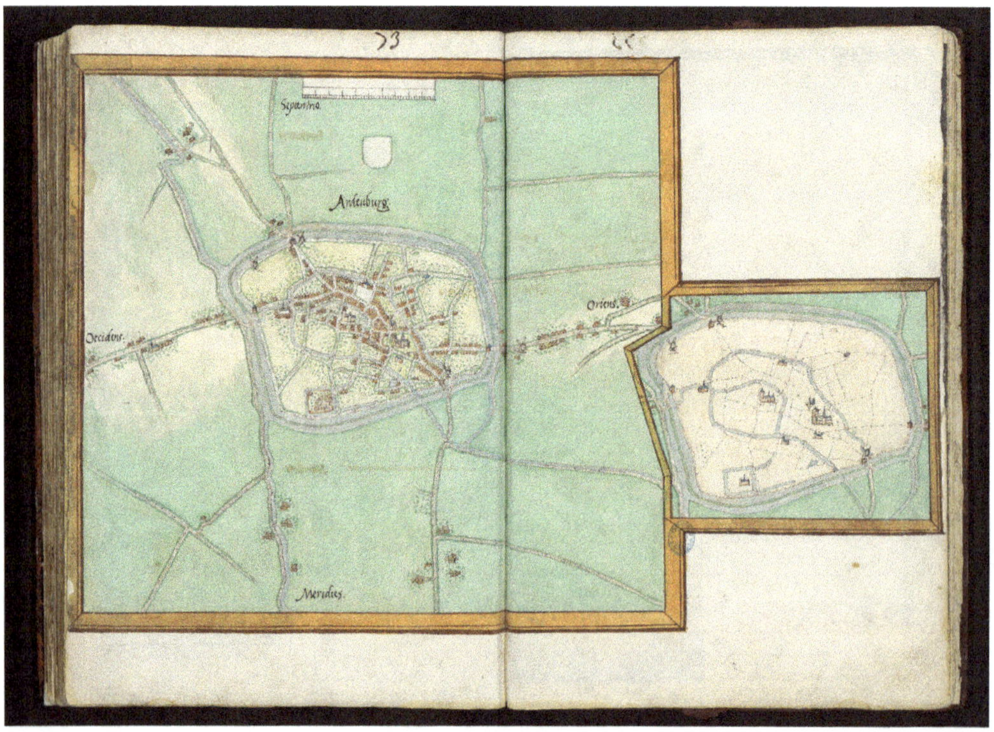

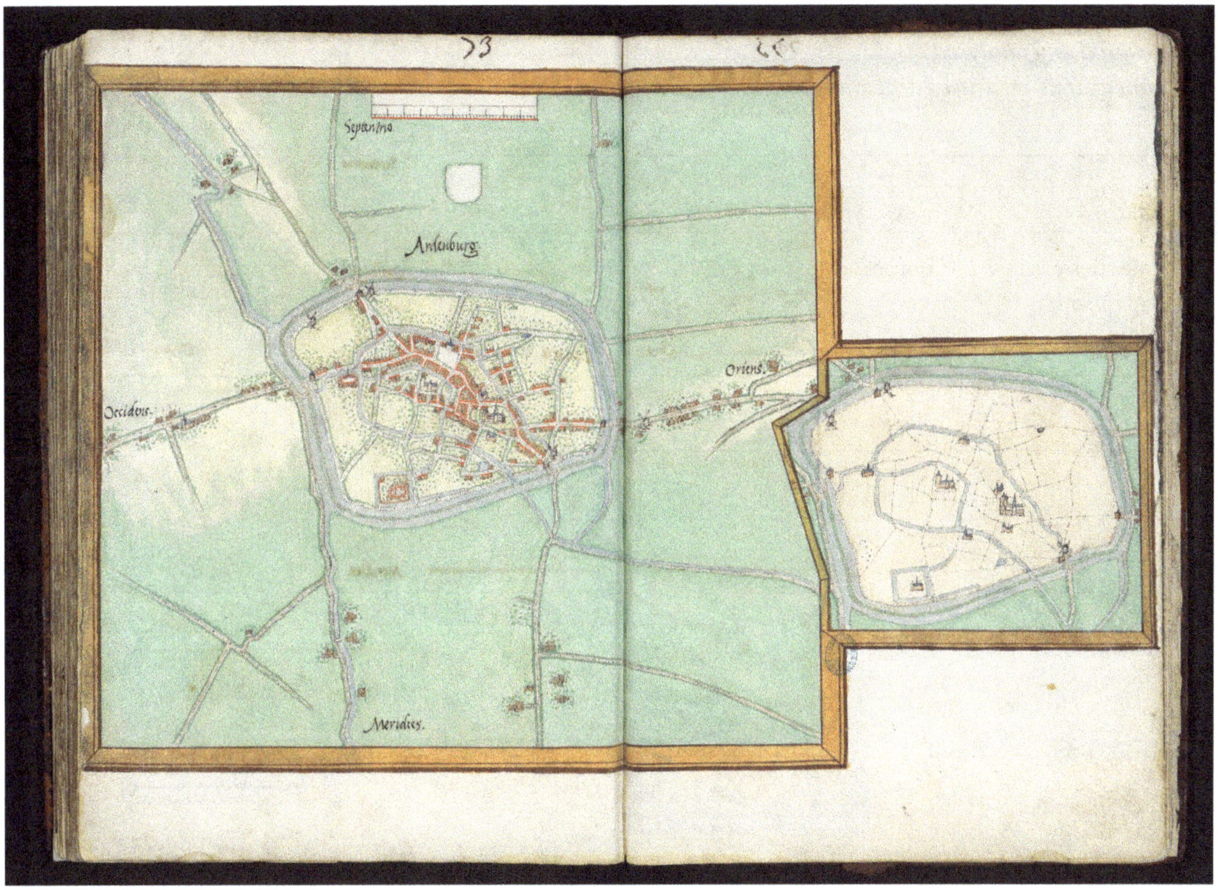
Septentrio
Ardenburg
Occidens.
Oriens.
Meridies.

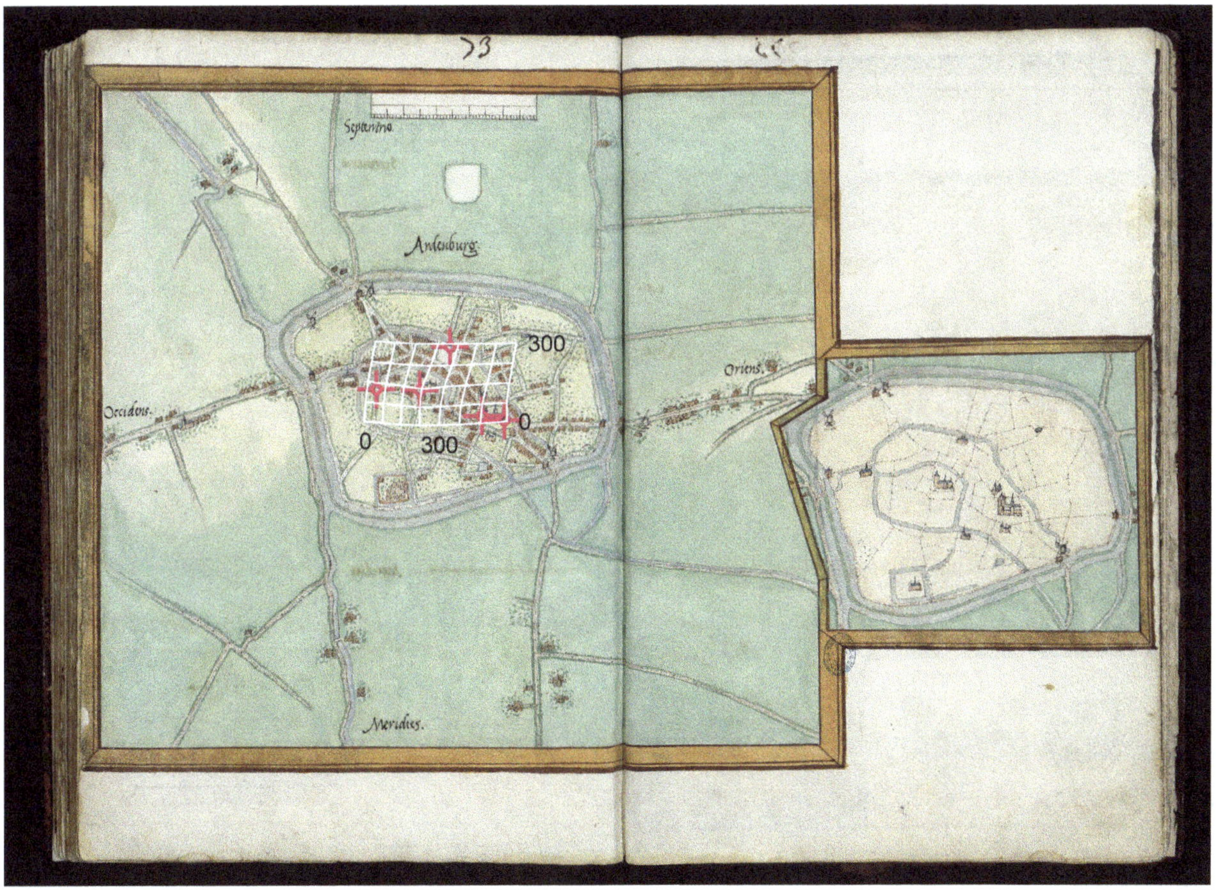
Septentrio
Ardenburg
300
Oriens.
Occidens.
0
0
300
Meridies.

3. Aarschot

Tables 3.3 (measured and analysed).
Figures 3.3 (unmeasured, measured and analysed).

Measured							
Pixels	Scale stick original map (pixels)	Houses	Street sides	Houses per street side	Inhabitants (RF 4.0)	Inhabitants (RF 5.5)	Inhabitants (RF 7.0)
3317	Missing	582	75	7.77	2330	3204	4077

Analysed	
September 5, 2023 1:36:32 PM	
Computation with 5 linked points	
(*italics* points are less reliable):	1. Onze-Lieve-Vrouwekerk
	2. Begijnhof
	3. Aurelianustoren
	4. Sint-Rochustoren
	5. Vestingstraat
Reference points (Reliability 0–5)	5
Transformation with 3 parameters	
Scale	1:8410
Rotation	15 °
Mean position error	18m

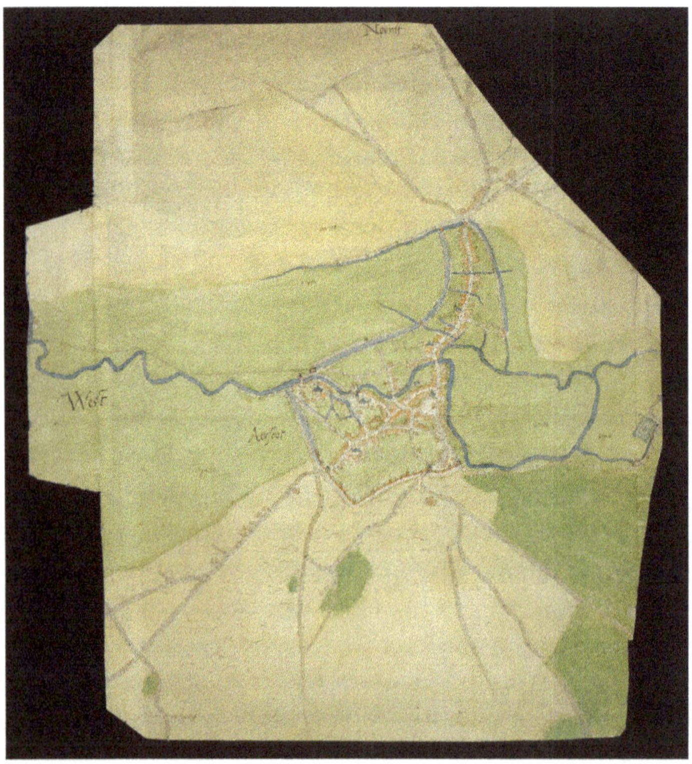

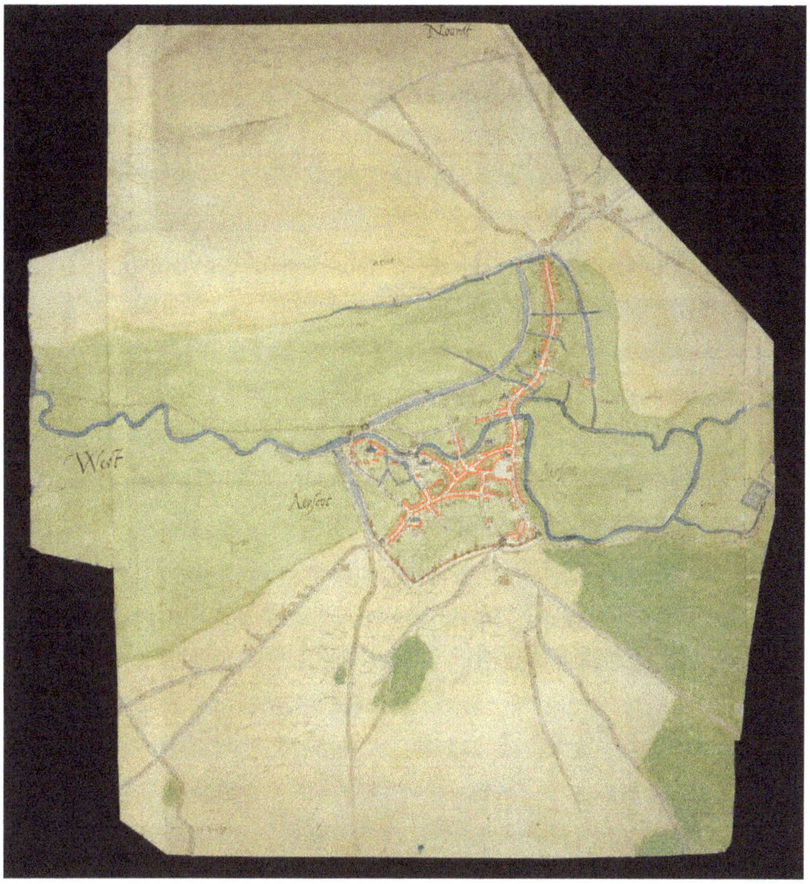
West

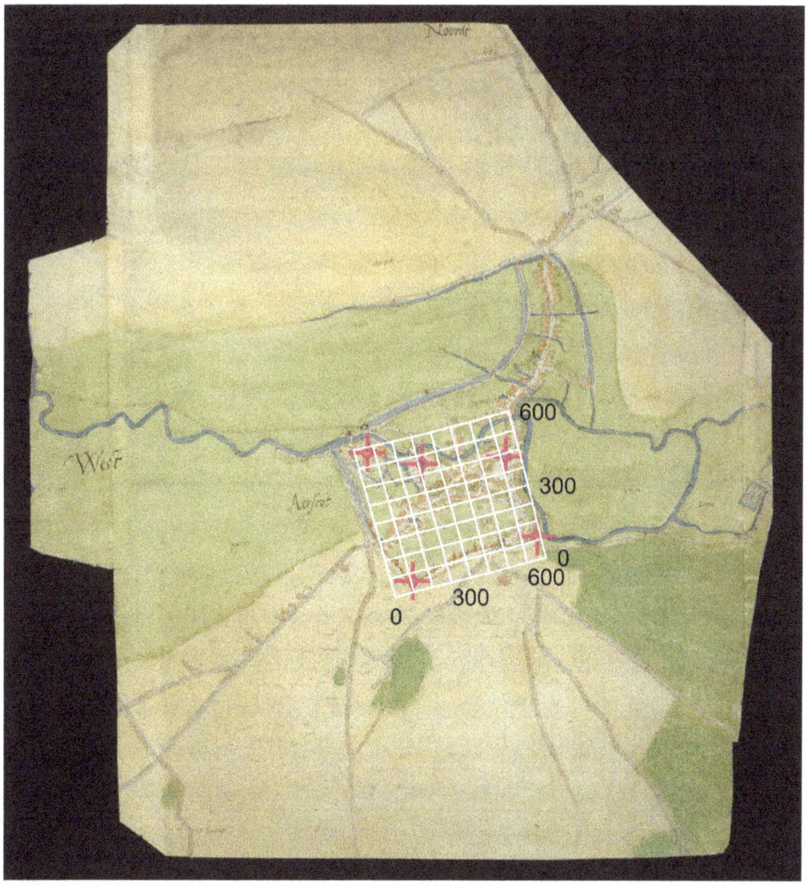
West
600
300
0
600
300
0

4. Aire-sur-la-Lys

Tables 3.4 (measured and analysed).
Figures 3.4 (unmeasured, measured and analysed).

Measured

Pixels	Scale stick original map (pixels)	Houses	Street sides	Houses per street side	Inhabitants (RF 4.0)	Inhabitants (RF 5.5)	Inhabitants (RF 7.0)
5784	1035	1016	126	8.06	4063	5586	7110

Analysed

February 11, 2023 6:06:57 PM	
Computation with 5 linked points	
(*italics* points are less reliable):	1. Église Saint Pierre 2. La Lys (ouest) *3. La Lys (Est)* *4. Mordyck* 5. Belfort
Reference points (Reliability 0–5)	3
Transformation with 3 parameters	
Scale	1:7800
Rotation	9 °
Mean position error	27m

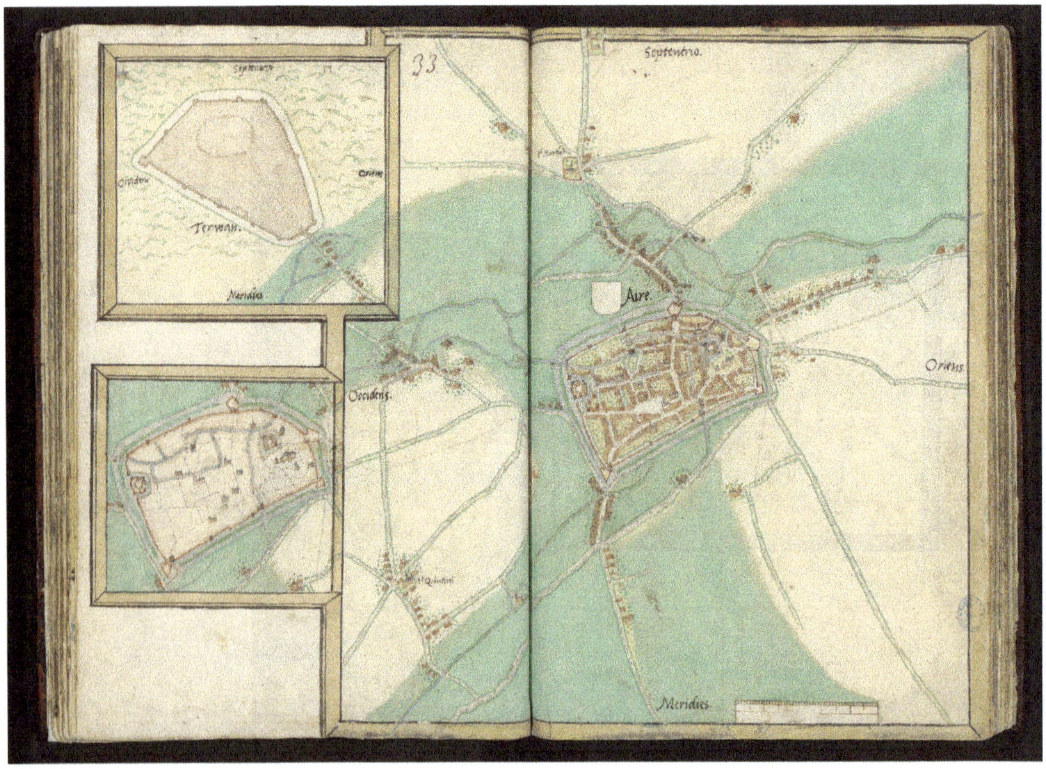

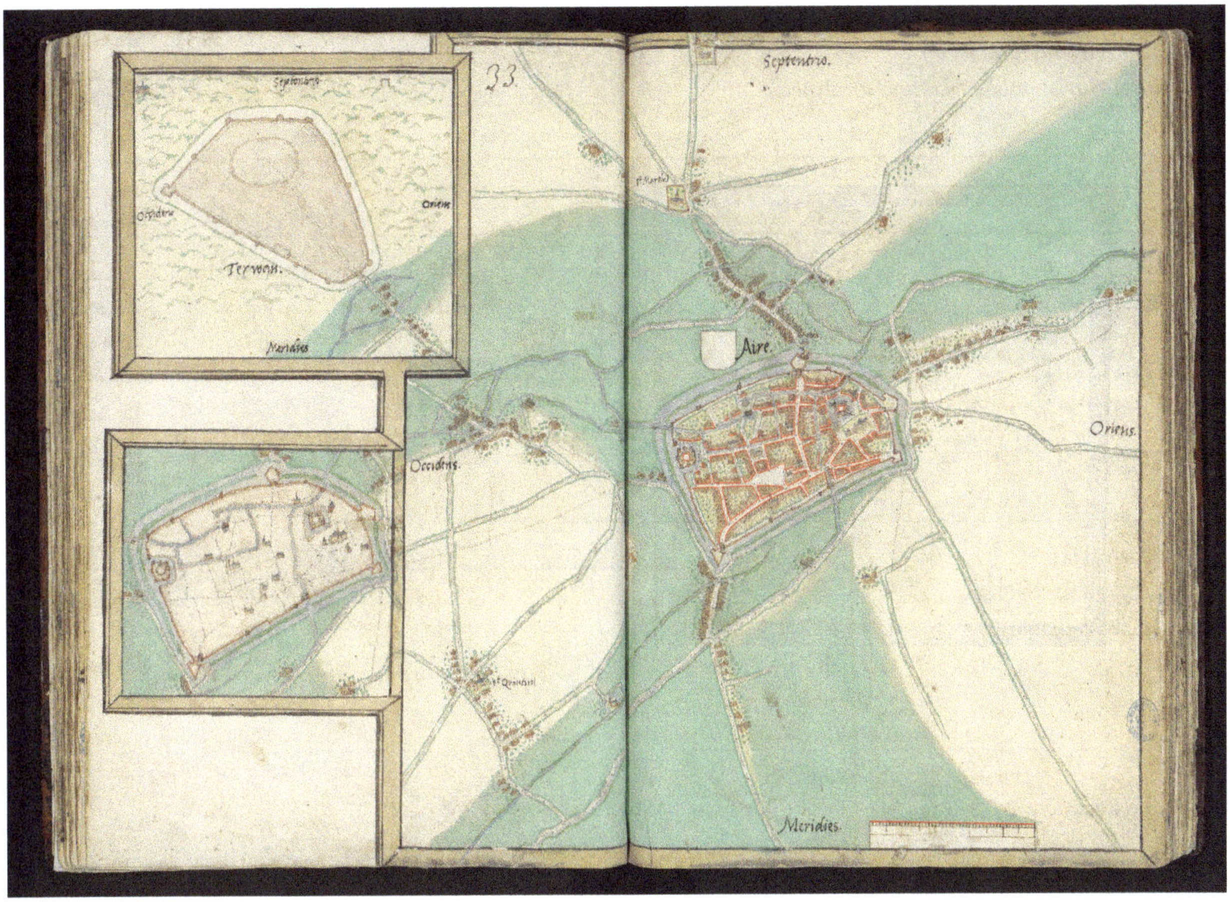
33.
Septentrio.
Aire.
Oriens.
Occidens.
Meridies.
Terwan.

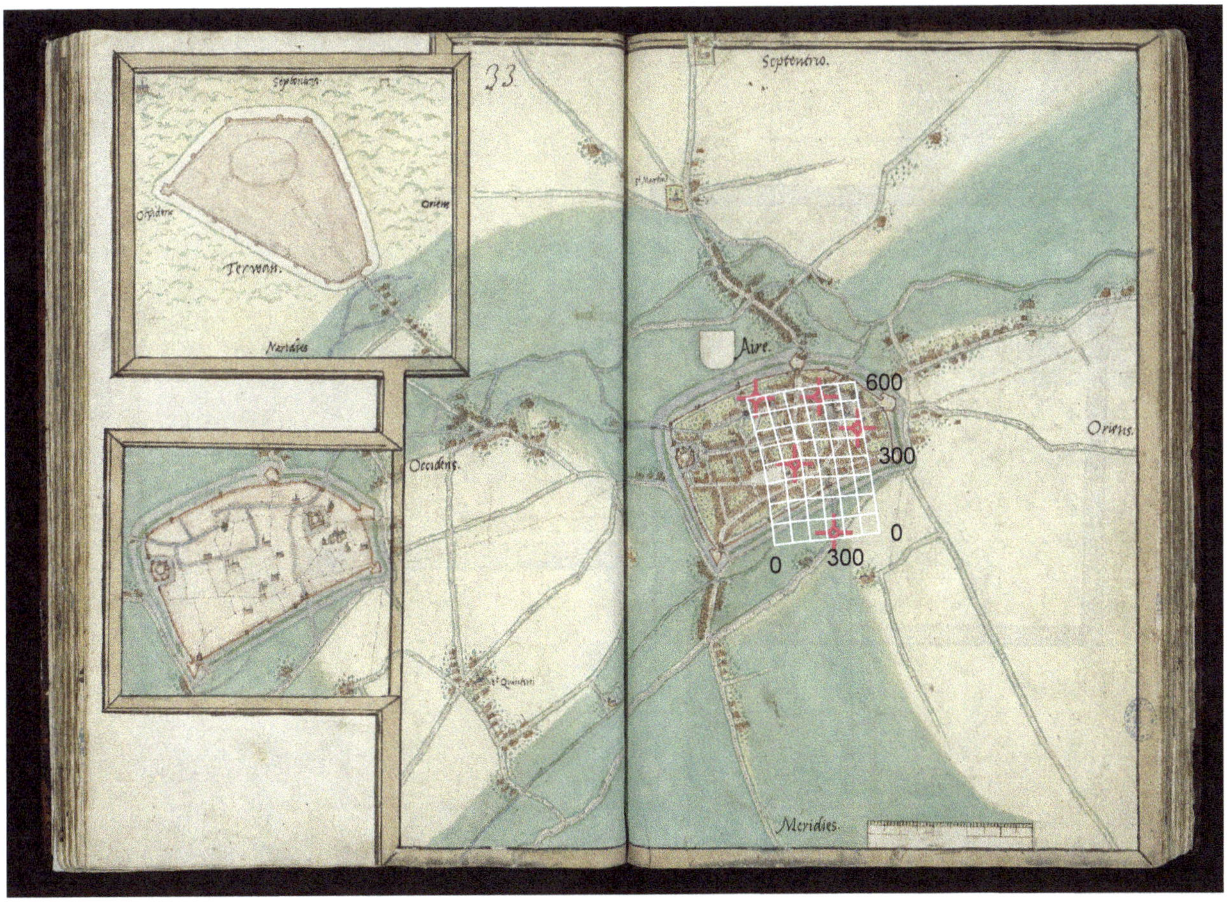
33.
Septentrio.
Aire.
600
300
0
0
300
Oriens.
Occidens.
Meridies.

5. Alkmaar

Tables 3.5 (measured and analysed).
Figures 3.5 (unmeasured, measured and analysed).

Measured

Pixels	Scale stick original map (pixels)	Houses	Street sides	Houses per street side	Inhabitants (RF 4.0)	Inhabitants (RF 5.5)	Inhabitants (RF 7.0)
7578	1006	1331	182	7.31	5323	7319	9315

Analysed

February 19, 2023 10:57:24 AM	
Computation with 5 linked points	
(*italics* points are less reliable):	1. Grote Sint-Laurenskerk 2. Stadhuis 3. Voordam 4. Bierkade 5. Bierkade
Reference points (Reliability 0–5)	5
Transformation with 3 parameters	
Scale	1:8260
Rotation	16 °
Mean position error	10m

40.
Septentrio
Alckmaer
Oriens.
Meridies.

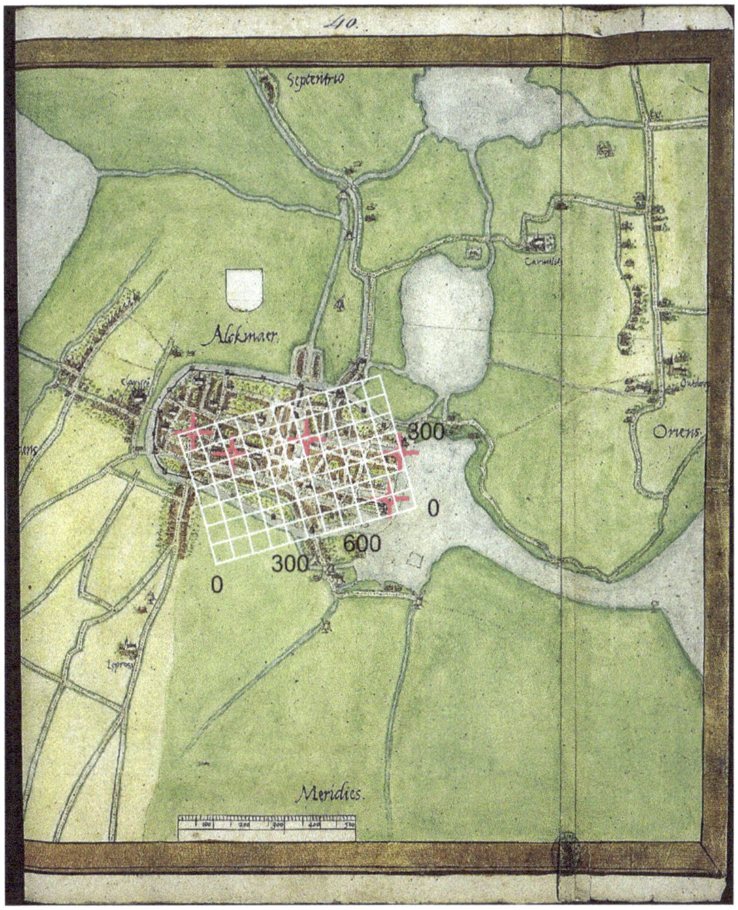
40.
Septentrio
Alckmaer
Oriens.
Meridies.
300
0
600
300
0

6. Almelo

Tables 3.6 (measured and analysed).
Figures 3.6 (unmeasured, measured and analysed).

Measured							
Pixels	Scale stick original map (pixels)	Houses	Street sides	Houses per street side	Inhabitants (RF 4.0)	Inhabitants (RF 5.5)	Inhabitants (RF 7.0)
1336	950	235	34	6.90	938	1290	1642

Analysed	
October 15, 2023 11:11:54 AM	
Computation with 5 linked points	
(*italics* points are less reliable):	1. Grote kerk
	2. Het Kolkje
	3. Burcht
	4. Gravenallee
	5. Gravenallee
Reference points (Reliability 0–5)	5
Transformation with 3 parameters	
Scale	1:8260
Rotation	10 °
Mean position error	9m

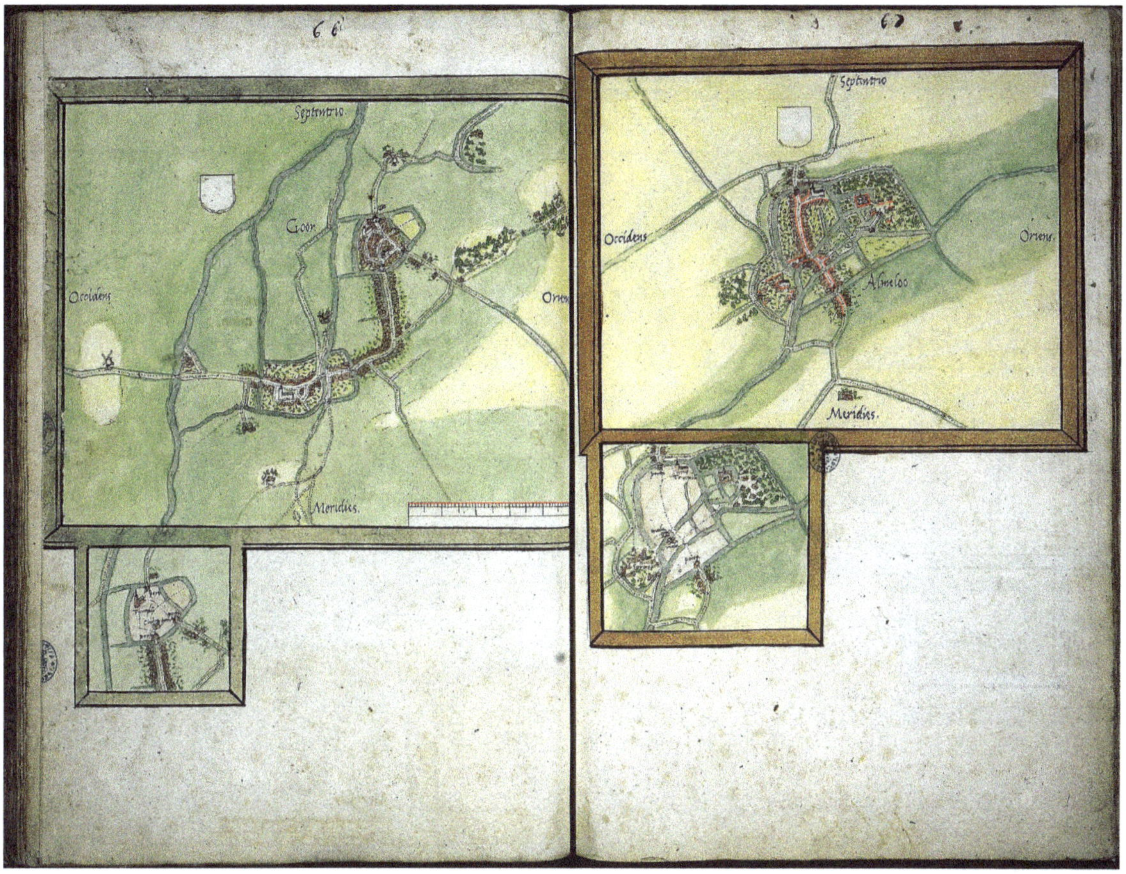
Septentrio
Goor
Occidens
Meridies
Septentrio
Occidens
Oriens
Almeloo
Meridies

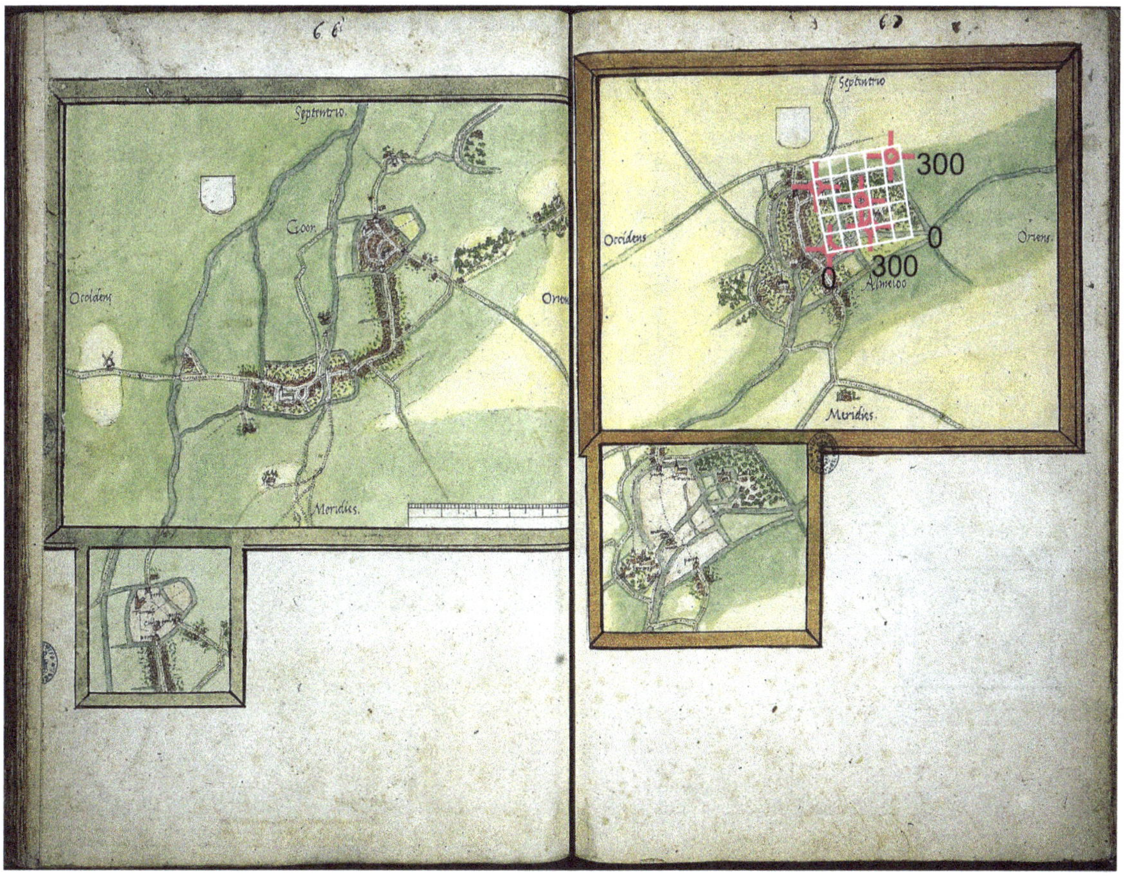
Septentrio
Goor
Occidens
Meridies
Septentrio
300
0
300
0
Occidens
Oriens
Almeloo
Meridies

7. Amersfoort

Tables 3.7 (measured and analysed).
Figures 3.7 (unmeasured, measured and analysed).

Measured							
Pixels	Scale stick original map (pixels)	Houses	Street sides	Houses per street side	Inhabitants (RF 4.0)	Inhabitants (RF 5.5)	Inhabitants (RF 7.0)
11,909	992	2091	263	7.95	8365	11,502	14,639

Analysed	
February 20, 2023 2:11:53 PM	
Computation with 5 linked points	
(*italics* points are less reliable):	1. Sint-Joriskerk
	2. Zuidsingel
	3. Beek
	4. Langestraat
	5. Grote Spui
Reference points (Reliability 0–5)	5
Transformation with 3 parameters	
Scale	1:8540
Rotation	9 °
Mean position error	11m

Septentrio.
Amersfoort
Oriens
Meridies

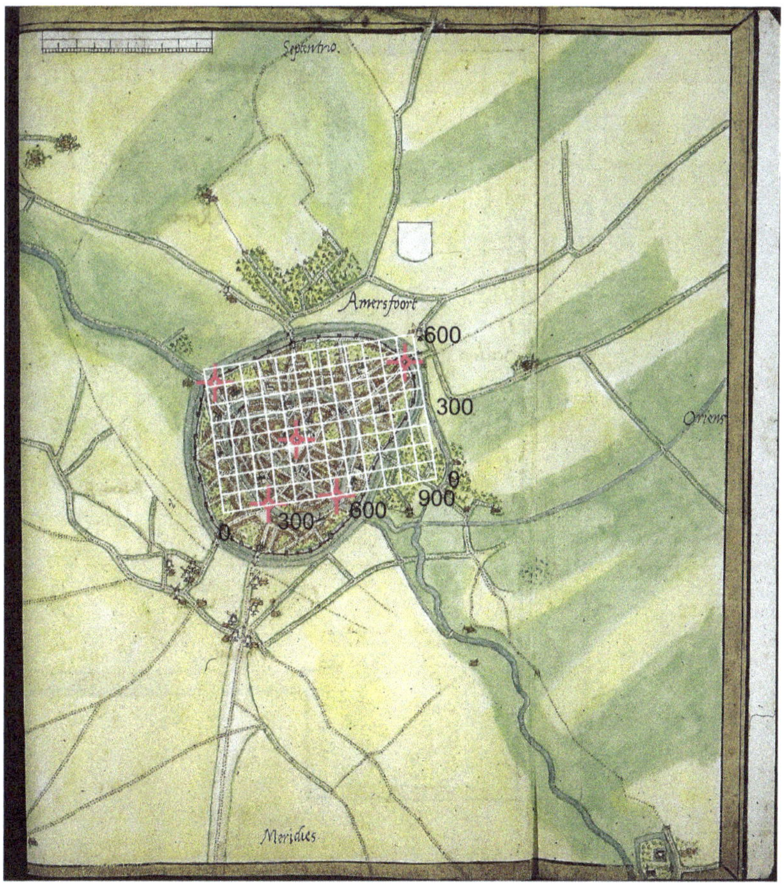
Septentrio.
Amersfoort
600
300
0
900
Oriens
0
300
600
Meridies

8. Amsterdam

Tables 3.8 (measured and analysed).
Figures 3.8 (unmeasured, measured and analysed).

Measured							
Pixels	Scale stick original map (pixels)	Houses	Street sides	Houses per street side	Inhabitants (RF 4.0)	Inhabitants (RF 5.5)	Inhabitants (RF 7.0)
18,009	1001	3162	548	5.77	12,650	17,393	22.137

Analysed	
February 20, 2023 8:55:34 PM	
Computation with 5 linked points	
(*italics* points are less reliable):	1. Oude kerk 2. Nieuwe kerk 3. Waag 4. Damrak 5. Amstel
Reference points (Reliability 0–5)	5
Transformation with 3 parameters	
Scale	1:8260
Rotation	14 °
Mean position error	14m

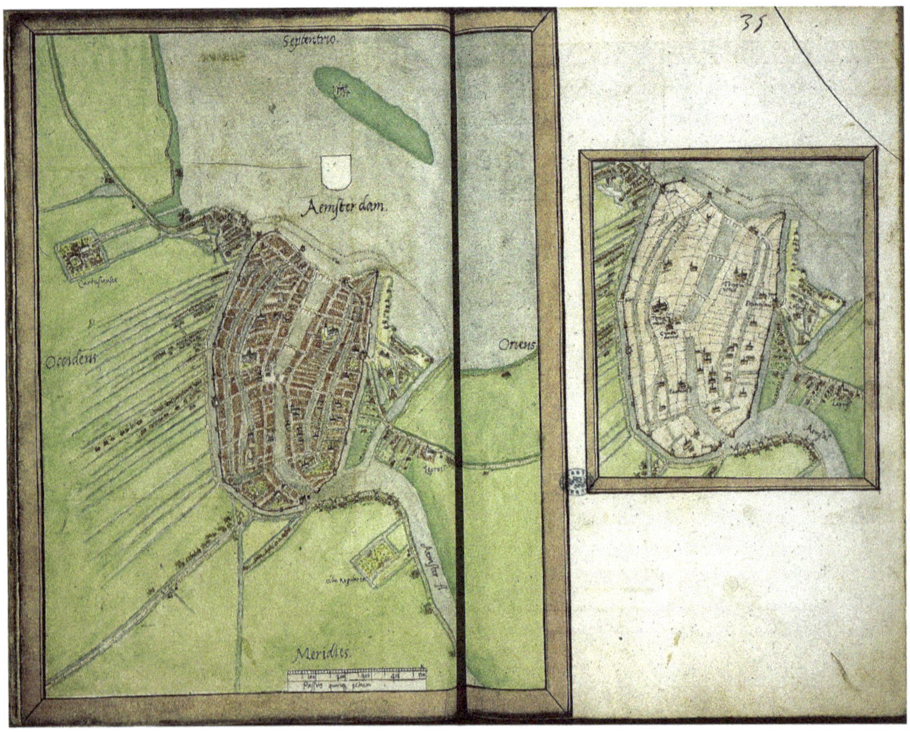

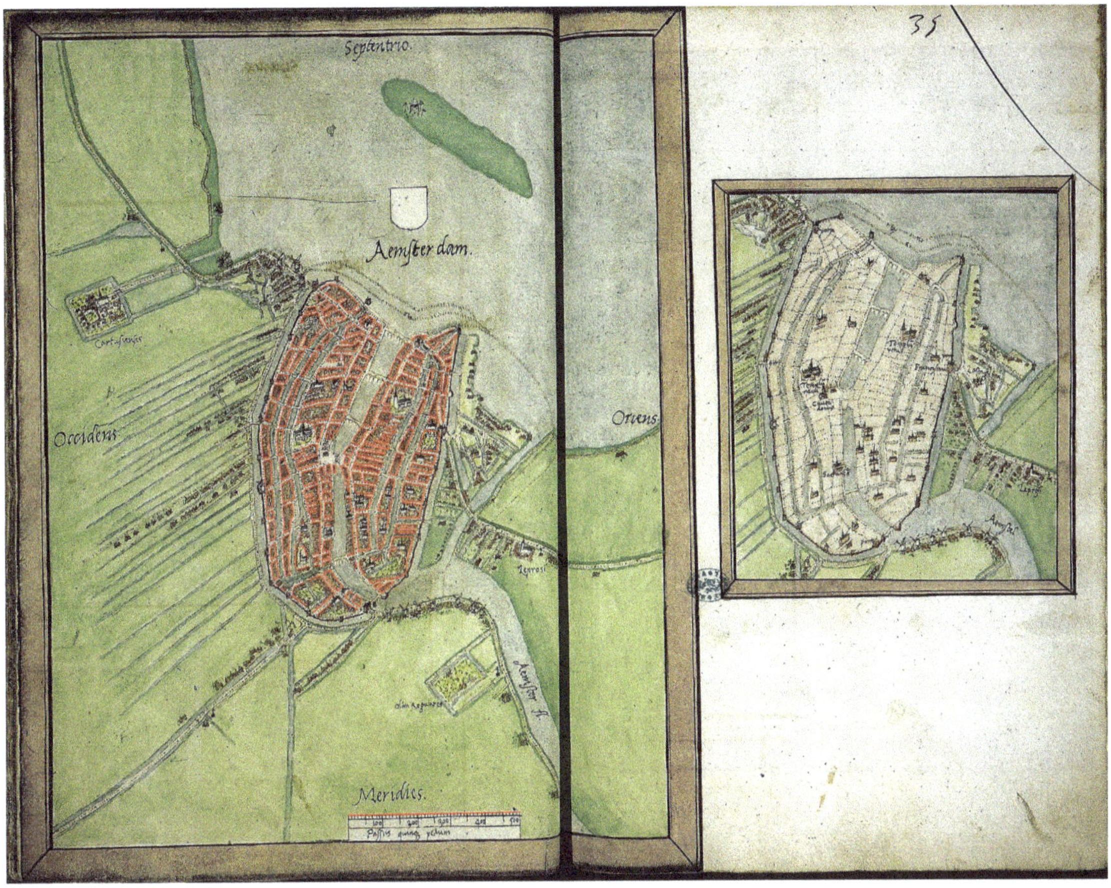
Septentrio.
Aemster dam.
Occidens.
Oriens.
Meridies.
35

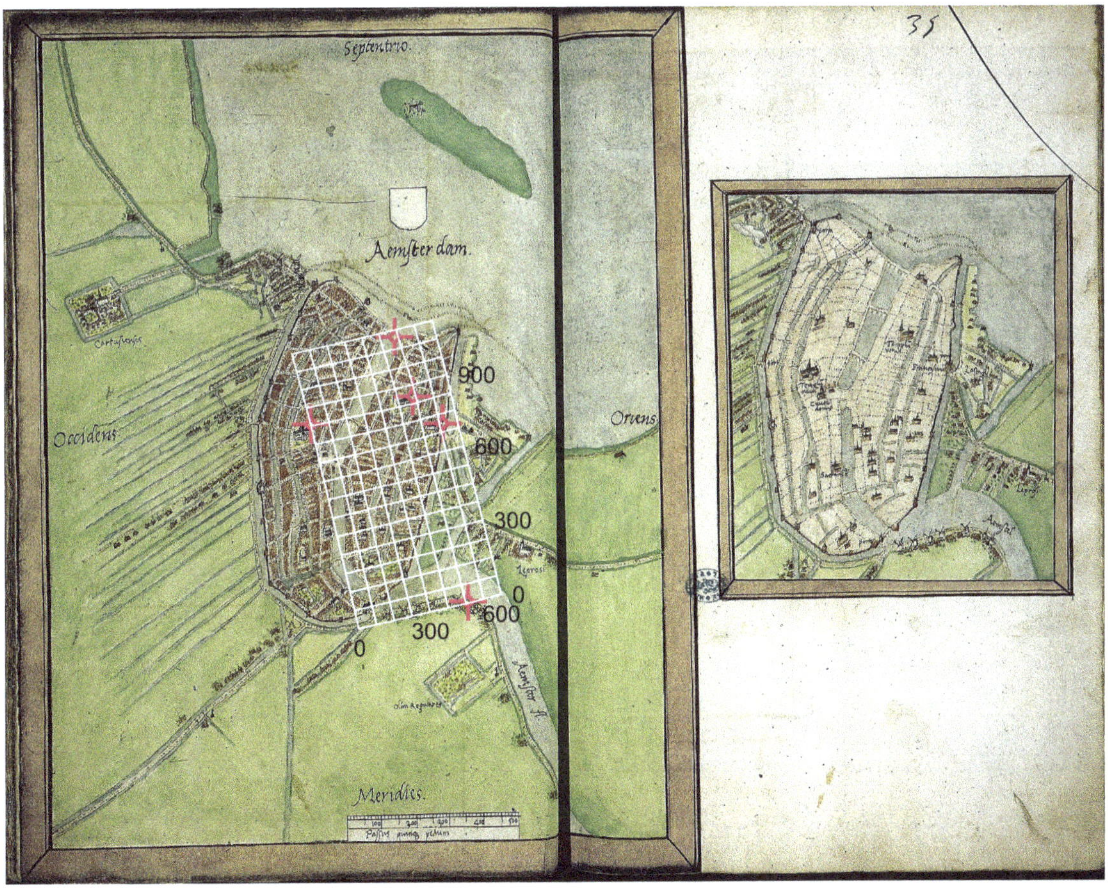
Septentrio.
Aemster dam.
Occidens.
Oriens.
Meridies.
35
900
600
300
0
0
300
600

9. Appingedam

Tables 3.9 (measured and analysed).
Figures 3.9 (unmeasured, measured and analysed).

Measured							
Pixels	Scale stick original map (pixels)	Houses	Street sides	Houses per street side	Inhabitants (RF 4.0)	Inhabitants (RF 5.5)	Inhabitants (RF 7.0)
2973	935	522	77	6.78	2088	2871	3654

Analysed	
February 20, 2023 4:55:48 PM	
Computation with 5 linked points	
(*italics* points are less reliable):	1. Nicolaikerk 2. Fivelkade 3. Hamweg 4. Jan Bronsweg 5. Annastraat
Reference points (Reliability 0–5)	5
Transformation with 3 parameters	
Scale	1:8150
Rotation	10 °
Mean position error	24m

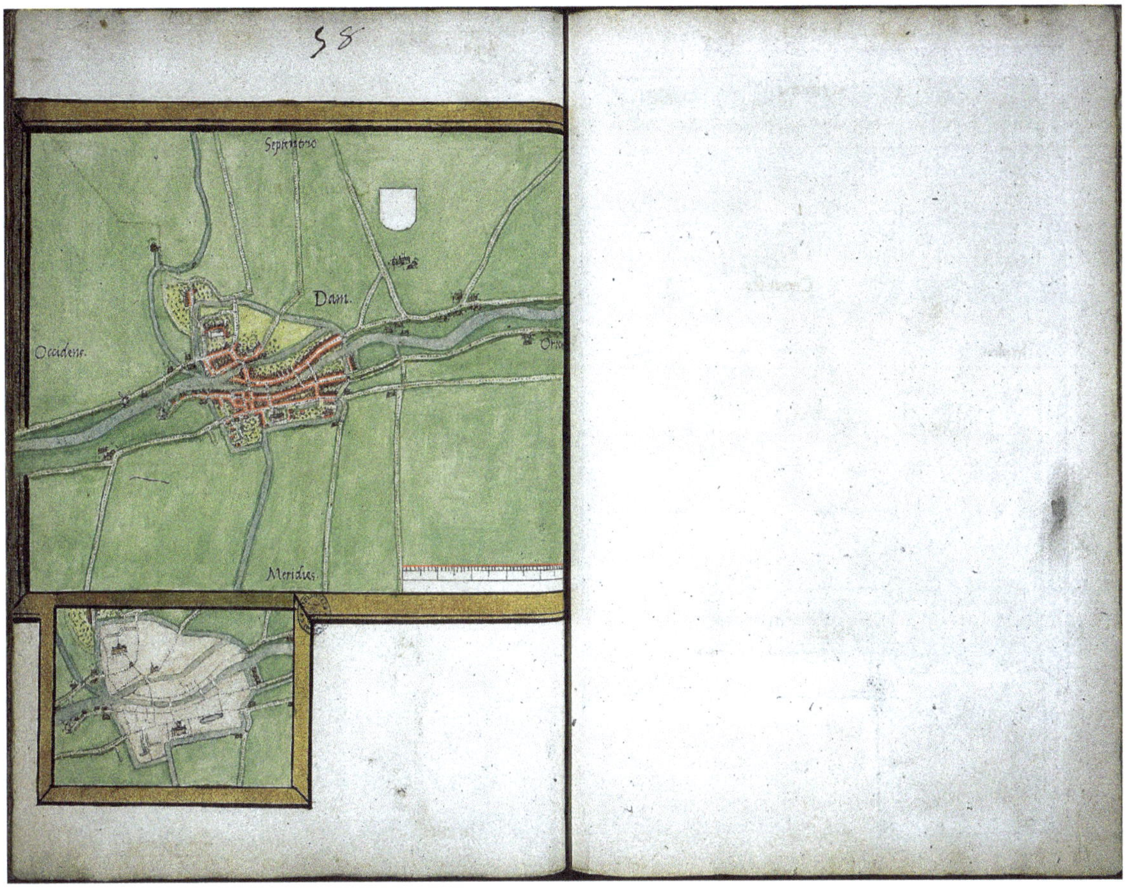
Dam.
Occidens.
Meridies.

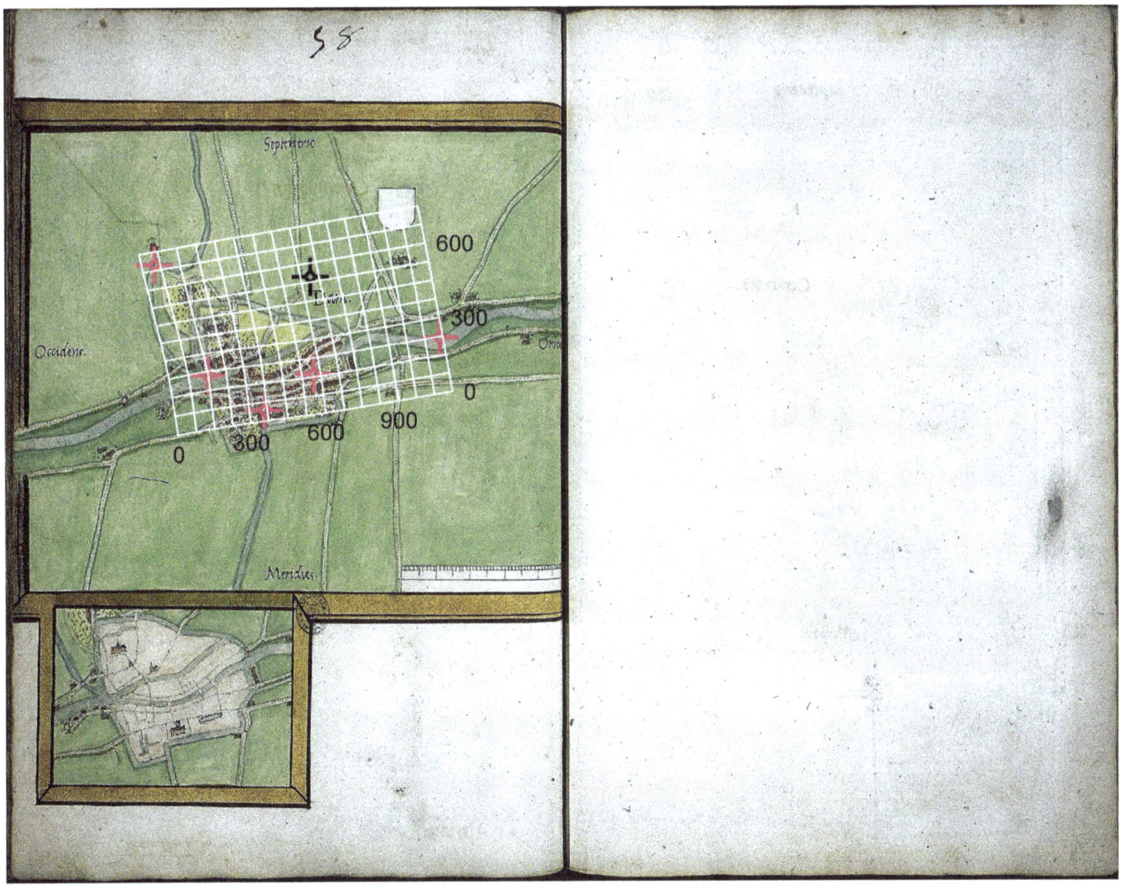
600
300
0
900
600
300
0
Occidens.
Meridies.

10. Arlon

Tables 3.10 measured and analysed).
Figures 3.10 (unmeasured, measured and analysed).

Measured							
Pixels	Scale stick original map (pixels)	Houses	Street sides	Houses per street side	Inhabitants (RF 4.0)	Inhabitants (RF 5.5)	Inhabitants (RF 7.0)
3156	missing	554	122	4.54	2217	3048	3879

Analysed	
September 5, 2023 9:19:15 PM	
Computation with 5 linked points	
(*italics* points are less reliable):	1. Sint-Donatuskerk
	2. Grand Place
	3. Grand Rue
	4. Marché au Legumes
	5. Rue des Remparts
Reference points (Reliability 0–5)	5
Transformation with 3 parameters	
Scale	1:6840
Rotation	9 °
Mean position error	8m

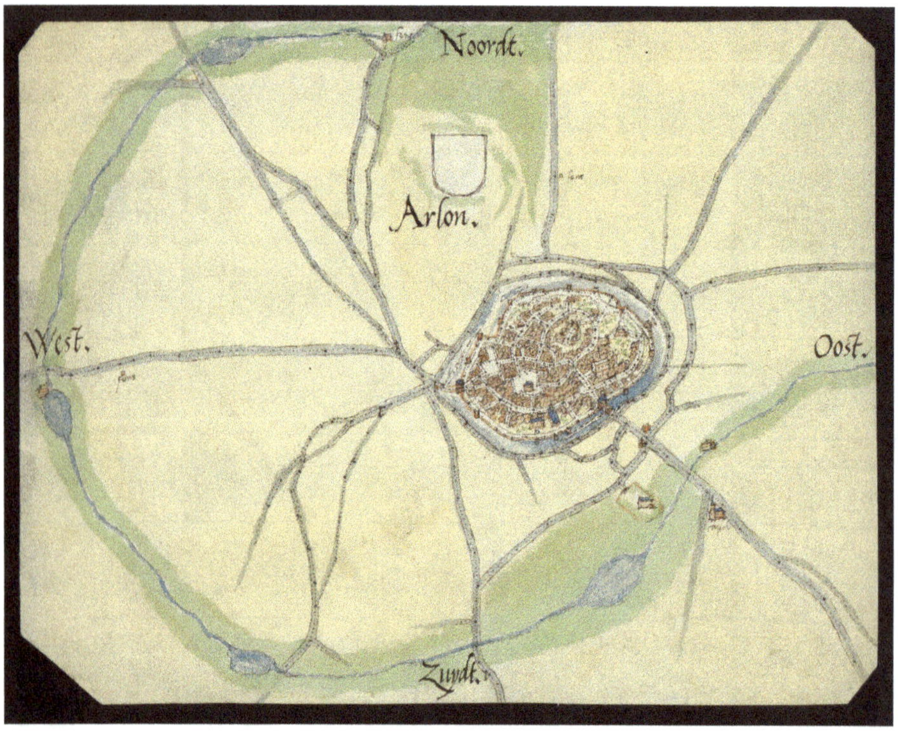

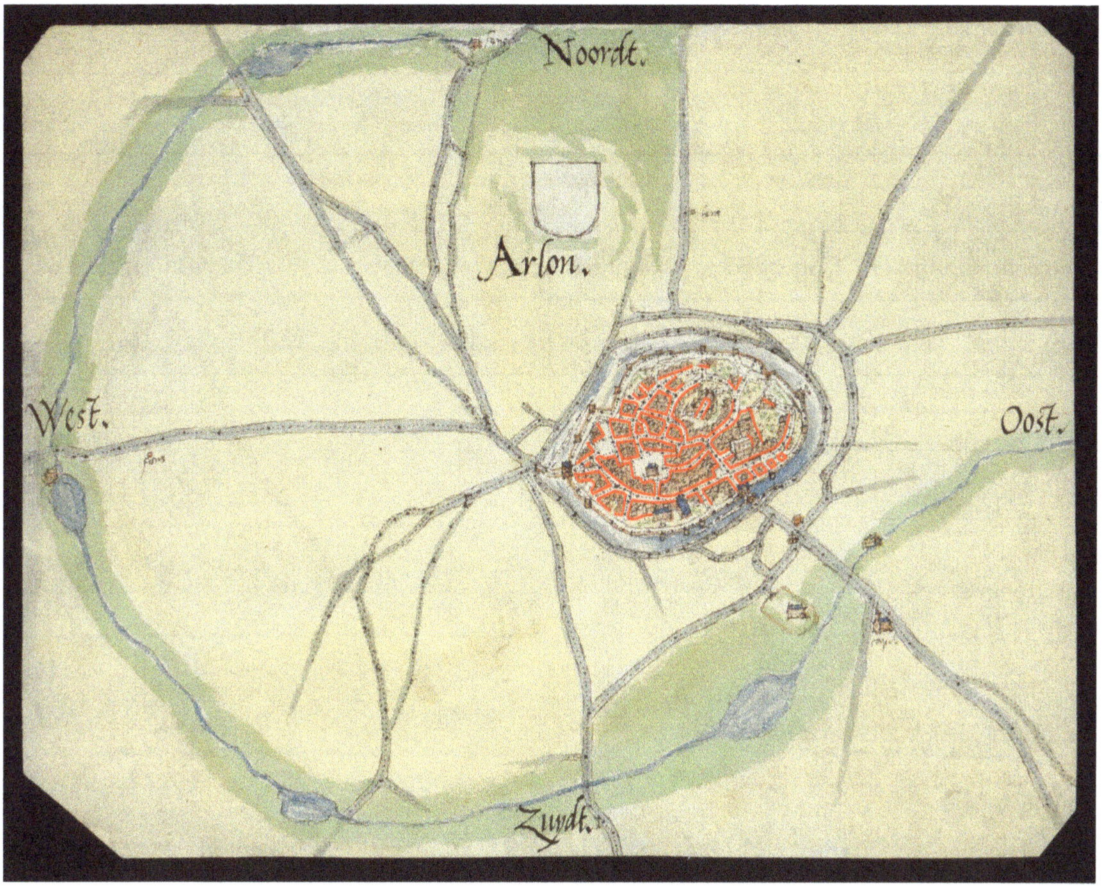
Noordt.
Arlon.
West.
Oost.
Zuydt.

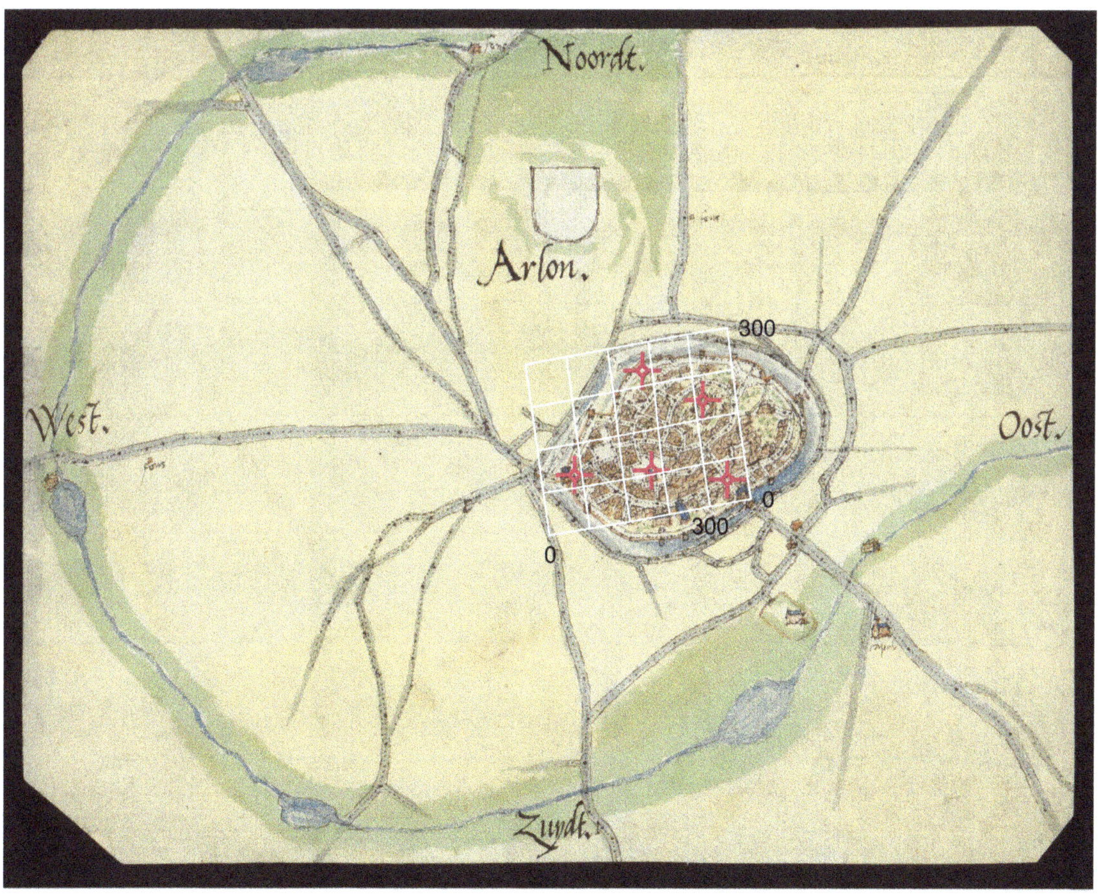
Noordt.
Arlon.
West.
Oost.
Zuydt.
300
0
300
0

11. Armentieres

Tables 3.11 (measured and analysed).
Figures 3.11 (unmeasured, measured and analysed).

Measured							
Pixels	Scale stick original map (pixels)	Houses	Street sides	Houses per street side	Inhabitants (RF 4.0)	Inhabitants (RF 5.5)	Inhabitants (RF 7.0)
7629	1027	1340	145	9.24	5359	7368	9378

Analysed	
October 15, 2023 4:47:56 PM	
Computation with 5 linked points	
(*italics* points are less reliable):	1. Sint-Vaast *2. Rue Jean Jaurès* *3. Rue de Dunkerque* *4. Rue des Fusilles* *5. Rue des Quais*
Reference points (Reliability 0–5)	1
Transformation with 3 parameters	
Scale	1:9410
Rotation	4°
Mean position error	53m

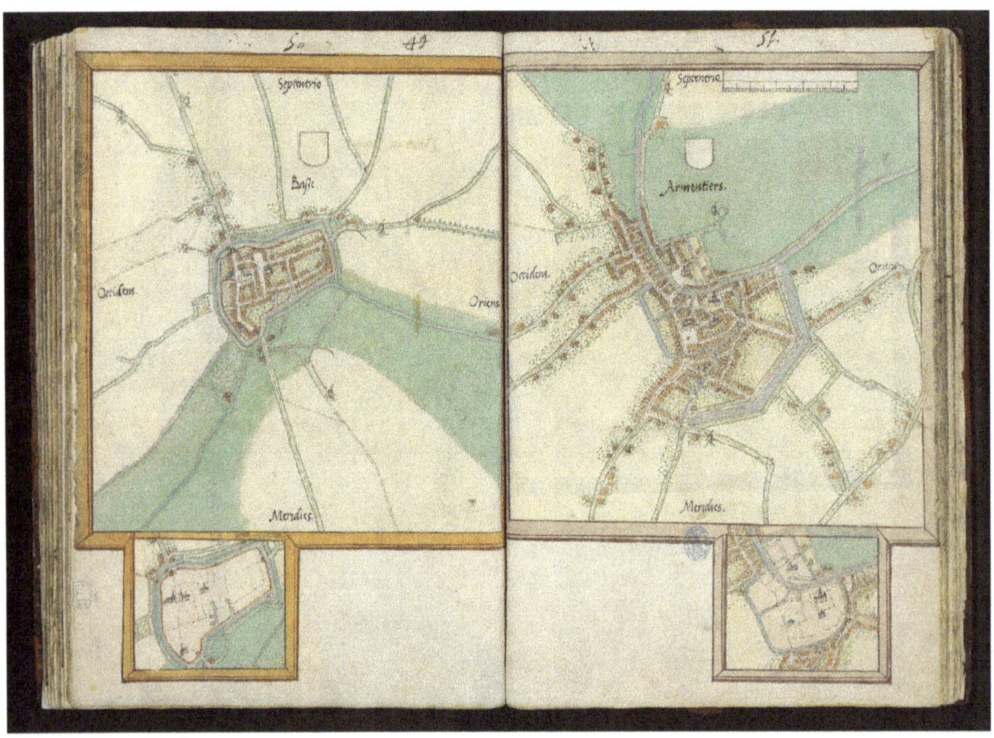

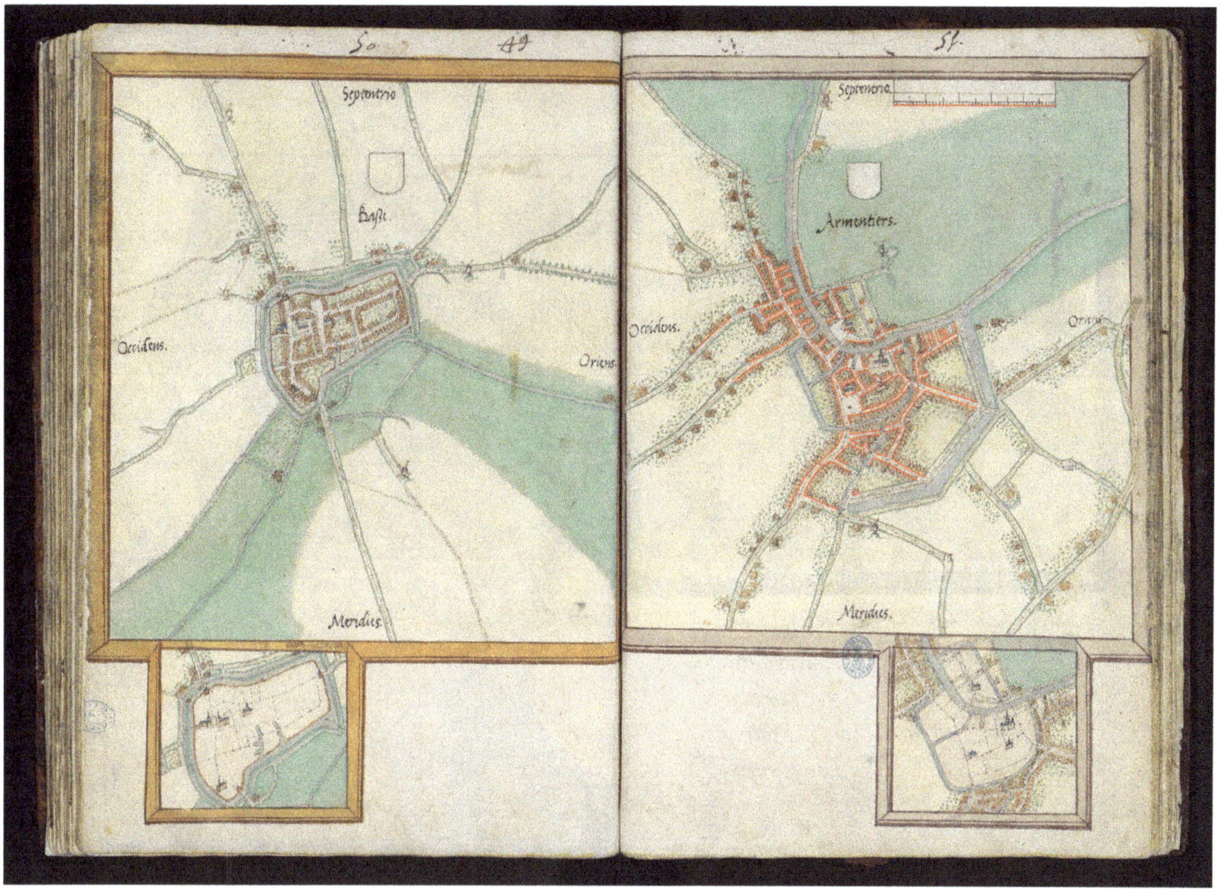
Septentrio
Occidens.
Oriens
Meridies.
Armentiers.

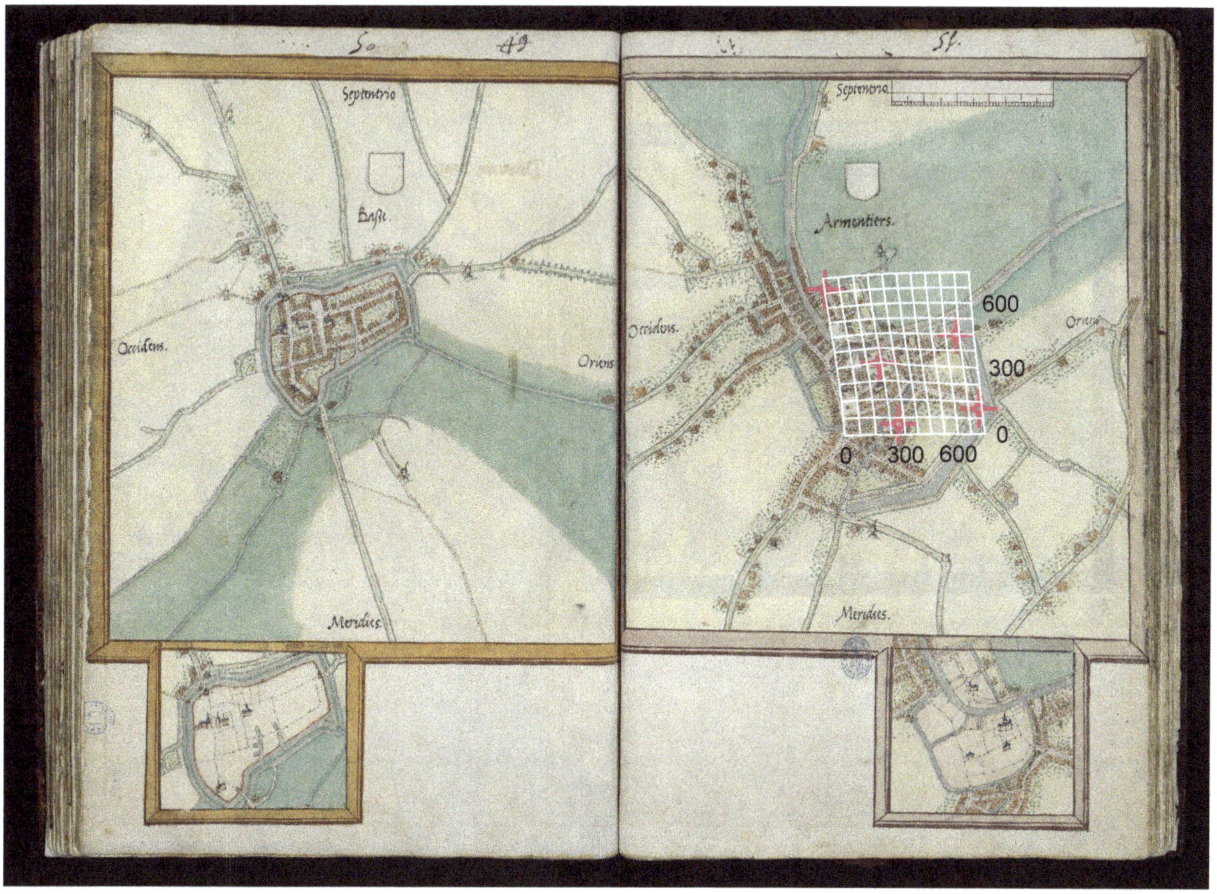
Septentrio
Occidens.
Oriens
Meridies.
Armentiers.
600
300
0
0
300
600

12. Arnemuiden

Tables 3.12 (measured and analysed).
Figures 3.12 (unmeasured, measured and analysed).

Measured

Pixels	Scale stick original map (pixels)	Houses	Street sides	Houses per street side	Inhabitants (RF 4.0)	Inhabitants (RF 5.5)	Inhabitants (RF 7.0)
1718	1008	302	27	11.17	1207	1659	2112

Analysed

January 24, 2023, 10:24:23 AM	
Computation with 5 linked points	
(*italics* points are less reliable):	*1. Molenweg*
	2. Lionstraat
	3. Marktstraat
	4. Walstraat
	5. Langstraat
Reference points (Reliability 0–5)	1
Transformation with 3 parameters	
Scale	1:8610
Rotation	10 °
Mean position error	29m

Armuyen.
Ortens.
4
© Biblioteca Nacional de España

Armuyen.
0 180 360 540 720 0
Ortens.
4

13. Arnhem

Tables 3.13 (measured and analysed).
Figures 3.13 (unmeasured, measured and analysed).

Measured

Pixels	Scale stick original map (pixels)	Houses	Street sides	Houses per street side	Inhabitants (RF 4.0)	Inhabitants (RF 5.5)	Inhabitants (RF 7.0)
6855	1009	1204	190	6.34	4815	6621	8426

Analysed

February 21, 2023 7:49:56 PM	
Computation with 5 linked points	
(*italics* points are less reliable):	1. Sint-Walburgisbasiliek 2. Sabelspoort 3. Sint-Eusebiuskerk 4. Janspoort 5. Rijnpoort
Reference points (Reliability 0–5)	5
Transformation with 3 parameters	
Scale	1:8700
Rotation	16 °
Mean position error	18m

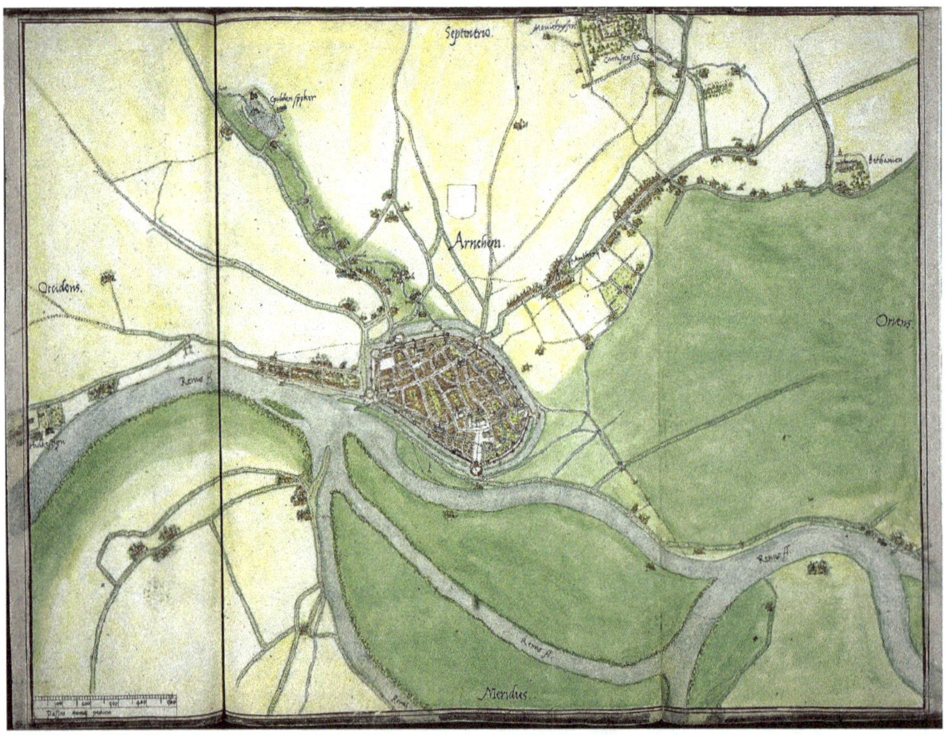

Septentrio.
Arnhem
Occidens.
Oriens.
Renus fl.
Meridies.

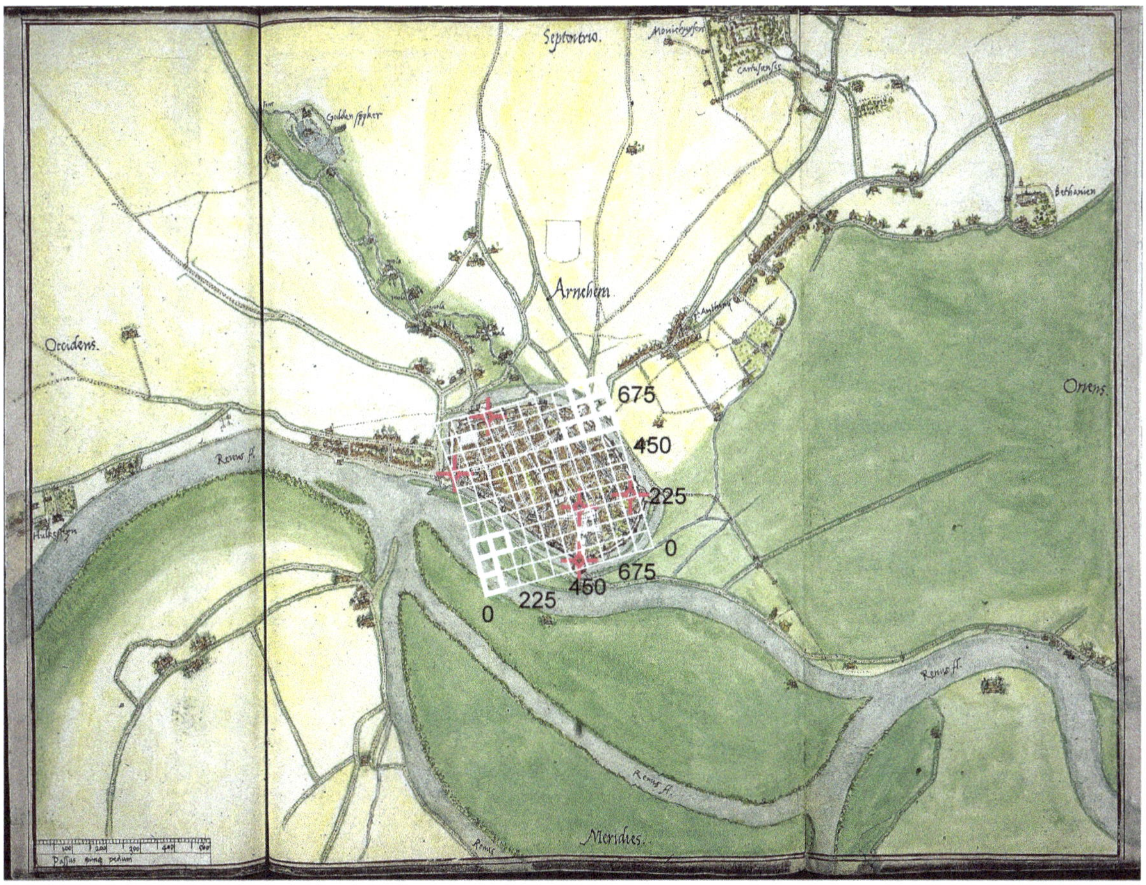
Septentrio.
Arnhem
Occidens.
Oriens.
Renus fl.
Meridies.
675
450
225
0
0
225
450
675

14. Arras
Tables 3.14 (measured and analysed).
Figures 3.14 (unmeasured, measured and analysed).

Measured

Pixels	Scale stick original map (pixels)	Houses	Street sides	Houses per street side	Inhabitants (RF 4.0)	Inhabitants (RF 5.5)	Inhabitants (RF 7.0)
21,938	897	3852	573	6.72	15,409	21,188	26,966

Analysed

February 21, 2023 11:21:34 PM	
Computation with 5 linked points	
(*italics* points are less reliable):	1. Cathedrale
	2. Bastion de Roeux
	3. La Scarpe
	4. Belfort
	5. Grande Place
Reference points (Reliability 0–5)	5
Transformation with 3 parameters	
Scale	1:8920
Rotation	8 °
Mean position error	30m

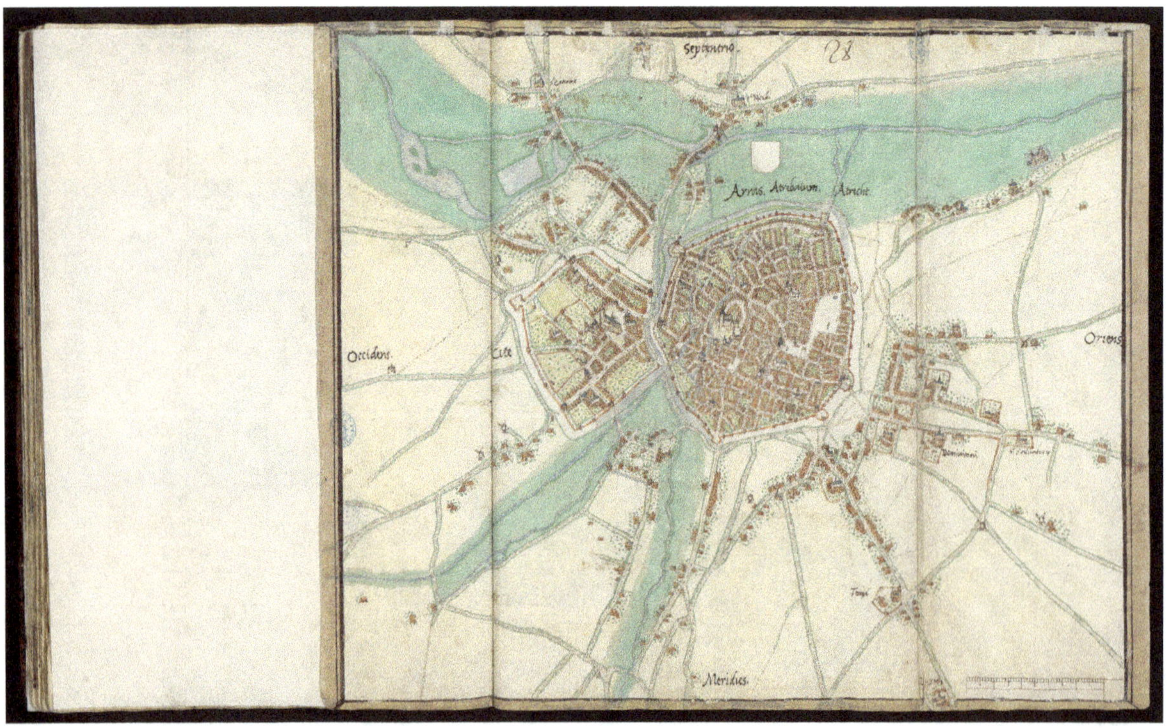

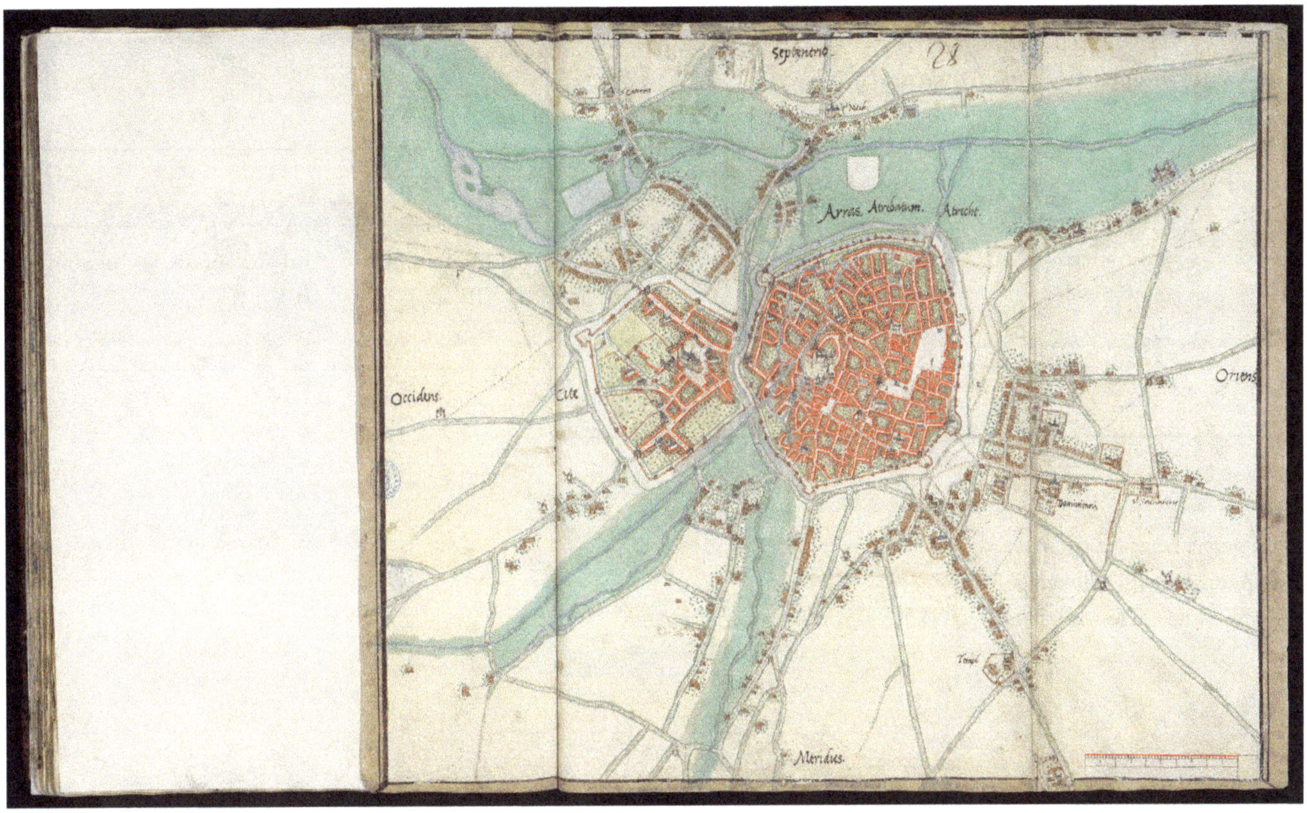
Septentrio
28
Arras. Atrebatum. Atrecht.
Occidens
Cite
Oriens
Meridies

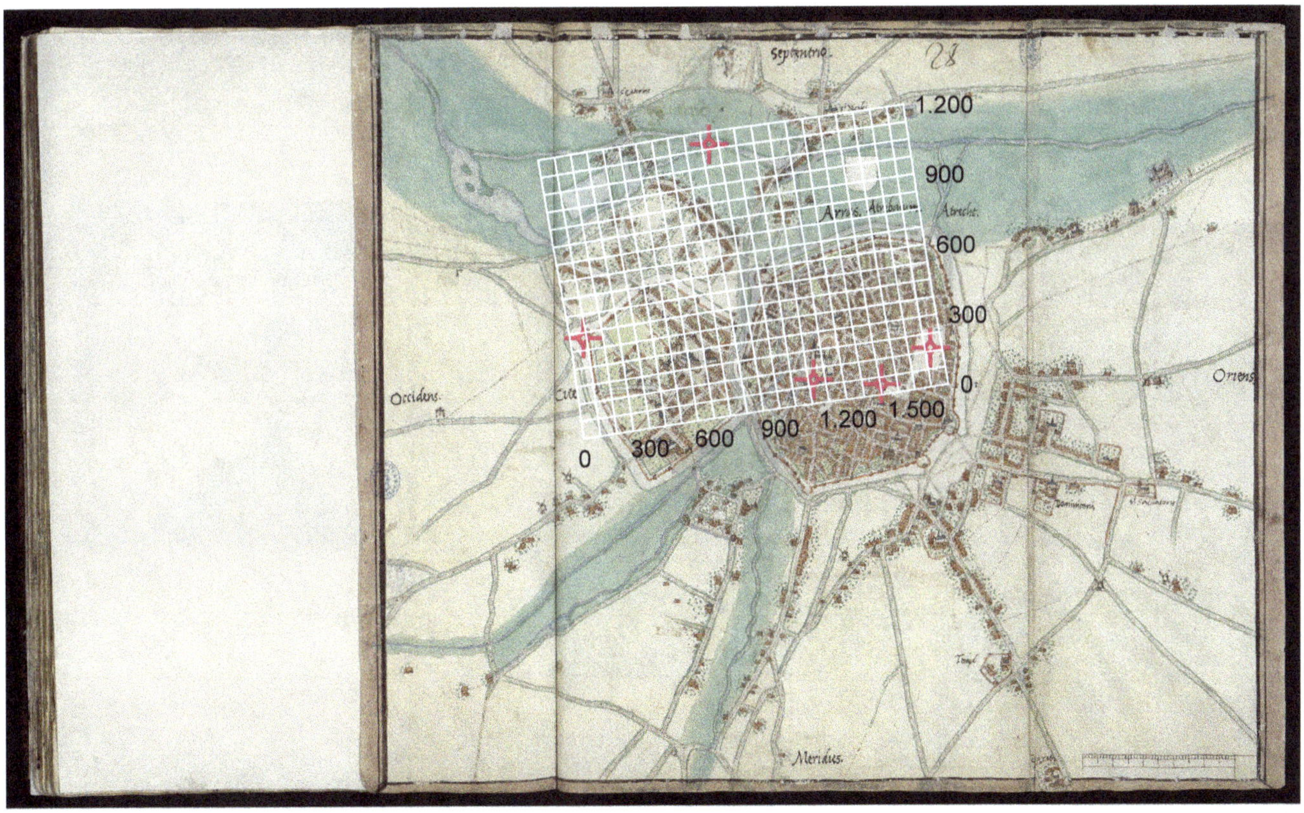
Septentrio
28
1.200
900
600
300
0
Occidens
Cite
Oriens
0
300
600
900
1.200
1.500
Meridies

15. Asperen

Tables 3.15 (measured and analysed).
Figures 3.15 (unmeasured, measured and analysed).

Measured							
Pixels	Scale stick original map (pixels)	Houses	Street sides	Houses per street side	Inhabitants (RF 4.0)	Inhabitants (RF 5.5)	Inhabitants (RF 7.0)
1739	1007	305	39	7.83	1221	1680	2138

Analysed	
February 22, 2023 4:24:35 PM	
Computation with 5 linked points	
(*italics* points are less reliable):	1. Sint-Catharinakerk
	2. Molenwal
	3. Stadswal
	4. Achterstraat
	5. Waterpoor
Reference points (Reliability 0–5)	5
Transformation with 3 parameters	
Scale	1:8040
Rotation	12 °
Mean position error	32m

Leerdam.
Asperen.
Occidens.
Oriens.
Huckelem.
Meridies

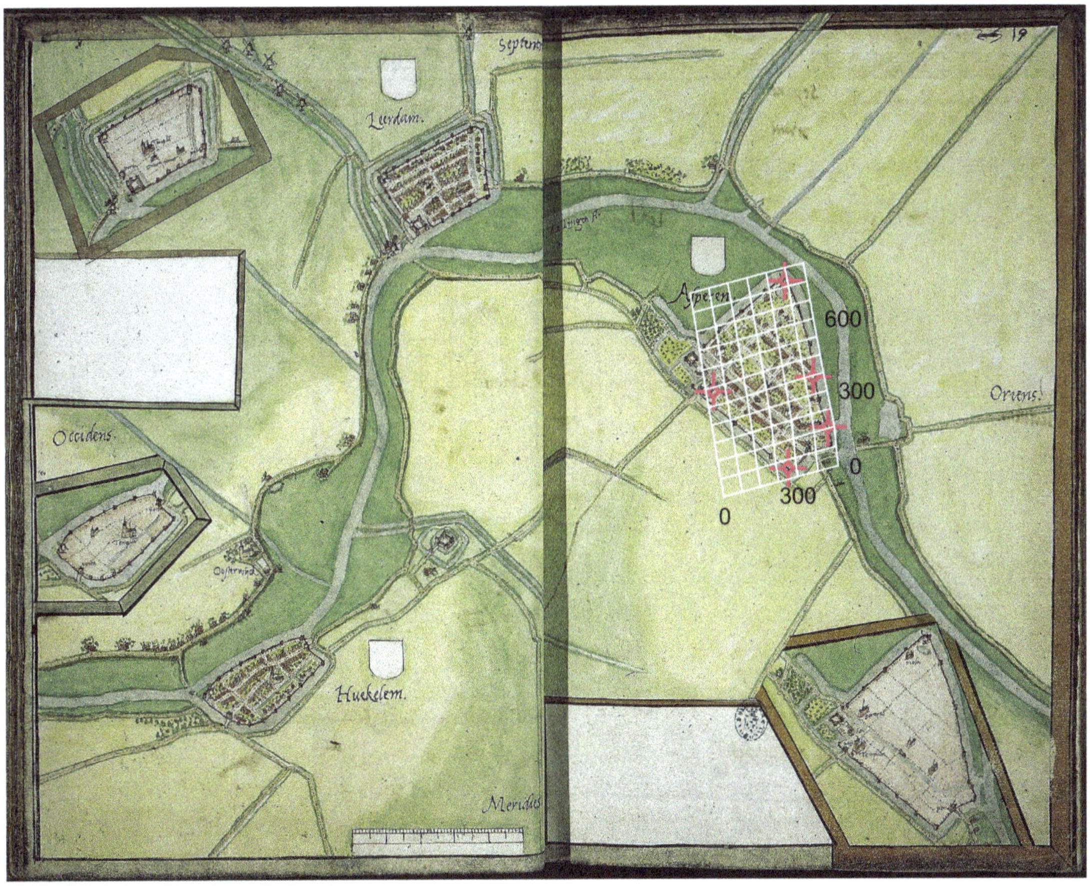
Leerdam.
Asperen.
600
300
0
300
0
Occidens.
Oriens.
Huckelem.
Meridies

16. Ath

Tables 3.16 (measured and analysed).
Figures 3.16 (unmeasured, measured and analysed).

Measured

Pixels	Scale stick original map (pixels)	Houses	Street sides	Houses per street side	Inhabitants (RF 4.0)	Inhabitants (RF 5.5)	Inhabitants (RF 7.0)
5581	1018	990	140	7.07	3960	5445	6930

Analysed

February 22, 2023 8:33:17 PM	
Computation with 5 linked points	
(*italics* points are less reliable):	1. Église Saint-Julien 2. Square Saint-Julien 3. Chateau Burbant 4. Grand Place 5. Rue du Noir Boeuf
Reference points (Reliability 0–5)	5
Transformation with 3 parameters	
Scale	1:7480
Rotation	11 °
Mean position error	16m

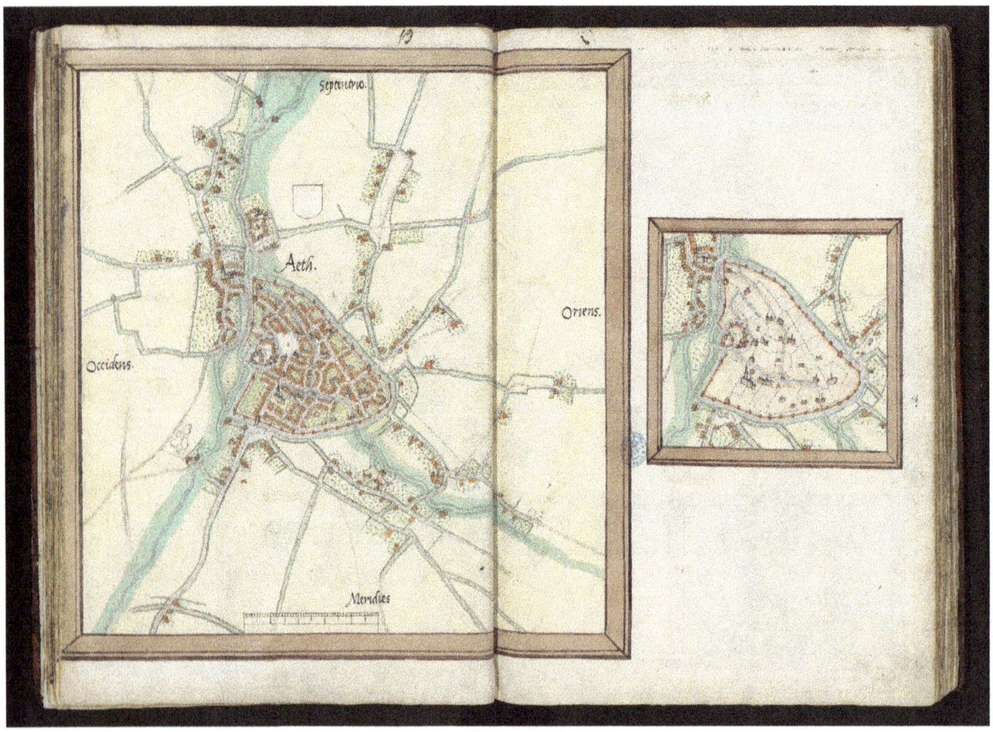

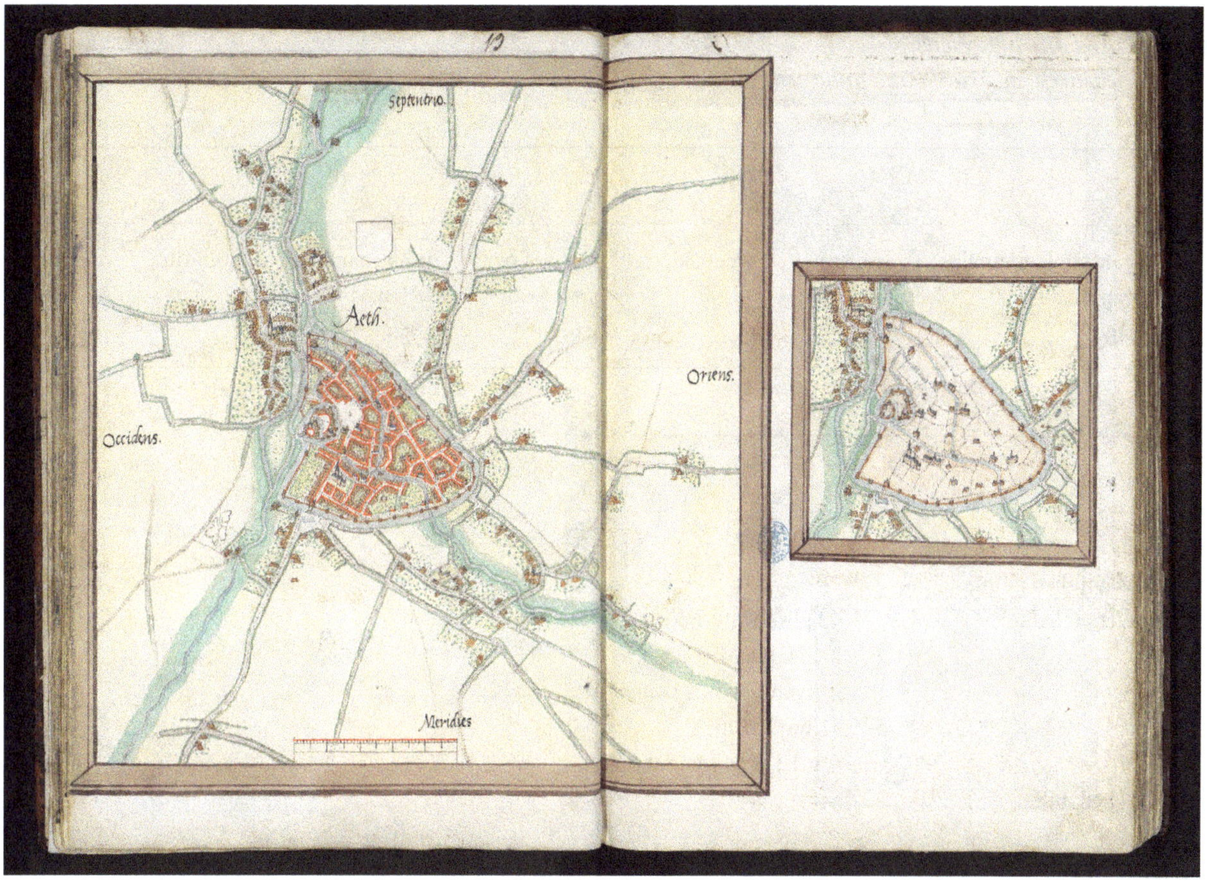
Septentrio.
Aeth.
Oriens.
Occidens.
Meridies

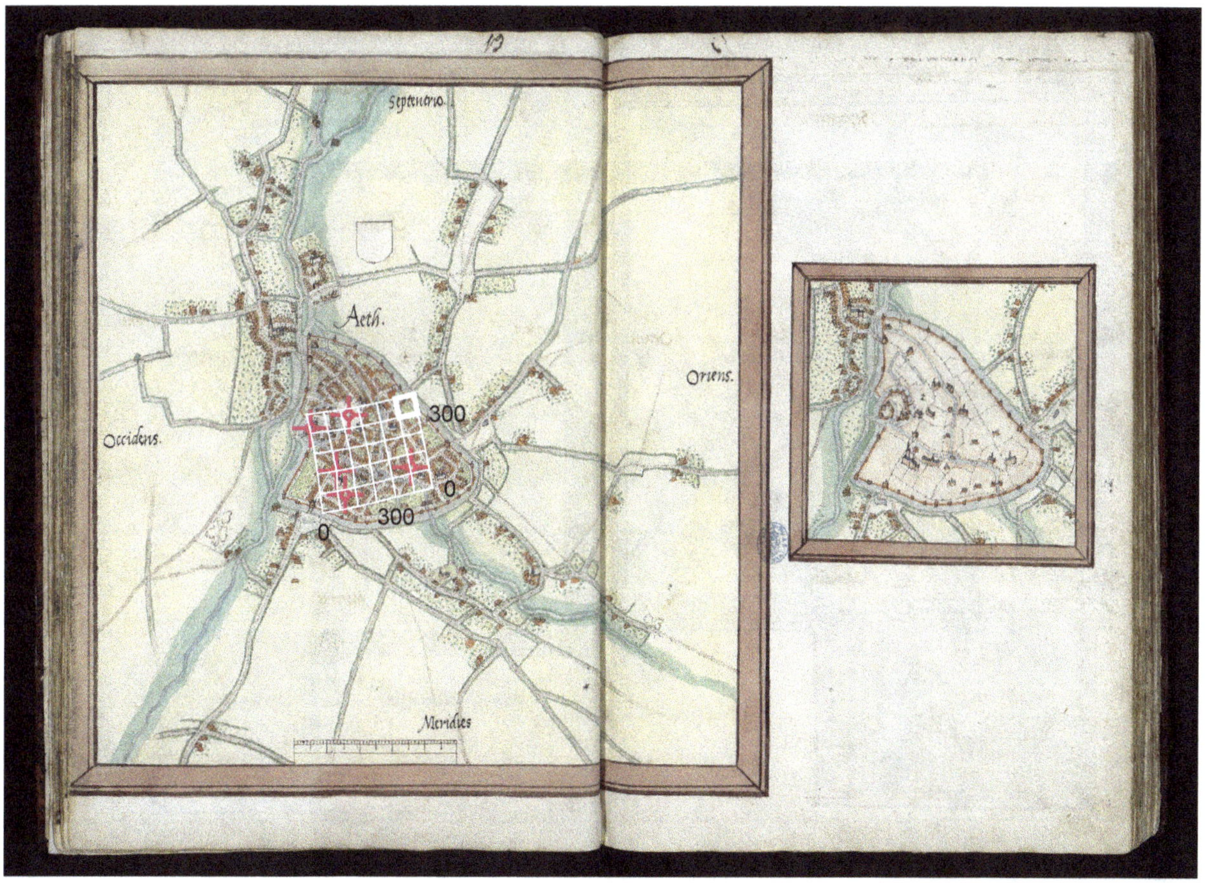
Septentrio.
Aeth.
Oriens.
Occidens.
300
0
300
0
Meridies

17. Avesnes-sur-Helpe
Tables 3.17 (measured and analysed).
Figures 3.17 (unmeasured, measured and analysed).

Measured							
Pixels	Scale stick original map (pixels)	Houses	Street sides	Houses per street side	Inhabitants (RF 4.0)	Inhabitants (RF 5.5)	Inhabitants (RF 7.0)
4061	1015	713	119	5.99	2852	3922	4992

Analysed	
February 22, 2023 10:36:10 PM	
Computation with 5 linked points	
(*italics* points are less reliable):	1. Église Saint-Nicolas 2. L'Helpe Majeur 3. Avenue Louis Loucheur 4. Rue Tassin 5. 2. L'Helpe Majeur (est)
Reference points (Reliability 0–5)	5
Transformation with 3 parameters	
Scale	1:7430
Rotation	11 °
Mean position error	16m

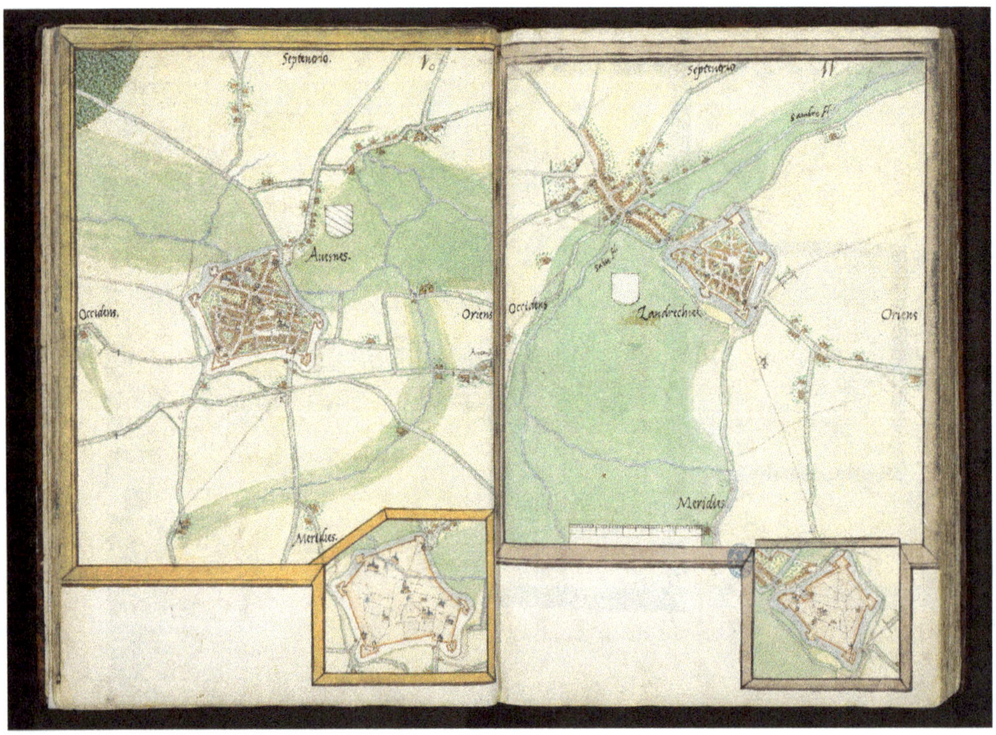

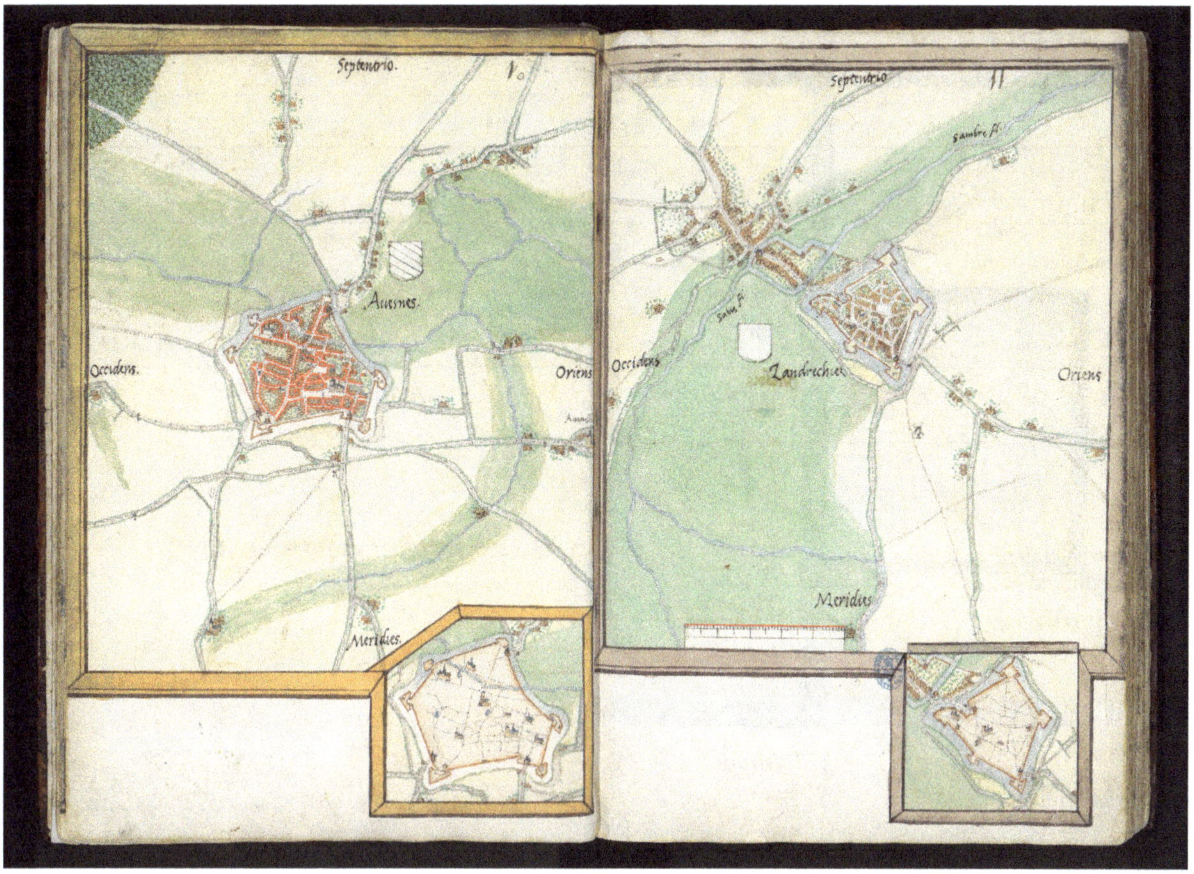
Septentrio.
10
Auesnes.
Occidens.
Oriens
Meridies.
Septentrio
11
Sambre fl.
Occidens
Landrechies
Oriens
Meridus

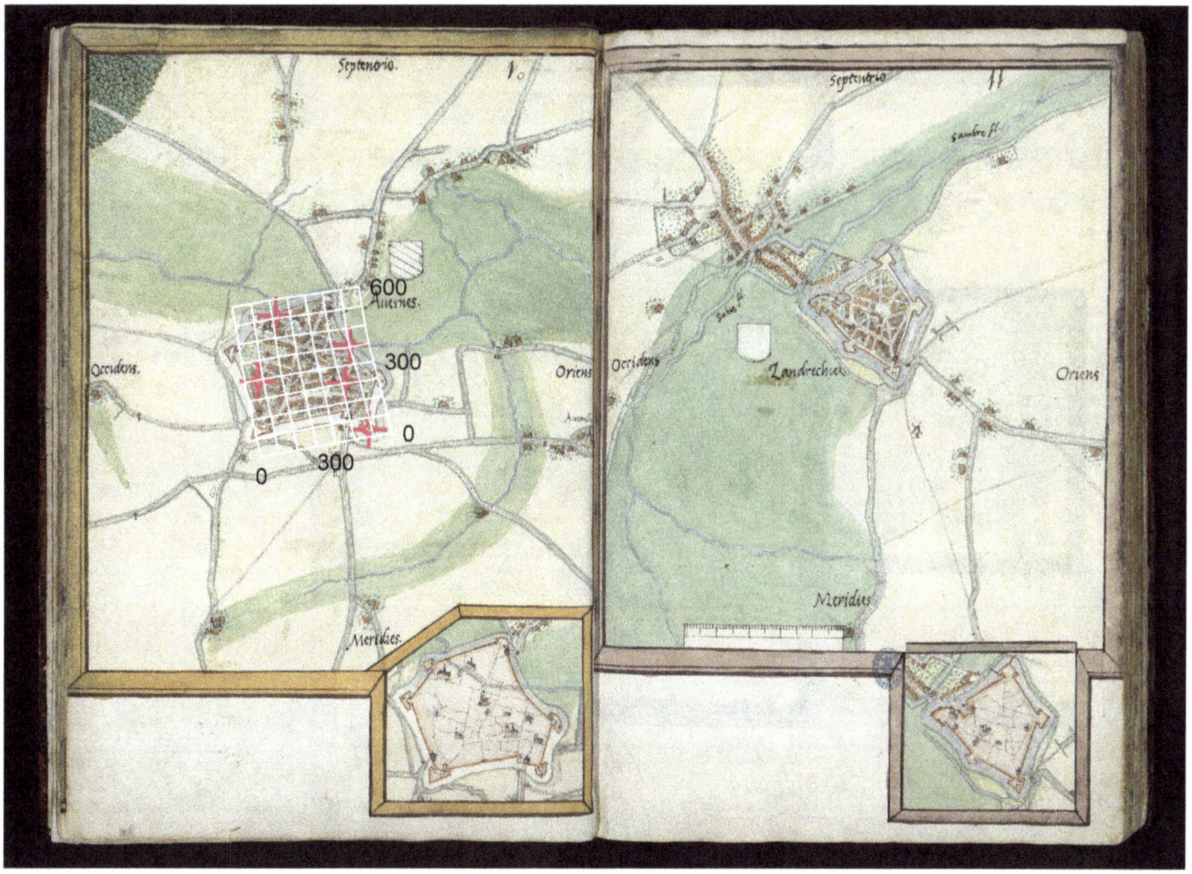
Septentrio.
10
600
Auesnes.
300
Occidens.
Oriens
0
0
300
Meridies.
Septentrio
11
Sambre fl.
Occidens
Landrechies
Oriens
Meridus

18. Axel

Tables 3.18 (measured and analysed).
Figures 3.18 (unmeasured, measured and analysed).

Measured

Pixels	Scale stick original map (pixels)	Houses	Street sides	Houses per street side	Inhabitants (RF 4.0)	Inhabitants (RF 5.5)	Inhabitants (RF 7.0)
6124	1019	1075	151	7.12	4301	5915	7528

Analysed

February 23, 2023 10:32:54 AM	
Computation with 5 linked points	
(*italics* points are less reliable):	*1. Markt*
	2. Molenstraat
	3. Pachtershof
	4. Kasteel
	5. Walstraat
Reference points (Reliability 0–5)	0
Transformation with 3 parameters	
Scale	1:8520
Rotation	5 °
Mean position error	21m

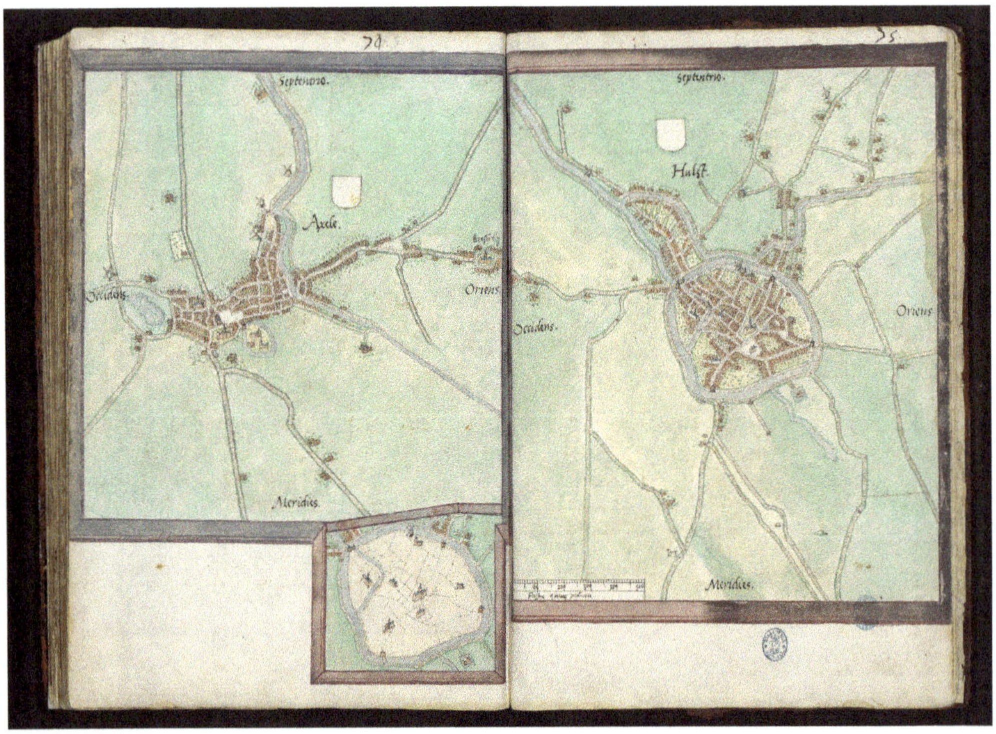

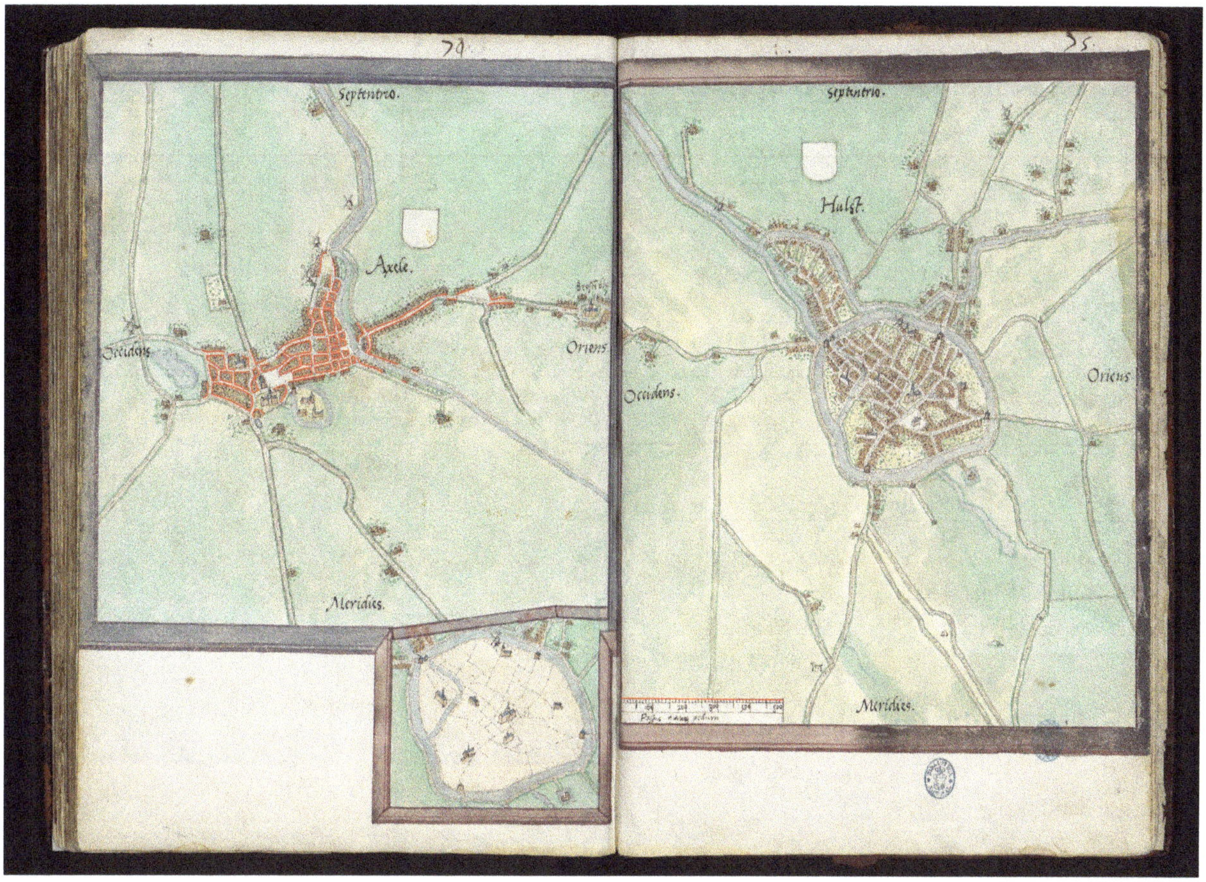
Septentrio.
Axele.
Occidens.
Oriens.
Meridies.
Hulst.
Septentrio.
Occidens.
Oriens.
Meridies.

Septentrio.
Axele.
Occidens.
Oriens.
Meridies.
Hulst.
Septentrio.
Occidens.
Oriens.
Meridies.

19. Bailleul

Tables 3.19 (measured and analysed).
Figures 3.19 (unmeasured, measured and analysed).

Measured							
Pixels	Scale stick original map (pixels)	Houses	Street sides	Houses per street side	Inhabitants (RF 4.0)	Inhabitants (RF 5.5)	Inhabitants (RF 7.0)
5246	1027	921	69	13.35	3685	5067	6448

Analysed	
February 24, 2023 9:00:19 PM	
Computation with 5 linked points	
(*italics* points are less reliable):	1. Sint-Vaaskerk 2. Rue de Lille 3. Rue de Cassel 4. Becque du Mont noir 5. Rue de Ypres
Reference points (Reliability 0–5)	5
Transformation with 3 parameters	
Scale	1:7620
Rotation	6 °
Mean position error	24m

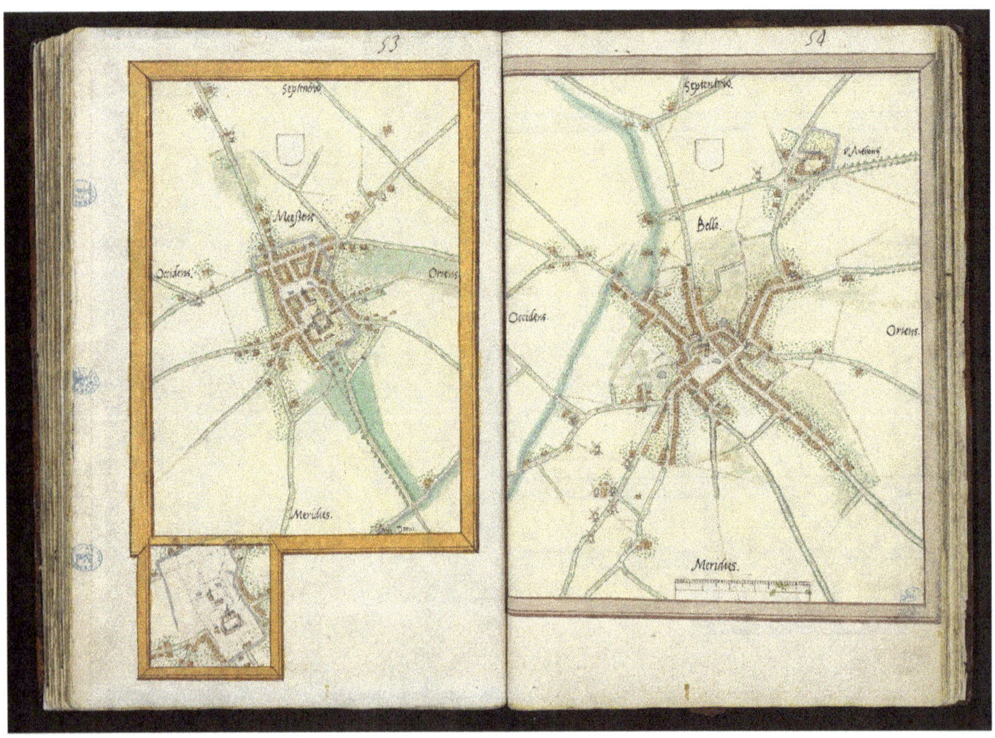

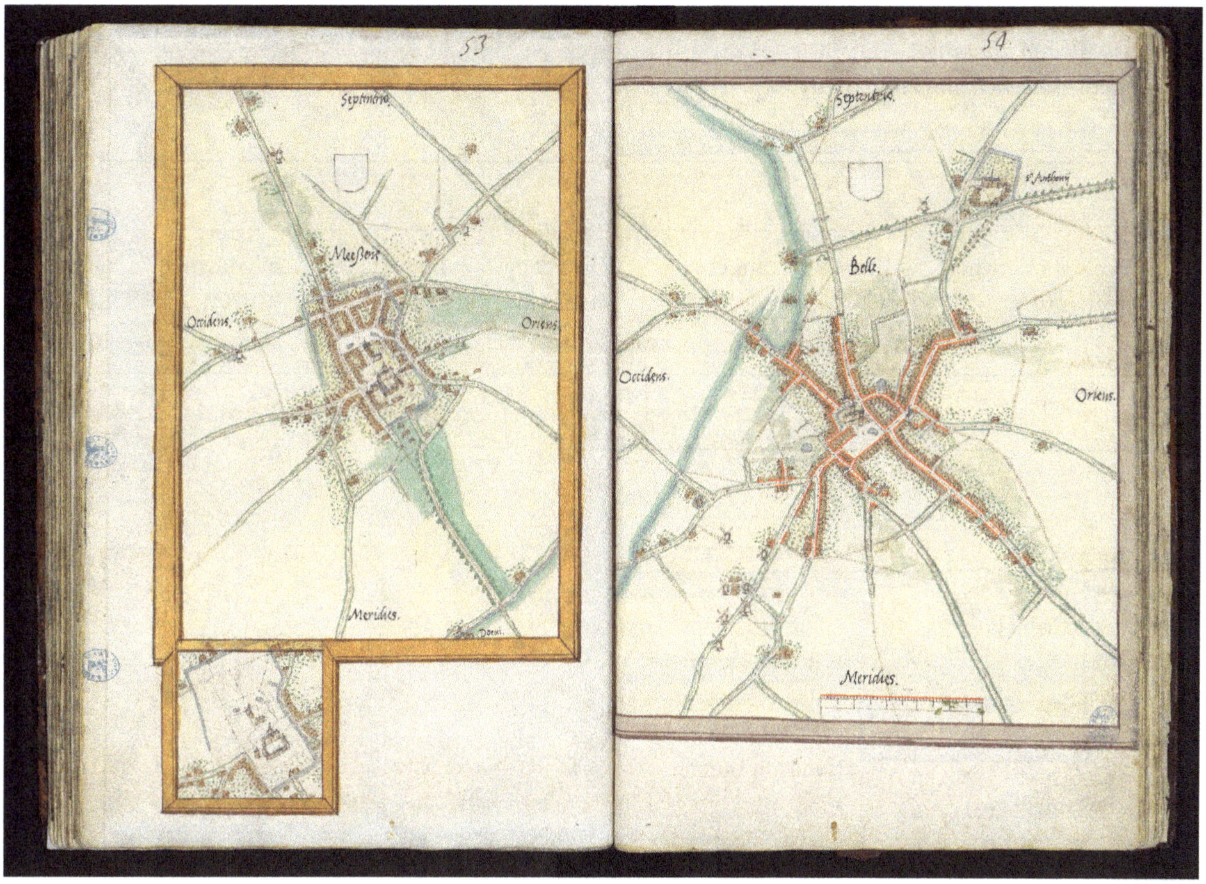
53
54
Septentrio
Occidens
Oriens
Meridies
Belle

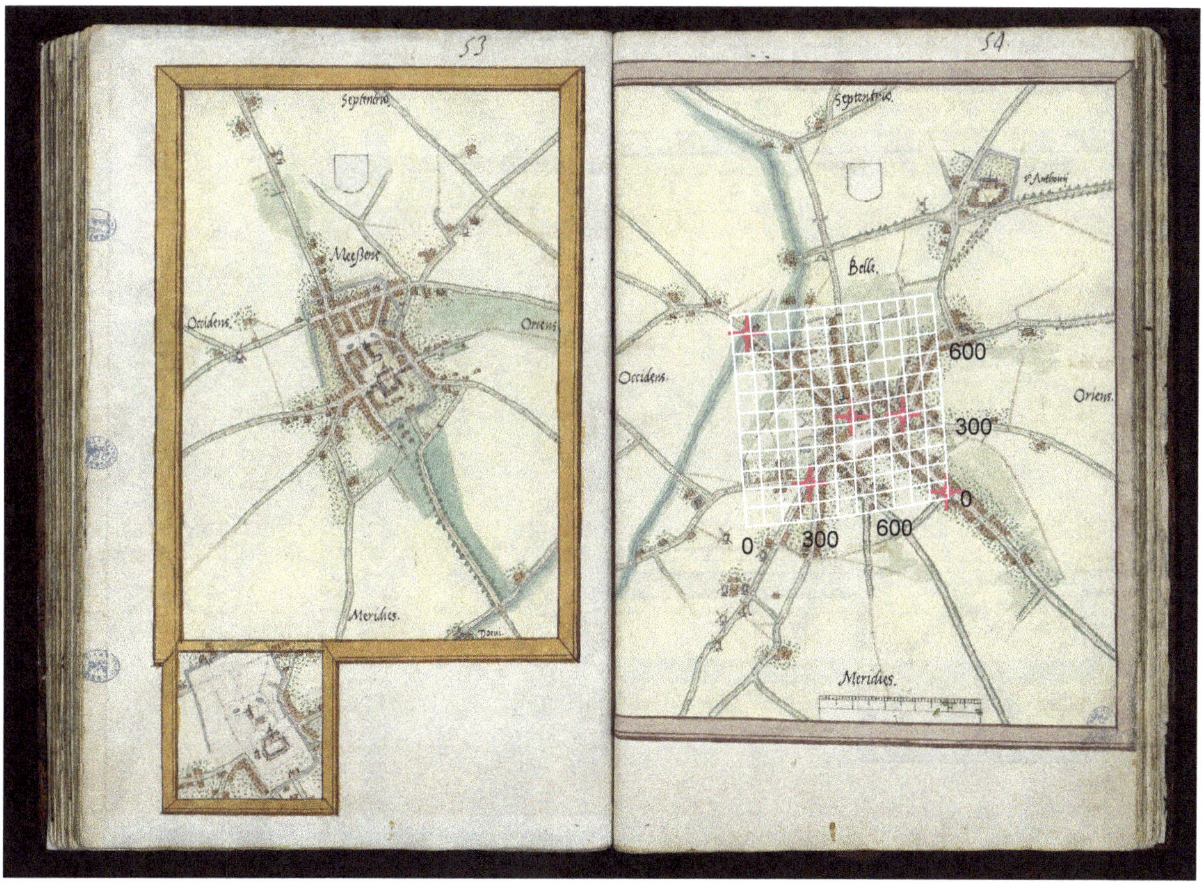
53
54
Septentrio
Occidens
Oriens
Meridies
Belle
600
300
0
0
300
600

20. Bapaume
Tables 3.20 (measured and analysed).
Figures 3.20 (unmeasured, measured and analysed).

Measured							
Pixels	Scale stick original map (pixels)	Houses	Street sides	Houses per street side	Inhabitants (RF 4.0)	Inhabitants (RF 5.5)	Inhabitants (RF 7.0)
3215	1023	565	67	8.4	3258	3105	3952

Analysed	
February 24, 2023 11:15:08 PM Computation with 5 linked points (*italics* points are less reliable):	 1. Rue de la République 2. Rue de Douai 3. Rue dus bessol 4. Eglise Saint-Nicolas 5. Rue du Donjon
Reference points (Reliability 0–5)	5
Transformation with 3 parameters	
Scale	1:8310
Rotation	12 °
Mean position error	9m

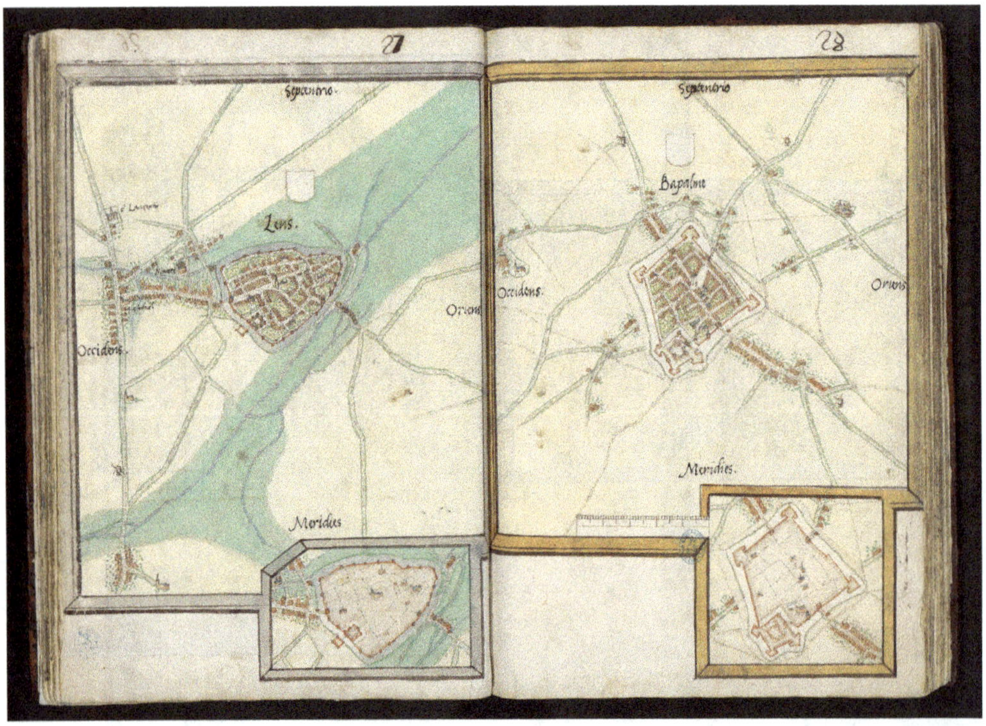

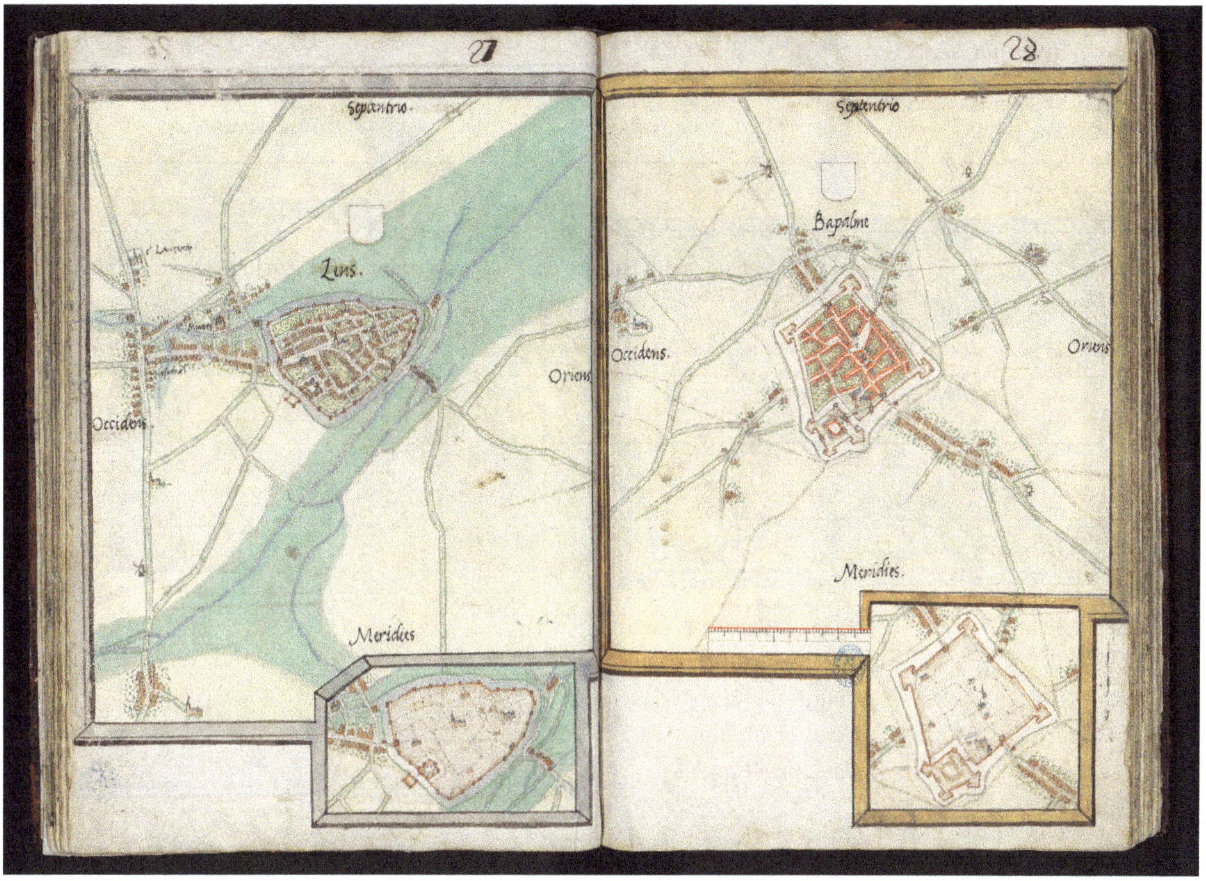
27
28
Septentrio
Lens
Occidens
Oriens
Meridies
Septentrio
Bapalme
Occidens
Oriens
Meridies

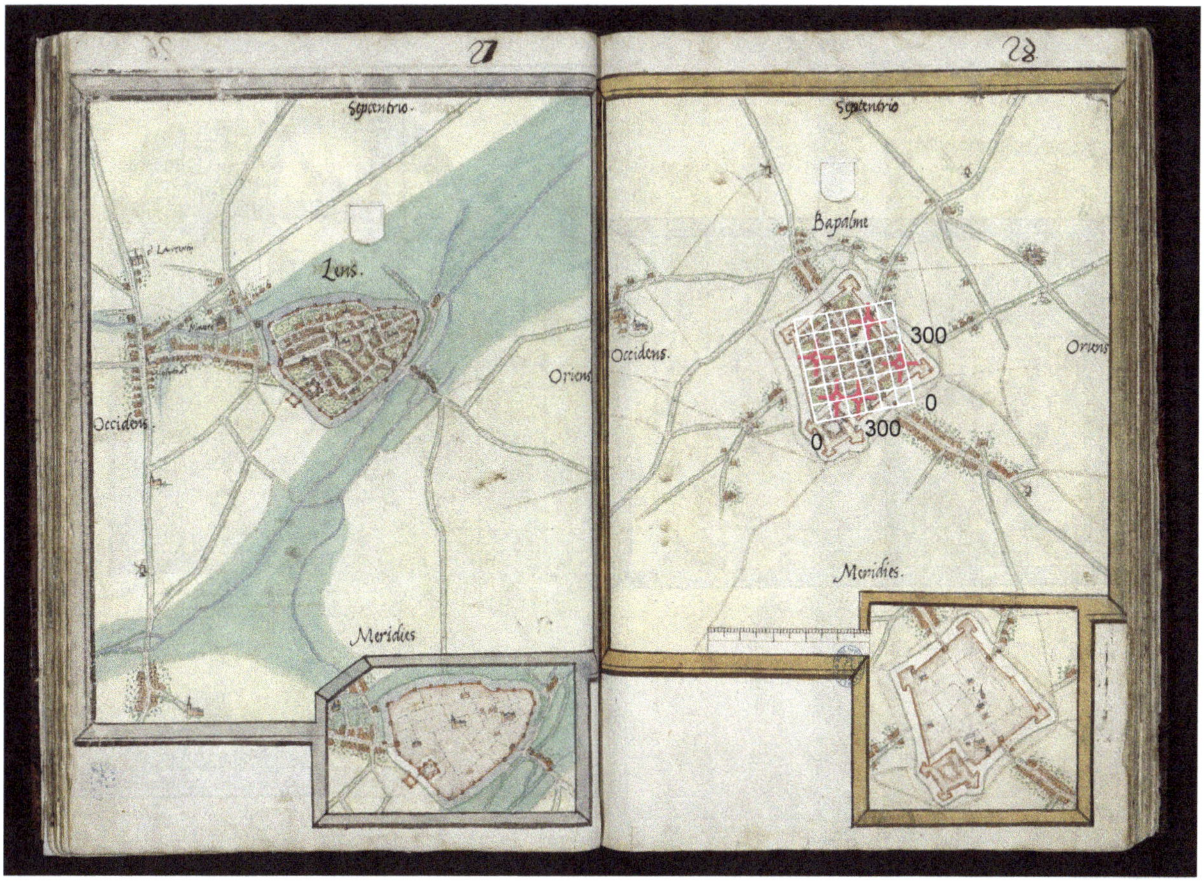
27
28
Septentrio
Lens
Occidens
Oriens
Meridies
Septentrio
Bapalme
300
0
0
300
Occidens
Oriens
Meridies

21. Bastogne
Tables 3.21 (measured and analysed).
Figures 3.21 (unmeasured, measured and analysed).

Measured							
Pixels	Scale stick original map (pixels)	Houses	Street sides	Houses per street side	Inhabitants (RF 4.0)	Inhabitants (RF 5.5)	Inhabitants (RF 7.0)
2671	missing	469	87	5.39	1876	2580	3283

Analysed	
September 5, 2023 10:58:46 PM	
Computation with 5 linked points	
(*italics* points are less reliable):	1. Rue des Remparts (sud) *2. Rue des Remparts (nord)* *3. Rue de la Halle* *4. Rue des Jardins* *5. Rue Porte-Haute*
Reference points (Reliability 0–5)	1
Transformation with 3 parameters	
Scale	1:7240
Rotation	10 °
Mean position error	36m

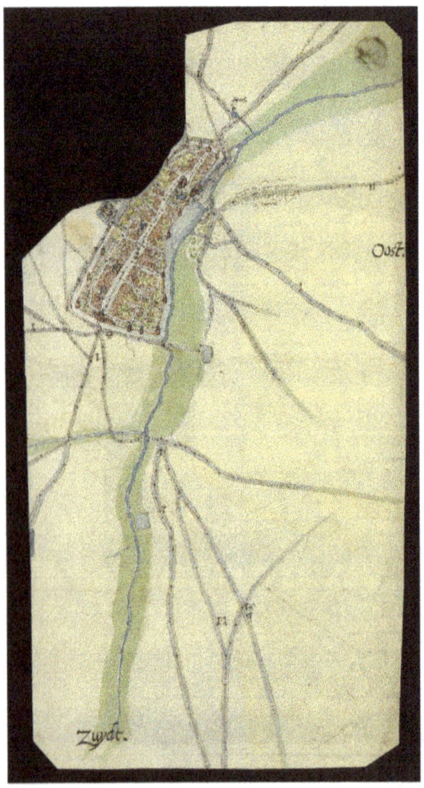

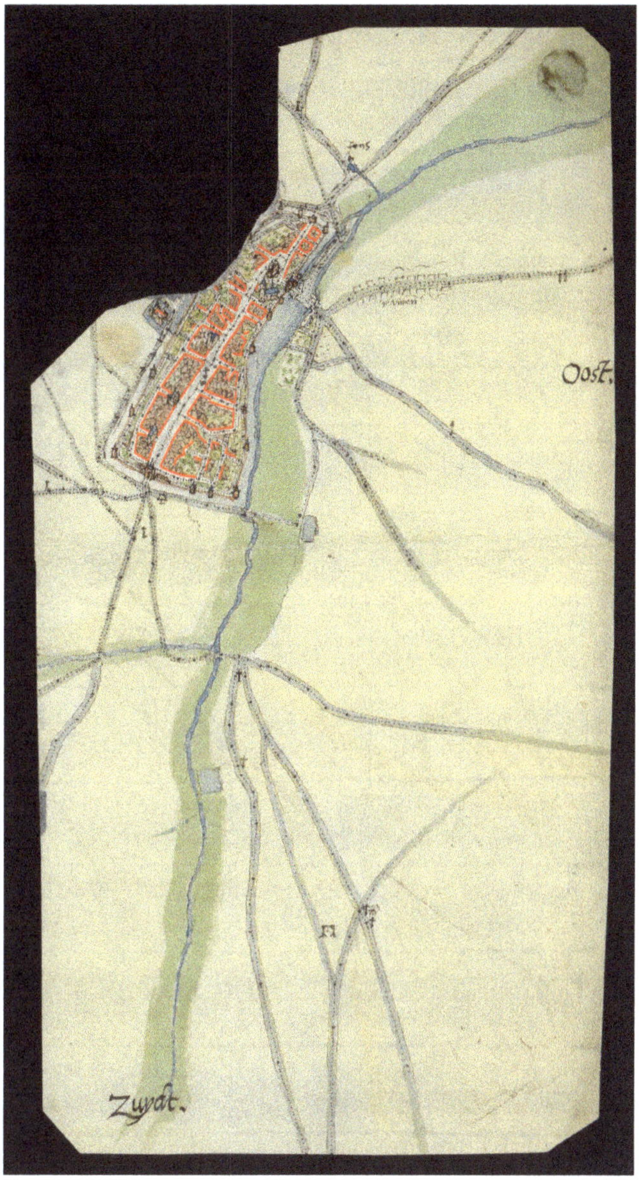
Oost.
Zuydt.

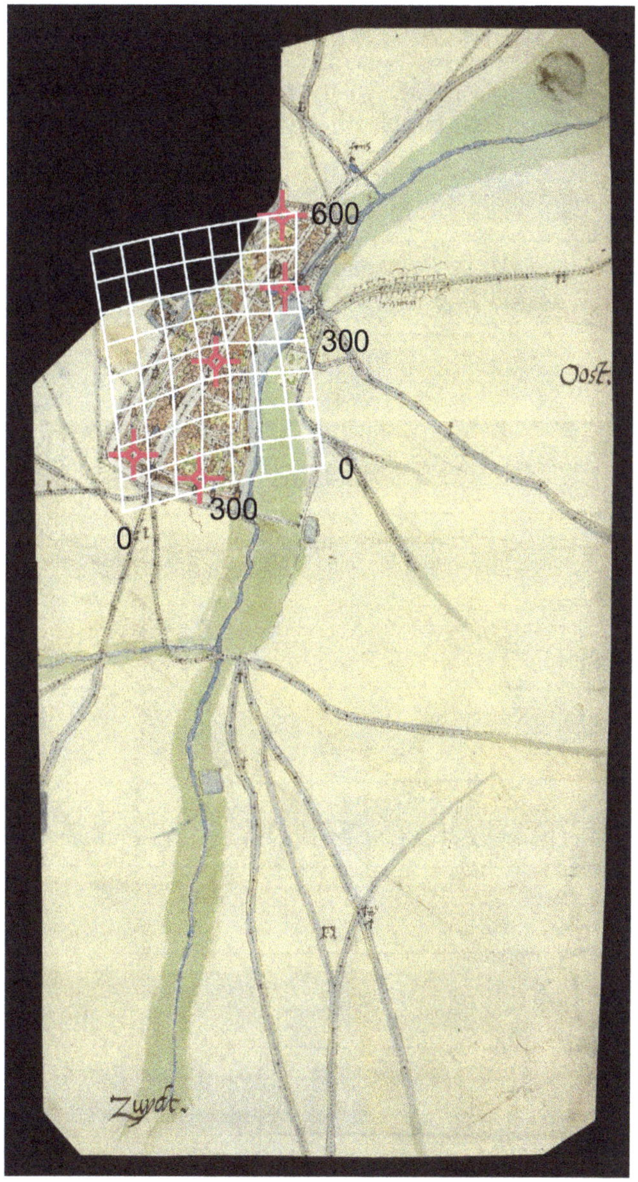
600
300
0
300
0
Oost.
Zuydt.

22. Bavay
Tables 3.22 (measured and analysed).
Figures 3.22 (unmeasured, measured and analysed).

Measured

Pixels	Scale stick original map (pixels)	Houses	Street sides	Houses per street side	Inhabitants (RF 4.0)	Inhabitants (RF 5.5)	Inhabitants (RF 7.0)
1666	1004	293	42	6.97	1170	1609	2048

Analysed

February 26, 2023 5:40:09 PM	
Computation with 5 linked points	
(*italics* points are less reliable):	1. Notre-Dame de lÁssomption 2. Rempart 1 *3. Rempart 2* 4. Rempart 3 5. Rempart 4
Reference points (Reliability 0–5)	4
Transformation with 3 parameters	
Scale	1:8270
Rotation	16 °
Mean position error	24m

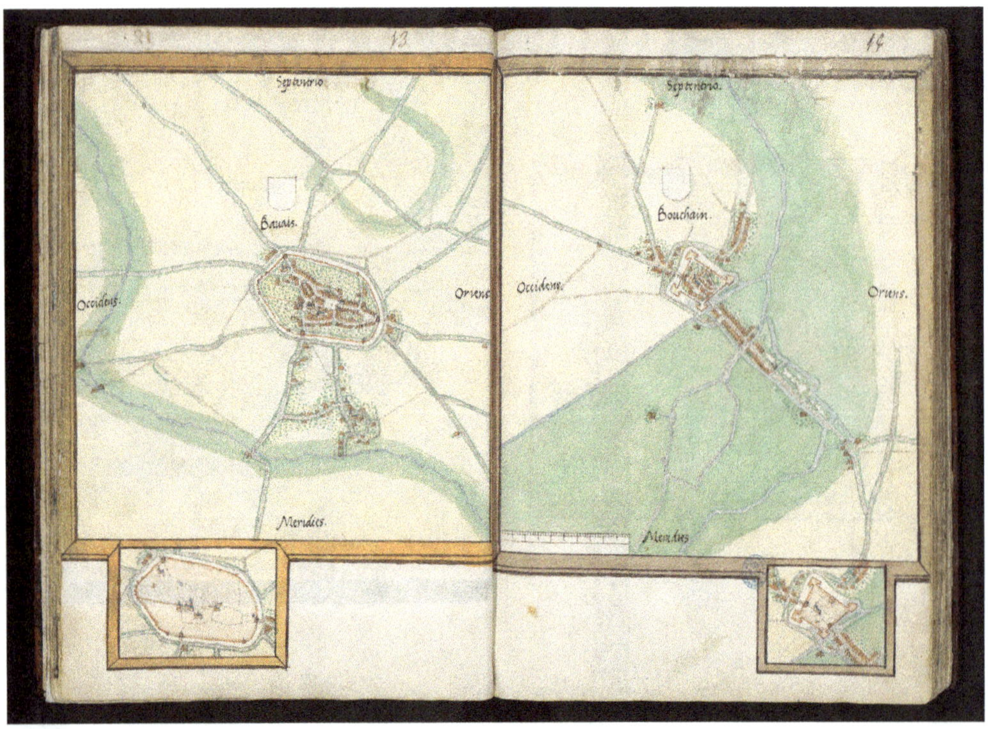

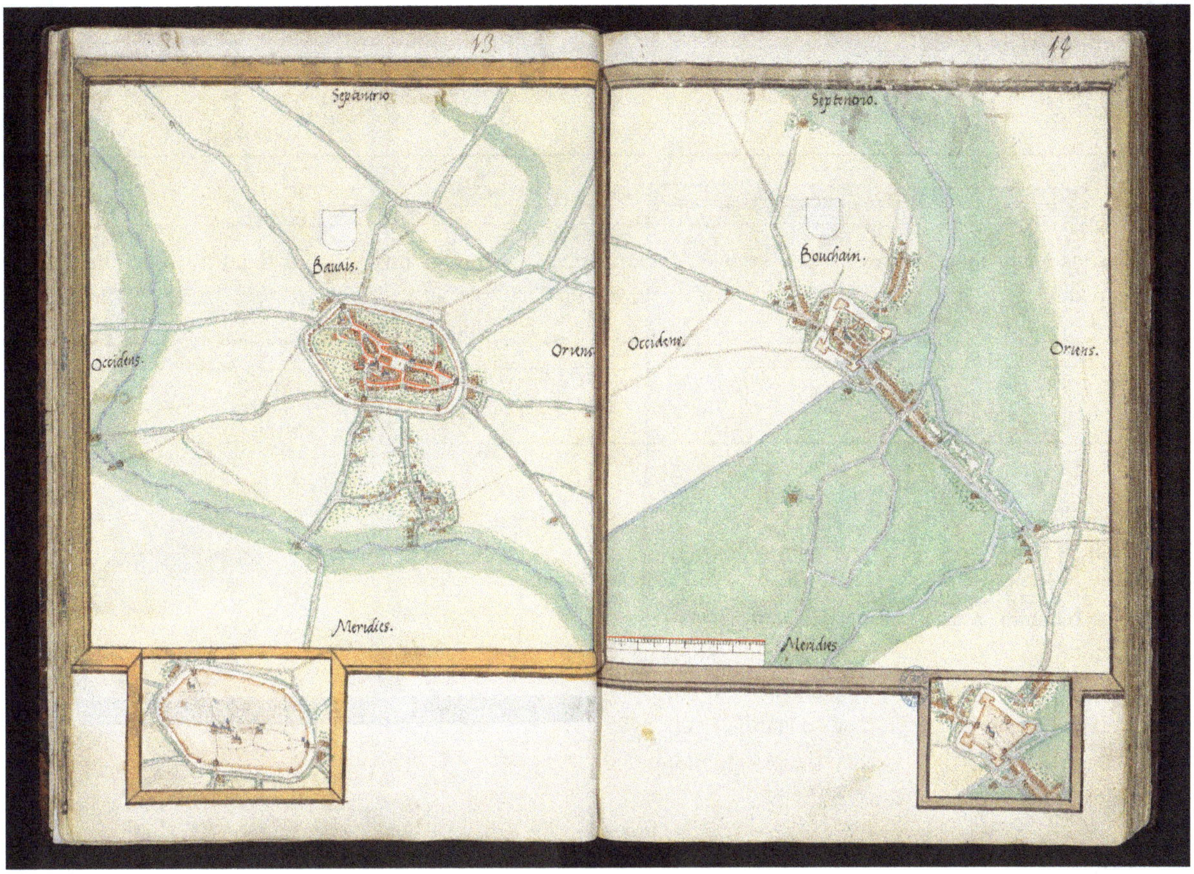
13
14
Septentrio
Bauais.
Occidens.
Oriens
Meridies.
Septentrio.
Bouchain.
Occidens.
Oriens.
Meridies

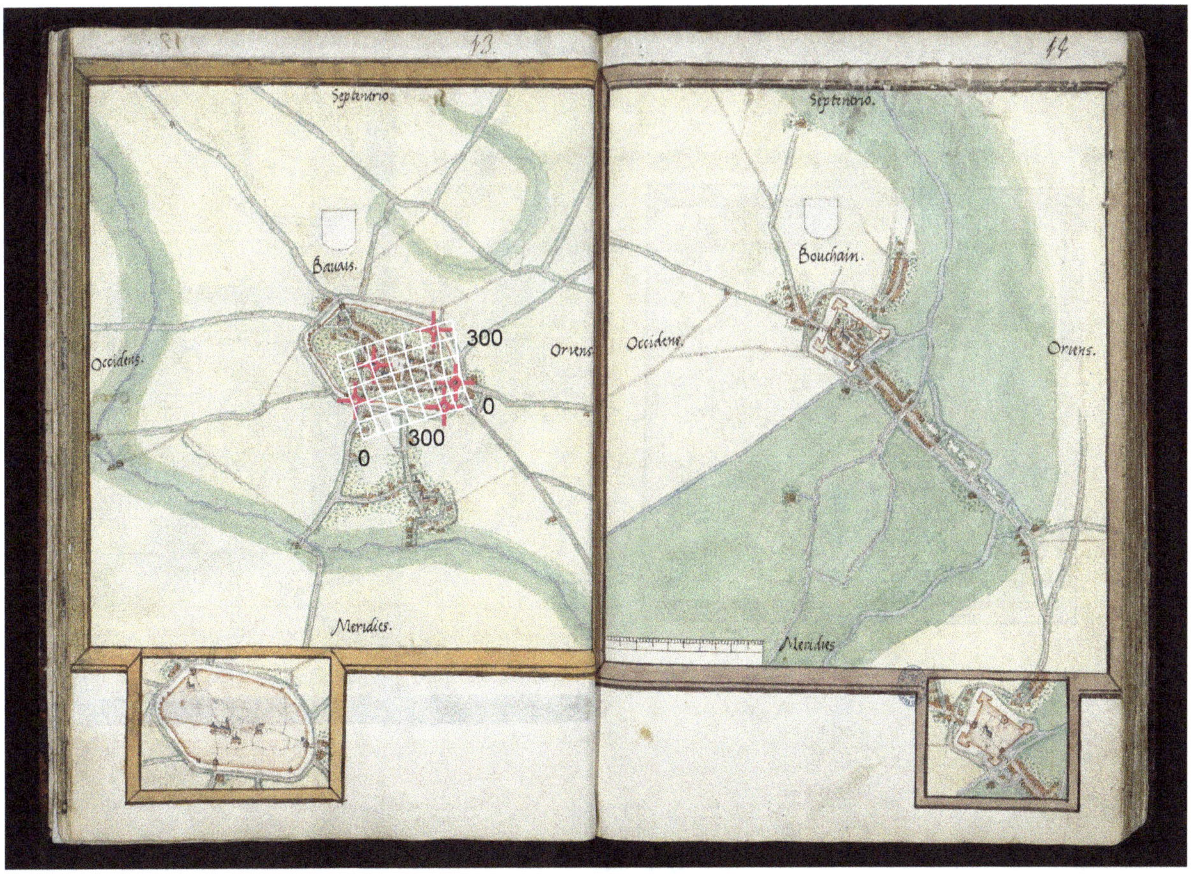
13
14
Septentrio
Bauais.
300
0
300
0
Occidens.
Oriens
Meridies.
Septentrio.
Bouchain.
Occidens.
Oriens.
Meridies

23. Beaumont

Tables 3.23 (measured and analysed).
Figures 3.23 (unmeasured, measured and analysed).

Measured							
Pixels	Scale stick original map (pixels)	Houses	Streetsides	Houses per street side	Inhabitants (RF 4.0)	Inhabitants (RF 5.5)	Inhabitants (RF 7.0)
3252	1018	571	97	5.89	2284	3141	3997

Analysed	
February 27, 2023 7:48:19 PM	
Computation with 5 linked points	
(*italics* points are less reliable):	*1. La Hantes (Ouest)*
	2. Église Saint-Servais
	3. Rue de la Régence
	4. La Hantes (sud)
	5. Tour Salamandre
Reference points (Reliability 1–5)	4
Transformation with 3 parameters	
Scale	1:8220
Rotation	11 °
Mean position error	22m

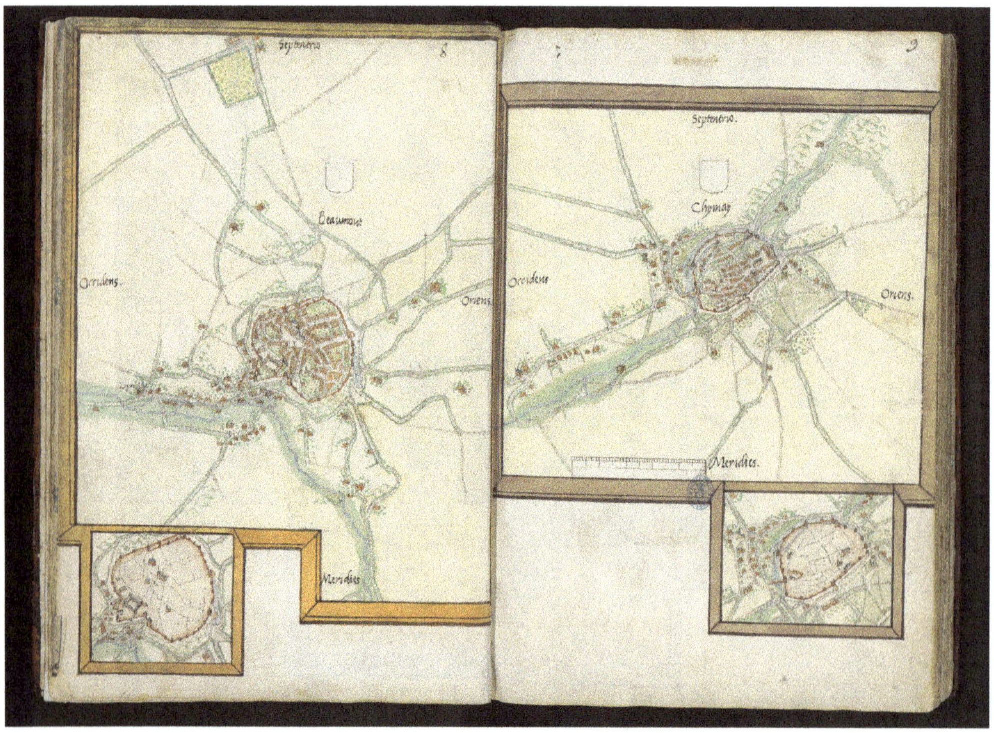

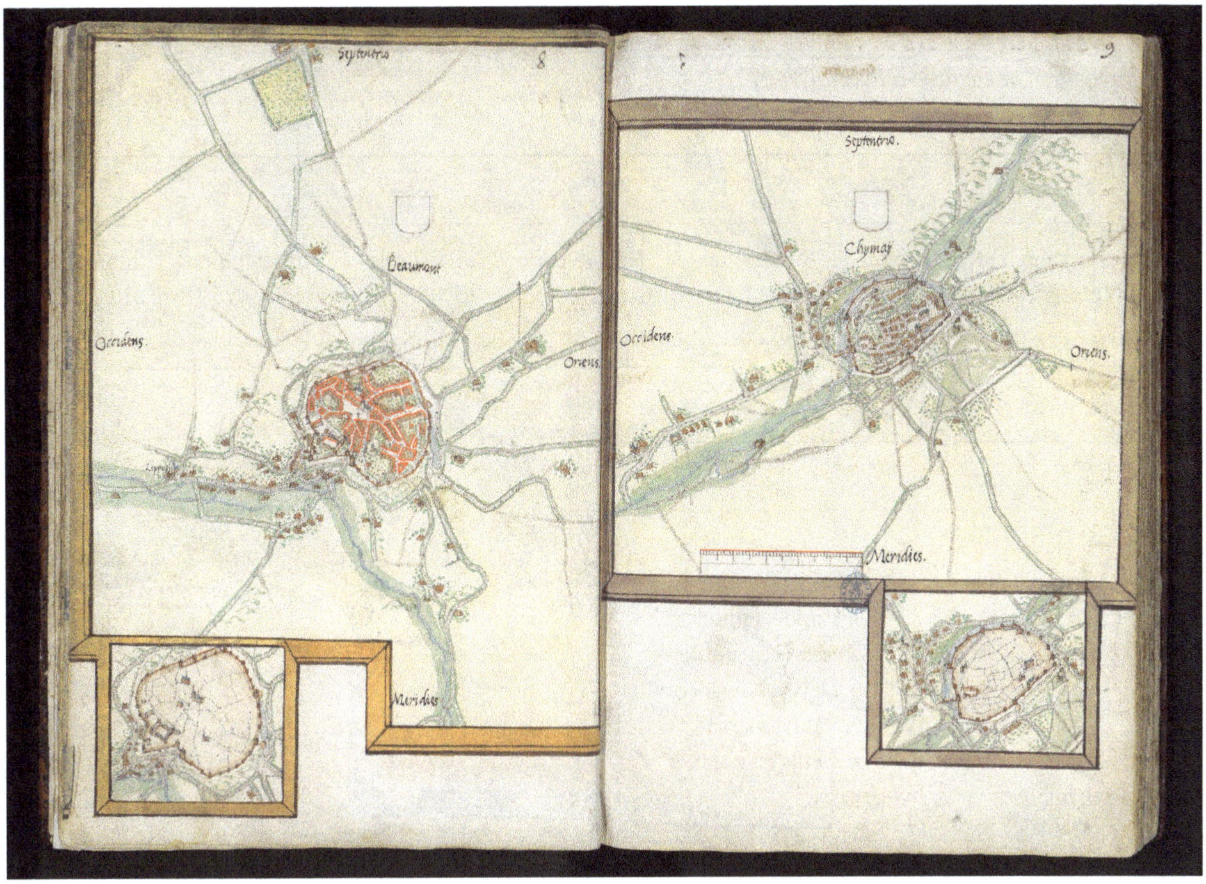
Septentrio
Beaumont
Occidens
Oriens
Meridies
Septentrio
Chymay
Occidens
Oriens
Meridies

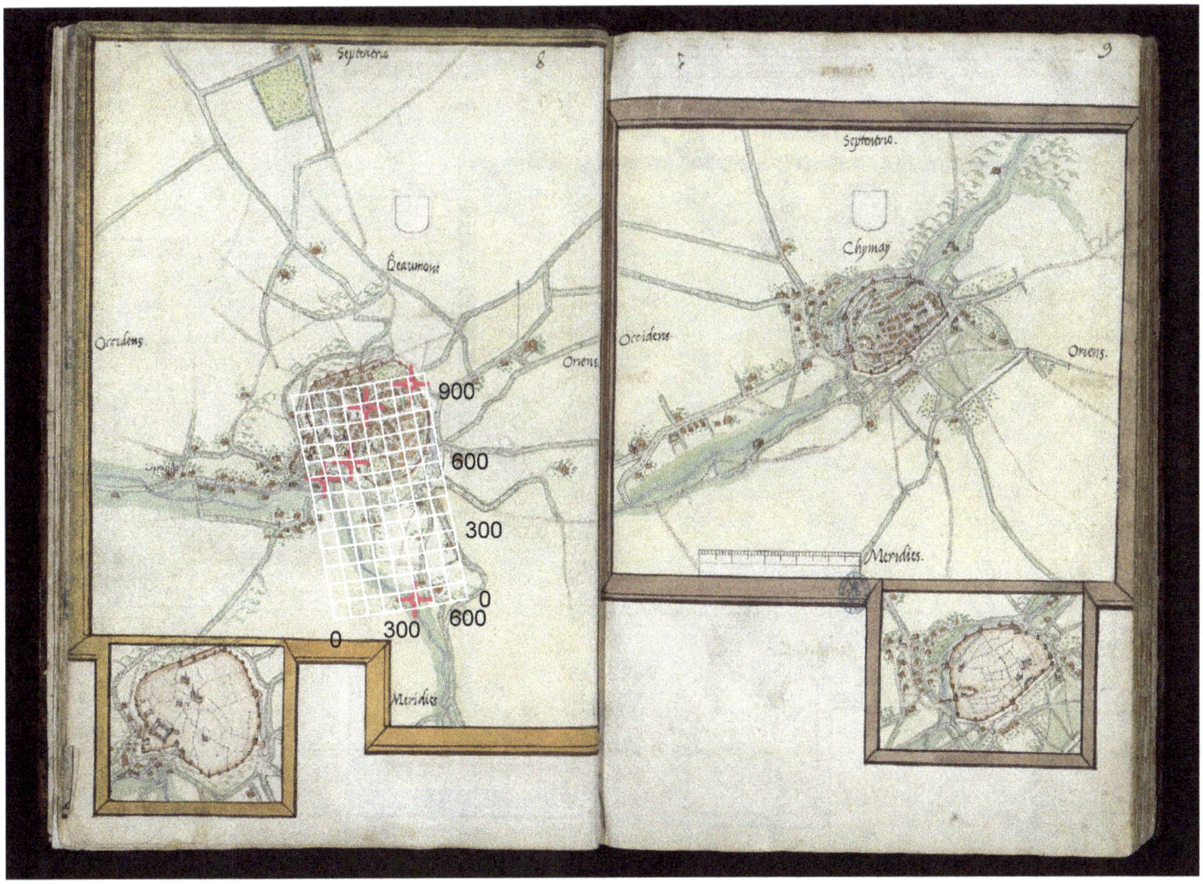
Septentrio
Beaumont
Occidens
Oriens
900
600
300
0
0
300
600
Meridies
Septentrio
Chymay
Occidens
Oriens
Meridies

24. Bergen op Zoom

Tables 3.24 (measured and analysed).
Figures 3.24 (unmeasured, measured and analysed).

Measured							
Pixels	Scale stick original map (pixels)	Houses	Street sides	Houses per street side	Inhabitants (RF 4.0)	Inhabitants (RF 5.5)	Inhabitants (RF 7.0)
9342	0	1640	264	6.21	6562	9023	11,483

Analysed	
September 6, 2023 11:43:12 AM	
Computation with 5 linked points	
(*italics* points are less reliable):	1. Markiezenhof
	2. Sint-Gertrudiskerk
	3. Lievevrouwepoort
	4. Wouwsestraat
	5. Getijdenmolen
Reference points (Reliability 0–5)	5
Transformation with 3 parameters	
Scale	1:8740
Rotation	20 °
Mean position error	18m

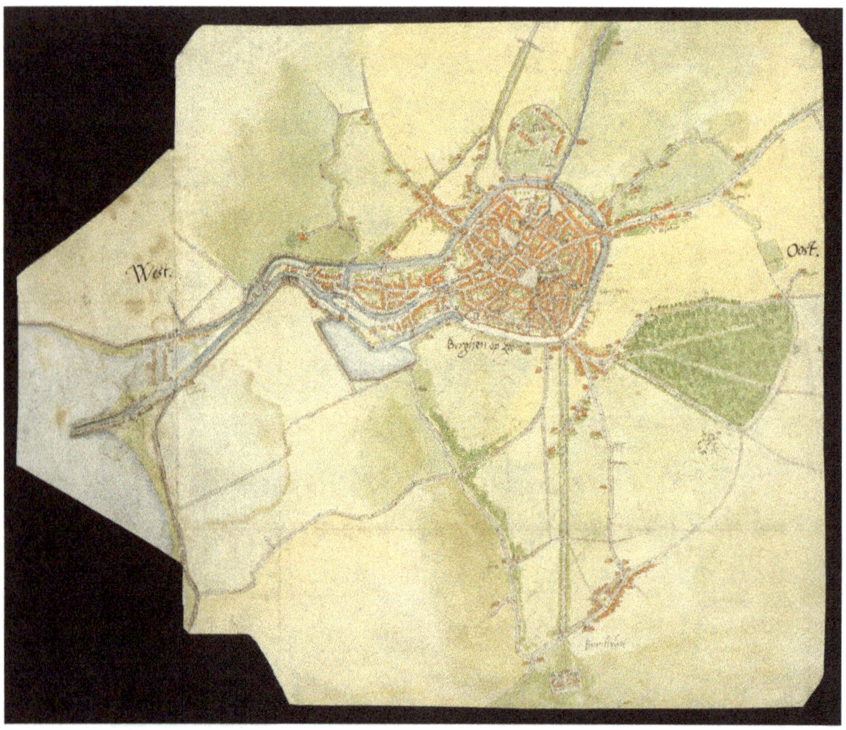

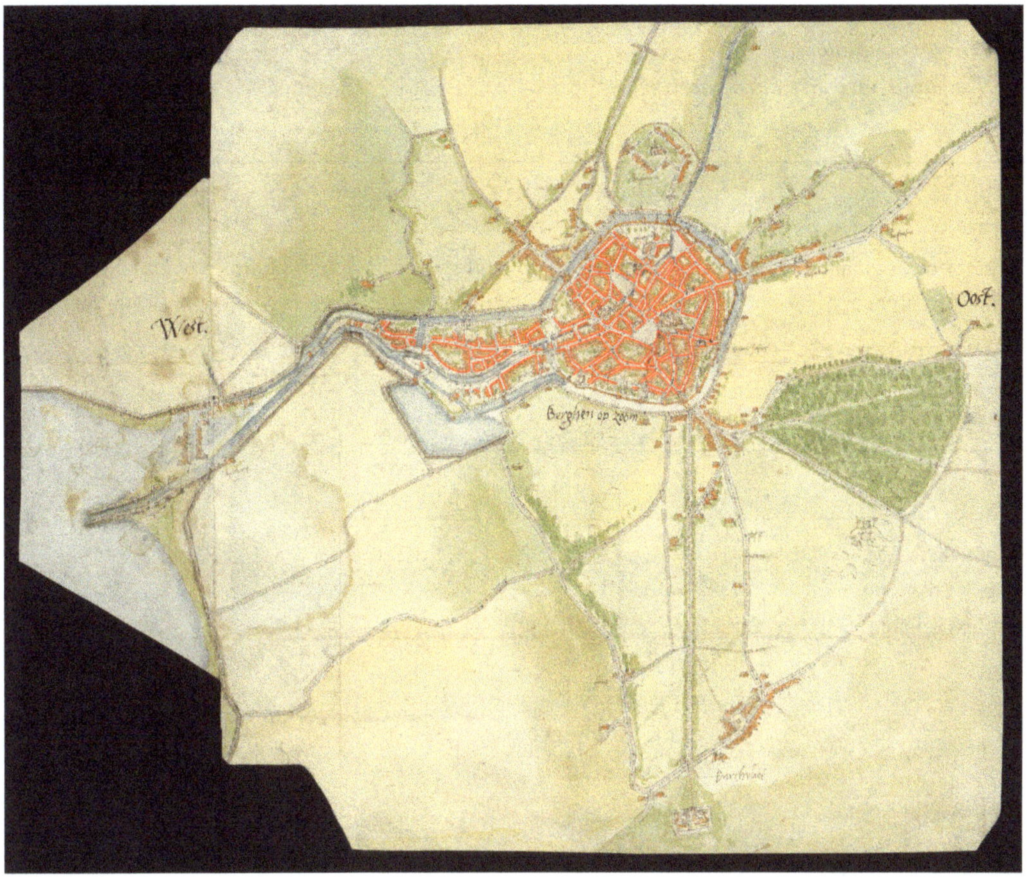
West.
Oost.
Berghen op Zoom

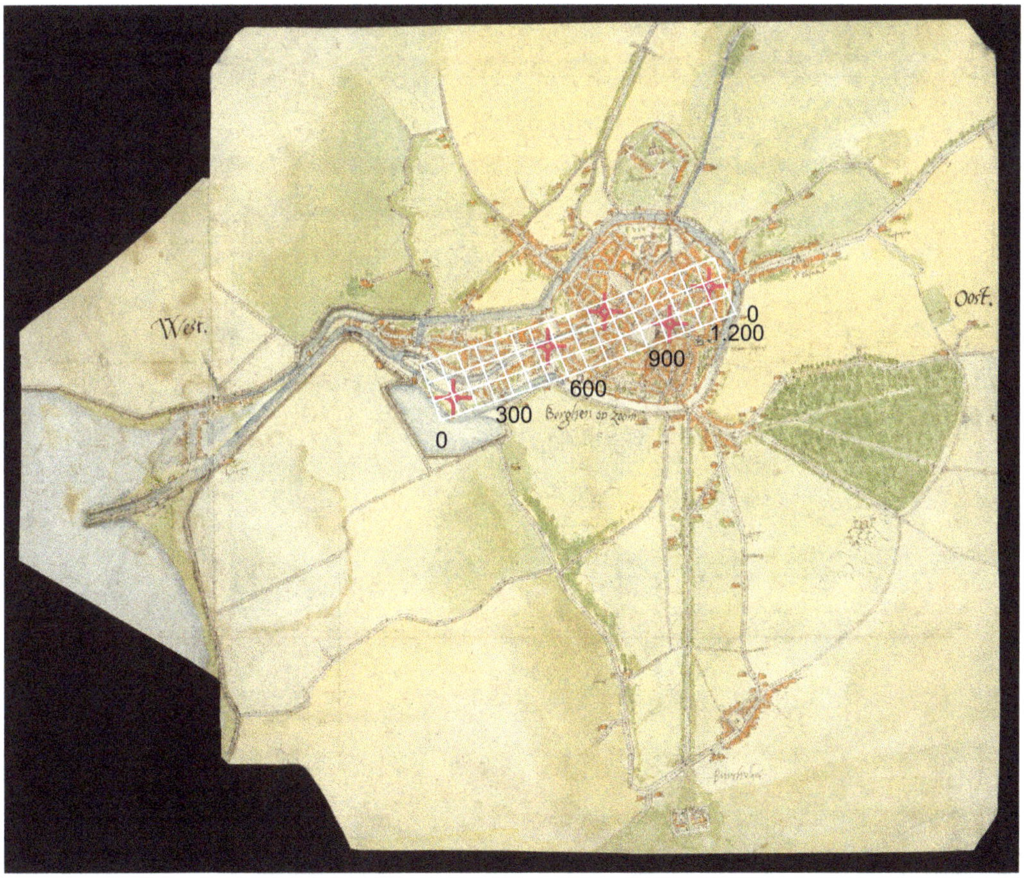
West.
Oost.
Berghen op Zoom
0
300
600
900
1.200
0

25. Bergues

Tables 3.25 (measured and analysed).
Figures 3.25 (unmeasured, measured and analysed).

Measured							
Pixels	Scale stick original map (pixels)	Houses	Street sides	Houses per street side	Inhabitants (RF 4.0)	Inhabitants (RF 5.5)	Inhabitants (RF 7.0)
5993	1034	1052	133	7.91	4209	5788	7367

Analysed	
February 28, 2023 11:34:02 AM	
Computation with 5 linked points	
(*italics* points are less reliable):	1. Église Saint-Martin
	2. Porte de Bierre
	3. Tour des Fauw-Monnayeurs
	4. Porte d'Hondschoote
	5. Porte de Cassel
Reference points (Reliability 0–5)	5
Transformation with 3 parameters	
Scale	1:7860
Rotation	8 °
Mean position error	20m

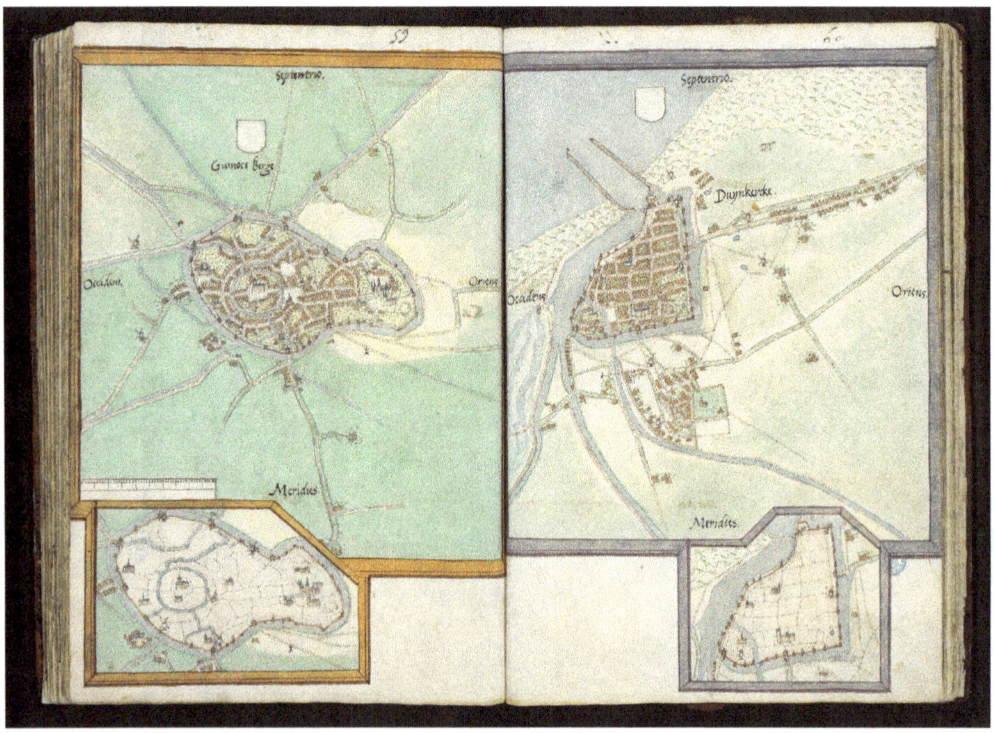

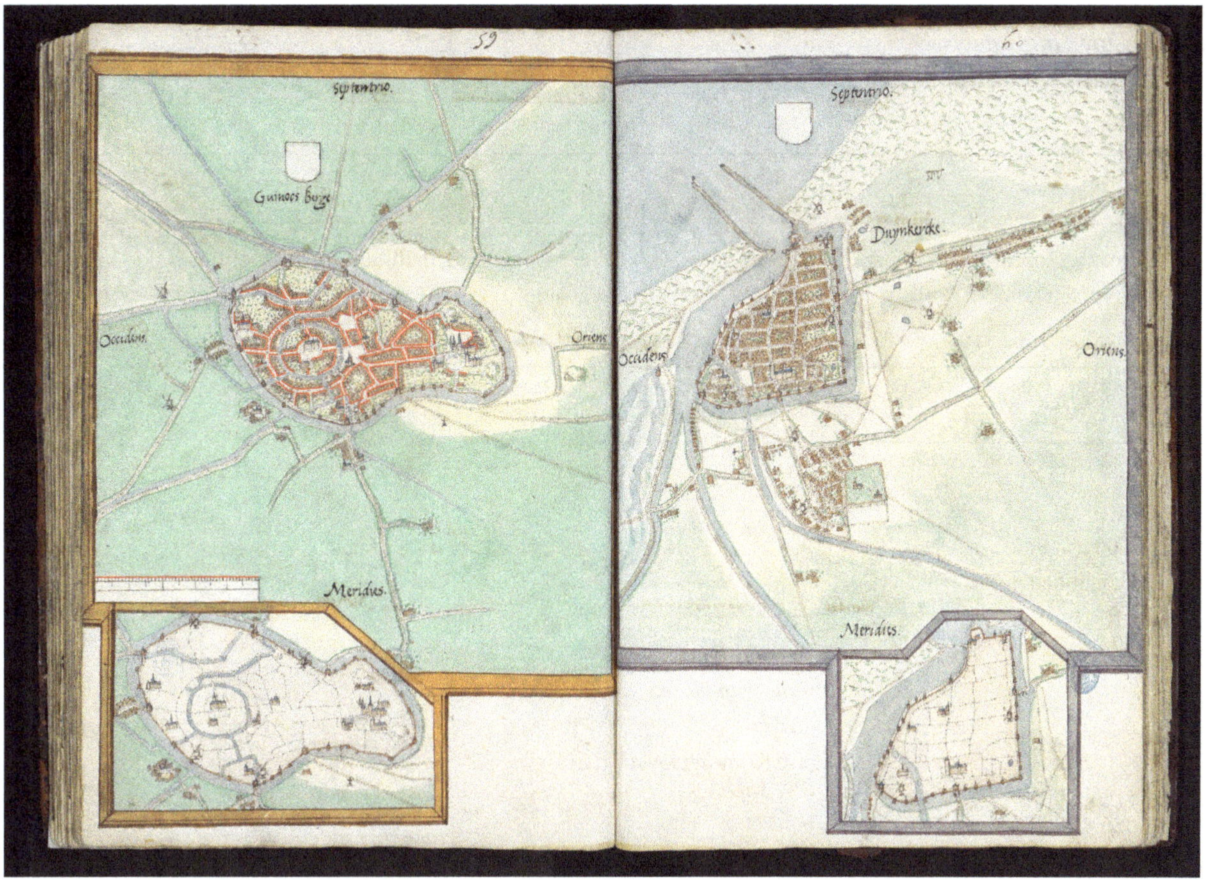
59
60
Septentrio.
Septentrio.
Occidens.
Oriens
Occidens
Oriens.
Duynkercke.
Meridies.
Meridies.

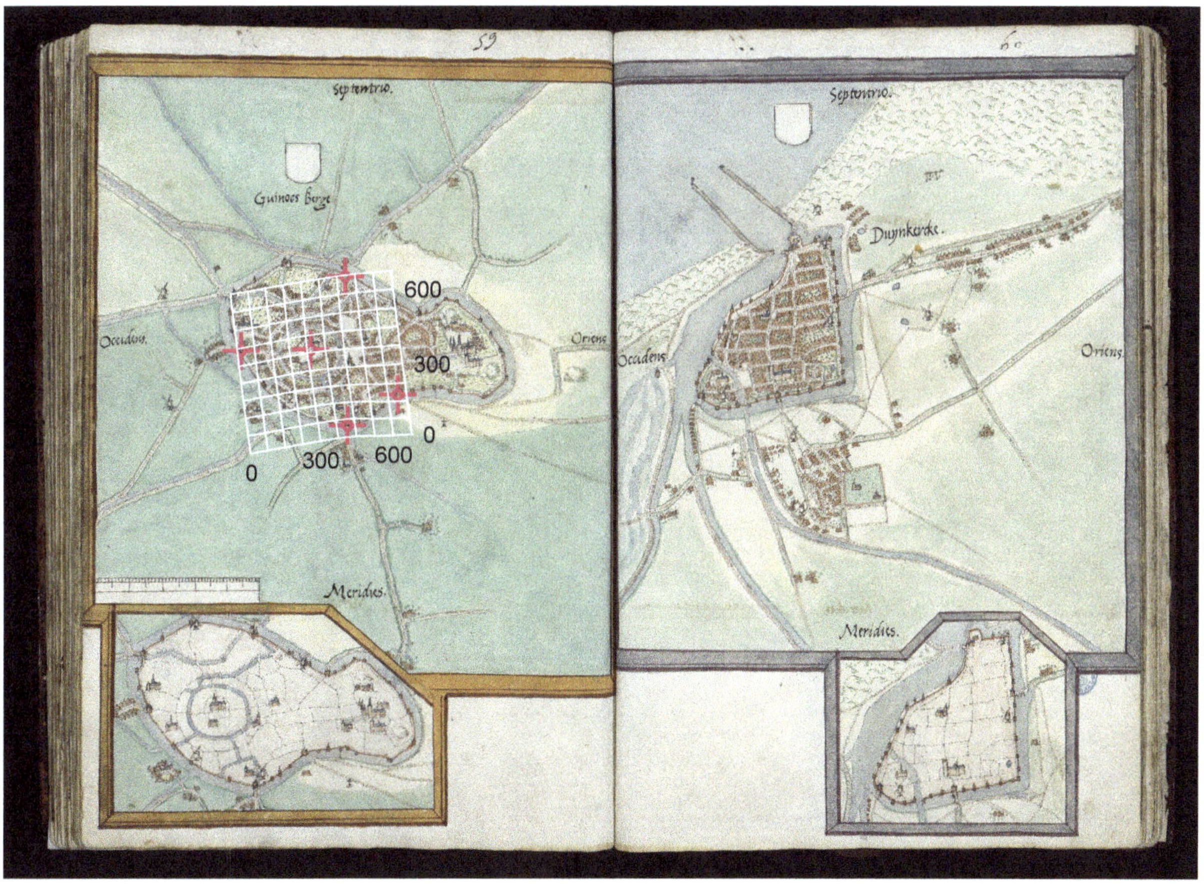
59
60
Septentrio.
Septentrio.
600
300
0
0
300
600
Occidens.
Oriens
Occidens
Oriens.
Duynkercke.
Meridies.
Meridies.

26. Béthune

Tables 3.26 (measured and analysed).
Figures 3.26 (unmeasured, measured and analysed).

Measured

Pixels	Scale stick original map (pixels)	Houses	Street sides	Houses per street side	Inhabitants (RF 4.0)	Inhabitants (RF 5.5)	Inhabitants (RF 7.0)
4567	1021	802	114	7.03	3208	4411	5614

Analysed

February 28, 2023 6:13:56 PM	
Computation with 5 linked points	
(*italics* points are less reliable):	*1. Beffroi*
	2. Tour Saint-Ignace
	3. Avenue Jean Jaures
	4. La Lawe
	5. Église Saint-Servaast
Reference points (Reliability 0–5)	1
Transformation with 3 parameters	
Scale	1:8770
Rotation	9 °
Mean position error	31m

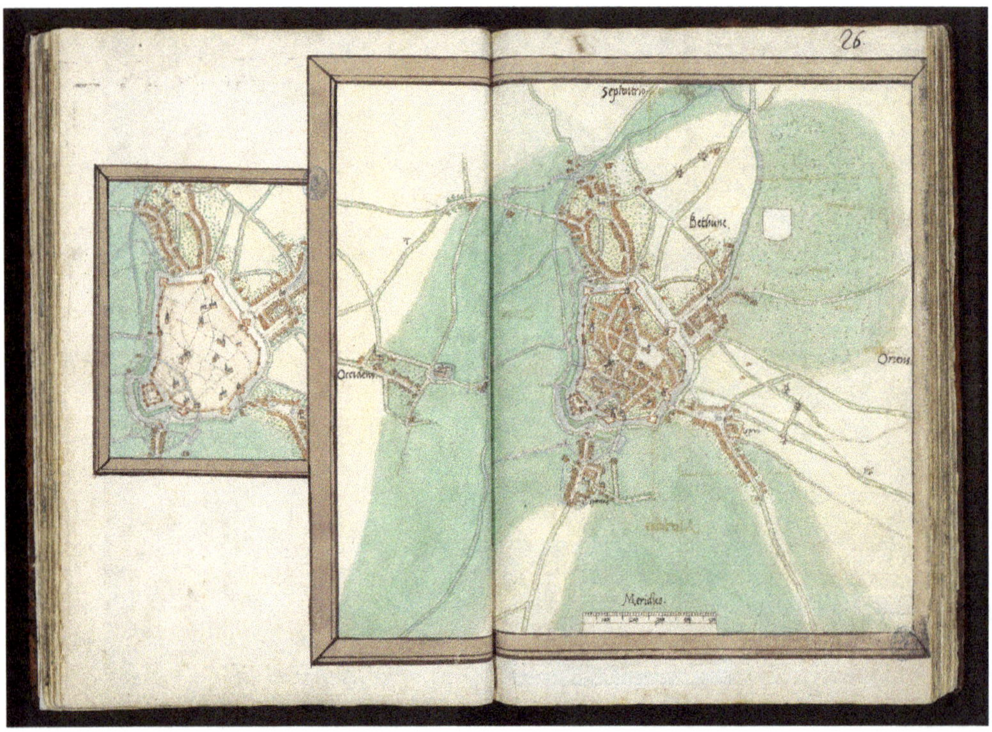

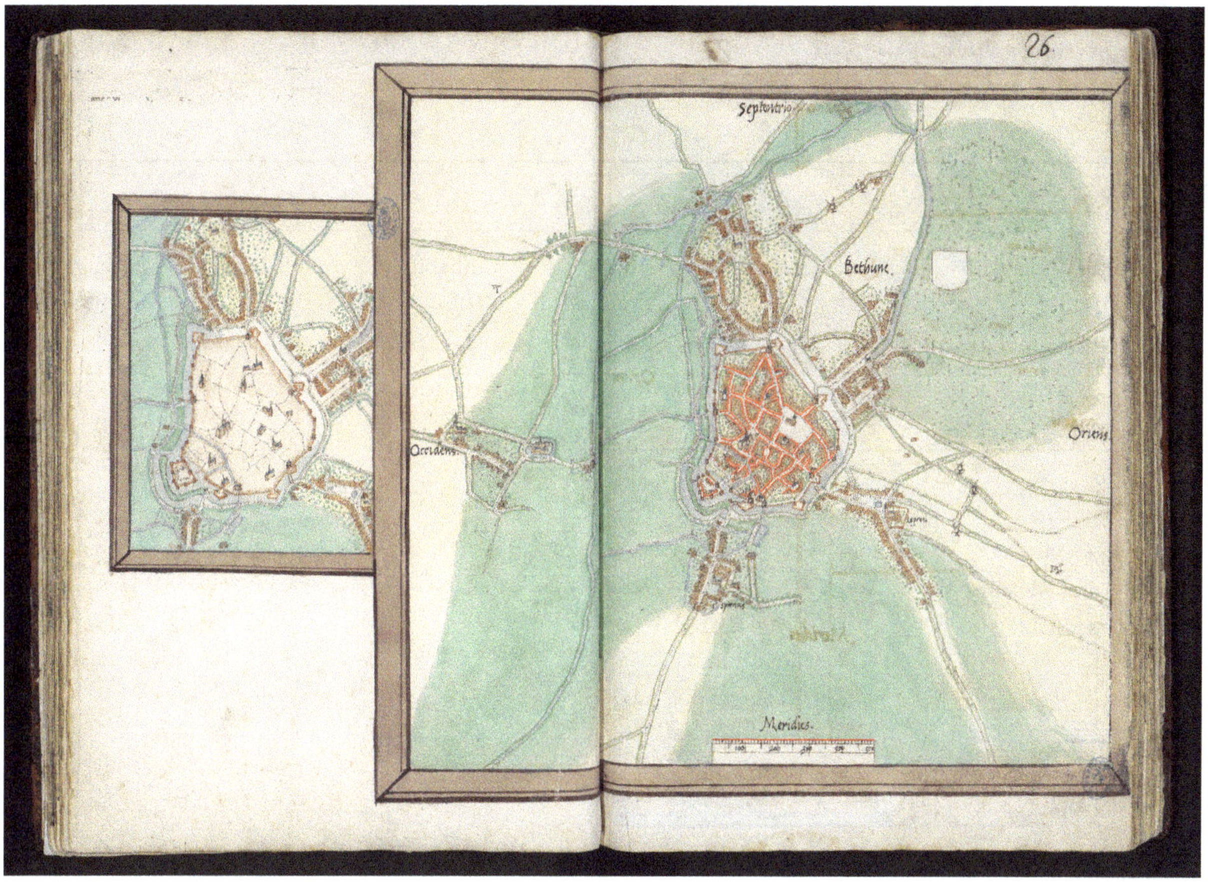
26
Septentrio
Bethune
Occidens
Oriens
Meridies

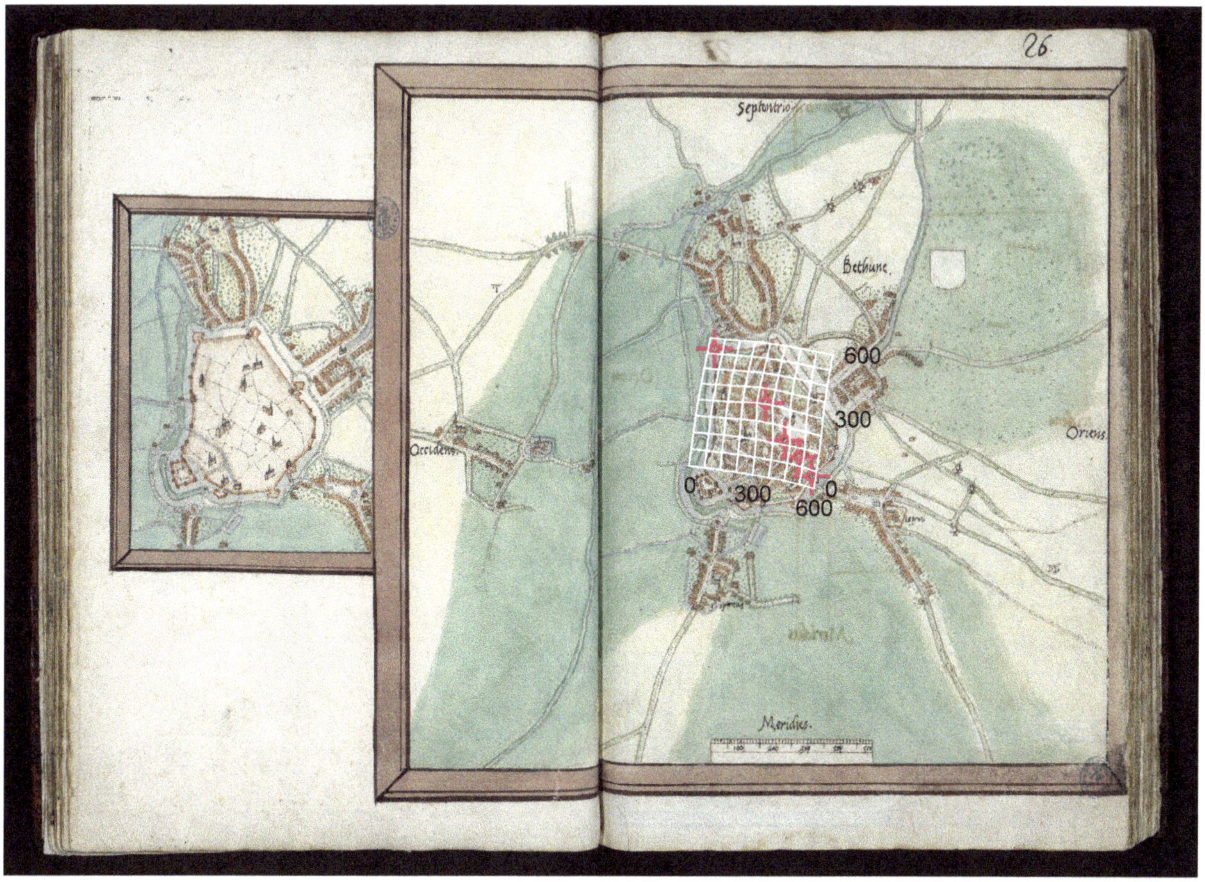
26
Septentrio
Bethune
600
300
Occidens
Oriens
0
300
600
0
Meridies

27. Beverwijk
Tables 3.27 (measured and analysed).
Figures 3.27 (unmeasured, measured and analysed).

Measured							
Pixels	Scale stick original map (pixels)	Houses	Street sides	Houses per street side	Inhabitants (RF 4.0)	Inhabitants (RF 5.5)	Inhabitants (RF 7.0)
2215	1006	389	47	8.28	1556	2139	2723

Analysed	
March 1, 2023 11:09:44 AM	
Computation with 5 linked points	
(*italics* points are less reliable):	1. Grote Kerk 2. Breestraat 3. Koningsstraat 4. Koningsstraat 5. Breestraat
Reference points (Reliability 0–5)	5
Transformation with 3 parameters	
Scale	1:9020
Rotation	12 °
Mean position error	15m

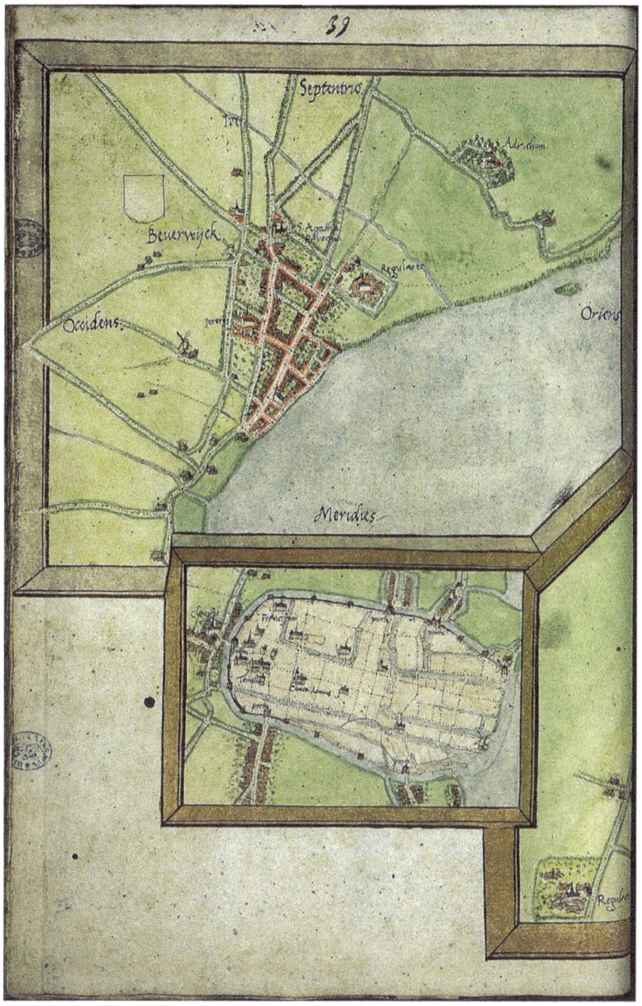
39
Septentrio
Beuerwijck
Occidens.
Oriens
Meridies

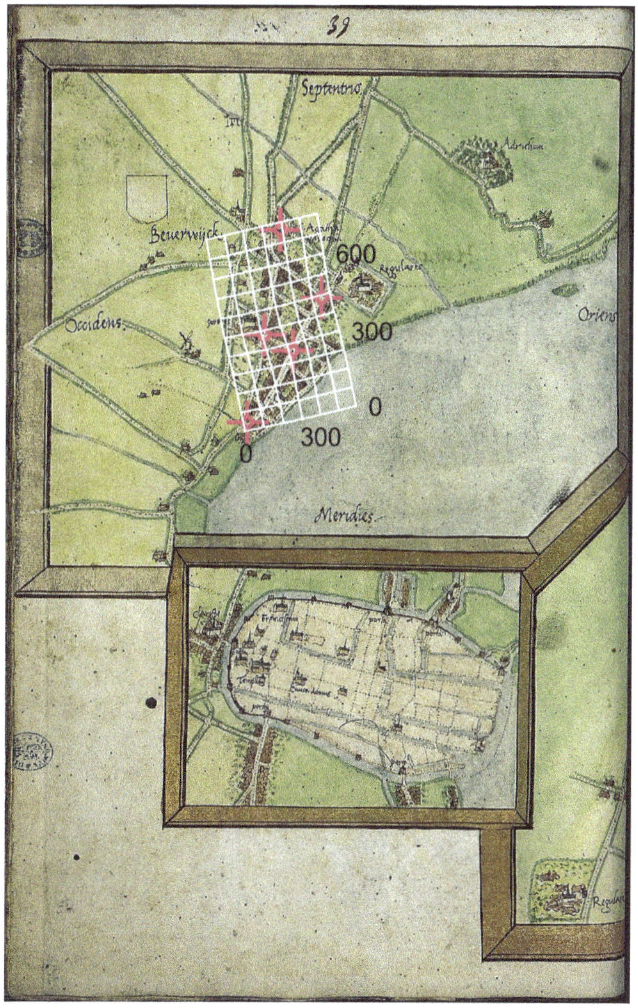
39
Septentrio
Beuerwijck
Occidens.
Oriens
600
300
0
300
0
Meridies

28. Binche
Tables 3.28 (measured and analysed).
Figures 3.28 (unmeasured, measured and analysed).

Measured							
Pixels	Scale stick original map (pixels)	Houses	Street sides	Houses per street side	Inhabitants (RF 4.0)	Inhabitants (RF 5.5)	Inhabitants (RF 7.0)
4911	1038	862	150	5.75	3449	4743	6037

Analysed	
March 1, 2023 1:27:02 PM Computation with 5 linked points (*italics* points are less reliable):	1. Église Saint-Ursmer 2. Beffroi 3. Remparts 4. Remparts 5. Remparts
Reference points (Reliability 0–5)	5
Transformation with 3 parameters	
Scale	1:7310
Rotation	10 °
Mean position error	11m

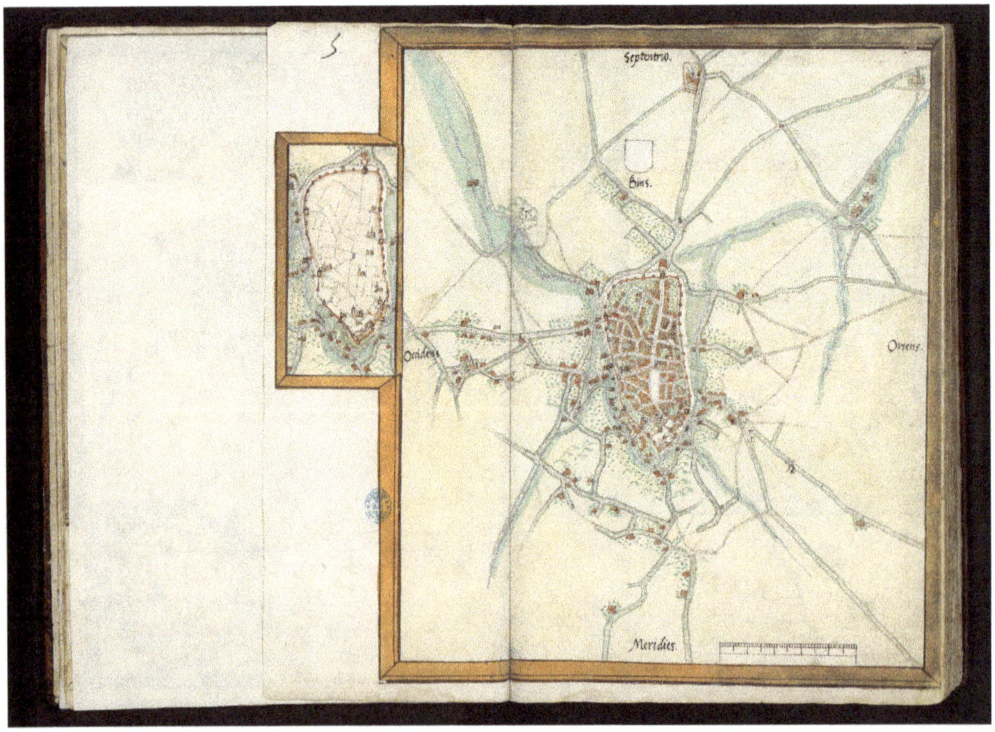

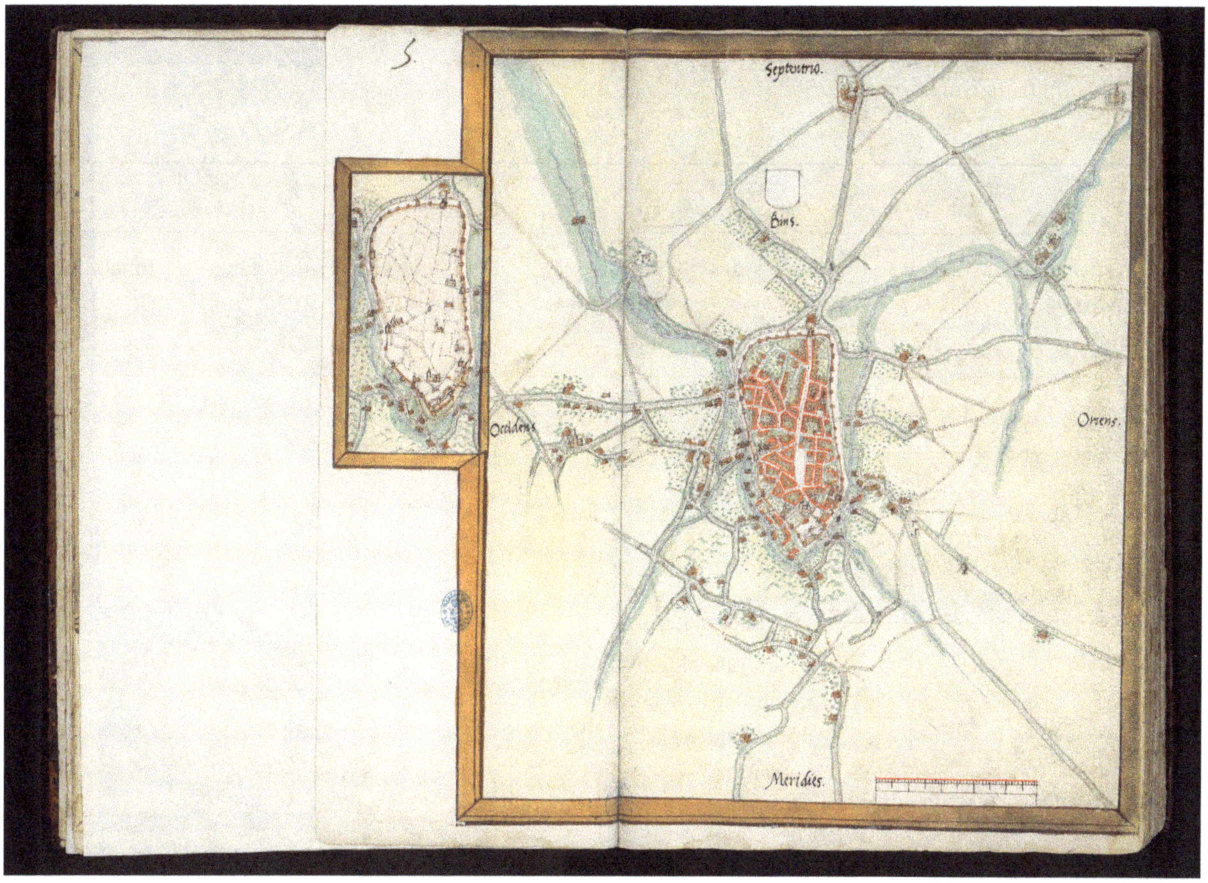
5.
Septentrio.
Bins.
Occidens
Oriens.
Meridies.

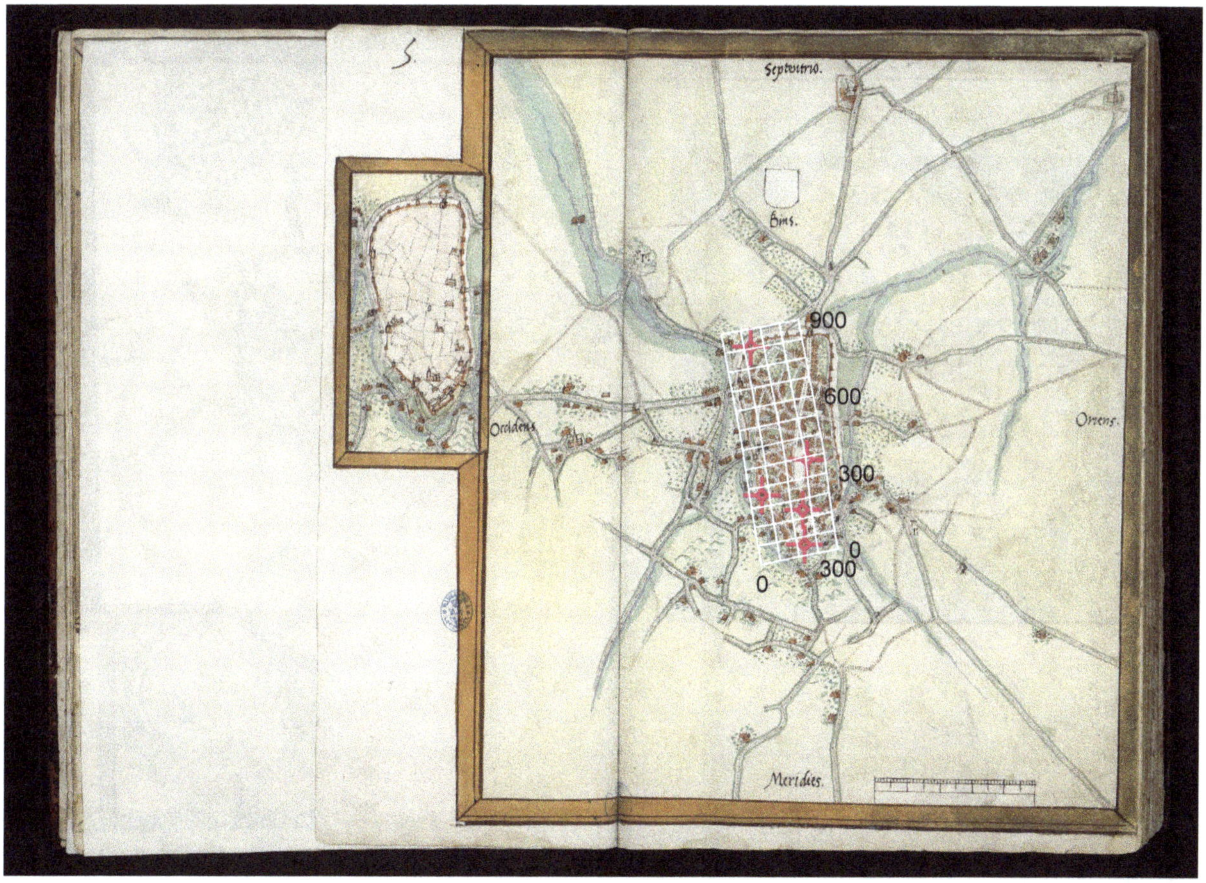
5.
Septentrio.
Bins.
900
600
Occidens
Oriens.
300
0
300
0
Meridies.

29. Bitburg

Tables 3.29 (measured and analysed).
Figures 3.29 (unmeasured, measured and analysed).

Measured							
Pixels	Scale stick original map (pixels)	Houses	Street sides	Houses per street side	Inhabitants (RF 4.0)	Inhabitants (RF 5.5)	Inhabitants (RF 7.0)
2072	missing	364	75	4.85	1455	2001	2547

Analysed	
September 8, 2023 9:43:05 AM	
Computation with 5 linked points	
(*italics* points are less reliable):	1. Hauptstrasse süd 2. Liebfrauenkirch 3. Hauptstrasse norden 4. Schliezgasse *5. Am Markt*
Reference points (Reliability 0–5)	4
Transformation with 3 parameters	
Scale	1:7260
Rotation	9 °
Mean position error	21m

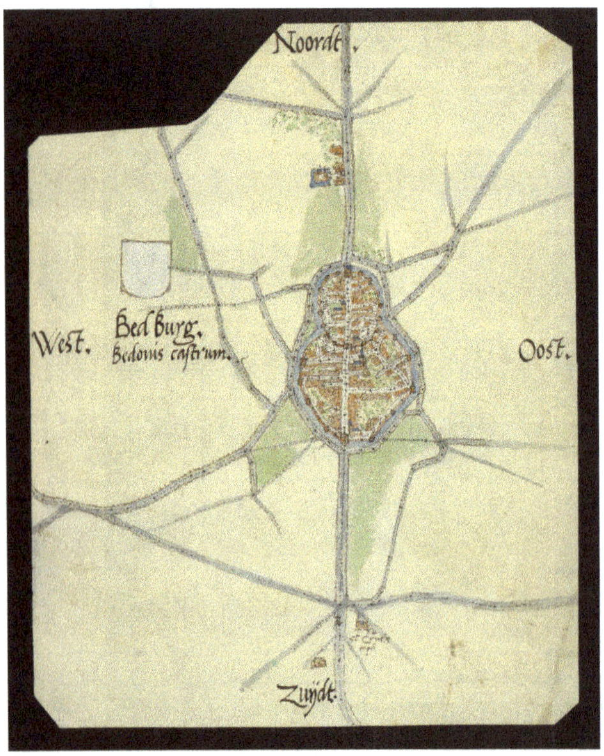

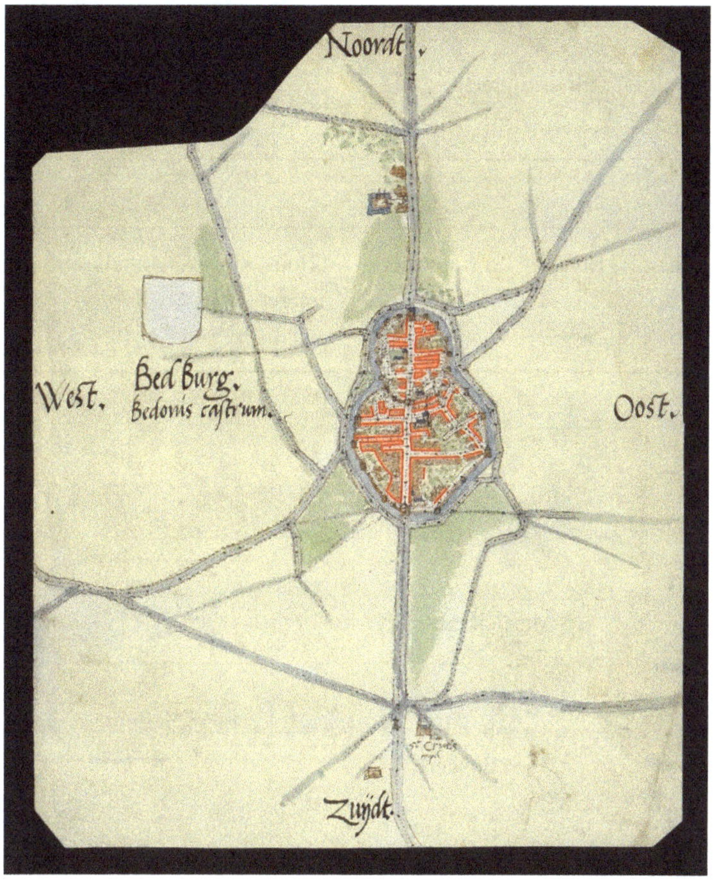
Noordt.
West.
Bed Burg.
Bedonis castrum.
Oost.
Zuijdt.

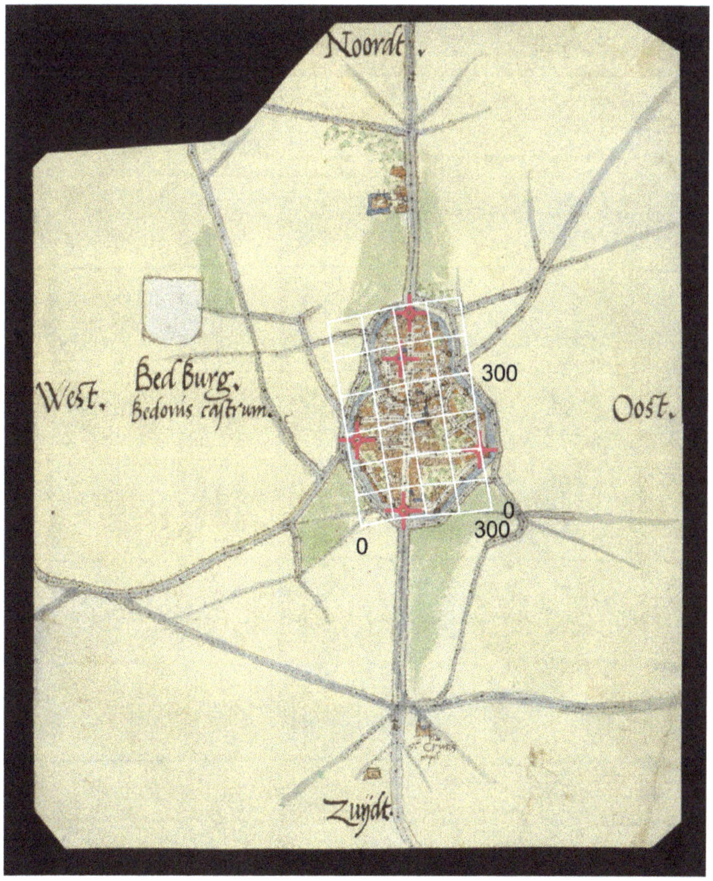
Noordt.
West.
Bed Burg.
Bedonis castrum.
Oost.
Zuijdt.
300
0
0
300

30. Bolsward

Tables 3.30 (measured and analysed).
Figures 3.30 (unmeasured, measured and analysed).

Measured							
Pixels	Scale stick original map (pixels)	Houses	Street sides	Houses per street side	Inhabitants (RF 4.0)	Inhabitants (RF 5.5)	Inhabitants (RF 7.0)
3682	missing	647	94	6.88	2586	3556	4526

Analysed	
March 1, 2023 4:40:59 PM	
Computation with 5 linked points	
(*italics* points are less reliable):	1. Martinikerk 2. Broerekerk 3. Marktstraat 4. Laag Bolwerk 5. Laag Bolwerk
Reference points (Reliability 0–5)	5
Transformation with 3 parameters	
Scale	1:8220
Rotation	12 °
Mean position error	12m

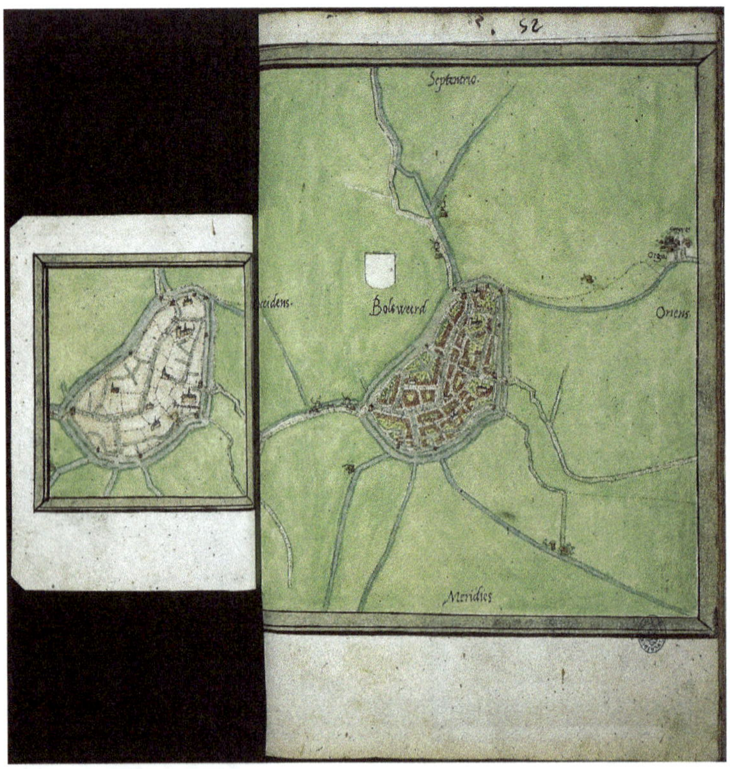

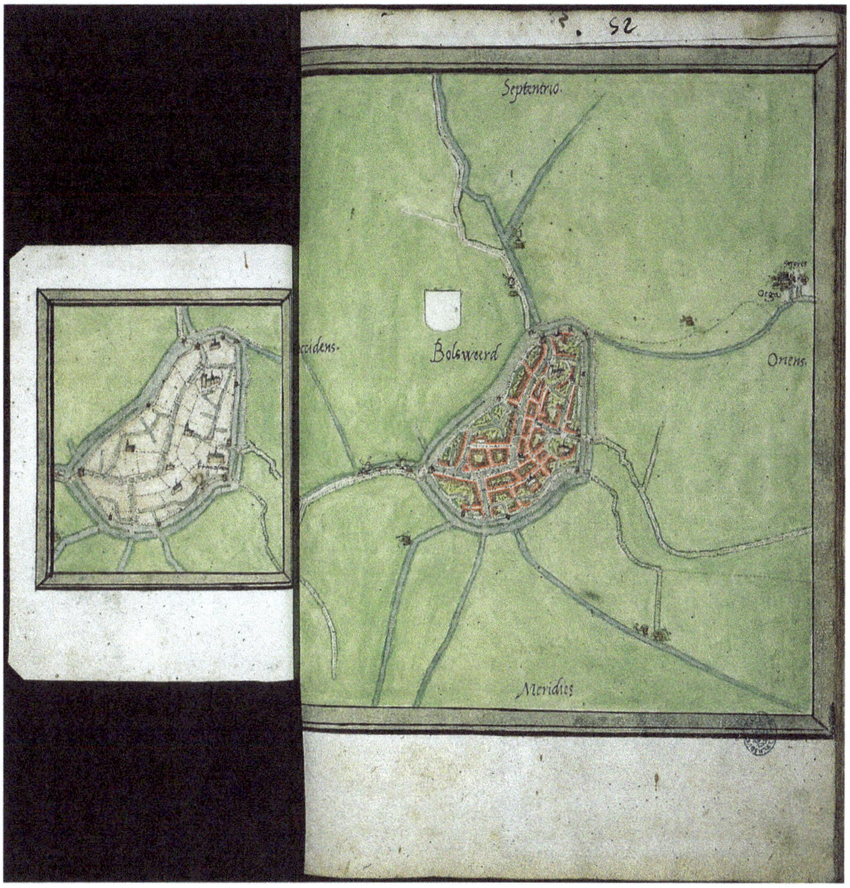
Septentrio.
Bolswerd
Occidens.
Oriens.
Meridies

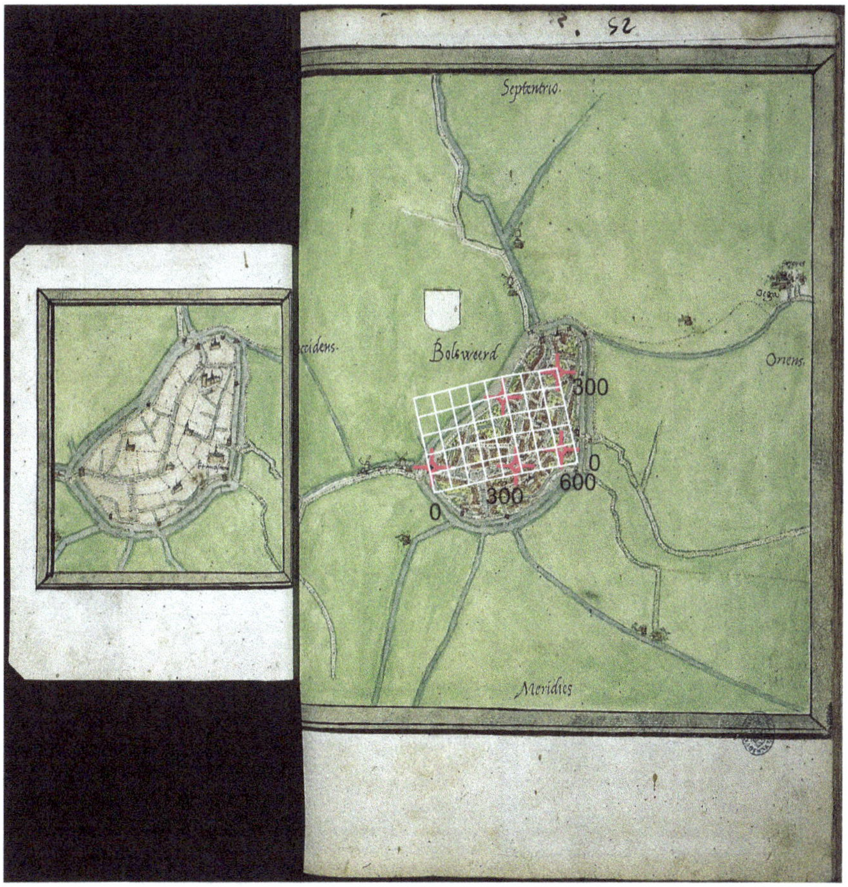
Septentrio.
Bolswerd
Oriens.
Meridies
300
0
600
300
0

31. Bouchain

Tables 3.31 (measured and analysed).
Figures 3.31 (unmeasured, measured and analysed).

Measured							
Pixels	Scale stick original map (pixels)	Houses	Street sides	Houses per street side	Inhabitants (RF 4.0)	Inhabitants (RF 5.5)	Inhabitants (RF 7.0)
615	1005	108	30	3.60	432	594	756

Analysed	
March 2, 2023 6:09:54 PM	
Computation with 5 linked points	
(*italics* points are less reliable):	1. Église Saint-Quentin
	2. Bastion des Forges
	3. Rue d'Ostrevant
	4. Tour d'Ostrevant
	5. Rue Léon Piérard
Reference points (Reliability 0–5)	4
Transformation with 3 parameters	
Scale	1:7110
Rotation	6 °
Mean position error	12m

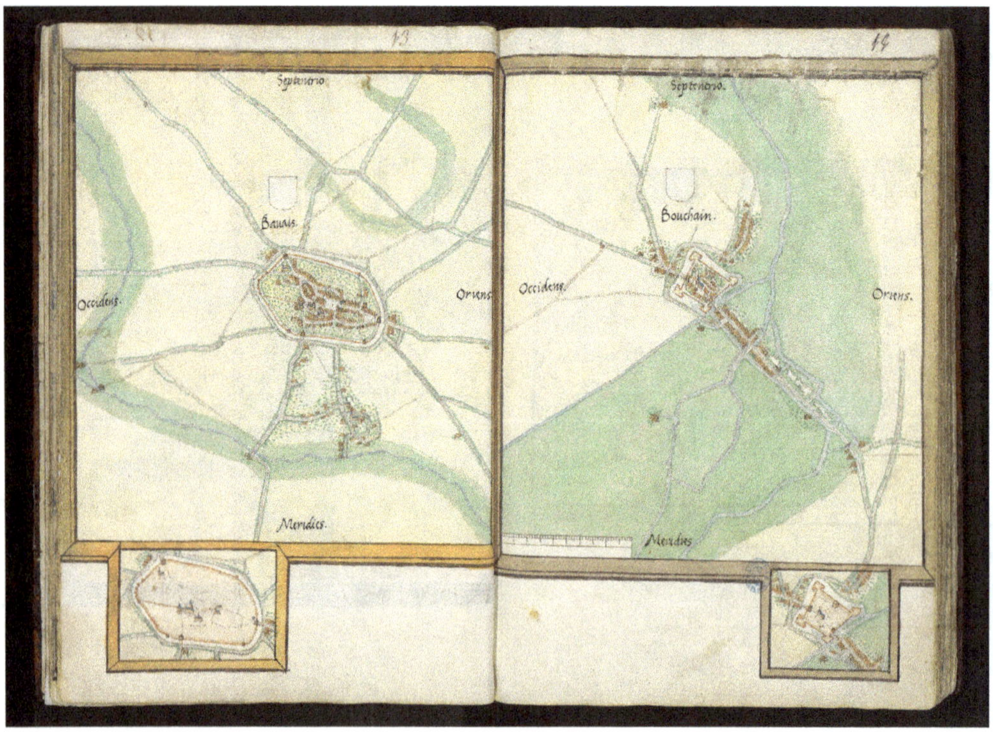

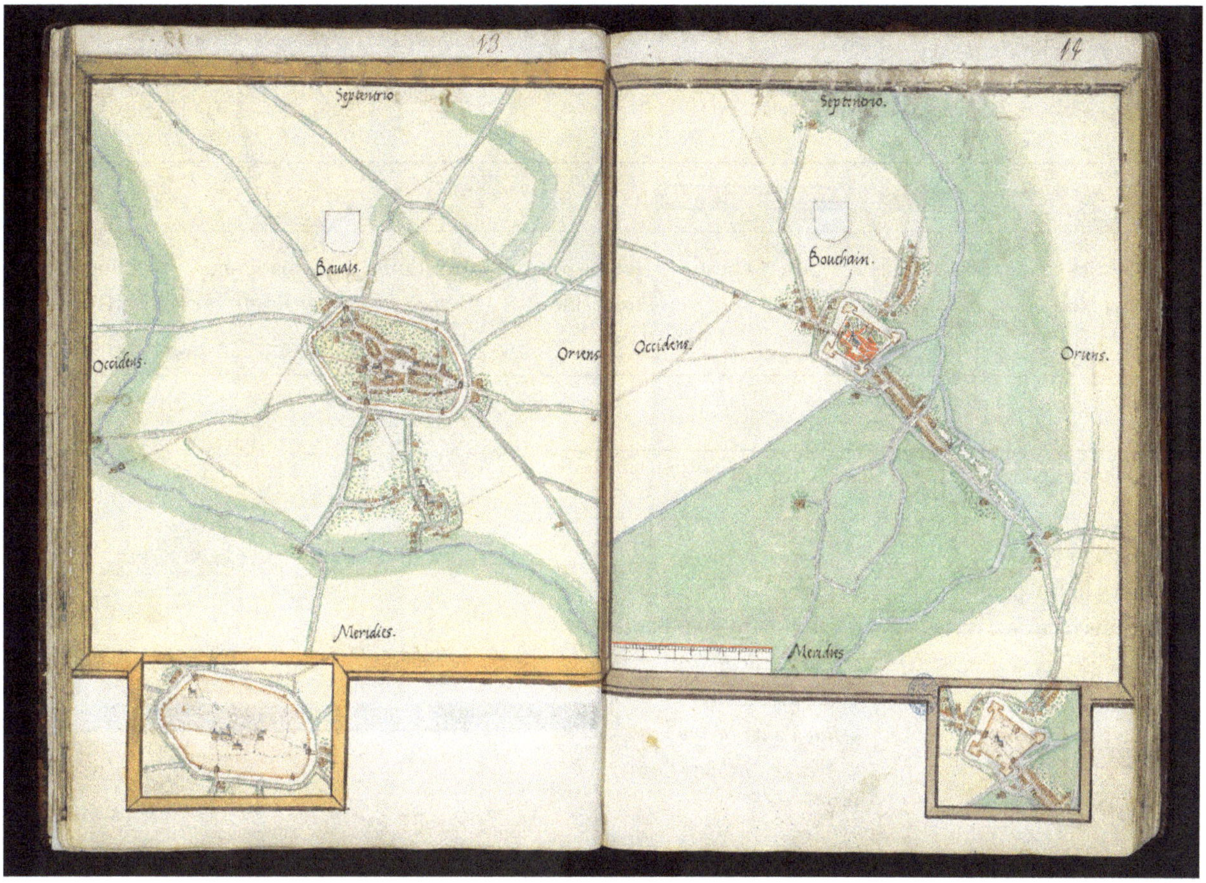
13.
14
Septentrio
Bauais.
Occidens.
Oriens
Meridies.
Septentrio.
Bouchain.
Occidens.
Oriens.
Meridies

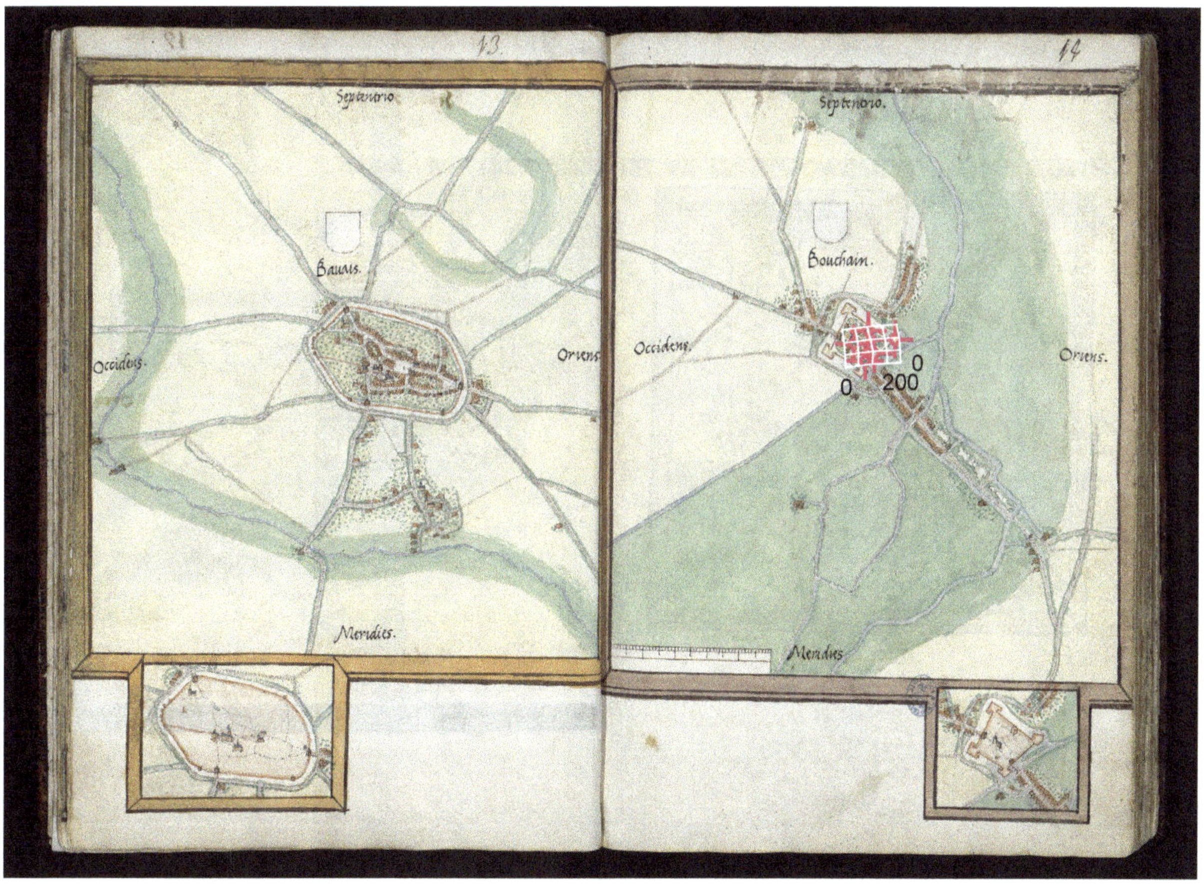
13.
14
Septentrio
Bauais.
Occidens.
Oriens
Meridies.
Septentrio.
Bouchain.
Occidens.
Oriens.
0
0
200
Meridies

32. Bourbourg
Tables 3.32 (measured and analysed).
Figures 3.32 (unmeasured, measured and analysed).

Measured							
Pixels	Scale stick original map (pixels)	Houses	Street sides	Houses per street side	Inhabitants (RF 4.0)	Inhabitants (RF 5.5)	Inhabitants (RF 7.0)
2338	1010	411	51	8.05	1642	2258	2874

Analysed	
March 26, 2023 12:20:19 PM	
Computation with 5 linked points	
(*italics* points are less reliable):	1. Église Saint-Jean-Baptiste
	2. Rue des Remparts
	3. Rue de l'Abbay
	4. Rue Paul Machy
	5. Avenue Anthony Caro
Reference points (Reliability 0–5)	1
Transformation with 3 parameters	
Scale	1:7710
Rotation	8 °
Mean position error	22m

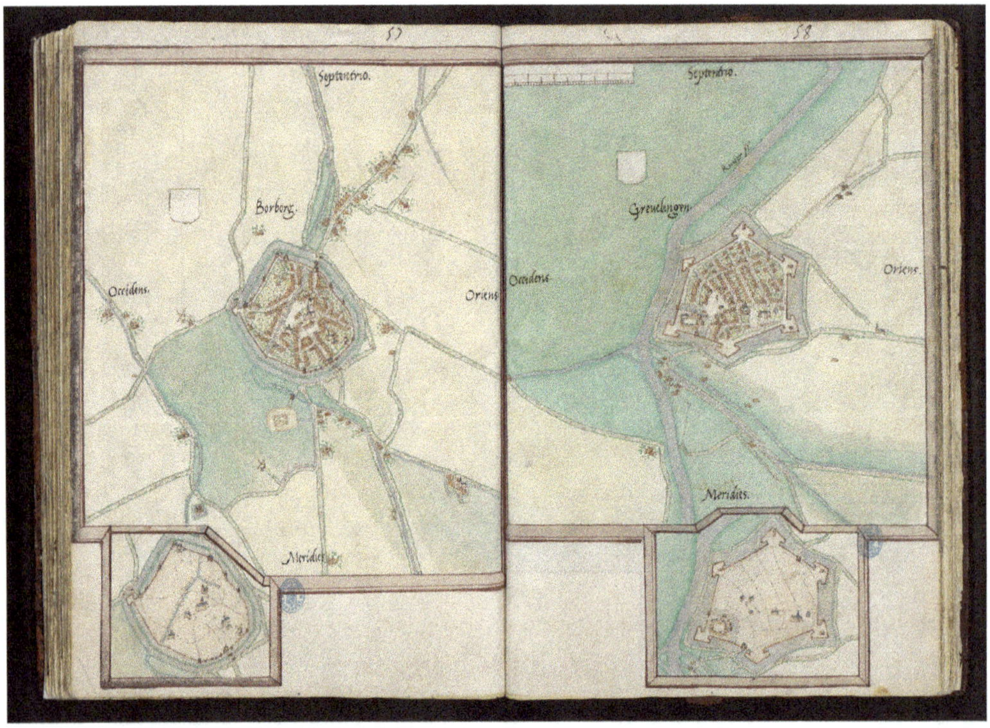

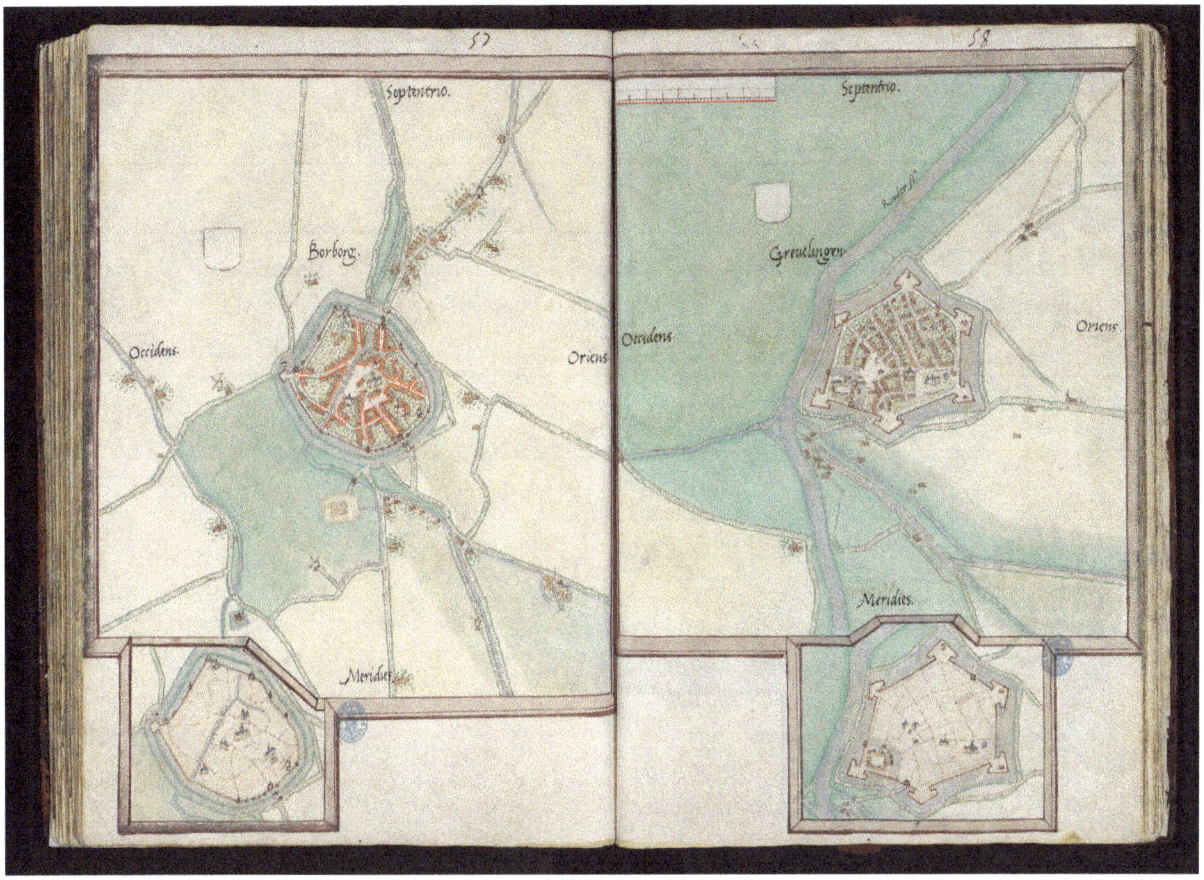
57
58
Septentrio.
Borborg.
Occidens.
Oriens
Meridies.
Septentrio.
Greuelingen
Occidens.
Oriens.
Meridies.

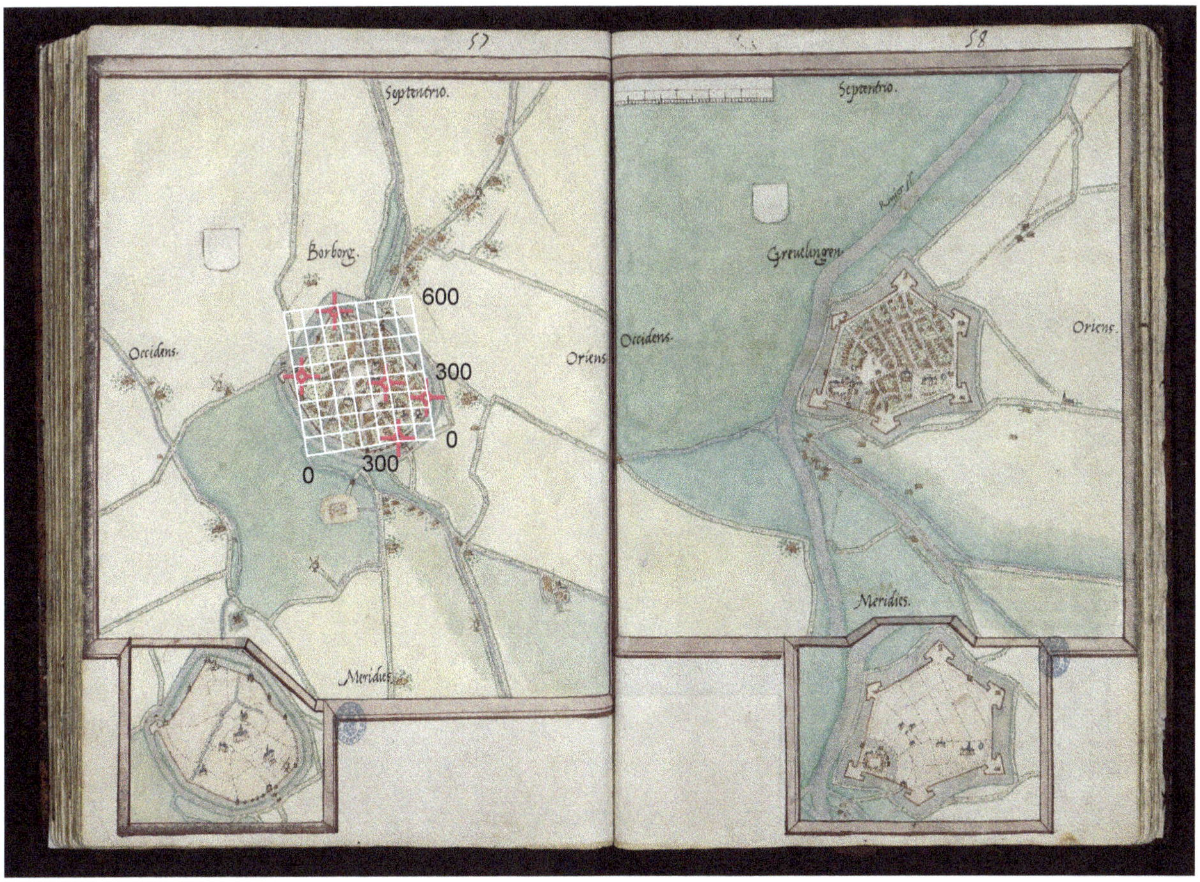
57
58
Septentrio.
Borborg.
600
300
0
0
300
Occidens.
Oriens
Meridies.
Septentrio.
Greuelingen
Occidens.
Oriens.
Meridies.

33. Bouvignes-sur-Meuse

Tables 3.33 (measured and analysed).
Figures 3.33 (unmeasured, measured and analysed).

Measured							
Pixels	Scale stick original map (pixels)	Houses	Street sides	Houses per street side	Inhabitants (RF 4.0)	Inhabitants (RF 5.5)	Inhabitants (RF 7.0)
2505	missing	440	77	5.71	1760	2419	3079

Analysed	
September 10, 2023 12:58:14 PM	
Computation with 5 linked points	
(*italics* points are less reliable):	1. Église Saint-Lambert 2. Porte Chevalier 3. Château de Crèvecoeur 4. Rue de Béguines 5. Rue du Fourneau
Reference points (Reliability 0–5)	5
Transformation with 3 parameters	
Scale	1:7190
Rotation	15 °
Mean position error	12m

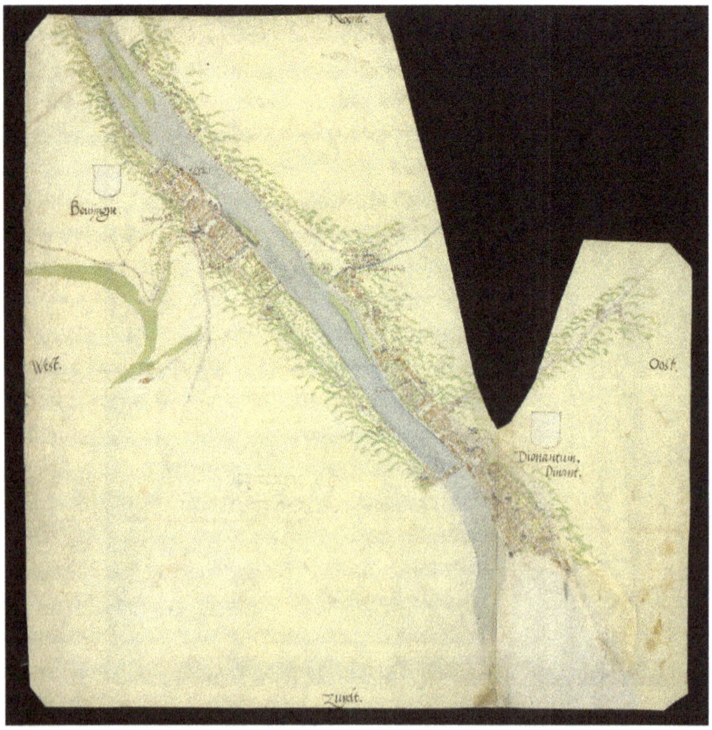

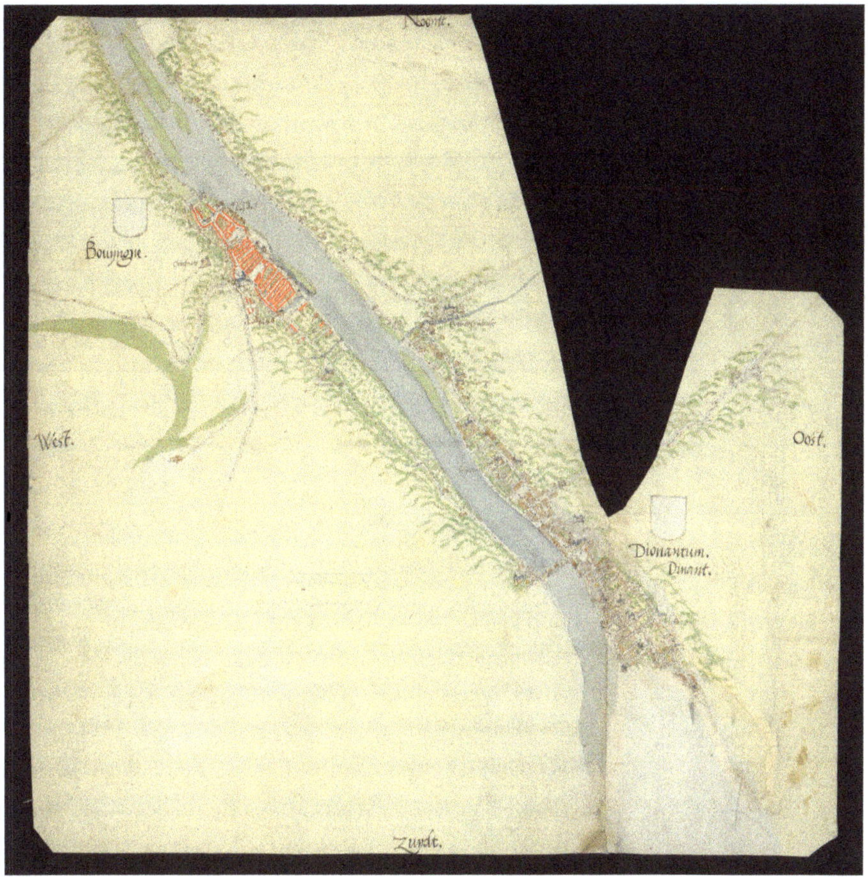
Bouvigne.
West.
Oost.
Dionantum.
Dinant.
Zuydt.

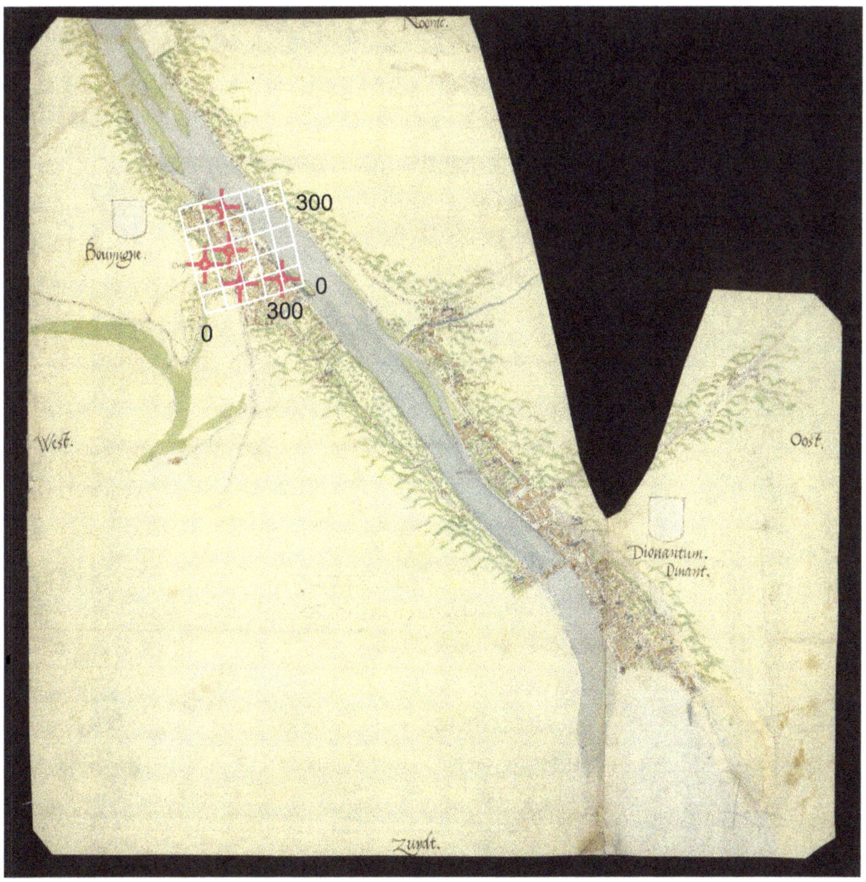
300
Bouvigne.
0
300
0
West.
Oost.
Dionantum.
Dinant.
Zuydt.

34. Braine-le-Comte

Tables 3.34 (measured and analysed).
Figures 3.34 (unmeasured, measured and analysed).

Measured

Pixels	Scale stick original map (pixels)	Houses	Street sides	Houses per street side	Inhabitants (RF 4.0)	Inhabitants (RF 5.5)	Inhabitants (RF 7.0)
2069	1034	363	69	5.27	1453	1998	2543

Analysed

February 24, 2023 10:05:19 PM	
Computation with 5 linked points	
(*italics* points are less reliable):	1. Grand Place
	2. Rue Samson
	3. Rue des Digues
	4. Rue d'Eglise
	5. Rue Edouard Etienne
Reference points (Reliability 0–5)	2
Transformation with 3 parameters	
Scale	1:6870
Rotation	2 °
Mean position error	16m

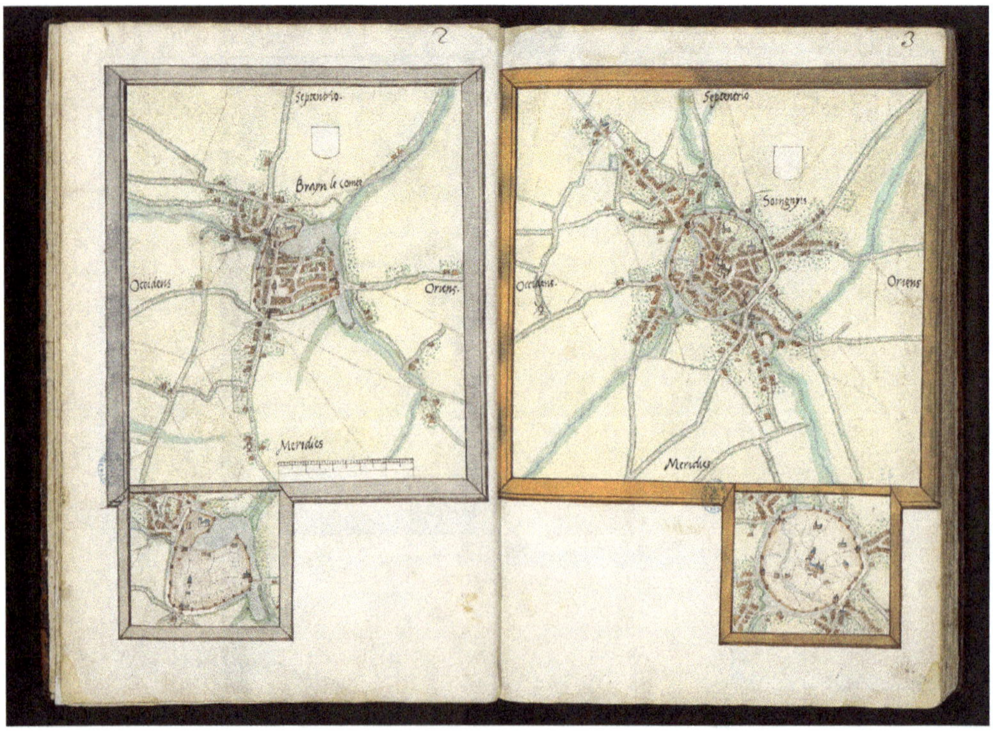

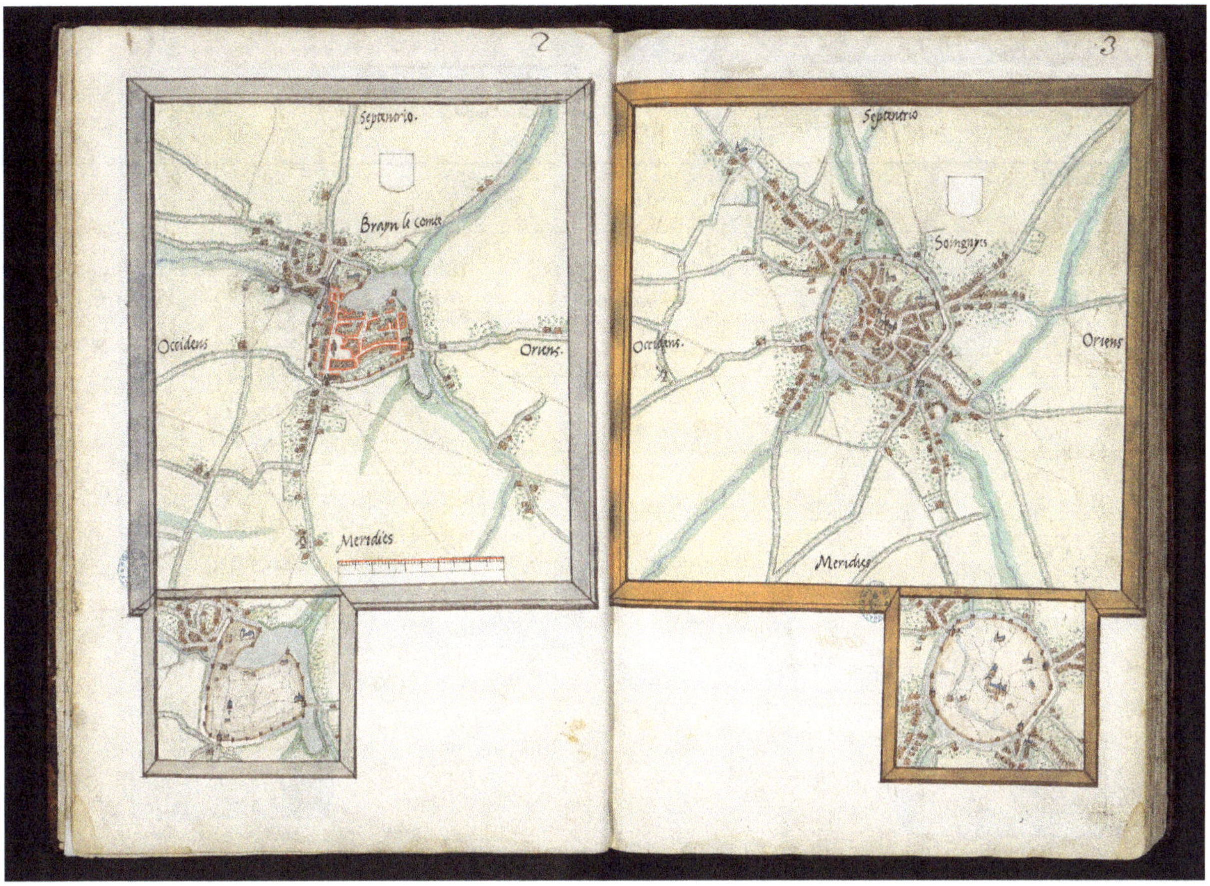
Occidens
Oriens
Meridies
Occidens
Oriens

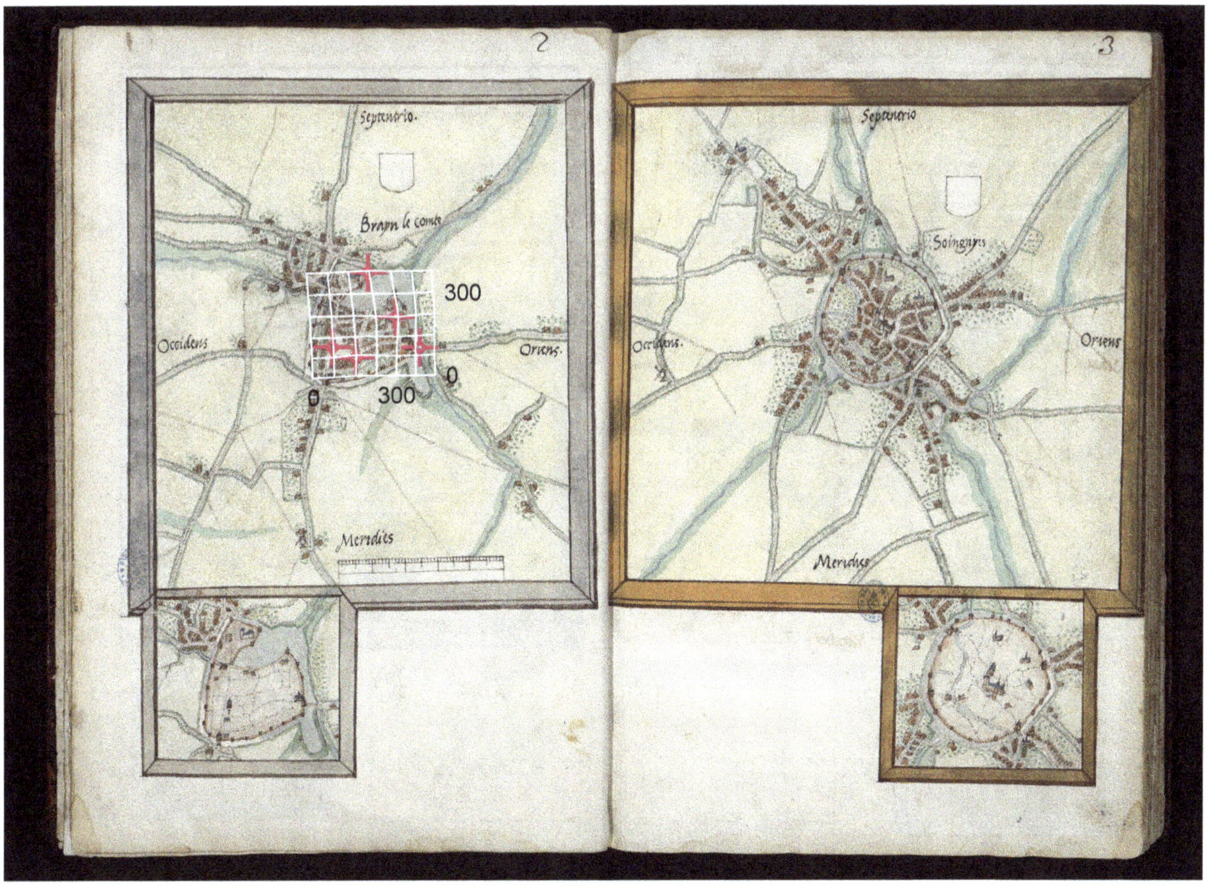
300
0
300
0
Occidens
Oriens
Meridies
Oriens

35. Bredevoort

Tables 3.35 (measured and analysed).
Figures 3.35 (unmeasured, measured and analysed).

Measured

Pixels	Scale stick original map (pixels)	Houses	Street sides	Houses per street side	Inhabitants (RF 4.0)	Inhabitants (RF 5.5)	Inhabitants (RF 7.0)
622	1007	109	15	7.28	437	601	765

Analysed

March 4, 2023 10:14:35 AM	
Computation with 5 linked points	
(*italics* points are less reliable):	1. Sint-Joriskerk 2. *Markt* 3. *Landstraat* 4. *Landstraat* 5. *Landstraat*
Reference points (Reliability 0–5)	1
Transformation with 3 parameters	
Scale	1:8810
Rotation	5 °
Mean position error	15m

Septentrio.
Doetecum.
Occidens.
Oriens.
Meridies
Septentrio.
Occidens.
Oriens
Meridies

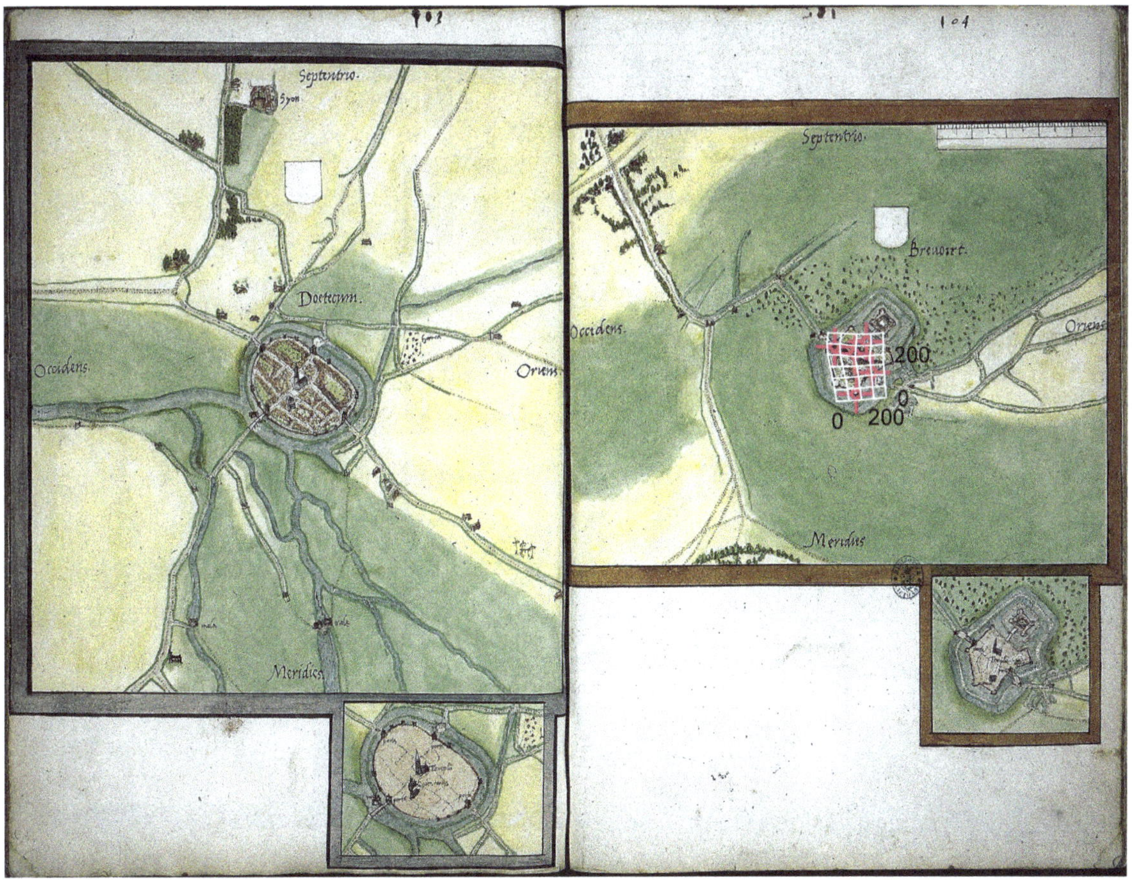
Septentrio.
Doetecum.
Occidens.
Oriens.
Meridies
Septentrio.
Occidens.
Oriens
200
0
0
200
Meridies

36. Brielle

Tables 3.36 (measured and analysed).
Figures 3.36 (unmeasured, measured and analysed).

Measured							
Pixels	Scale stick original map (pixels)	Houses	Street sides	Houses per street side	Inhabitants (RF 4.0)	Inhabitants (RF 5.5)	Inhabitants (RF 7.0)
3973	998	698	112	6.23	2791	3837	4884

Analysed	
March 6, 2023 12:53:09 PM	
Computation with 5 linked points	
(*italics* points are less reliable):	1. Sint-Catherijnekerk 2. *Turfkade* 3. Langestraat 4. *Venkelstraat* 5. Maarland Zuidzijde
Reference points (Reliability 0–5)	3
Transformation with 3 parameters	
Scale	1:8570
Rotation	16 °
Mean position error	18m

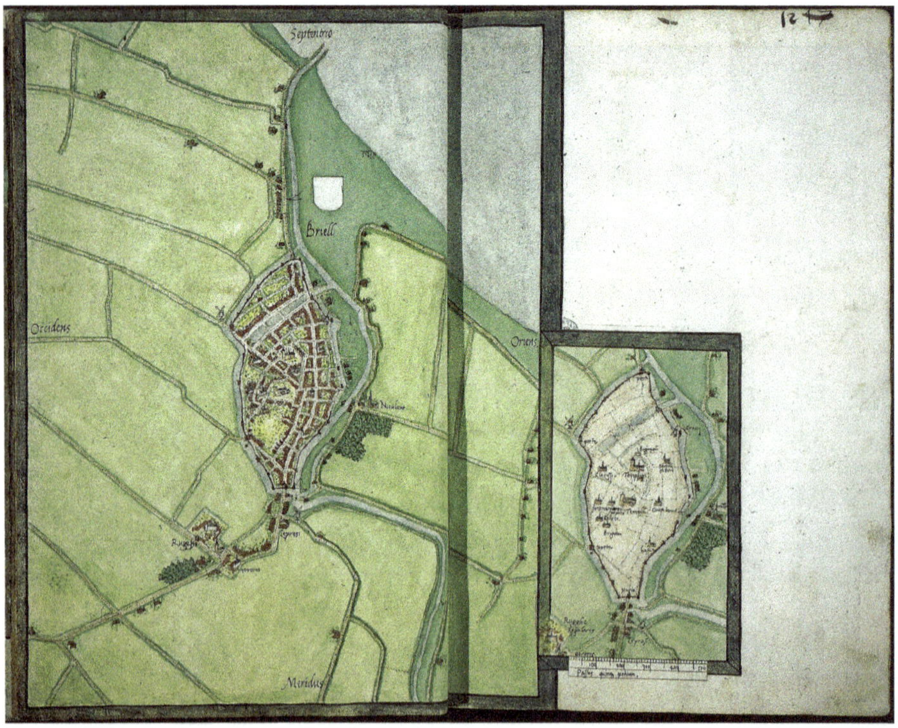

Septentrio
Briell
Occidens
Oriens
Meridus

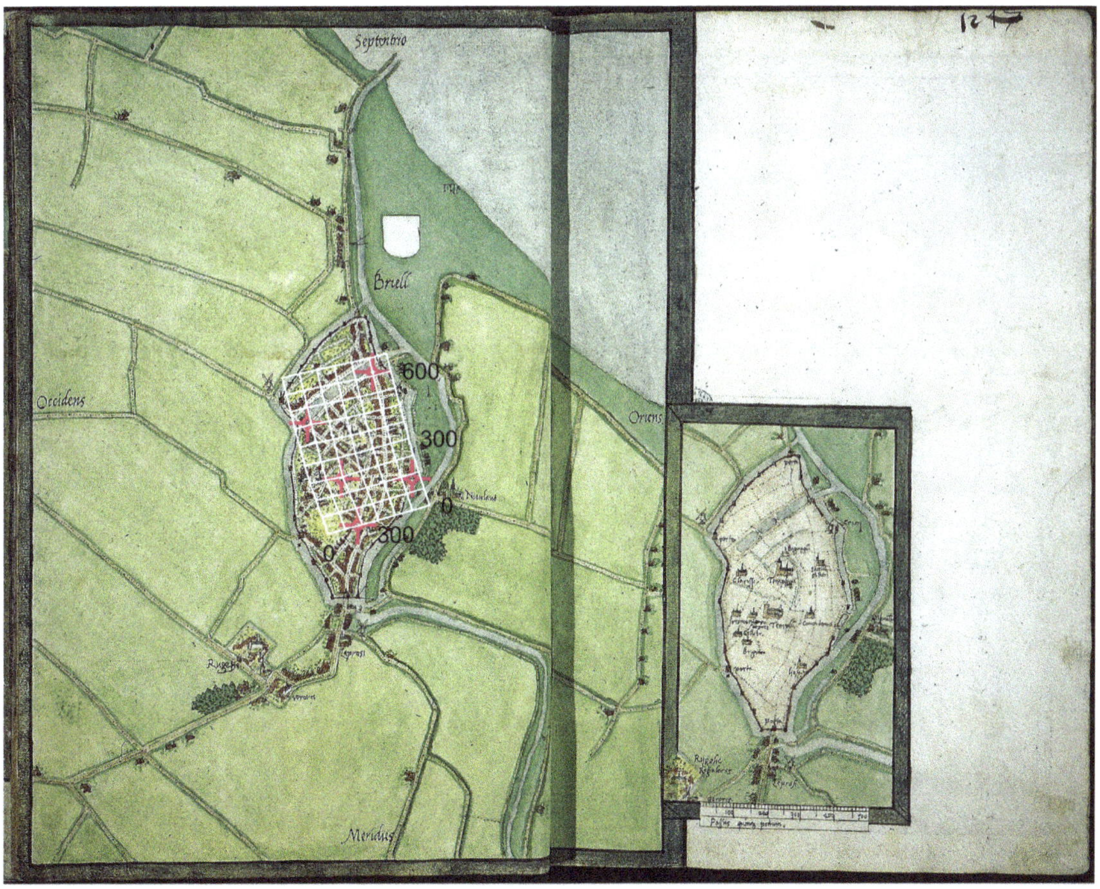
Septentrio
Briell
Occidens
Oriens
600
300
0
0
300
Meridus

37. Brouwershaven
Tables 3.37 (measured and analysed).
Figures 3.37 (unmeasured, measured and analysed).

Measured							
Pixels	Scale stick original map (pixels)	Houses	Street sides	Houses per street side	Inhabitants (RF 4.0)	Inhabitants (RF 5.5)	Inhabitants (RF 7.0)
2758	999	484	51	9.50	1937	2664	3390

Analysed	
January 23, 2023, 11:24:18 AM	
Computation with 5 linked points	
(*italics* points are less reliable):	1. Kerkpad 2. Markt 3. Haven Noordzijde 4. Drilhoek *5. Kromwal*
Reference points (Reliability 0–5)	4
Transformation with 3 parameters	
Scale	1:9150
Rotation	9 °
Mean position error	34m

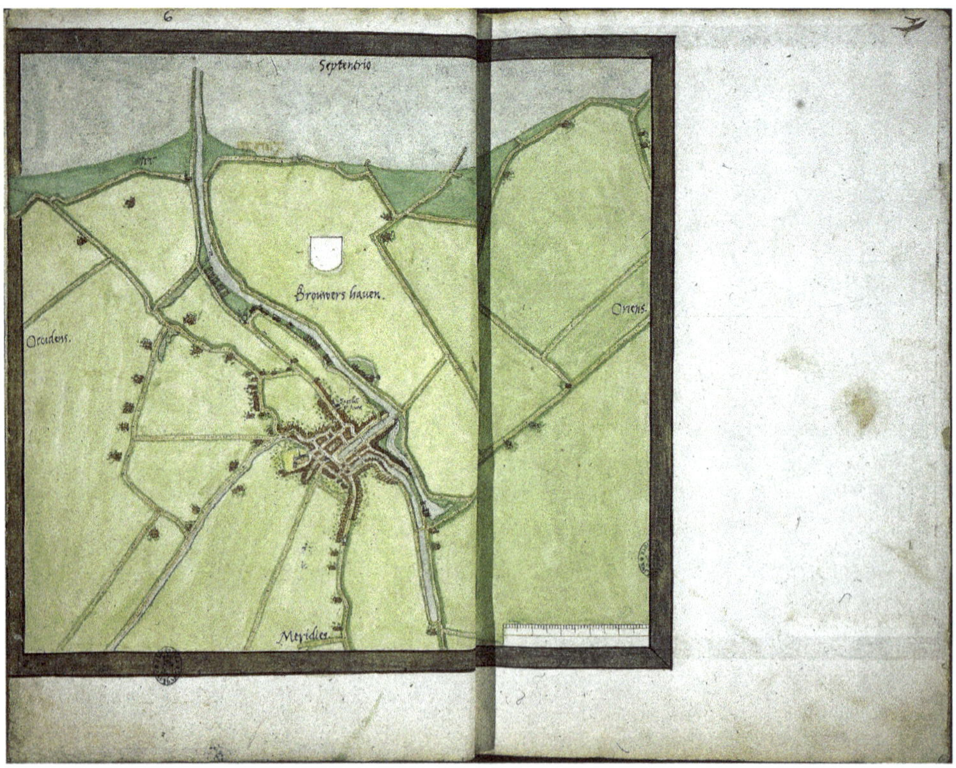

Septentrio
Brouwers hauen
Occidens
Oriens
Meridies

Septentrio
Brouwers hauen
Occidens
Oriens
Meridies
300
0
0
300

38. Brugge

Tables 3.38 (measured and analysed).
Figures 3.38 (unmeasured, measured and analysed).

Measured							
Pixels	Scale stick original map (pixels)	Houses	Street sides	Houses per street side	Inhabitants (RF 4.0)	Inhabitants (RF 5.5)	Inhabitants (RF 7.0)
51,916	1357	9116	1125	8.10	36,466	50.140	63,815

Analysed	
March 10, 2023 7:31:23 PM	
Computation with 5 linked points	
(*italics* points are less reliable):	1. Belfort 2. Ezelspoort 3. Kruisport 4. Gentpoort 5. Smedenpoort
Reference points (Reliability 0–5)	5
Transformation with 3 parameters	
Scale	1:7.930
Rotation	10 °
Mean Position Error	**35m**

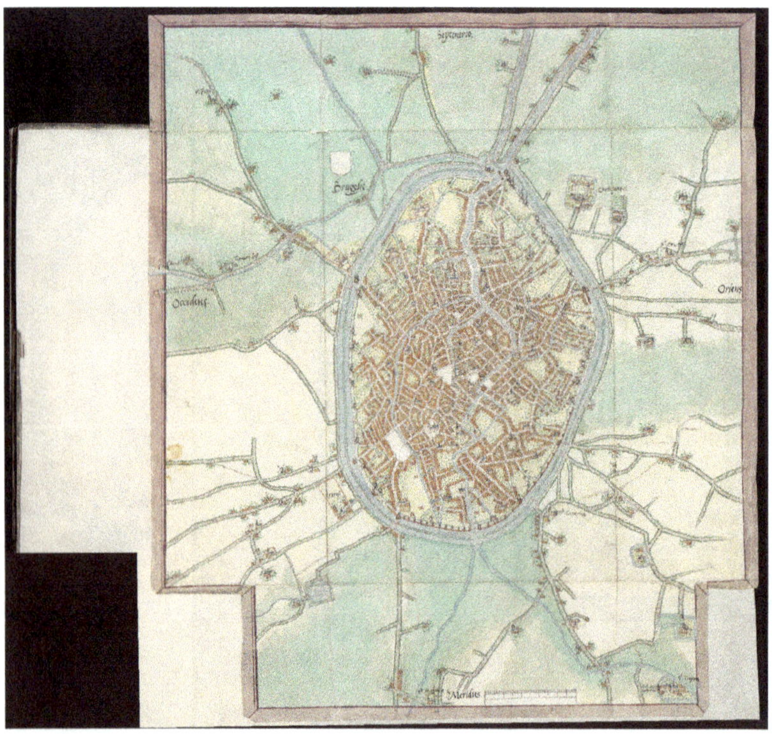

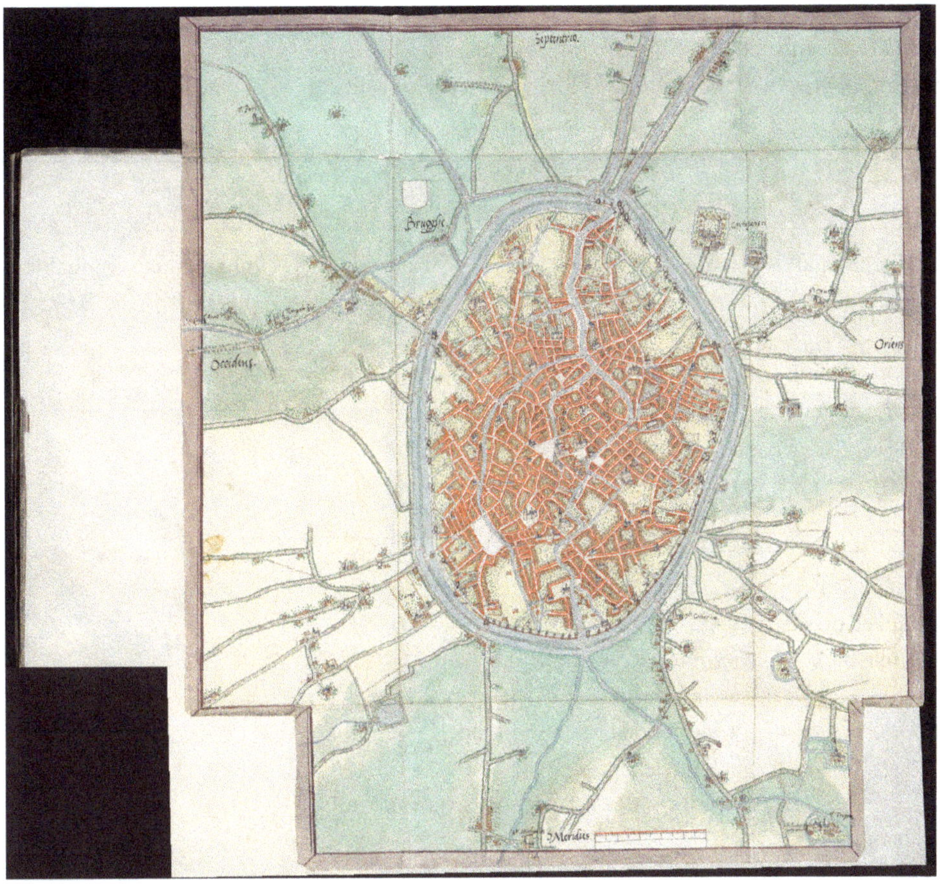
Septentrio.
Brugghe
Occidens
Oriens
Meridies

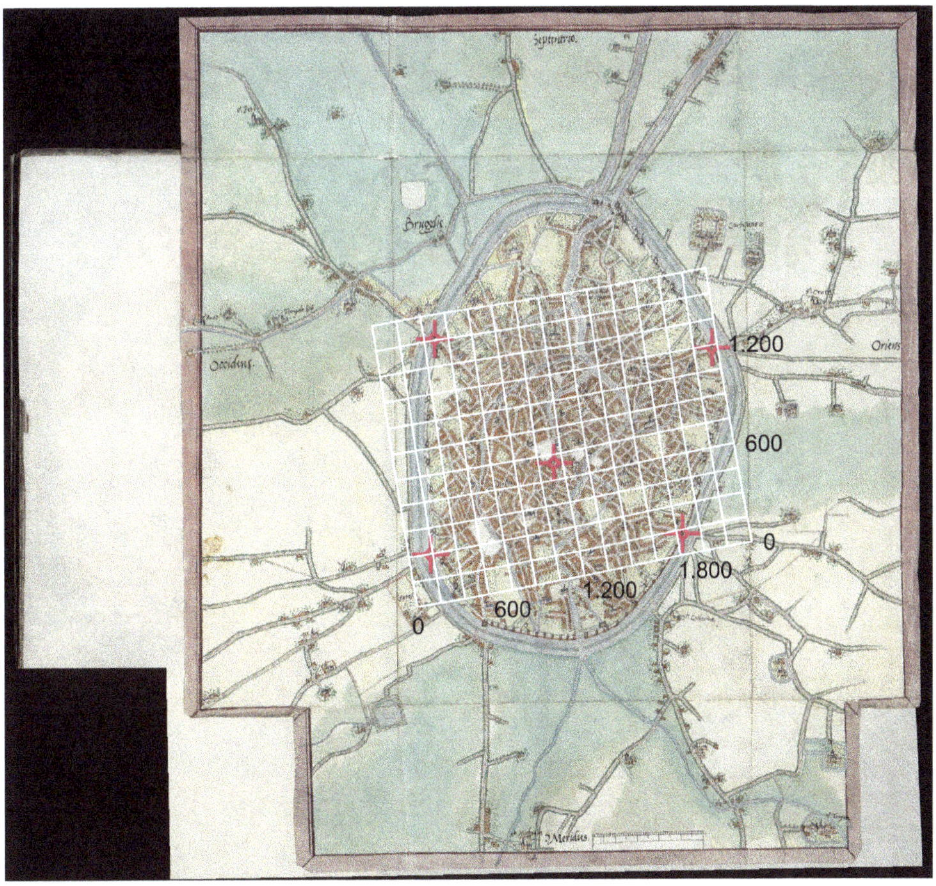
Septentrio.
Brugghe
Occidens
Oriens
1.200
600
0
1.800
1.200
600
0
Meridies

39. Brussel

Tables 3.39 (*measured and analysed*).
Figures 3.39 (*unmeasured, measured and analysed*).

Measured

Pixels	Scale stick original map (pixels)	Houses	Street sides	Houses per street side	Inhabitants (RF 4.0)	Inhabitants (RF 5.5)	Inhabitants (RF 7.0)
35,753	missing	6278	767	8.19	25,113	34,530	43,948

Analysed

September 13, 2023 8:36:18 PM	
Computation with 5 linked points	
(*italics* points are less reliable):	1. Grote Markt 2. Hallepoort 3. Kathedraal van Sint-Michiel 4. Kapellekerk 5. Onze-Lieve-Vrouwe-ter-Zavelkerk
Reference points (Reliability 0–5)	5
Transformation with 3 parameters	
Scale	1:8940
Rotation	15 °
Mean position error	16m

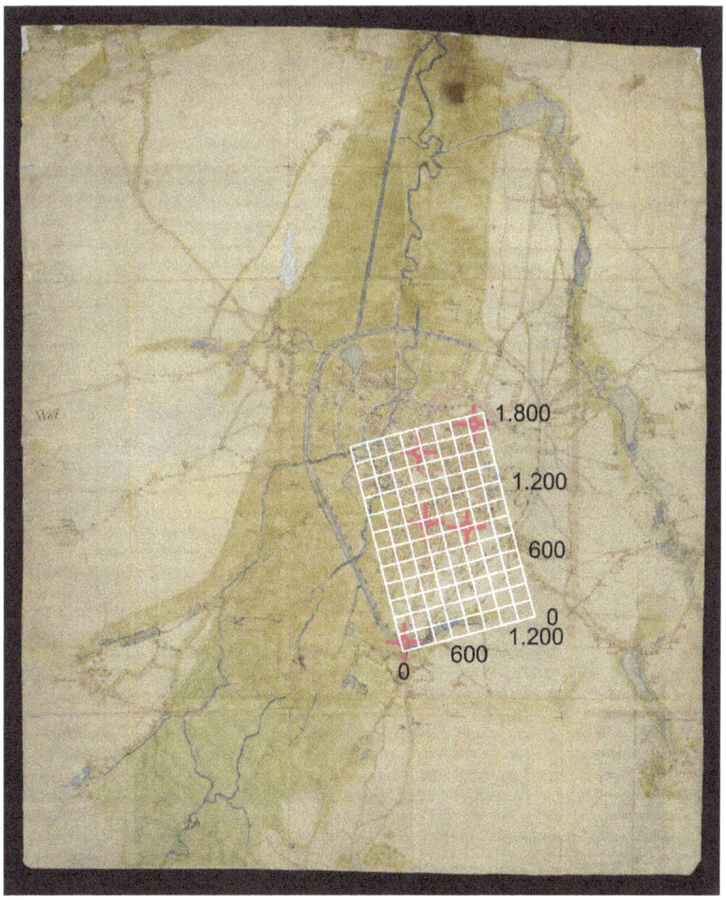
1.800
1.200
600
0
1.200
600
0

40. Buren

Tables 3.40 (measured and analysed).
Figures 3.40 (unmeasured, measured and analysed).

Measured

Pixels	Scale stick original map (pixels)	Houses	Street sides	Houses per street side	Inhabitants (RF 4.0)	Inhabitants (RF 5.5)	Inhabitants (RF 7.0)
1923	992	338	44	7.67	1351	1857	2364

Analysed

March 11, 2023 10:24:17 AM	
Computation with 5 linked points	
(italics points are less reliable):	1. Sint-Lambertuskerk 2. Culemborgse Poort 3. Prins van Oranje (molen) 4. Kornewal 5. Weeshuiswal
Reference points (Reliability 0–5)	5
Transformation with 3 parameters	
Scale	1:7210
Rotation	11 °
Mean position error	12m

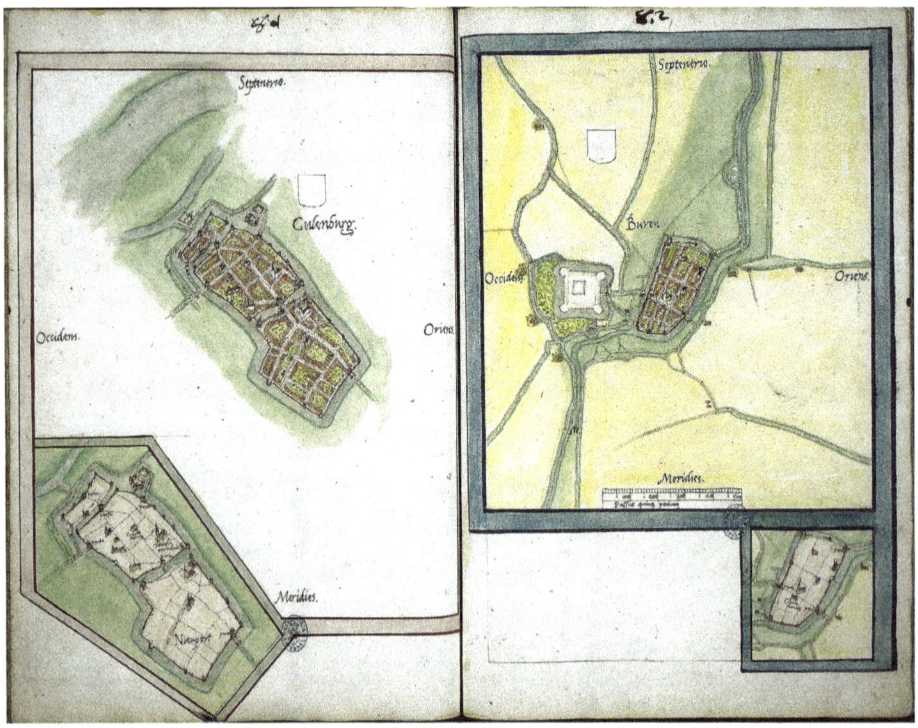

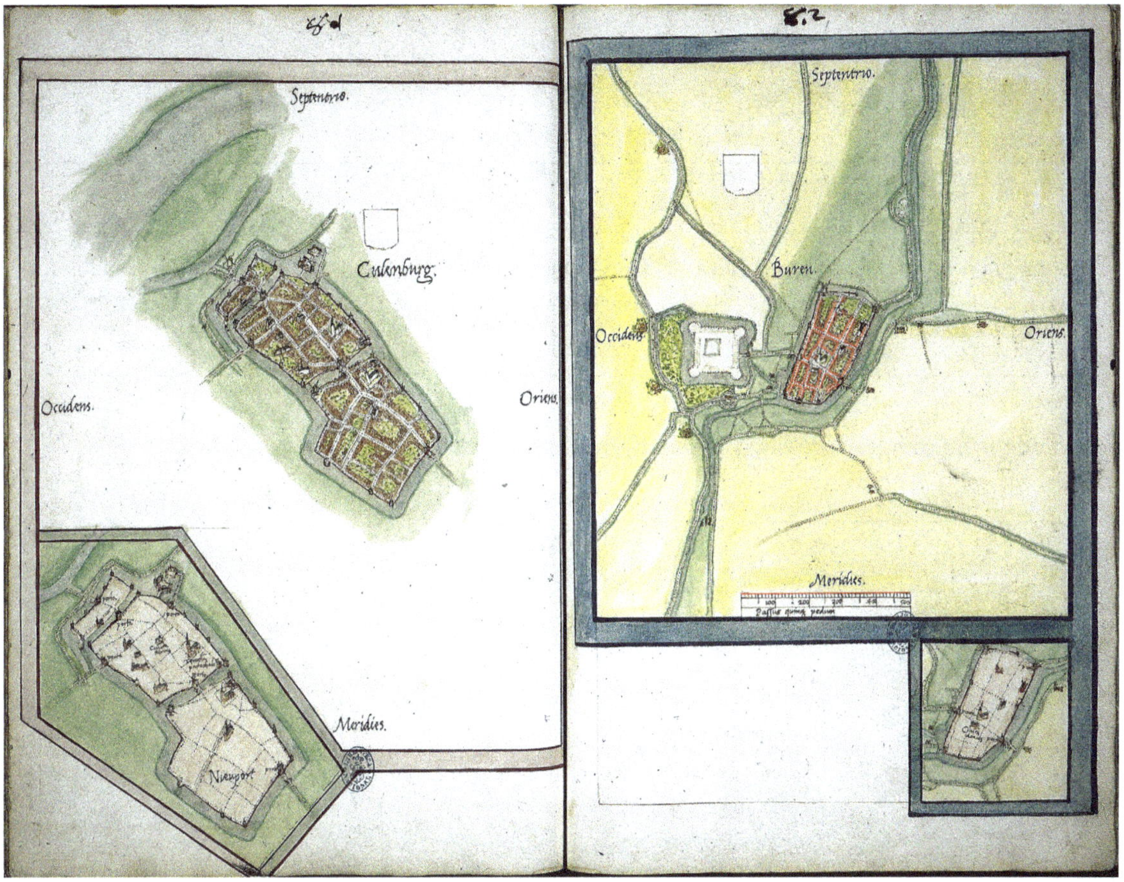
Septentrio.
Culenburg.
Occidens.
Oriens.
Meridies.
Nieupoort
Septentrio.
Buren.
Occidens.
Oriens.
Meridies.

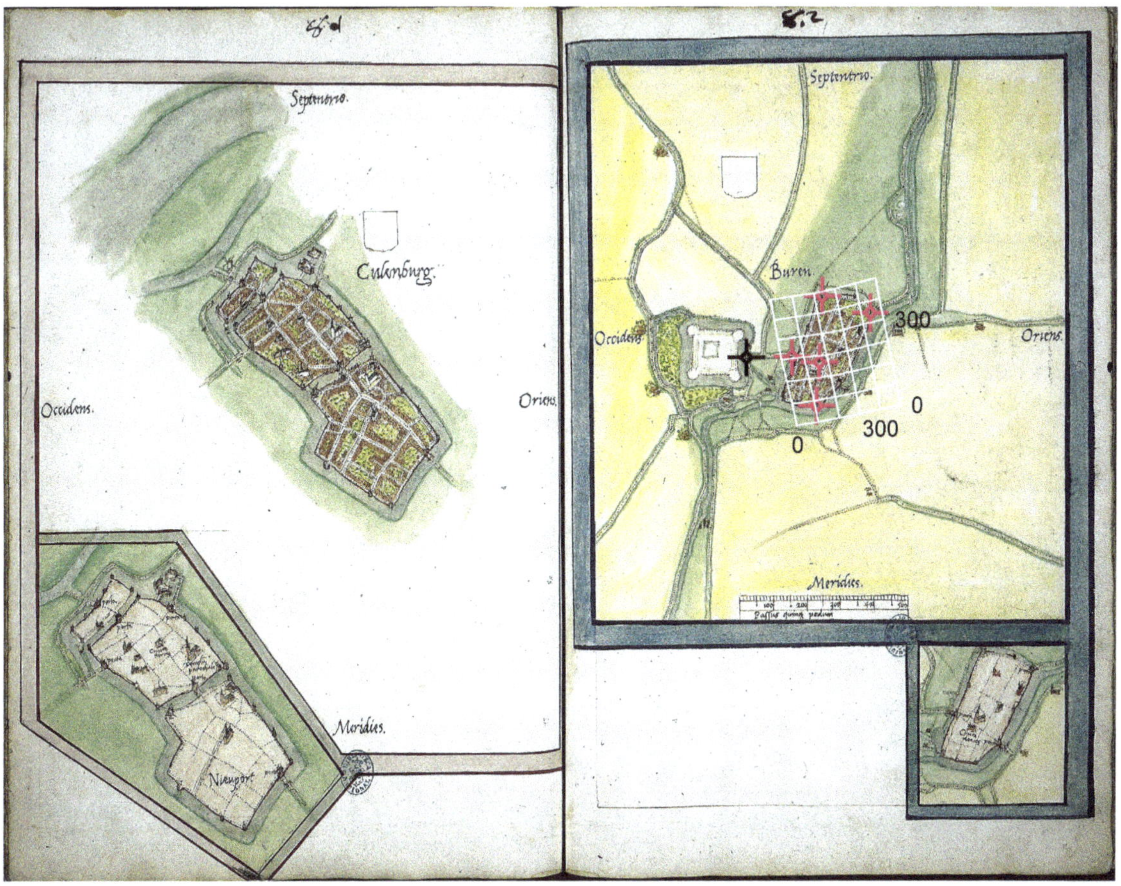
Septentrio.
Culenburg.
Occidens.
Oriens.
Meridies.
Nieupoort
Septentrio.
Buren.
300
Occidens.
Oriens.
0
300
0
Meridies.

41. Cambrai

Tables 3.41 (measured and analysed).
Figures 3.41 (unmeasured, measured and analysed).

Measured							
Pixels	Scale stick original map (pixels)	Houses	Street sides	Houses per street side	Inhabitants (RF 4.0)	Inhabitants (RF 5.5)	Inhabitants (RF 7.0)
–	1387	–	–	–	–	–	–

In this map de streets and the buildings of the town are not drawn.

Analysed	
March 11, 2023 11:13:07 AM	
Computation with 5 linked points	
(*italics* points are less reliable):	1. Citadel
	2. Porte de Paris
	3. Chateau de Selles
	4. Rue de Solesmes
	5. Tour des Arquets
Reference points (Reliability 0–5)	1
Transformation with 3 parameters	
Scale	1:8160
Rotation	12 °
Mean position error	50m

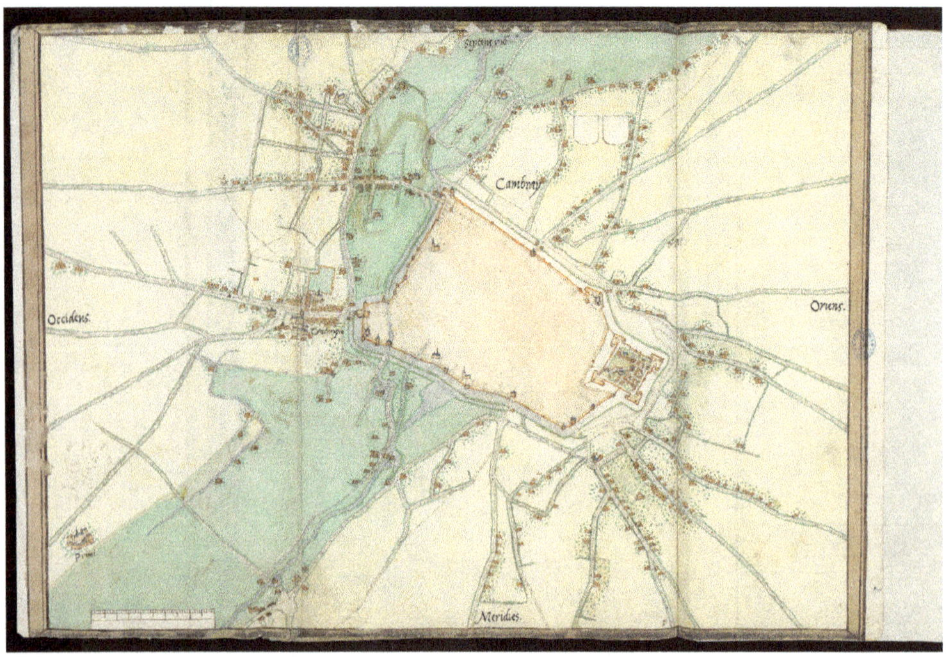

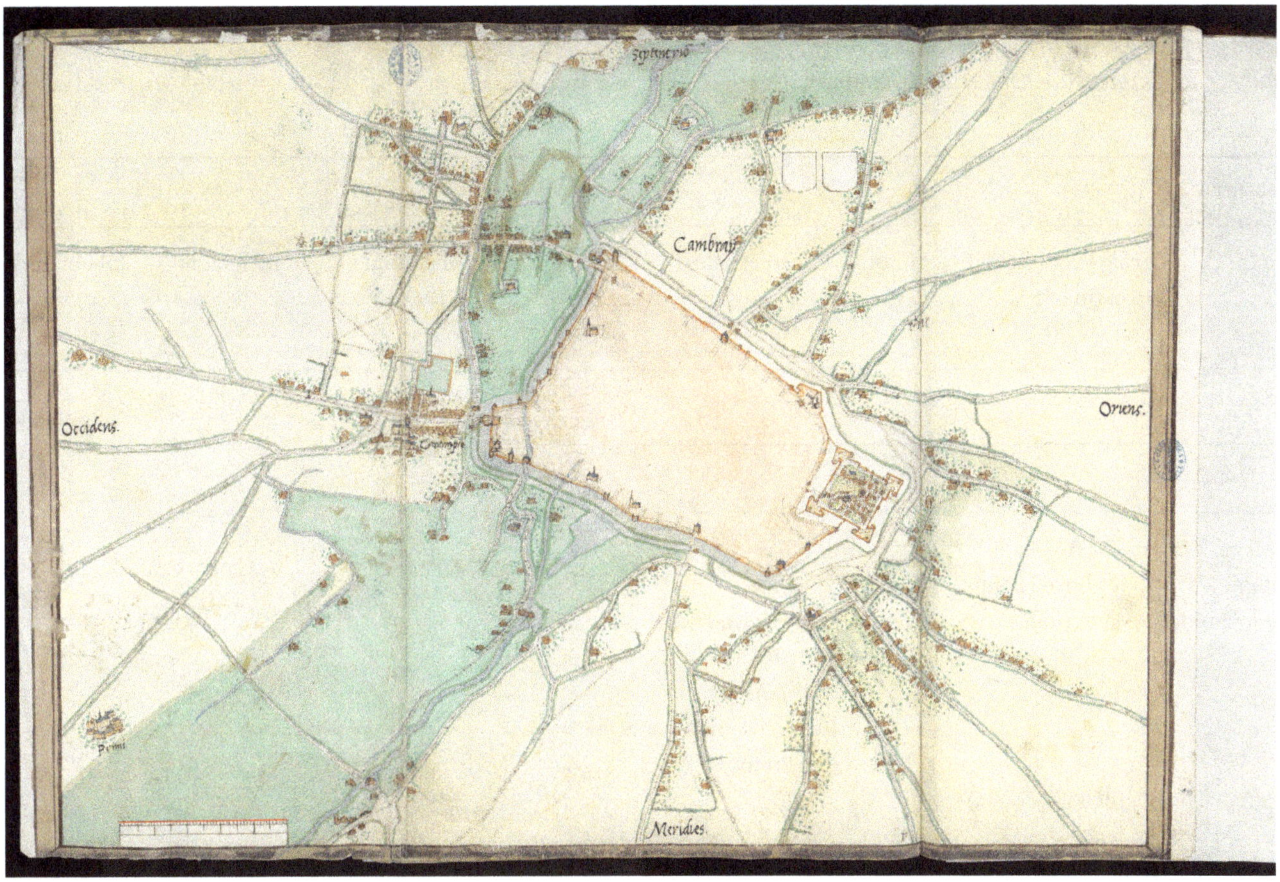
Septentrio
Cambray
Occidens.
Oriens.
Meridies.

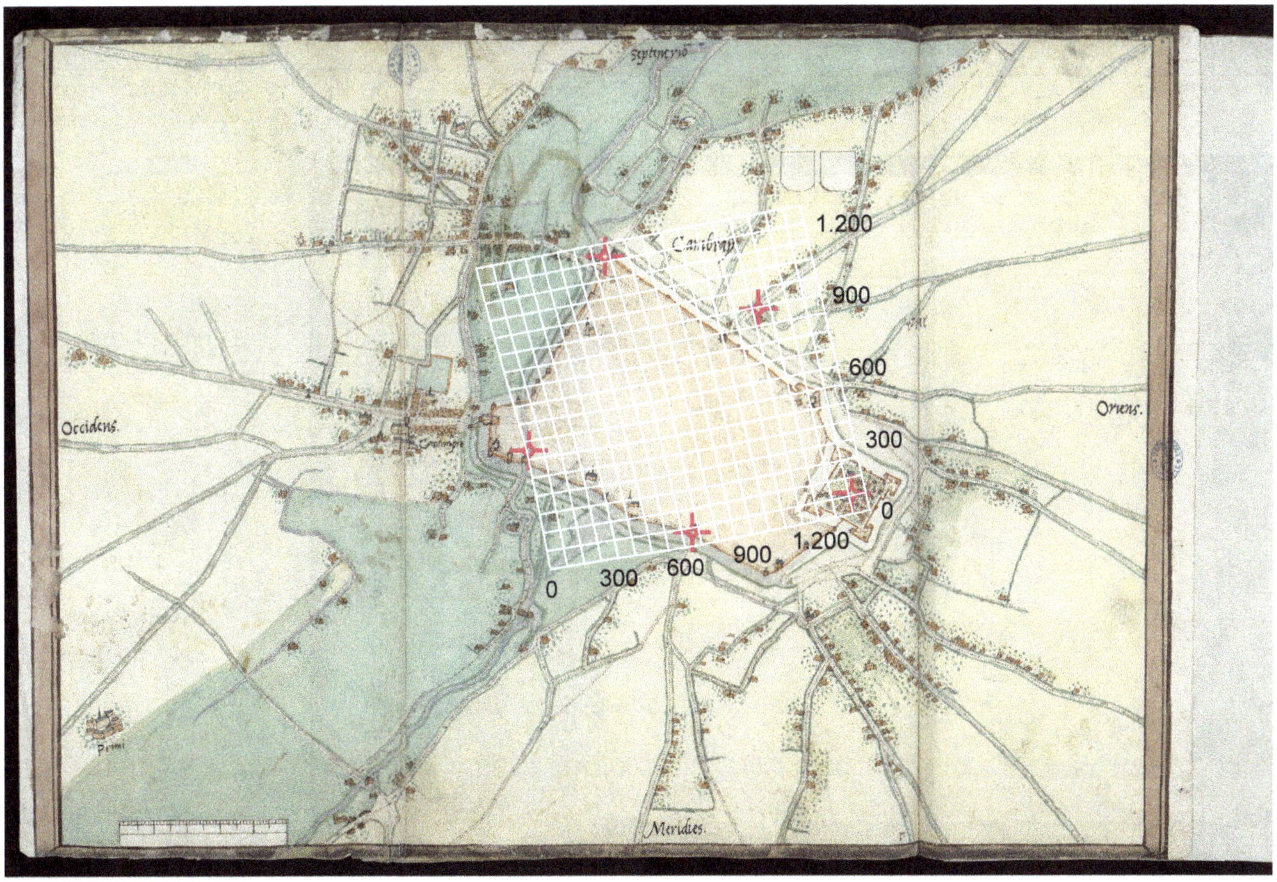
Septentrio
Cambray
Occidens.
Oriens.
Meridies.
1.200
900
600
300
0
0
300
600
900
1.200

42. Cassel
Tables 3.42 (measured and analysed).
Figures 3.42 (unmeasured, measured and analysed).

Measured

Pixels	Scale stick original map (pixels)	Houses	Street sides	Houses per street side	Inhabitants (RF 4.0)	Inhabitants (RF 5.5)	Inhabitants (RF 7.0)
2923	1021	513	69	7.44	2053	2823	3593

Analysed

April 30, 2023 11:19:04 AM	
Computation with 5 linked points	
(*italics* points are less reliable):	1. Église Notre-Dame 2. Porte d'Aire *3. Rue Saint-Nicolas* 4. Rue Constant Moeneclaey 5. Rue Profonde
Reference points (Reliability 0–5)	4
Transformation with 3 parameters	
Scale	1:7870
Rotation	4 °
Mean position error	27m

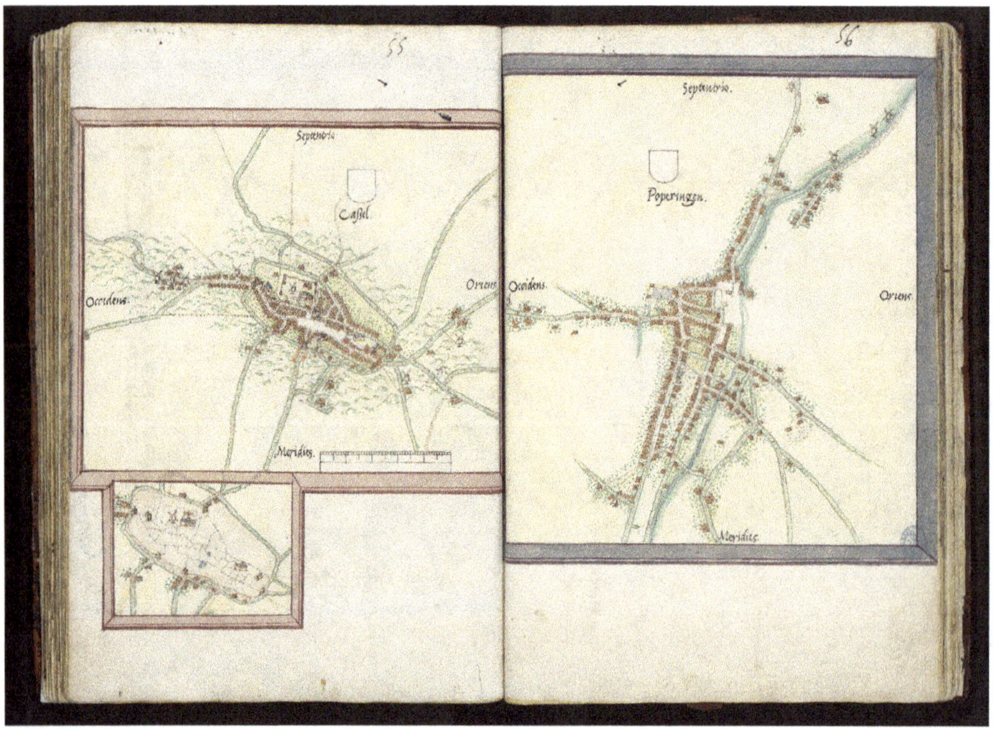

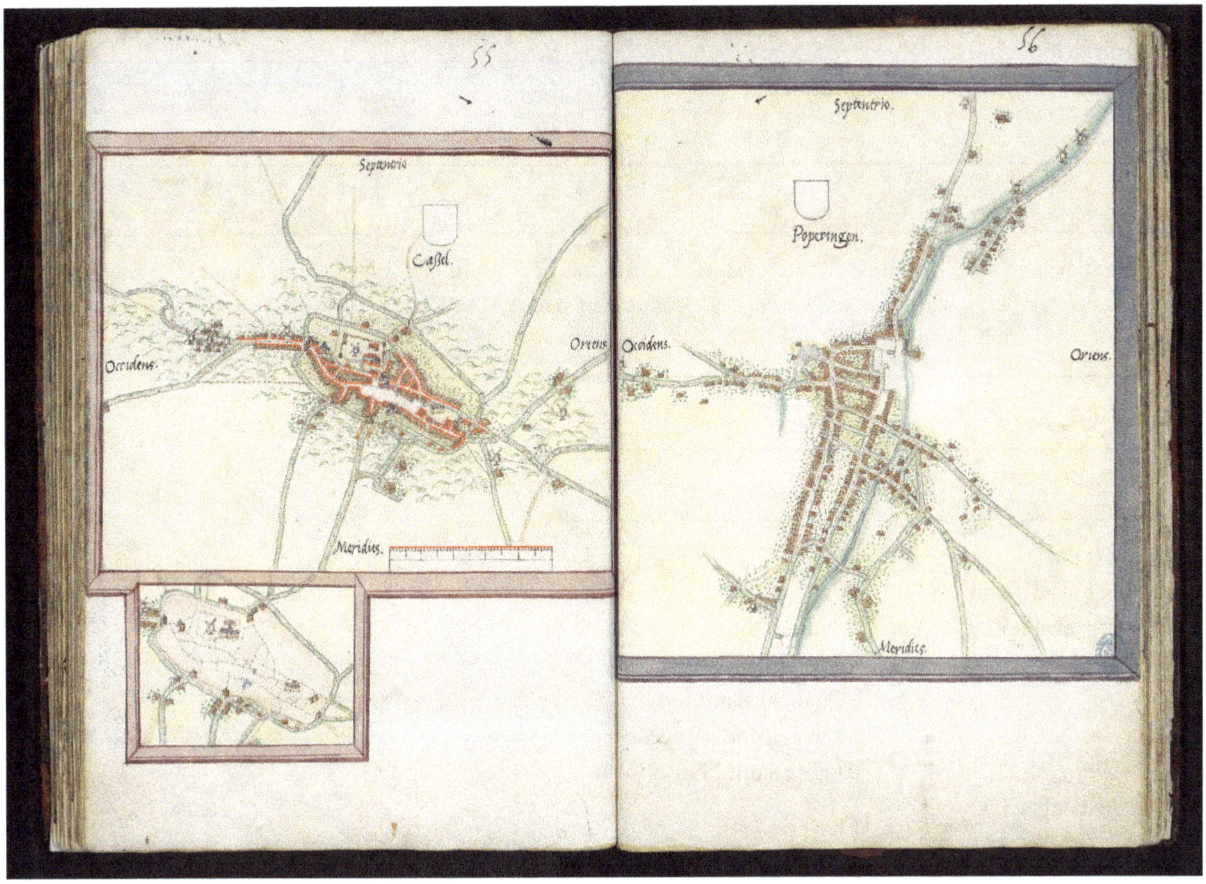
55
56
Septentrio
Caßel.
Occidens.
Oriens.
Meridies.
Septentrio.
Poperingen.
Occidens.
Oriens.
Meridies.

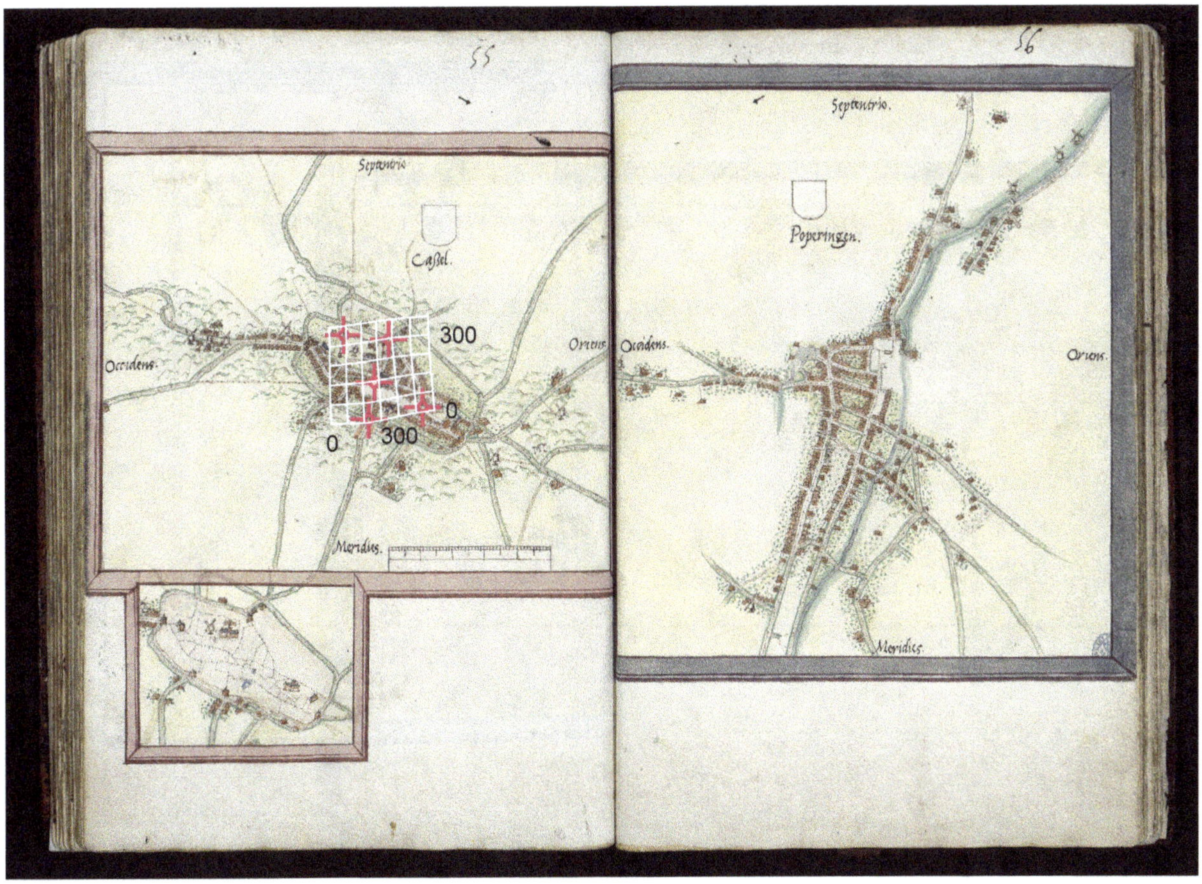
55
56
Septentrio
Caßel.
300
Occidens.
Oriens.
0
0
300
Meridies.
Septentrio.
Poperingen.
Occidens.
Oriens.
Meridies.

43. Chièvres

Tables 3.43 (measured and analysed).
Figures 3.43 (unmeasured, measured and analysed).

Measured

Pixels	Scale stick original map (pixels)	Houses	Street sides	Houses per street side	Inhabitants (RF 4.0)	Inhabitants (RF 5.5)	Inhabitants (RF 7.0)
1104	1030	194	26	7.46	775	1066	1357

Analysed

March 13, 2023 3:03:21 PM	
Computation with 5 linked points	
(*italics* points are less reliable):	*1. Tour de Grave*
	2. La Petite Hunelle
	3. Chapelle Saint Jean
	4. Chapelle de la Ladrerie
	5. Église Saint-Martin
Reference points (Reliability 0–5)	3
Transformation with 3 parameters	
Scale	1:7510
Rotation	9 °
Mean position error	32m

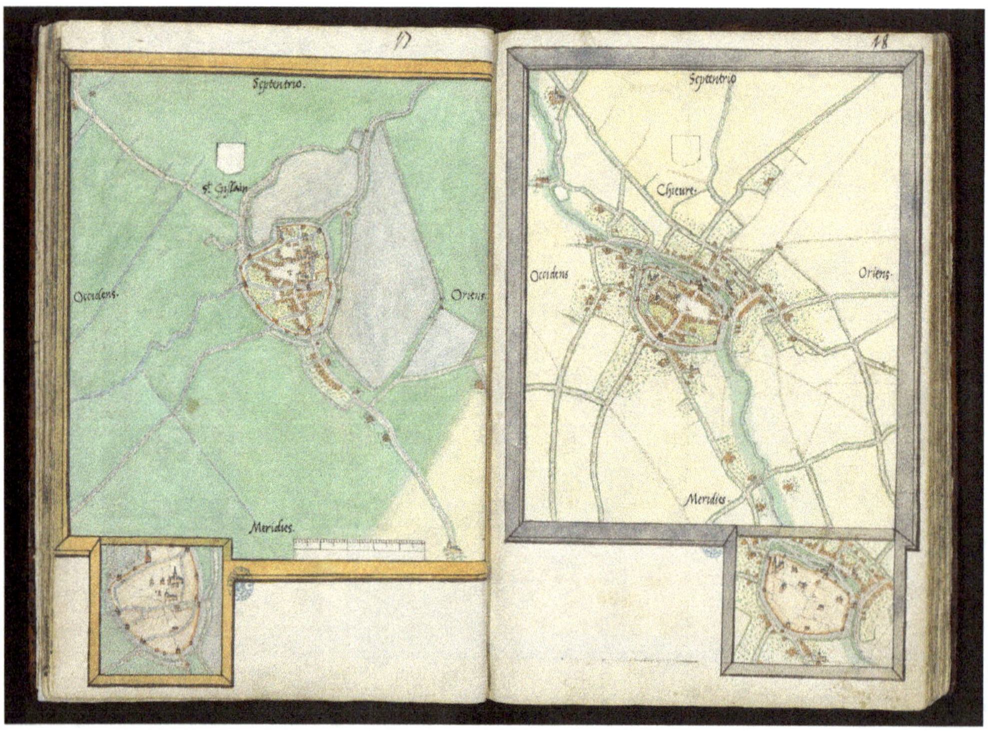

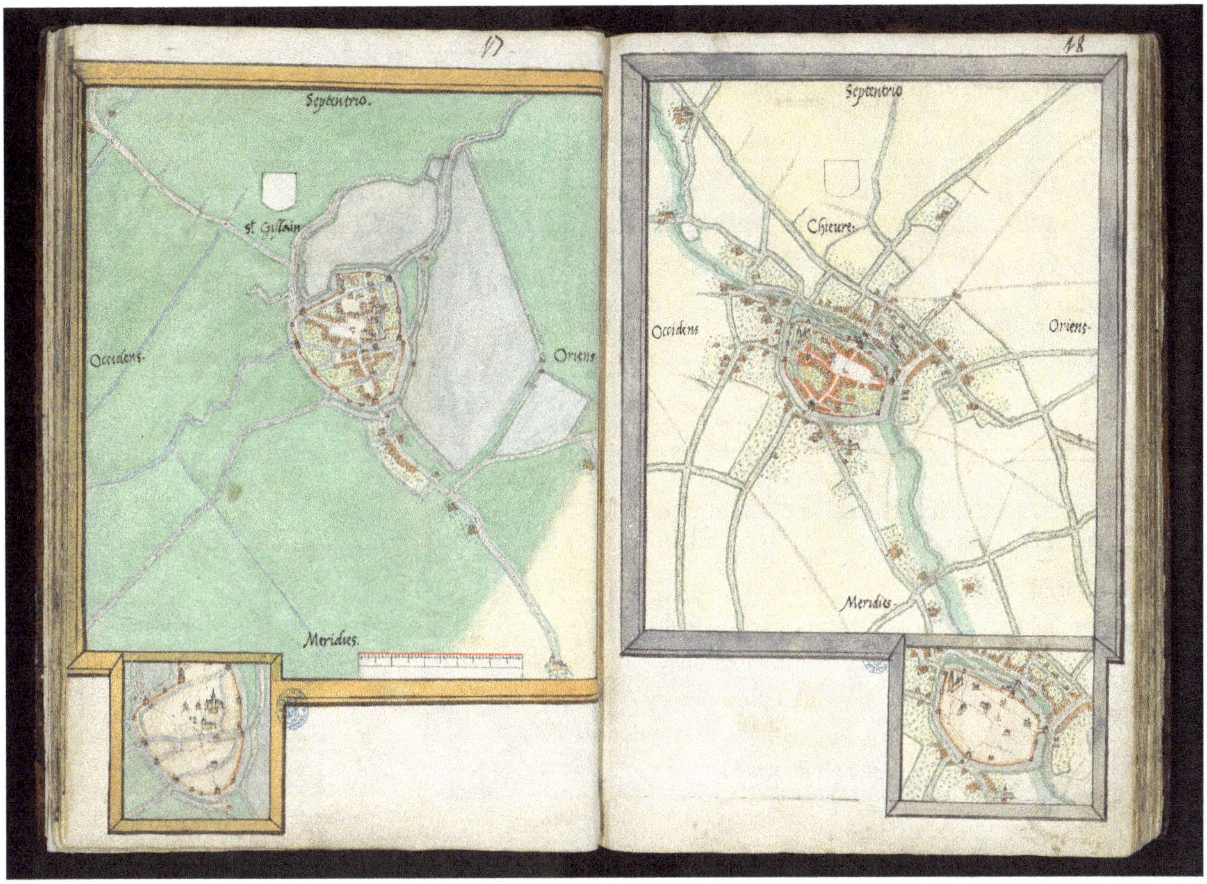
17
Septentrio.
St. Gislain
Occidens.
Oriens.
Meridies.
18
Septentrio
Chieure
Occidens
Oriens
Meridies

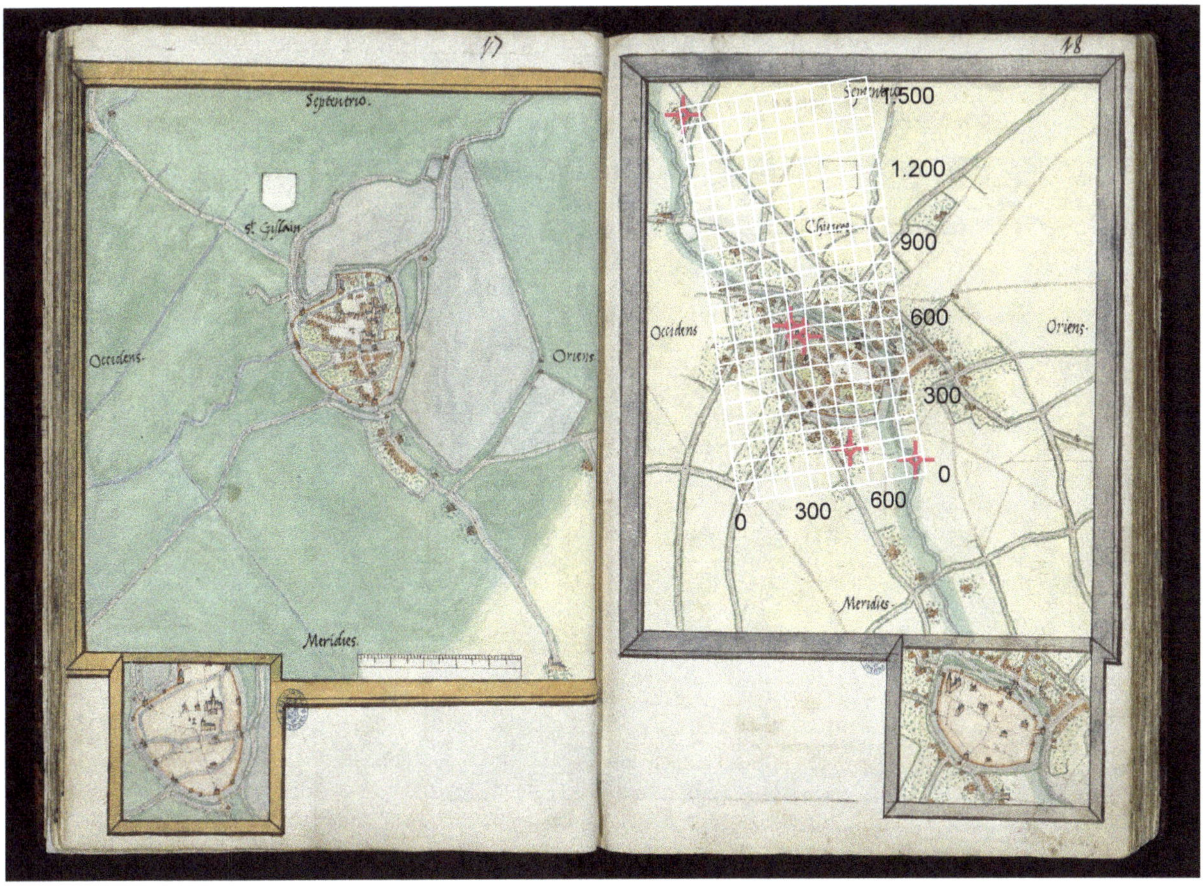
17
Septentrio.
St. Gislain
Occidens.
Oriens.
Meridies.
18
1.500
1.200
900
600
300
0
0
300
600
Occidens
Oriens
Meridies

44. Chimay
Tables 3.44 (measured and analysed).
Figures 3.44 (unmeasured, measured and analysed).

Measured							
Pixels	Scale stick original map (pixels)	Houses	Street sides	Houses per street side	Inhabitants (RF 4.0)	Inhabitants (RF 5.5)	Inhabitants (RF 7.0)
1912	1018	336	93	3.61	1343	1847	2351

Analysed	
March 14, 2023 11:00:31 AM	
Computation with 5 linked points	
(*italics* points are less reliable):	1. Saints Pierre et Paul
	2. Chateau de Chimay
	3. Vieille Tour
	4. Notre-Dame de Miséricorde
	5. Eau Blanche
Reference points (Reliability 0–5)	4
Transformation with 3 parameters	
Scale	1:6800
Rotation	13 °
Mean position error	21m

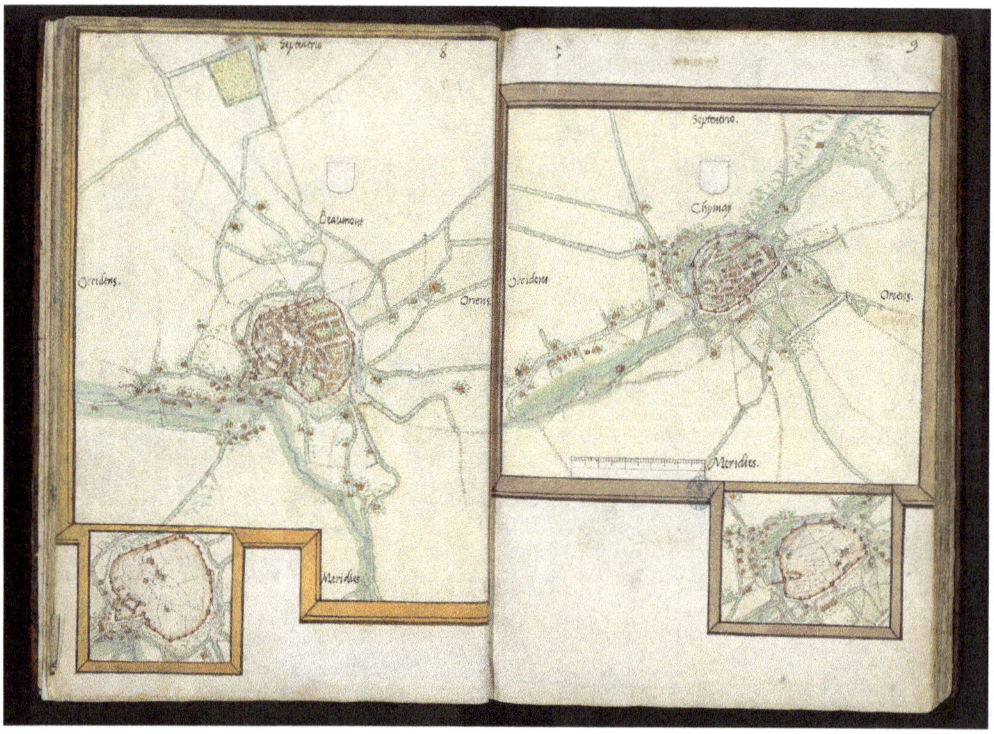

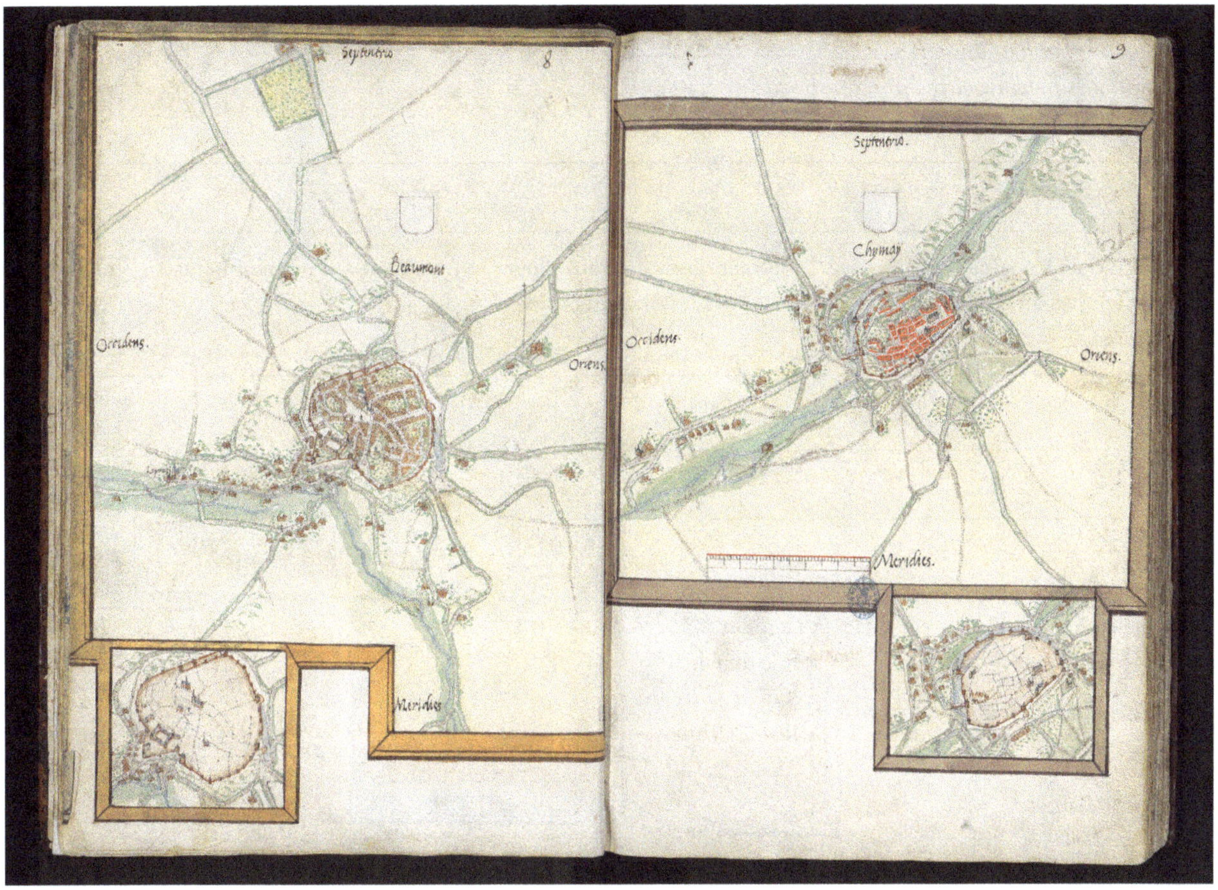
Septentrio
8
Beaumont
Occidens
Oriens
Meridies
9
Septentrio
Chymay
Occidens
Oriens
Meridies

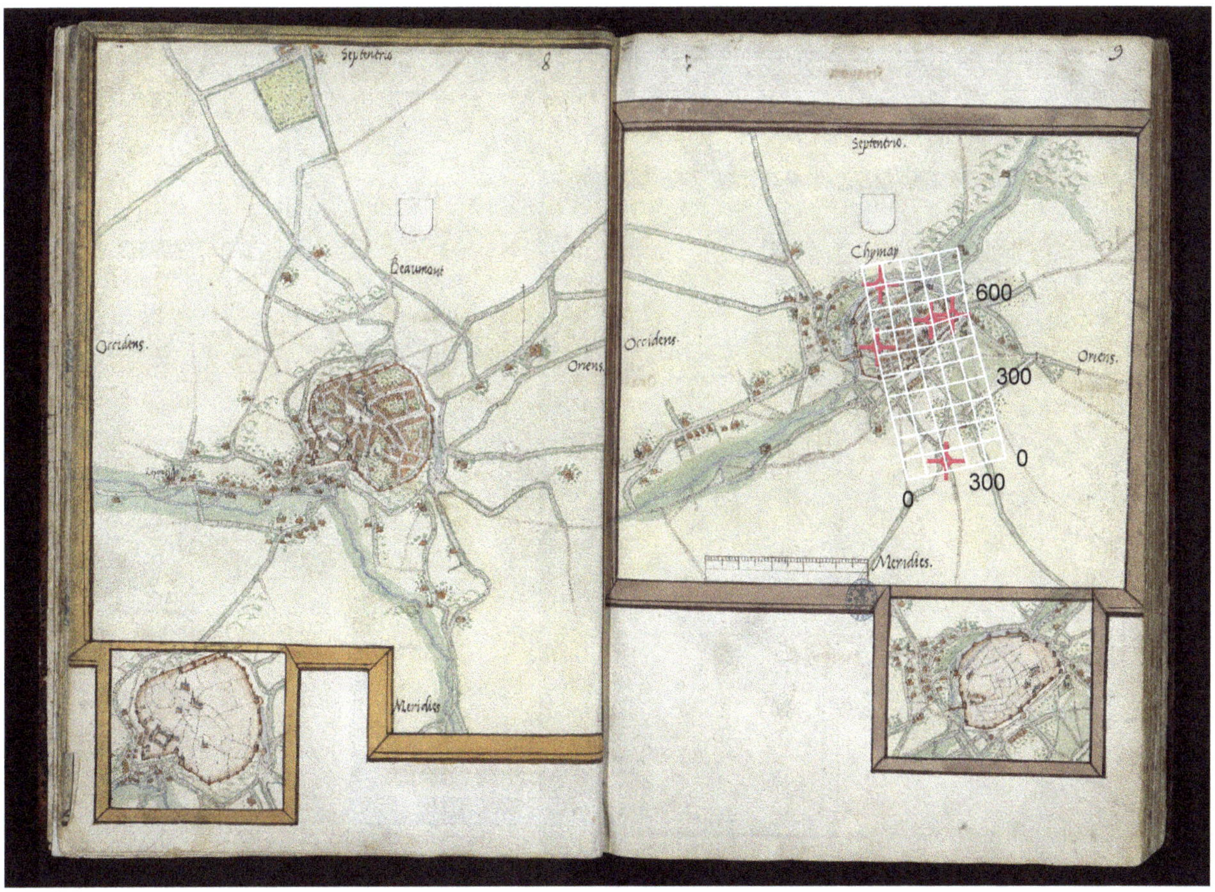
Septentrio
8
Beaumont
Occidens
Oriens
Meridies
9
Septentrio
Chymay
600
300
0
300
0
Occidens
Oriens
Meridies

45. Chiny

Tables 3.45 (measured and analysed).
Figures 3.45 (unmeasured, measured and analysed).

Measured								
Pixels	Scale stick original map (pixels)	Houses	Street sides	Houses per street side	Inhabitants (RF 4.0)	Inhabitants (RF 5.5)	Inhabitants (RF 7.0)
810	missing	142	19	7.49	569	782	995

Analysed	
September 10, 2023 5:05:21 PM	
Computation with 5 linked points	
(*italics* points are less reliable):	*1. La Noue*
	2. Rue du Fort
	3. Rue de Lorrène
	4. Rue de Briga
	5. Rue de la Goulette
Reference points (Reliability 0–5)	4
Transformation with 3 parameters	
Scale	1:7.080
Rotation	11 °
Mean position error	23m

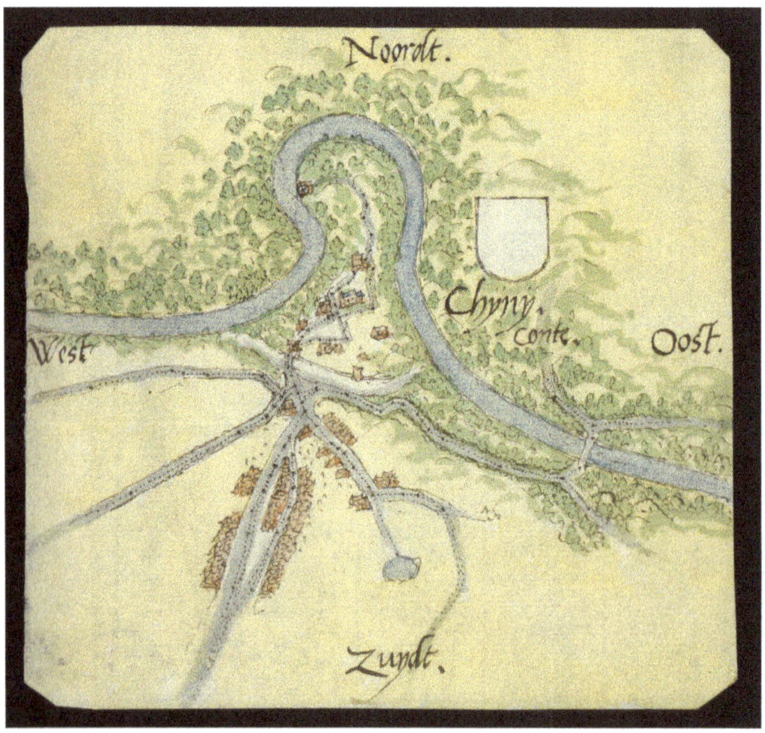

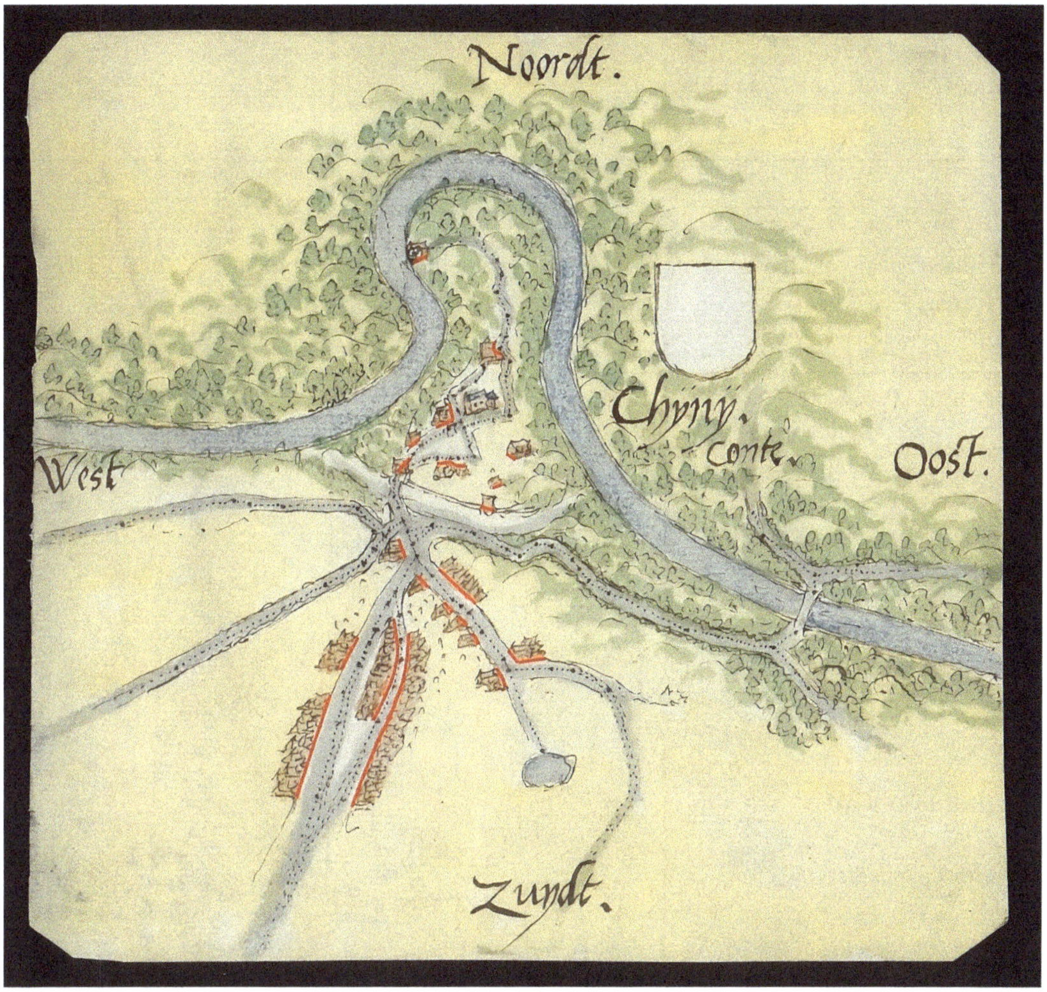
Noordt.
Chyny.
Conte.
West
Oost.
Zuydt.

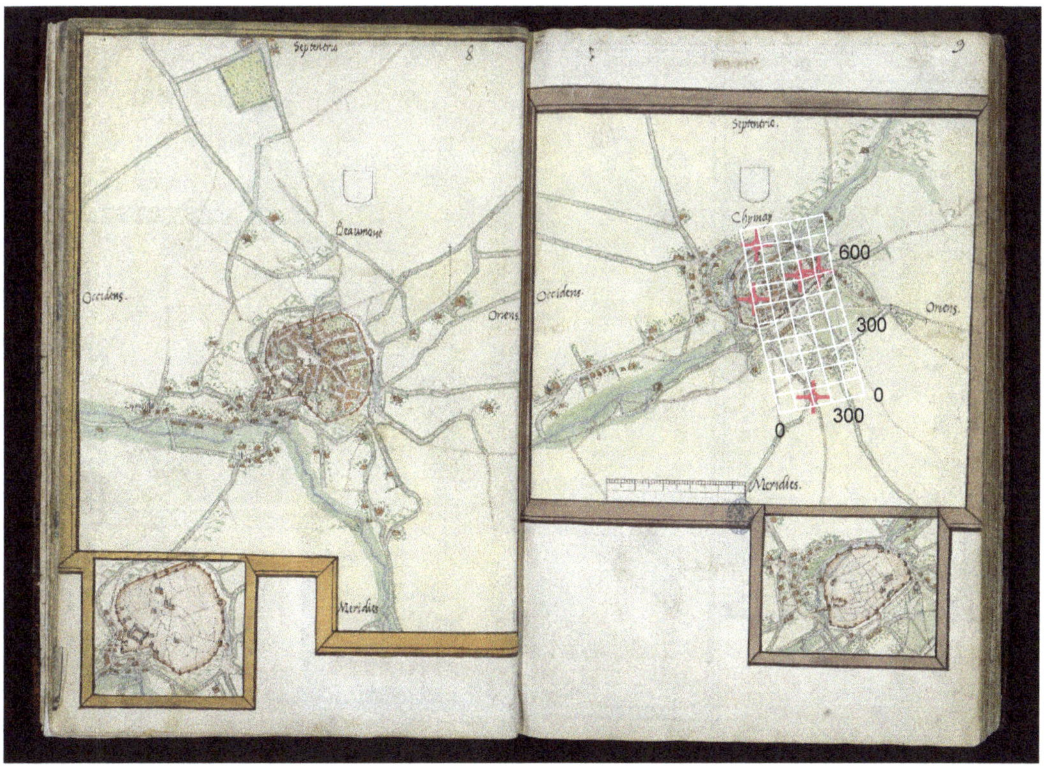
600
300
0
300
0

46. Coevorden

Tables 3.46 (measured and analysed).
Figures 3.46 (unmeasured, measured and analysed).

Measured							
Pixels	Scale stick original map (pixels)	Houses	Street sides	Houses per street side	Inhabitants (RF 4.0)	Inhabitants (RF 5.5)	Inhabitants (RF 7.0)
1202	1018	211	31	6.81	844	1161	1478

Analysed	
March 15, 2023 4:15:18 PM	
Computation with 5 linked points	
(*italics* points are less reliable):	1. Hervormde Kerk 2. Kasteel Coevorden *3. Rijnsestraat* *4. Gracht* *5. Gracht*
Reference points (Reliability 0–5)	2
Transformation with 3 parameters	
Scale	1:9680
Rotation	17 °
Mean position error	29m

59
Septentrio.
Coeuerden.
Occidens
Oriens
Meridies

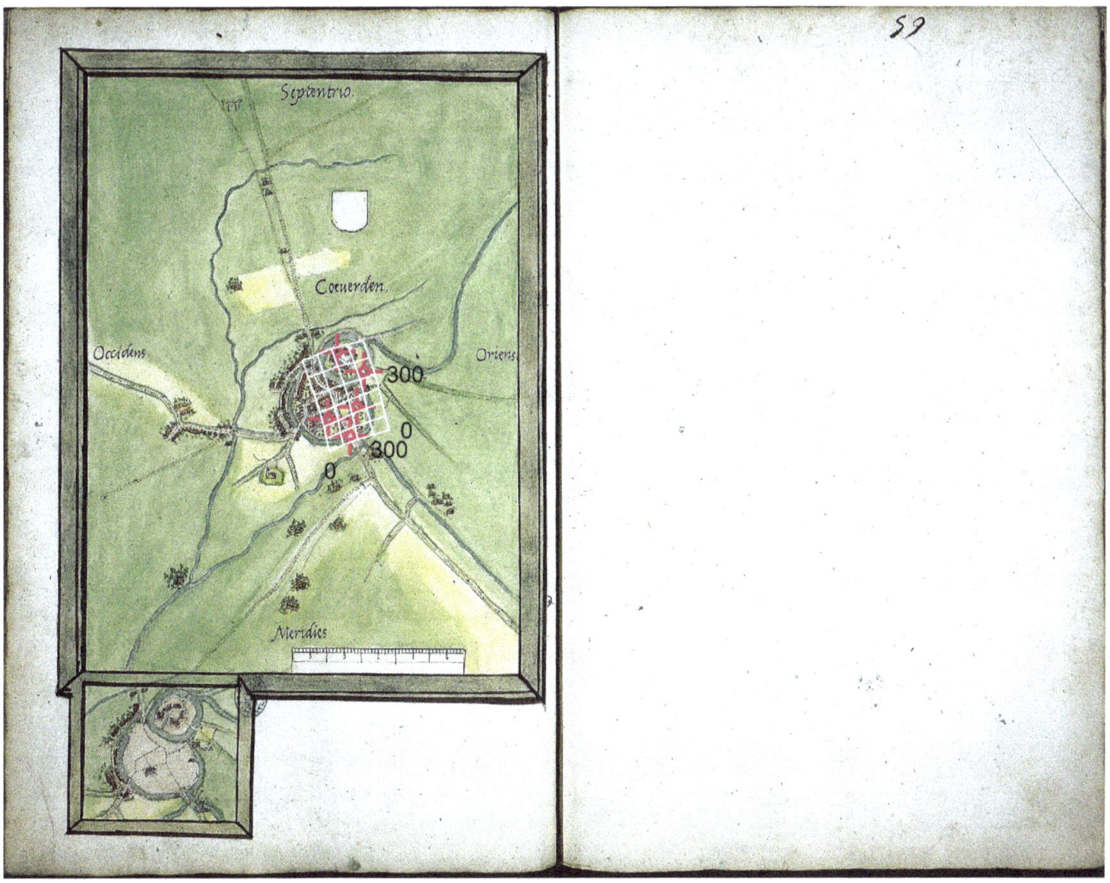
59
Septentrio.
Coeuerden.
Occidens
Oriens
300
0
300
0
Meridies

47. Comines

Tables 3.47 (measured and analysed).
Figures 3.47 (unmeasured, measured and analysed).

Measured

Pixels	Scale stick original map (pixels)	Houses	Street sides	Houses per street side	Inhabitants (RF 4.0)	Inhabitants (RF 5.5)	Inhabitants (RF 7.0)
4010	1014	704	69	10.21	2817	3873	4929

Analysed

October 11, 2023 10:05:18 AM	
Computation with 5 linked points	
(*italics* points are less reliable):	1. Église Saint-Chrysole
	2. Belfort
	3. Rue des Hirondelles
	4. Rue des Arts
	5. Rue du Chateau
Reference points (Reliability 0–5)	4
Transformation with 3 parameters	
Scale	1:7.080
Rotation	11 °
Mean position error	22m

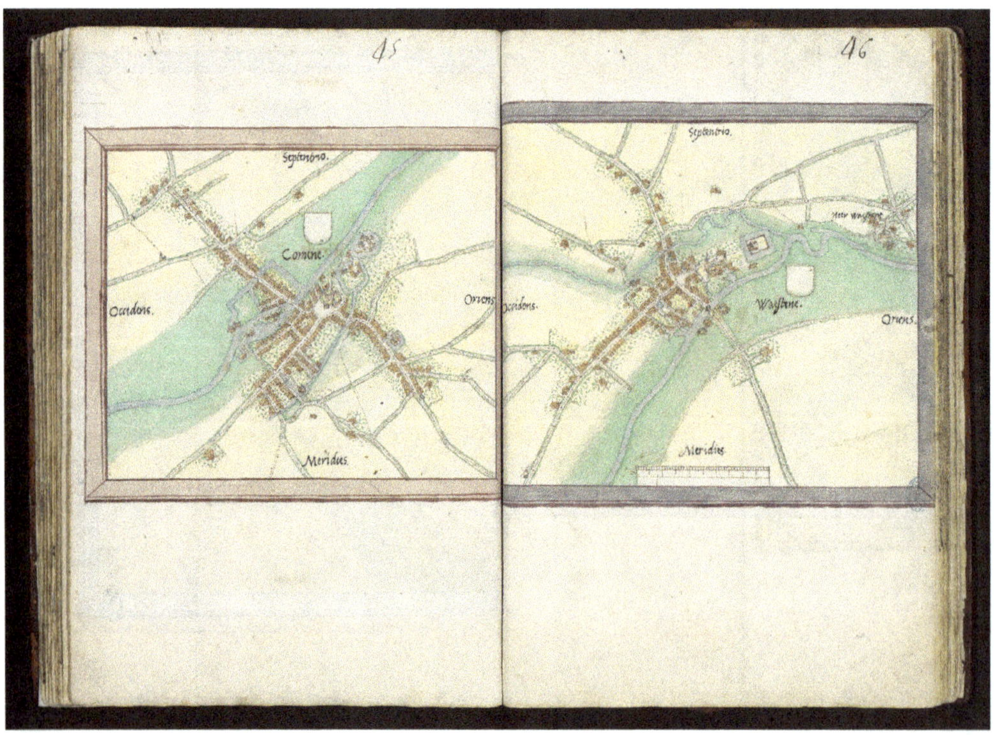

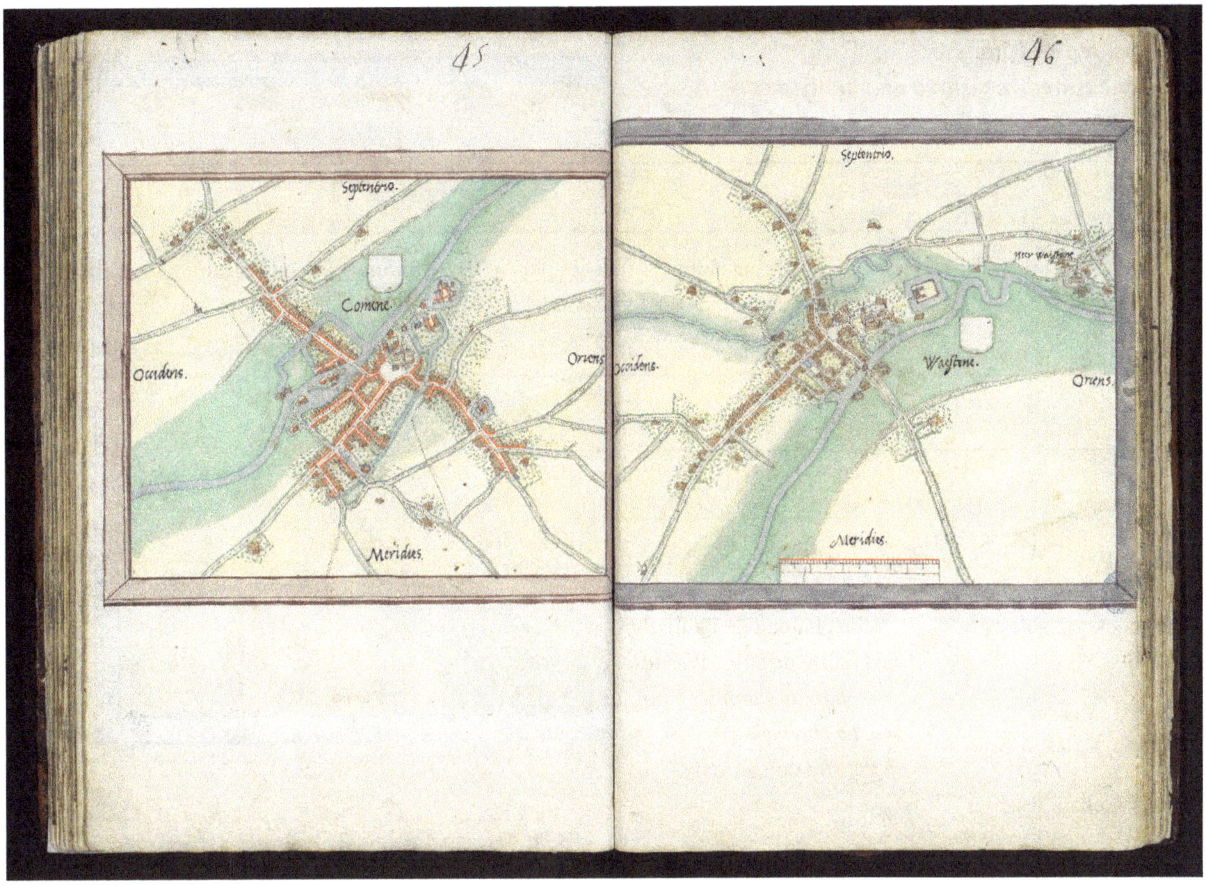
45
46
Septentrio.
Comine.
Occidens.
Oriens
Meridies.
Septentrio.
Occidens.
Waestene.
Oriens.
Meridies.

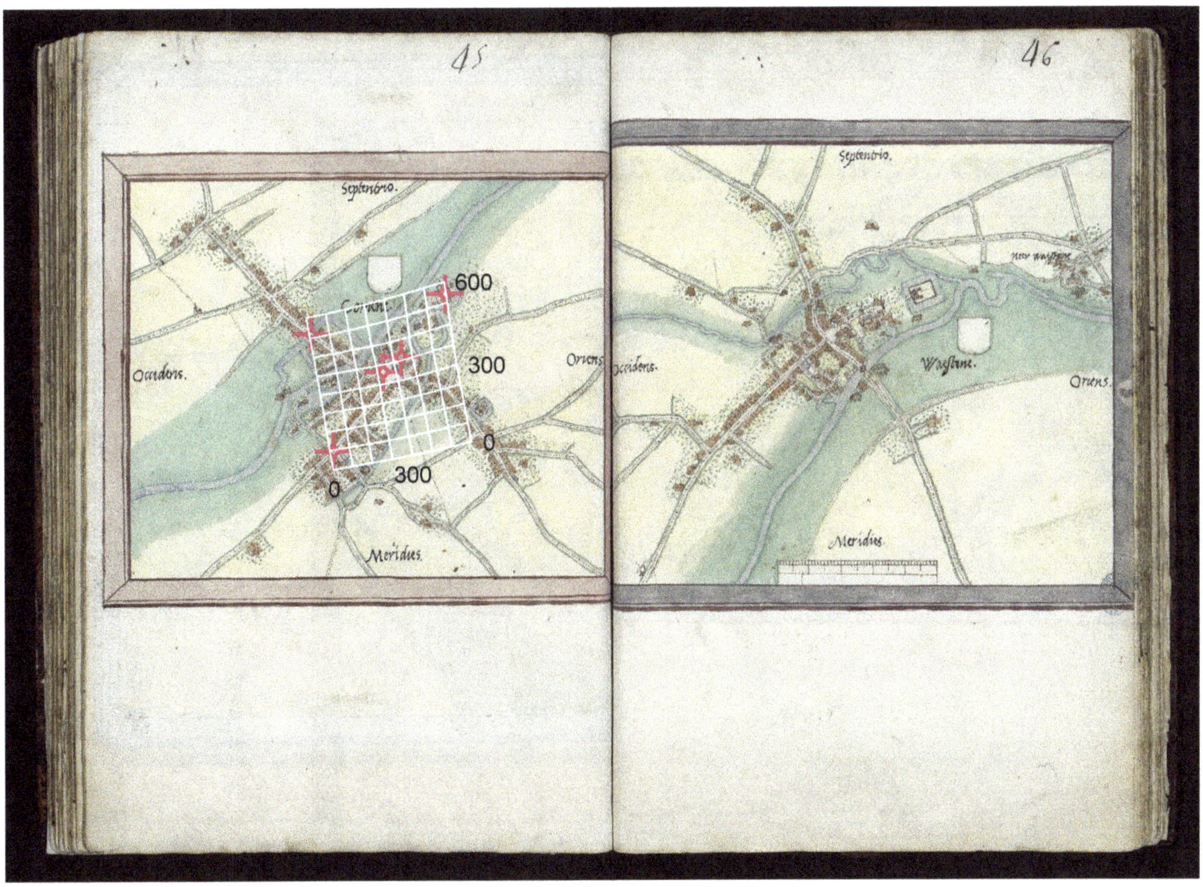
45
46
Septentrio.
600
300
0
300
0
Occidens.
Oriens
Meridies.
Septentrio.
Occidens.
Waestene.
Oriens.
Meridies.

48. Conde-sur-l'Escaut

Tables 3.48 (measured and analysed).
Figures 3.48 (unmeasured, measured and analysed).

Measured

Pixels	Scale stick original map (pixels)	Houses	Street sides	Houses per street side	Inhabitants (RF 4.0)	Inhabitants (RF 5.5)	Inhabitants (RF 7.0)
3536	1032	621	82	7.57	2484	3415	4346

Analysed

January 23, 2023 2:41:07 PM	
Computation with 5 linked points	
(*italics* points are less reliable):	1. Le chateau de Bailleul
	2. Église de Saint-Wasnon
	3. Chateau Comptes Hainaut
	4. La Hayne
	5. Porte de Tournai
Reference points (Reliability 0–5)	3
Transformation with 3 parameters	
Scale	1:7610
Rotation	7 °
Mean position error	19m

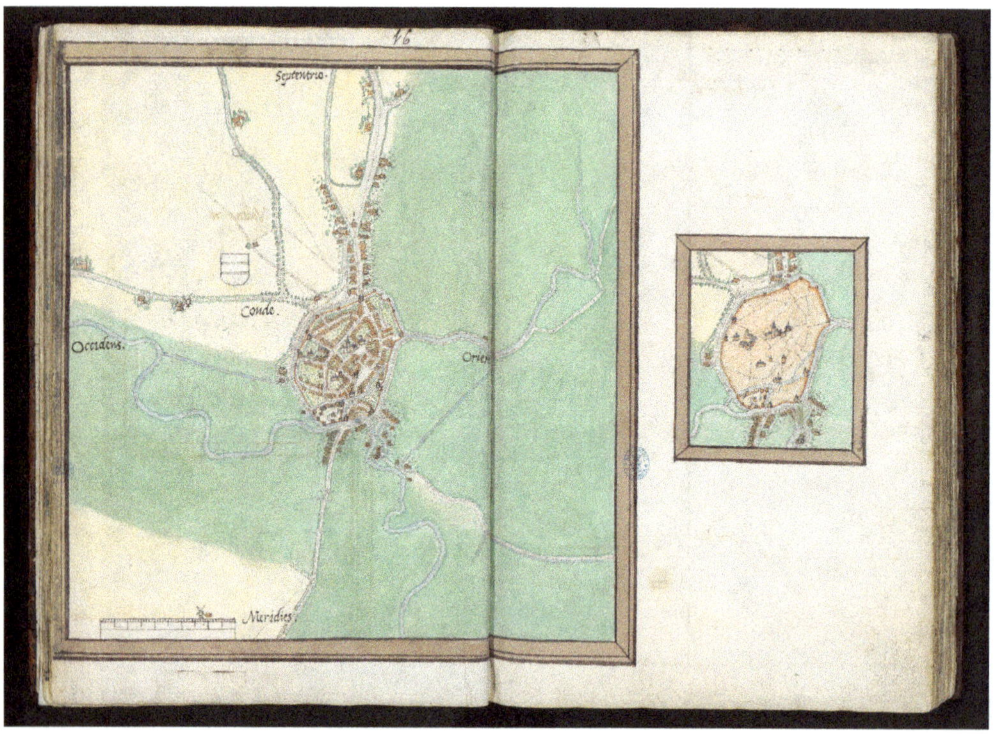

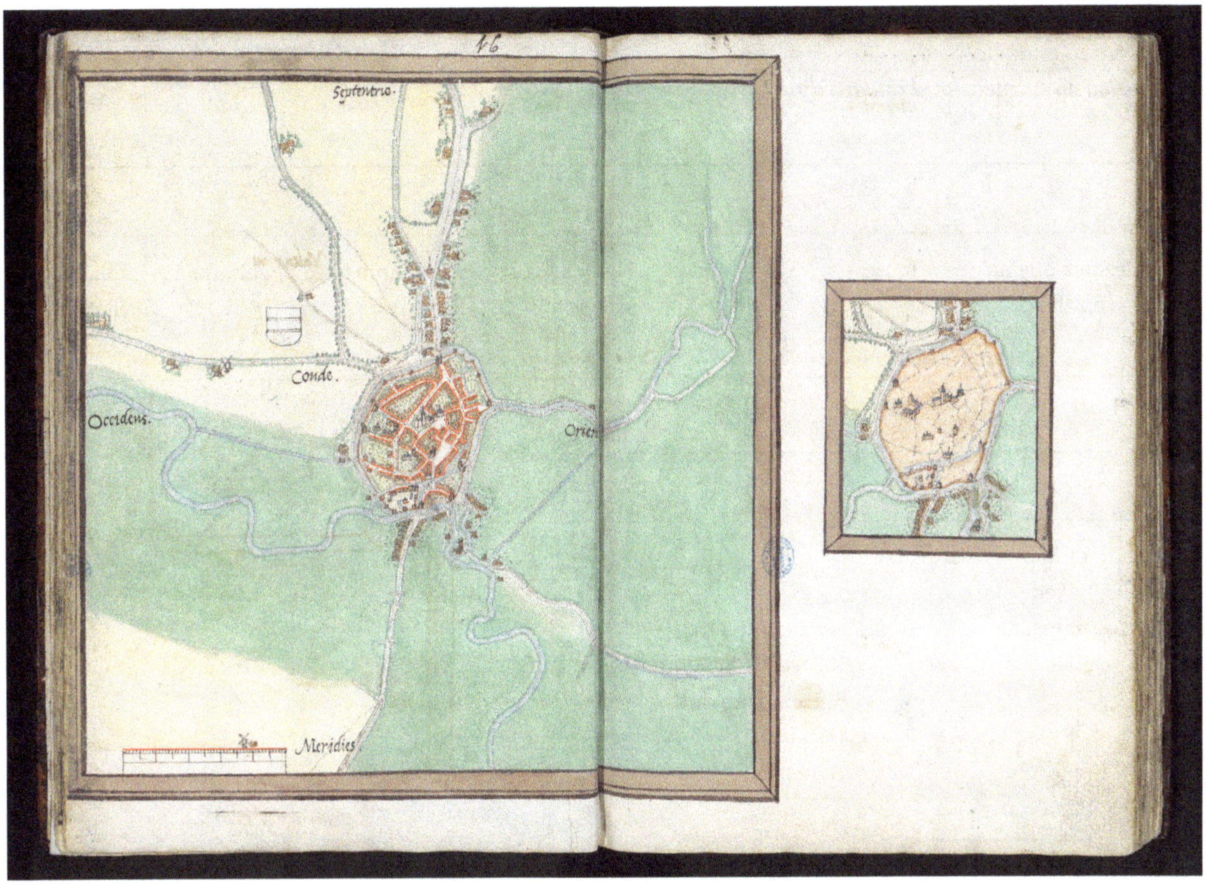
16
Septentrio.
Conde.
Occidens.
Meridies

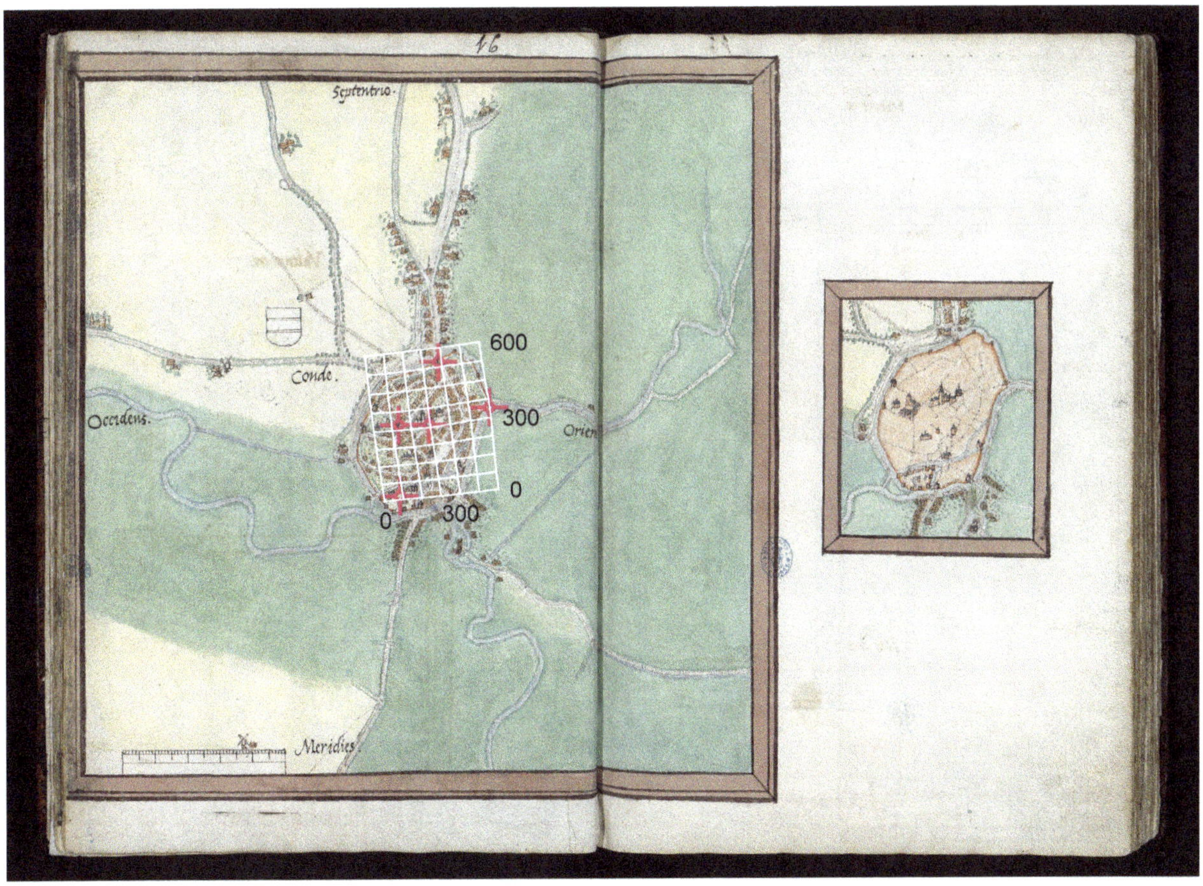
16
Septentrio.
Conde.
Occidens.
600
300
0
0
300
Meridies

49. Culemborg
Tables 3.49 (measured and analysed).
Figures 3.49 (unmeasured, measured and analysed).

Measured							
Pixels	Scale stick original map (pixels)	Houses	Street sides	Houses per street side	Inhabitants (RF 4.0)	Inhabitants (RF 5.5)	Inhabitants (RF 7.0)
5626	991	988	112	8.82	3952	5434	6916

Analysed	
March 17, 2023 6:00:01 PM	
Computation with 5 linked points	
(*italics* points are less reliable):	1. Lanxmeerpoort
	2. Kasteel Culemborg
	3. Sint-Barbarakerk
	4. Oosterwal
	5. Zuiderwal
Reference points (Reliability 0–5)	4
Transformation with 3 parameters	
Scale	1:7850
Rotation	16 °
Mean position error	26m

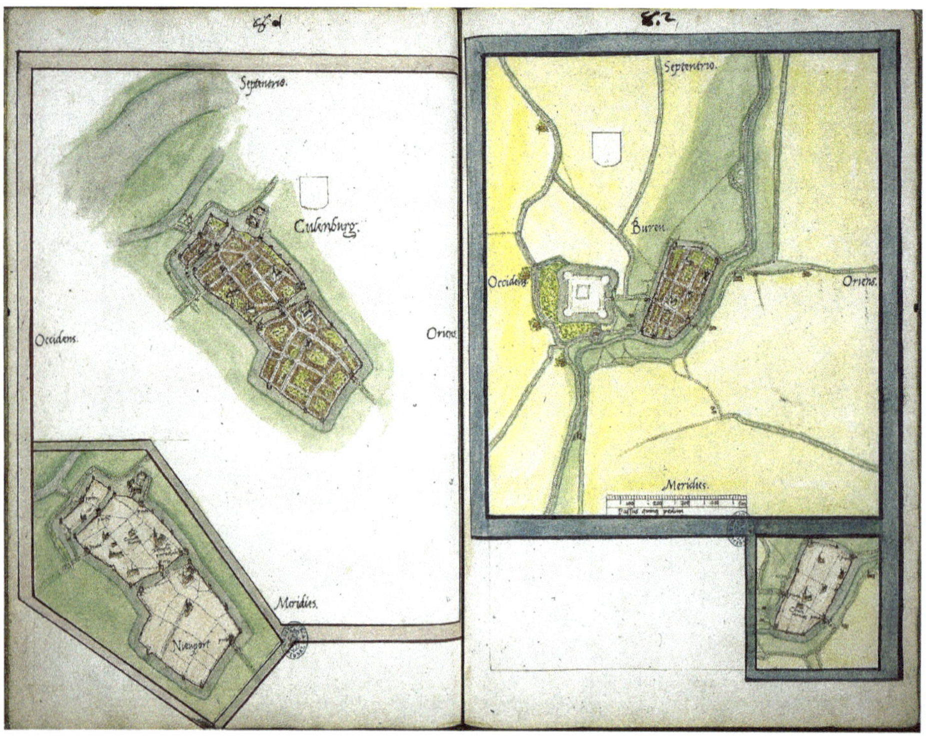

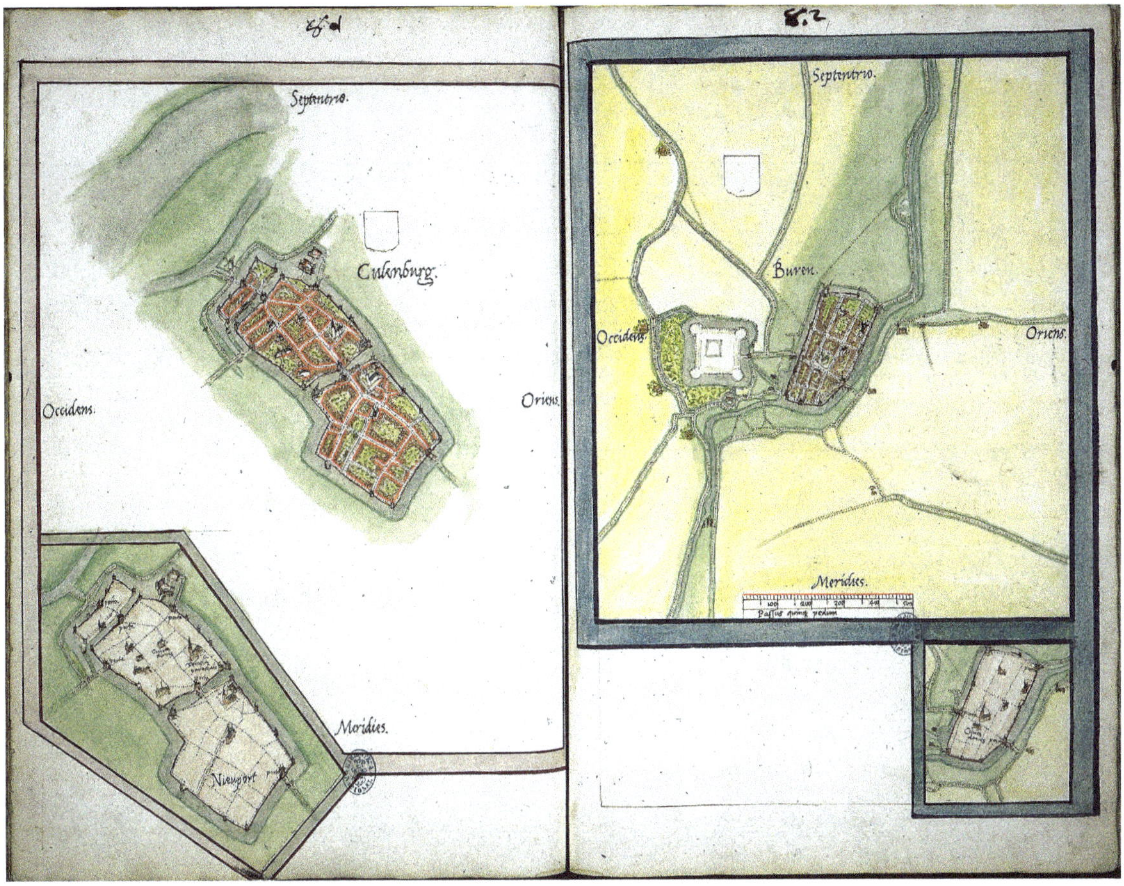
Septentrio.
Culenburg.
Occidens.
Oriens.
Meridies.
Nieuport
Septentrio.
Buren.
Occidens.
Oriens.
Meridies.

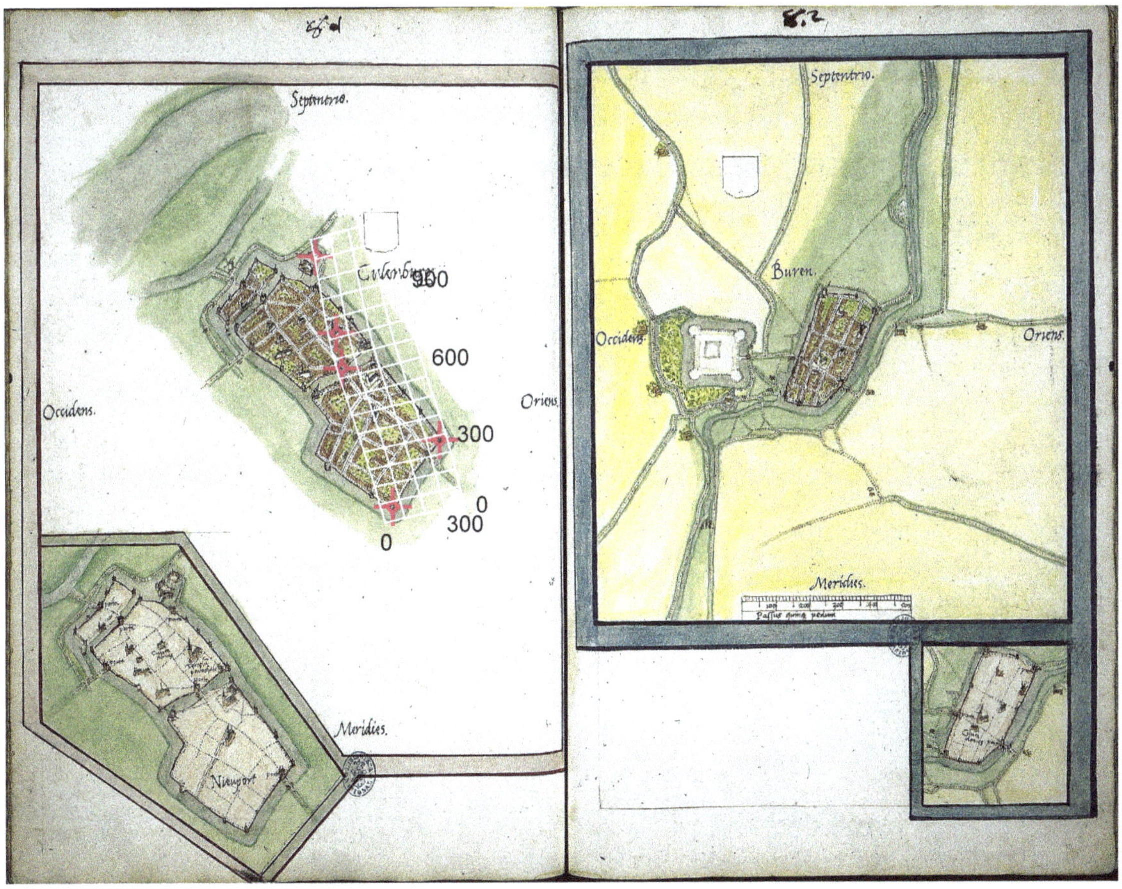
Septentrio.
Culenburg.
900
600
300
0
300
0
Occidens.
Oriens.
Meridies.
Nieuport
Septentrio.
Buren.
Occidens.
Oriens.
Meridies.

50. Damme

Tables 3.50 (measured and analysed).
Figures 3.50 (unmeasured, measured and analysed).

Measured

Pixels	Scale stick original map (pixels)	Houses	Street sides	Houses per street side	Inhabitants (RF 4.0)	Inhabitants (RF 5.5)	Inhabitants (RF 7.0)
2257	1018	396	76	5.21	1585	2180	2775

Analysed

March 20, 2023 9:36:41 AM	
Computation with 5 linked points	
(*italics* points are less reliable):	1. OLV-Hemelvaartkerk
	2. Stadhuis
	3. Corneliestraat
	4. Burgstraat
	5. Sint-Jansstraat
Reference points (Reliability 0–5)	2
Transformation with 3 parameters	
Scale	1:7730
Rotation	4 °
Mean position error	30m

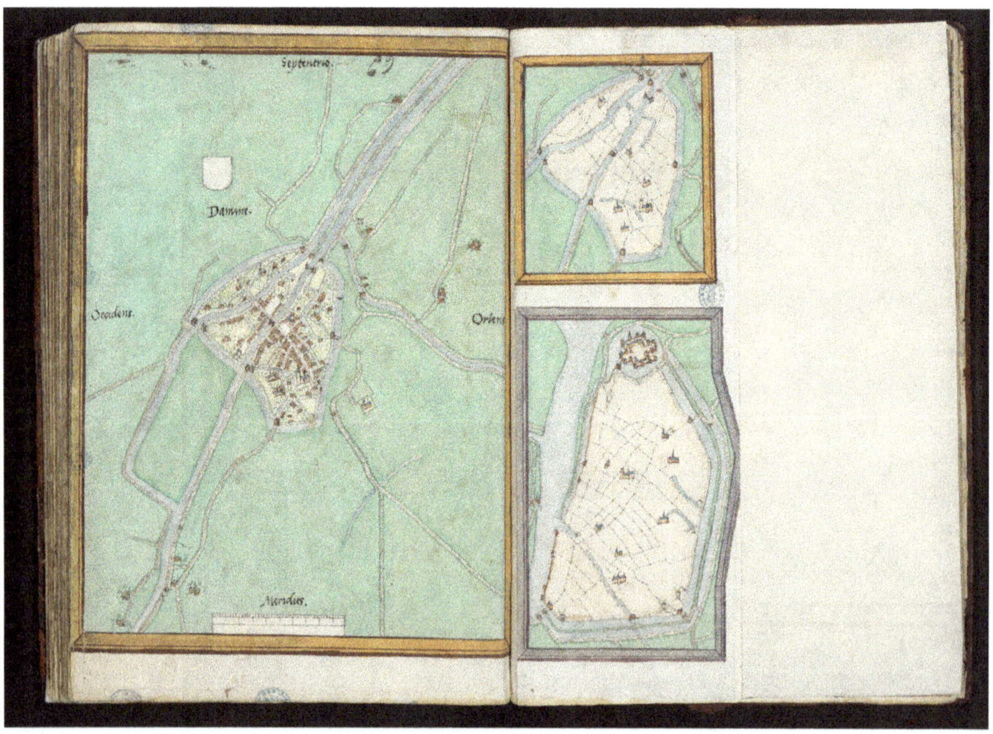

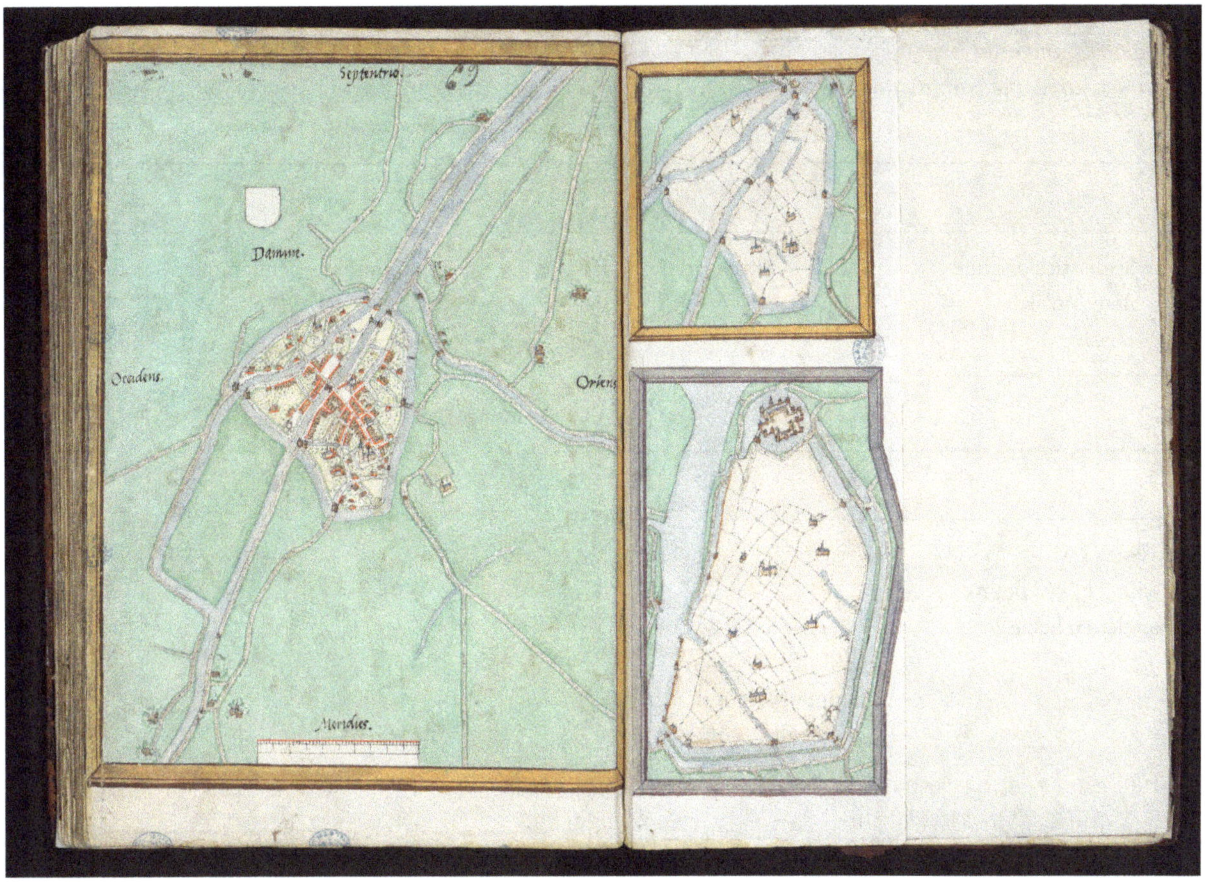
Septentrio.
Damme.
Occidens.
Oriens.
Meridies.

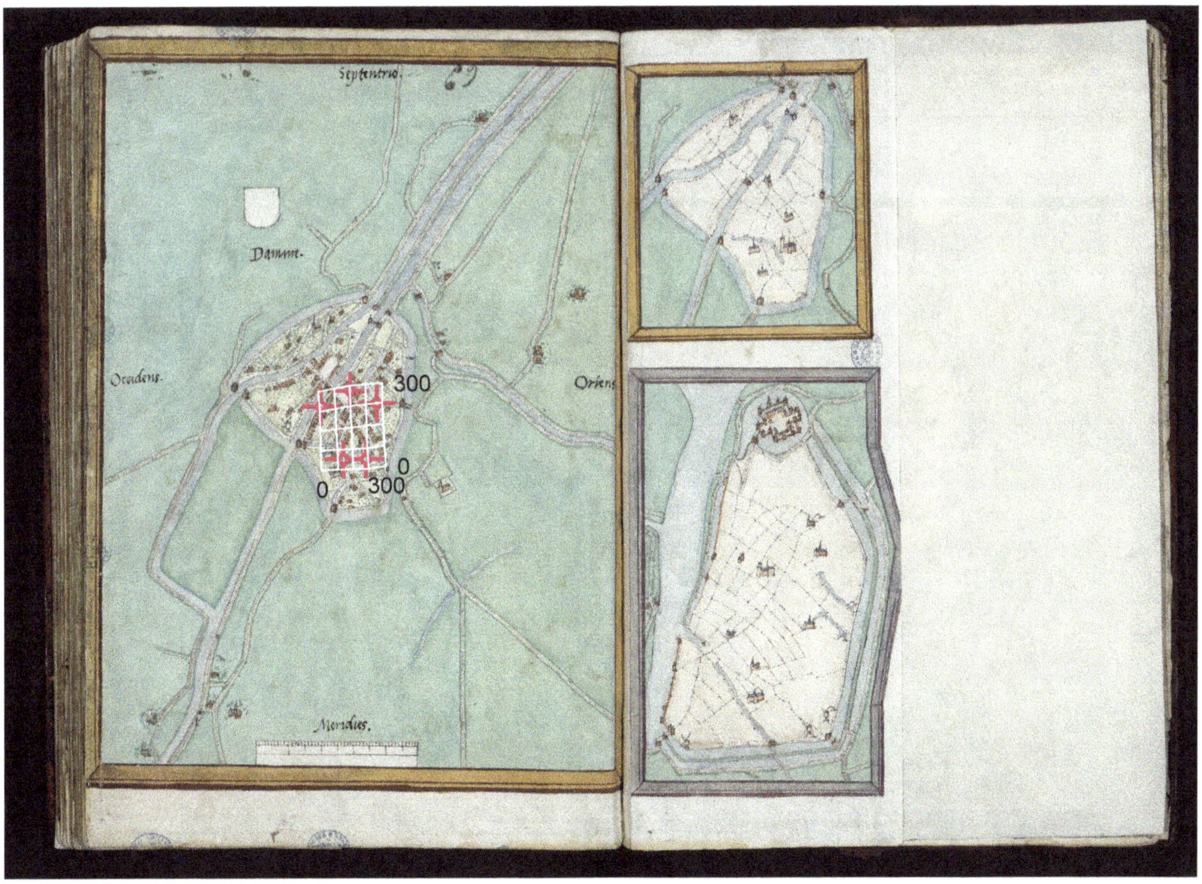
Septentrio.
Damme.
Occidens.
Oriens.
300
0
0
300
Meridies.

51. Damvillers
Tables 3.51 (measured and analysed).
Figures 3.51 (unmeasured, measured and analysed).

Measured							
Pixels	Scale stick original map (pixels)	Houses	Street sides	Houses per street side	Inhabitants (RF 4.0)	Inhabitants (RF 5.5)	Inhabitants (RF 7.0)
1582	missing	278	51	5.45	1111	1528	1945

Analysed	
April 18, 2023 9:47:22 AM	
Computation with 5 linked points	
(*italics* points are less reliable):	1. Église Saint-Maurice 2. Rue du Bastion Sainte-Marguérite 3. Rue des Tanneries 4. Place Liégois 5. Rue Carnot
Reference points (Reliability 0–5)	5
Transformation with 3 parameters	
Scale	1:7450
Rotation	12 °
Mean position error	10m

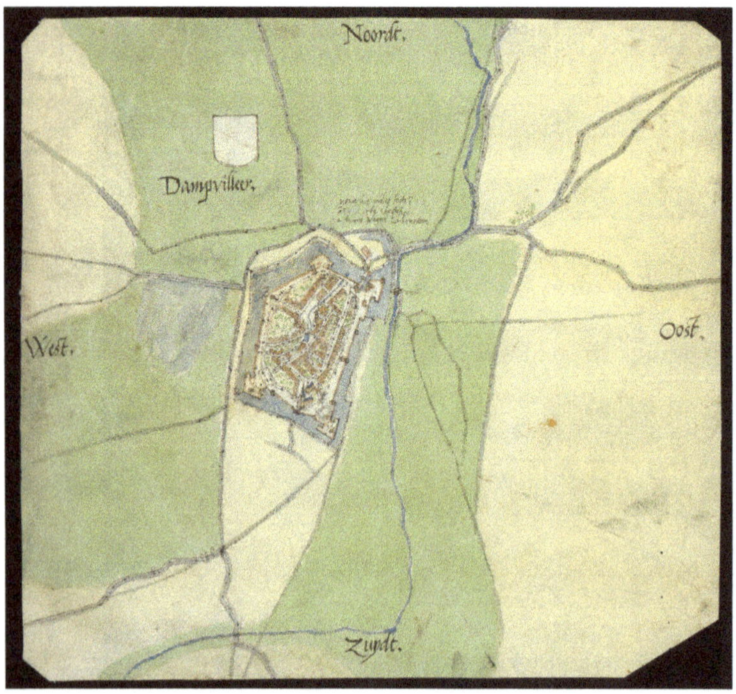

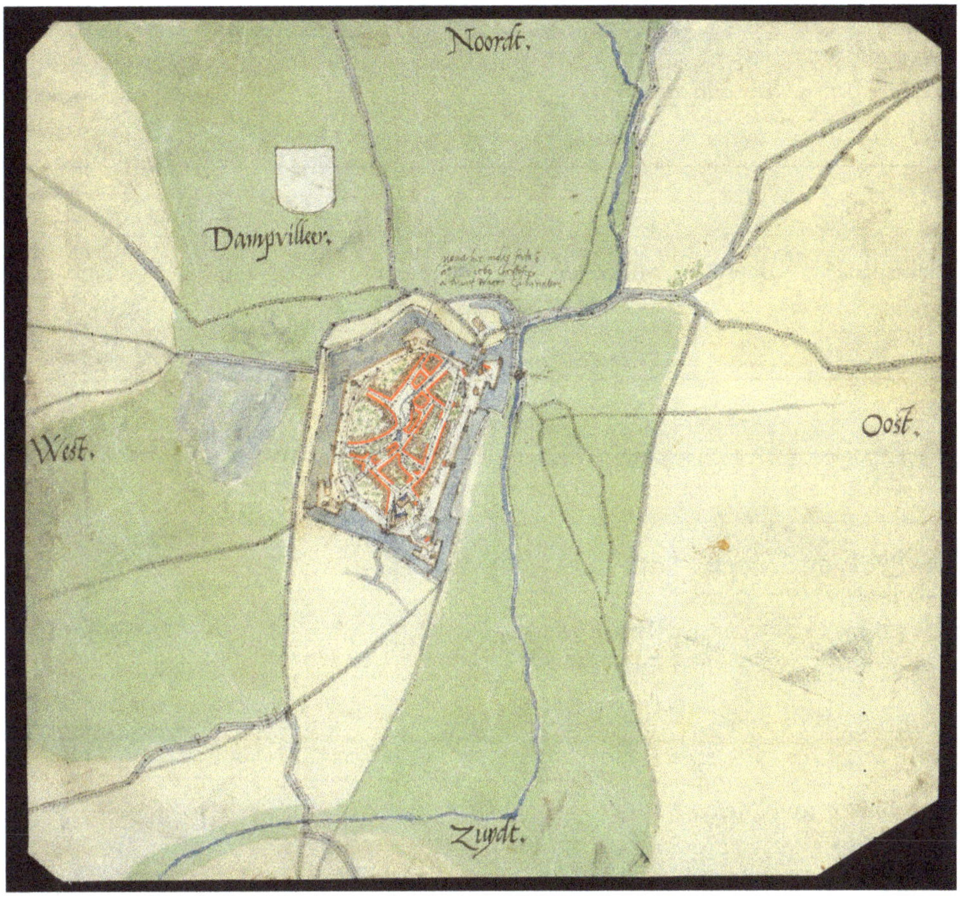
Noordt.
Dampvilleer.
West.
Oost.
Zuydt.

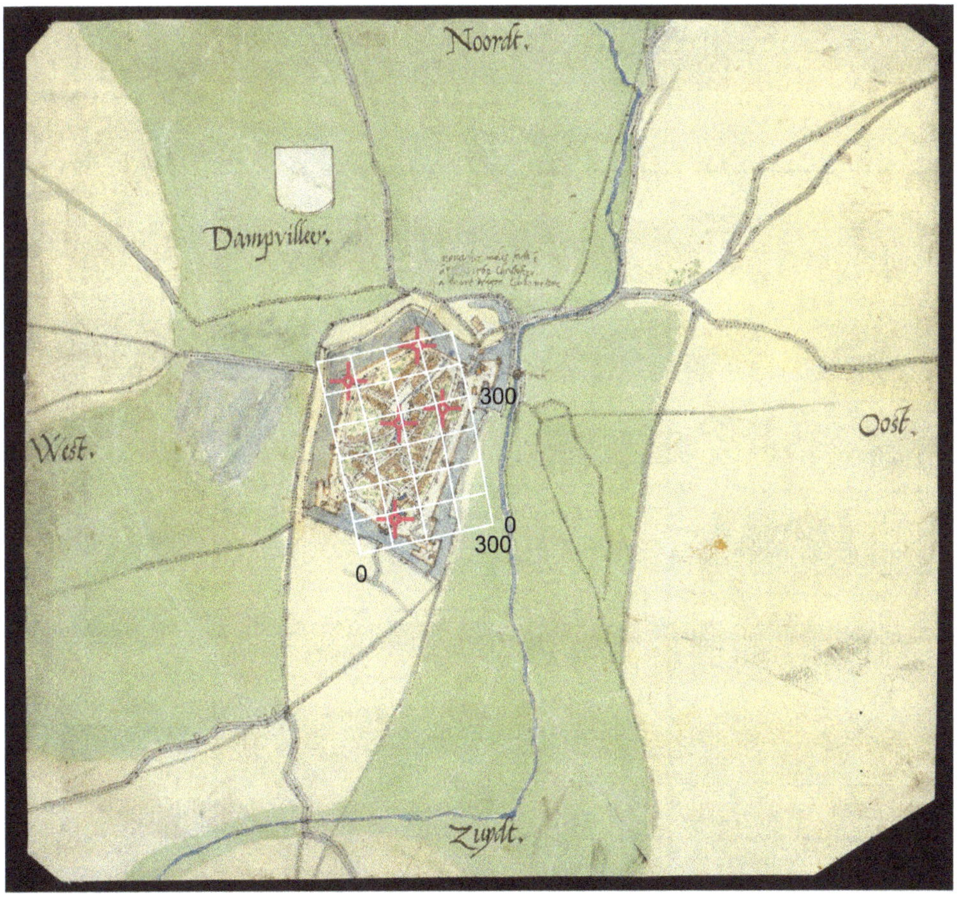
Noordt.
Dampvilleer.
West.
Oost.
Zuydt.
300
0
300
0

52. Delden

Tables 3.52 (measured and analysed).
Figures 3.52 (unmeasured, measured and analysed).

Measured							
Pixels	Scale stick original map (pixels)	Houses	Street sides	Houses per street side	Inhabitants (RF 4.0)	Inhabitants (RF 5.5)	Inhabitants (RF 7.0)
1736	1005	305	48	6.35	1220	1677	2134

Analysed	
March 21, 2023 3:35:59 PM	
Computation with 5 linked points	
(*italics* points are less reliable):	1. Oude Blasiuskerk
	2. Twickel
	3. Langestraat
	4. Noordwal
	5. Meijlingstraat
Reference points (Reliability 0–5)	2
Transformation with 3 parameters	
Scale	1:8100
Rotation	7 °
Mean position error	23m

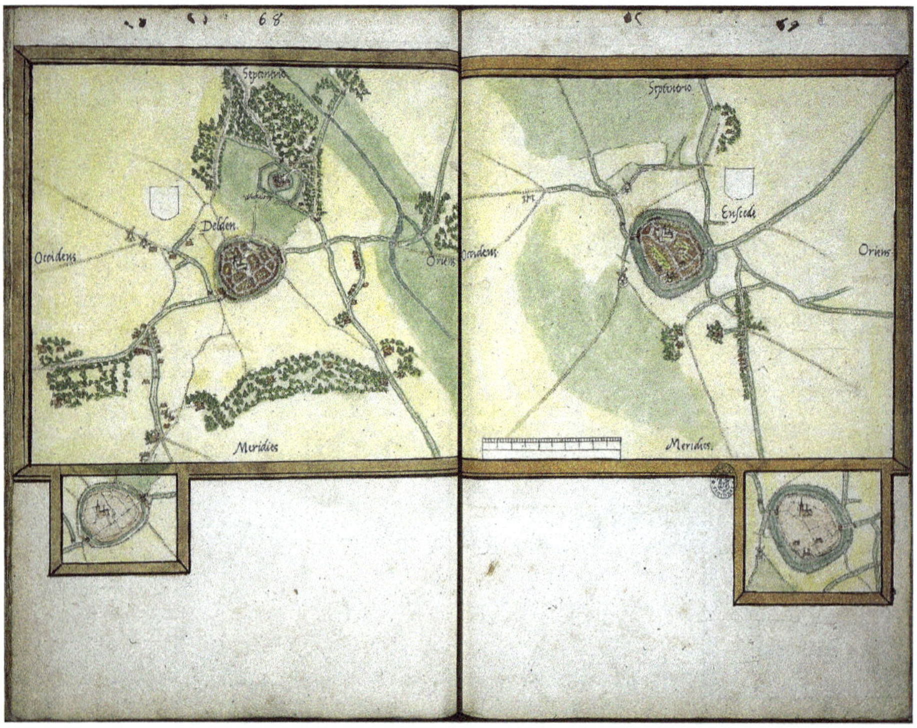

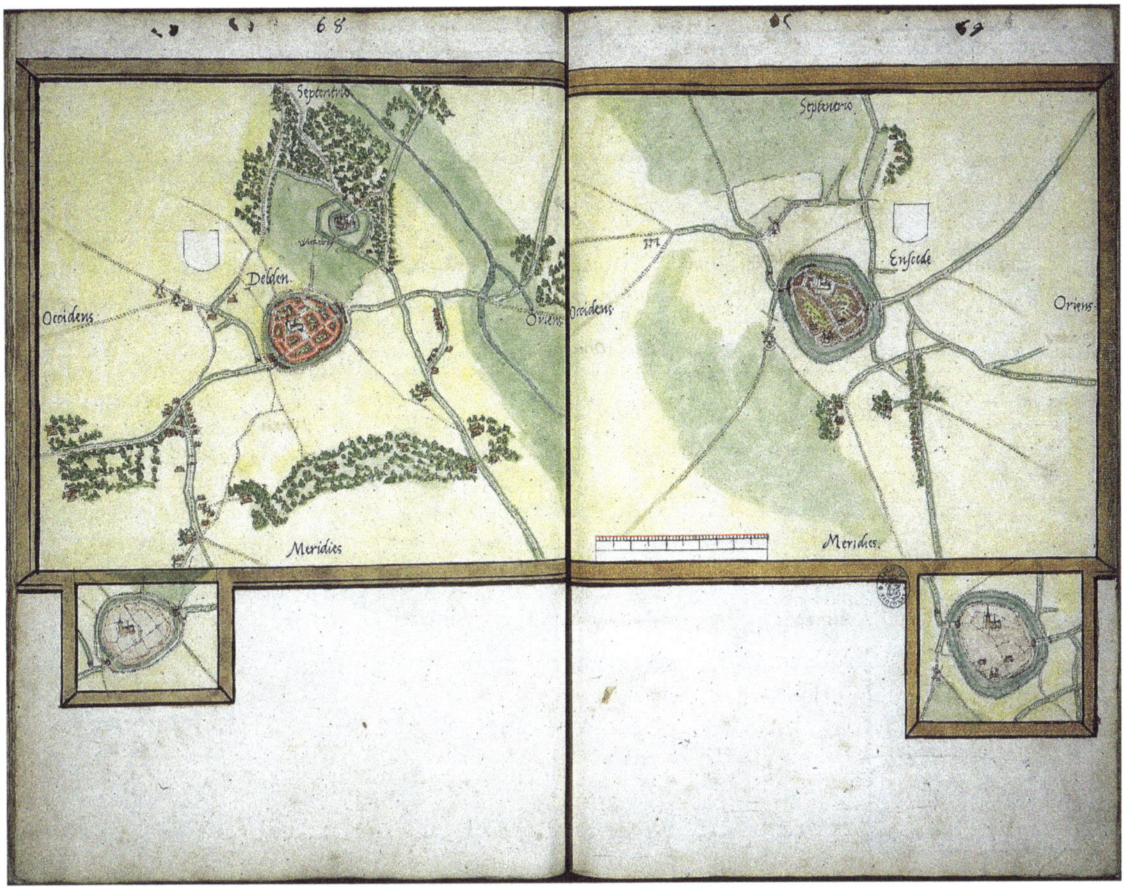
Septentrio
Delden
Occidens
Oriens
Meridies
Enschede

Septentrio
600
300
0
300
0
Occidens
Oriens
Meridies
Enschede

53. Delft

Tables 3.53 (measured and analysed).
Figures 3.53 (unmeasured, measured and analysed).

Measured							
Pixels	Scale stick original map (pixels)	Houses	Street sides	Houses per street side	Inhabitants (RF 4.0)	Inhabitants (RF 5.5)	Inhabitants (RF 7.0)
12,839	1010	2255	234	9.63	9018	12,400	15,782

Analysed	
March 21, 2023 8:07:30 PM	
Computation with 5 linked points	
(*italics* points are less reliable):	1. Lutherse Kerk
	2. Nieuwe Kerk
	3. Oude Kerk
	4. Korte Geer
	5. Oostplantsoen
Reference points (Reliability 0–5)	5
Transformation with 3 parameters	
Scale	1:8400
Rotation	9 °
Mean position error	19m

Septentrio.
Delft.
Oriens.
Meridies.
100
200
300
400
500

Septentrio.
1.200
900
600
300
0
0
300
600
Oriens.
Meridies.

54. Den Haag

Tables 3.54 (measured and analysed).
Figures 3.54 (unmeasured, measured and analysed).

Measured							
Pixels	Scale stick original map (pixels)	Houses	Street sides	Houses per street side	Inhabitants (RF 4.0)	Inhabitants (RF 5.5)	Inhabitants (RF 7.0)
9506	1013	1669	177	9.43	6677	9181	11,685

Analysed	
March 26, 2023 11:14:04 PM	
Computation with 5 linked points	
(*italics* points are less reliable):	1. Grote Kerk 2. Hofvijver 3. Ridderzaal 4. Gevangenpoort 5. Spuistraat
Reference points (Reliability 0–5)	5
Transformation with 3 parameters	
Scale	1:8550
Rotation	11 °
Mean position error	14m

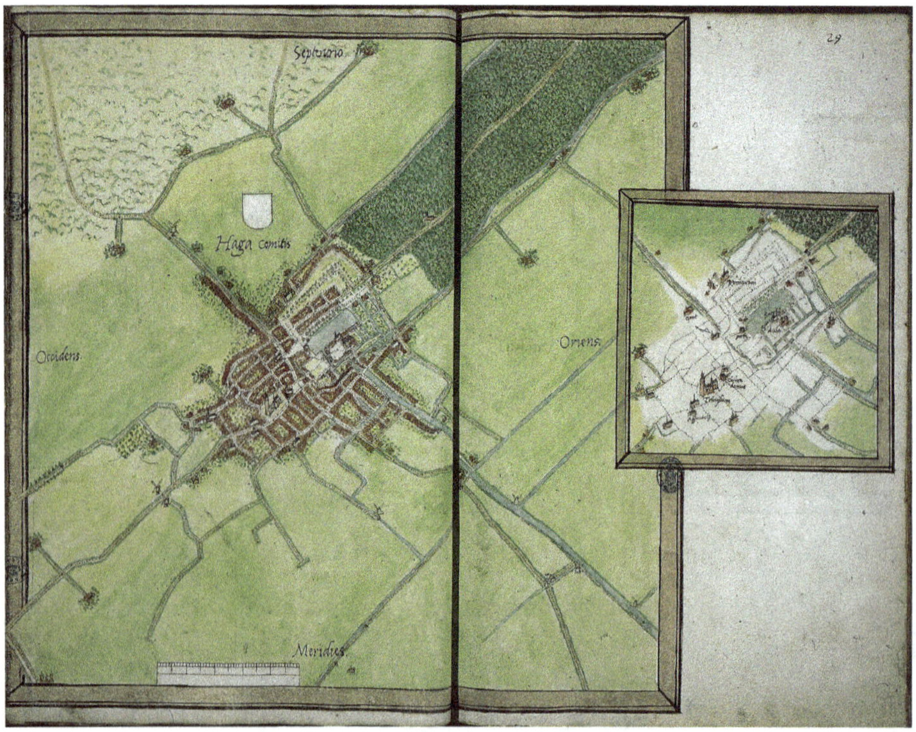

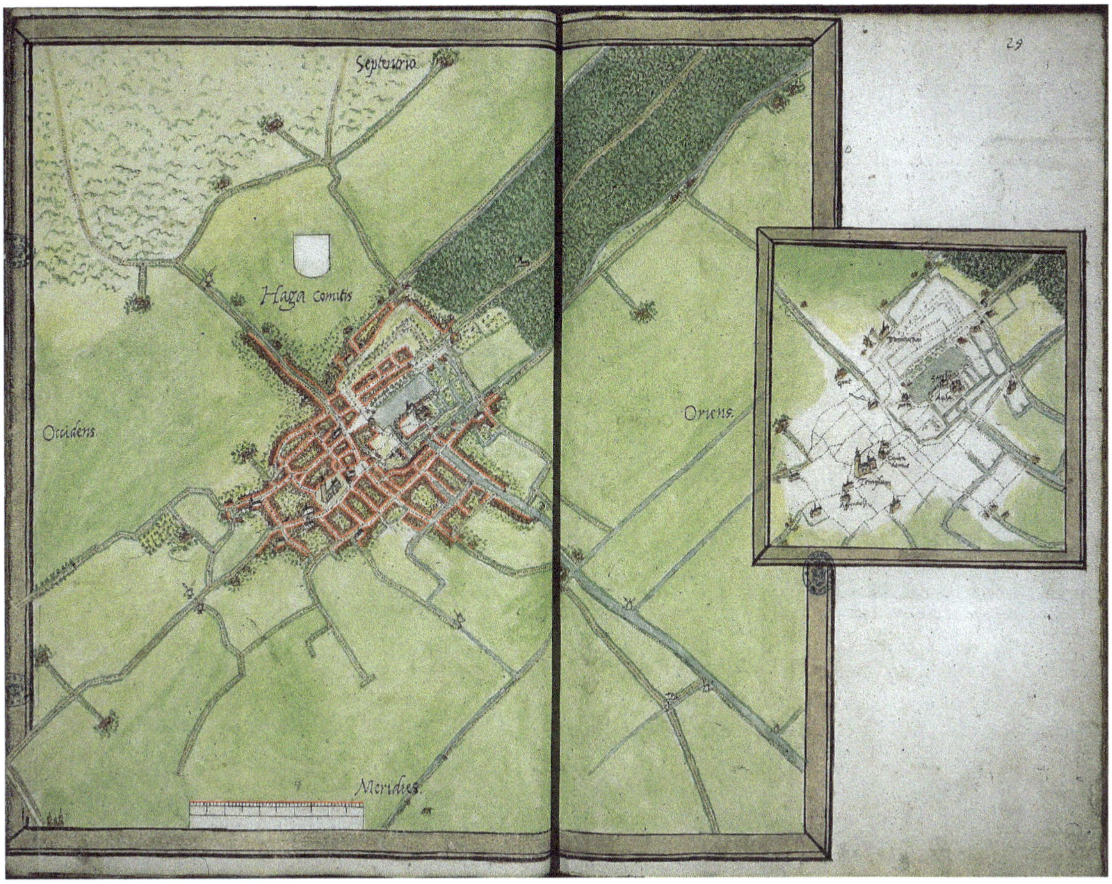
Septentrio
Haga comitis
Occidens
Oriens
Meridies
29

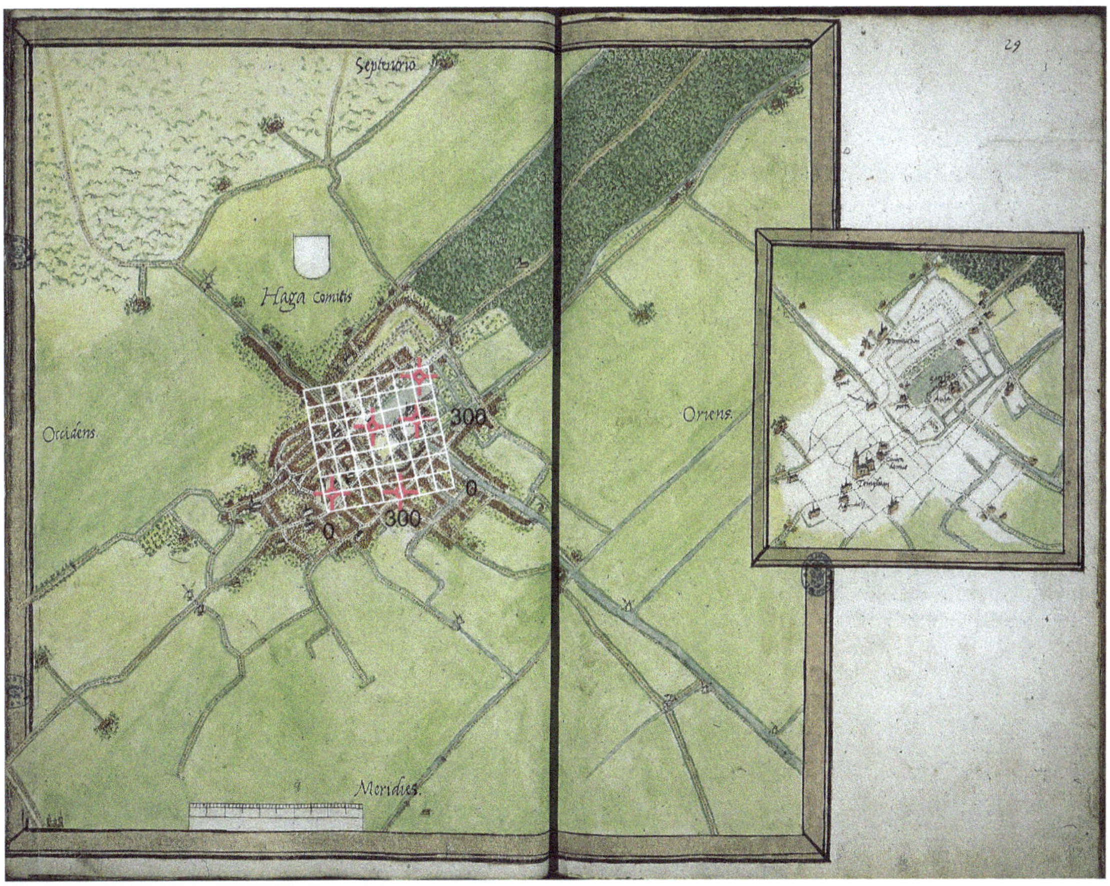
Septentrio
Haga comitis
Occidens
Oriens
Meridies
300
0
0
300
29

55. Dendermonde

Tables 3.55 (measured and analysed).
Figures 3.55 (unmeasured, measured and analysed).

Measured							
Pixels	Scale stick original map (pixels)	Houses	Street sides	Houses per street side	Inhabitants (RF 4.0)	Inhabitants (RF 5.5)	Inhabitants (RF 7.0)
5767	1030	1013	126	8.04	4051	5570	7089

Analysed	
March 28, 2023 11:59:00 AM	
Computation with 5 linked points	
(*italics* points are less reliable):	1. Sint-Gillis-Binnenkerk
	2. Onze-Lieve-Vrouwekerk
	3. Kasteelstraat
	4. Belfort
	5. Begijnhof
Reference points (Reliability 0–5)	**3**
Transformation with 3 parameters	
Scale	1:7880
Rotation	10 °
Mean position error	33m

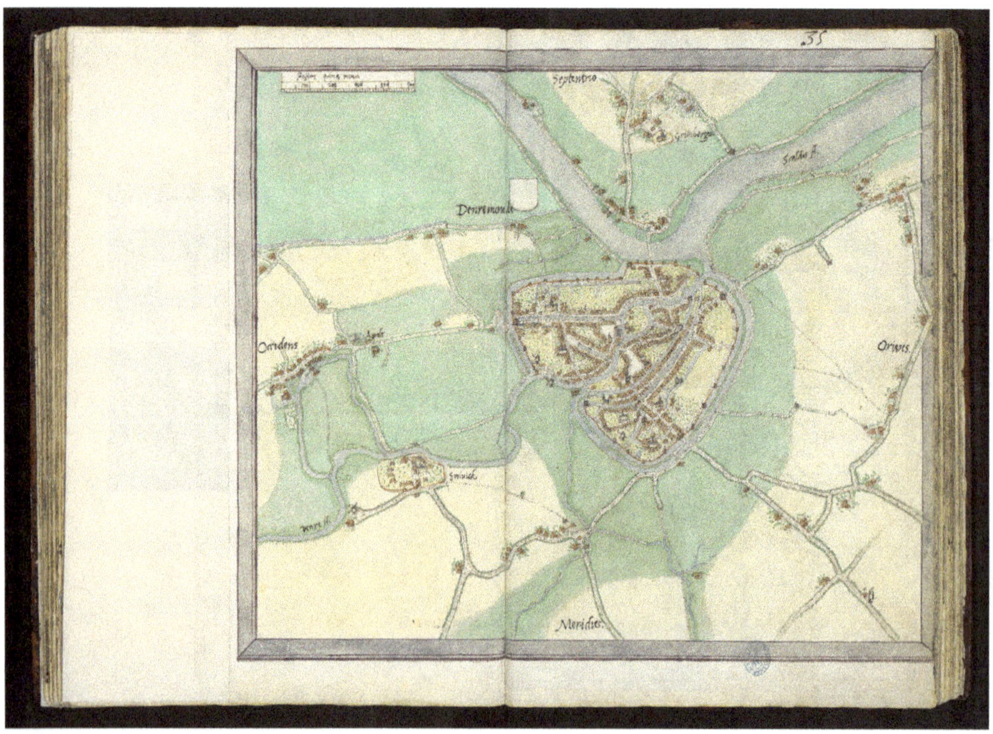

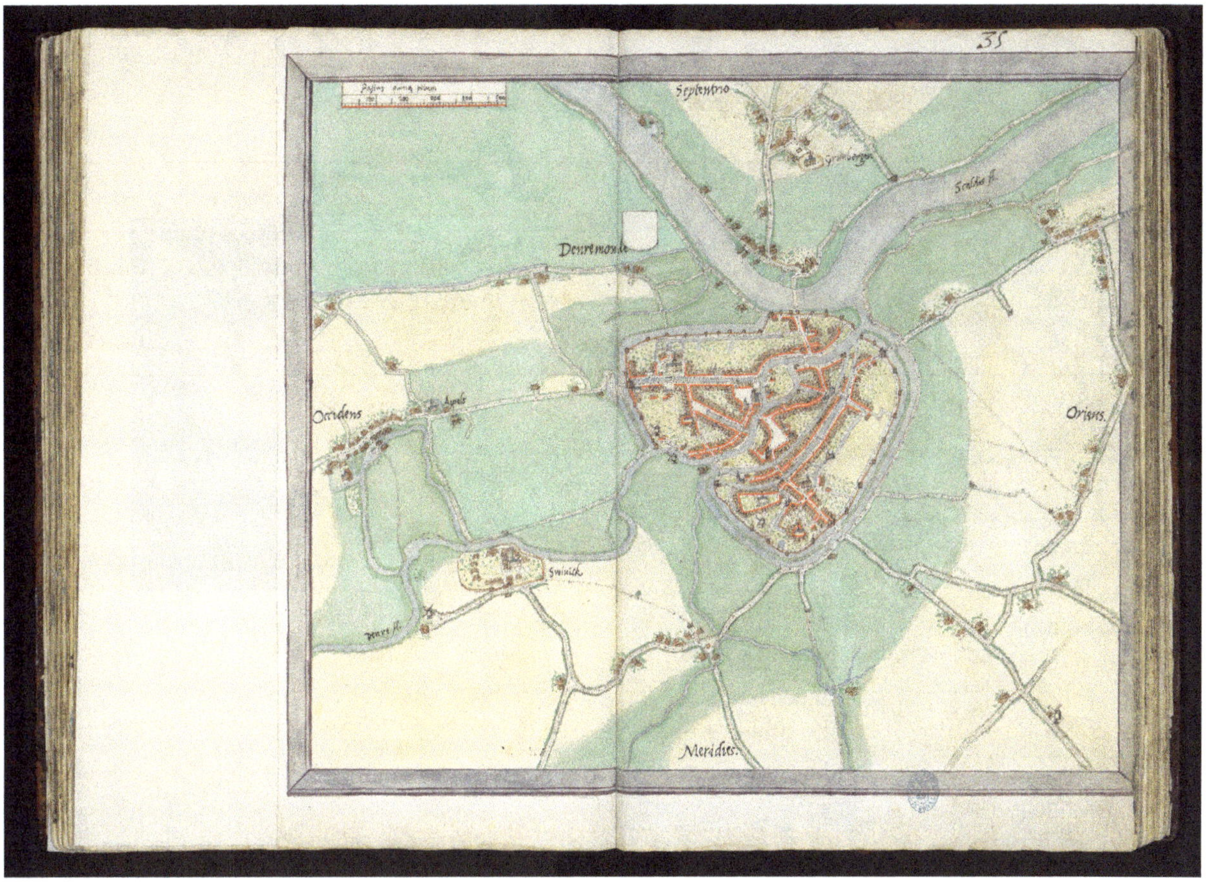
35
Septentrio
Occidens
Oriens.
Meridies.

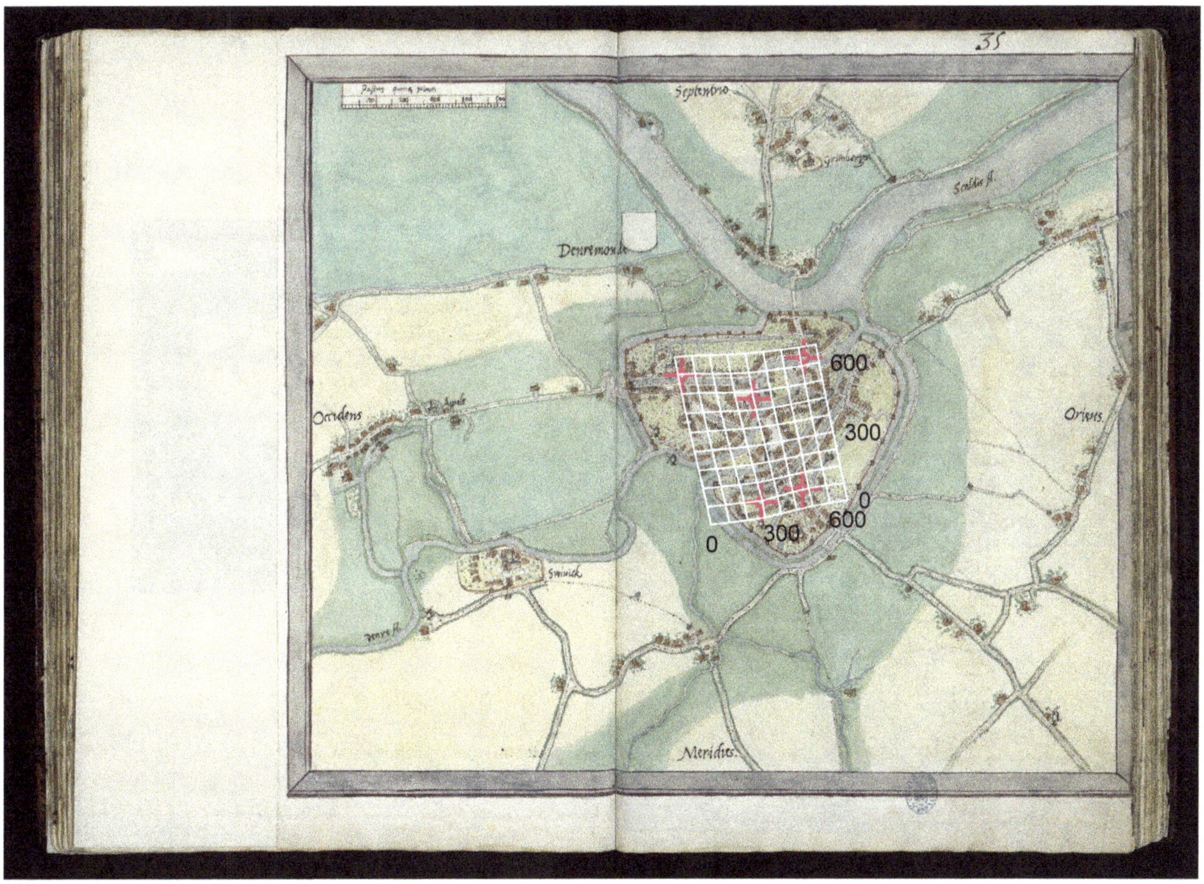
35
Septentrio
Occidens
Oriens.
Meridies.
600
300
0
0
300
600

56. Deventer

Tables 3.56 (measured and analysed).
Figures 3.56 (unmeasured, measured and analysed).

Measured							
Pixels	Scale stick original map (pixels)	Houses	Street sides	Houses per street side	Inhabitants (RF 4.0)	Inhabitants (RF 5.5)	Inhabitants (RF 7.0)
8563	997	1504	194	7.75	6015	8270	10,525

Analysed	
March 29, 2023 5:29:31 PM	
Computation with 5 linked points	
(*italics* points are less reliable):	1. Lebuinuskerk
	2. Bergkerk
	3. Broederenkerk
	4. De Waag
	5. Smedenstraat
Reference points (Reliability 0–5)	5
Transformation with 3 parameters	
Scale	1:8480
Rotation	7 °
Mean position error	18m

Septentrio.
Deuenter
Dauentria.
Oriens.
Meridies.

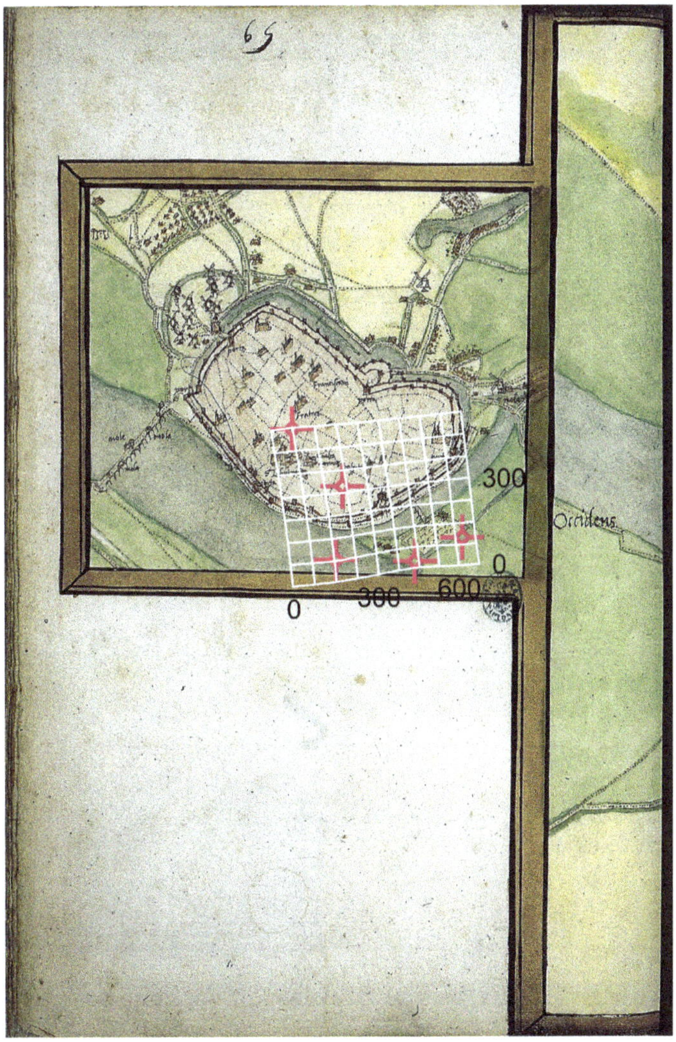
65
300
0
0
300
600
Occidens.

57. Diksmuide

Tables 3.57 (measured and analysed).
Figures 3.57 (unmeasured, measured and analysed).

Measured							
Pixels	Scale stick original map (pixels)	Houses	Street sides	Houses per street side	Inhabitants (RF 4.0)	Inhabitants (RF 5.5)	Inhabitants (RF 7.0)
4317	1023	758	86	8.81	3032	4169	5306

Analysed	
March 30, 2023 2:31:51 PM	
Computation with 5 linked points	
(*italics* points are less reliable):	1. Sint-Niklaaskerk 2. Belfort 3. Begijnhof 4. Sint-Rochuskapel *5. Stationsstraat*
Reference points (Reliability 0–5)	4
Transformation with 3 parameters	
Scale	1:7170
Rotation	8°
Mean position error	22m

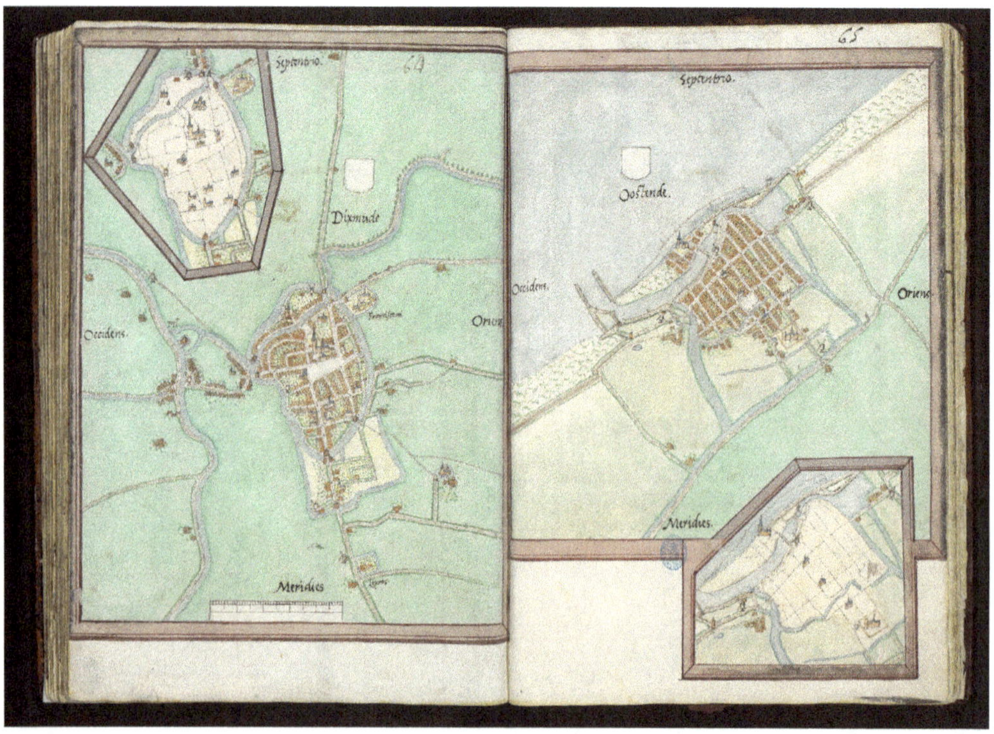

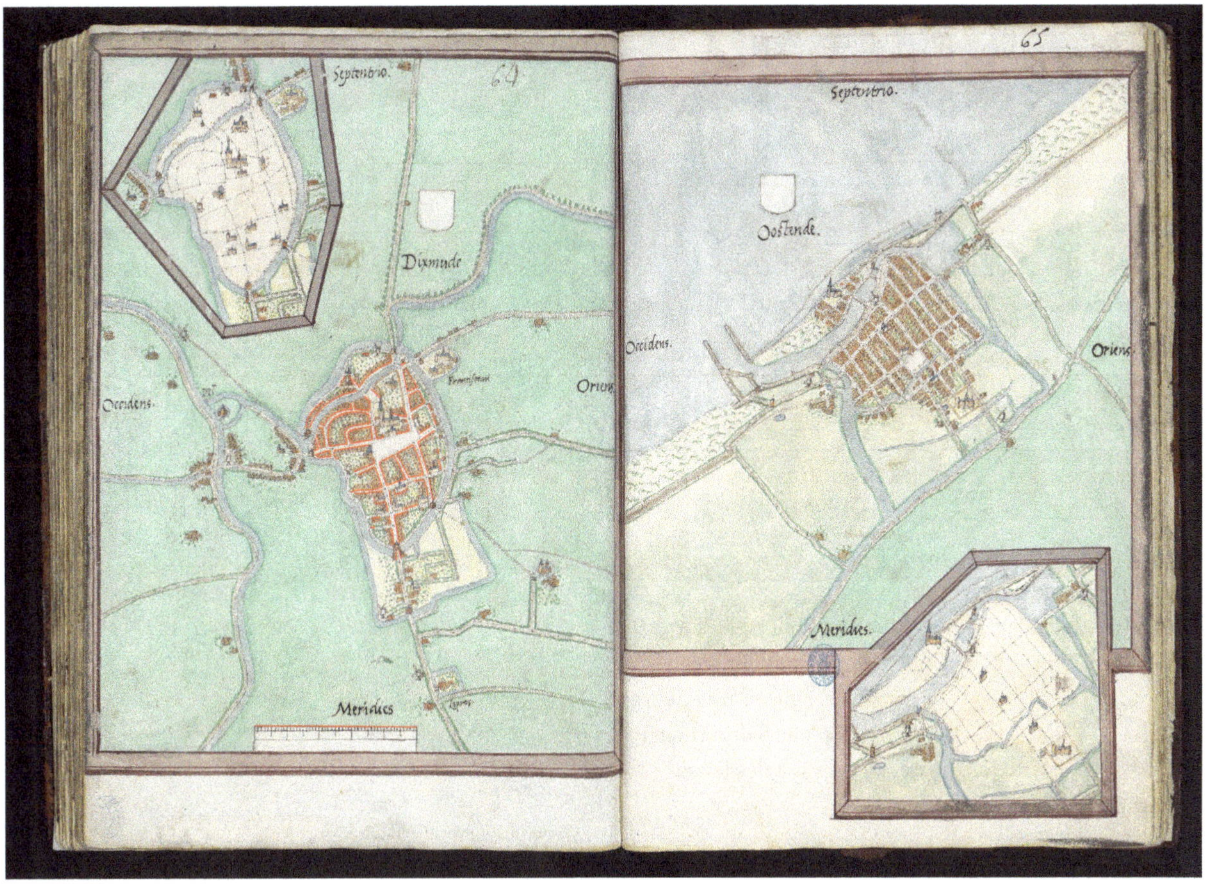
64
Septentrio.
Dixmude
Occidens.
Oriens
Meridies
65
Septentrio.
Oostende.
Occidens.
Oriens
Meridies.

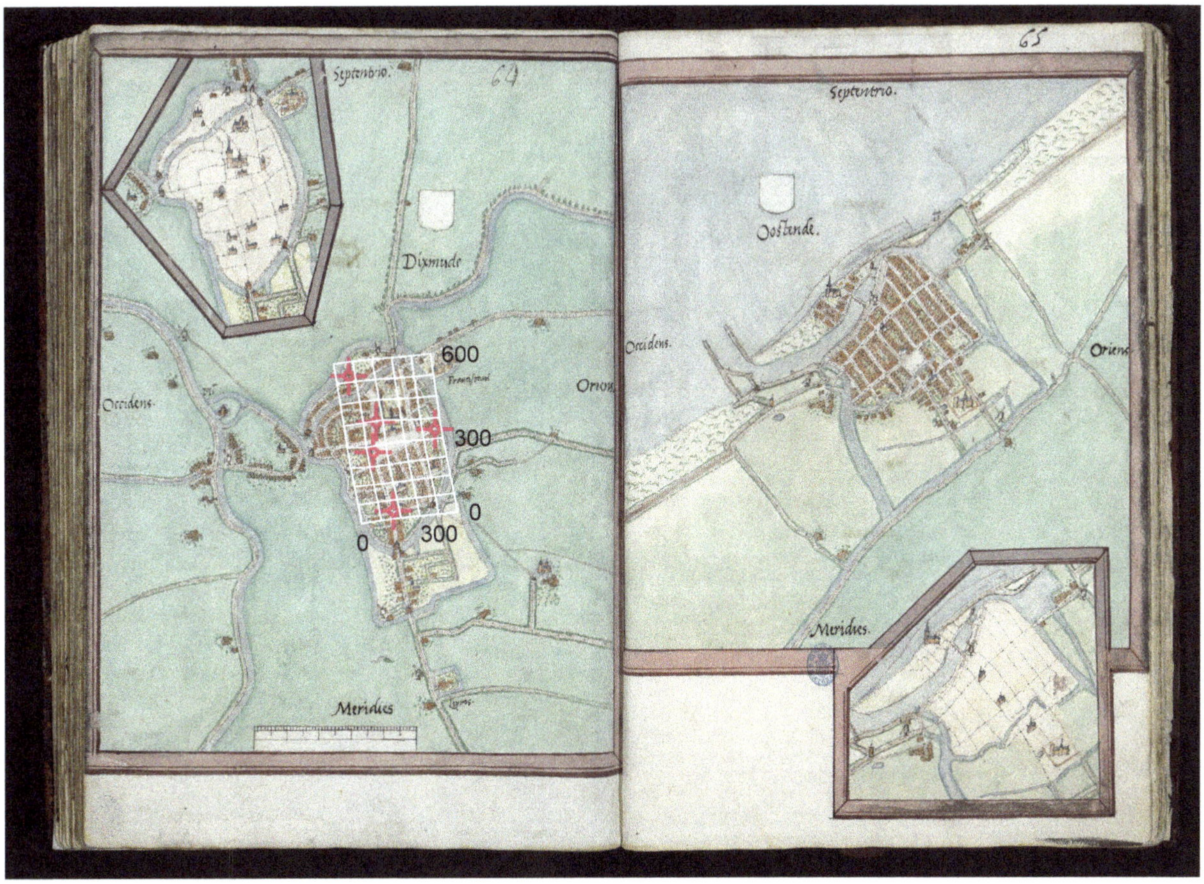
64
Septentrio.
Dixmude
600
300
0
0
300
Occidens.
Oriens
Meridies
65
Septentrio.
Oostende.
Occidens.
Oriens
Meridies.

58. Dinant

Tables 3.58 (measured and analysed).
Figures 3.58 (unmeasured, measured and analysed).

Measured							
Pixels	Scale stick original map (pixels)	Houses	Street sides	Houses per street side	Inhabitants (RF 4.0)	Inhabitants (RF 5.5)	Inhabitants (RF 7.0)
4287	missing	753	163	4.62	3011	4140	5269

Analysed	
September 18, 2023 10:50:02 AM	
Computation with 5 linked points	
(*italics* points are less reliable):	1. Collégiale Notre-Dame 2. Citadelle 3. Église Saint-Georges 4. Rue Saint-Michel 5. Rue des Fossés
Reference points (Reliability 0–5)	5
Transformation with 3 parameters	
Scale	1:7720
Rotation	10 °
Mean position error	14m

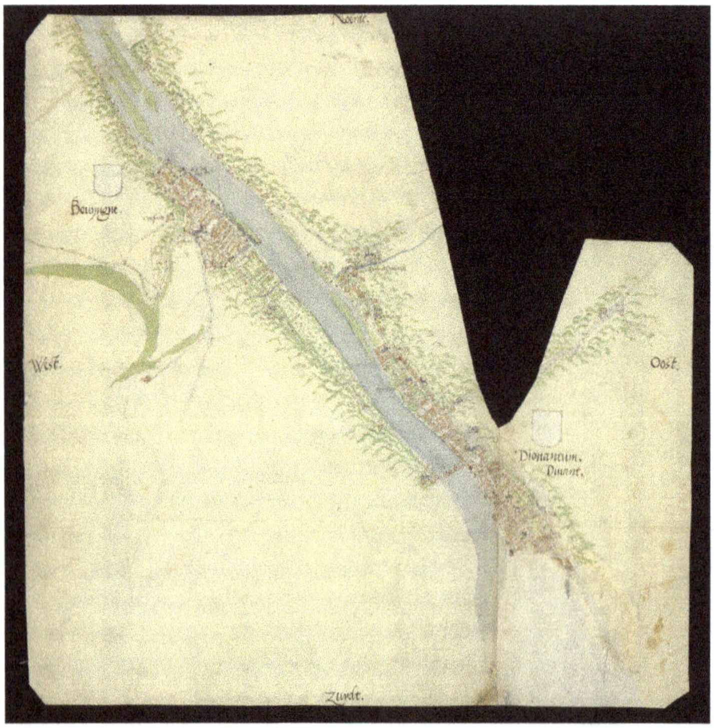

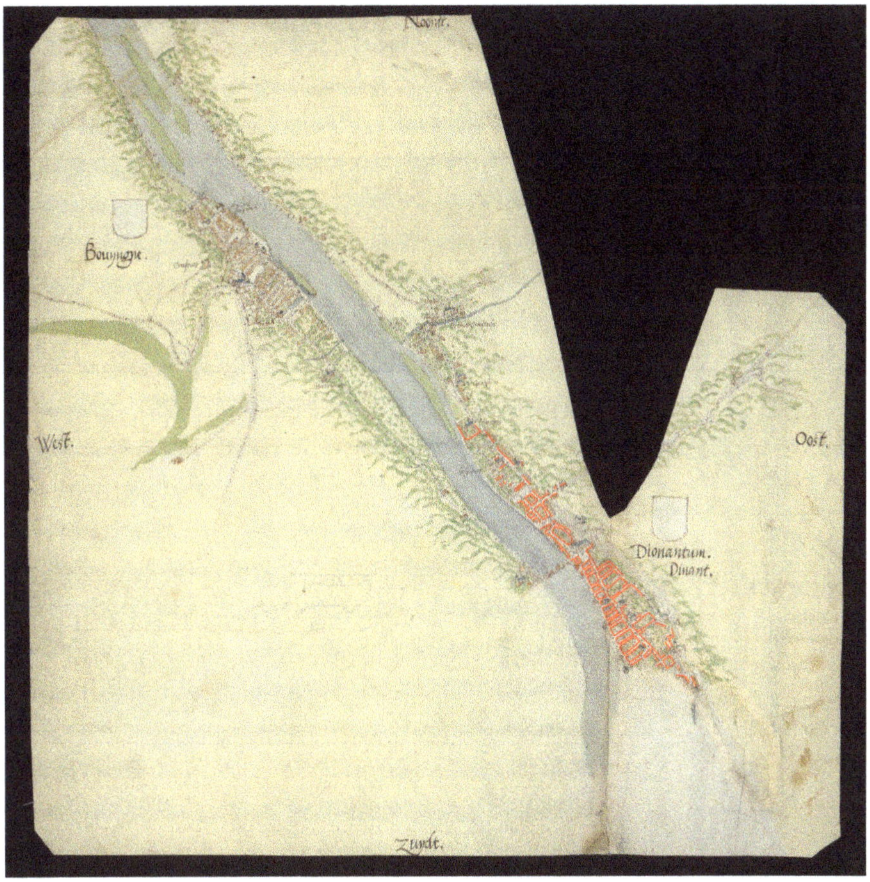
Noordt.
Bouvigne.
West.
Oost.
Dionantum.
Dinant.
Zuydt.

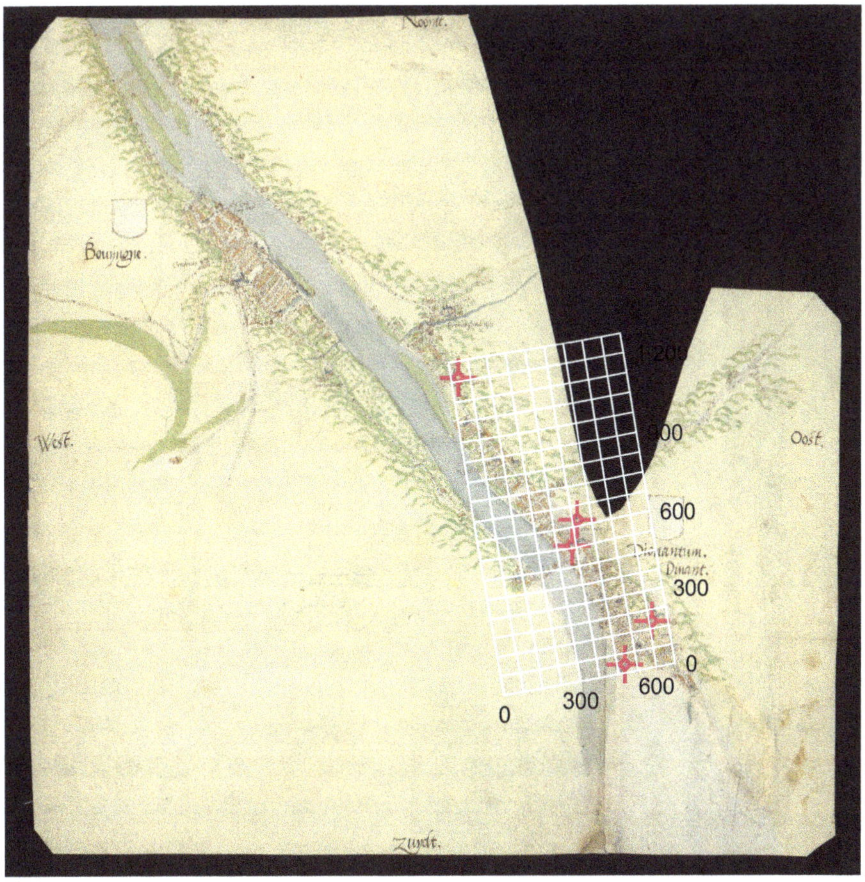
Noordt.
Bouvigne.
West.
Oost.
Dinant.
1200
900
600
300
0
0
300
600
Zuydt.

59. Doesburg

Tables 3.59 (measured and analysed).
Figures 3.59 (unmeasured, measured and analysed).

Measured

Pixels	Scale stick original map (pixels)	Houses	Street sides	Houses per street side	Inhabitants (RF 4.0)	Inhabitants (RF 5.5)	Inhabitants (RF 7.0)
4771	1000	838	72	11.64	3351	4608	5865

Analysed

March 31, 2023 11:38:35 AM	
Computation with 5 linked points	
(*italics* points are less reliable):	1. Martinikerk 2. Gasthuiskerk 3. Veerpoortwal 4. Lindewal 5. Ooipoort
Reference points (Reliability 0–5)	5
Transformation with 3 parameters	
Scale	1:8580
Rotation	10 °
Mean position error	10m

Septentrio
Occidens.
Doesburg.
Oriens.
Drempt
Meridies.

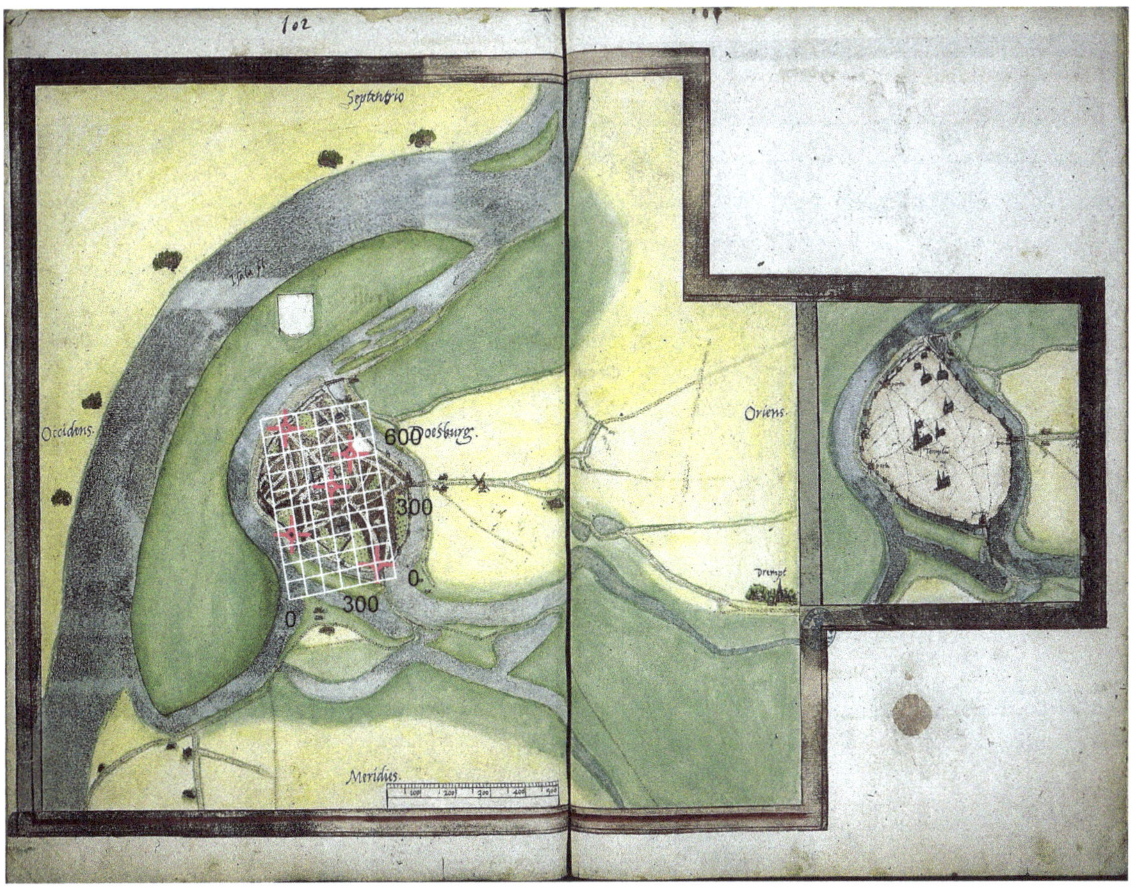
Septentrio
Occidens.
600
300
0
0
300
Oriens.
Drempt
Meridies.

60. Doetinchem

Tables 3.60 (measured and analysed).
Figures 3.60 (unmeasured, measured and analysed).

Measured							
Pixels	Scale stick original map (pixels)	Houses	Street sides	Houses per street side	Inhabitants (RF 4.0)	Inhabitants (RF 5.5)	Inhabitants (RF 7.0)
2294	1007	403	47	8.57	1612	2216	2820

Analysed	
March 31, 2023 12:30:03 PM	
Computation with 5 linked points	
(*italics* points are less reliable):	1. Catharinakerk
	2. Hamburgerstraat
	3. Waterstraat
	4. Walstraat
	5. Nieuwstad
Reference points (Reliability 0–5)	5
Transformation with 3 parameters	
Scale	1:8610
Rotation	10 °
Mean position error	18m

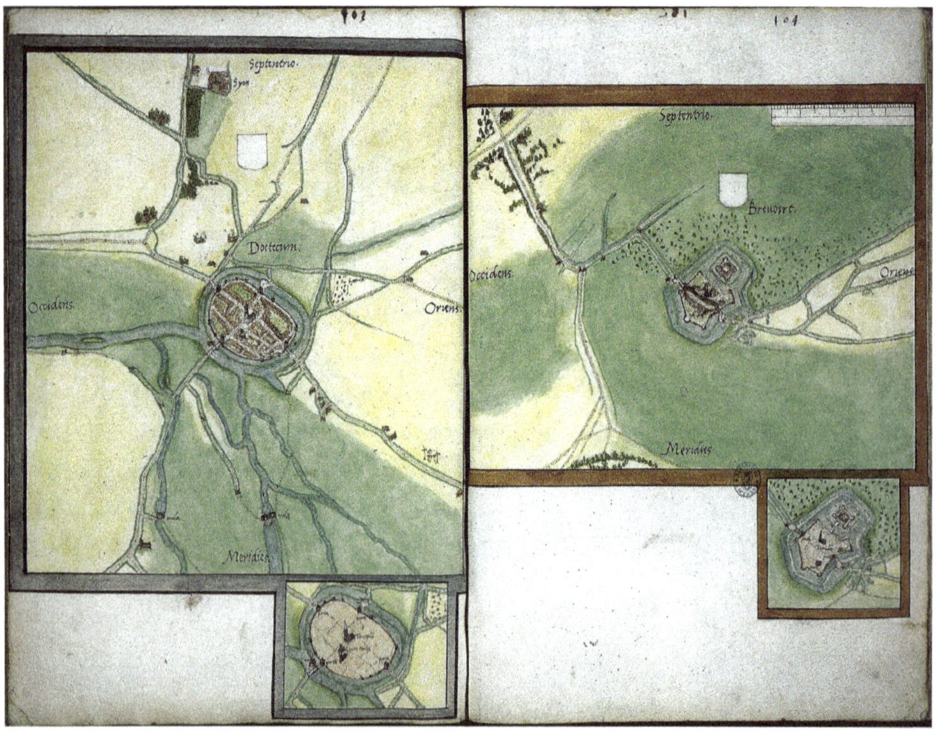

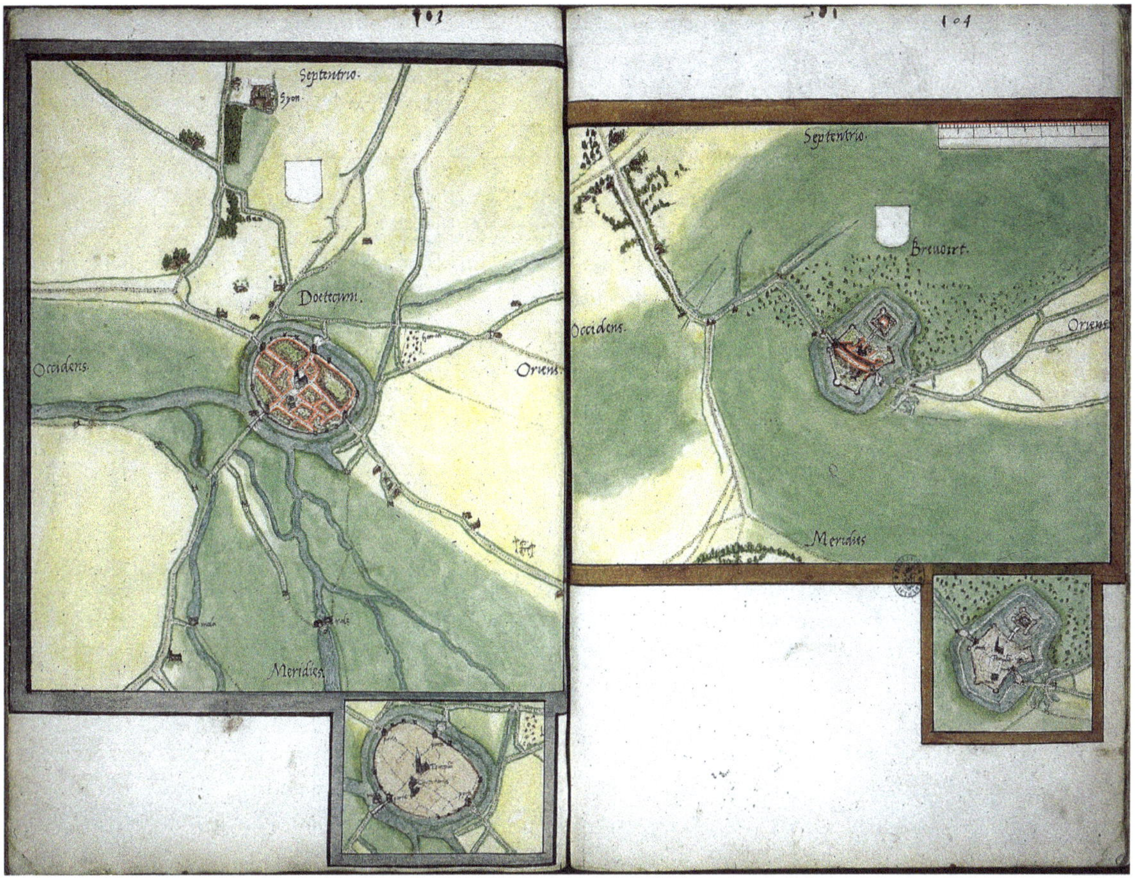
Septentrio
Occidens
Oriens
Meridies
Septentrio
Occidens
Meridies

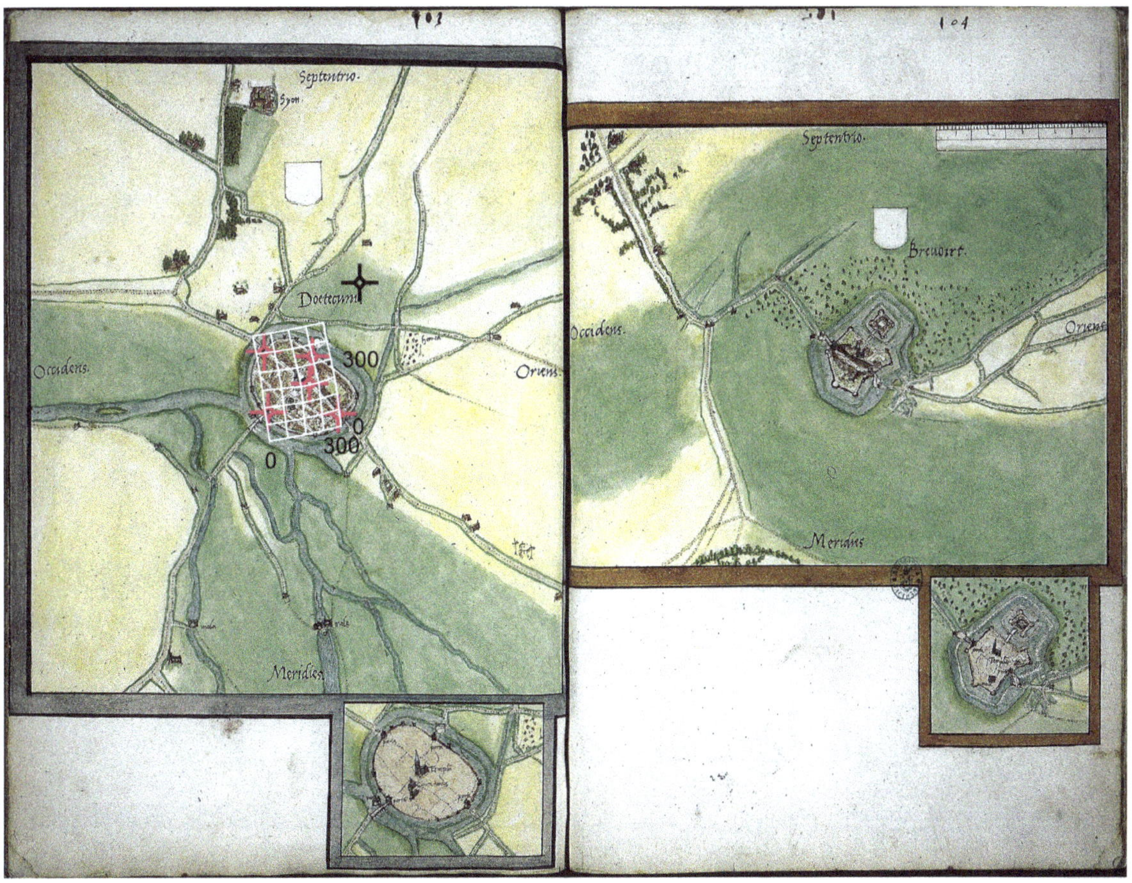
Septentrio
Occidens
Oriens
Meridies
300
0
300
0
Septentrio
Occidens
Meridies

61. Dokkum

Tables 3.61 (measured and analysed).
Figures 3.61 (unmeasured, measured and analysed).

Measured

Pixels	Scale stick original map (pixels)	Houses	Street sides	Houses per street side	Inhabitants (RF 4.0)	Inhabitants (RF 5.5)	Inhabitants (RF 7.0)
3939	1004	692	144	4.80	2767	3804	4841

Analysed

March 31, 2023 2:57:57 PM	
Computation with 5 linked points	
(*italics* points are less reliable):	1. Sint-Martinuskerk 2. Nauwstraat 3. Koornmarkt 4. Hanspoort 5. Diepswal
Reference points (Reliability 0–5)	5
Transformation with 3 parameters	
Scale	1:8370
Rotation	12 °
Mean position error	7m

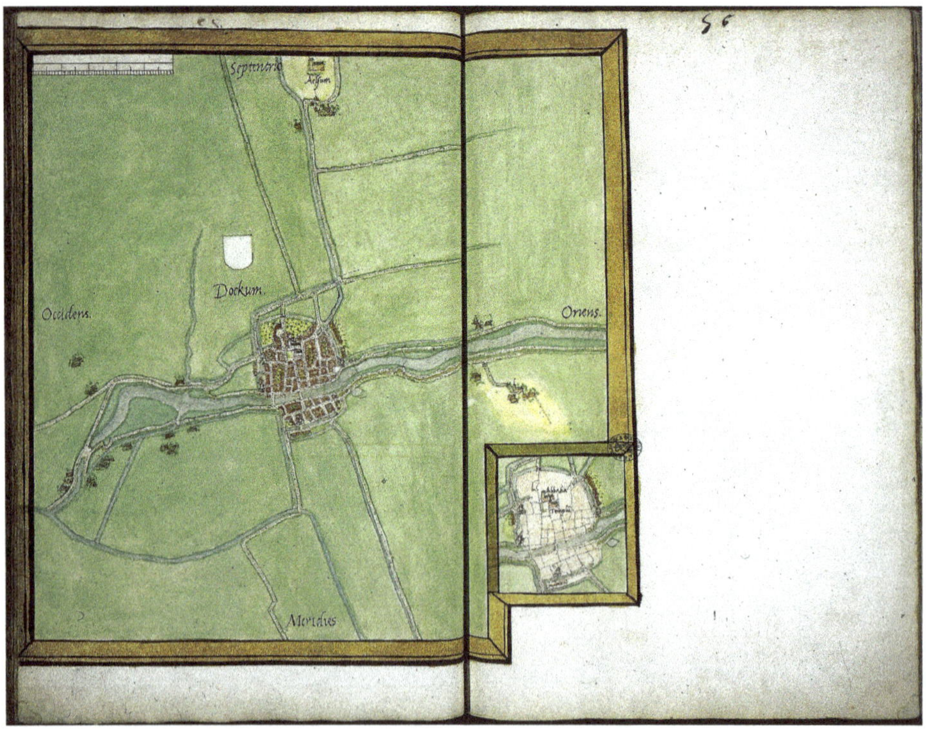

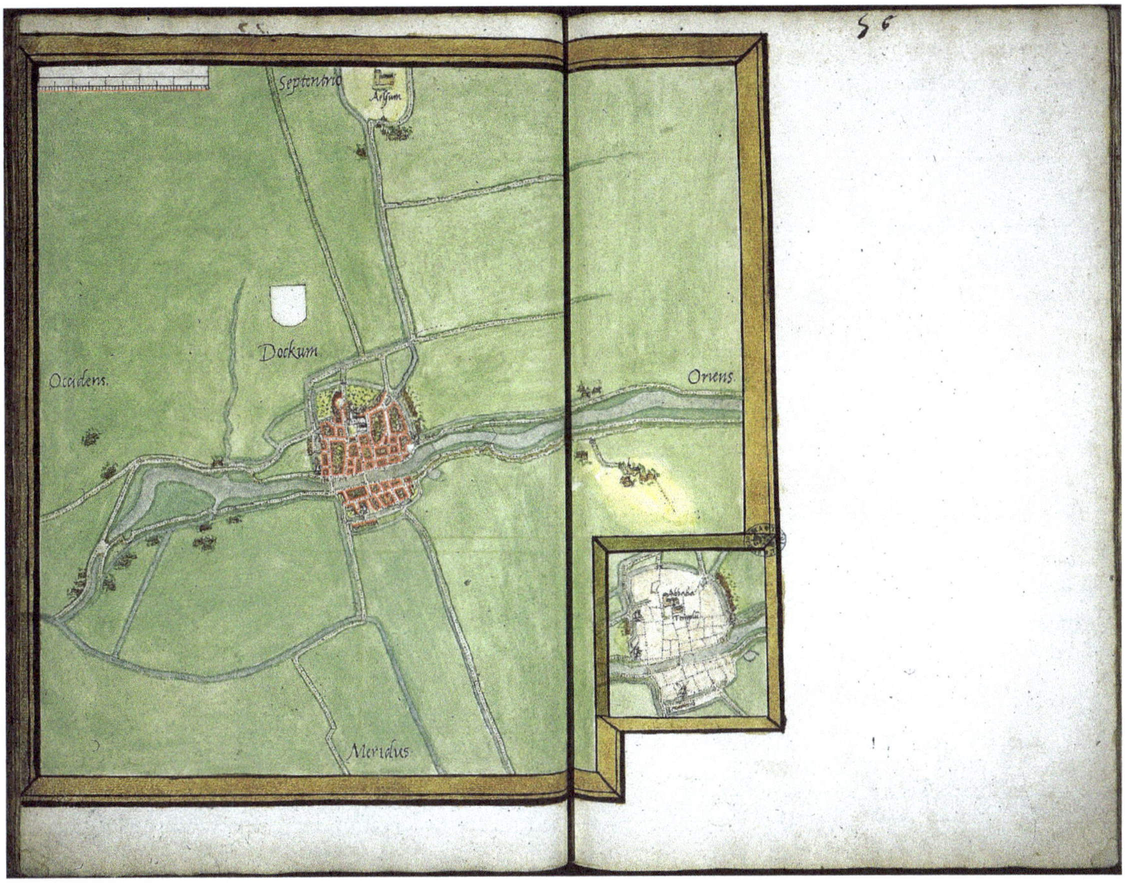
Septentrio
Dockum
Occidens
Oriens

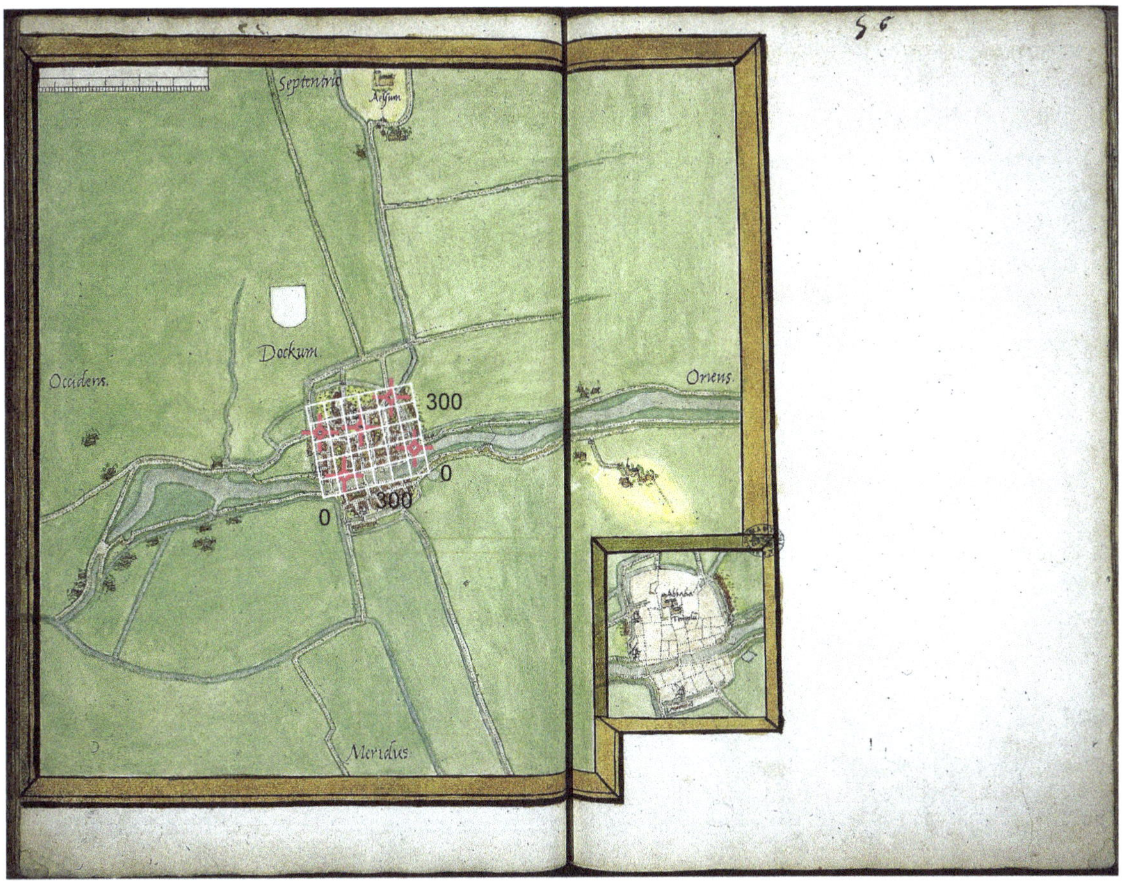
Septentrio
Dockum
Occidens
Oriens
300
0
0
300

62. Dordrecht

Tables 3.62 (measured and analysed).
Figures 3.62 (unmeasured, measured and analysed).

Measured							
Pixels	Scale stick original map (pixels)	Houses	Street sides	Houses per street side	Inhabitants (RF 4.0)	Inhabitants (RF 5.5)	Inhabitants (RF 7.0)
9812	1035	1723	284	6.07	6892	9476	12,060

Analysed	
April 1, 2023 3:31:27 PM	
Computation with 5 linked points	
(*italics* points are less reliable):	1. Grote Kerk
	2. Stadhuis
	3. Augustijnenkerk
	4. Groothoofdspoort
	5. Vriesebrug
Reference points (Reliability 0–5)	5
Transformation with 3 parameters	
Scale	1:8630
Rotation	11°
Mean position error	15m

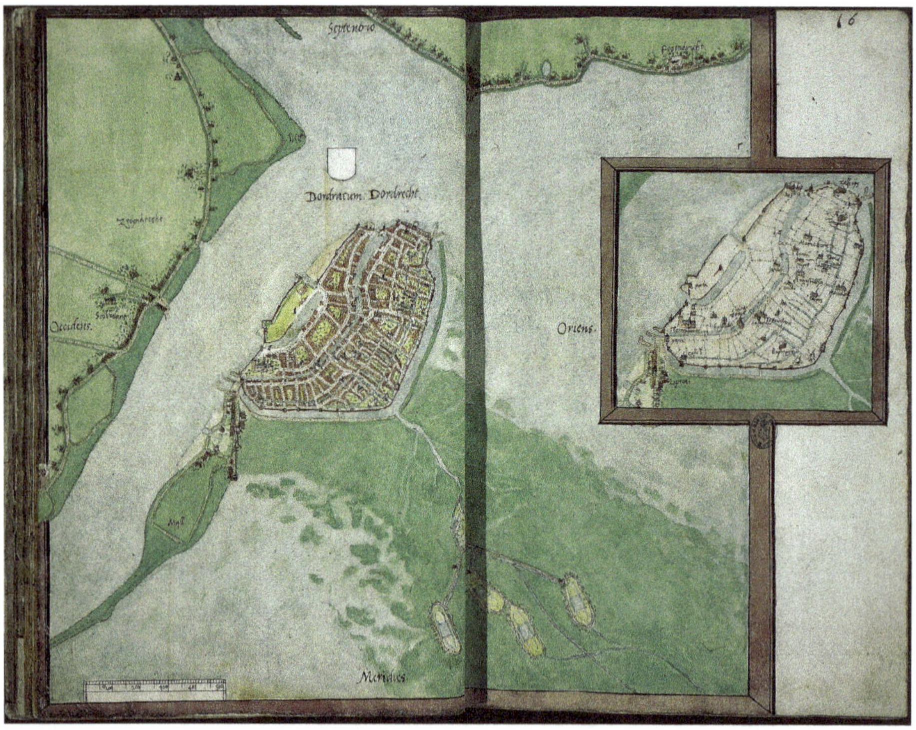

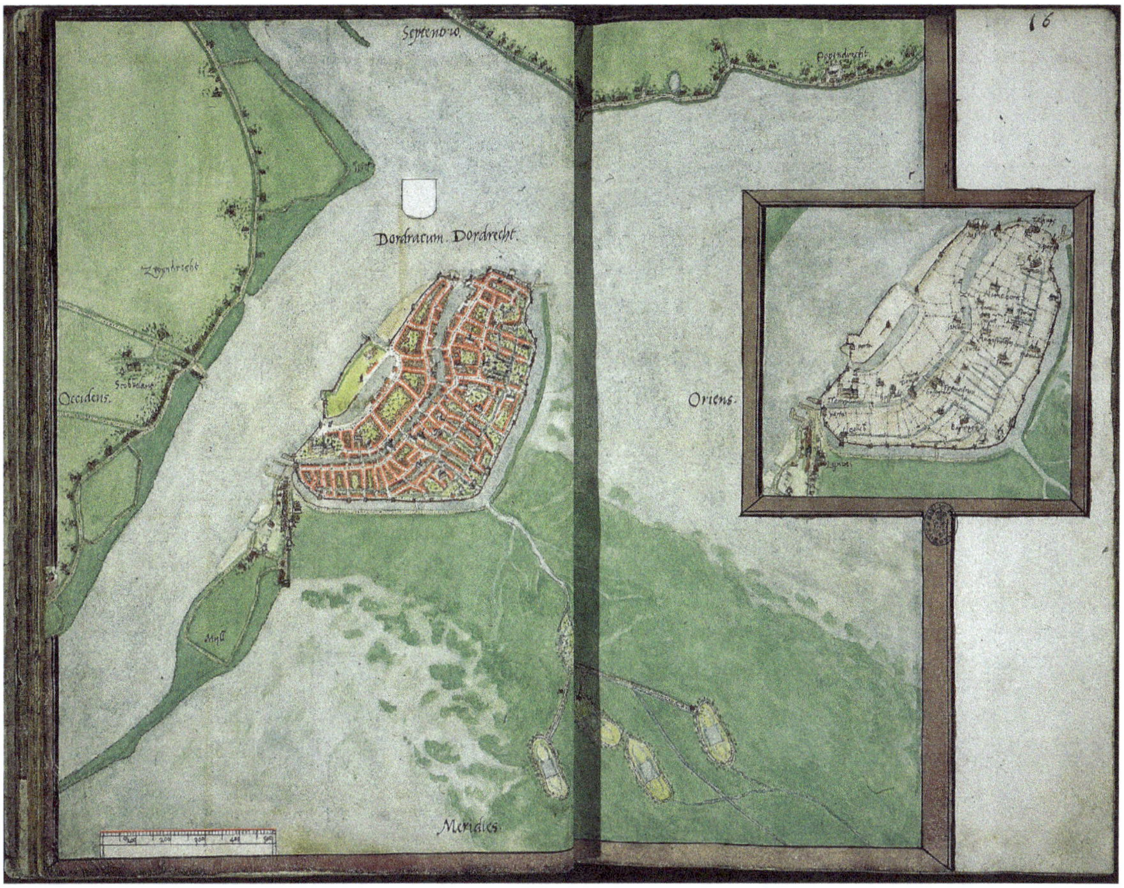
Septentrio
Dordracum Dordrecht
Occidens.
Oriens.
Meridies.

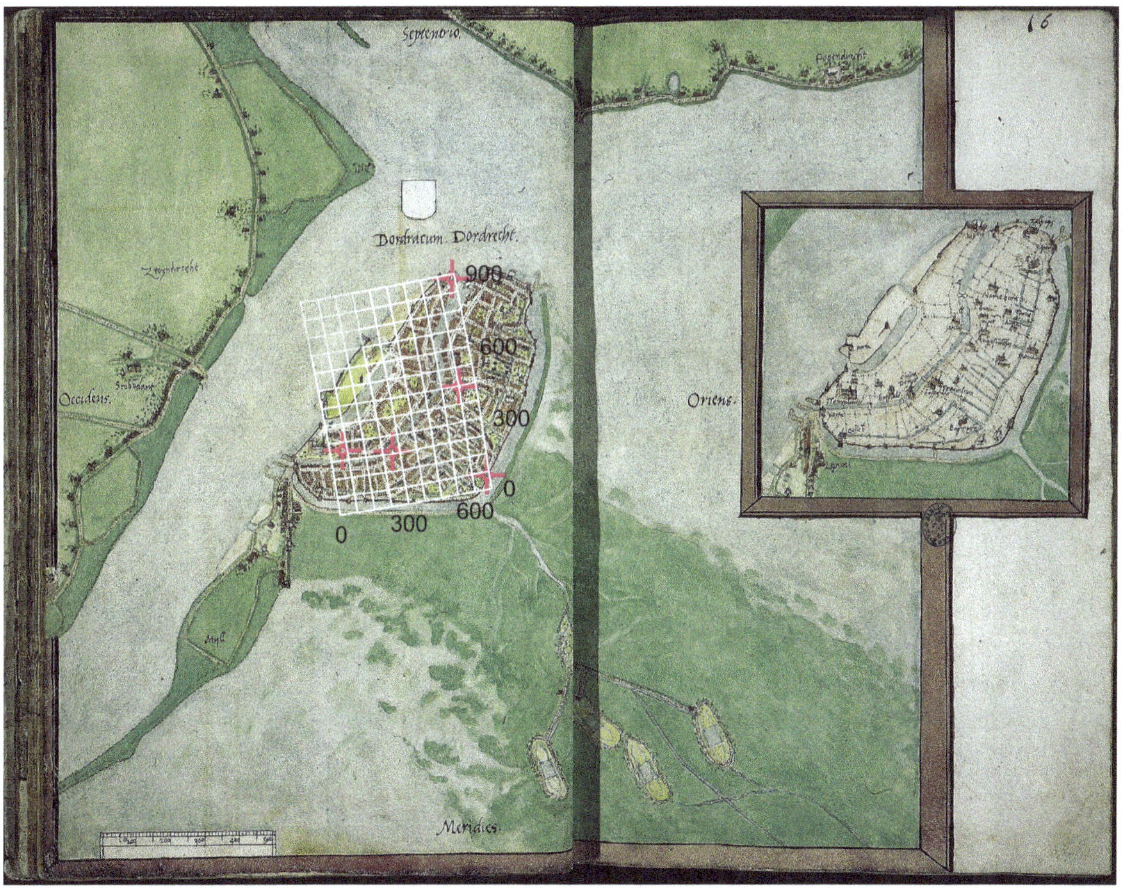
Septentrio
Dordracum Dordrecht
900
600
300
0
0
300
600
Occidens.
Oriens.
Meridies.

63. Douai

Tables 3.63 (measured and analysed).
Figures 3.63 (unmeasured, measured and analysed).

Measured

Pixels	Scale stick original map (pixels)	Houses	Street sides	Houses per street side	Inhabitants (RF 4.0)	Inhabitants (RF 5.5)	Inhabitants (RF 7.0)
20,077	869	3526	426	8.28	14,102	19,390	24,678

Analysed

April 2, 2023 10:20:17 AM	
Computation with 5 linked points	
(*italics* points are less reliable):	1. Saint-Pierre de Douai 2. Église de Notre Dame 3. Belfort 4. La Scarpe *5. La Scarpe*
Reference points (Reliability 0–5)	4
Transformation with 3 parameters	
Scale	1:9070
Rotation	14 °
Mean position error	22m

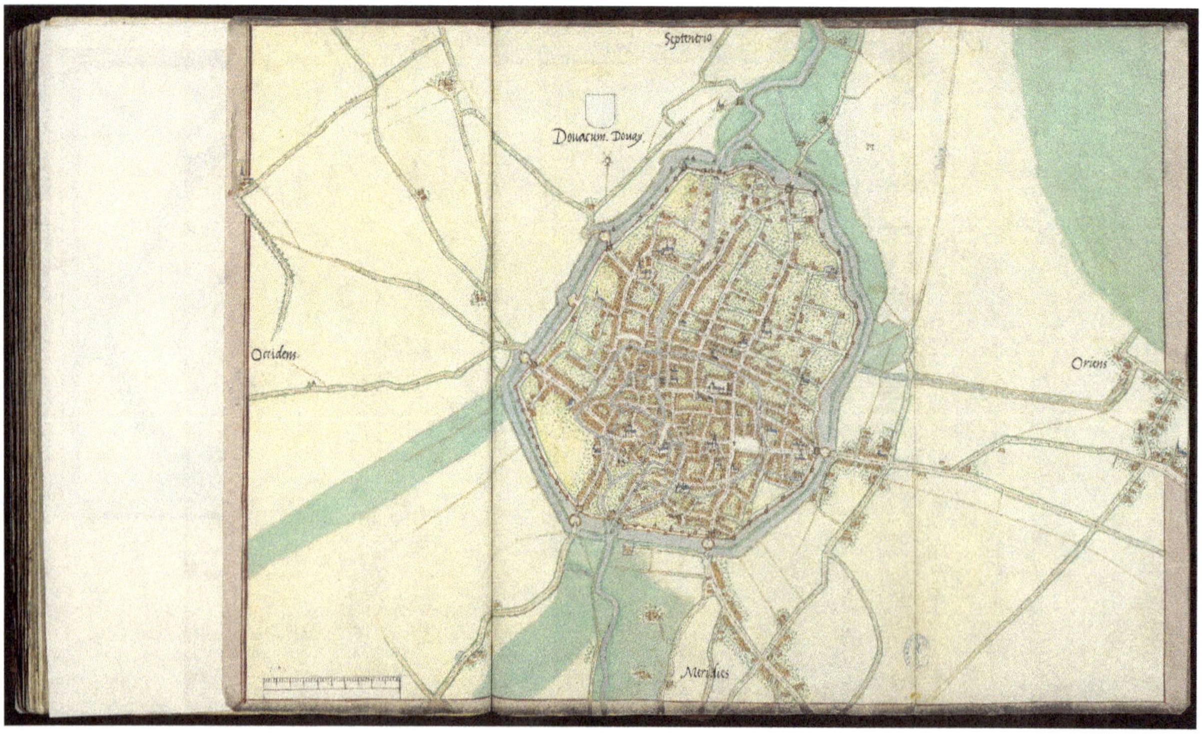

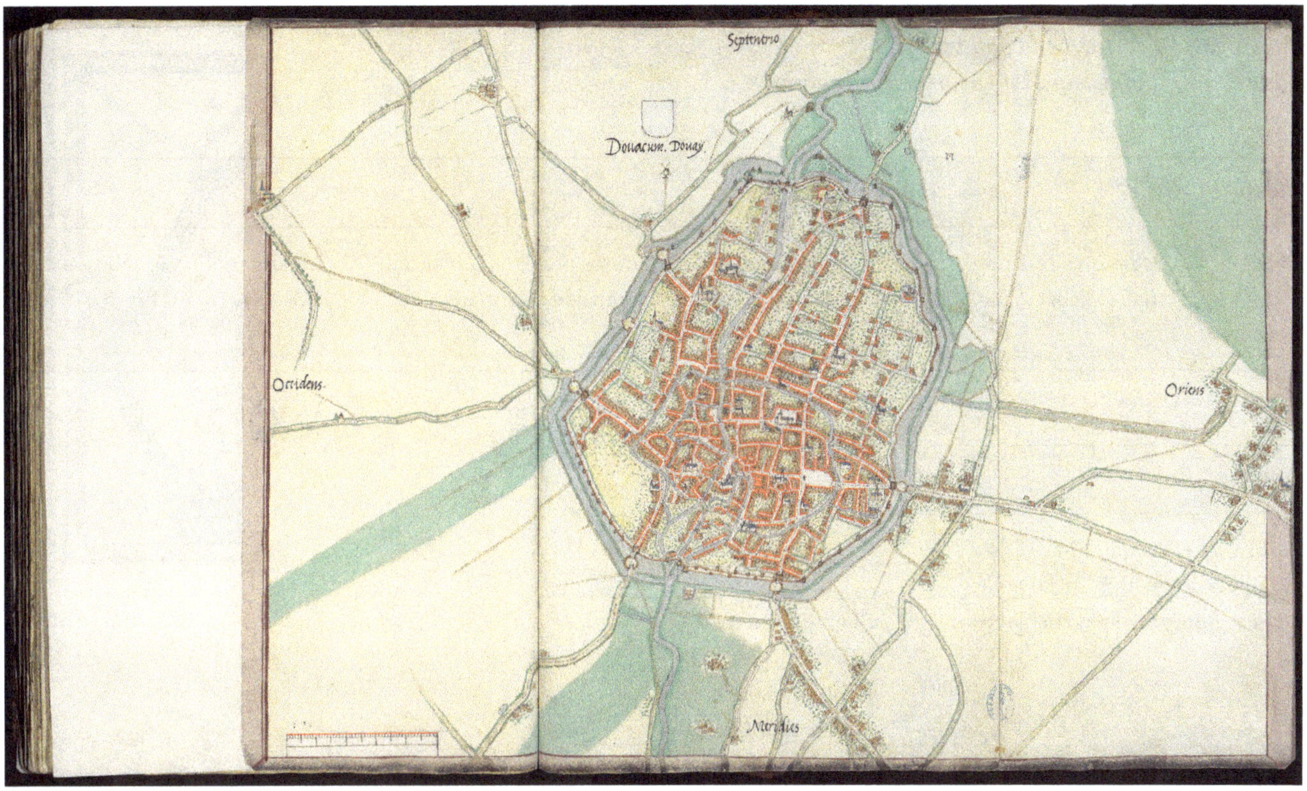
Septentrio
Douacum. Douay.
Occidens.
Oriens
Meridies

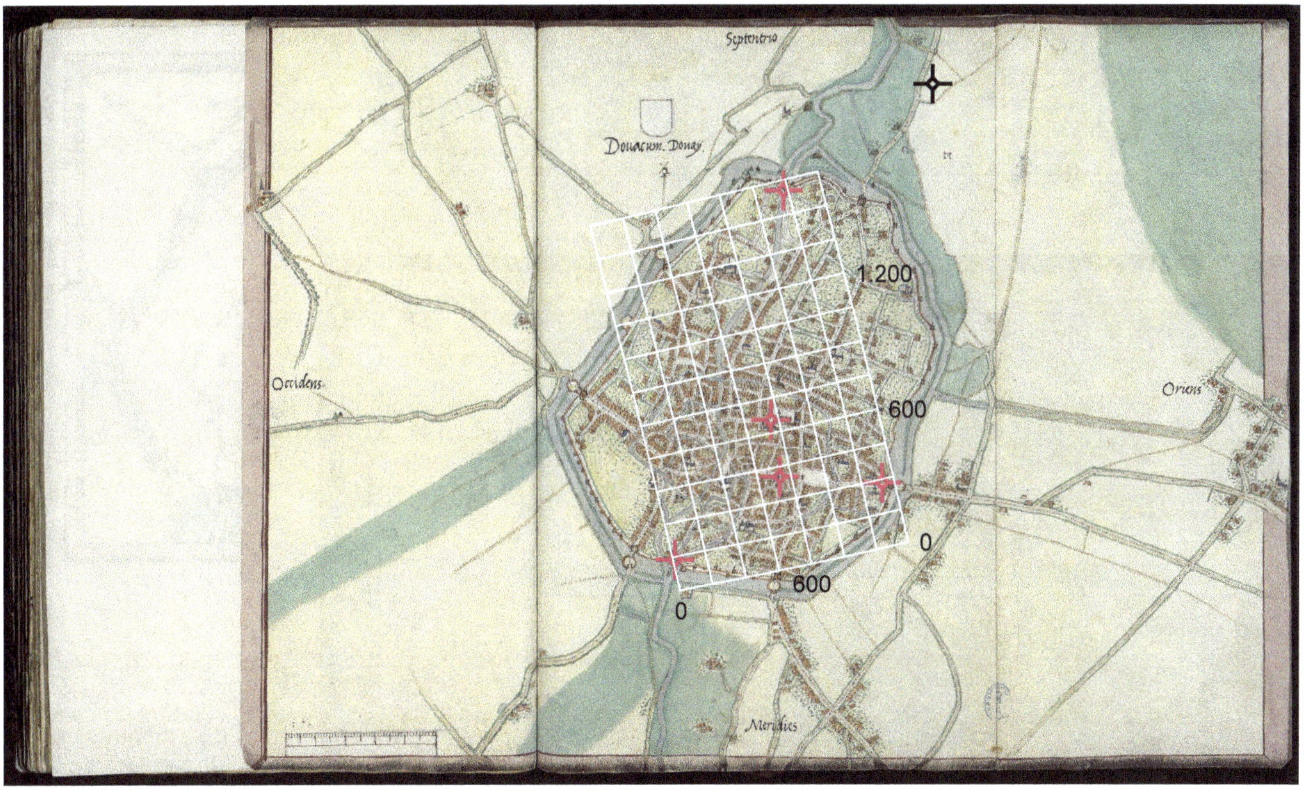
Septentrio
Douacum. Douay.
Occidens.
Oriens
Meridies
1.200
600
0
600
0

64. Dunkerque

Tables 3.64 (measured and analysed).
Figures 3.64 (unmeasured, measured and analysed).

Measured							
Pixels	Scale stick original map (pixels)	Houses	Street sides	Houses per street side	Inhabitants (RF 4.0)	Inhabitants (RF 5.5)	Inhabitants (RF 7.0)
5021	1034	882	111	7.94	3527	4849	6171

Analysed	
April 2, 2023 12:44:16 PM	
Computation with 5 linked points	
(*italics* points are less reliable):	1. Église Saint-Eloi
	2. Belfort
	3. Tour du Leughenaer
	4. Quai de la Citadelle
	5. Rue du président Poincarré
Reference points (Reliability 0–5)	3
Transformation with 3 parameters	
Scale	1:7800
Rotation	7 °
Mean position error	13m

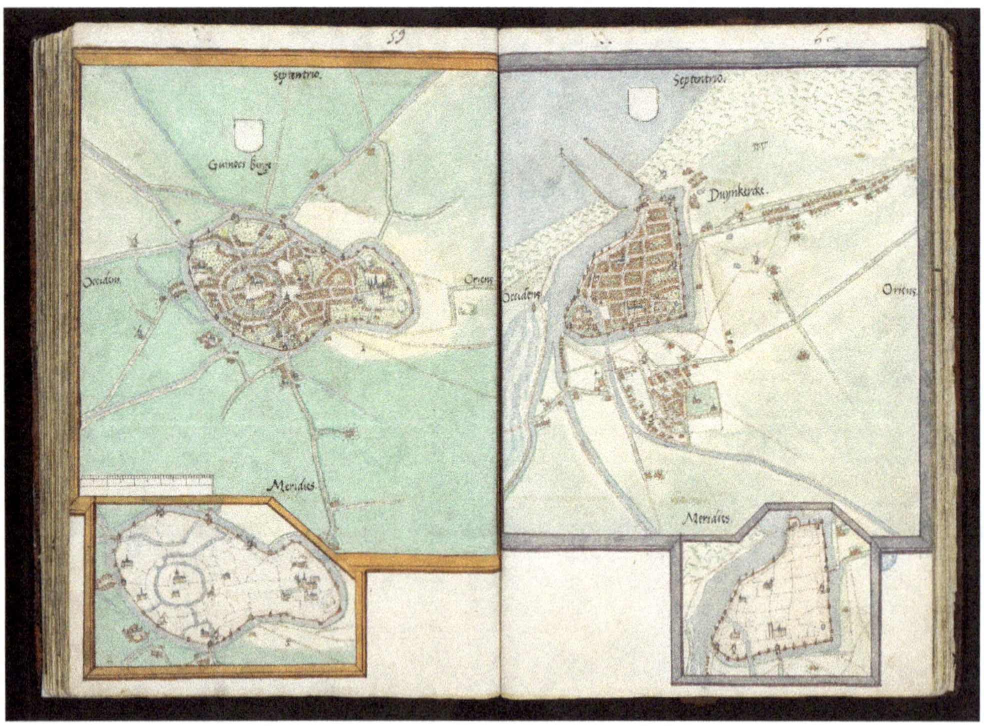

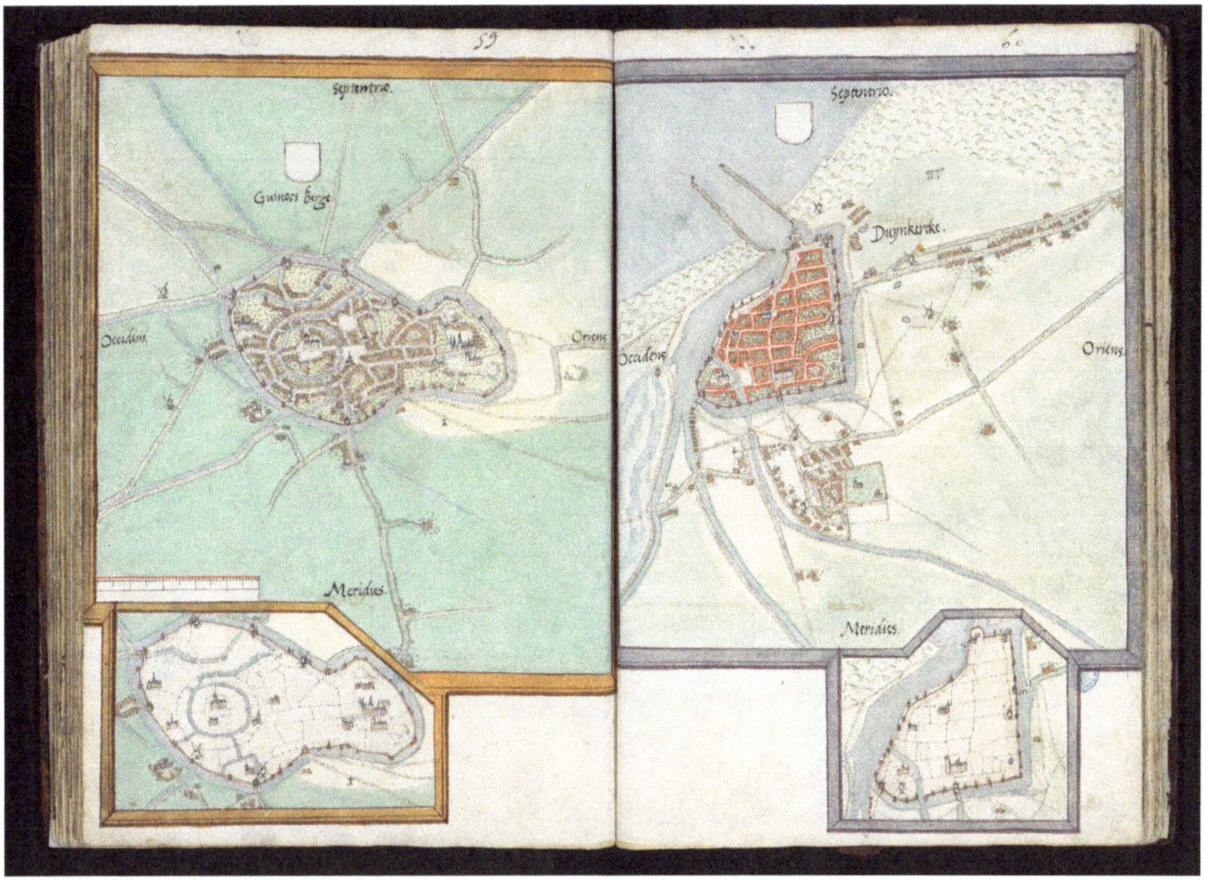
59
60
Septentrio.
Occidens
Oriens
Meridies.
Septentrio.
Duynkercke.
Occidens
Oriens
Meridies.

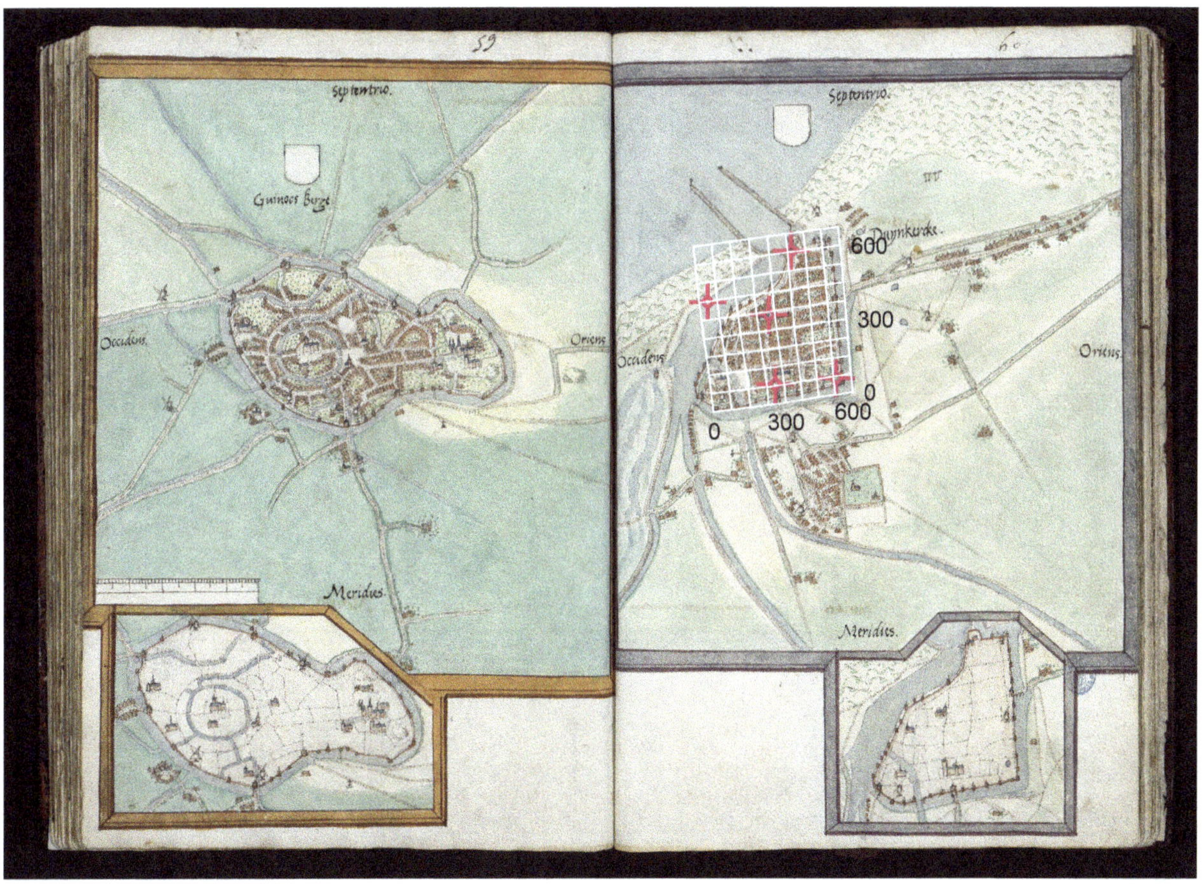
59
60
Septentrio.
Occidens
Oriens
Meridies.
Septentrio.
Occidens
Oriens
Meridies.
600
300
0
0
300
600

65. Echt

Tables 3.65 (measured and analysed).
Figures 3.65 (unmeasured, measured and analysed).

Measured							
Pixels	Scale stick original map (pixels)	Houses	Street sides	Houses per street side	Inhabitants (RF 4.0)	Inhabitants (RF 5.5)	Inhabitants (RF 7.0)
1914	1008	336	46	7.31	1345	1849	2353

Analysed	
April 3, 2023 12:09:21 PM	
Computation with 5 linked points	
(*italics* points are less reliable):	1. Sint-Landricuskerk 2. Zuiderpoort 3. Huis Verduynen 4. Echtermolenbeek 5. Wijnstraat
Reference points (Reliability 0–5)	5
Transformation with 3 parameters	
Scale	1:8680
Rotation	13 °
Mean position error	10m

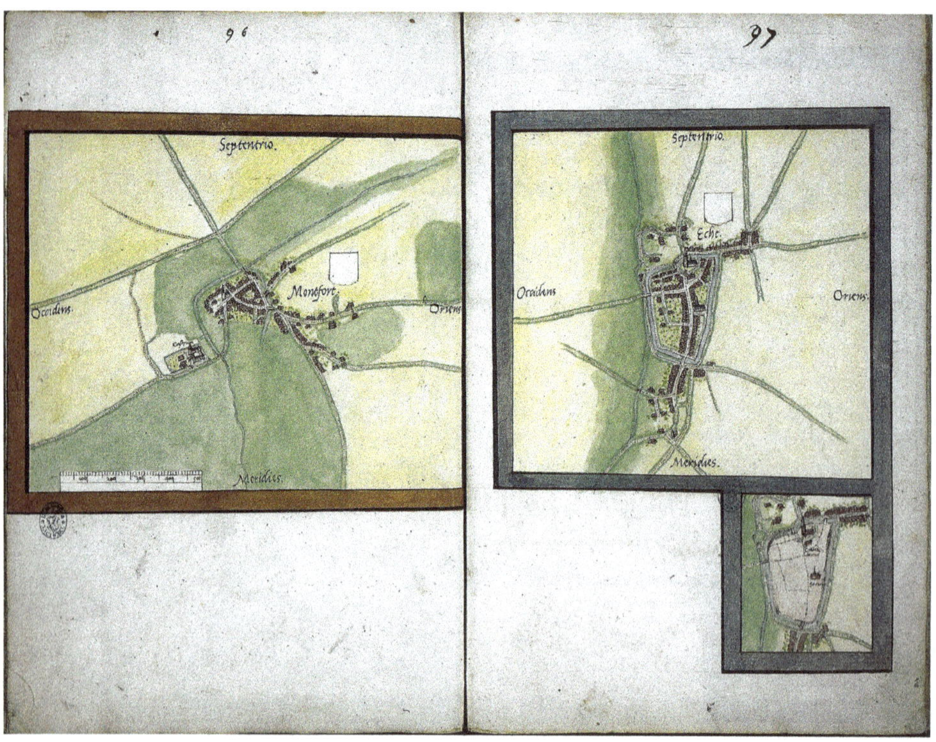

96
97
Septentrio.
Montfort
Occidens.
Meridies.
Septentrio.
Echt.
Occidens
Oriens.
Meridies.

96
97
Septentrio.
Montfort
Occidens.
Meridies.
Septentrio.
600
300
0
300
0
Occidens
Oriens.
Meridies.

66. Echternach

Tables 3.66 (measured and analysed).
Figures 3.66 (unmeasured, measured and analysed).

Measured							
Pixels	Scale stick original map (pixels)	Houses	Street sides	Houses per street side	Inhabitants (RF 4.0)	Inhabitants (RF 5.5)	Inhabitants (RF 7.0)
4752	missing	834	145	5.75	3337	4589	5841

Analysed	
April 18, 2023 9:47:22 AM	
Computation with 5 linked points	
(*italics* points are less reliable):	1. Basilique 2. Église Saints-Pierre-et Paul 3. Rue Neuve 4. Rue des Remparts 5. Rue des Redoutes
Reference points (Reliability 0–5)	5
Transformation with 3 parameters	
Scale	1:7300
Rotation	6 °
Mean position error	17m

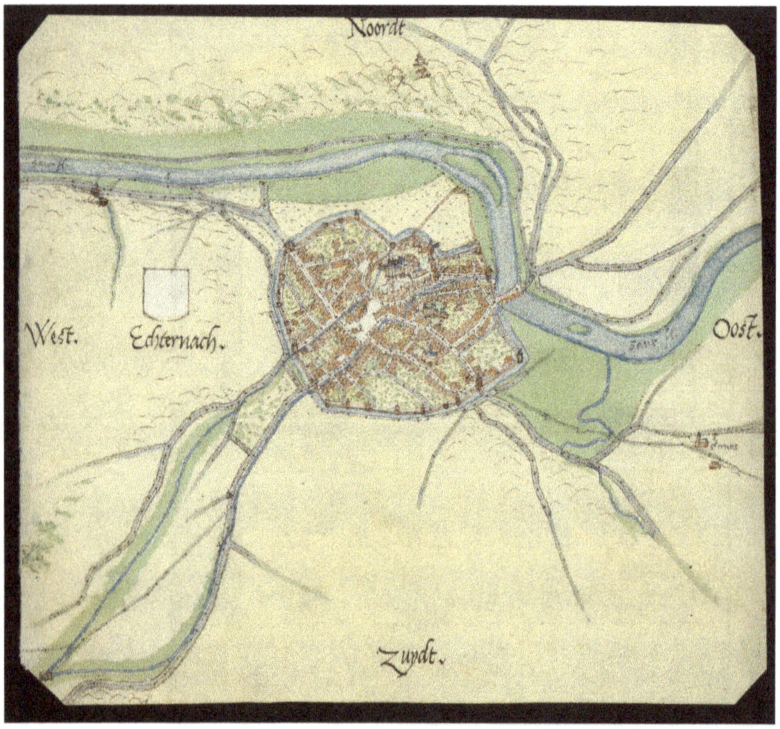

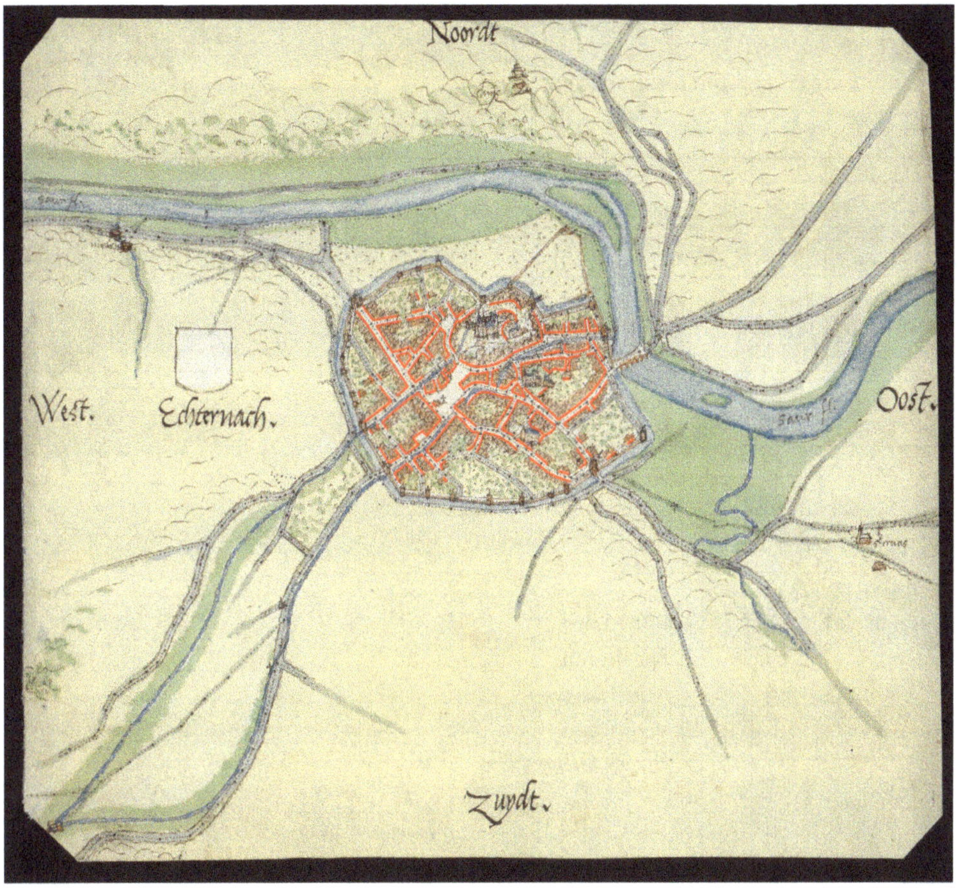
Noordt
West.
Echternach.
Oost.
Zuydt.

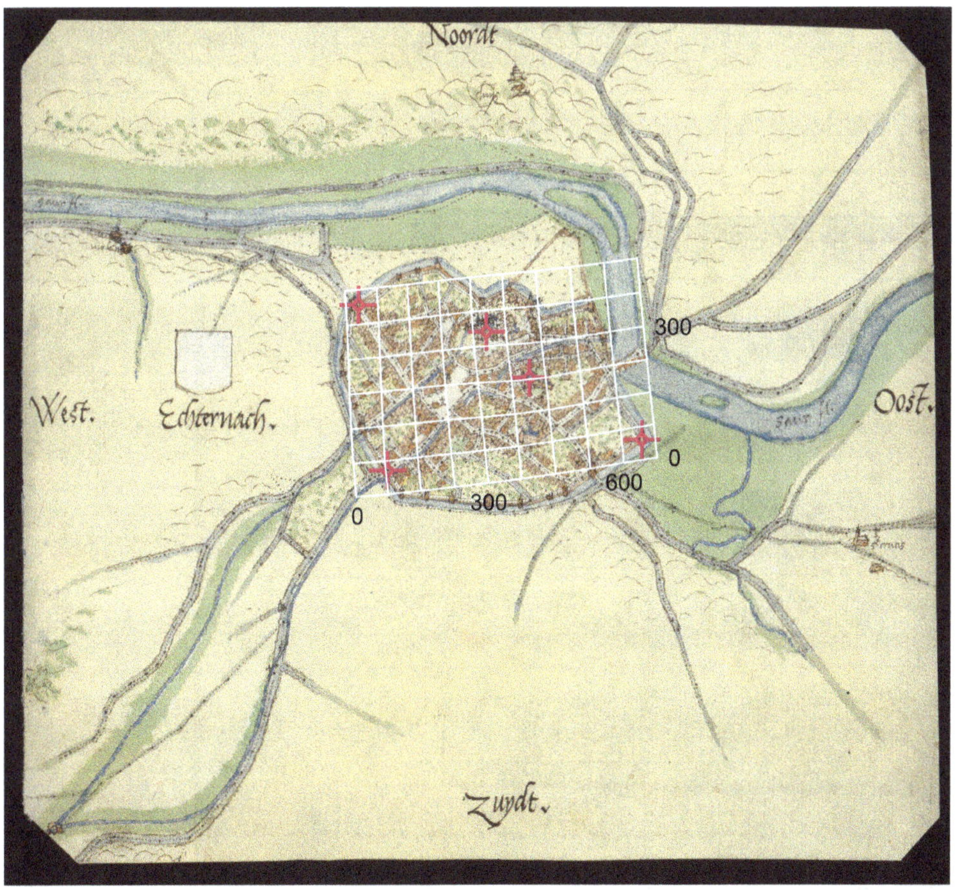
Noordt
West.
Echternach.
300
Oost.
0
600
300
0
Zuydt.

67. Edam

Tables 3.67 (measured and analysed).
Figures 3.67 (unmeasured, measured and analysed).

Measured							
Pixels	Scale stick original map (pixels)	Houses	Street sides	Houses per street side	Inhabitants (RF 4.0)	Inhabitants (RF 5.5)	Inhabitants (RF 7.0)
3185	1010	559	85	6.58	2237	3076	3915

Analysed	
April 4, 2023 10:33:32 AM	
Computation with 5 linked points	
(*italics* points are less reliable):	1. Grote Kerk
	2. Speeltoren
	3. Achterhaven
	4. Voorhaven
	5. Nieuwehaven
Reference points (Reliability 0–5)	**3**
Transformation with 3 parameters	
Scale	1:8530
Rotation	10 °
Mean position error	26m

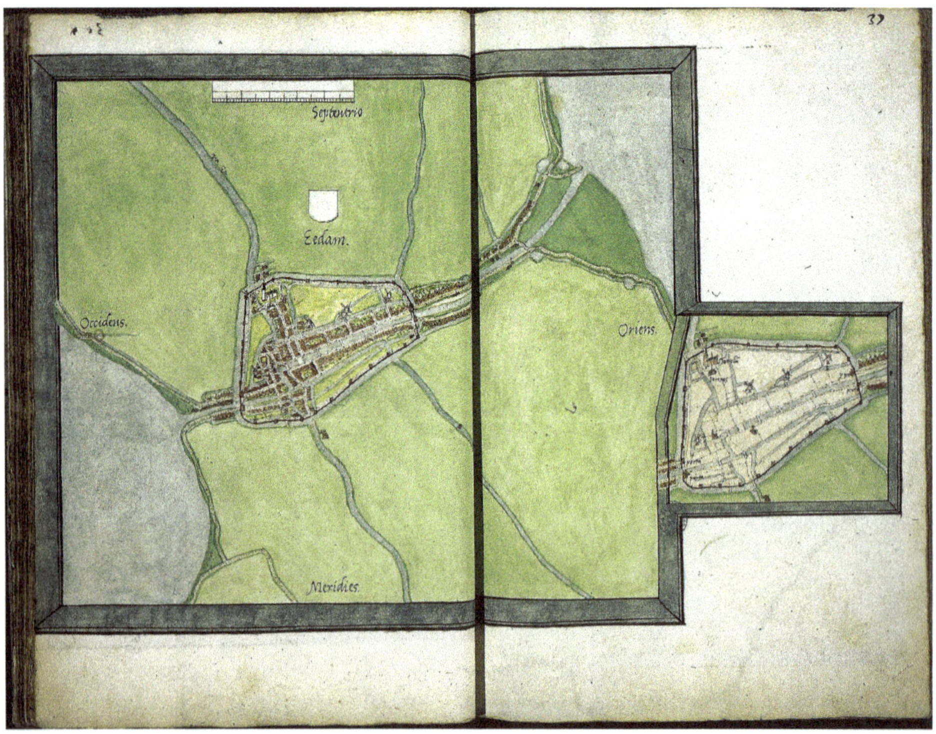

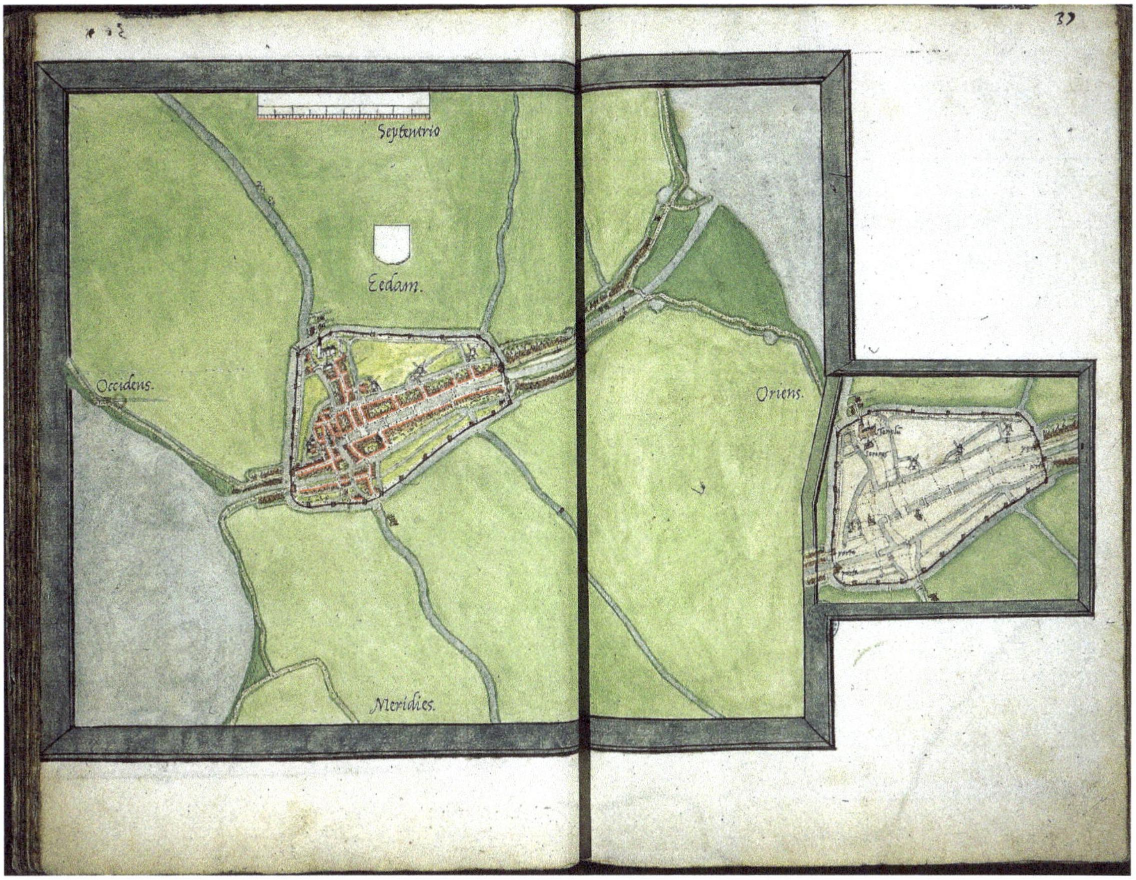
Septentrio
Eedam.
Occidens.
Oriens.
Meridies.

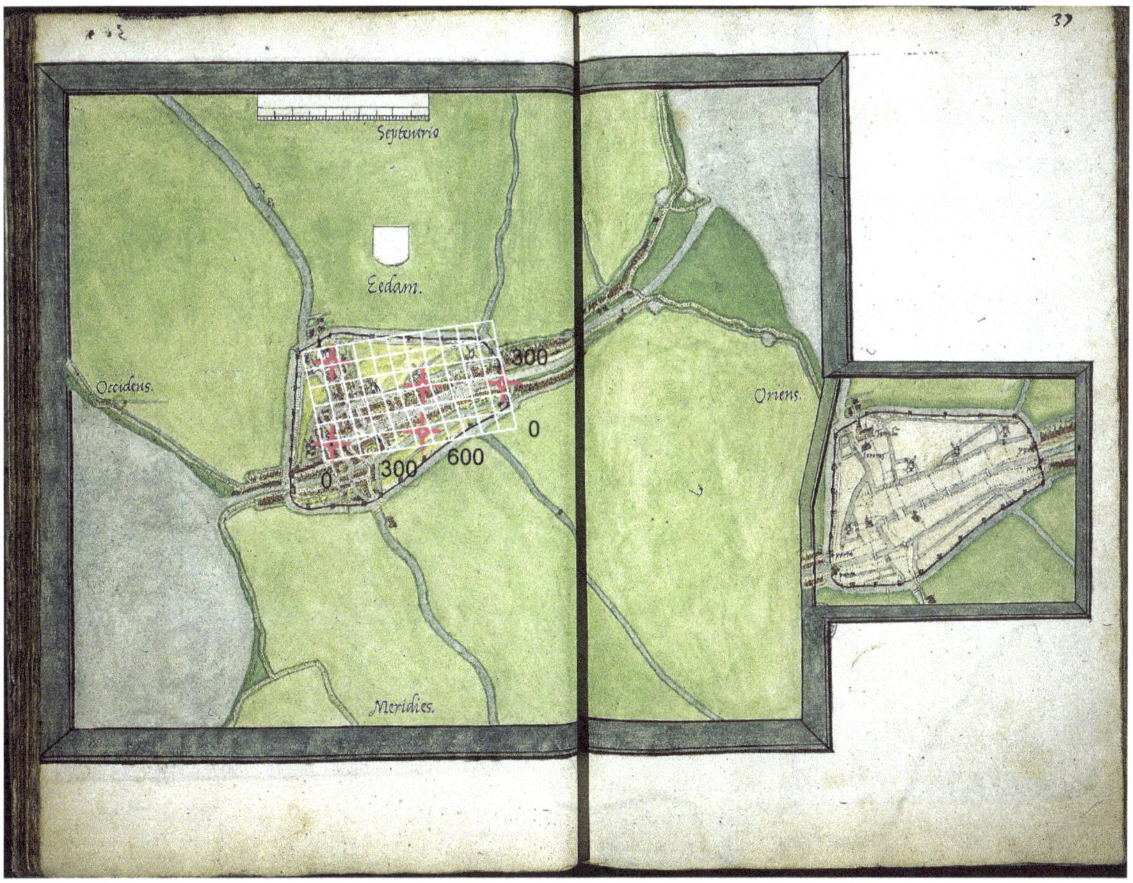
Septentrio
Eedam.
Occidens.
Oriens.
Meridies.
300
0
0
300
600

68. Eindhoven

Tables 3.68 (measured and analysed).
Figures 3.68 (unmeasured, measured and analysed).

Measured							
Pixels	Scale stick original map (pixels)	Houses	Street sides	Houses per street side	Inhabitants (RF 4.0)	Inhabitants (RF 5.5)	Inhabitants (RF 7.0)
2080	missing	365	49	7.45	1461	2009	2557

Analysed	
September 20, 2023 8:46:35 AM	
Computation with 5 linked points	
(*italics* points are less reliable):	1. Markt
	2. Vestdijk
	3. Wal
	4. Demer
	5. Vrijstraat
Reference points (Reliability 0–5)	5
Transformation with 3 parameters	
Scale	1:8710
Rotation	7 °
Mean position error	17m

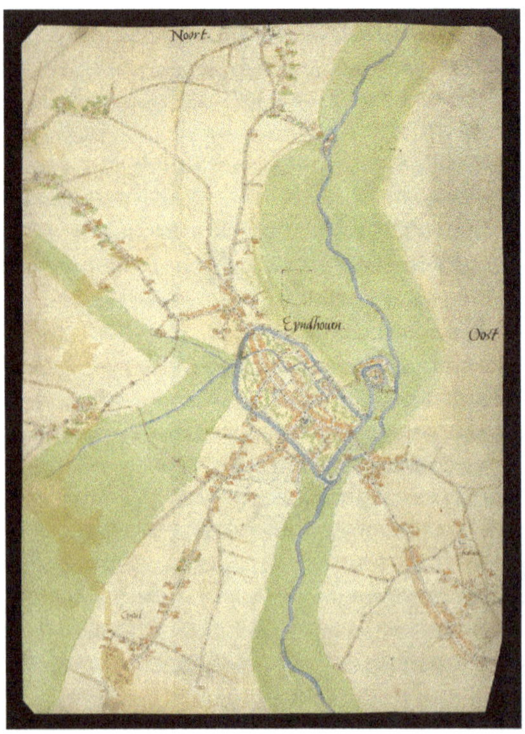

Noort.
Eyndhouen.
Oost

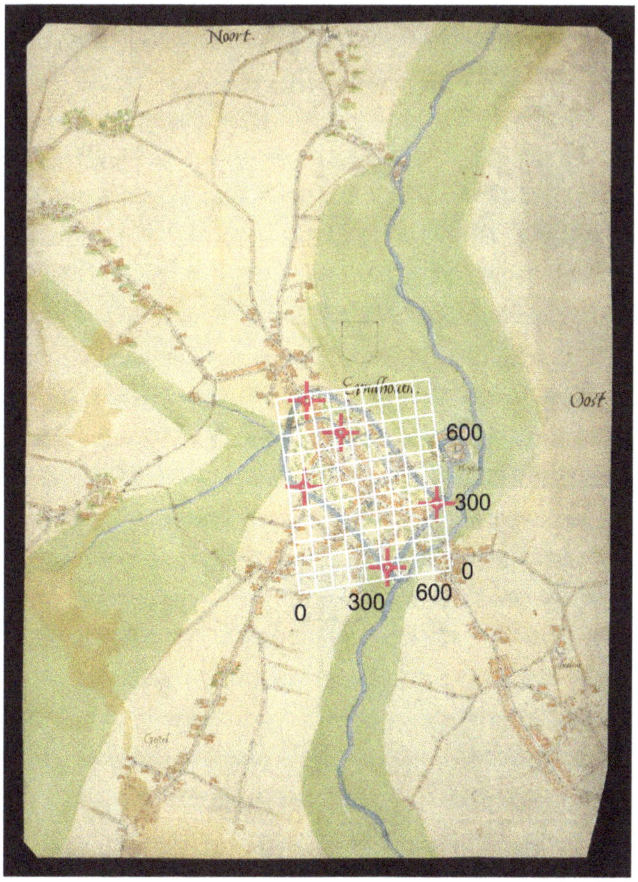
Noort.
Oost
600
300
0
0
300
600

69. Elburg

Tables 3.69 (measured and analysed).
Figures 3.69 (unmeasured, measured and analysed).

Measured							
Pixels	Scale stick original map (pixels)	Houses	Street sides	Houses per street side	Inhabitants (RF 4.0)	Inhabitants (RF 5.5)	Inhabitants (RF 7.0)
3124	1012	549	116	4.73	2194	3017	3840

Analysed	
April 6, 2023 11:14:15 AM	
Computation with 5 linked points	
(*italics* points are less reliable):	1. Noorderwalstraat
	2. Westerwalstraat
	3. Sint-Nicolaaskerk
	4. Stadsbeek
	5. Eiermarkt
Reference points (Reliability 0–5)	4
Transformation with 3 parameters	
Scale	1:8050
Rotation	11 °
Mean position error	16m

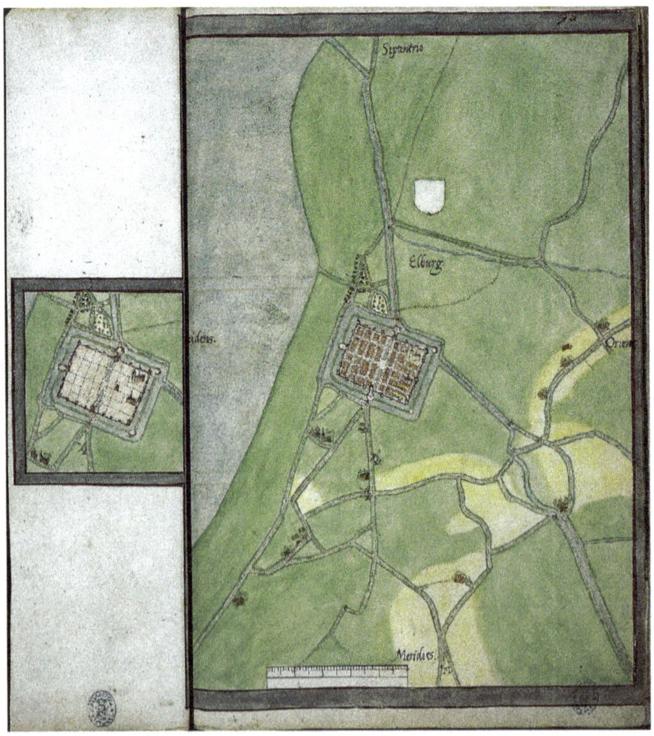

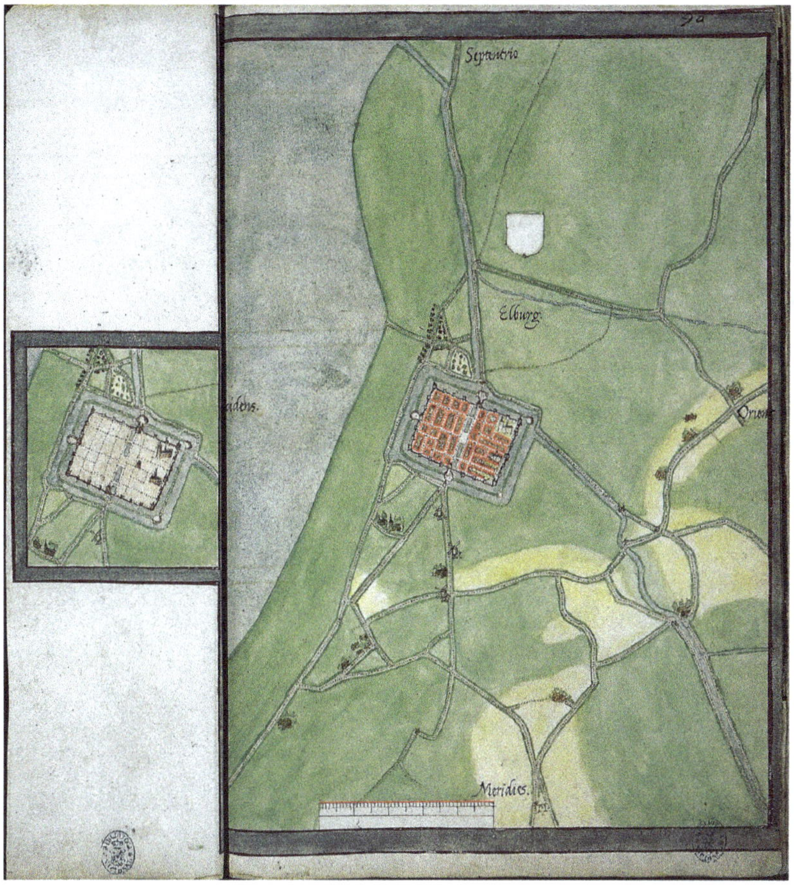
Septentrio
Elburg
Meridies

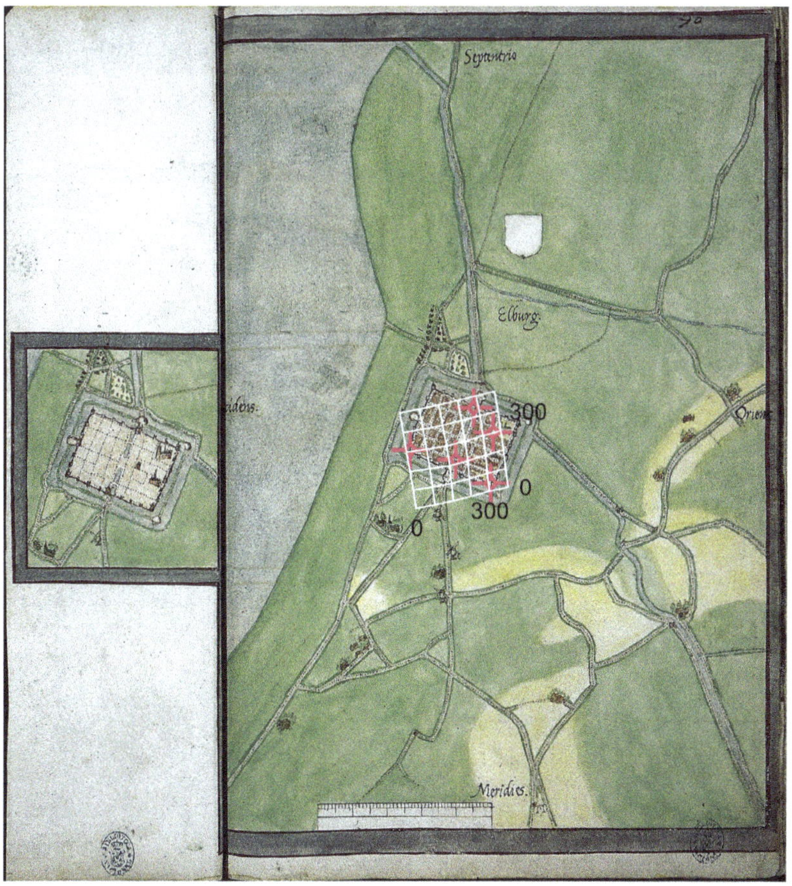
Septentrio
Elburg
300
0
300
0
Meridies

70. Enghien

Tables 3.70 (measured and analysed).
Figures 3.70 (unmeasured, measured and analysed).

Measured

Pixels	Scale stick original map (pixels)	Houses	Street sides	Houses per street side	Inhabitants (RF 4.0)	Inhabitants (RF 5.5)	Inhabitants (RF 7.0)
6142	1032	1079	118	9.14	4314	5932	7550

Analysed

April 6, 2023 8:04:35 PM	
Computation with 5 linked points	
(*italics* points are less reliable):	1. Eglise Saint-Nicolas
	2. Rue d'Hoves
	3. Etang du miroir
	4. Rue du sac
	5. Boulevard d'Arenberg
Reference points (Reliability 0–5)	**3**
Transformation with 3 parameters	
Scale	1:7760
Rotation	9 °
Mean position error	34m

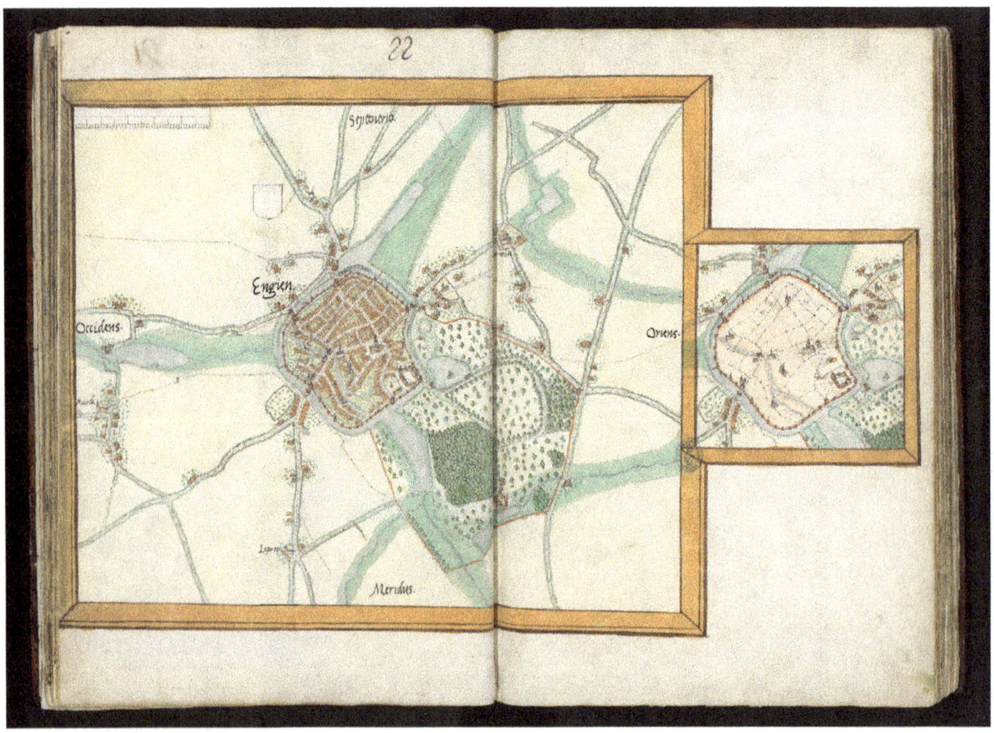

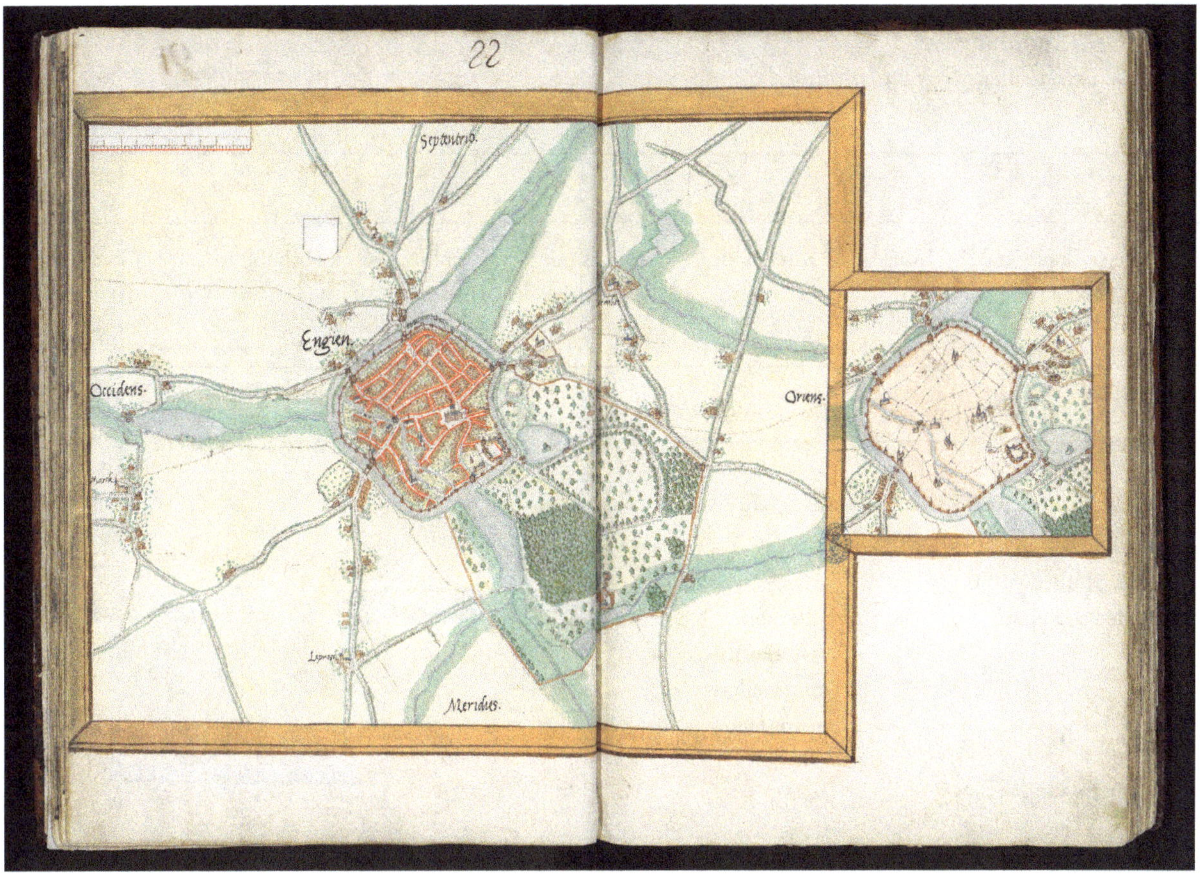
22
Septentrio.
Engien.
Occidens.
Oriens.
Meridies.

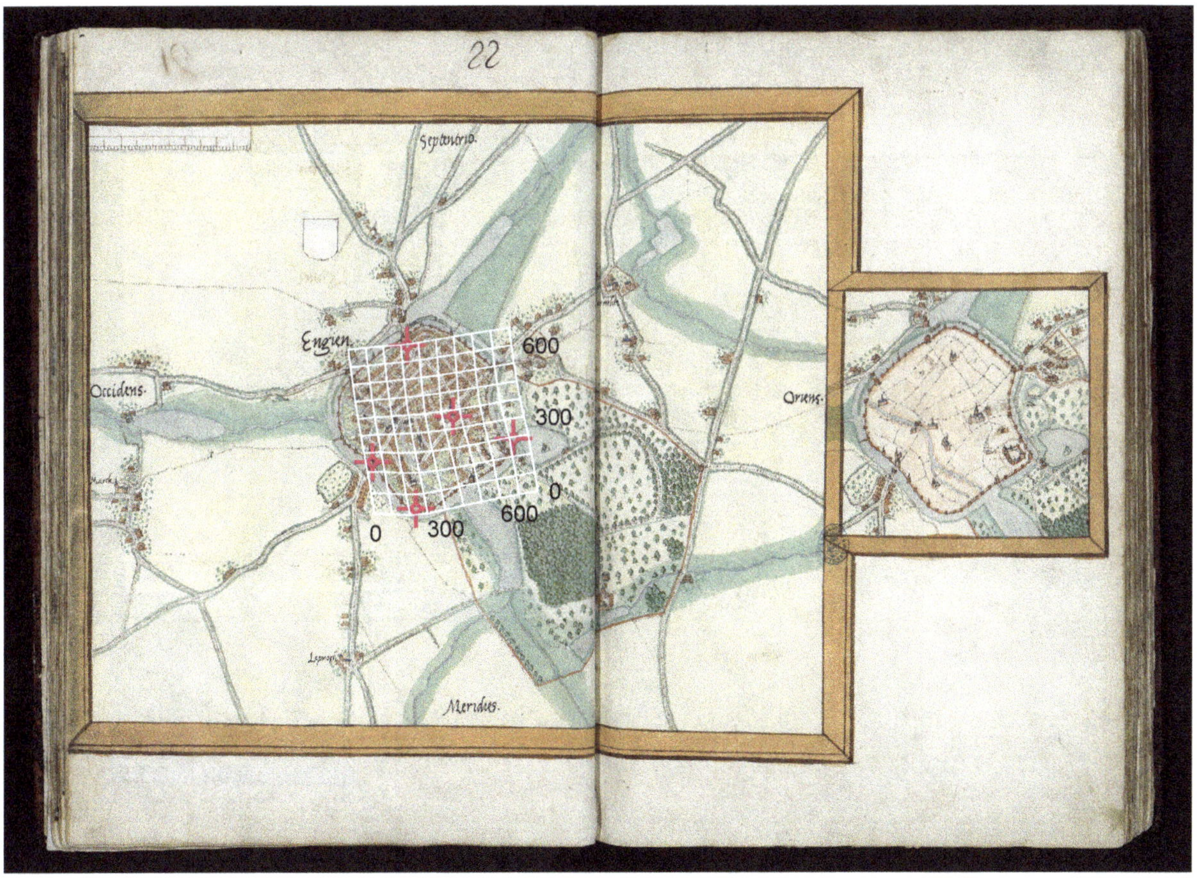
22
Septentrio.
Engien.
600
300
0
Occidens.
Oriens.
0
300
600
Meridies.

71. Enkhuizen

Tables 3.71 (measured and analysed).
Figures 3.71 (unmeasured, measured and analysed).

Measured							
Pixels	Scale stick original map (pixels)	Houses	Street sides	Houses per street side	Inhabitants (RF 4.0)	Inhabitants (RF 5.5)	Inhabitants (RF 7.0)
7165	1001	1258	217	5.80	5033	6920	8807

Analysed	
April 9, 2023 11:10:47 PM	
Computation with 5 linked points	
(*italics* points are less reliable):	1. Zuiderkerk
	2. Westerkerk
	3. Drommedaris
	4. Spaans leger
	5. Verlaat
Reference points (Reliability 0–5)	5
Transformation with 3 parameters	
Scale	1:8500
Rotation	8°
Mean position error	16m

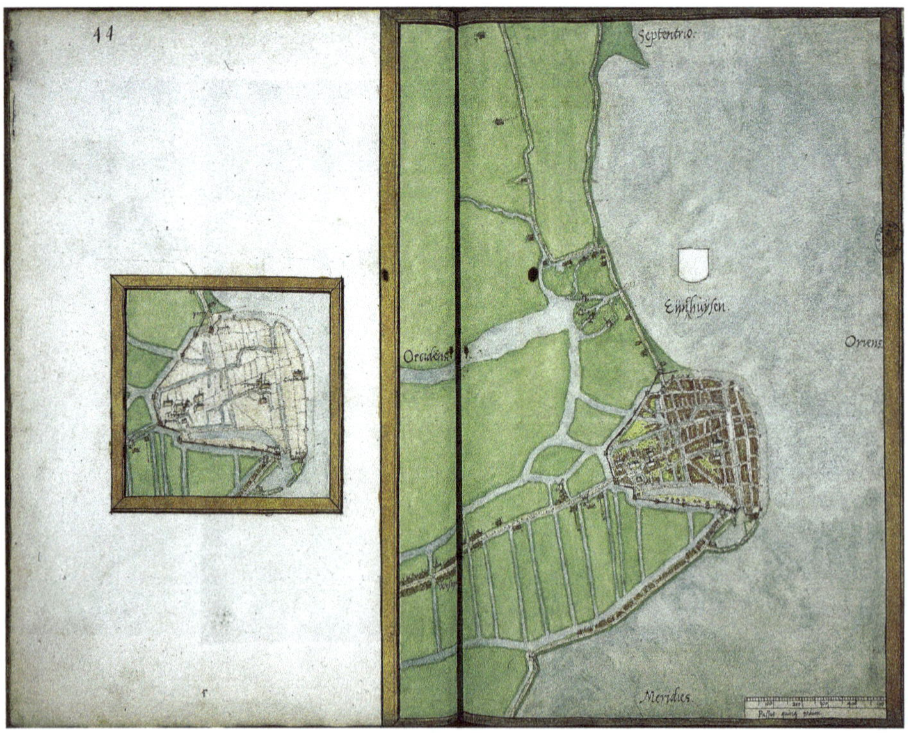

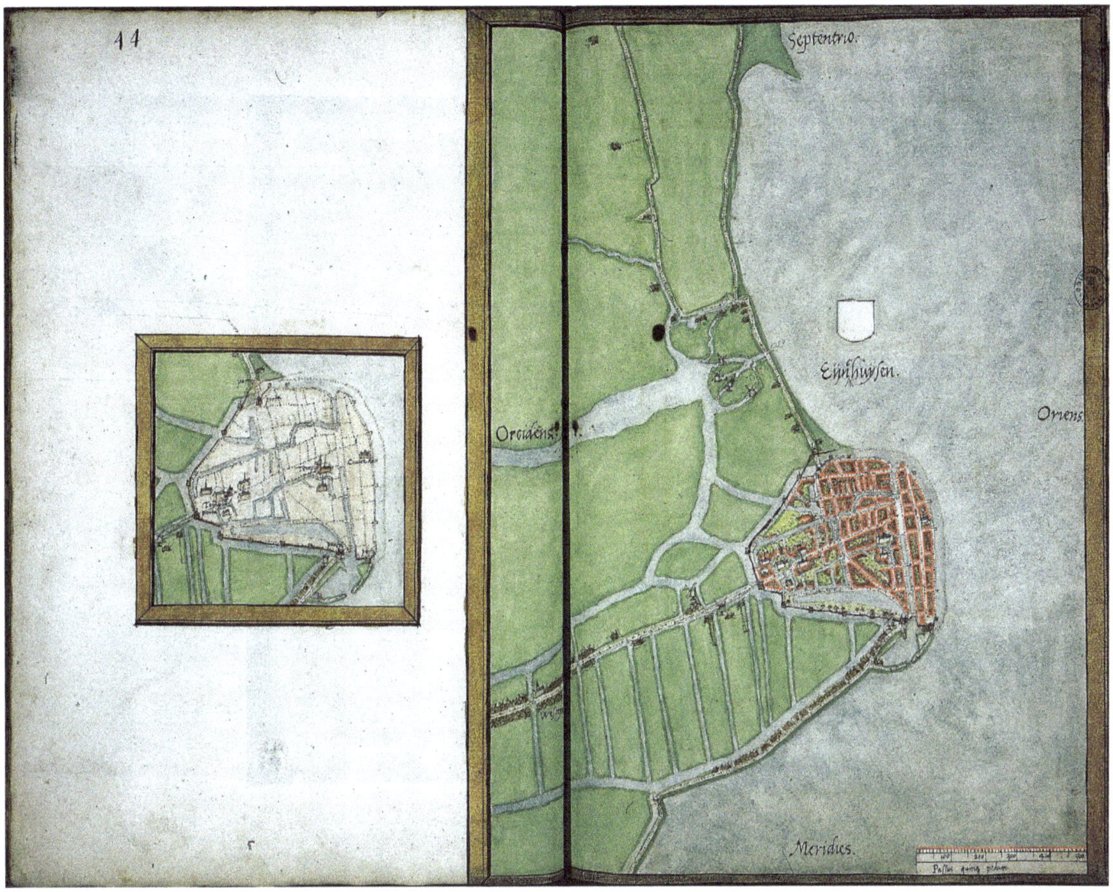
Septentrio.
Oriens
Meridies.

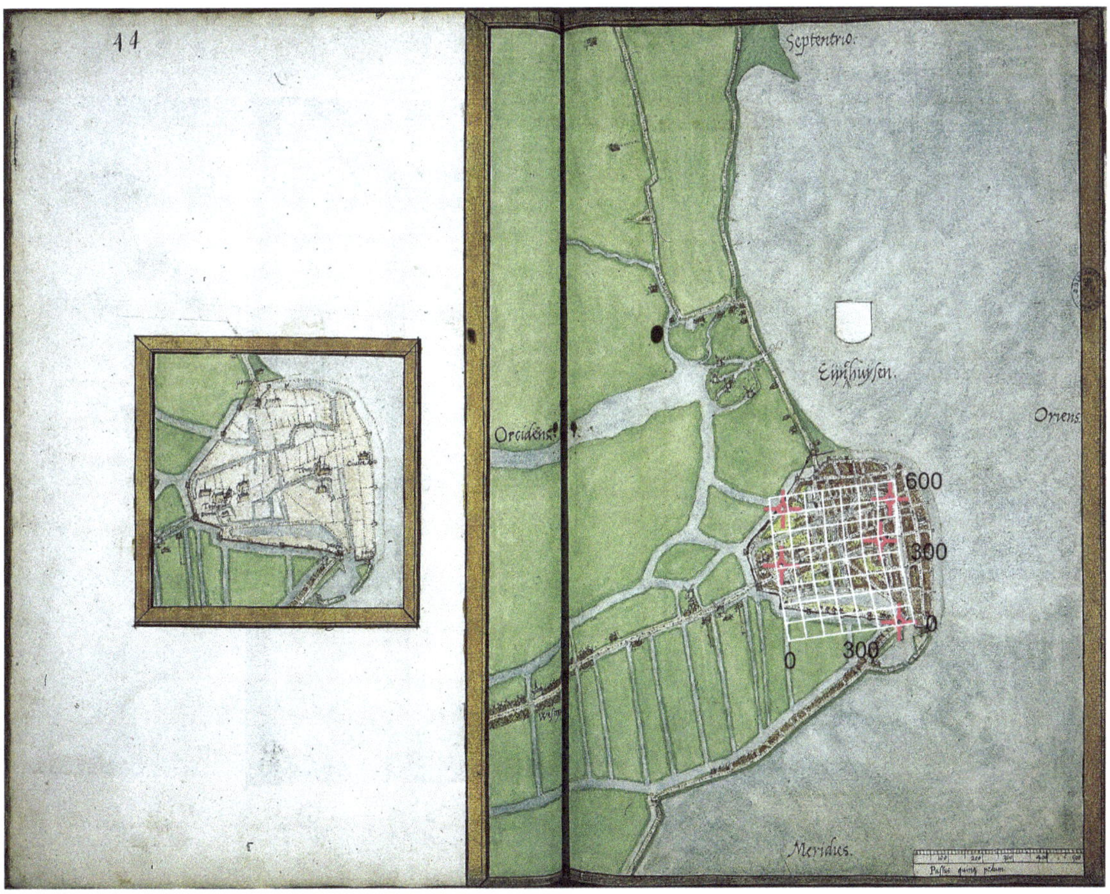
Septentrio.
Oriens
600
300
0
0
300
Meridies.

72. Enschede

Tables 3.72 (measured and analysed).
Figures 3.72 (unmeasured, measured and analysed).

Measured							
Pixels	Scale stick original map (pixels)	Houses	Street sides	Houses per street side	Inhabitants (RF 4.0)	Inhabitants (RF 5.5)	Inhabitants (RF 7.0)
1704	998	299	37	8.09	1197	1646	2095

Analysed	
April 11, 2023 5:00:32 PM	
Computation with 5 linked points	
(*italics* points are less reliable):	1. Grote Kerk
	2. Windbrugstraat
	3. Langestraat
	4. Martktstraat
	5. Haverstraatpassage
Reference points (Reliability 0–5)	5
Transformation with 3 parameters	
Scale	1:8300
Rotation	14 °
Mean position error	9m

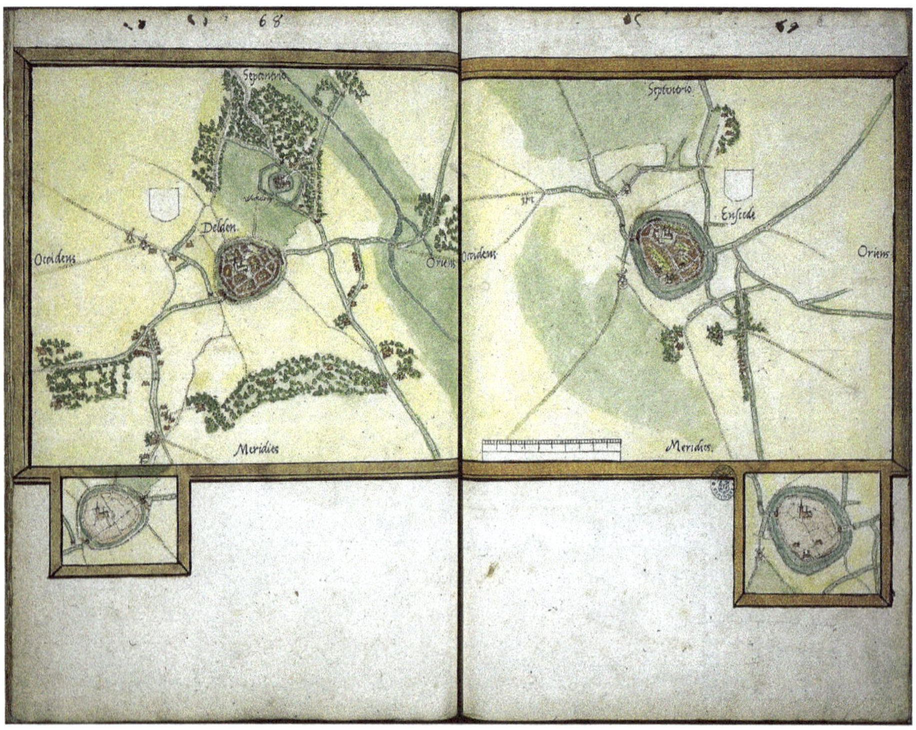

68
Septentrio
Delden
Occidens
Oriens
Meridies
Septentrio
Ensede
Occidens
Oriens
Meridies

68
Septentrio
Delden
Occidens
Oriens
Meridies
Septentrio
300
0
300
0
Occidens
Oriens
Meridies

73. Erkelenz

Tables 3.73 (measured and analysed).
Figures 3.73 (unmeasured, measured and analysed).

Measured

Pixels	Scale stick original map (pixels)	Houses	Street sides	Houses per street side	Inhabitants (RF 4.0)	Inhabitants (RF 5.5)	Inhabitants (RF 7.0)
2048	1009	360	47	7.65	1439	1978	2517

Analysed

April 15, 2023 10:24:43 AM	
Computation with 5 linked points	
(*italics* points are less reliable):	1. St. Lambertus
	2. Hexenturm
	3. Kölner Strasse
	4. Aachener Strasse
	5. Westpromenade
Reference points (Reliability 0–5)	5
Transformation with 3 parameters	
Scale	1:8370
Rotation	11 °
Mean position error	15m

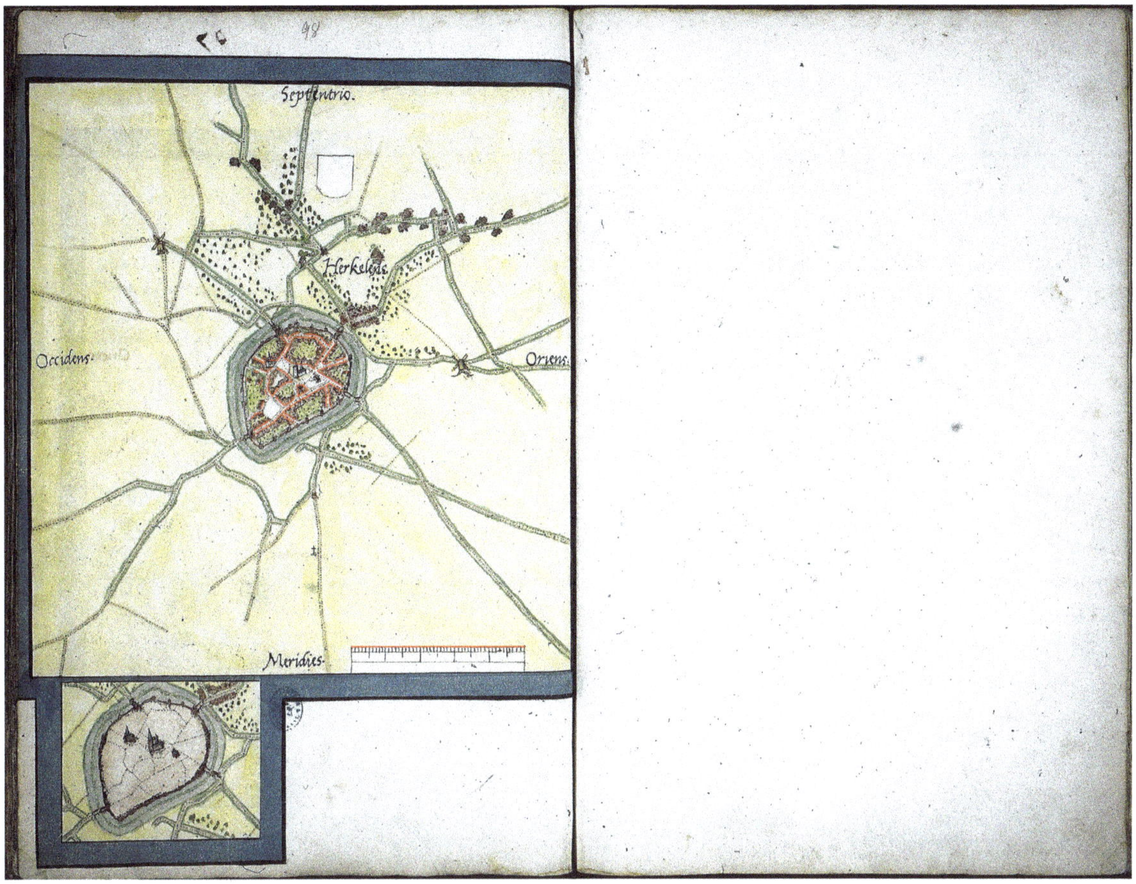
Septentrio.
Occidens.
Oriens.
Meridies.

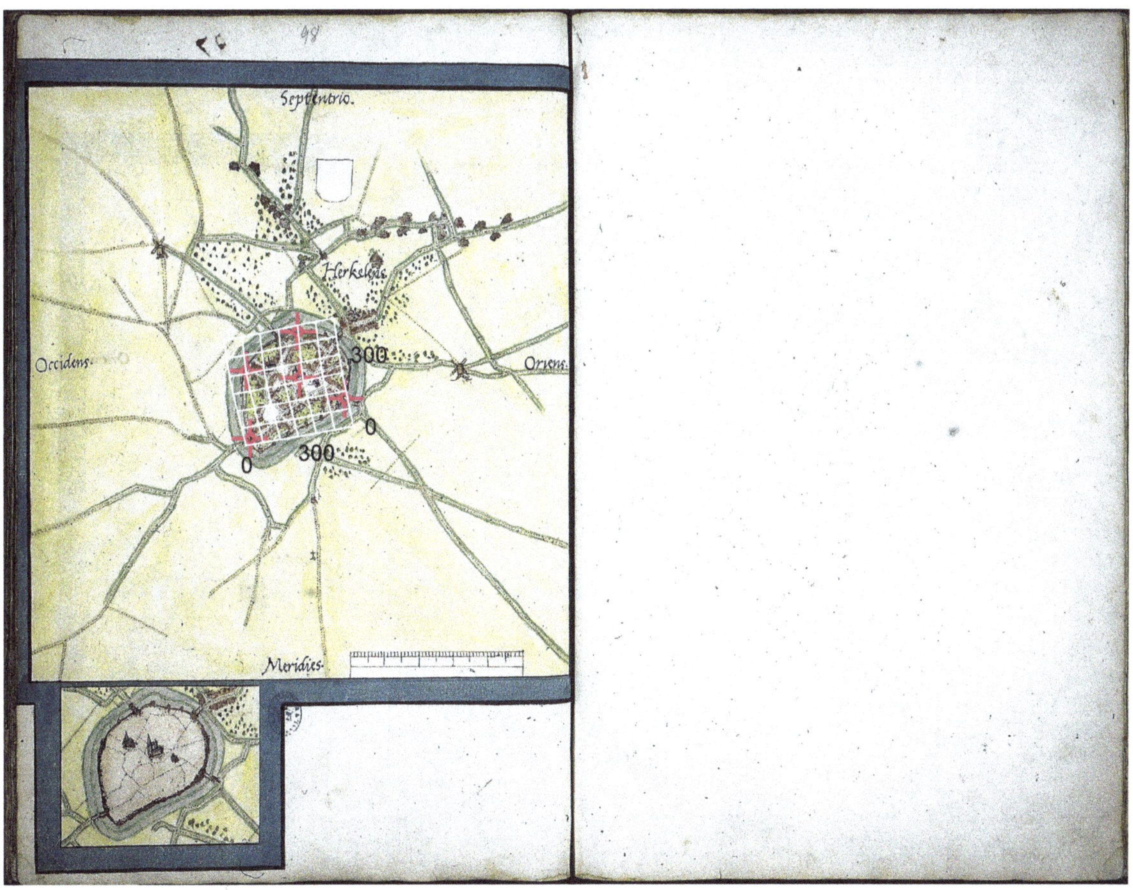
Septentrio.
Occidens.
Oriens.
300
0
0
300
Meridies.

74. Franeker

Tables 3.74 (measured and analysed).
Figures 3.74 (unmeasured, measured and analysed).

Measured							
Pixels	Scale stick original map (pixels)	Houses	Street sides	Houses per street side	Inhabitants (RF 4.0)	Inhabitants (RF 5.5)	Inhabitants (RF 7.0)
3331	1003	585	78	7.50	2340	3217	4094

Analysed	
April 15, 2023 12:36:45 PM	
Computation with 5 linked points	
(*italics* points are less reliable):	1. Martinikerk 2. Noord 3. *Westerbolwerk* 4. Godsacker 5. Noorderbolwerk
Reference points (Reliability 0–5)	4
Transformation with 3 parameters	
Scale	1:7970
Rotation	6 °
Mean position error	20m

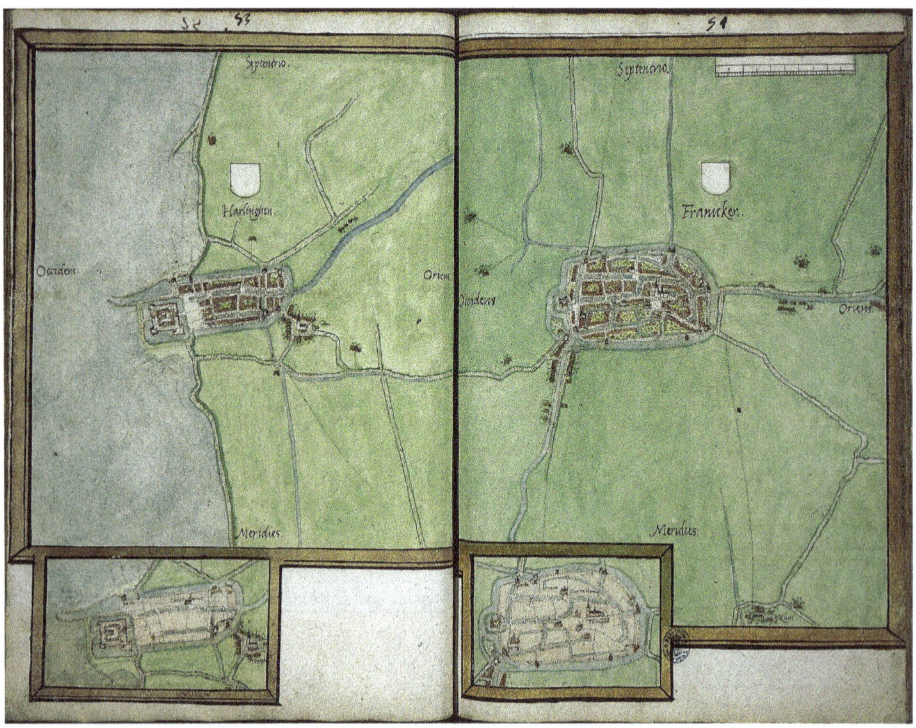

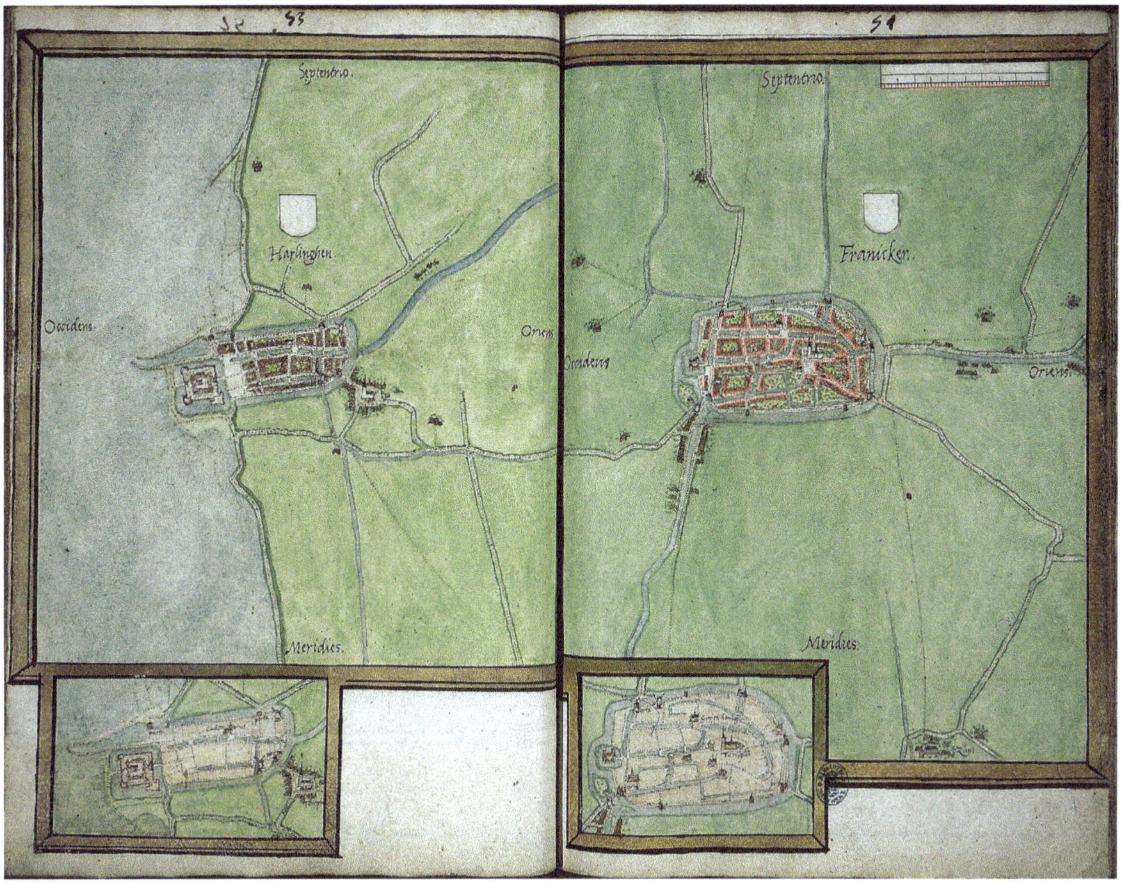
Septentrio.
Harlinghen.
Occidens.
Oriens.
Meridies.
Septentrio.
Franicker.
Occidens
Oriens.
Meridies.

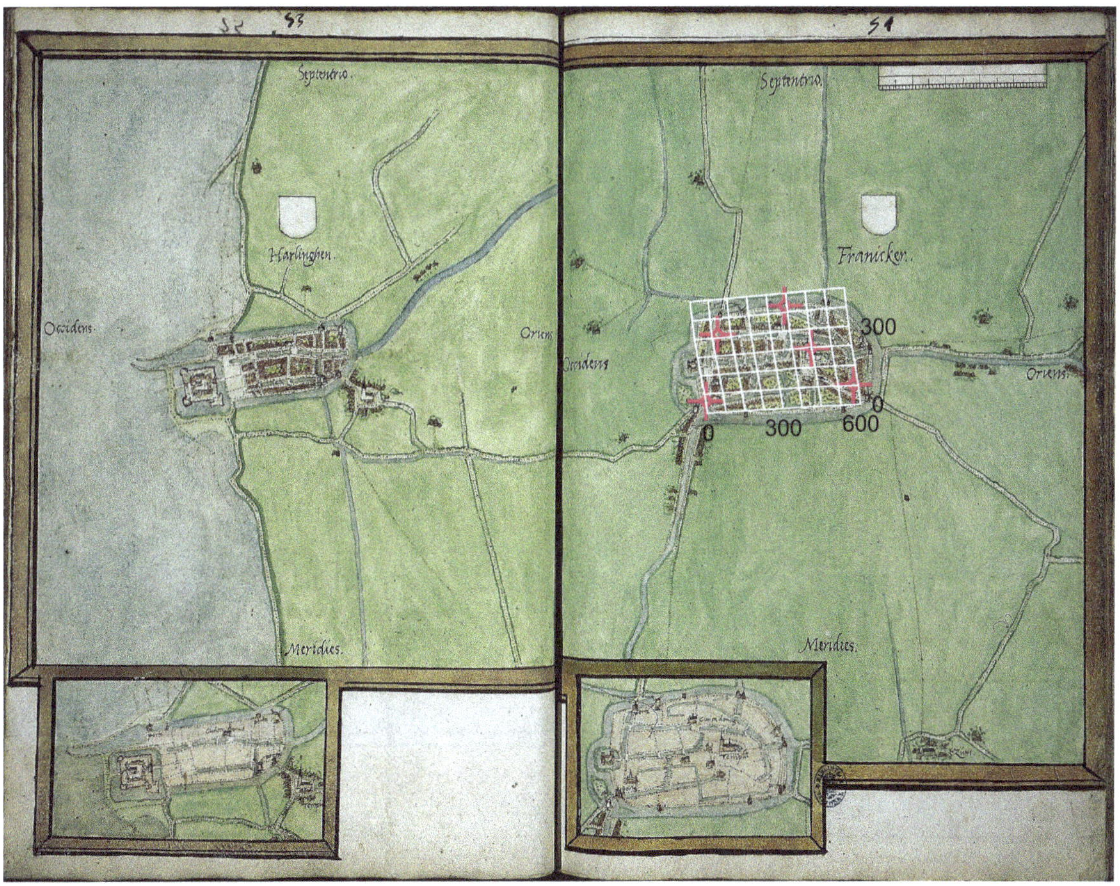
Septentrio.
Harlinghen.
Occidens.
Oriens.
Meridies.
Septentrio.
Franicker.
300
0
0
300
600
Occidens
Oriens.
Meridies.

75. Geertruidenberg

Tables 3.75 (measured and analysed).
Figures 3.75 (unmeasured, measured and analysed).

Measured

Pixels	Scale stick original map (pixels)	Houses	Street sides	Houses per street side	Inhabitants (RF 4.0)	Inhabitants (RF 5.5)	Inhabitants (RF 7.0)
2434	1007	427	49	8.72	1710	2351	2992

Analysed

April 15, 2023 12:54:36 PM	
Computation with 5 linked points	
(*italics* points are less reliable):	1. Geertruidskerk
	2. Raadhuis
	3. Koestraat
	4. Venestraat
	5. Haven
Reference points (Reliability 0–5)	5
Transformation with 3 parameters	
Scale	1:8630
Rotation	8 °
Mean position error	10m

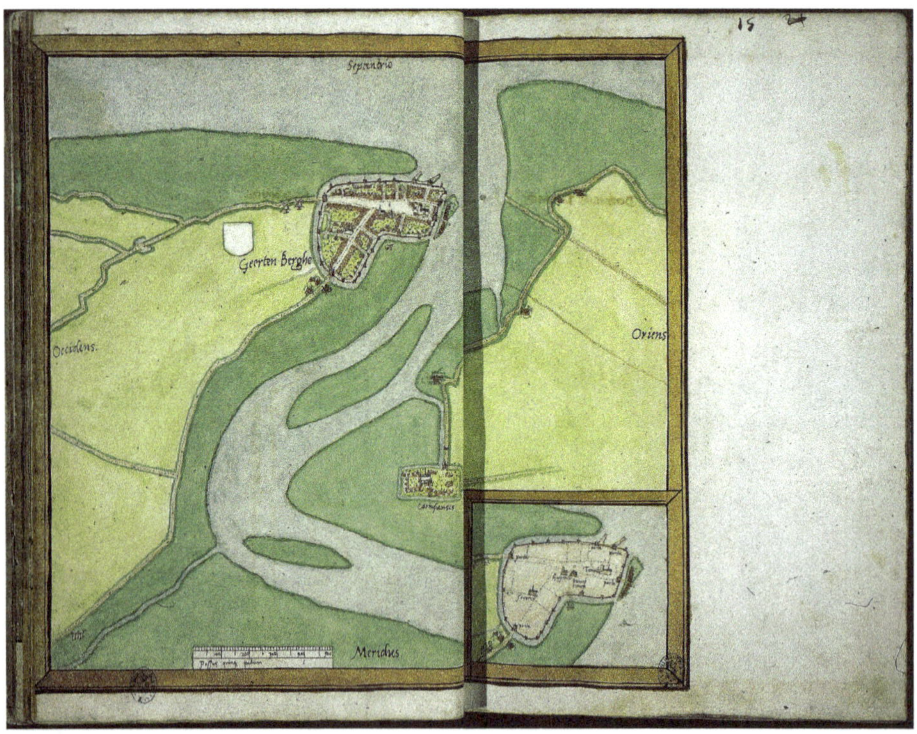

Septentrio
Geerten Berghe
Occidens.
Oriens

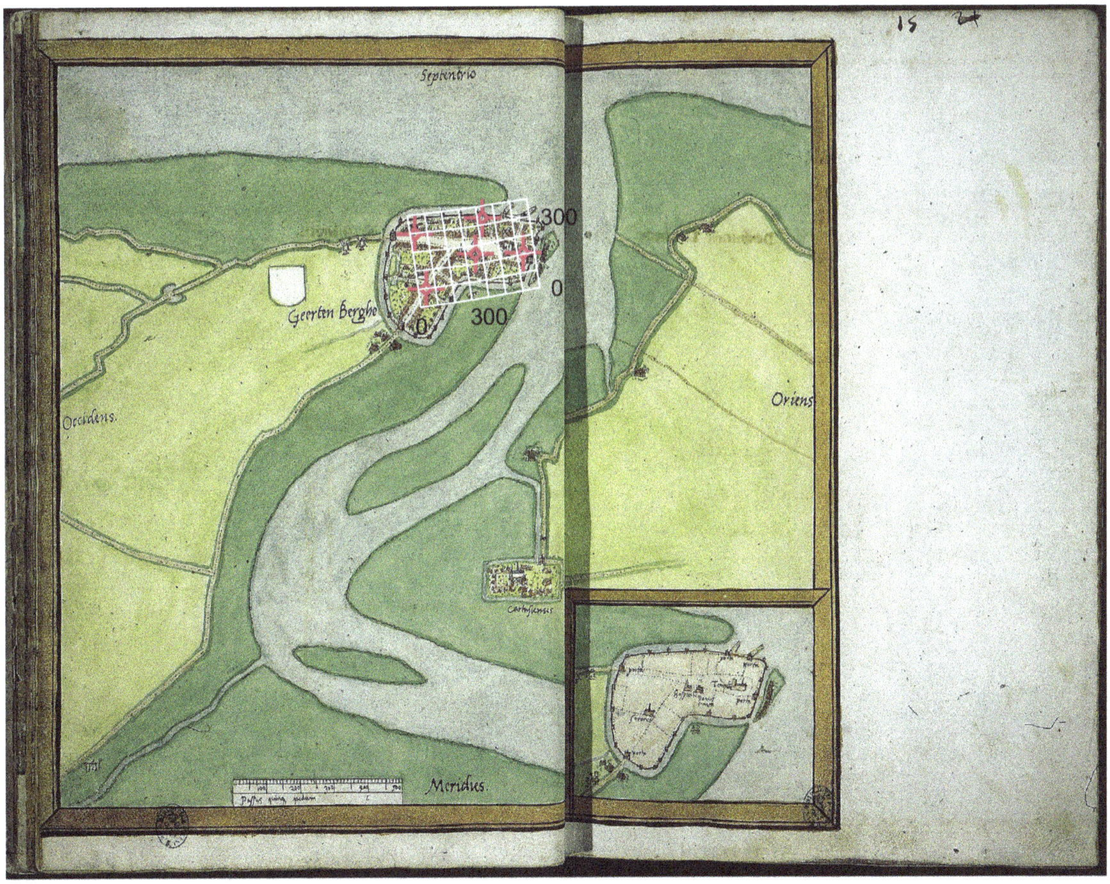
Septentrio
300
0
300
0
Geerten Berghe
Occidens.
Oriens

76. Geervliet

Tables 3.76 (measured and analysed).
Figures 3.76 (unmeasured, measured and analysed).

Measured							
Pixels	Scale stick original map (pixels)	Houses	Street sides	Houses per street side	Inhabitants (RF 4.0)	Inhabitants (RF 5.5)	Inhabitants (RF 7.0)
795	missing	140	41	3.40	559	768	977

Analysed	
April 17, 2023 9:45:21 AM	
Computation with 5 linked points	
(*italics* points are less reliable):	1. Onze Lieve Vrouw 2. Noorddijk 3. Bernisse 4. Oude Singel 5. Oude Singel
Reference points (Reliability 0–5)	5
Transformation with 3 parameters	
Scale	1:8540
Rotation	15 °
Mean position error	15m

13.
Septentrio.
Geervliet.
Occidens.
Meridies.

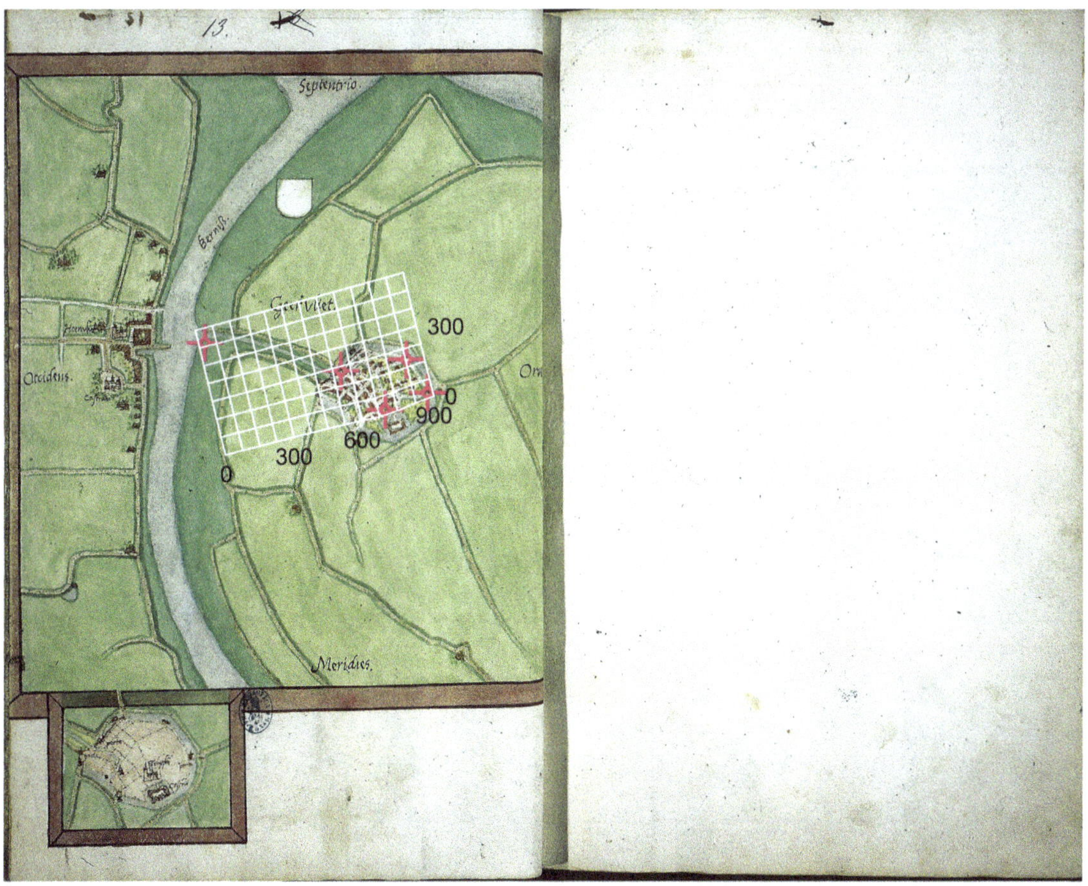
13.
Septentrio.
Geervliet.
Occidens.
Meridies.
300
0
900
0
300
600

77. Geldern

Tables 3.77 (measured and analysed).
Figures 3.77 (unmeasured, measured and analysed).

Measured							
Pixels	Scale stick original map (pixels)	Houses	Street sides	Houses per street side	Inhabitants (RF 4.0)	Inhabitants (RF 5.5)	Inhabitants (RF 7.0)
2813	missing	494	47	10.51	1976	2717	3458

Analysed	
April 17, 2023 8:15:11 PM	
Computation with 5 linked points	
(*italics* points are less reliable):	1. Sankt Maria Magdalena 2. Geldertor 3. Issumertor 4. Westwall 5. Nordwall
Reference points (Reliability 0–5)	5
Transformation with 3 parameters	
Scale	1:8520
Rotation	11 °
Mean position error	9m

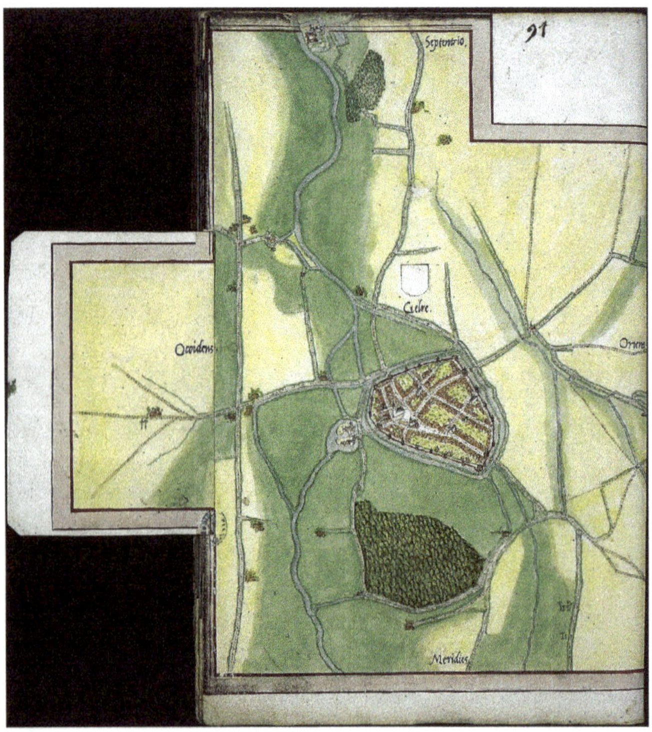

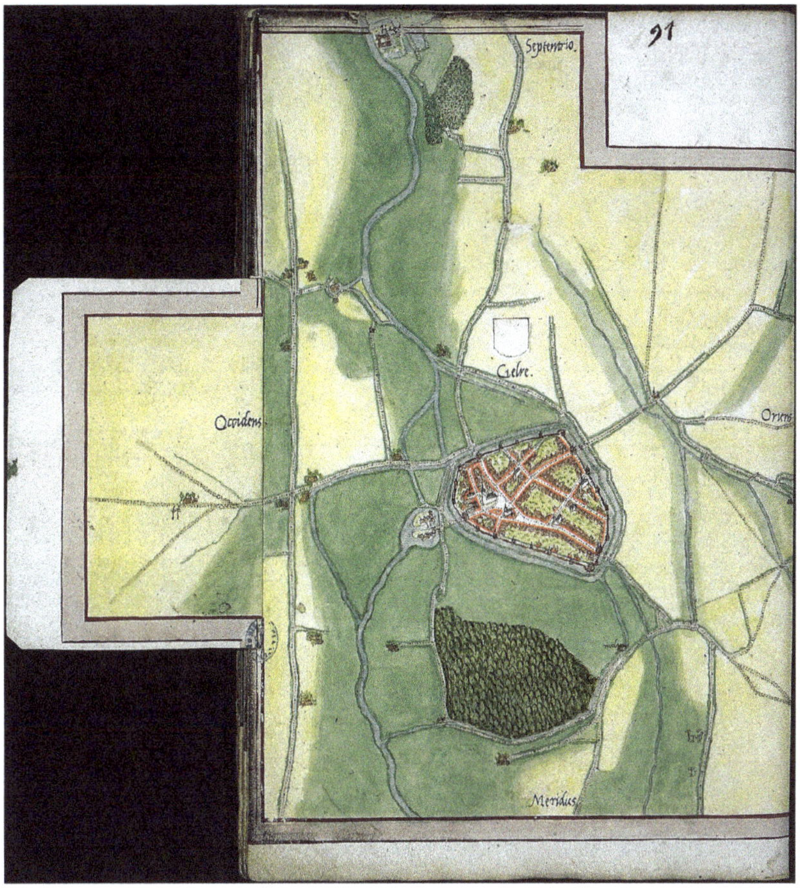
91
Septentrio.
Gelre.
Occidens
Meridies

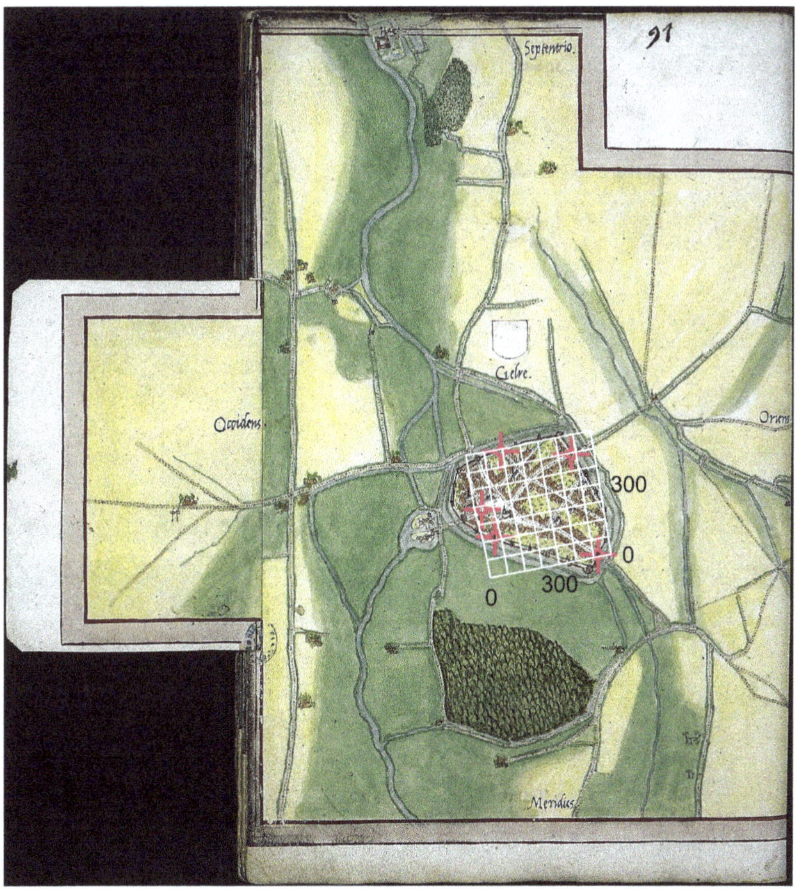
91
Septentrio.
Gelre.
Occidens
300
0
0
300
Meridies

78. Gembloux

Tables 3.78 (measured and analysed).
Figures 3.78 (unmeasured, measured and analysed).

Measured

Pixels	Scale stick original map (pixels)	Houses	Street sides	Houses per street side	Inhabitants (RF 4.0)	Inhabitants (RF 5.5)	Inhabitants (RF 7.0)
1815	missing	319	65	4.90	1275	1753	2231

Analysed

September 20, 2023 3:03:56 PM	
Computation with 5 linked points	
(*italics* points are less reliable):	1. Beffroi du Gembloux
	2. Grand Rue
	3. Rue Gustave Docq
	4. Rue Sigebert
	5. Rue Albert
Reference points (Reliability 0–5)	5
Transformation with 3 parameters	
Scale	1:8170
Rotation	6 °
Mean position error	15m

Noordt.
Gemblacum.
Gemblours.
West.
Oost.
Zuijdt.

Noordt.
Gemblacum.
Gemblours.
West.
Oost.
Zuijdt.
300
0
300
0

79. Genemuiden

Tables 3.79 (measured and analysed).
Figures 3.79 (unmeasured, measured and analysed).

Measured							
Pixels	Scale stick original map (pixels)	Houses	Street sides	Houses per street side	Inhabitants (RF 4.0)	Inhabitants (RF 5.5)	Inhabitants (RF 7.0)
1544	993	271	33	8.22	1084	1491	1898

Analysed	
April 18, 2023 9:47:22 AM	
Computation with 5 linked points	
(*italics* points are less reliable):	*1. Jaagpad*
	2. Blokhuis
	3. Zwarte Water
	4. Langestraat
	5. Langestraat
Reference points (Reliability 0–5)	1
Transformation with 3 parameters	
Scale	1:8270
Rotation	12 °
Mean position error	42m

76
Septentrio
Occidens
Geelmuyden
Oriens
Meridies
Septentrio
77
Vollenhove
Oriens
Meridies

76
Septentrio
1.200
900
600
300
0
0
300
Occidens
Geelmuyden
Oriens
Meridies
Septentrio
77
Vollenhove
Oriens
Meridies

80. Gent

Tables 3.80 (measured and analysed).
Figures 3.80 (unmeasured, measured and analysed).

Measured

Pixels	Scale stick original map (pixels)	Houses	Street sides	Houses per street side	Inhabitants (RF 4.0)	Inhabitants (RF 5.5)	Inhabitants (RF 7.0)
56,662	1359	9950	1286	7.74	39,799	54,724	69,649

Analysed

January 23, 2023 6:21:30 PM	
Computation with 5 linked points	
(*italics* points are less reliable):	1. Leie-bocht
	2. Sint-Baafsplein
	3. Kuipgat
	4. Lievekaai
	5. De Krook
Reference points (Reliability 0–5)	5
Transformation with 3 parameters	
Scale	1:8040
Rotation	9 °
Mean position error	34m

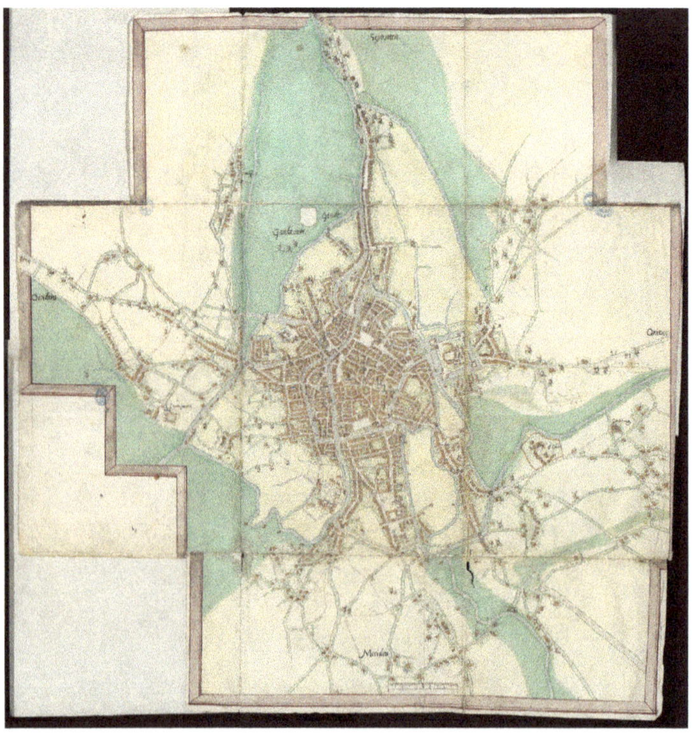

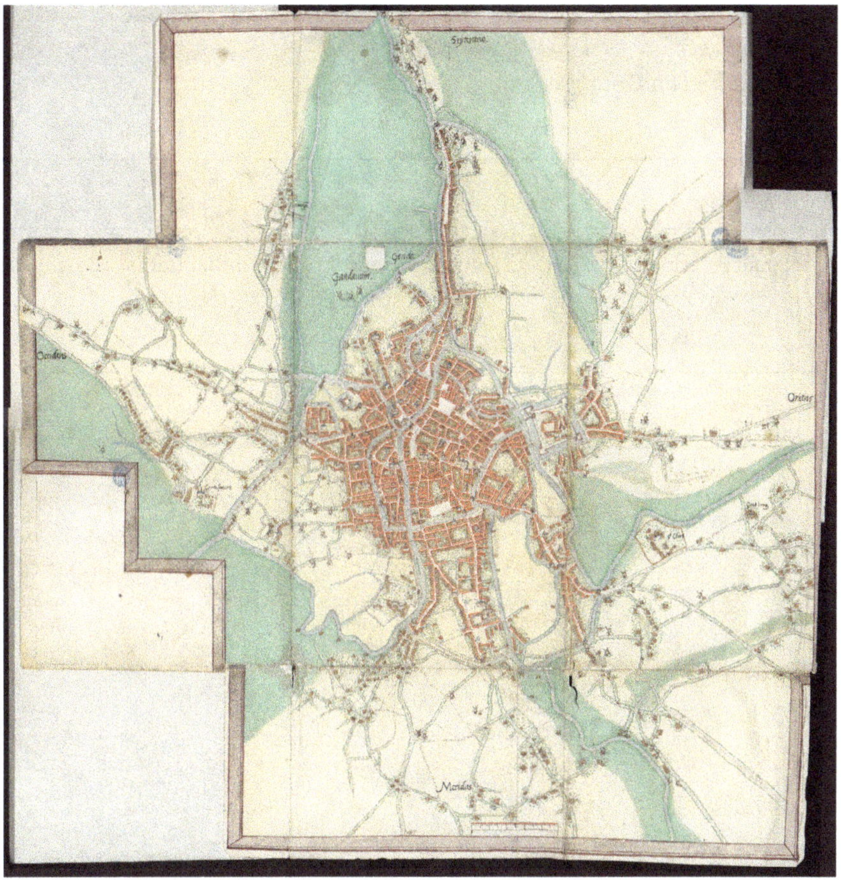

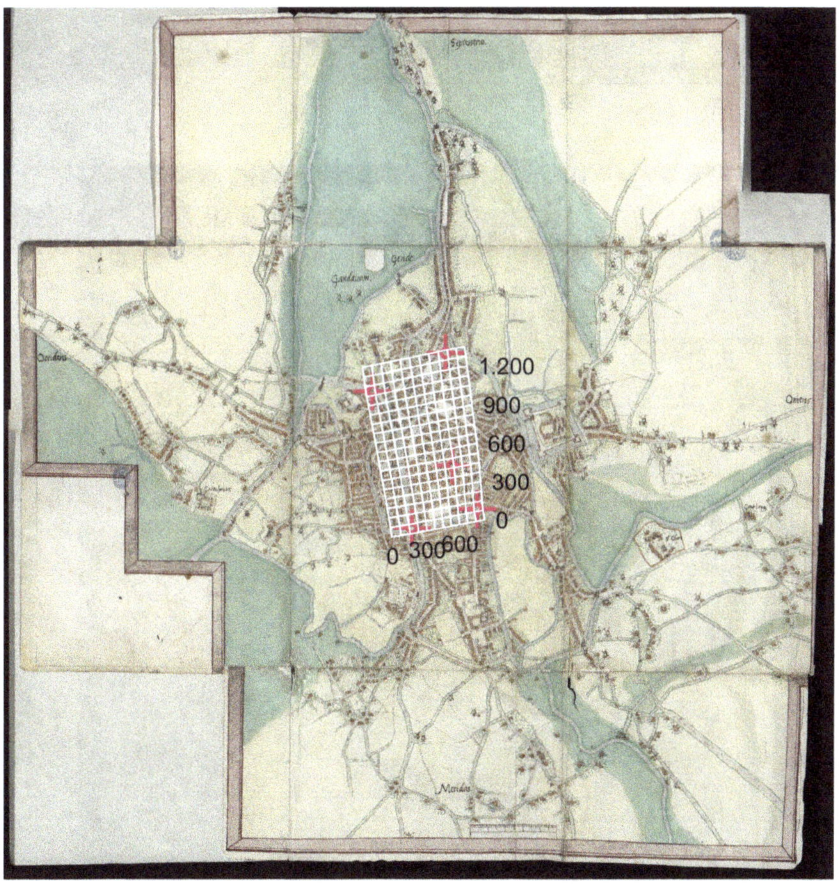
1.200
900
600
300
0
0 300 600

81. Geraardsbergen

Tables 3.81 (measured and analysed).
Figures 3.81 (unmeasured, measured and analysed).

Measured							
Pixels	Scale stick original map (pixels)	Houses	Street sides	Houses per street side	Inhabitants (RF 4.0)	Inhabitants (RF 5.5)	Inhabitants (RF 7.0)
6703	1023	1177	161	7.31	4708	6474	8240

Analysed	
April 19, 2023 5:41:51 PM	
Computation with 5 linked points	
(*italics* points are less reliable):	*1. Markt*
	2. Abdijpoort
	3. Kapelmuur
	4. Karmelieten
	5. Dender
Reference points (Reliability 0–5)	4
Transformation with 3 parameters	
Scale	1:7740
Rotation	8 °
Mean position error	25m

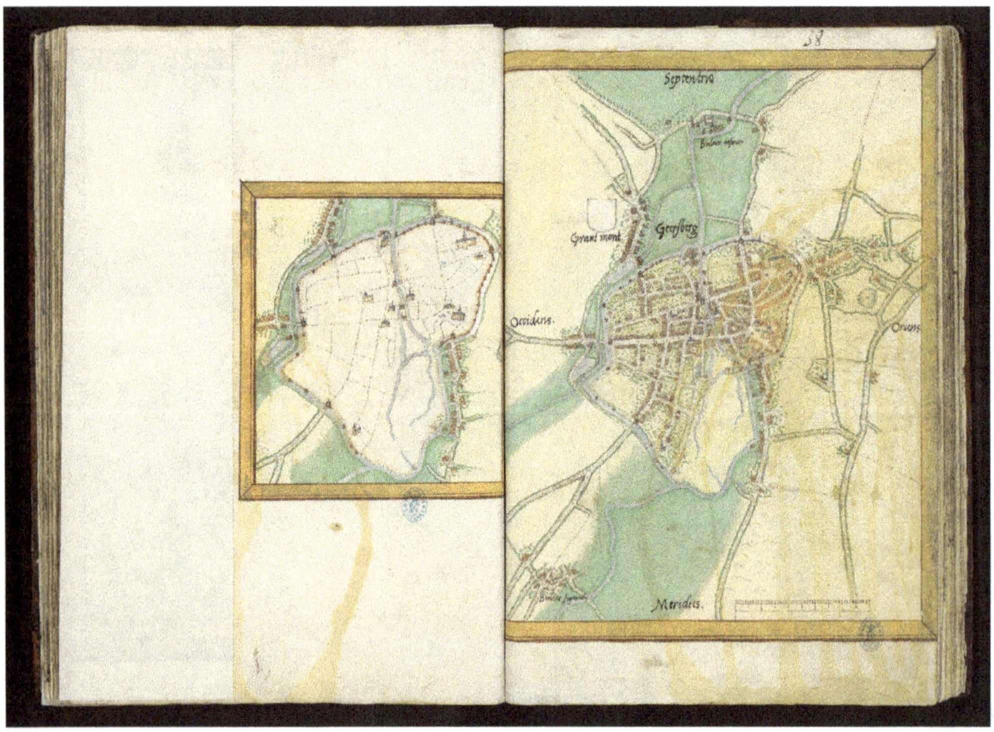

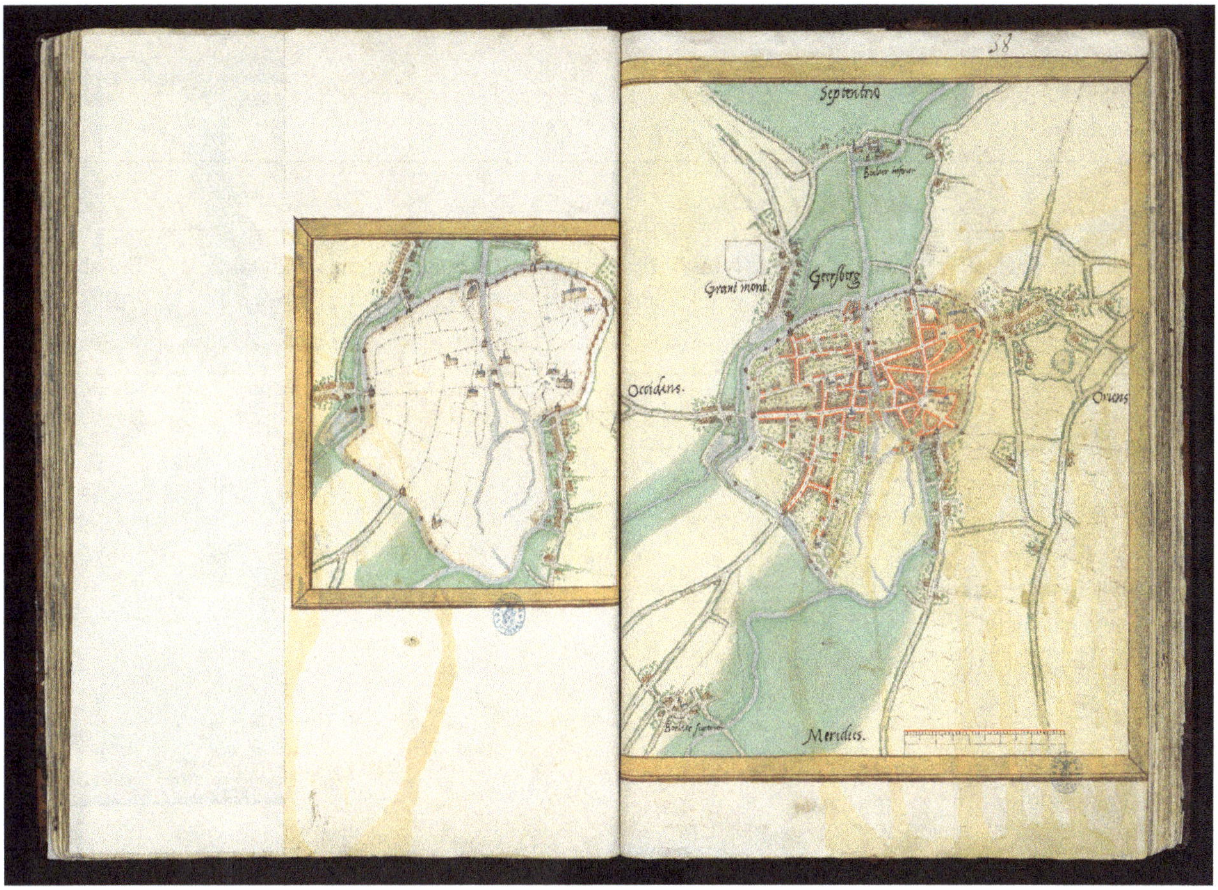
Septentrio
Occidens.
Meridies.

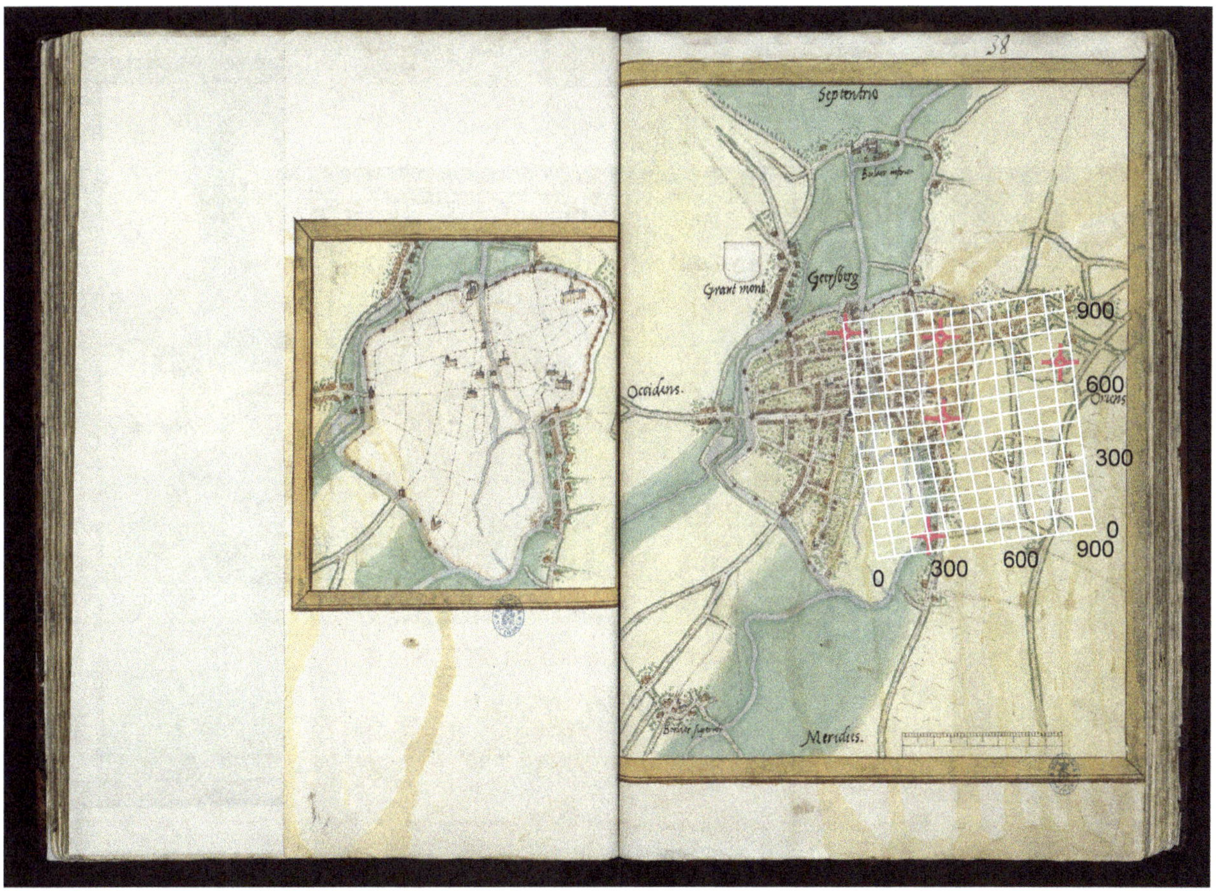
Septentrio
Occidens.
Meridies.
900
600
300
0
0
300
600
900

82. Gistel

Tables 3.82 (measured and analysed).
Figures 3.82 (unmeasured, measured and analysed).

Measured							
Pixels	Scale stick original map (pixels)	Houses	Street sides	Houses per street side	Inhabitants (RF 4.0)	Inhabitants (RF 5.5)	Inhabitants (RF 7.0)
2058	1021	361	53	6.82	1446	1988	2530

Analysed	
June 30, 2023 5:31:24 PM	
Computation with 5 linked points	
(*italics* points are less reliable):	1. Onze-Lieve-Vrouwekerk 2. Bruidstraat 3. Vaartsraat 4. Hoogstraat 5. Stationsstraat
Reference points (Reliability 0–5)	5
Transformation with 3 parameters	
Scale	1:8050
Rotation	9 °
Mean position error	18m

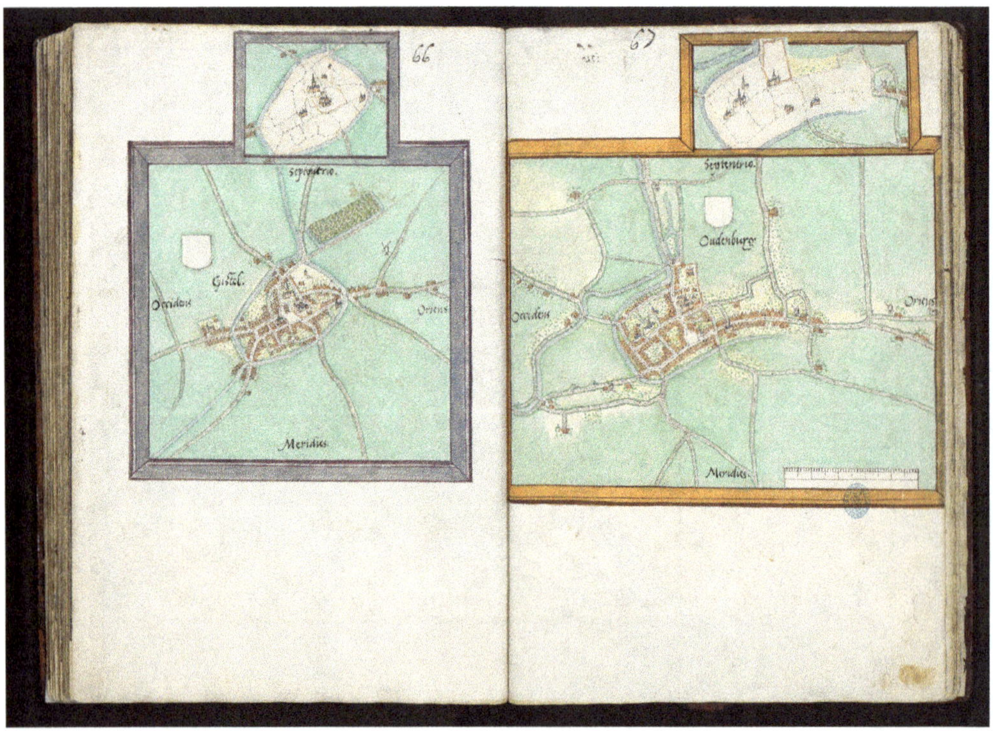

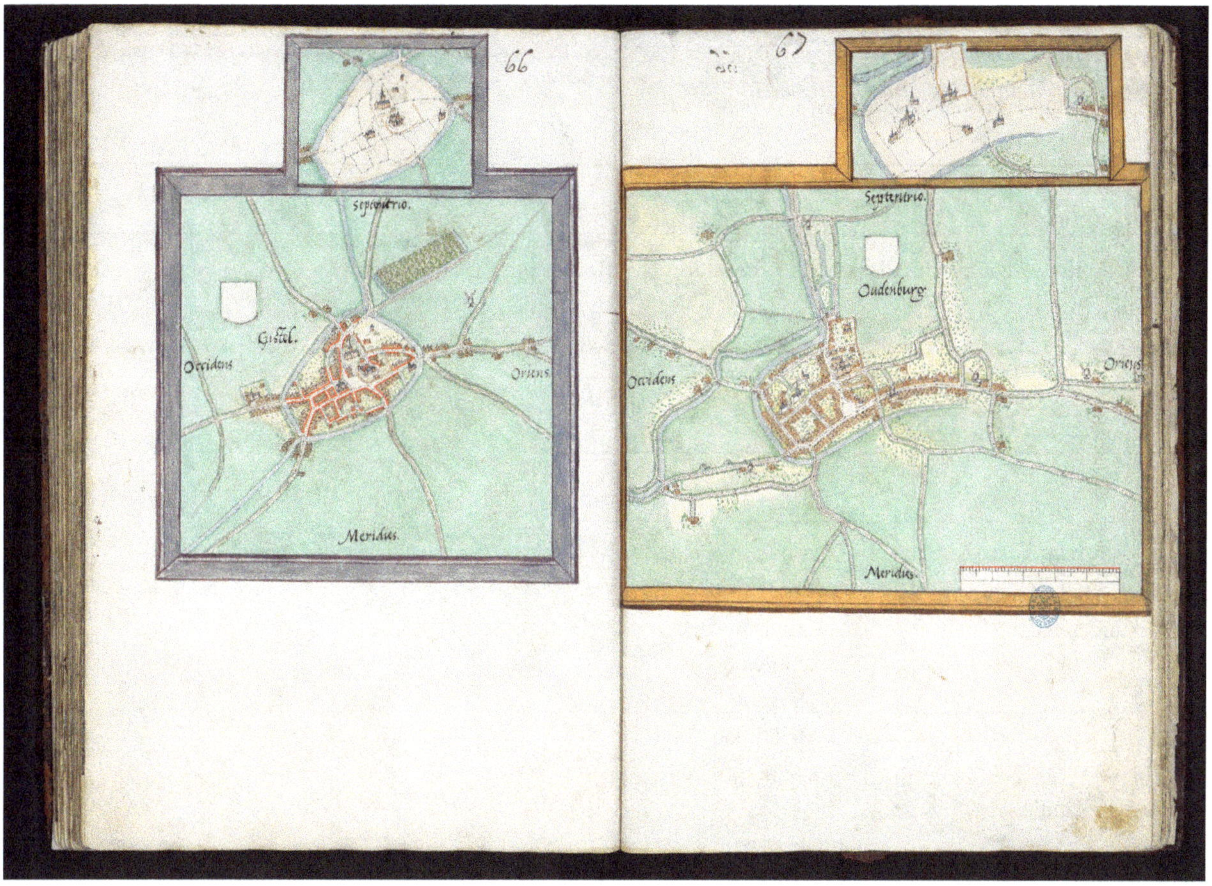
66
67
Septentrio.
Gistel.
Occidens
Oriens
Meridies.
Oudenburg.

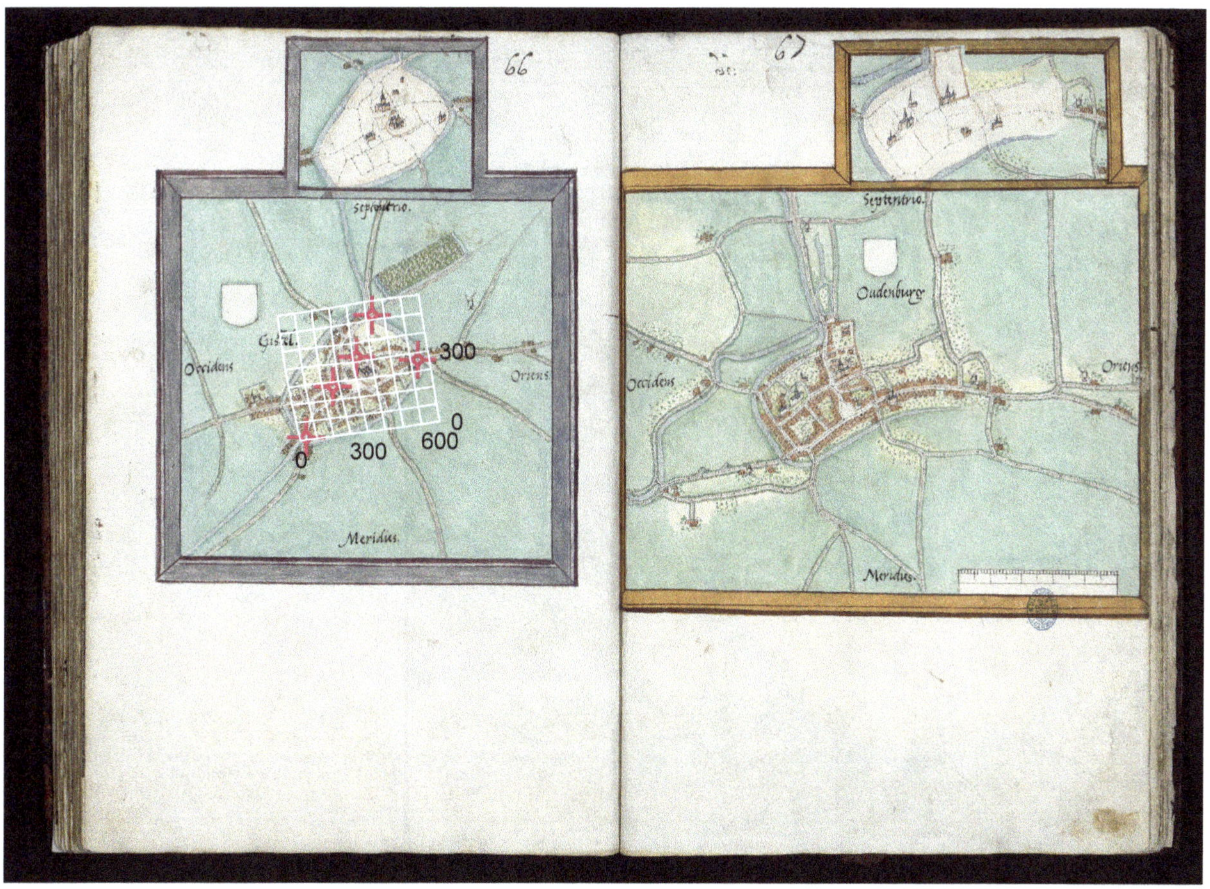
66
67
Septentrio.
Gistel.
Occidens
Oriens
Meridies.
Oudenburg.
300
0
600
0
300
0

83. Goedereede

Tables 3.83 (measured and analysed).
Figures 3.83 (unmeasured, measured and analysed).

Measured							
Pixels	Scale stick original map (pixels)	Houses	Street sides	Houses per street side	Inhabitants (RF 4.0)	Inhabitants (RF 5.5)	Inhabitants (RF 7.0)
1373	1001	241	47	5.13	964	1326	1688

Analysed	
April 20, 2023 11:21:22 AM	
Computation with 5 linked points	
(*italics* points are less reliable):	1. Catharinakerk 2. Windvang 3. Markt 4. Melkdijkje 5. Doelweg
Reference points (Reliability 0–5)	5
Transformation with 3 parameters	
Scale	1:7970
Rotation	13 °
Mean position error	11m

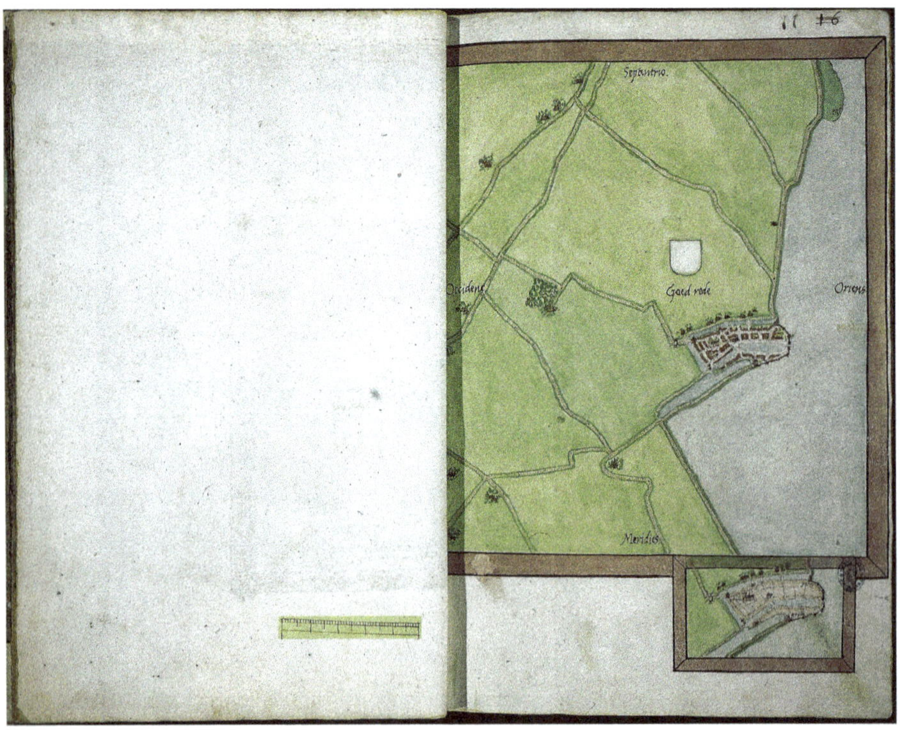

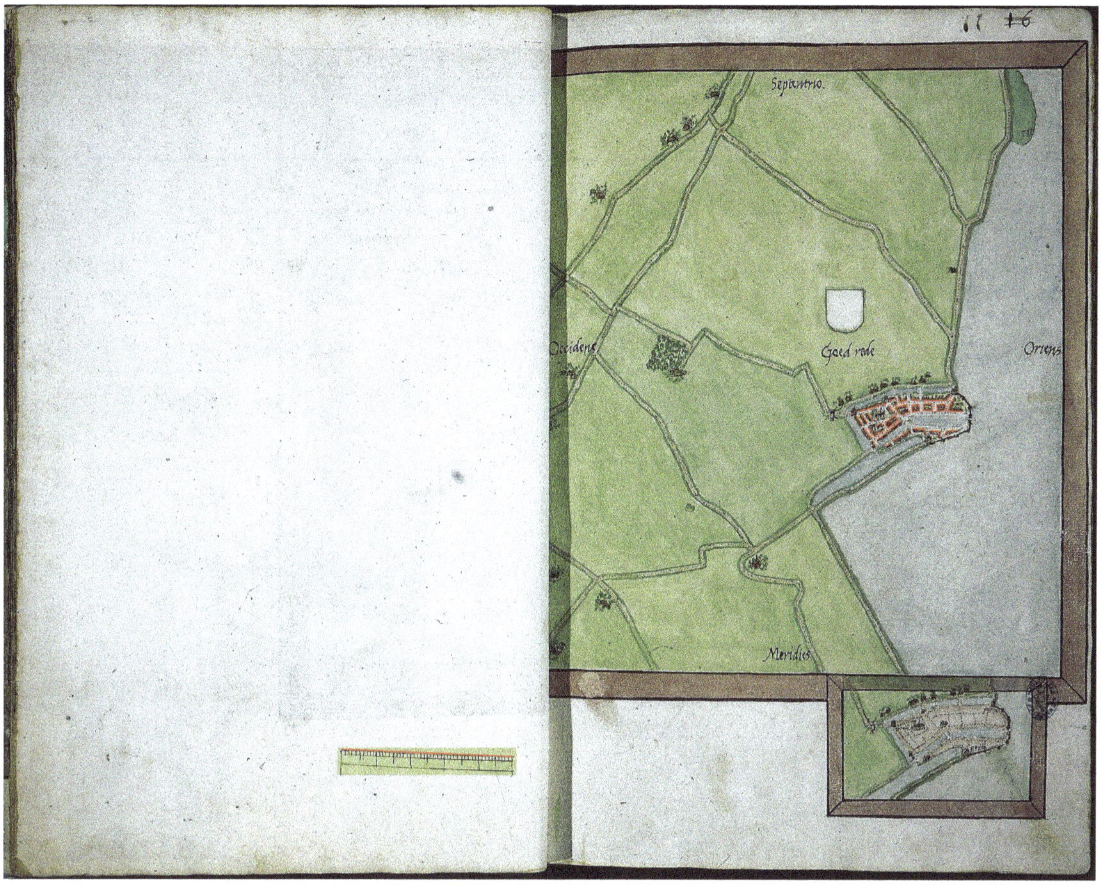
Septentrio.
Occidens
Oriens
Meridies

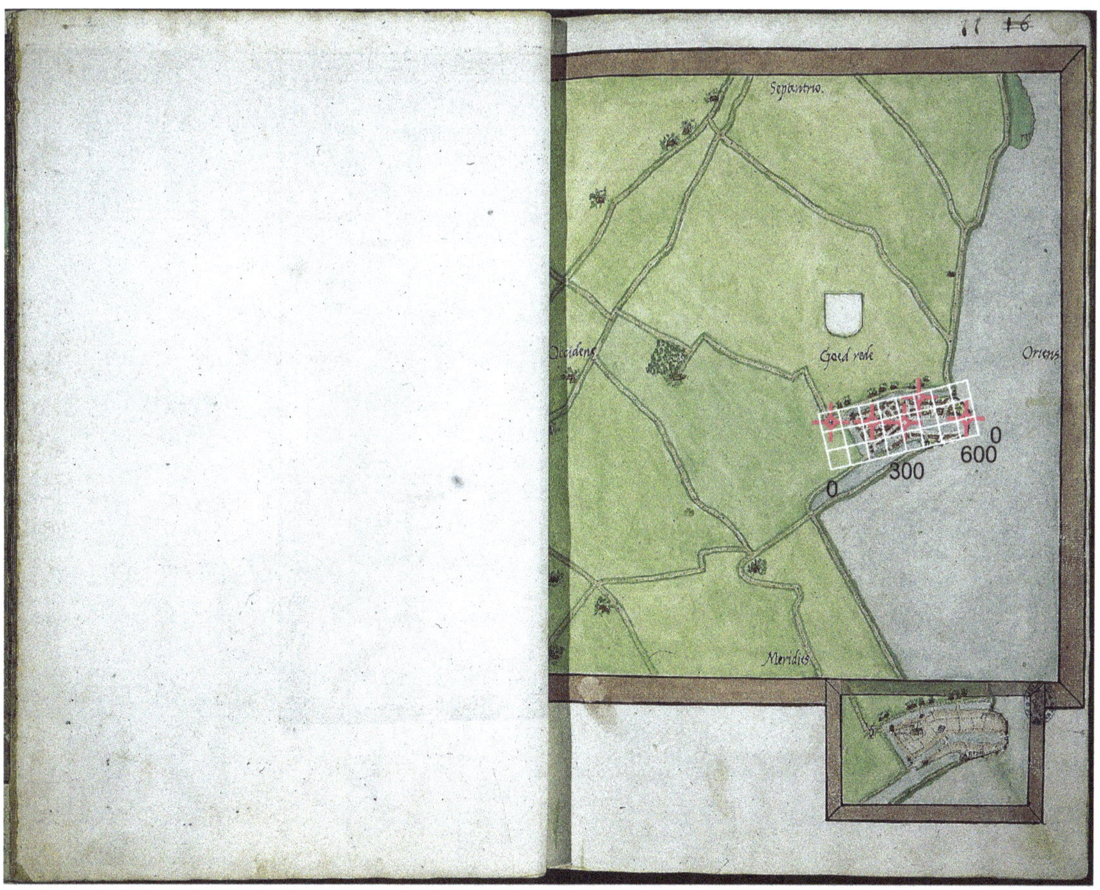
Septentrio.
Occidens
Oriens
Meridies
0
300
600
0

84. Goes

Tables 3.84 (measured and analysed).
Figures 3.84 (unmeasured, measured and analysed).

Measured							
Pixels	Scale stick original map (pixels)	Houses	Street sides	Houses per street side	Inhabitants (RF 4.0)	Inhabitants (RF 5.5)	Inhabitants (RF 7.0)
3188	995	560	59	9.49	2239	3079	3919

Analysed	
January 23, 2023 5:04:07 PM	
Computation with 5 linked points (*italics* points are less reliable):	1. Grote Kerk 2. Grote markt 3. Bierkade 4. Oostwal *5. Ravelijn de Grenadier*
Reference points (Reliability 0–5)	4
Transformation with 3 parameters	
Scale	1:9350
Rotation	15 °
Mean position error	26m

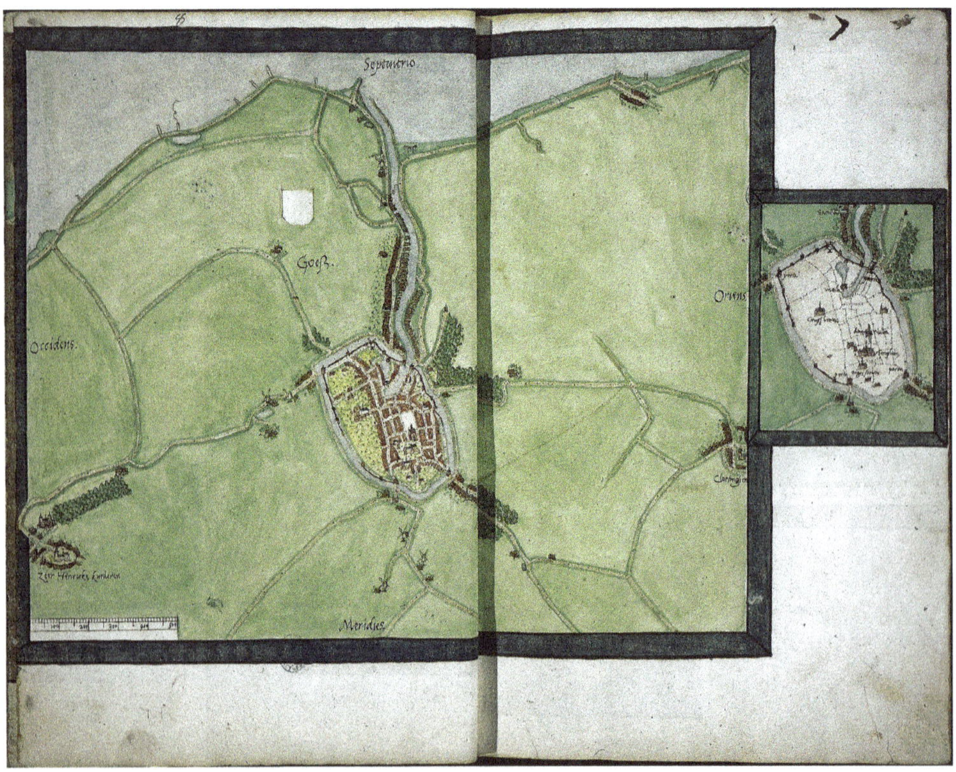

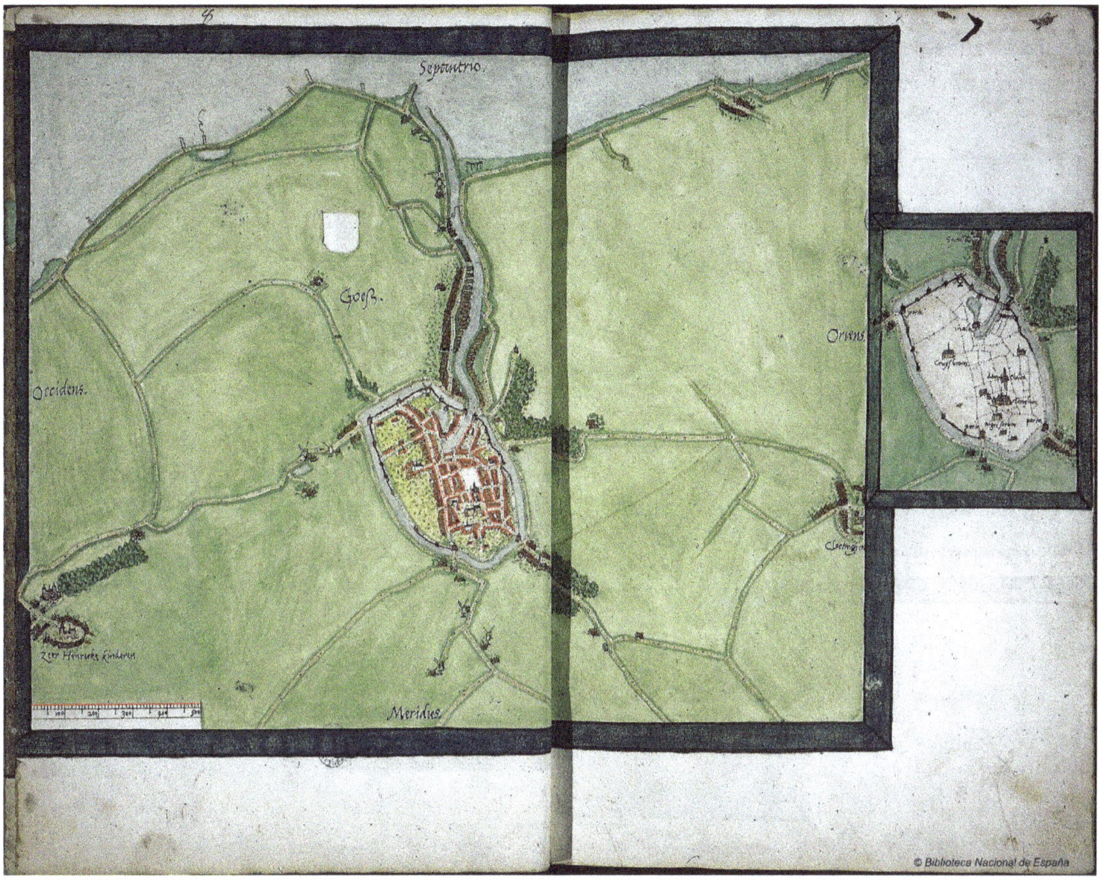
Septentrio.
Goeß.
Occidens.
Oriens
Meridies.
© Biblioteca Nacional de España

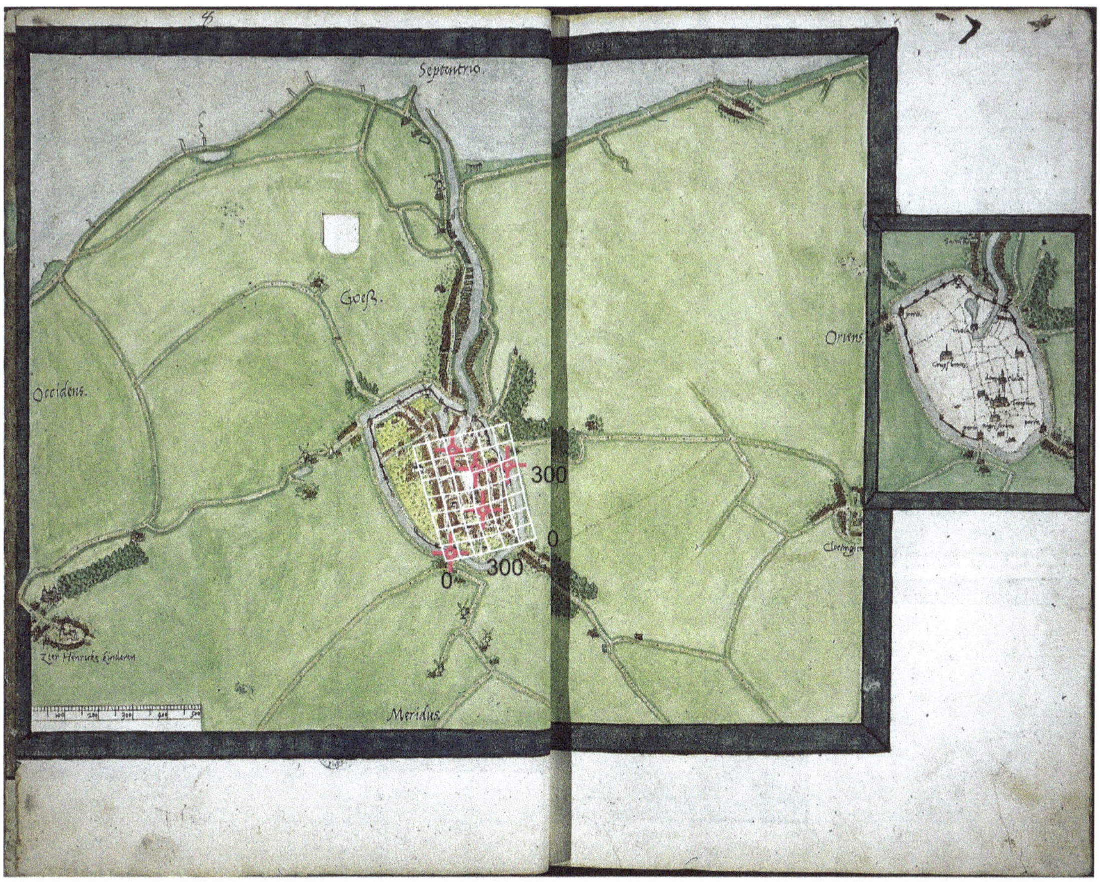
Septentrio.
Goeß.
Occidens.
Oriens
Meridies.
300
0
300
0

85. Goor

Tables 3.85 (measured and analysed).
Figures 3.85 (unmeasured, measured and analysed).

Measured							
Pixels	Scale stick original map (pixels)	Houses	Street sides	Houses per street side	Inhabitants (RF 4.0)	Inhabitants (RF 5.5)	Inhabitants (RF 7.0)
2042	950	359	37	9.69	1434	1972	2510

Analysed	
April 18, 2023 9:47:22 AM	
Computation with 5 linked points (*italics* points are less reliable):	1. Hofkerk 2. Kerkstraat 3. Voorstraat 4. Grotestraat 5. Bleekstraat
Reference points (Reliability 0–5)	5
Transformation with 3 parameters	
Scale	1:7830
Rotation	13 °
Mean position error	16m

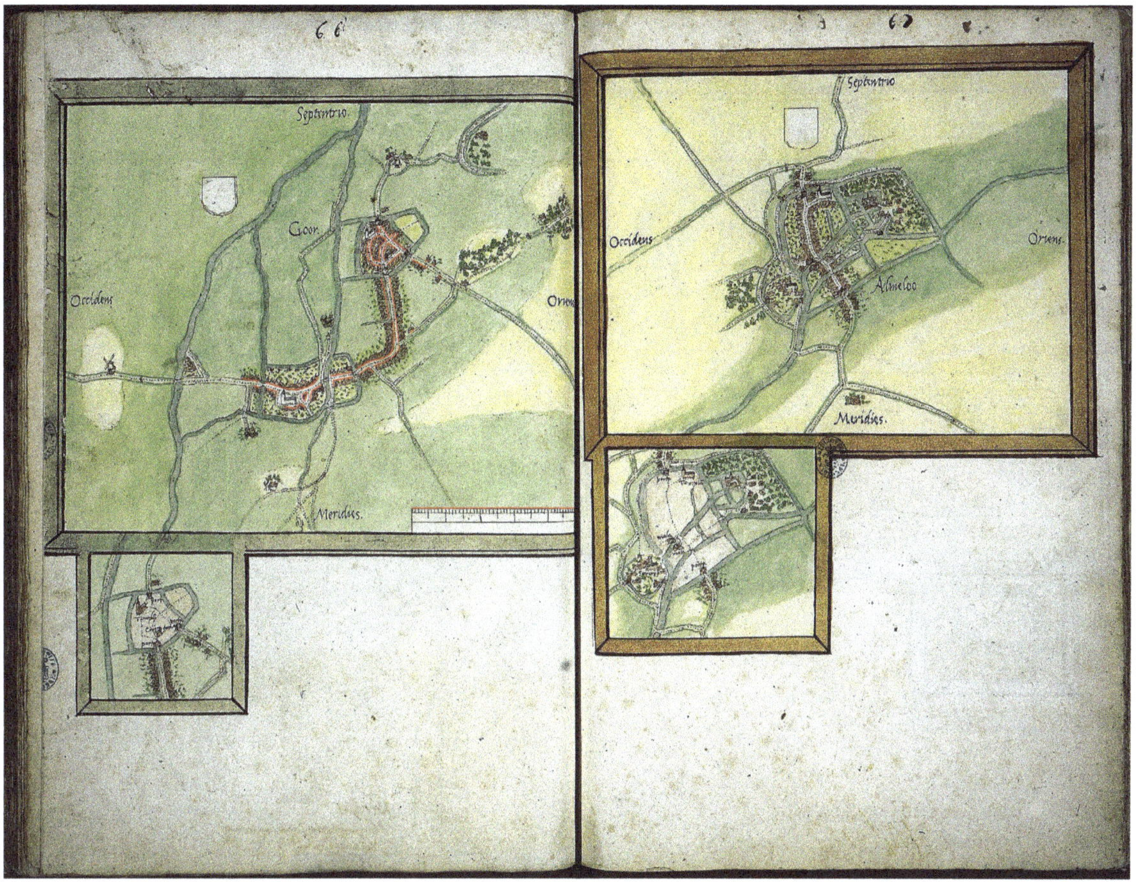
Septentrio
Goor
Occidens
Oriens
Meridies.
Septentrio
Occidens
Oriens
Almeloo
Meridies.

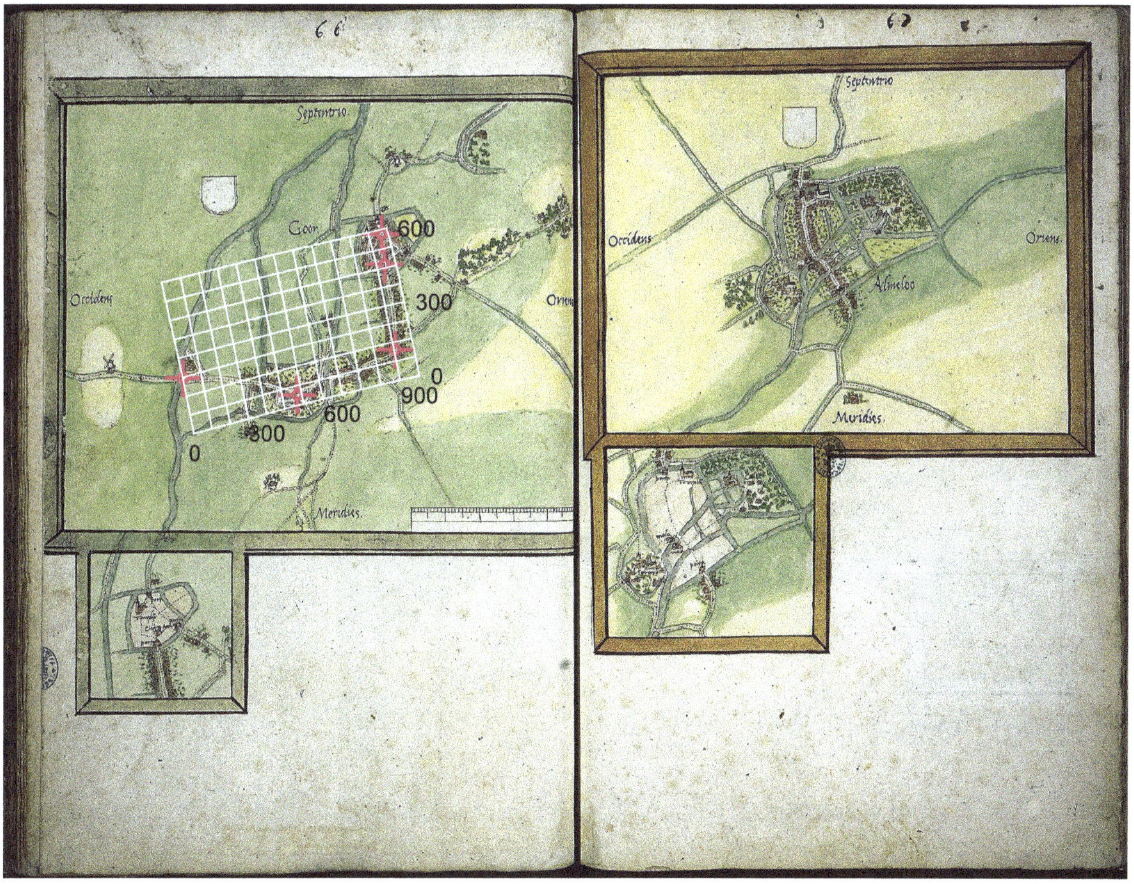
Septentrio
Goor
600
300
0
900
600
300
0
Occidens
Oriens
Meridies.
Septentrio
Occidens
Oriens
Almeloo
Meridies.

86. Gorinchem

Tables 3.86 (measured and analysed).
Figures 3.86 (unmeasured, measured and analysed).

Measured

Pixels	Scale stick original map (pixels)	Houses	Street sides	Houses per street side	Inhabitants (RF 4.0)	Inhabitants (RF 5.5)	Inhabitants (RF 7.0)
5829	998	1024	152	6.73	4.095	5630	7165

Analysed

April 18, 2023 9:47:22 AM	
Computation with 5 linked points (*italics* points are less reliable):	1. Grote Kerk 2. Kapeltoren 3. Burgstraat 4. Eind 5. Korenbrug
Reference points (Reliability 0–5)	5
Transformation with 3 parameters	
Scale	1:7780
Rotation	9 °
Mean position error	18m

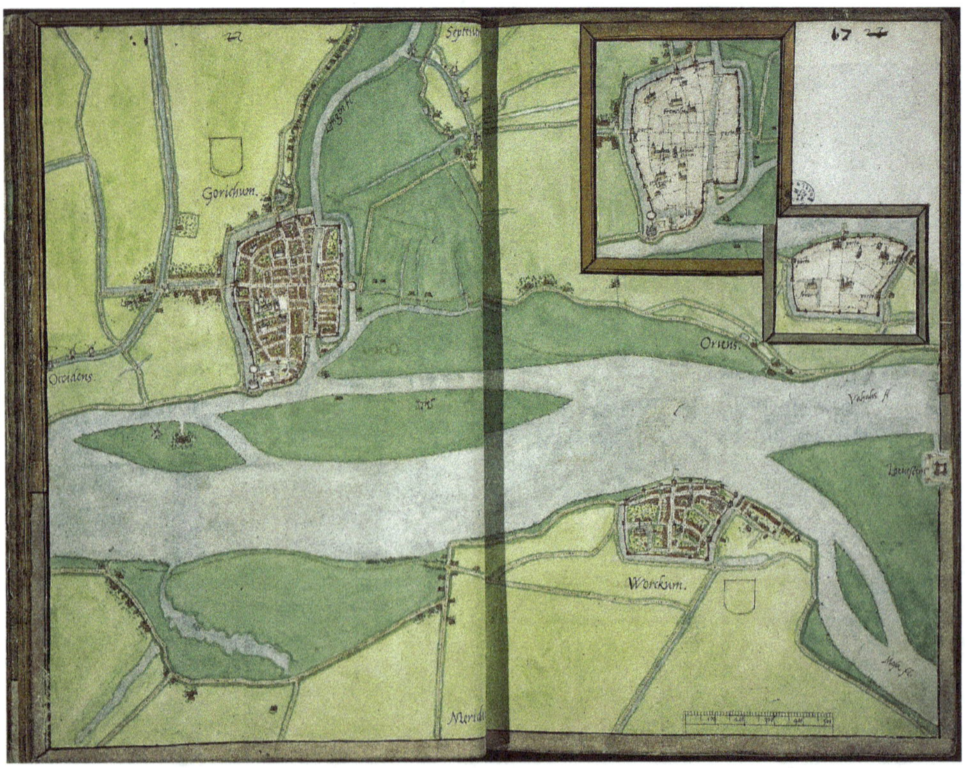

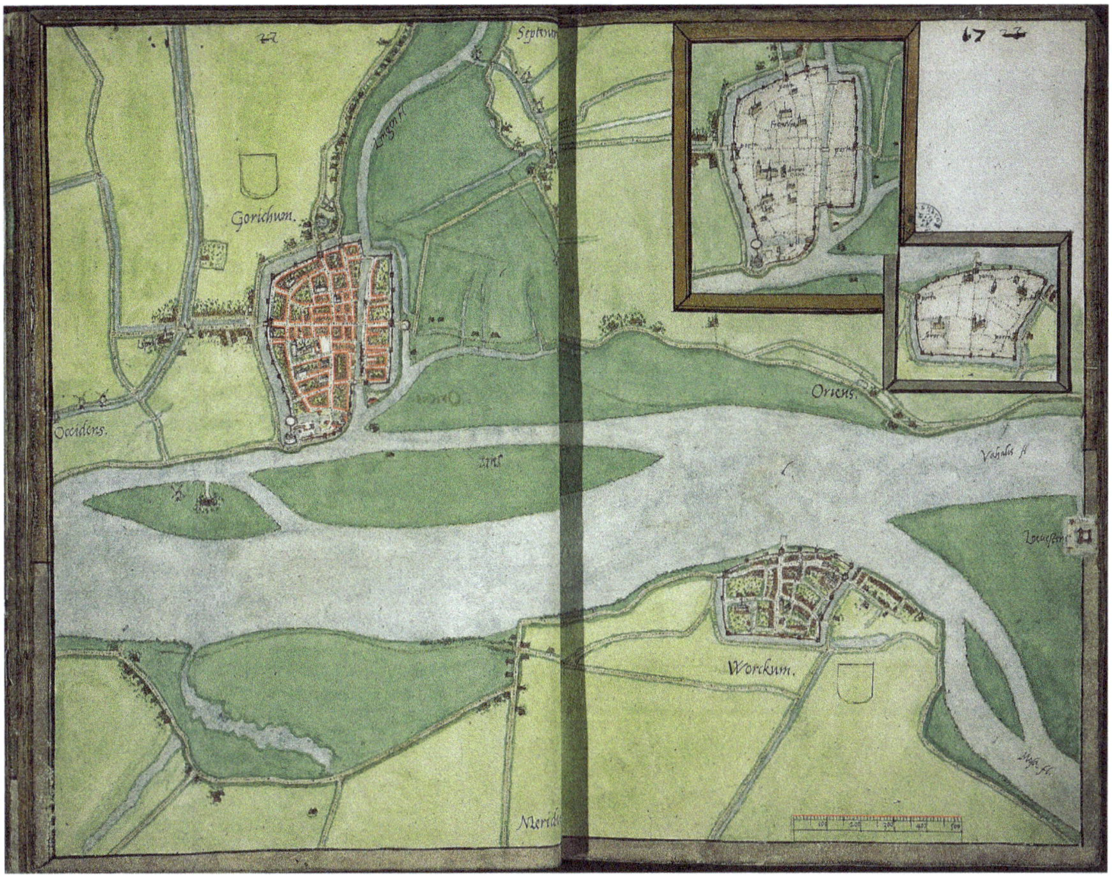
Gorichum.
Occidens.
Oriens.
Worckum.

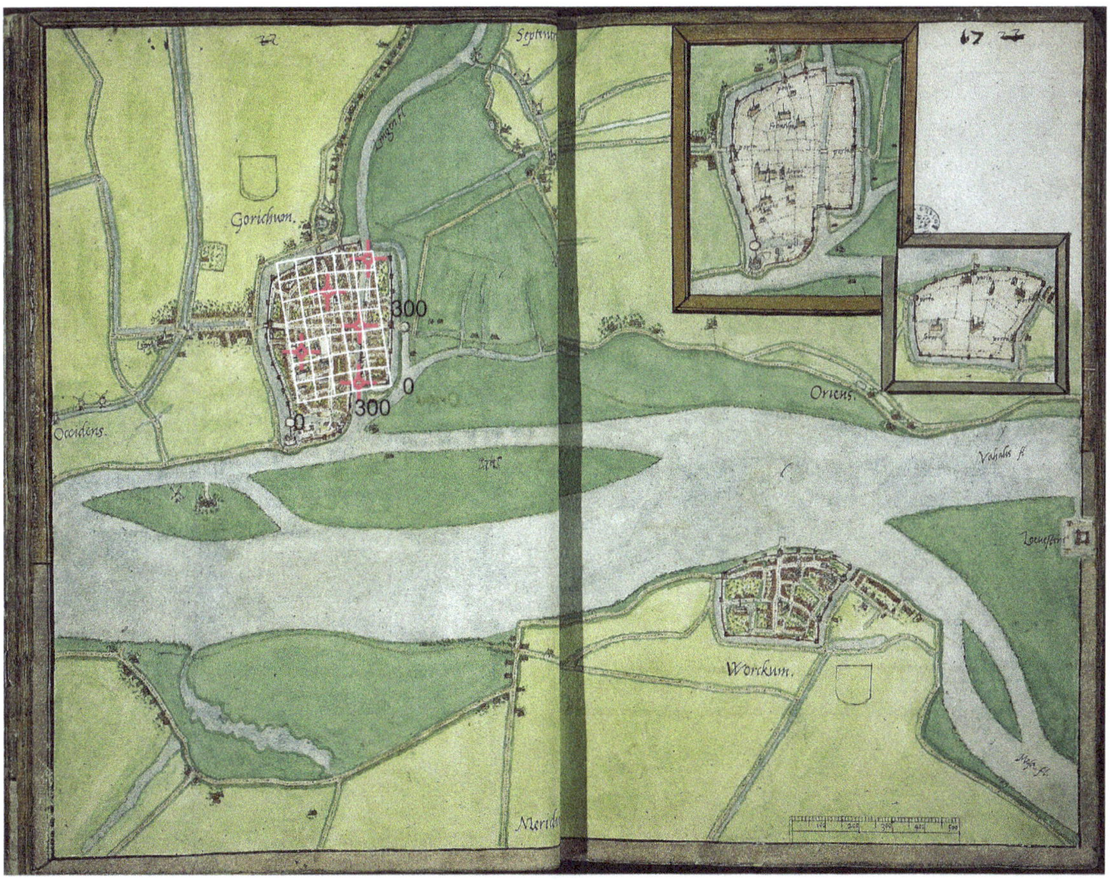
Gorichum.
300
0
300
0
Occidens.
Oriens.
Worckum.

87. Gouda

Tables 3.87 (measured and analysed).
Figures 3.87 (unmeasured, measured and analysed).

Measured							
Pixels	Scale stick original map (pixels)	Houses	Street sides	Houses per street side	Inhabitants (RF 4.0)	Inhabitants (RF 5.5)	Inhabitants (RF 7.0)
9903	1001	1739	230	7.56	6956	9564	12,172

Analysed	
April 26, 2023 11:16:32 AM	
Computation with 5 linked points (*italics* points are less reliable):	1. Stadhuis 2. Sint-Janskerk 3. Kieuwebrug 4. Oosthaven 5. Vest
Reference points (Reliability 0–5)	5
Transformation with 3 parameters	
Scale	1:7890
Rotation	10 °
Mean position error	14m

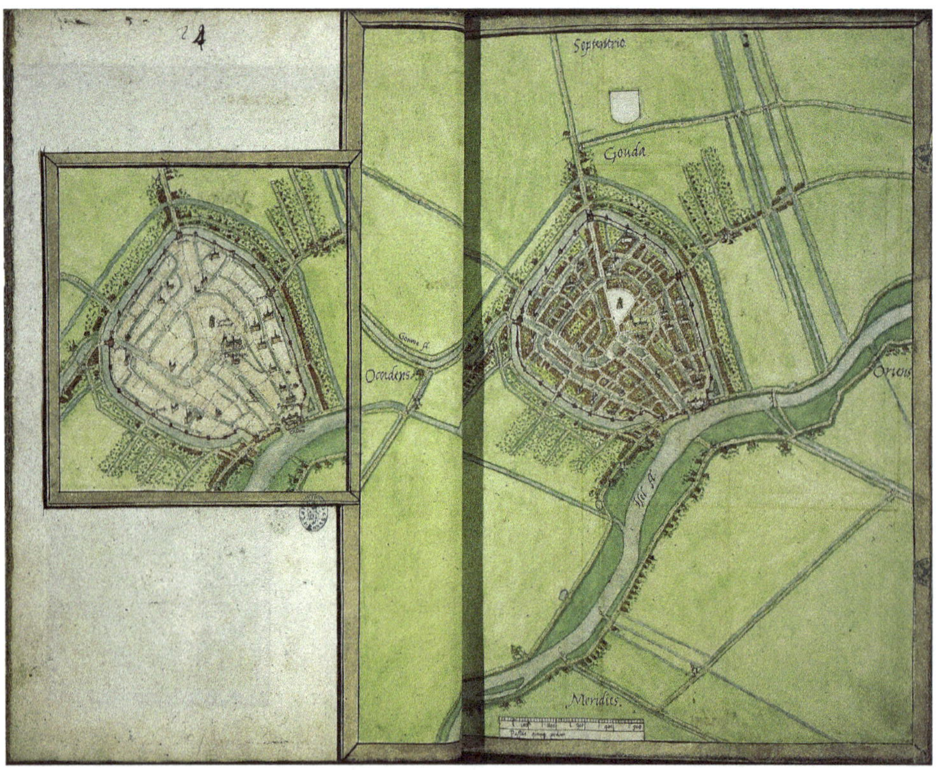

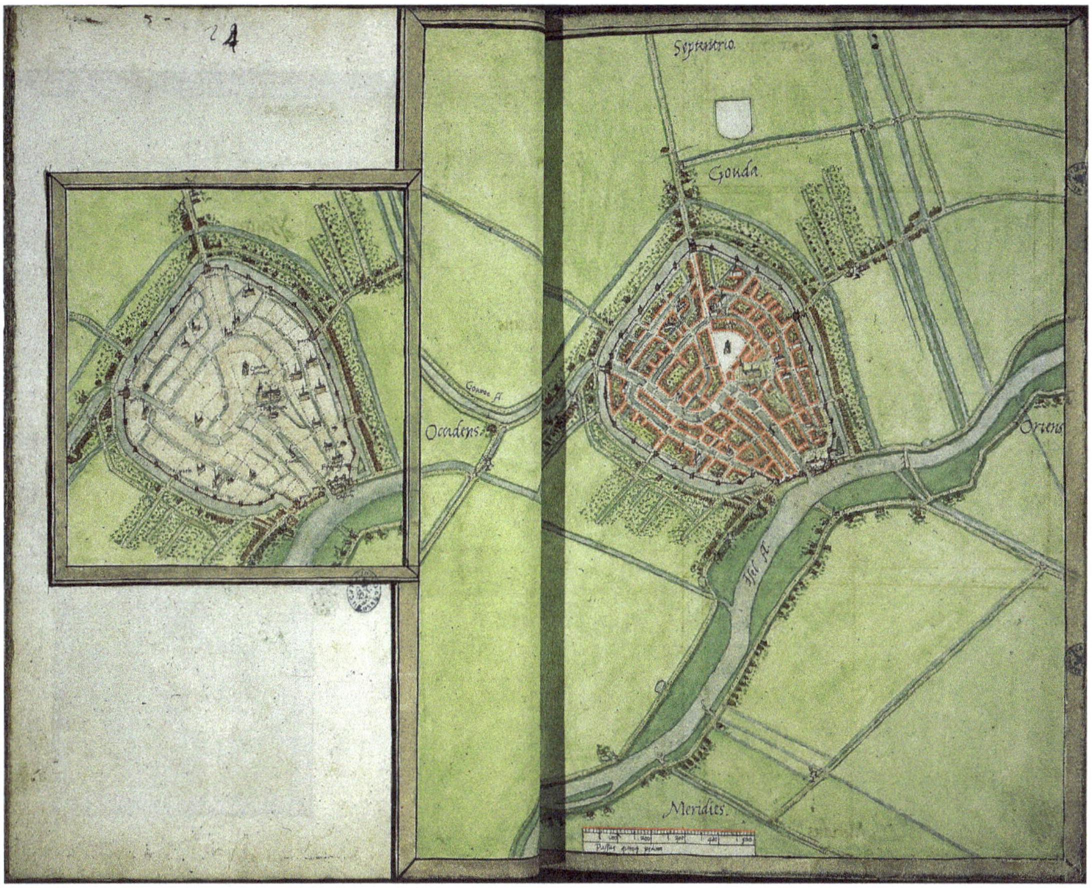
Septentrio
Gouda
Occidens
Oriens
Meridies

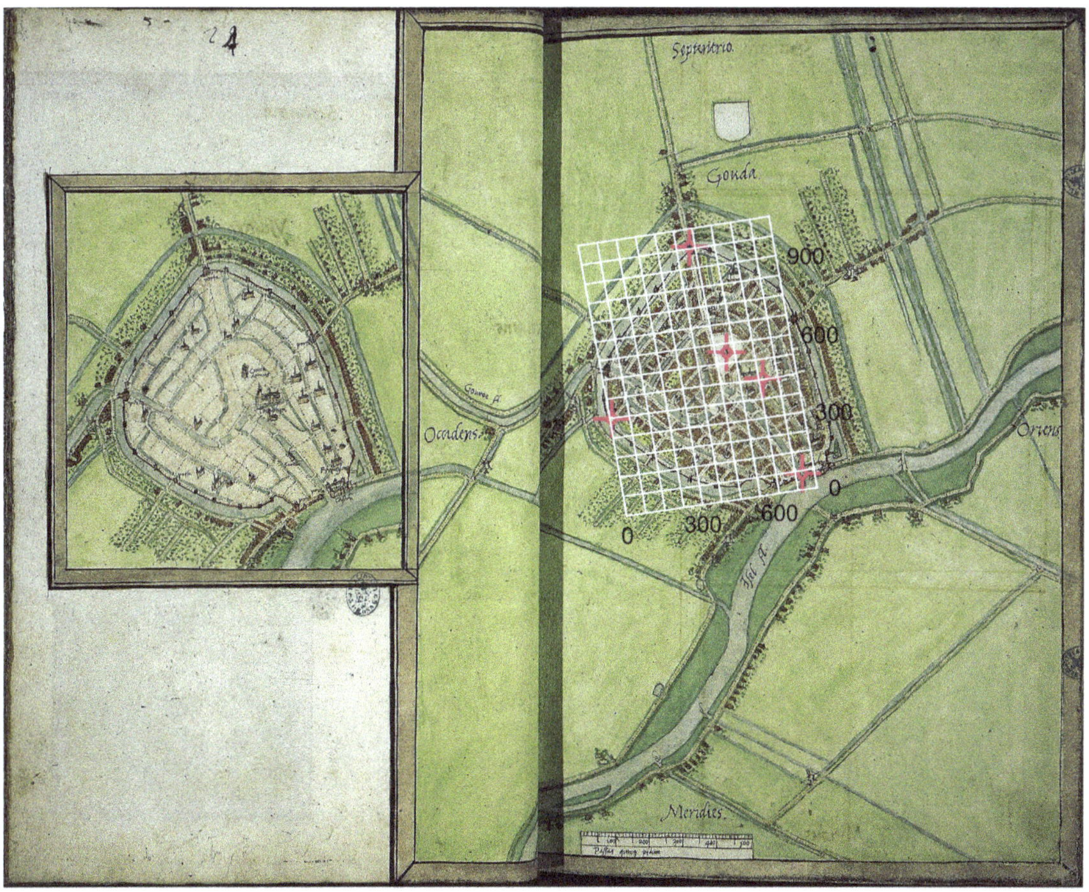
Septentrio
Gouda
900
600
300
0
Occidens
Oriens
0
300
600
Meridies

88. Gravelines

Tables 3.88 (measured and analysed).
Figures 3.88 (unmeasured, measured and analysed).

Measured

Pixels	Scale stick original map (pixels)	Houses	Street sides	Houses per street side	Inhabitants (RF 4.0)	Inhabitants (RF 5.5)	Inhabitants (RF 7.0)
2386	1010	419	55	7.62	1676	2304	2932

Analysed

April 29, 2023 5:49:53 PM	
Computation with 5 linked points (*italics* points are less reliable):	1. Bastion (ouest) 2. Porte de l 'Arsénal 3. Bastion (est) 4. Bastion (sud) 5. Bastion (nord)
Reference points (Reliability 0–5)	5
Transformation with 3 parameters	
Scale	1:8220
Rotation	12 °
Mean position error	17m

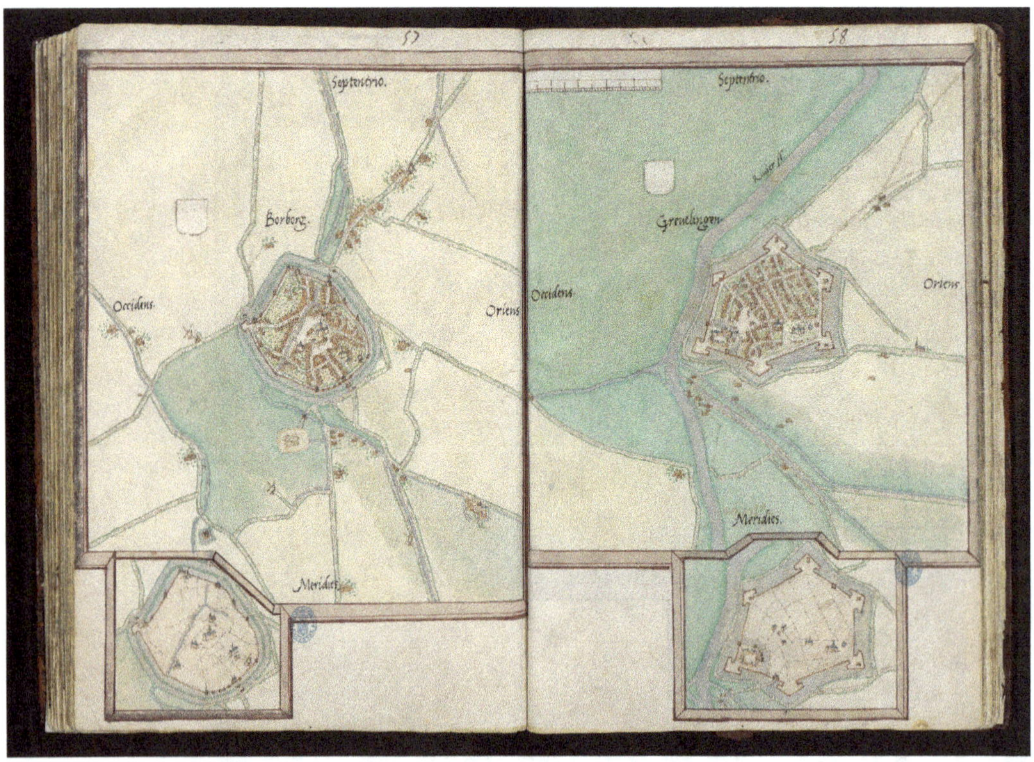

57
58
Septentrio.
Borborg.
Occidens.
Oriens.
Meridies.
Septentrio.
Greuelingen
Occidens.
Oriens.
Meridies.

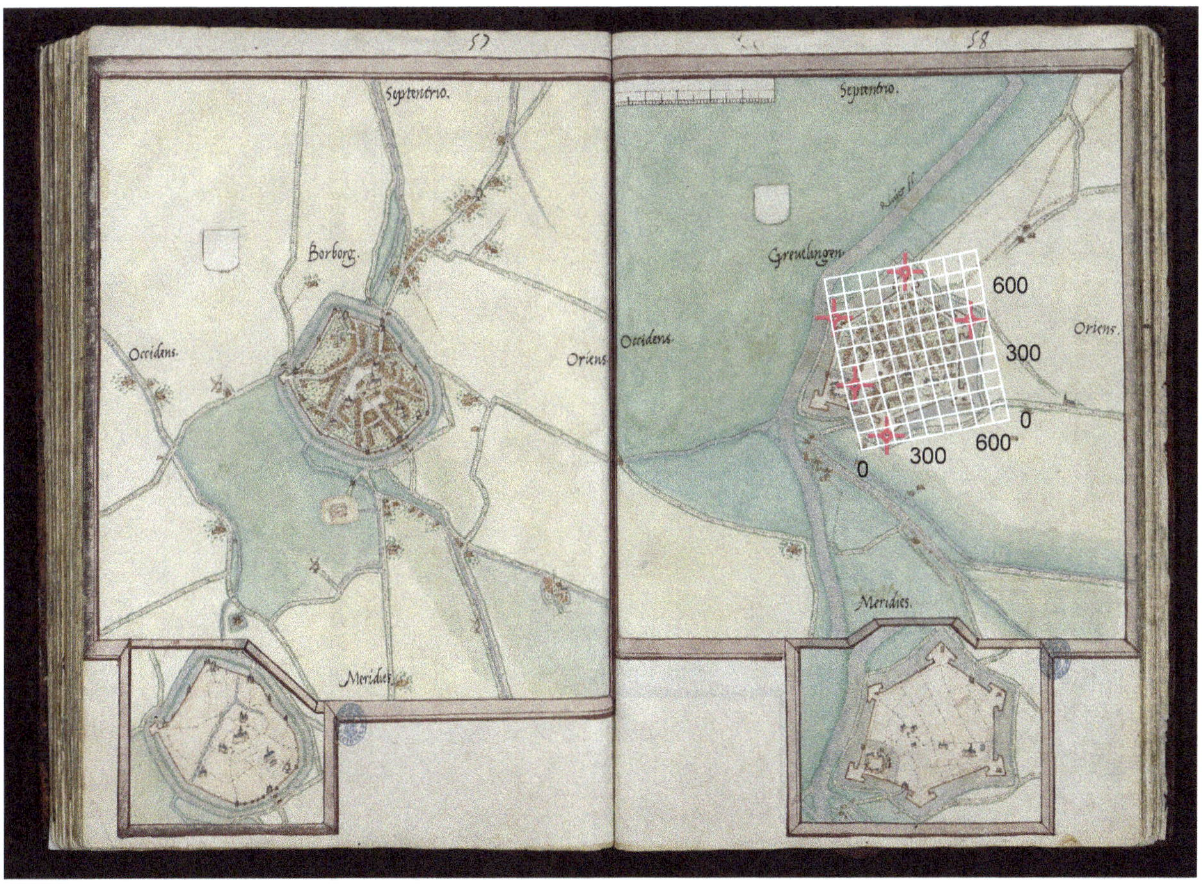
57
58
Septentrio.
Borborg.
Occidens.
Oriens.
Meridies.
Septentrio.
Greuelingen
Occidens.
Oriens.
Meridies.
600
300
0
0
300
600

89. Groenlo

Tables 3.89 (measured and analysed).
Figures 3.89 (unmeasured, measured and analysed).

Measured							
Pixels	Scale stick original map (pixels)	Houses	Street sides	Houses per street side	Inhabitants (RF 4.0)	Inhabitants (RF 5.5)	Inhabitants (RF 7.0)
2483	1021	436	59	7.39	1744	2398	3052

Analysed	
April 30, 2023 9:35:16 AM	
Computation with 5 linked points (*italics* points are less reliable):	1. Oude Calixtuskerk 2. Noteboomstraat 3. Nieuwstraat 4. Lepelstraat 5. Gasthuisstraat
Reference points (Reliability 0–5)	5
Transformation with 3 parameters	
Scale	1:8340
Rotation	12 °
Mean position error	13m

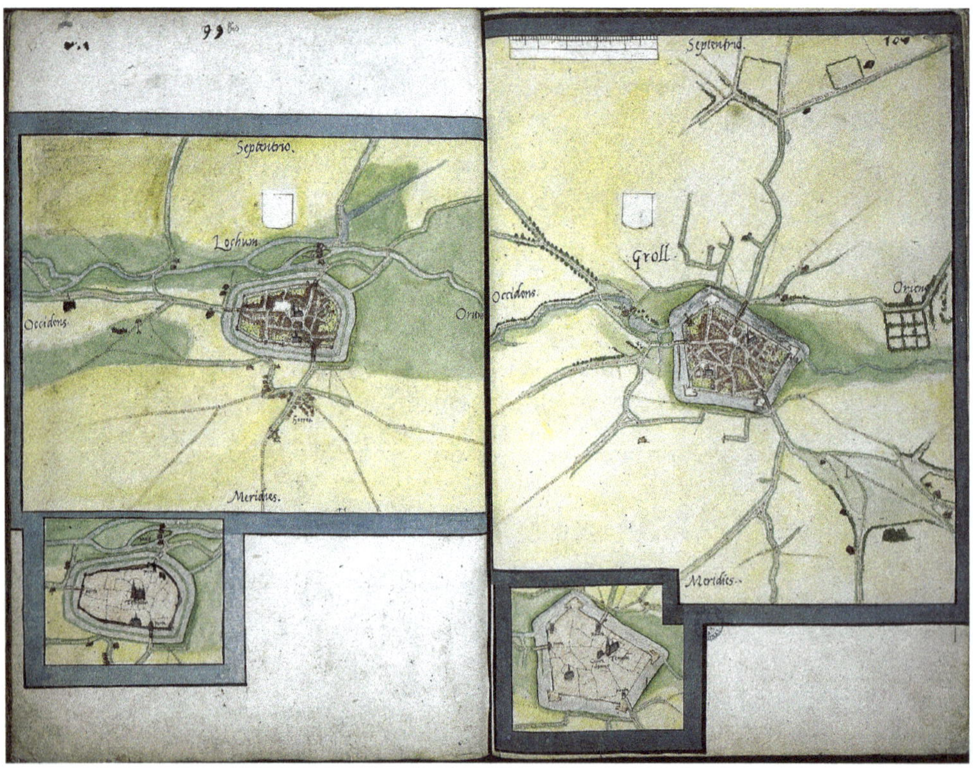

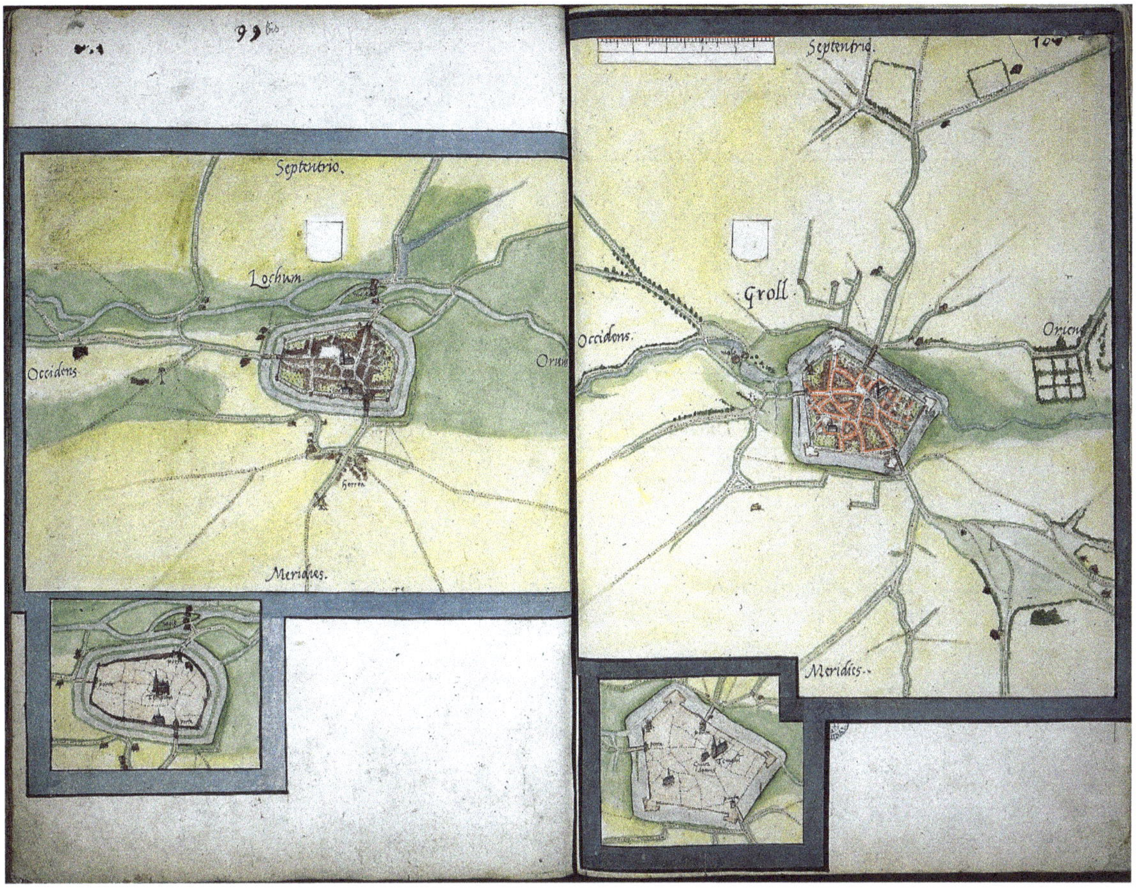
99 bis
Septentrio.
Lochum
Occidens
Meridies.
Septentrio.
Groll
Occidens.
Meridies.

99 bis
Septentrio.
Lochum
Occidens
Meridies.
Septentrio.
Groll
Occidens.
300
0
300
0
Meridies.

90. Groningen

Tables 3.90 (measured and analysed).
Figures 3.90 (unmeasured, measured and analysed).

Measured

Pixels	Scale stick original map (pixels)	Houses	Street sides	Houses per street side	Inhabitants (RF 4.0)	Inhabitants (RF 5.5)	Inhabitants (RF 7.0)
13,903	999	2441	309	7.90	9766	13,428	17,090

Analysed

May 2, 2023 11:10:12 AM	
Computation with 5 linked points (*italics* points are less reliable):	1. Martinikerk 2. Der Aa-kerk 3. Oude Boteringsestraat 4. Gedempte Zuiderdiep 5. *Noorderhaven*
Reference points (Reliability 0–5)	4
Transformation with 3 parameters	
Scale	1:8230
Rotation	11 °
Mean position error	24m

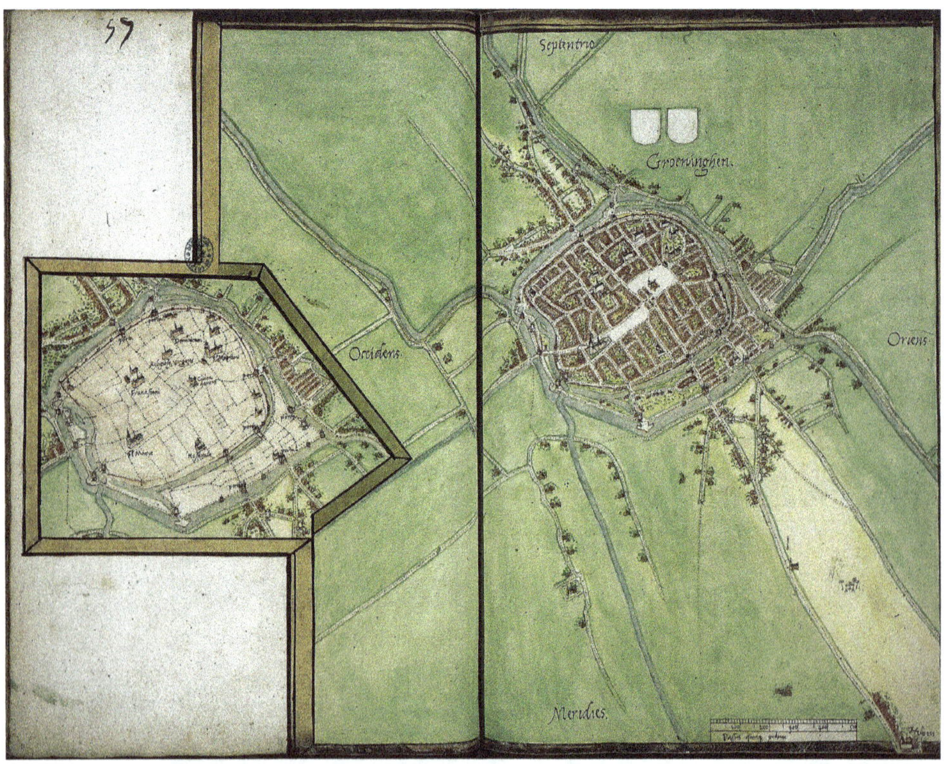

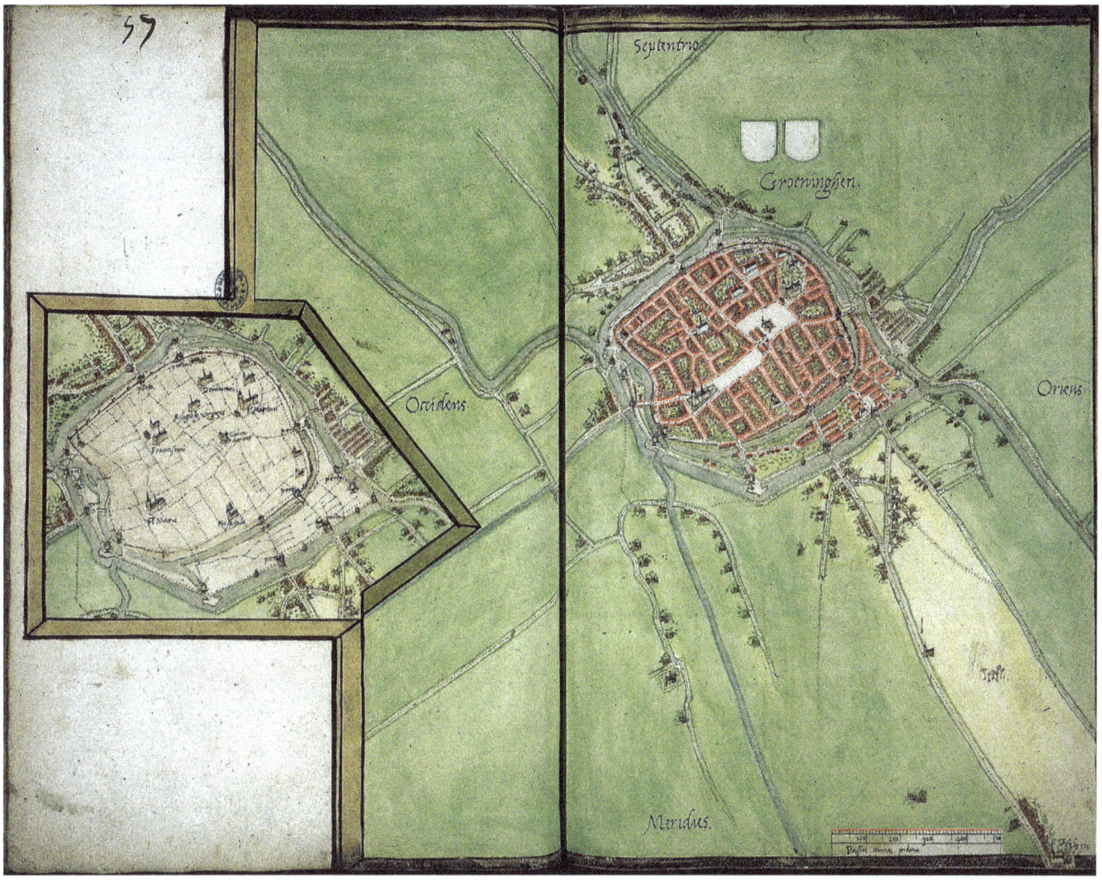
Septentrio
Groeninghen.
Occidens
Oriens
Meridies.

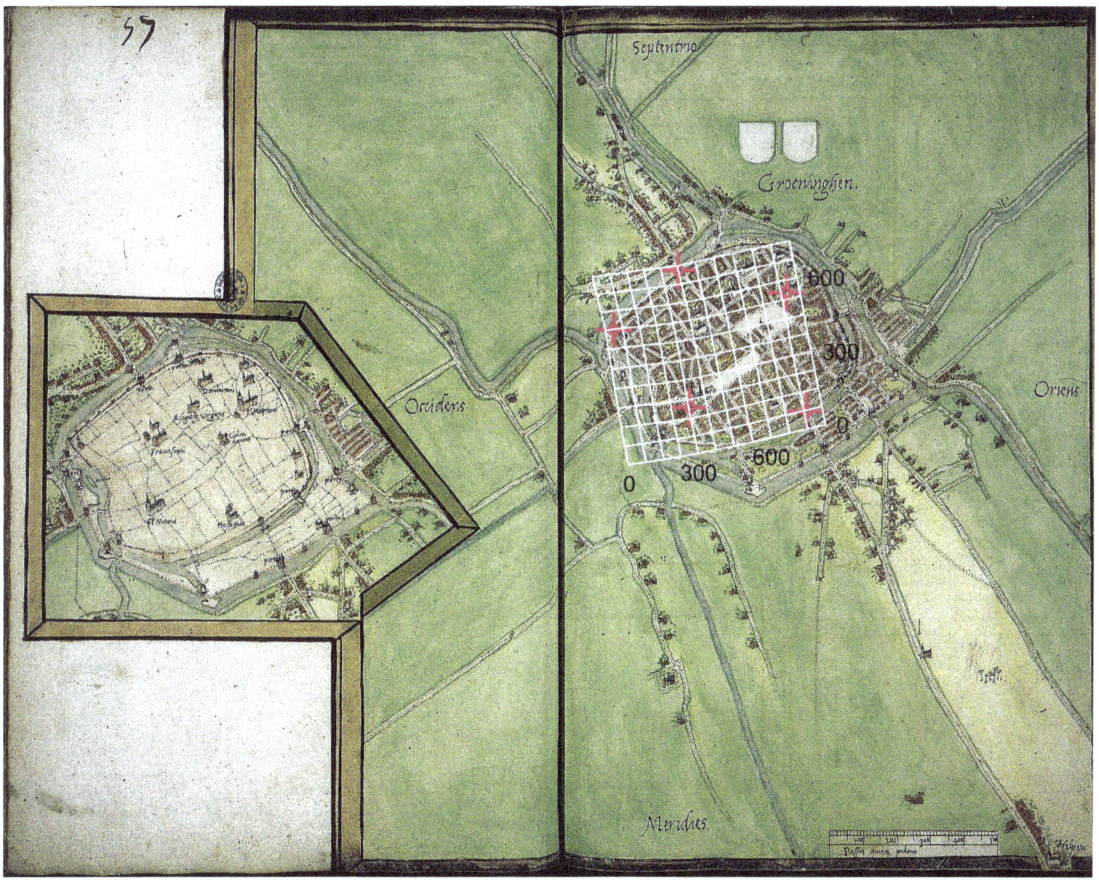
Septentrio
Groeninghen.
Occidens
Oriens
Meridies.
600
300
0
0
300
600

91. Grootebroek

Tables 3.91 (measured and analysed).
Figures 3.91 (unmeasured, measured and analysed).

Measured							
Pixels	Scale stick original map (pixels)	Houses	Street sides	Houses per street side	Inhabitants (RF 4.0)	Inhabitants (RF 5.5)	Inhabitants (RF 7.0)
4151	1004	716	36	19,9	2865	3939	5013

Analysed	
October 12, 2023 10:42:32 PM	
Computation with 5 linked points (*italics* points are less reliable):	*1. Toren Hervormde Kerk* *2. Bovenkarspel* *3. Lutjebroek* *4. Burgemeester Schrijnderlaan* *5. Richtershof*
Reference points (Reliability 0–5)	0
Transformation with 3 parameters	
Scale	1:7800
Rotation	10 °
Mean position error	73m

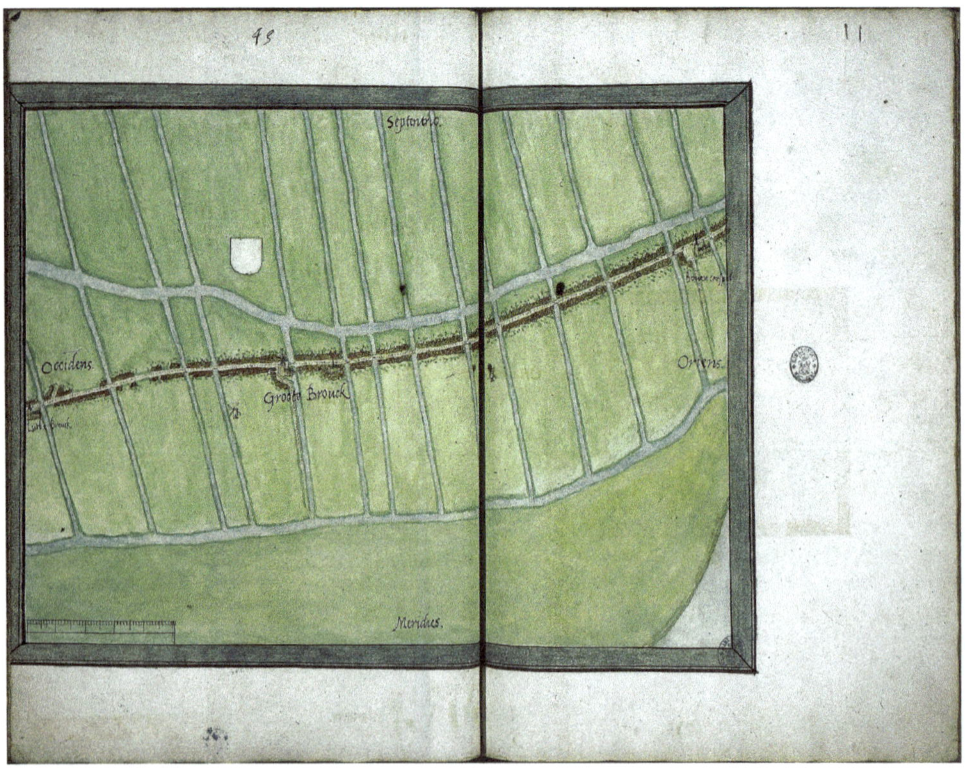

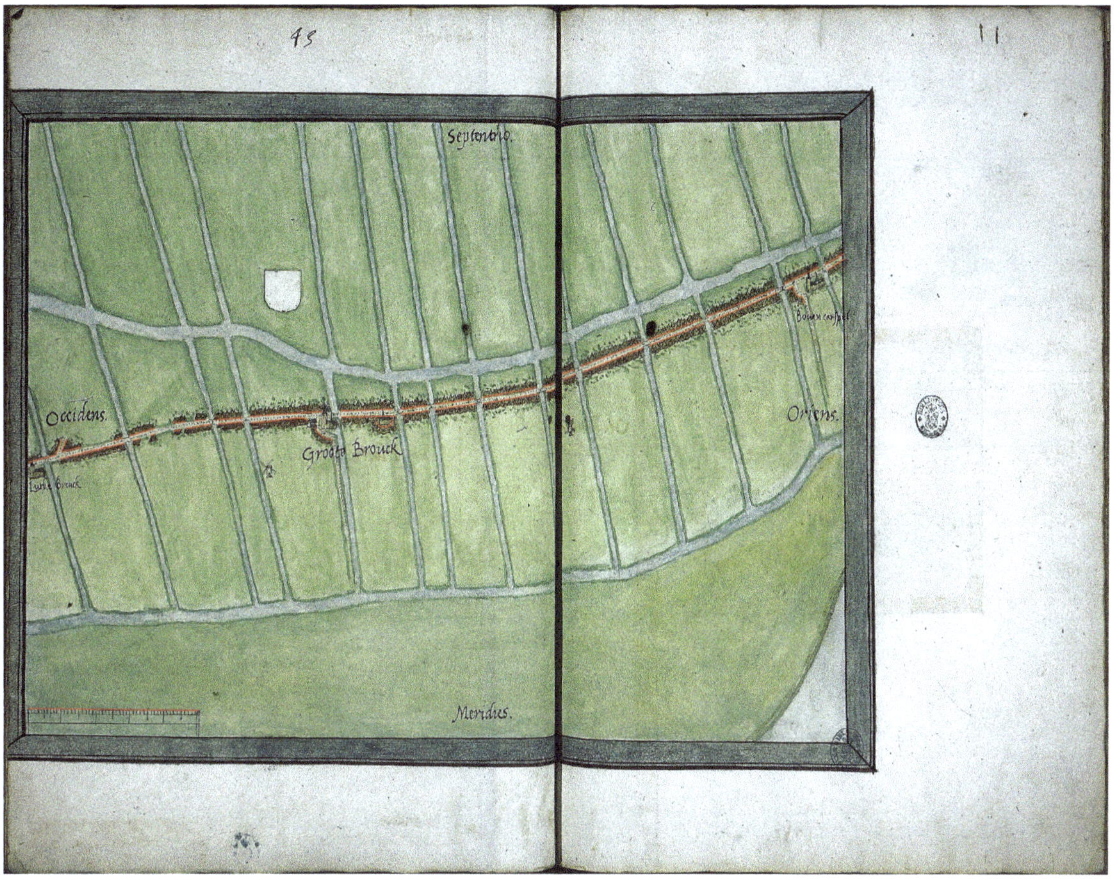
Septentrio.
Occidens.
Groote Brouck
Oriens.
Meridies.

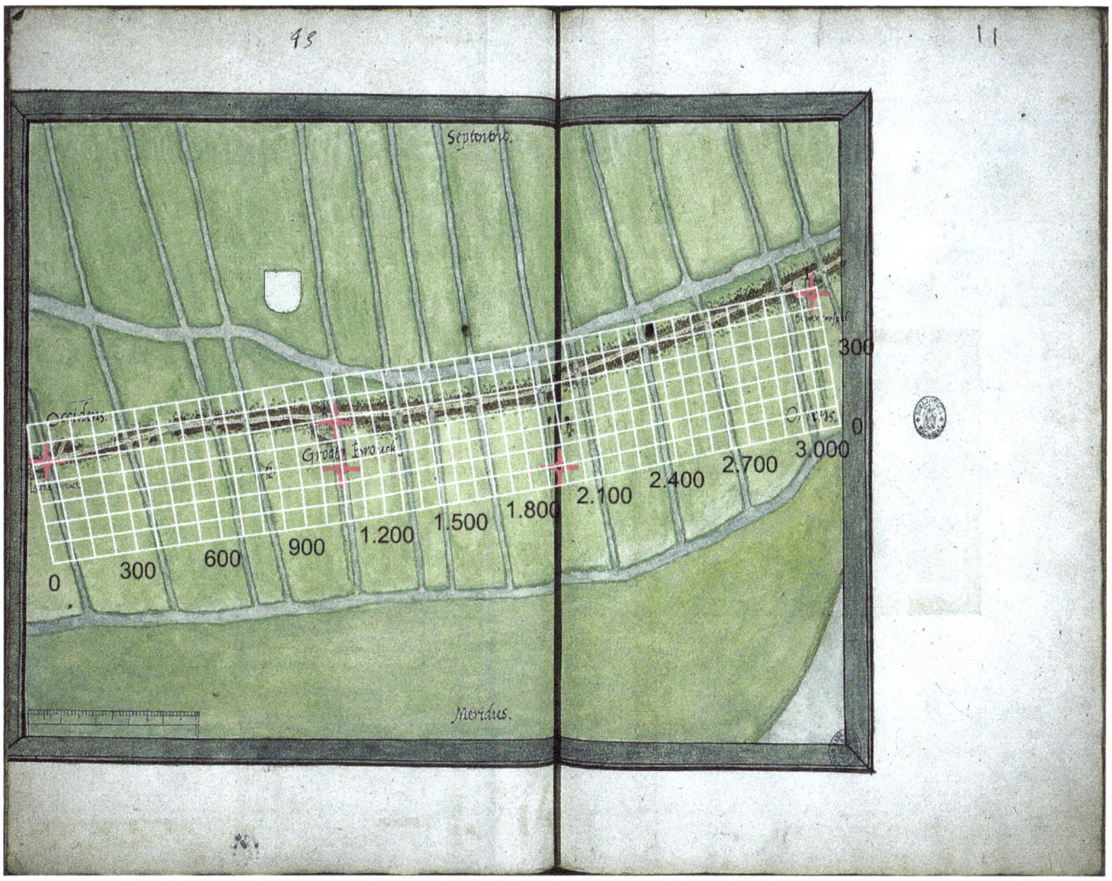
Septentrio.
Meridies.
0
300
600
900
1.200
1.500
1.800
2.100
2.400
2.700
3.000
0
300

92. Haarlem

Tables 3.92 (measured and analysed).
Figures 3.92 (unmeasured, measured and analysed).

Measured							
Pixels	Scale stick original map (pixels)	Houses	Street sides	Houses per street side	Inhabitants (RF 4.0)	Inhabitants (RF 5.5)	Inhabitants (RF 7.0)
18,845	1006	3309	430	7.70	13,237	18,201	23,165

Analysed	
May 6, 2023 3:01:24 PM	
Computation with 5 linked points (*italics* points are less reliable):	1. Sint-Bavokerk 2. Amsterdamse Poort 3. Gedempte Oude Gracht 4. Gasthuisvest *5. Nieuwe Gracht*
Reference points (Reliability 0–5)	4
Transformation with 3 parameters	
Scale	1:8390
Rotation	10 °
Mean position error	26m

Septentrio.
Haerlem.
Oriens.
Meridies.

Septentrio.
Haerlem.
Oriens.
Meridies.
900
600
300
0
0
300
600
900

93. Halle

Tables 3.93 (measured and analysed).
Figures 3.93 (unmeasured, measured and analysed).

Measured							
Pixels	Scale stick original map (pixels)	Houses	Street sides	Houses per street side	Inhabitants (RF 4.0)	Inhabitants (RF 5.5)	Inhabitants (RF 7.0)
2716	1029	477	82	5.82	1908	2623	3338

Analysed	
May 6, 2023 3:52:58 PM	
Computation with 5 linked points (*italics* points are less reliable):	1. Sint-Martinusbasiliek 2. Korte Vest 3. Arkenvest 4. Parklaan 5. Vestingstraat
Reference points (Reliability 0–5)	5
Transformation with 3 parameters	
Scale	1:7650
Rotation	10 °
Mean position error	9m

Septentrio
Halle
Occidens
Oriens
Meridies

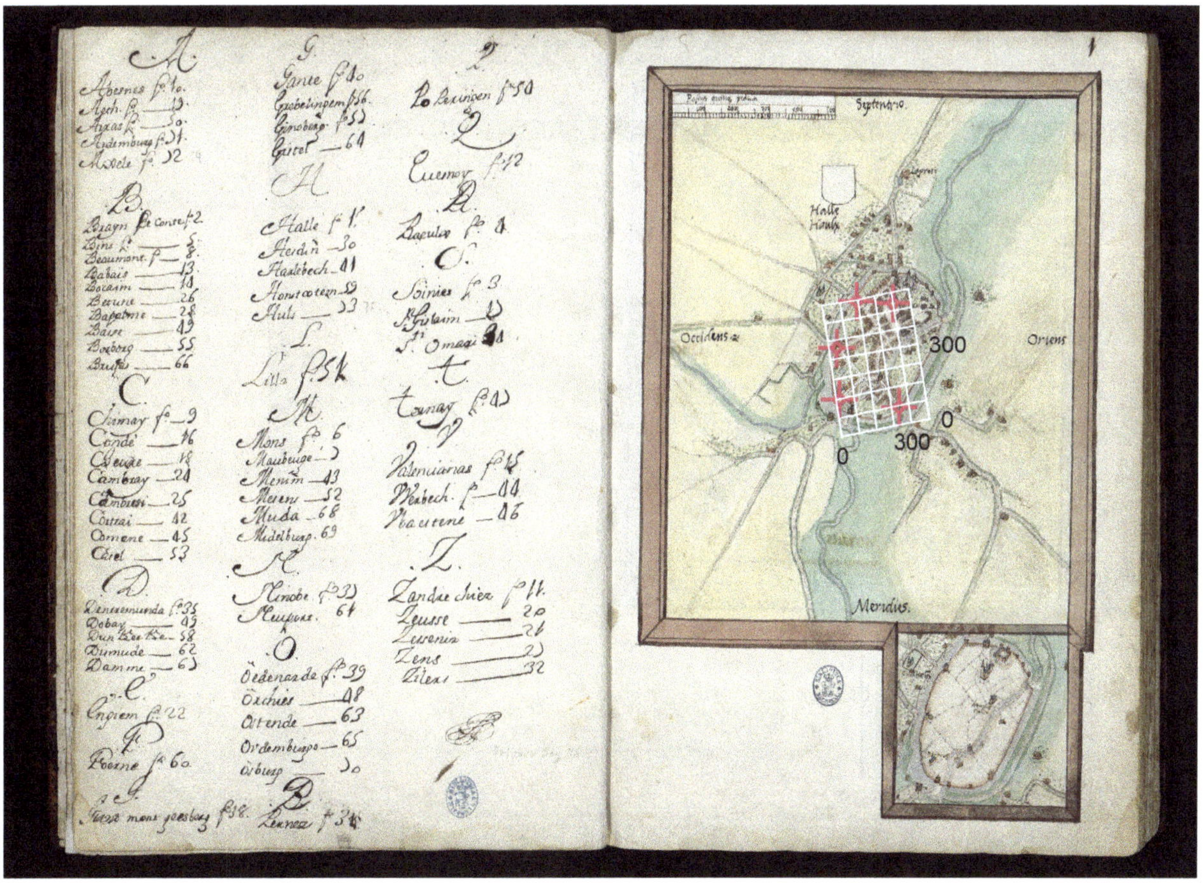
Septentrio
Halle
Occidens
300
0
300
0
Oriens
Meridies

94. Halluin

Tables 3.94 (measured and analysed).
Figures 3.94 (unmeasured, measured and analysed).

Measured							
Pixels	Scale stick original map (pixels)	Houses	Street sides	Houses per street side	Inhabitants (RF 4.0)	Inhabitants (RF 5.5)	Inhabitants (RF 7.0)
2377	1027	417	36	11.59	1670	2296	2922

Analysed	
October 15, 2023 6:56:57 PM	
Computation with 5 linked points (*italics* points are less reliable):	1. Église Saint-Hilaire *2. Rue de Lille* *3. Rue Jean Jaurès* *4. Rue Marthe Nollet* *5. Leie*
Reference points (Reliability 0–5)	1
Transformation with 3 parameters	
Scale	1:7460
Rotation	25 °
Mean position error	74m

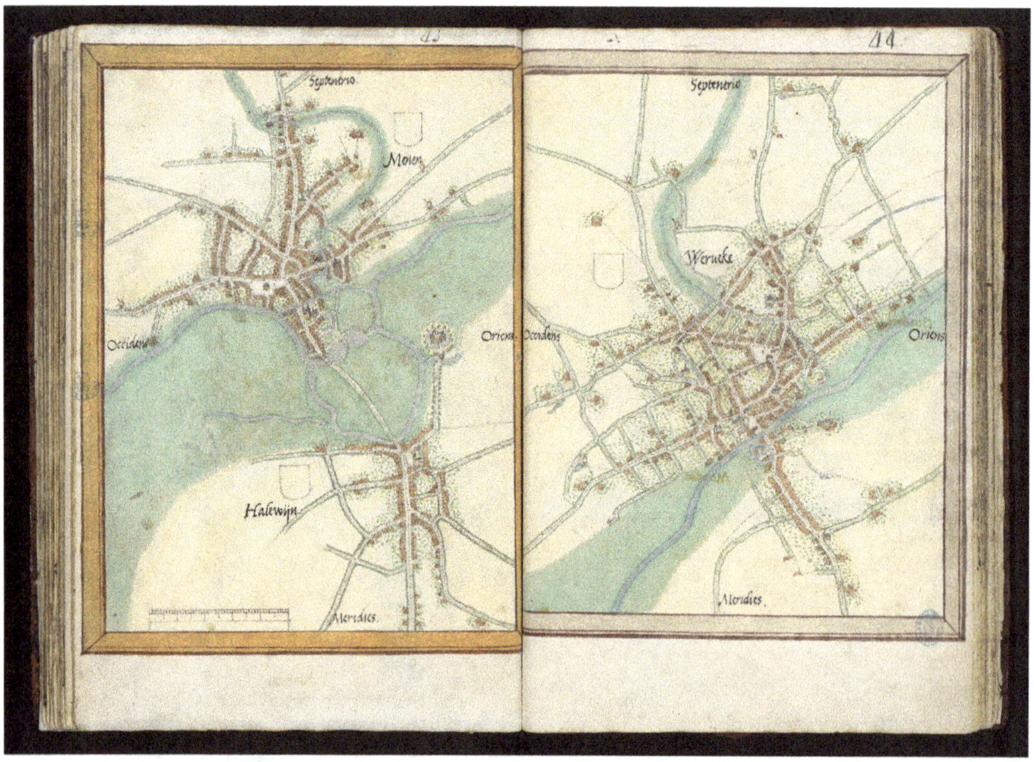

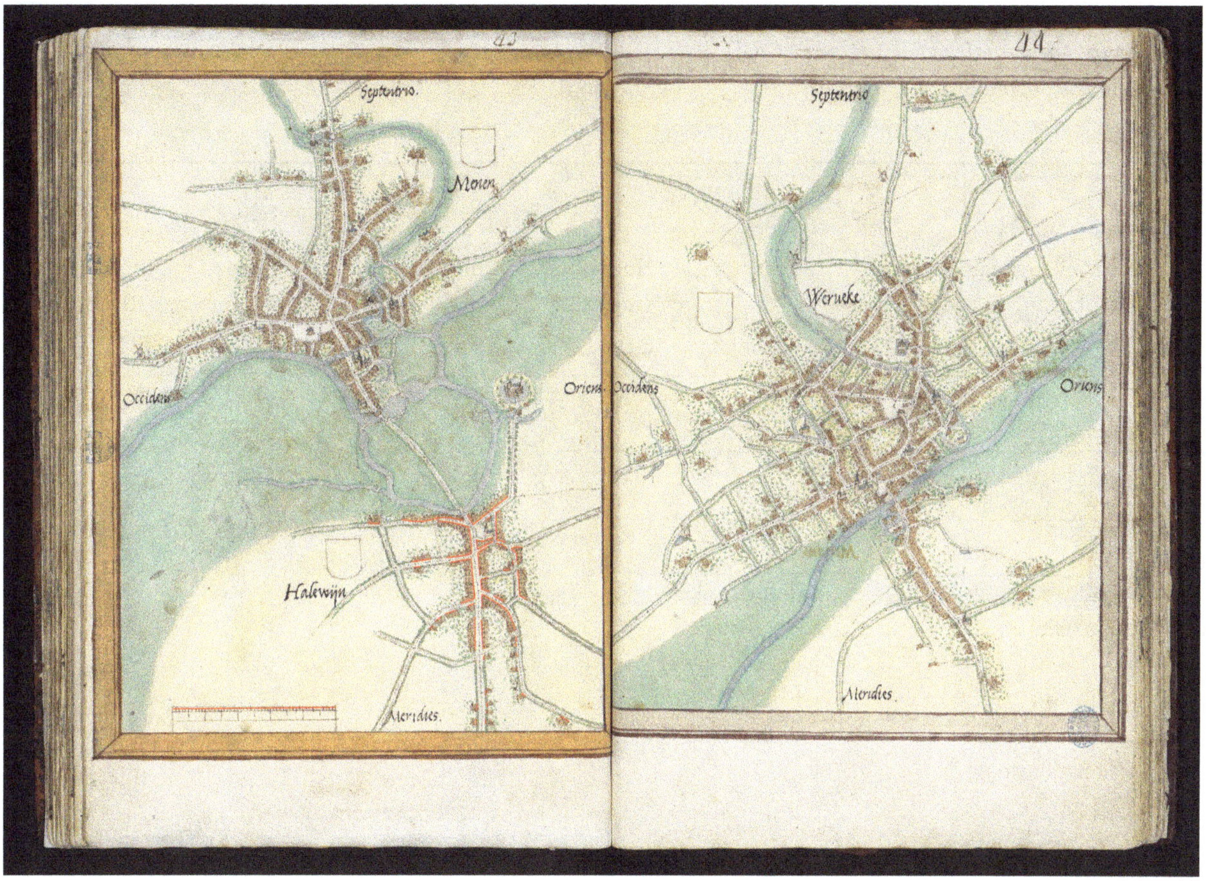
Septentrio.
Menen
Occidens
Oriens
Halewijn
Meridies.
Septentrio
Werueke
Occidens
Oriens
Meridies
44

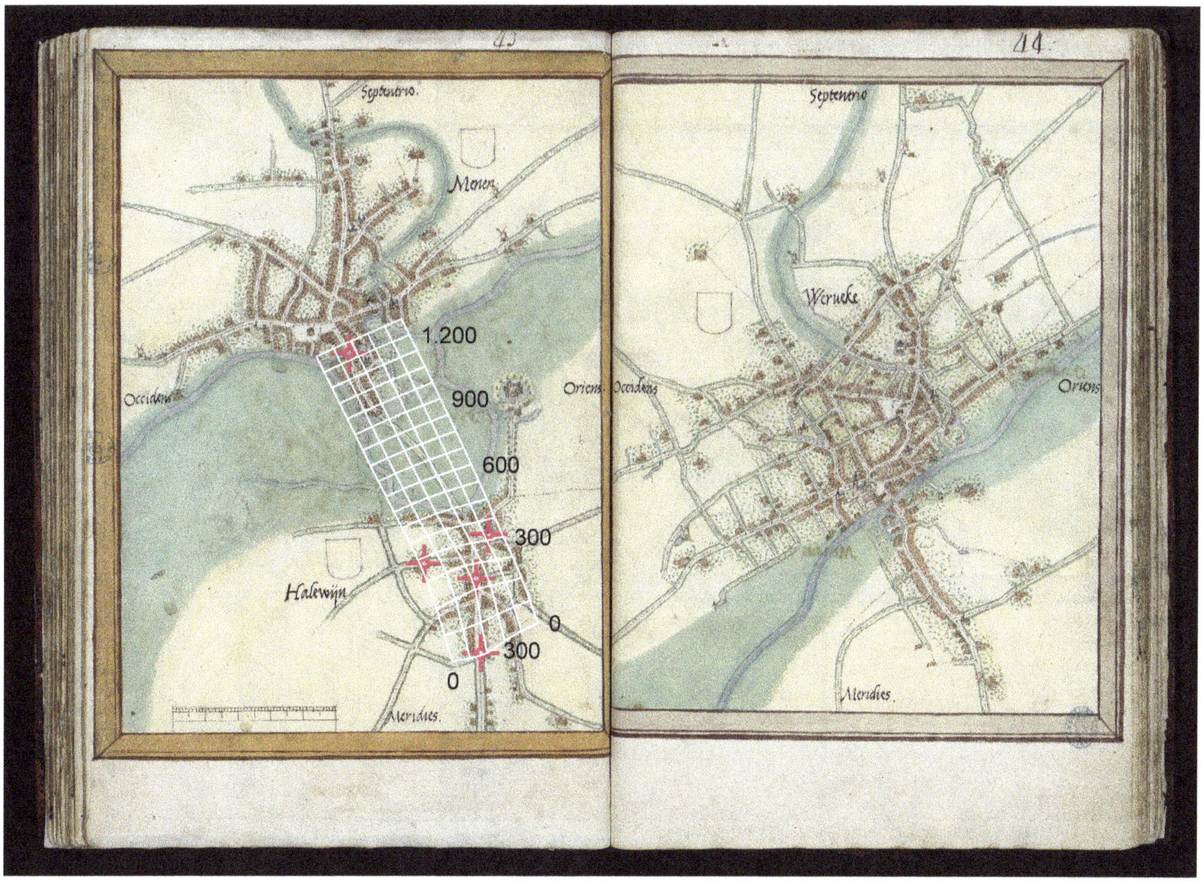
Septentrio.
Menen
1.200
900
600
300
0
300
0
Occidens
Oriens
Halewijn
Meridies.
Septentrio
Werueke
Occidens
Oriens
Meridies
44

95. Hardenberg

Tables 3.95 (measured and analysed).
Figures 3.95 (unmeasured, measured and analysed).

Measured							
Pixels	Scale stick original map (pixels)	Houses	Street sides	Houses per street side	Inhabitants (RF 4.0)	Inhabitants (RF 5.5)	Inhabitants (RF 7.0)
623	1003	109	21	5.21	438	602	766

Analysed	
May 9, 2023 12:49:53 PM	
Computation with 5 linked points (*italics* points are less reliable):	1. Stephanuskerk 2. Hoge Doelen 3. Kromme Steeg 4. Kruissteeg 5. Wilhelminaplein
Reference points (Reliability 0–5)	5
Transformation with 3 parameters	
Scale	1:8540
Rotation	13 °
Mean position error	9m

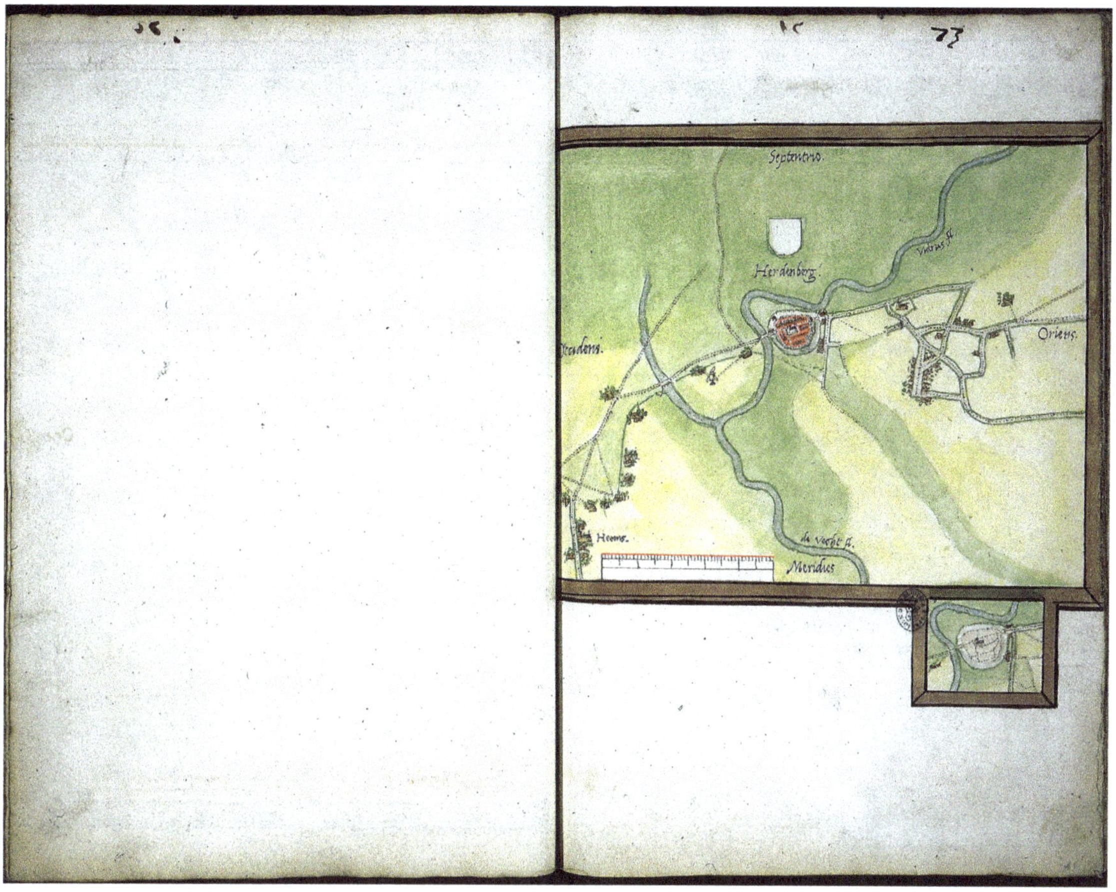
Septentrio.
Herdenberg.
Oriens.
Meridies

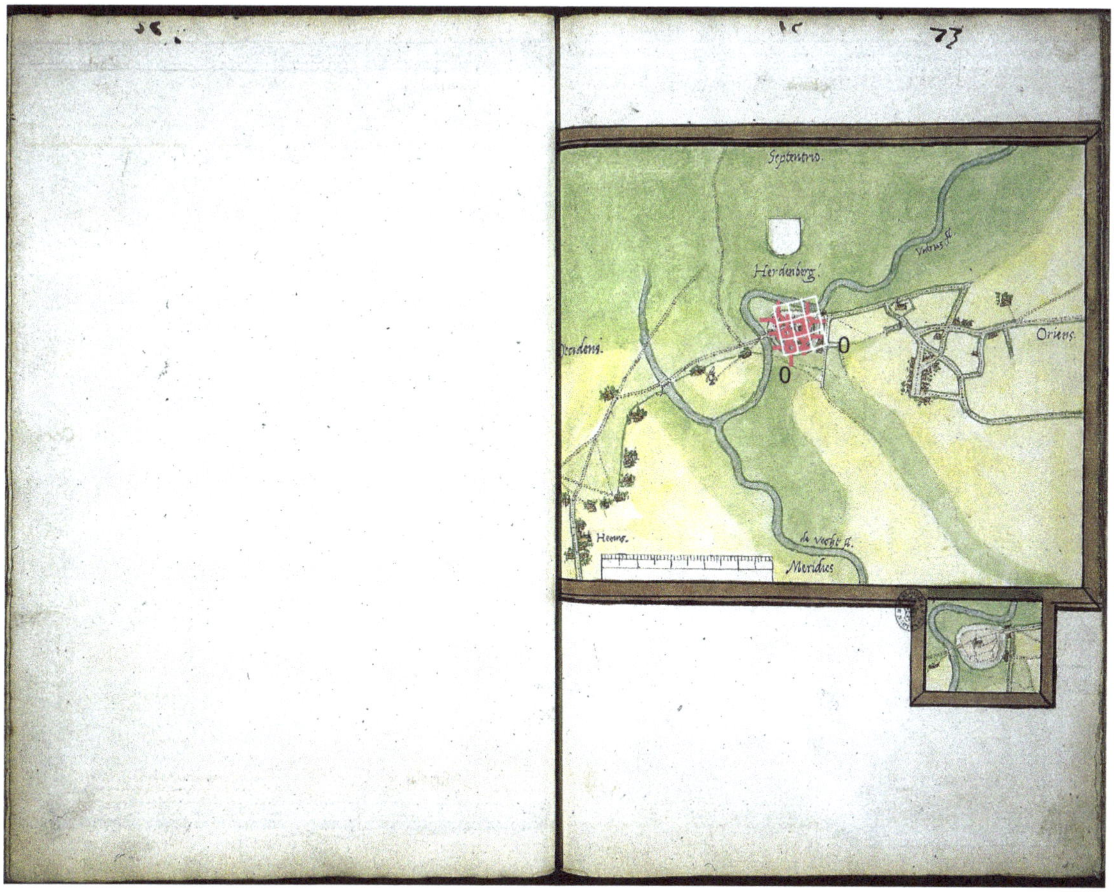
Septentrio.
Herdenberg.
Oriens.
Meridies

96. Harderwijk

Tables 3.96 (measured and analysed).
Figures 3.96 (unmeasured, measured and analysed).

Measured

Pixels	Scale stick original map (pixels)	Houses	Street sides	Houses per street side	Inhabitants (RF 4.0)	Inhabitants (RF 5.5)	Inhabitants (RF 7.0)
5460	missing	959	159	6.03	3835	5273	6711

Analysed

May 12, 2023 11:42:07 AM	
Computation with 5 linked points (*italics* points are less reliable):	1. Grote Kerk 2. Vischpoort 3. Smeepoort *4. Vissershaven* *5. Doelenstraat*
Reference points (Reliability 0–5)	3
Transformation with 3 parameters	
Scale	1:7660
Rotation	9 °
Mean position error	33m

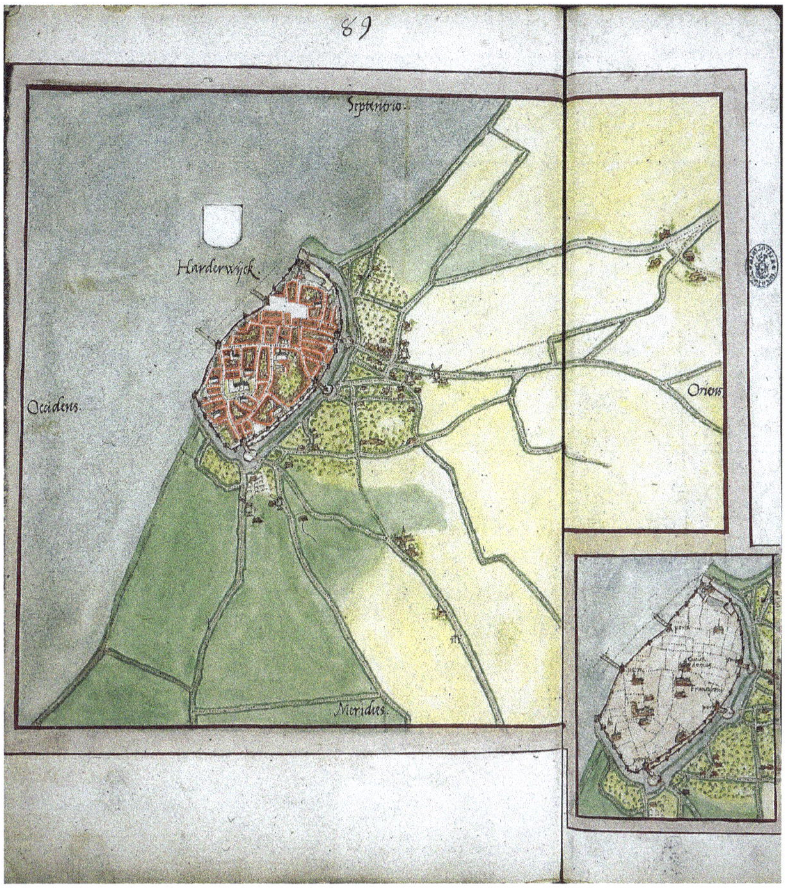
89
Septentrio
Harderwijck
Occidens
Oriens
Meridies

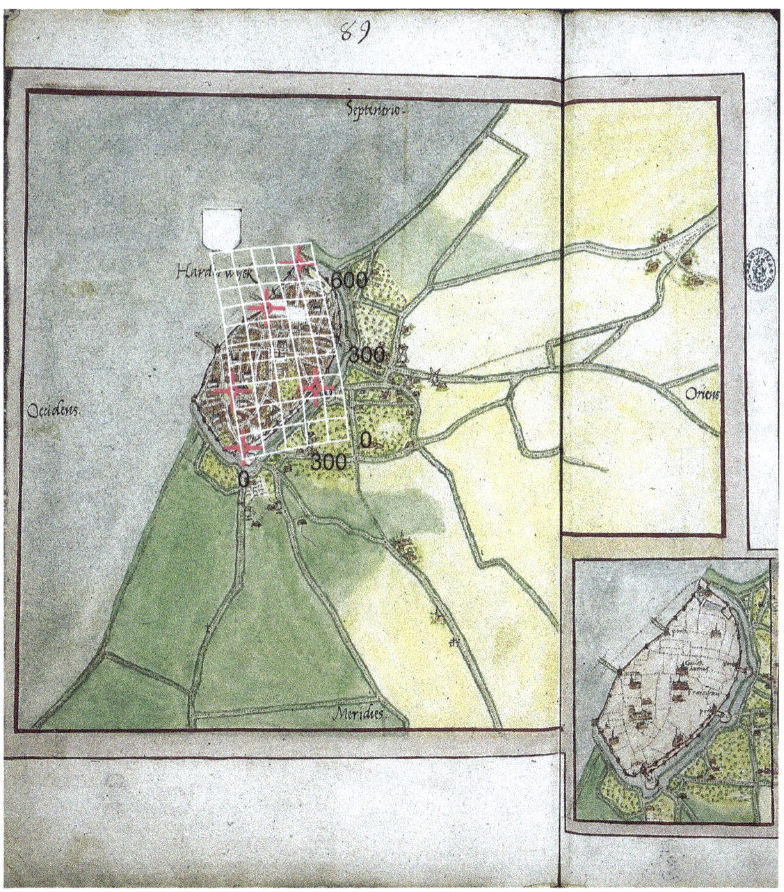
89
Septentrio
600
300
0
300
0
Occidens
Oriens
Meridies

97. Harelbeke

Tables 3.97 (measured and analysed).
Figures 3.97 (unmeasured, measured and analysed).

Measured							
Pixels	Scale stick original map (pixels)	Houses	Street sides	Houses per street side	Inhabitants (RF 4.0)	Inhabitants (RF 5.5)	Inhabitants (RF 7.0)
1598	1025	281	31	9.05	1122	1543	1964

Analysed	
May 13, 2023 11:24:05 AM	
Computation with 5 linked points (*italics* points are less reliable):	1. Sint-Salvatorkerk 2. Banmolen *3. Overleiestraat* 4. Hoge Brug *5. Marktstraat*
Reference points (Reliability 0–5)	**3**
Transformation with 3 parameters	
Scale	1:9440
Rotation	13 °
Mean position error	64m

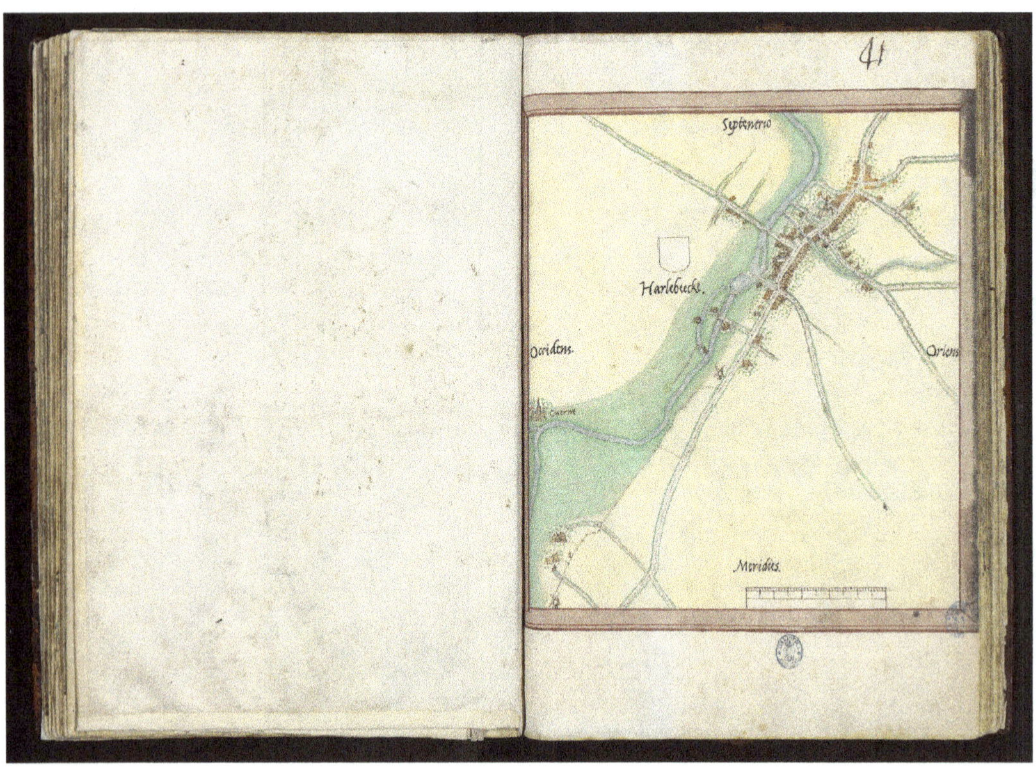

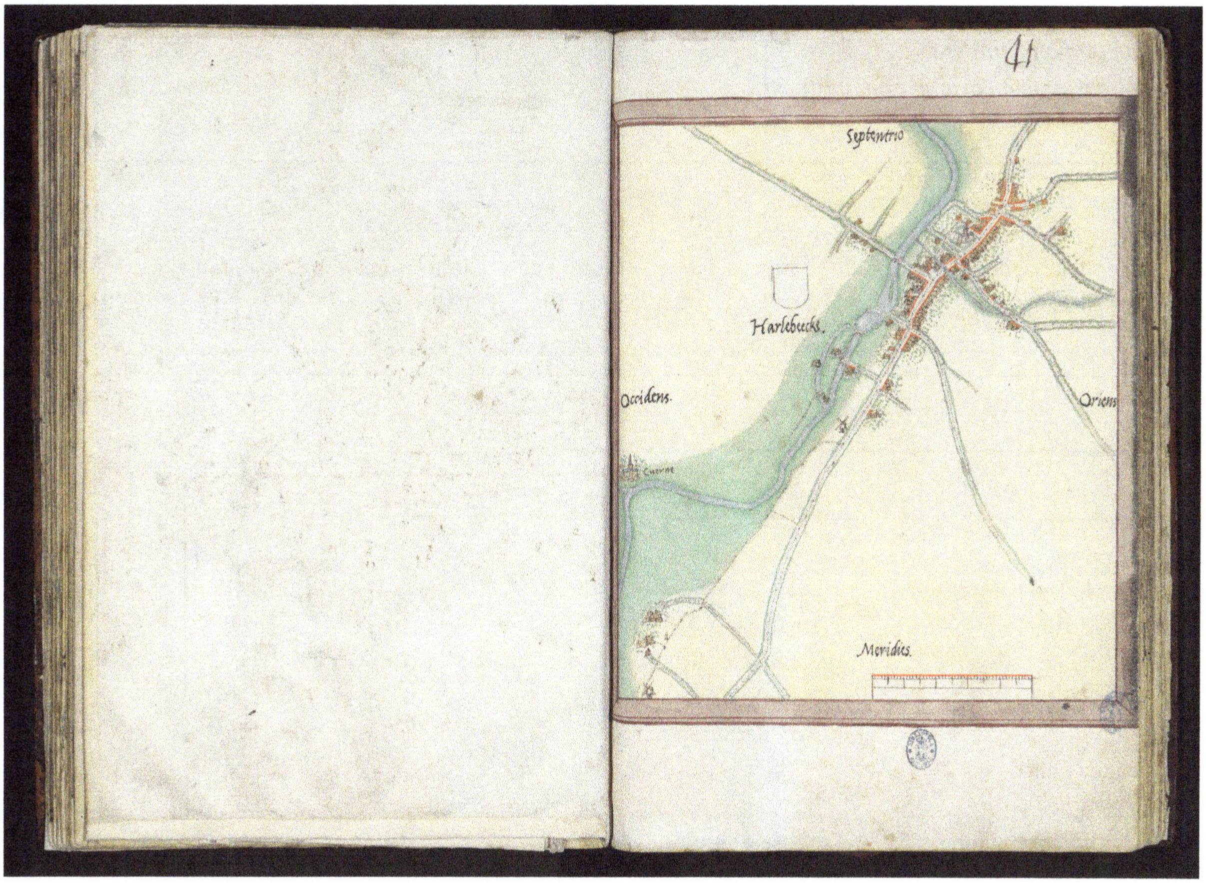
41
Septentrio
Harlebecks.
Occidens.
Oriens
Meridies.

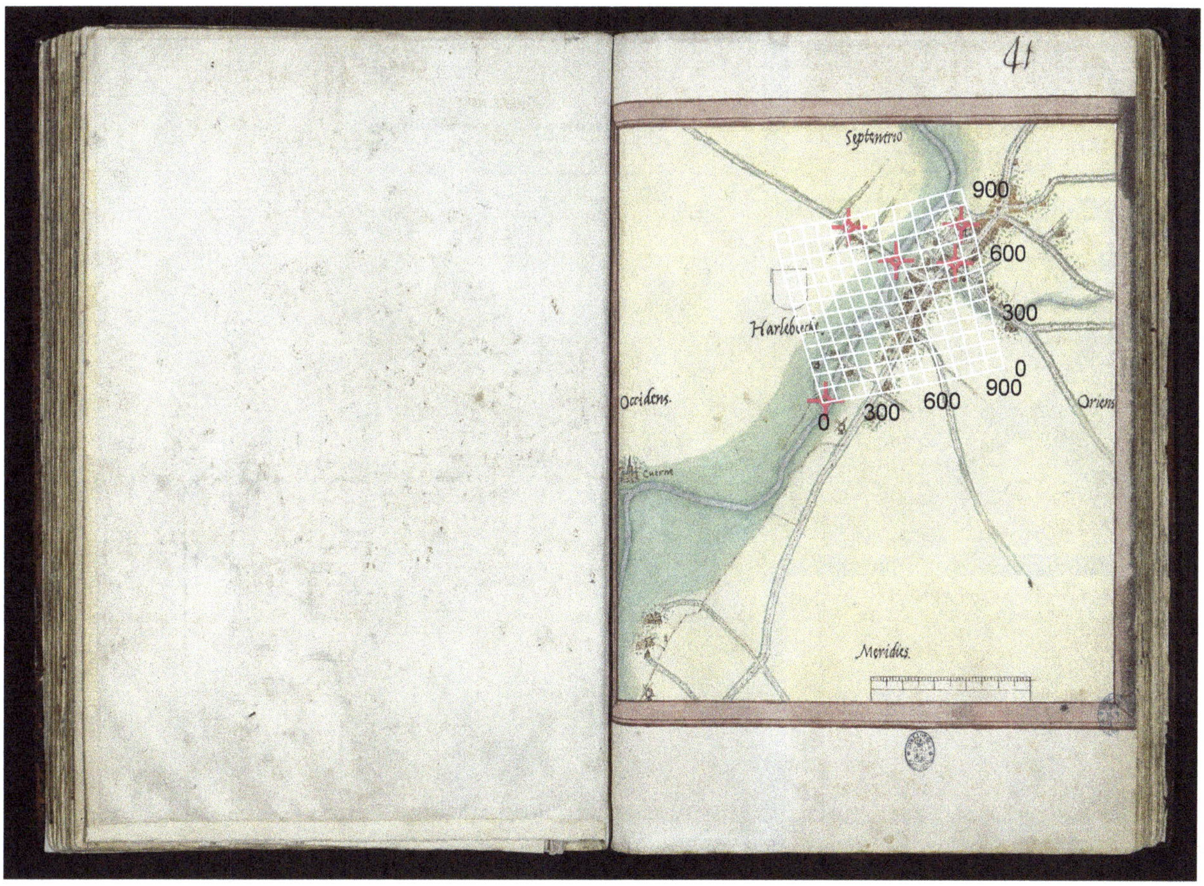
41
Septentrio
900
600
300
0
Harlebecke
Occidens.
0
300
600
900
Oriens
Meridies.

98. Harlingen

Tables 3.98 (measured and analysed).
Figures 3.98 (unmeasured, measured and analysed).

Measured							
Pixels	Scale stick original map (pixels)	Houses	Street sides	Houses per street side	Inhabitants (RF 4.0)	Inhabitants (RF 5.5)	Inhabitants (RF 7.0)
2855	1003	501	81	6.19	2005	2757	3509

Analysed	
May 14, 2023 10:15:02 PM	
Computation with 5 linked points (*italics* points are less reliable):	1. Grote Kerk 2. Heiligeweg 3. Schritsen 4. Noorderhaven 5. Noorderhaven
Reference points (Reliability 0–5)	5
Transformation with 3 parameters	
Scale	1:8760
Rotation	8 °
Mean position error	17m

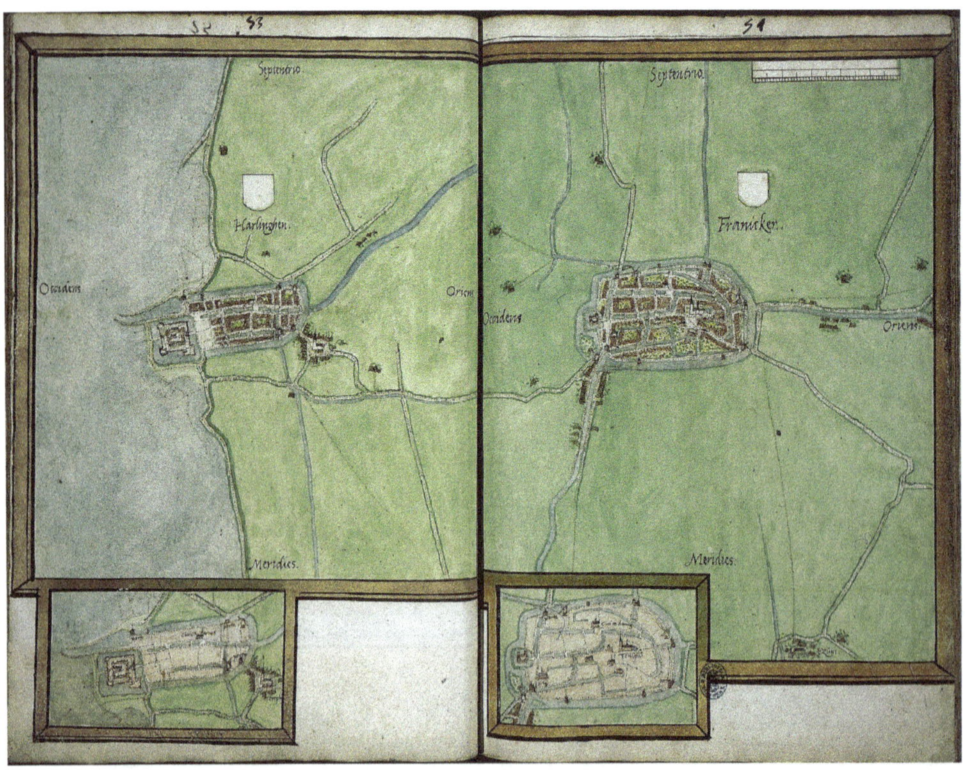

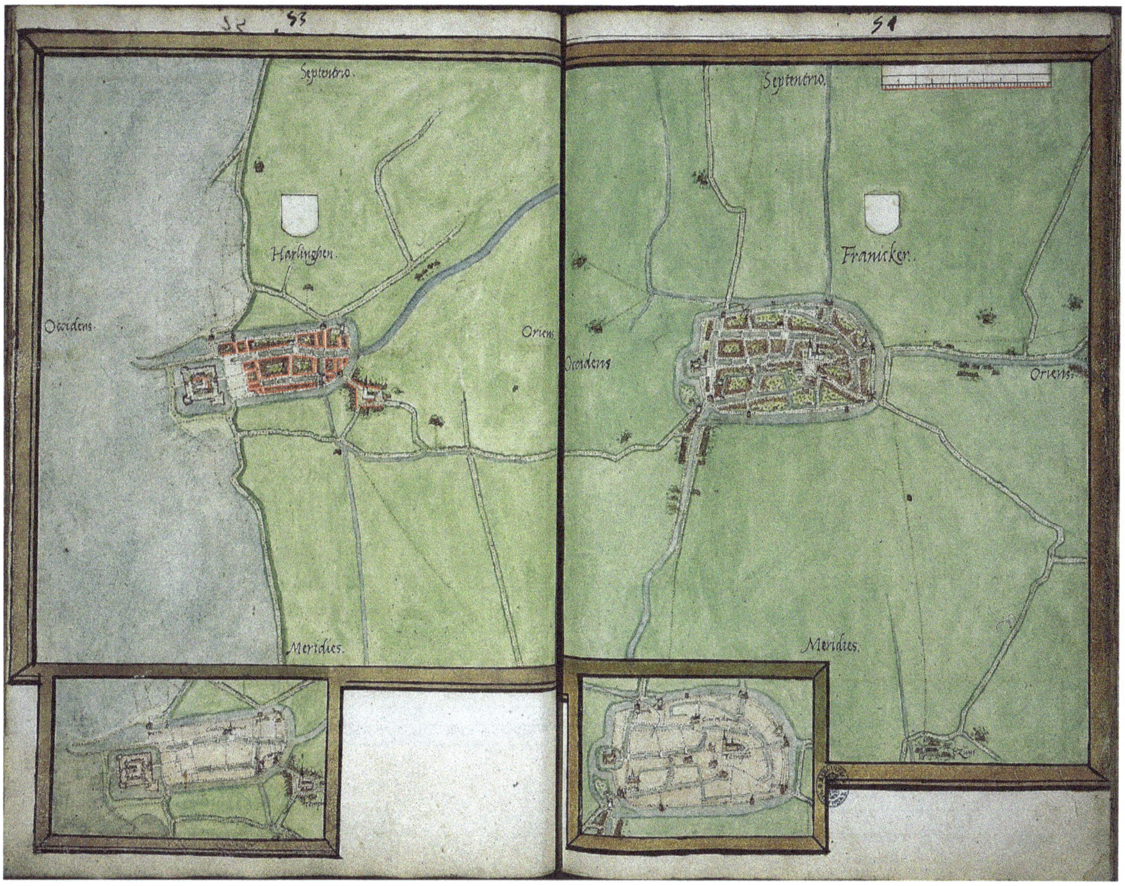
Septentrio.
Harlinghen.
Occidens.
Oriens.
Meridies.
Septentrio.
Franicker.
Occidens
Oriens.
Meridies.

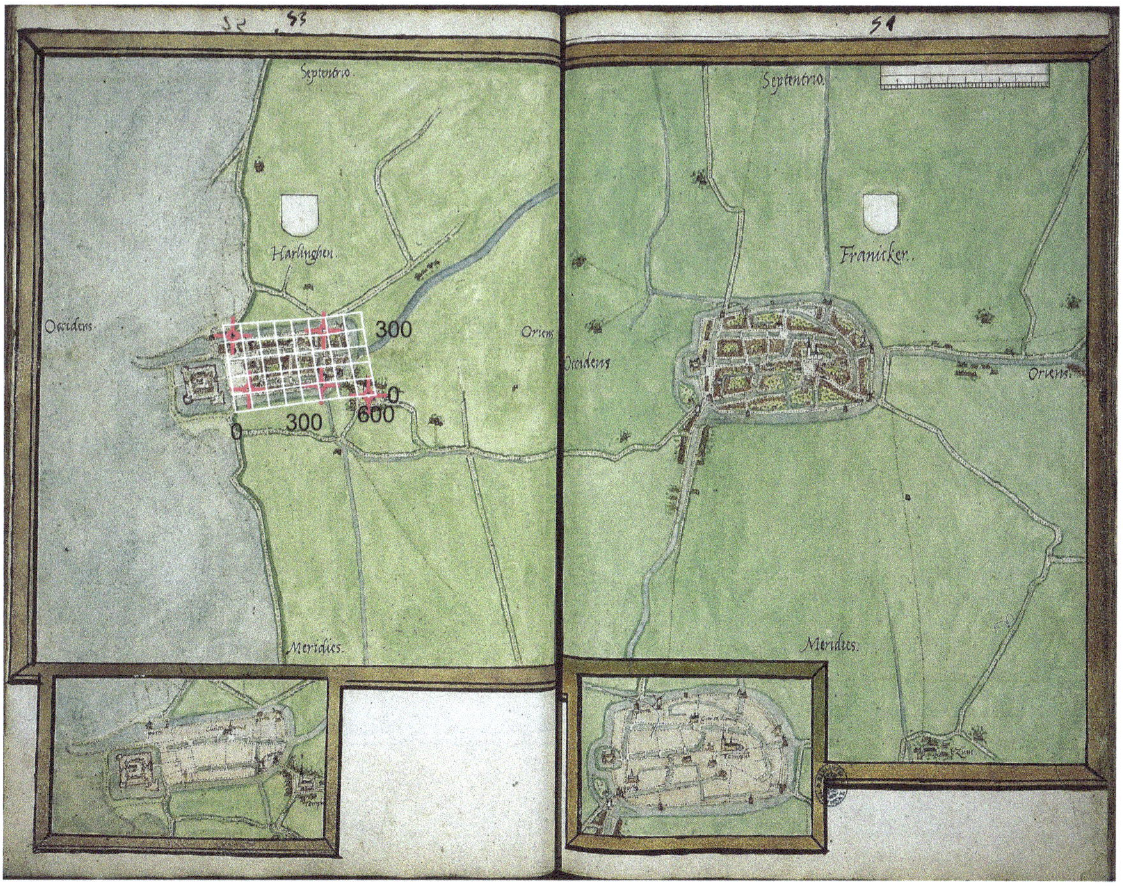
Septentrio.
Harlinghen.
Occidens.
Oriens.
300
0
600
0
300
Meridies.
Septentrio.
Franicker.
Occidens
Oriens.
Meridies.

99. Hasselt

Tables 3.99 (measured and analysed).
Figures 3.99 (unmeasured, measured and analysed).

Measured

Pixels	Scale stick original map (pixels)	Houses	Street sides	Houses per street side	Inhabitants (RF 4.0)	Inhabitants (RF 5.5)	Inhabitants (RF 7.0)
2759	998	484	65	7.45	1938	2665	3392

Analysed

May 16, 2023 9:17:28 AM	
Computation with 5 linked points (*italics* points are less reliable):	1. Sint-Stephanuskerk 2. Julianakade 3. Gasthuisstraat 4. Baangracht 5. Eikenlaan
Reference points (Reliability 0–5)	5
Transformation with 3 parameters	
Scale	1:8110
Rotation	12 °
Mean position error	18m

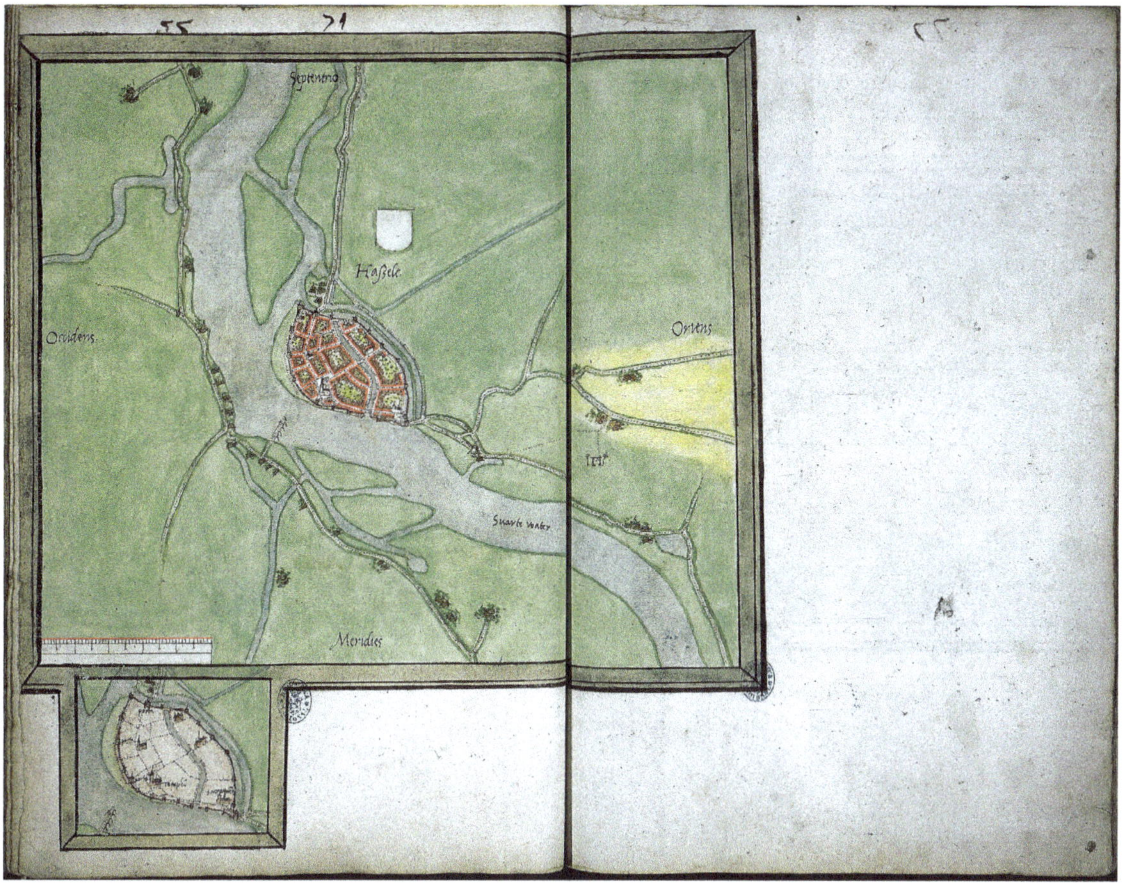
Haßele
Occidens
Oriens
Meridies

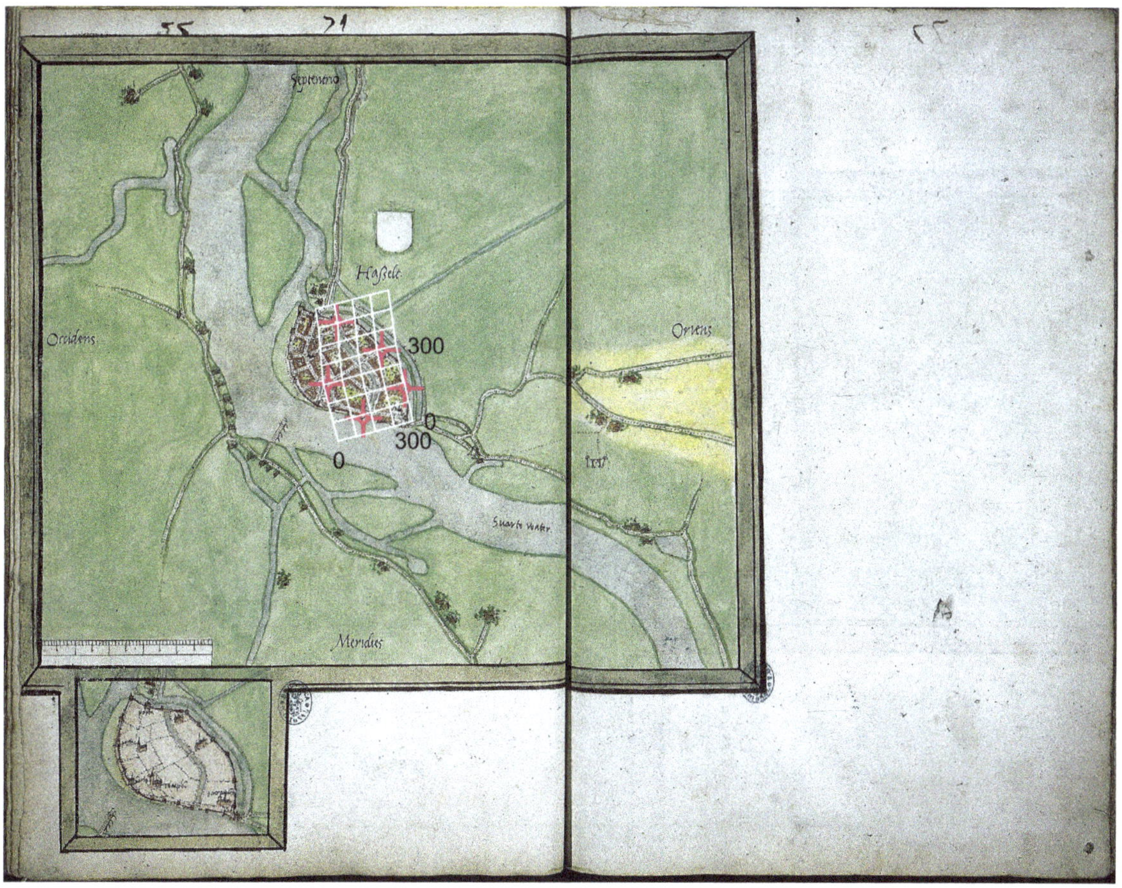
Haßele
Occidens
Oriens
300
0
300
0
Meridies

100. Hattem
Tables 3.100 (measured and analysed).
Figures 3.100 (unmeasured, measured and analysed).

Measured							
Pixels	Scale stick original map (pixels)	Houses	Street sides	Houses per street side	Inhabitants (RF 4.0)	Inhabitants (RF 5.5)	Inhabitants (RF 7.0)
1275	1012	224	35	6.40	895	1231	1567

Analysed	
April 18, 2023 9:47:22 AM	
Computation with 5 linked points (*italics* points are less reliable):	1. Andreaskerk 2. Dijkpoort 3. Zuidwal 4. Molencatense Beek 5. Ridderstraat
Reference points (Reliability 0–5)	5
Transformation with 3 parameters	
Scale	1:8620
Rotation	8 °
Mean position error	16m

88.
Septentrio.
Isel fl.
Hattem
Oriens
Meridies.

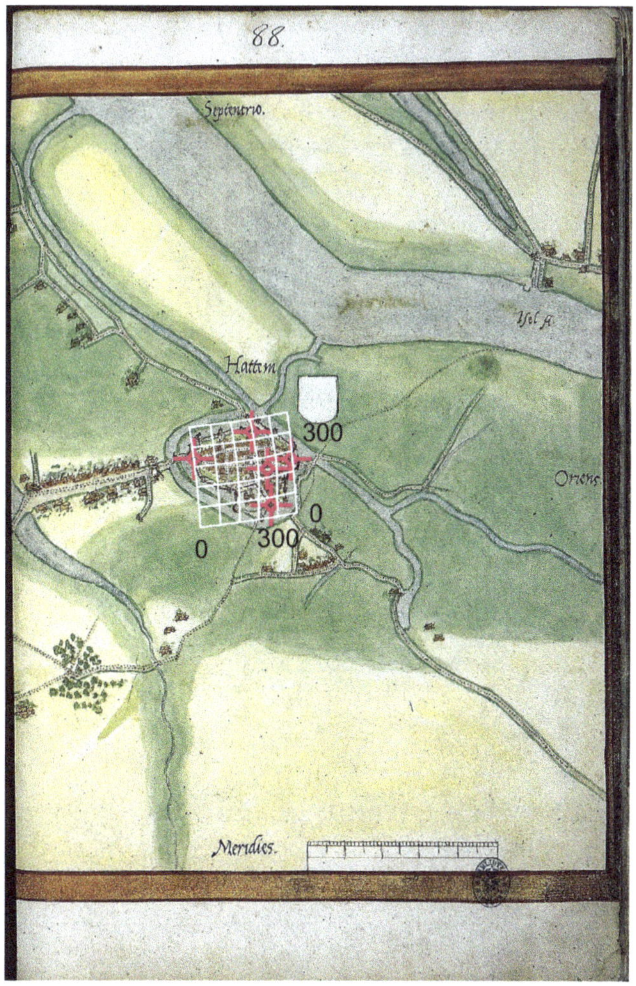
88.
Septentrio.
Isel fl.
Hattem
300
0
0
300
Oriens
Meridies.

101. Heerenberg, 's-

Tables 3.101 (measured and analysed).
Figures 3.101 (unmeasured, measured and analysed).

Measured

Pixels	Scale stick original map (pixels)	Houses	Street sides	Houses per street side	Inhabitants (RF 4.0)	Inhabitants (RF 5.5)	Inhabitants (RF 7.0)
1503	1011	264	29	9.10	1056	1452	1848

Analysed

May 18, 2023 1:12:49 PM	
Computation with 5 linked points (*italics* points are less reliable):	1. Sint Pancratiuskerk *2. Hof van Bergh* *3. Kellenstraat* *4. Walsteeg* *5. Marktstraat*
Reference points (Reliability 0–5)	1
Transformation with 3 parameters	
Scale	1:7280
Rotation	19 °
Mean position error	26m

Septentrio
s'Herē Berg.
Occidens.
Meridies

Septentrio
s'Herē Berg.
Occidens.
300
0
300
0
Meridies

102. Helmond

Tables 3.102 (measured and analysed).
Figures 3.102 (unmeasured, measured and analysed).

Measured							
Pixels	Scale stick original map (pixels)	Houses	Street sides	Houses per street side	Inhabitants (RF 4.0)	Inhabitants (RF 5.5)	Inhabitants (RF 7.0)
1878	missing	330	33	9.99	1319	1814	2309

Analysed	
September 20, 2023 4:07:29 PM	
Computation with 5 linked points (*italics* points are less reliable):	1. Kasteel Helmond 2. Sint-Lambertuskerk 3. Markt 4. Oude Aa 5. De Wiel
Reference points (Reliability 0–5)	5
Transformation with 3 parameters	
Scale	1:7840
Rotation	10 °
Mean position error	9m

Noordt.
Helmont.
West

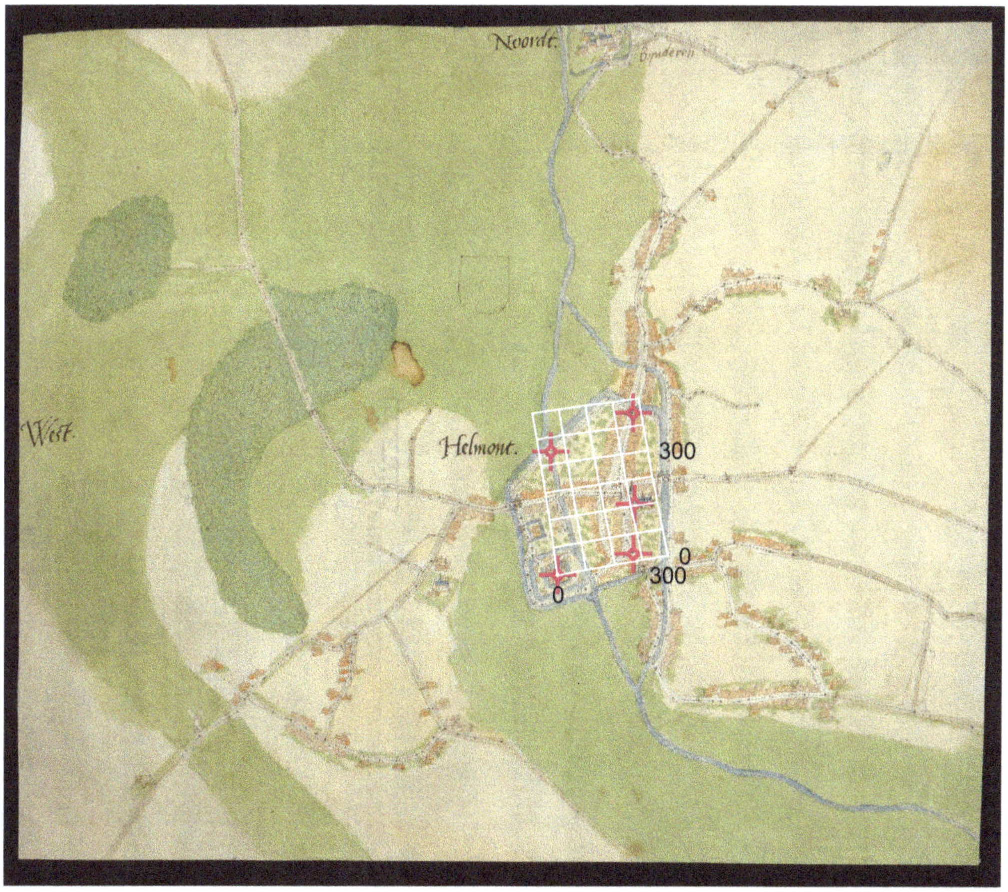
Noordt.
Helmont.
West
300
0
300
0

103. Herentals

Tables 3.103 (measured and analysed).
Figures 3.103 (unmeasured, measured and analysed).

Measured

Pixels	Scale stick original map (pixels)	Houses	Street sides	Houses per street side	Inhabitants (RF 4.0)	Inhabitants (RF 5.5)	Inhabitants (RF 7.0)
8799	missing	1545	173	8.93	6180	8498	10,816

Analysed

September 22, 2023 12:51:03 PM	
Computation with 5 linked points (*italics* points are less reliable):	1. Sint-Waltrudiskerk 2. Belfort 3. Zandpoort 4. Bovenpoort 5. Burchtstraat
Reference points (Reliability 0–5)	5
Transformation with 3 parameters	
Scale	1:8140
Rotation	12 °
Mean position error	17m

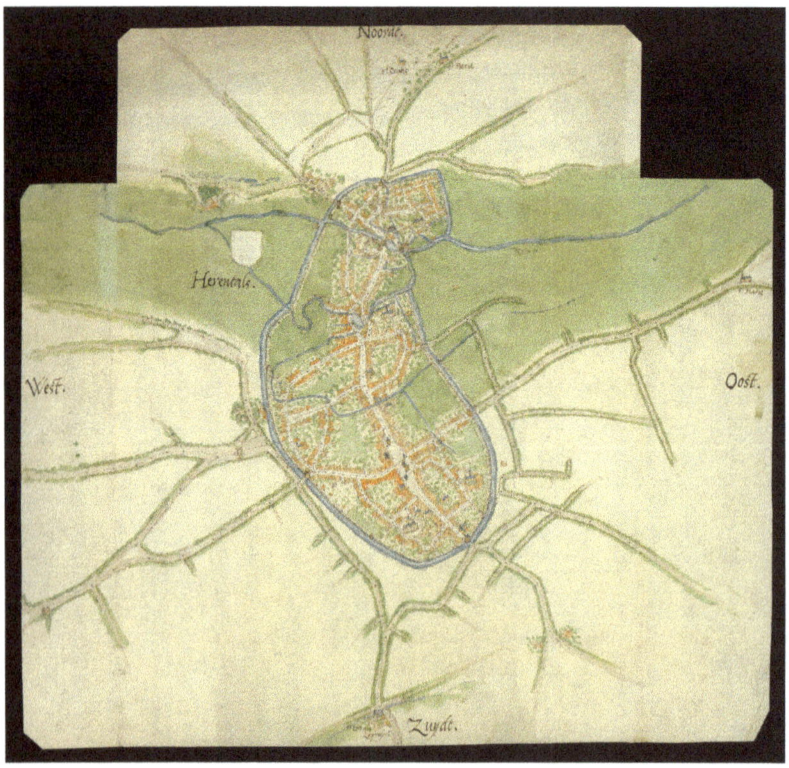

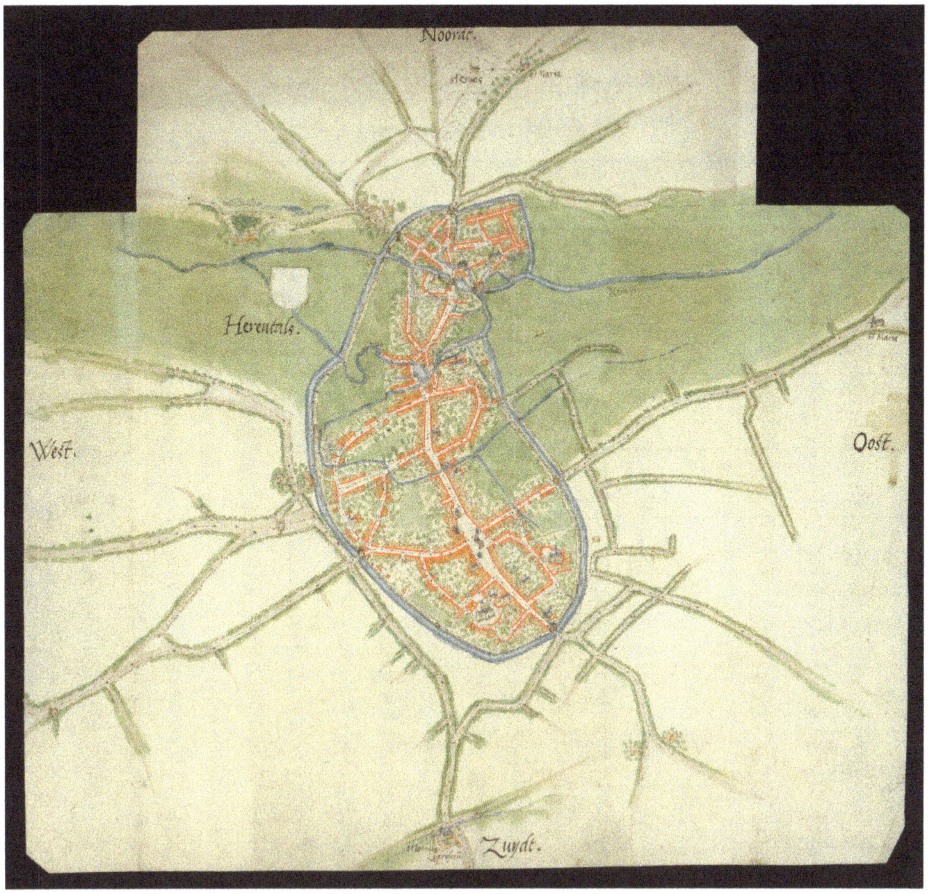
Noorde.
Herentals.
West.
Oost.
Zuydt.

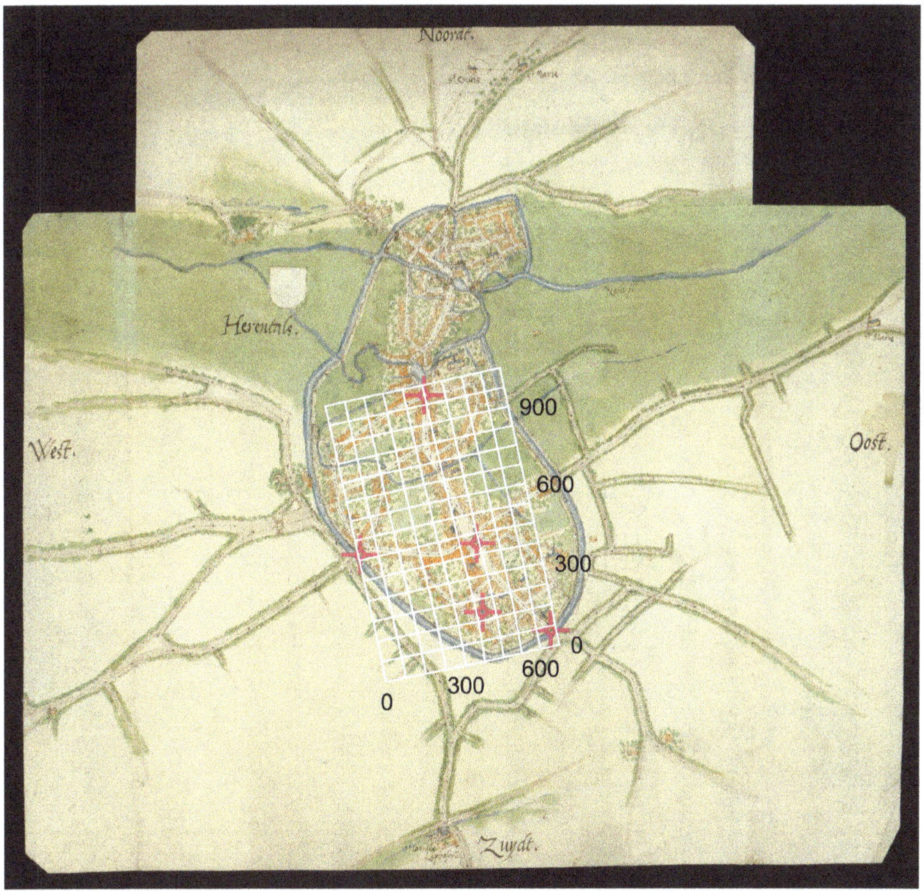
Noorde.
Herentals.
West.
Oost.
Zuydt.
900
600
300
0
600
300
0

104. Hertogenbosch, 's-

Tables 3.104 (measured and analysed).
Figures 3.104 (unmeasured, measured and analysed).

Measured							
Pixels	Scale stick original map (pixels)	Houses	Street sides	Houses per street side	Inhabitants (RF 4.0)	Inhabitants (RF 5.5)	Inhabitants (RF 7.0)
18,081	missing	3175	547	5.80	12,700	17,463	22,226

Analysed	
September 22, 2023 6:47:54 PM	
Computation with 5 linked points (*italics* points are less reliable):	1. Sint-Janskathedraal 2. Markt 3. Sint-Catharinakerk 4. Mariënburg 5. Orthenstraat
Reference points (Reliability 0–5)	5
Transformation with 3 parameters	
Scale	1:8670
Rotation	14 °
Mean position error	19m

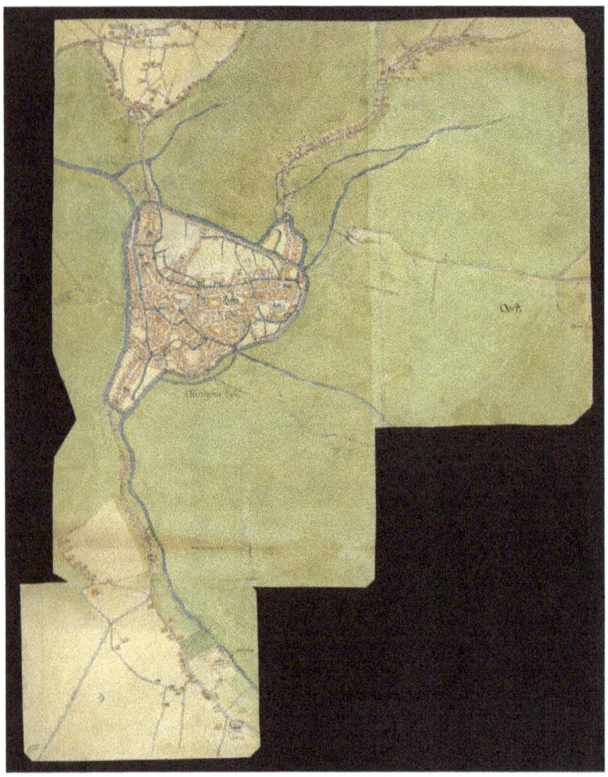

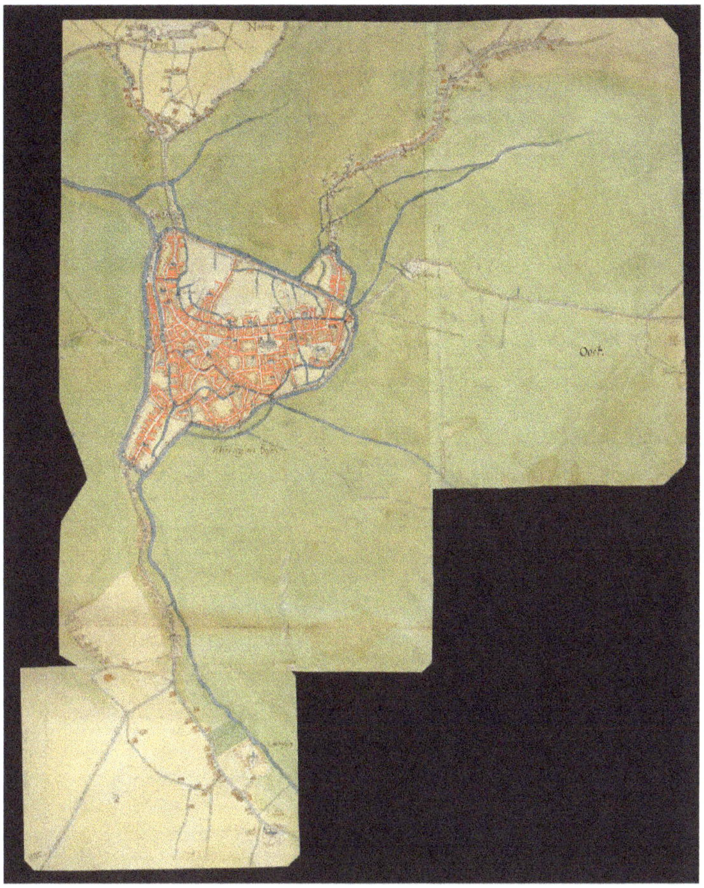

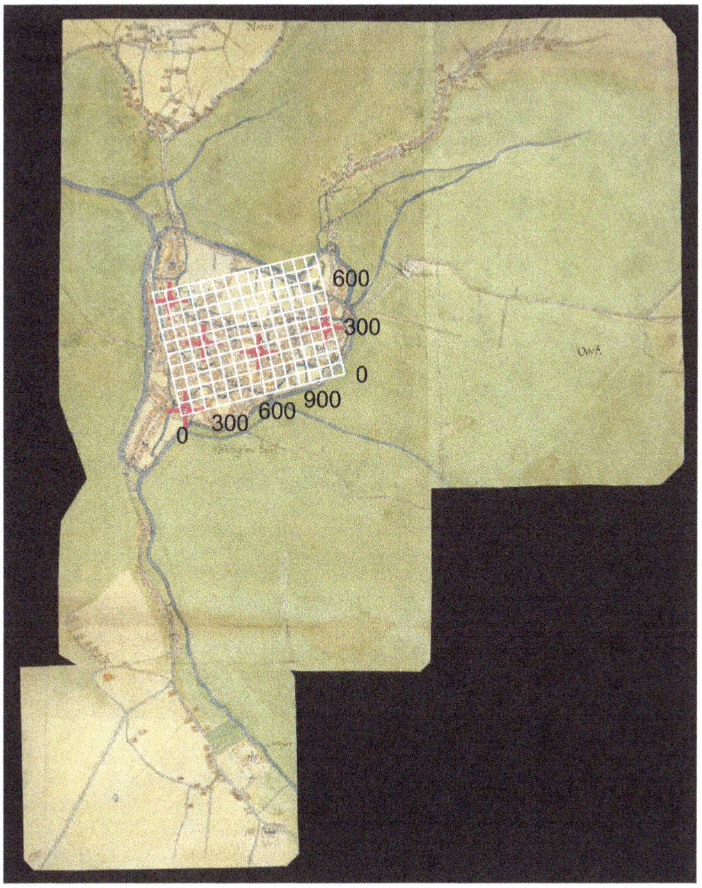
600
300
0
0
300
600
900

105. Herve
Tables 3.105 (measured and analysed).
Figures 3.105 (unmeasured, measured and analysed).

Measured

Pixels	Scale stick original map (pixels)	Houses	Street sides	Houses per street side	Inhabitants (RF 4.0)	Inhabitants (RF 5.5)	Inhabitants (RF 7.0)
1986	missing	349	52	6.71	1395	1918	2441

Analysed

September 22, 2023 7:19:08 PM	
Computation with 5 linked points (*italics* points are less reliable):	1. Église Saint-Jean-Baptiste 2. Rue Haute 3. Rue Jardon 4. Pré de Ville 5. Rue Chinville
Reference points (Reliability 0–5)	5
Transformation with 3 parameters	
Scale	1:7370
Rotation	15 °
Mean position error	12m

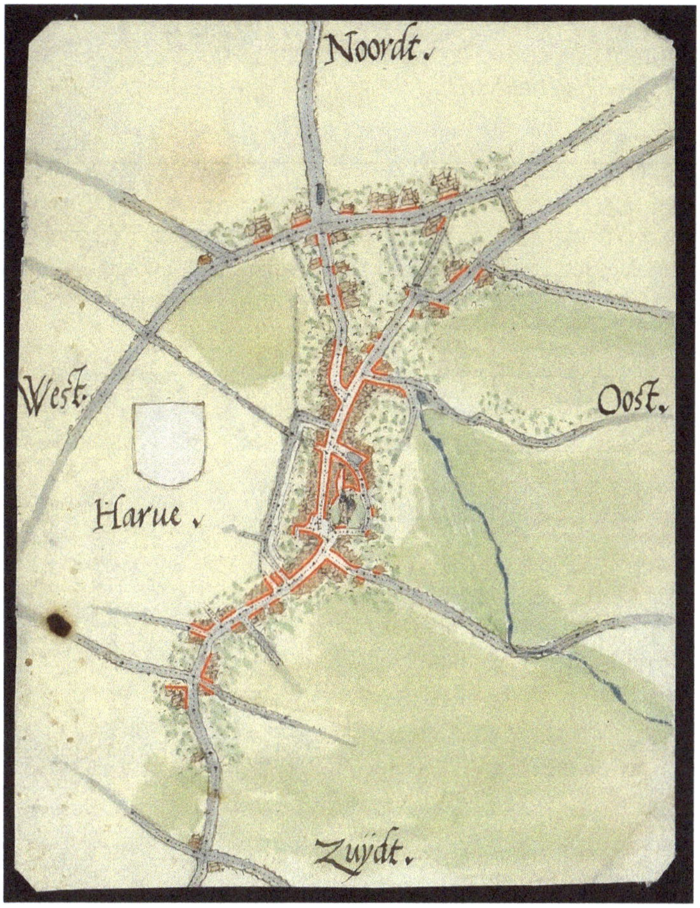
Noordt.
West.
Oost.
Harue.
Zuydt.

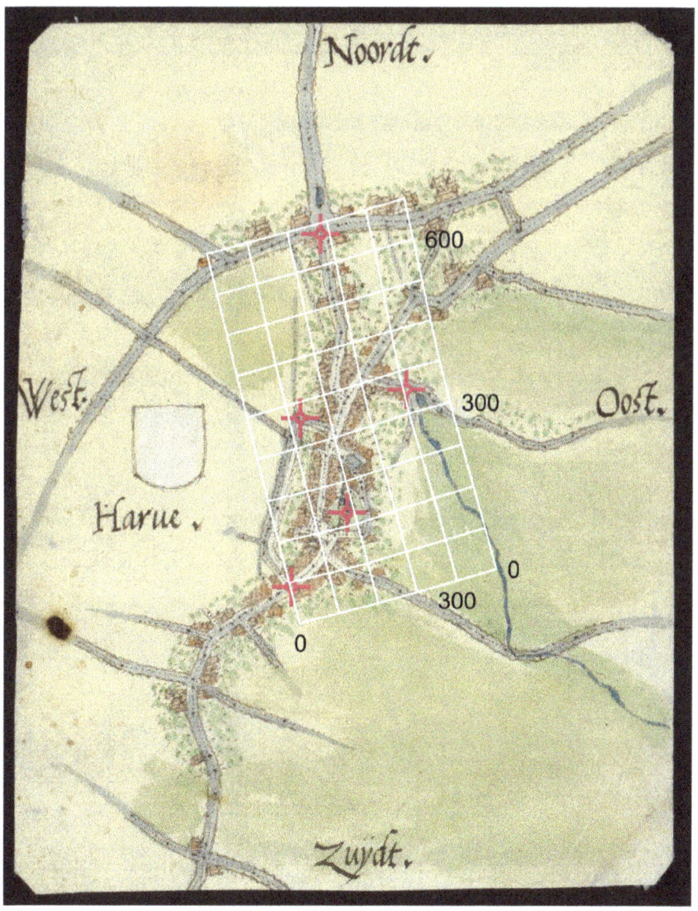
Noordt.
600
West.
300
Oost.
Harue.
0
300
0
Zuydt.

106. Herzogenrath

Tables 3.106 (measured and analysed).
Figures 3.106 (unmeasured, measured and analysed).

Measured

Pixels	Scale stick original map (pixels)	Houses	Street sides	Houses per street side	Inhabitants (RF 4.0)	Inhabitants (RF 5.5)	Inhabitants (RF 7.0)
1558	missing	274	47	5.82	1095	1505	1915

Analysed

September 22, 2023 8:09:43 PM	
Computation with 5 linked points (*italics* points are less reliable):	1. Rothenburg 2. Sankt Gertrud 3. Abdij 4. Kleikstrasse 5. Dammstrasse
Reference points (Reliability 0–5)	5
Transformation with 3 parameters	
Scale	1:7660
Rotation	11 °
Mean position error	6m

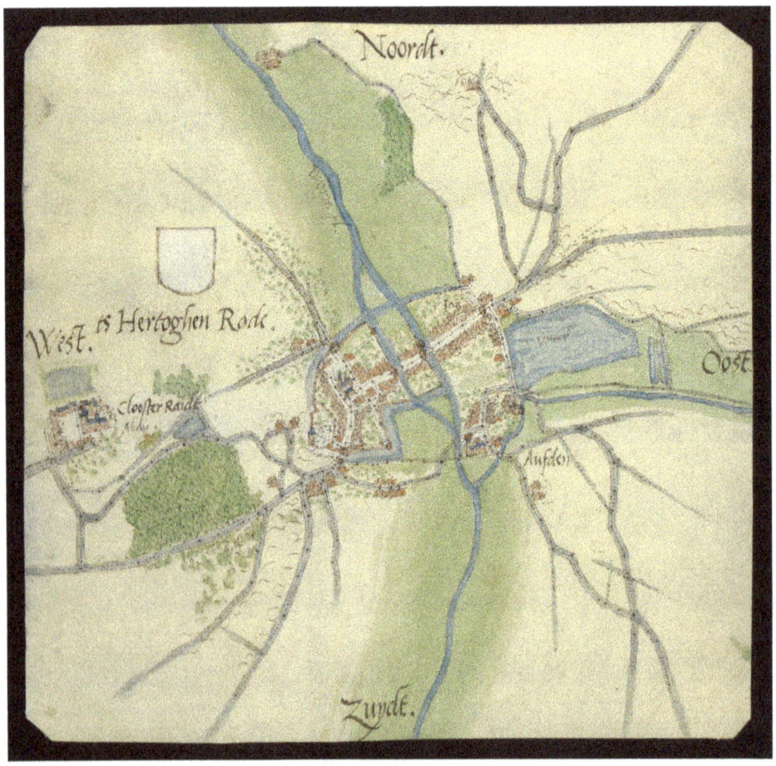

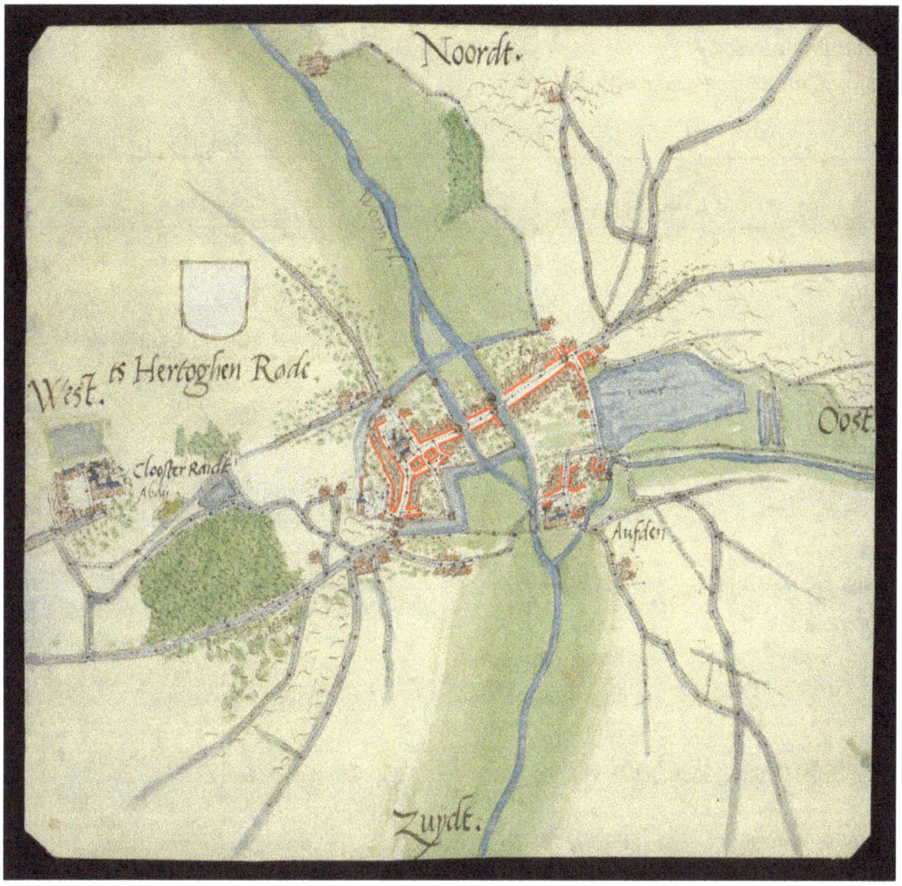
Noordt.
West.
ts Hertoghen Rode
Clooster Raide
Abdye
Aufden
Oost.
Zuydt.

Noordt.
West.
ts Hertoghen Rode
Clooster Raide
Oost.
300
0
900
600
300
0
Zuydt.

107. Hesdin

Tables 3.107 (measured and analysed).
Figures 3.107 (unmeasured, measured and analysed).

Measured

Pixels	Scale stick original map (pixels)	Houses	Street sides	Houses per street side	Inhabitants (RF 4.0)	Inhabitants (RF 5.5)	Inhabitants (RF 7.0)
1754	1026	308	62	4.97	1232	1694	2156

Analysed

May 18, 2023 7:39:03 PM	
Computation with 5 linked points (*italics* points are less reliable):	*1. La Canche midi* 2. La Canche ouest 3. La Canche est 4. Rue de Richelieu 5. Rue de la Paroisse
Reference points (Reliability 0–5)	4
Transformation with 3 parameters	
Scale	1:7760
Rotation	12 °
Mean position error	8m

30
Septentrio.
Occidens.
Hesdyn fert.
Oriens.
Meridies.
Hesdyn
Oriens
Meridies.

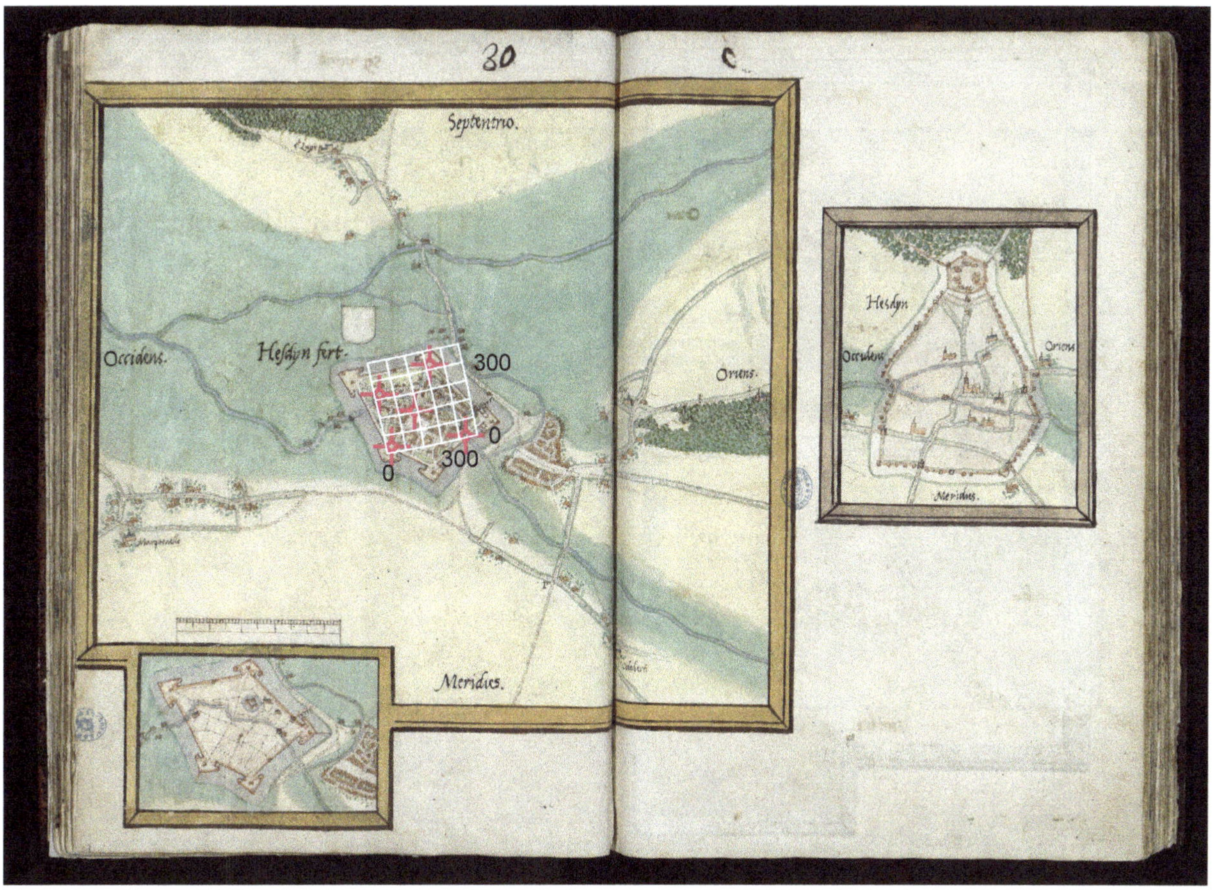
30
Septentrio.
Occidens.
Hesdyn fert.
300
0
0
300
Oriens.
Meridies.
Hesdyn
Oriens
Meridies.

108. Heukelum

Tables 3.108 (measured and analysed).
Figures 3.108 (unmeasured, measured and analysed).

Measured							
Pixels	Scale stick original map (pixels)	Houses	Street sides	Houses per street side	Inhabitants (RF 4.0)	Inhabitants (RF 5.5)	Inhabitants (RF 7.0)
953	1007	167	21	7.97	669	920	1171

Analysed	
May 21, 2023 10:04:09 PM	
Computation with 5 linked points (*italics* points are less reliable):	1. Hervormde Kerk 2. Kasteel Merckenburg 3. Rooyenburg 4. Kruissteiger *5. Voorstraat*
Reference points (Reliability 0–5)	4
Transformation with 3 parameters	
Scale	1:8720
Rotation	8 °
Mean position error	20m

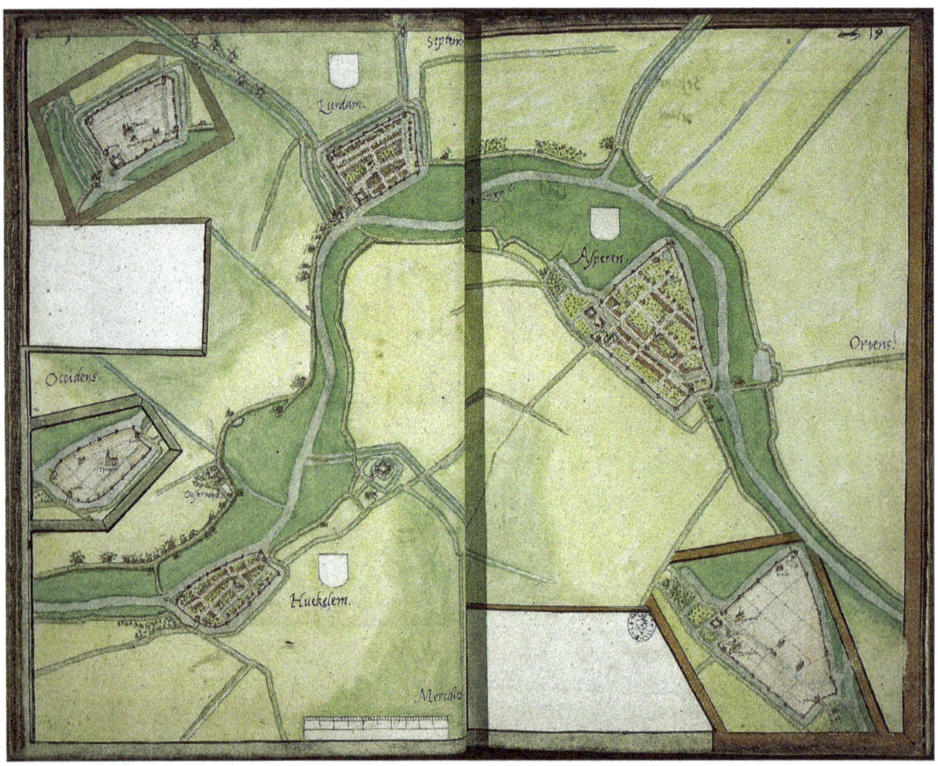

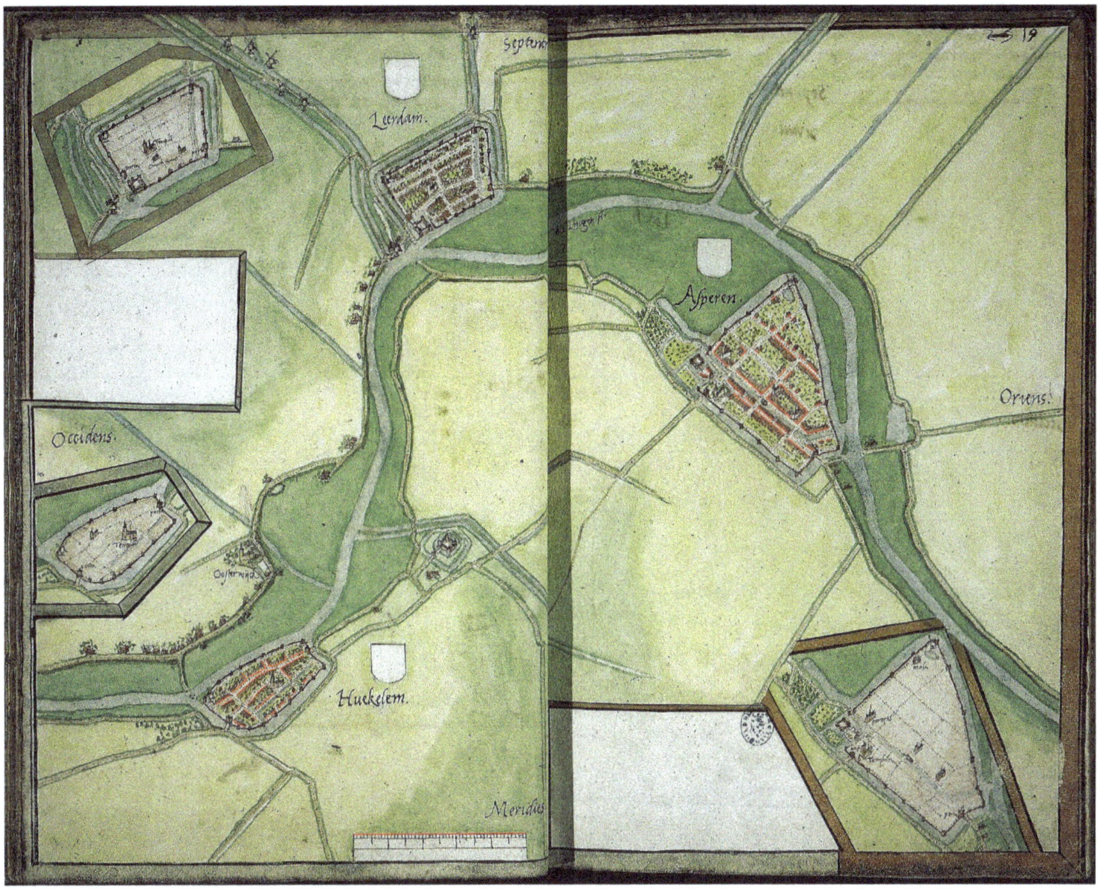
Lurdam.
Asperen
Occidens.
Oriens.
Huckelem.

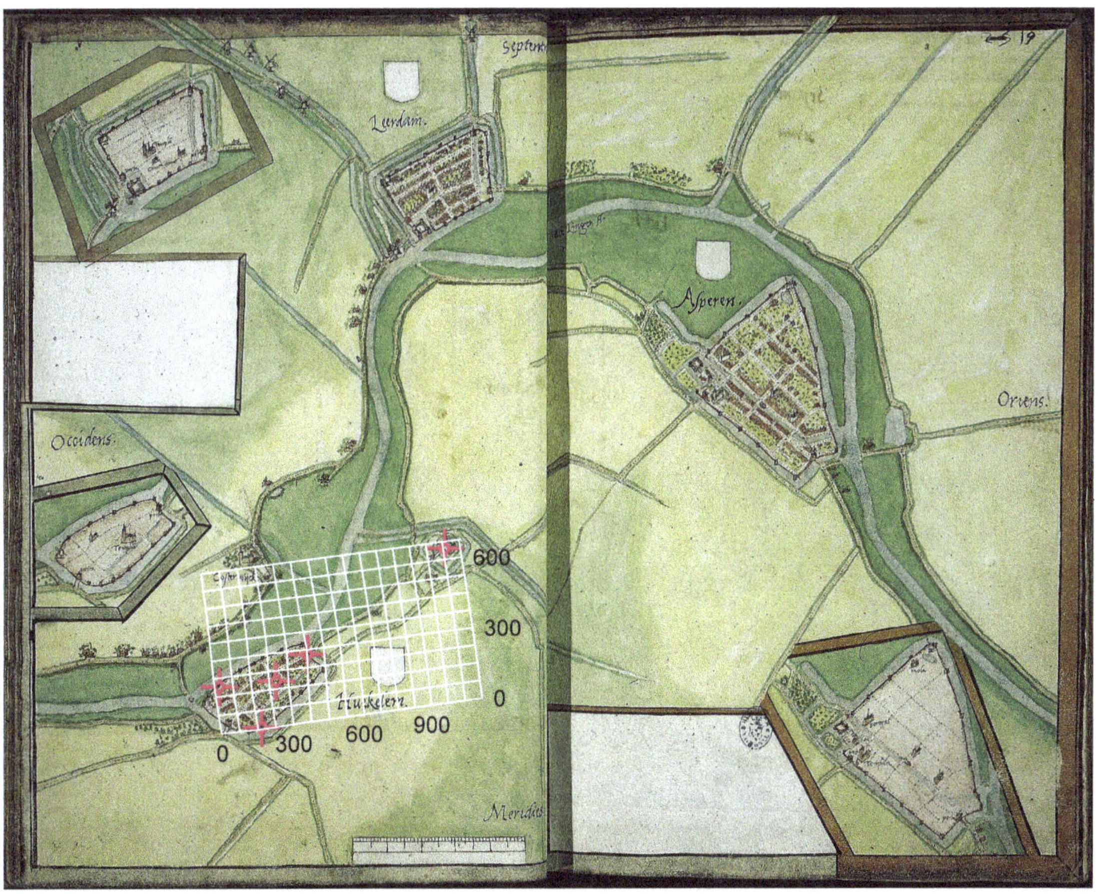
Lurdam.
Asperen
Occidens.
Oriens.
Huckelem.
600
300
0
0
300
600
900

109. Heusden

Tables 3.109 (measured and analysed).
Figures 3.109 (unmeasured, measured and analysed).

Measured							
Pixels	Scale stick original map (pixels)	Houses	Street sides	Houses per street side	Inhabitants (RF 4.0)	Inhabitants (RF 5.5)	Inhabitants (RF 7.0)
2563	1006	450	47	9.58	1800	2475	3150

Analysed	
May 22, 2023 9:06:21 AM	
Computation with 5 linked points (*italics* points are less reliable):	1. Sint-Catharinakerk 2. Demer 3. Molen 4. Putterstraat 5. Schramstraat
Reference points (Reliability 0–5)	5
Transformation with 3 parameters	
Scale	1:7710
Rotation	16 °
Mean position error	17m

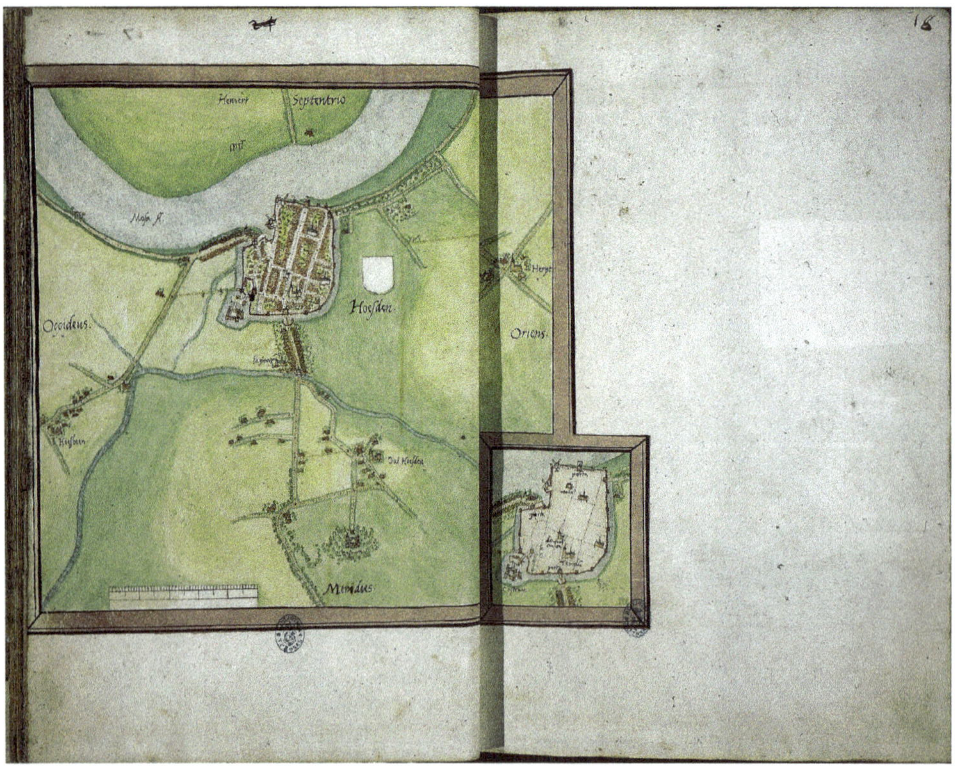

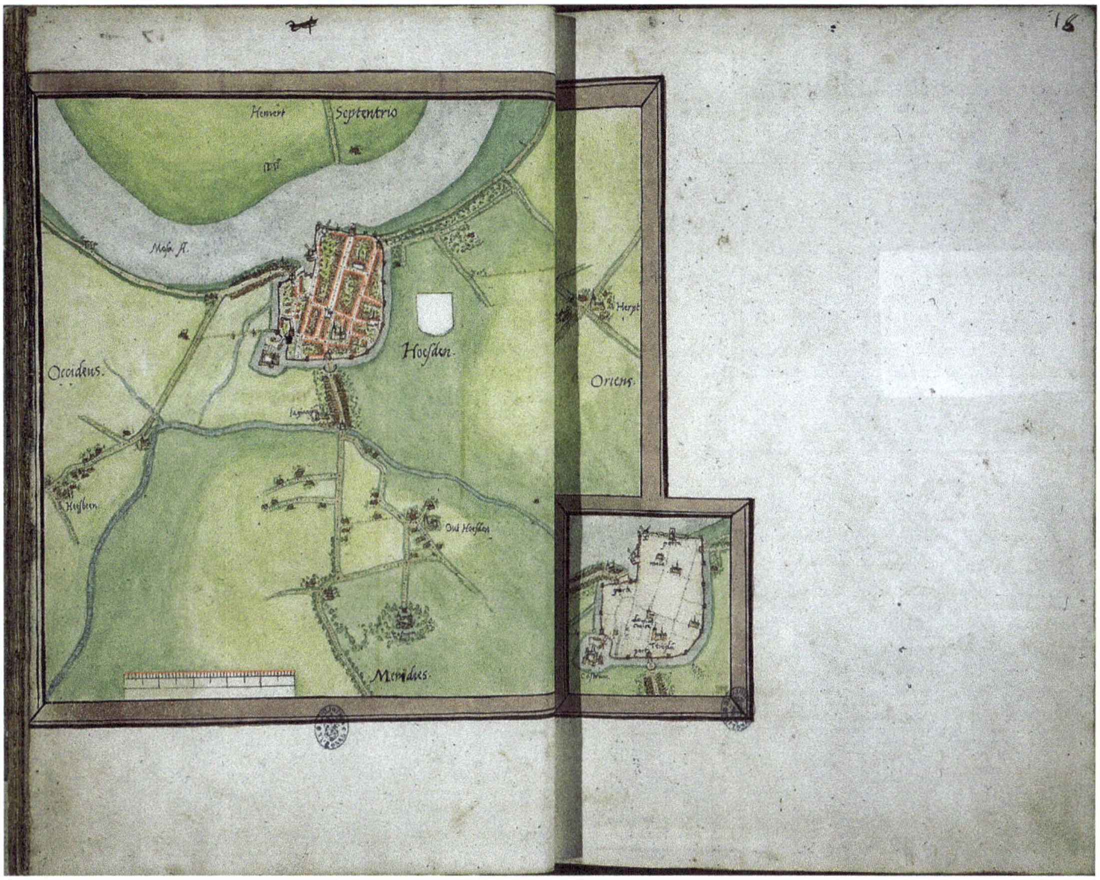
Septentrio
Occidens.
Oriens.
Meridies.

Septentrio
300
0
300
0
Occidens.
Oriens.
Meridies.

110. Hindeloopen

Tables 3.110 (measured and analysed).
Figures 3.110 (unmeasured, measured and analysed).

Measured							
Pixels	Scale stick original map (pixels)	Houses	Street sides	Houses per street side	Inhabitants (RF 4.0)	Inhabitants (RF 5.5)	Inhabitants (RF 7.0)
851	996	149	16	9.34	598	822	1046

Analysed	
May 22, 2023 3:34:24 PM	
Computation with 5 linked points (*italics* points are less reliable):	*1. 't Oost* *2. Nieuweweg* 3. Klinten 4. Dijkweg *5. Buren*
Reference points (Reliability 0–5)	2
Transformation with 3 parameters	
Scale	1:7750
Rotation	8 °
Mean position error	17m

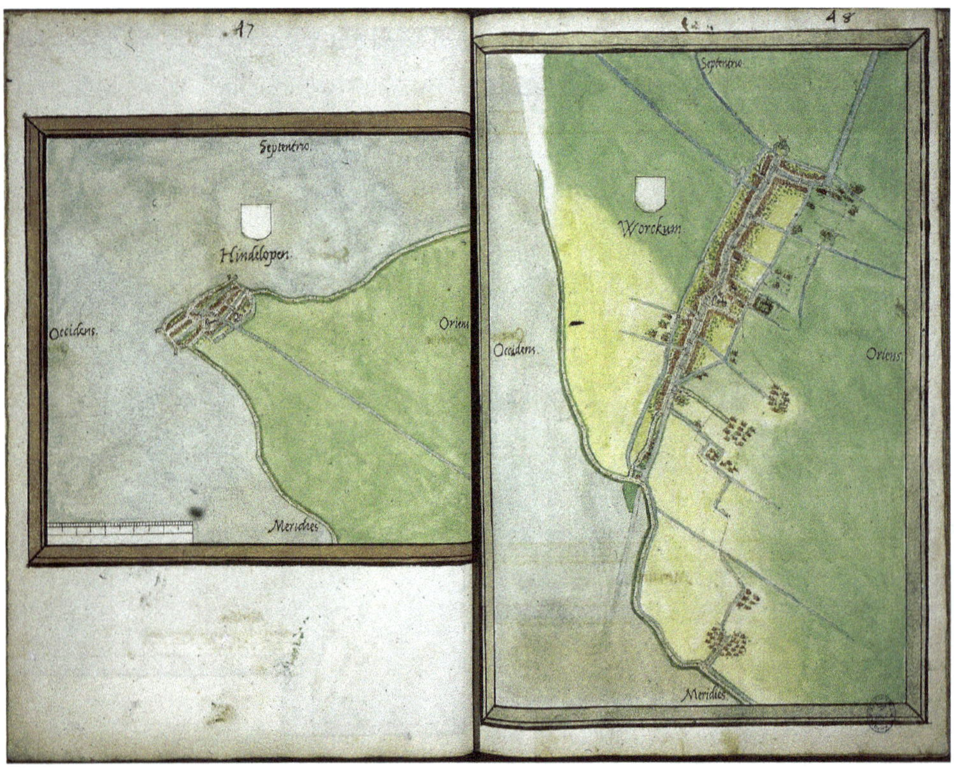

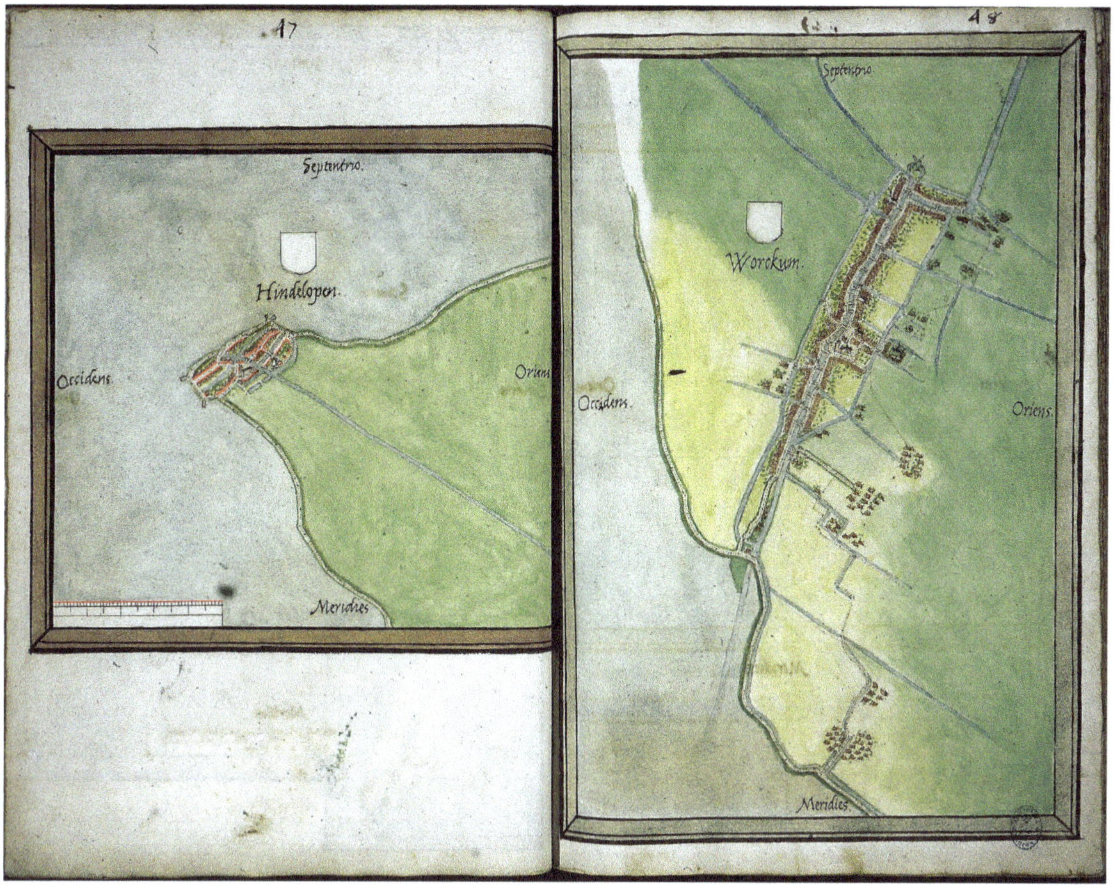
47
Septentrio.
Hindelopen.
Occidens.
Oriens
Meridies
48
Septentrio
Worckum.
Occidens.
Oriens.
Meridies

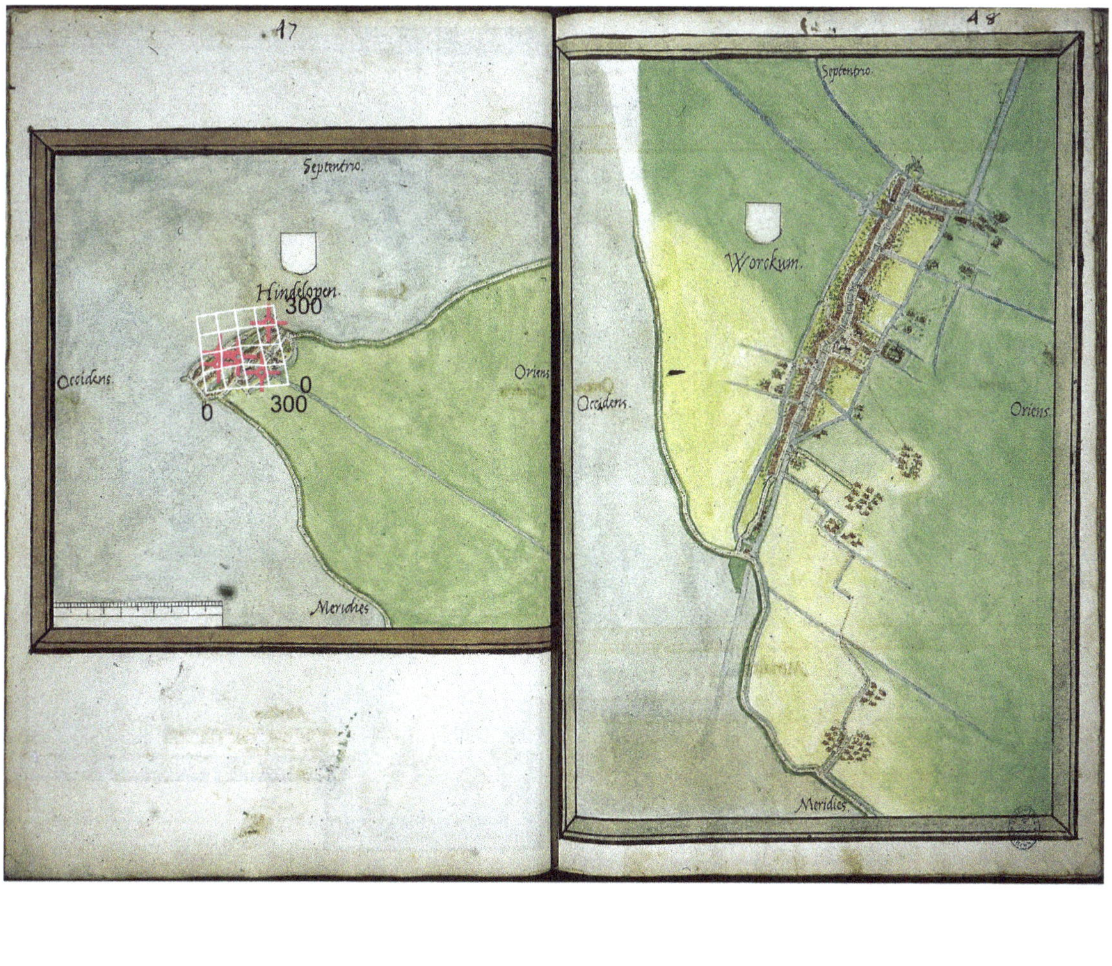
47
Septentrio.
Hindelopen.
300
0
0
300
Occidens.
Oriens
Meridies
48
Septentrio
Worckum.
Occidens.
Oriens.
Meridies

111. Hondschoote

Tables 3.111 (measured and analysed).
Figures 3.111 (unmeasured, measured and analysed).

Measured							
Pixels	Scale stick original map (pixels)	Houses	Street sides	Houses per street side	Inhabitants (RF 4.0)	Inhabitants (RF 5.5)	Inhabitants (RF 7.0)
6079	1026	1067	99	10.78	4270	5871	7472

Analysed	
May 22, 2023 4:56:29 PM	
Computation with 5 linked points (*italics* points are less reliable):	1. Église Saint-Vaast 2. Rue de la Cour 3. Rue de Saint Winoc 4. Rue de la liberation *5. Rue Coppens*
Reference points (Reliability 0–5)	4
Transformation with 3 parameters	
Scale	1:8240
Rotation	7 °
Mean position error	22m

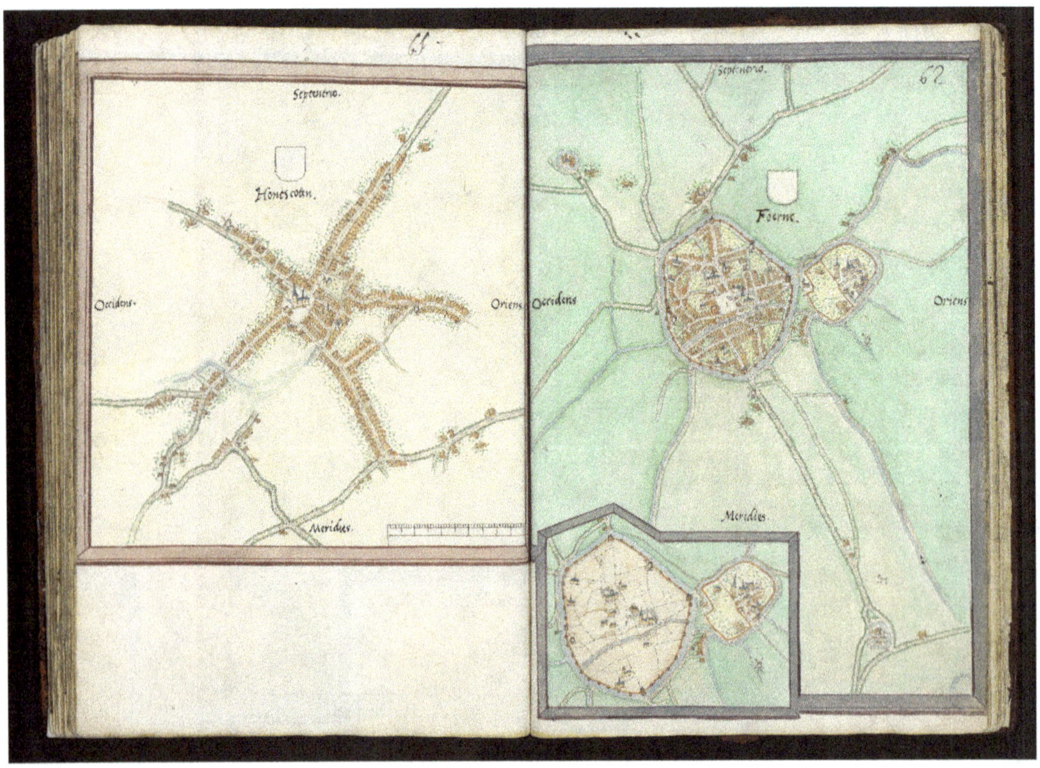

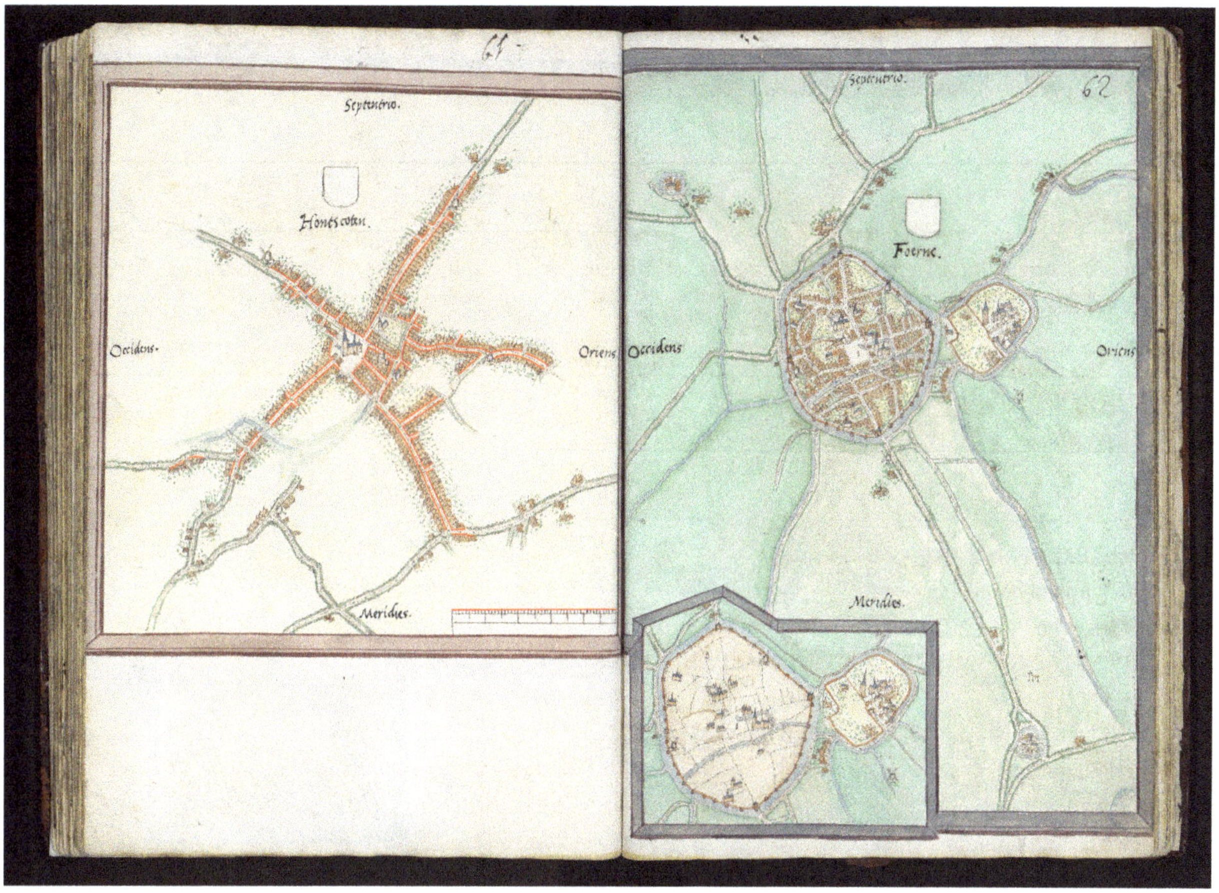
Septentrio.
Honts coten.
Occidens.
Oriens
Meridies.
Septentrio.
62
Foerne.
Occidens
Oriens
Meridies.

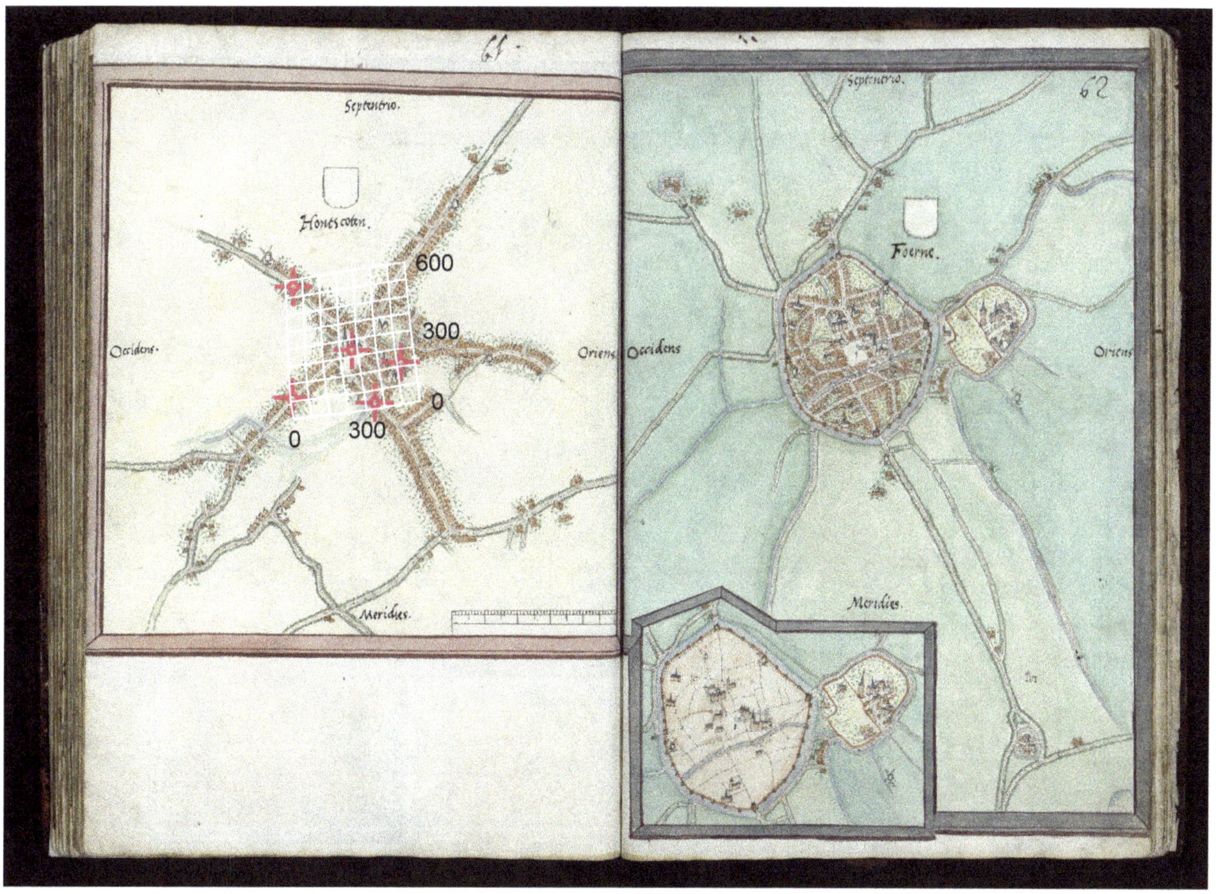
Septentrio.
Honts coten.
600
300
0
0
300
Occidens.
Oriens
Meridies.
Septentrio.
62
Foerne.
Occidens
Oriens
Meridies.

112. Hoorn

Tables 3.112 (measured and analysed).
Figures 3.112 (unmeasured, measured and analysed).

Measured							
Pixels	Scale stick original map (pixels)	Houses	Street sides	Houses per street side	Inhabitants (RF 4.0)	Inhabitants (RF 5.5)	Inhabitants (RF 7.0)
7811	1006	1372	212	6.47	5487	7544	9601

Analysed	
May 22, 2023 8:07:08 PM	
Computation with 5 linked points | (*italics* points are less reliable):	1. Grote Kerk 2. Hoofdtoren 3. Oosterkerk 4. Grote Oost 5. Westerdijk
Reference points (Reliability 0–5)	5
Transformation with 3 parameters	
Scale	1:8660
Rotation	10 °
Mean position error	17m

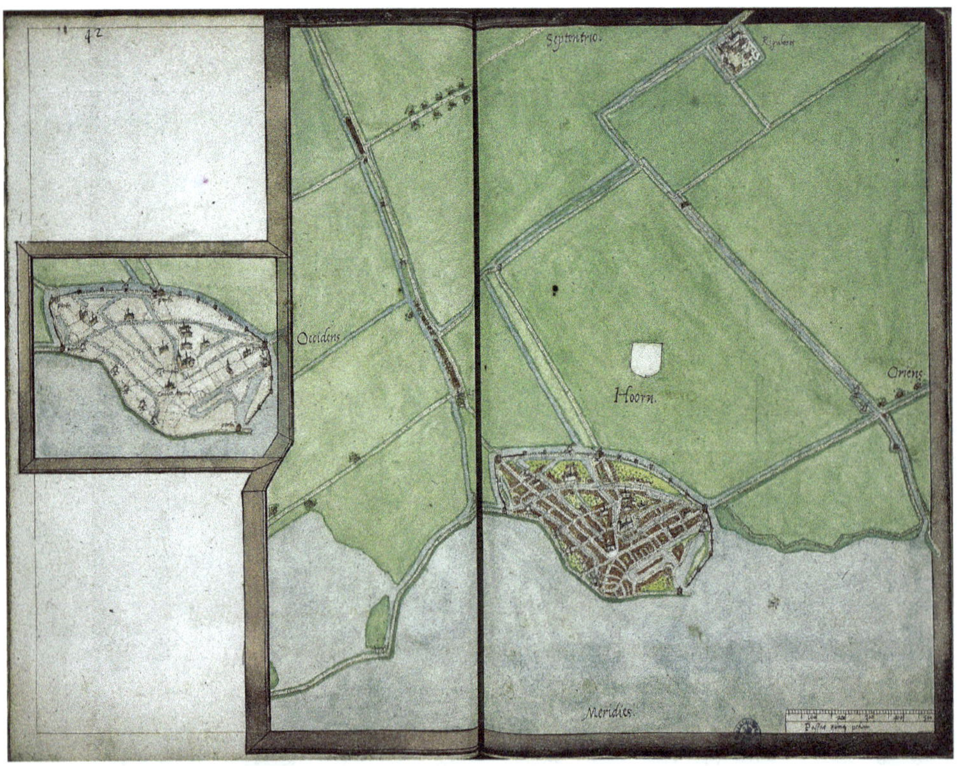

Septentrio.
Occidens
Hoorn.
Oriens
Meridies.

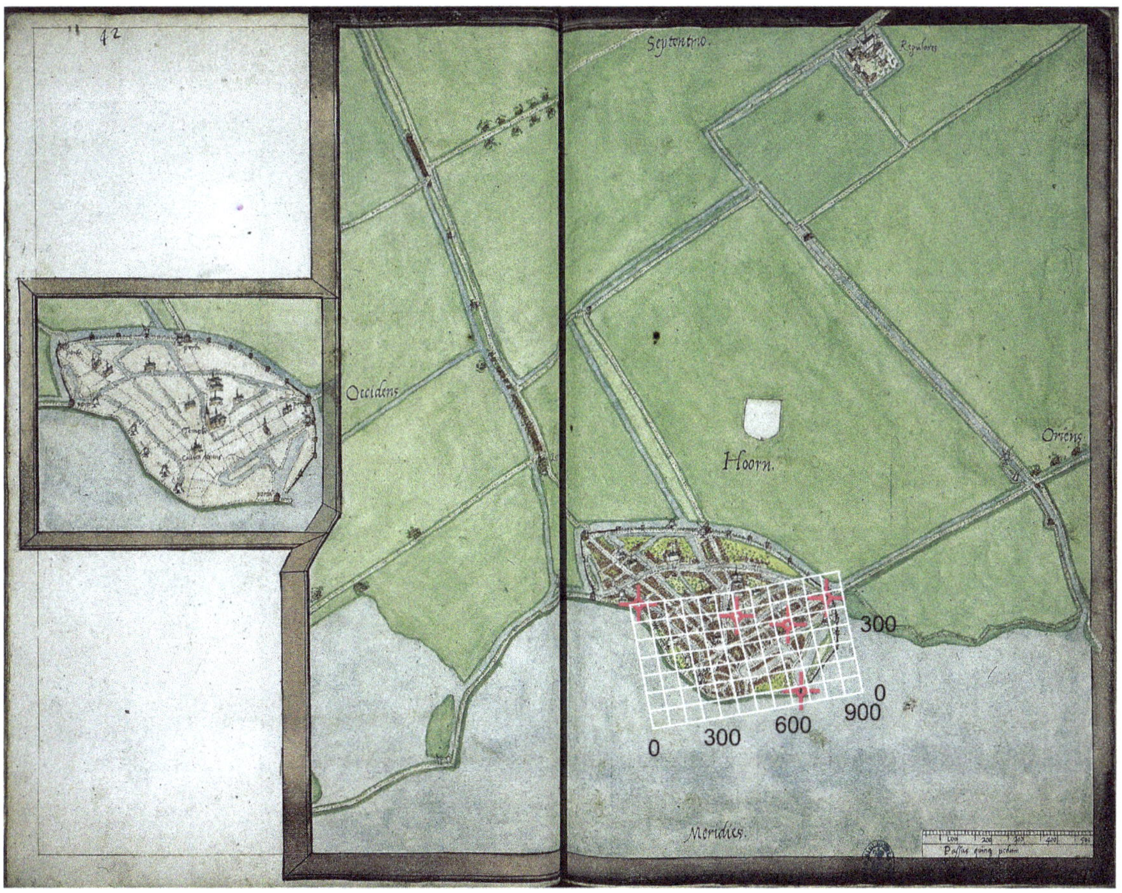
Septentrio.
Occidens
Hoorn.
Oriens
300
0
900
600
0
300
Meridies.

113. Hulst
Tables 3.113 (measured and analysed).
Figures 3.113 (unmeasured, measured and analysed).

Measured

Pixels	Scale stick original map (pixels)	Houses	Street sides	Houses per street side	Inhabitants (RF 4.0)	Inhabitants (RF 5.5)	Inhabitants (RF 7.0)
5031	1019	883	120	7.36	3534	4859	6184

Analysed

May 23, 2023 11:30:16 AM	
Computation with 5 linked points (*italics* points are less reliable):	1. Sint-Willibrordusbasiliek 2. Grote Markt 3. Vismarkt 4. Keldermanspoort 5. Paardenmarkt
Reference points (Reliability 0–5)	5
Transformation with 3 parameters	
Scale	1:8010
Rotation	8 °
Mean position error	11m

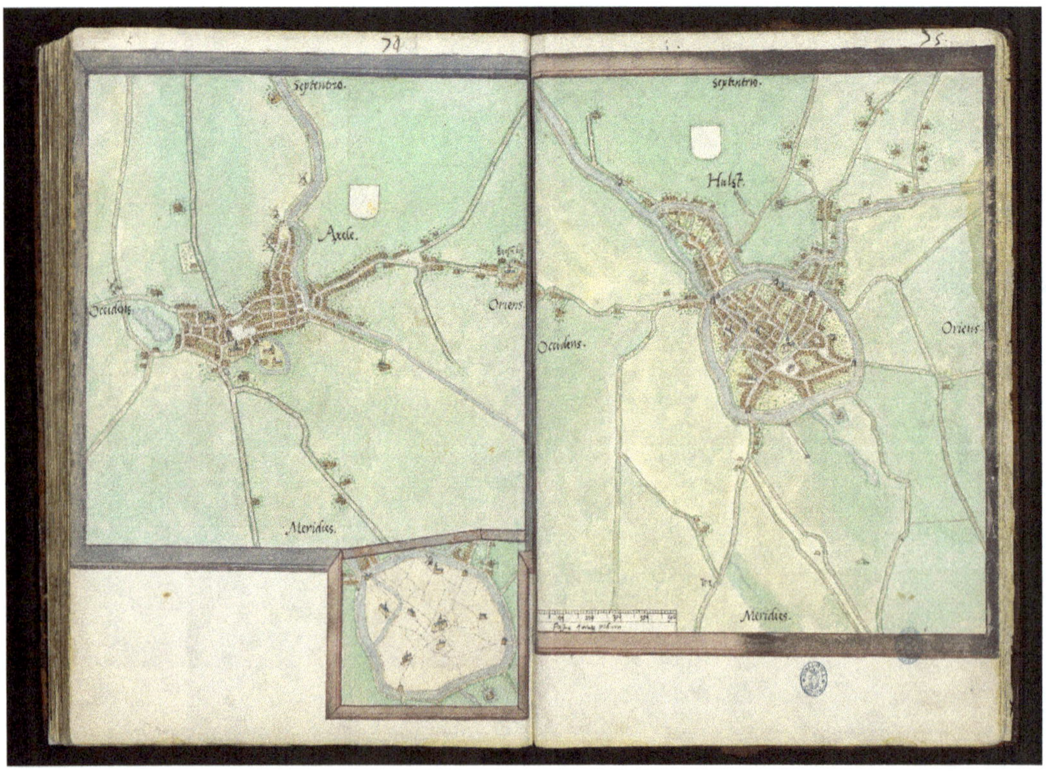

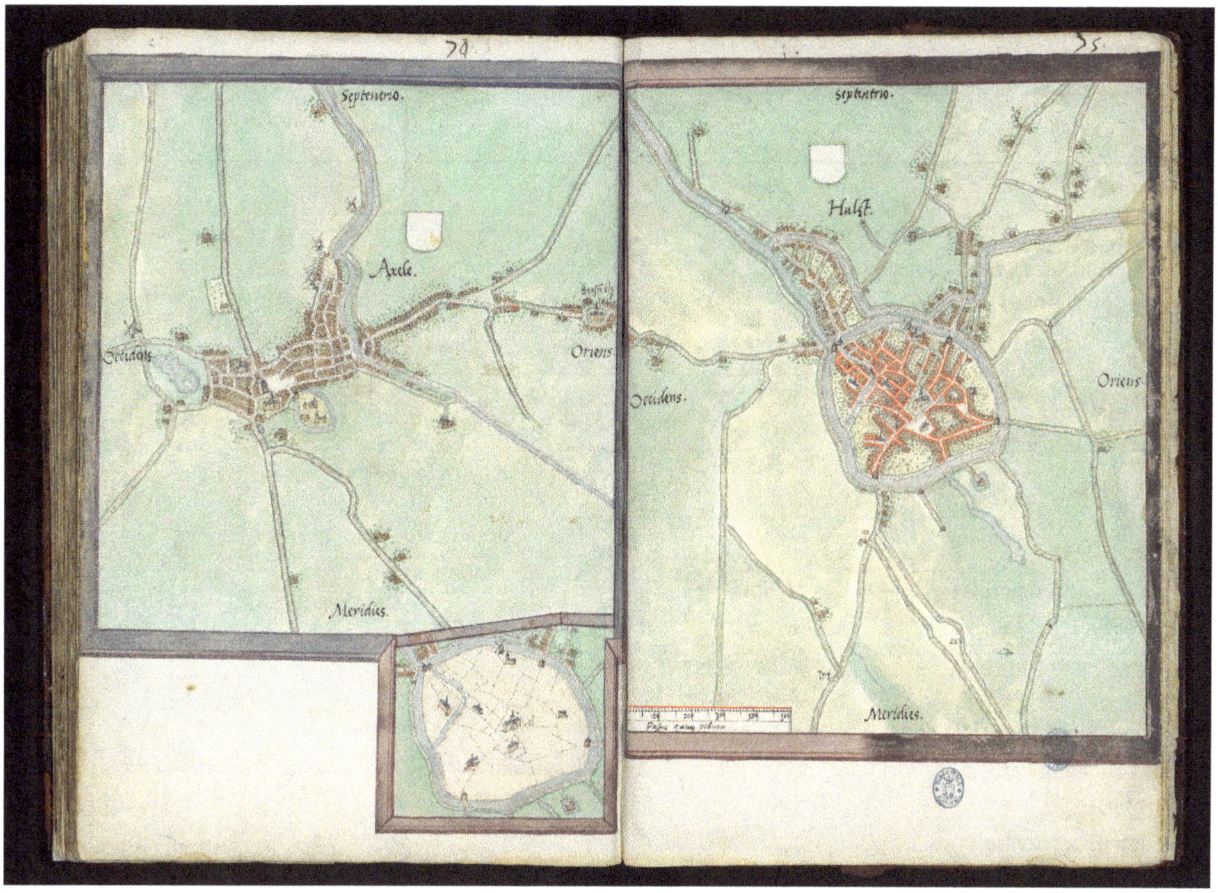
Septentrio.
Axele.
Occidens.
Oriens
Meridies.
Septentrio.
Hulst.
Occidens.
Oriens
Meridies.

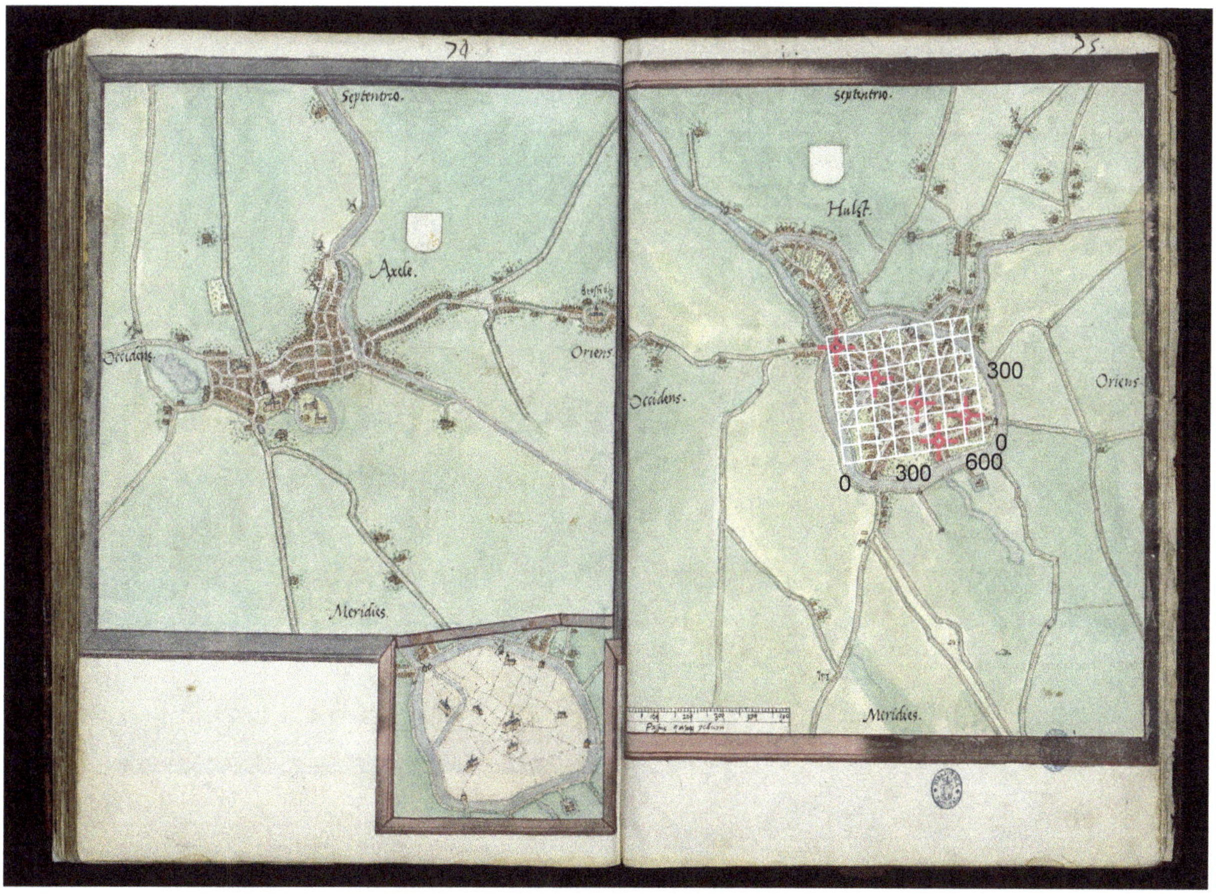
Septentrio.
Axele.
Occidens.
Oriens
Meridies.
Septentrio.
Hulst.
Occidens.
Oriens
300
0
0
300
600
Meridies.

114. IJlst

Tables 3.114 (measured and analysed).
Figures 3.114 (unmeasured, measured and analysed).

Measured							
Pixels	Scale stick original map (pixels)	Houses	Street sides	Houses per street side	Inhabitants (RF 4.0)	Inhabitants (RF 5.5)	Inhabitants (RF 7.0)
950	989	167	8	20.85	668	918	1168

Analysed	
May 23, 2023 12:13:50 PM	
Computation with 5 linked points (*italics* points are less reliable):	1. Eegracht 2. *Uilenburg* 3. Sudergoweg 4. Galamagracht 5. Kerkhofsteeg
Reference points (Reliability 0–5)	4
Transformation with 3 parameters	
Scale	1:8230
Rotation	7 °
Mean position error	23m

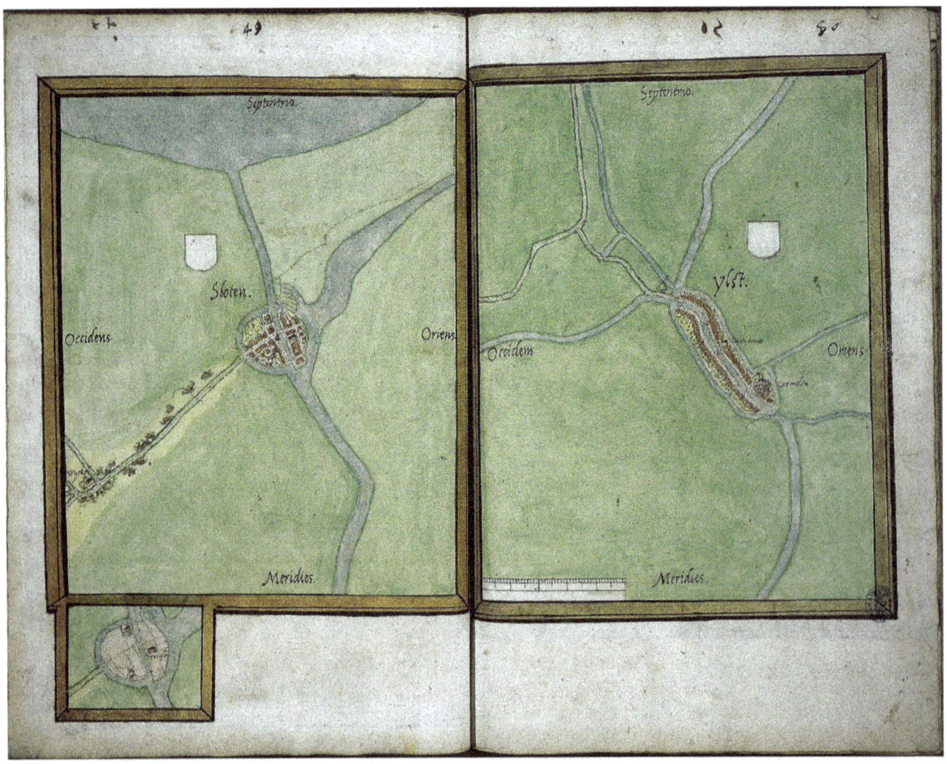

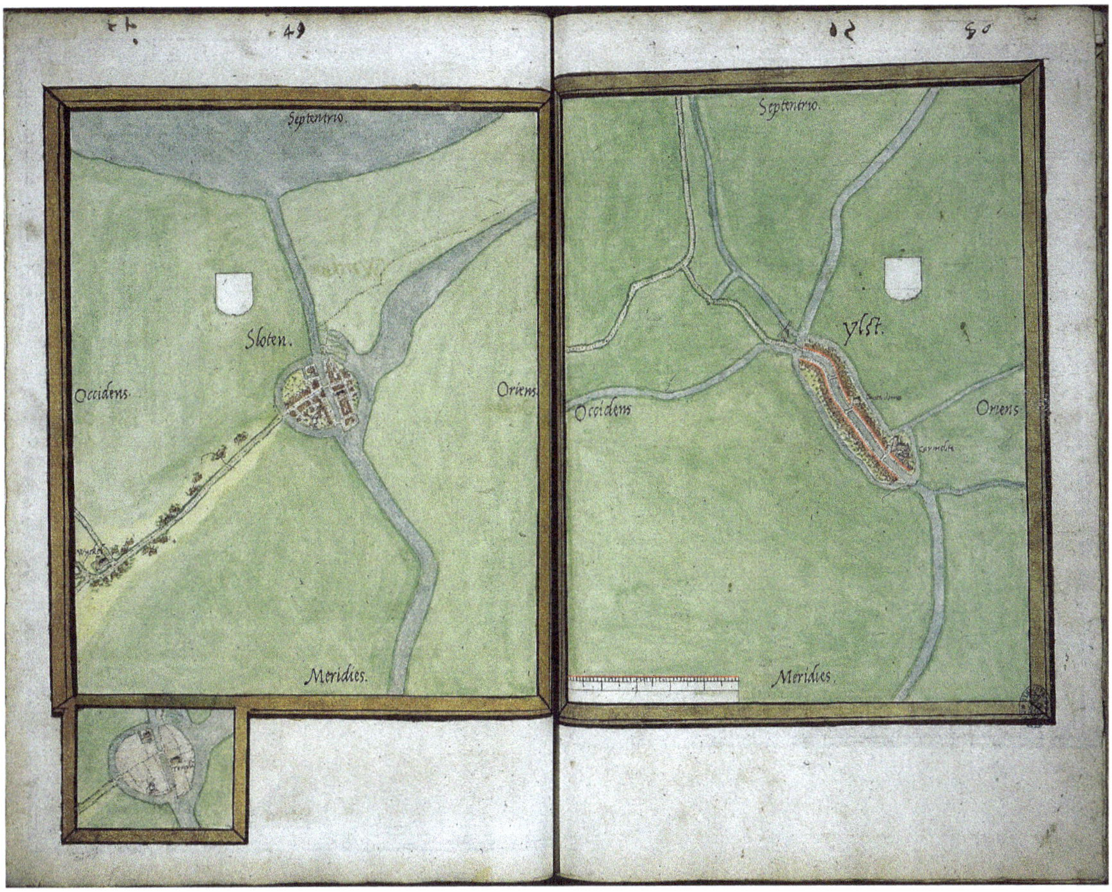
Septentrio.
Sloten.
Occidens.
Oriens.
Meridies.
Septentrio.
Ylst.
Occidens
Oriens
Meridies.

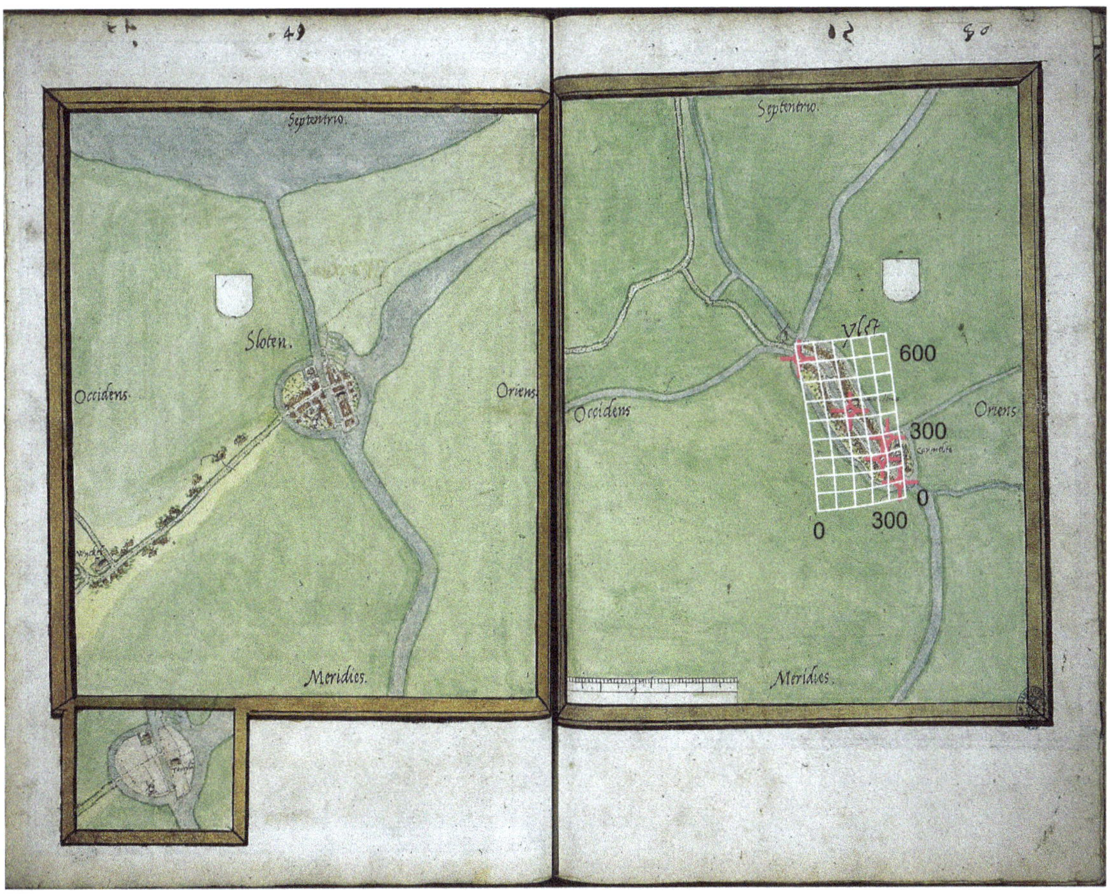
Septentrio.
Sloten.
Occidens.
Oriens.
Meridies.
Septentrio.
Ylst
600
Occidens
300
Oriens
0
0
300
Meridies.

115. IJsselstein

Tables 3.115 (measured and analysed).
Figures 3.115 (unmeasured, measured and analysed).

Measured							
Pixels	Scale stick original map (pixels)	Houses	Street sides	Houses per street side	Inhabitants (RF 4.0)	Inhabitants (RF 5.5)	Inhabitants (RF 7.0)
1924	995	338	40	8.45	1351	1858	2365

Analysed	
May 23, 2023 12:48:15 PM	
Computation with 5 linked points (*italics* points are less reliable):	1. Oude Nicolaaskerk 2. Oude Raadhuis 3. De Windotter 4. Sint Nicolaasstraat 5. Havenstraat
Reference points (Reliability 0–5)	5
Transformation with 3 parameters	
Scale	1:8220
Rotation	13 °
Mean position error	9m

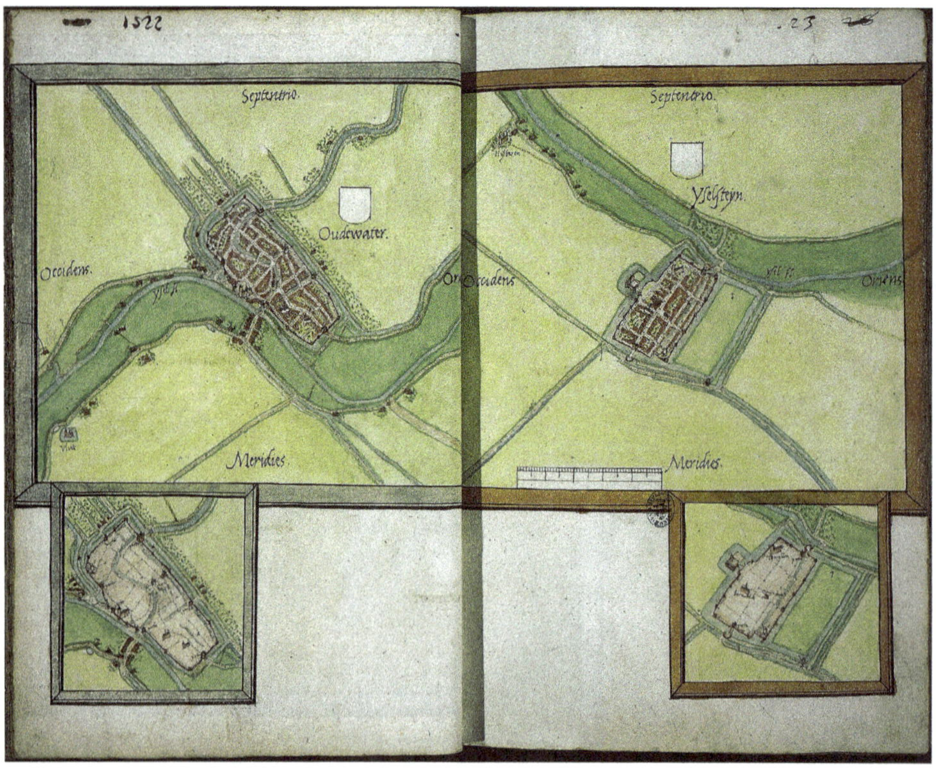

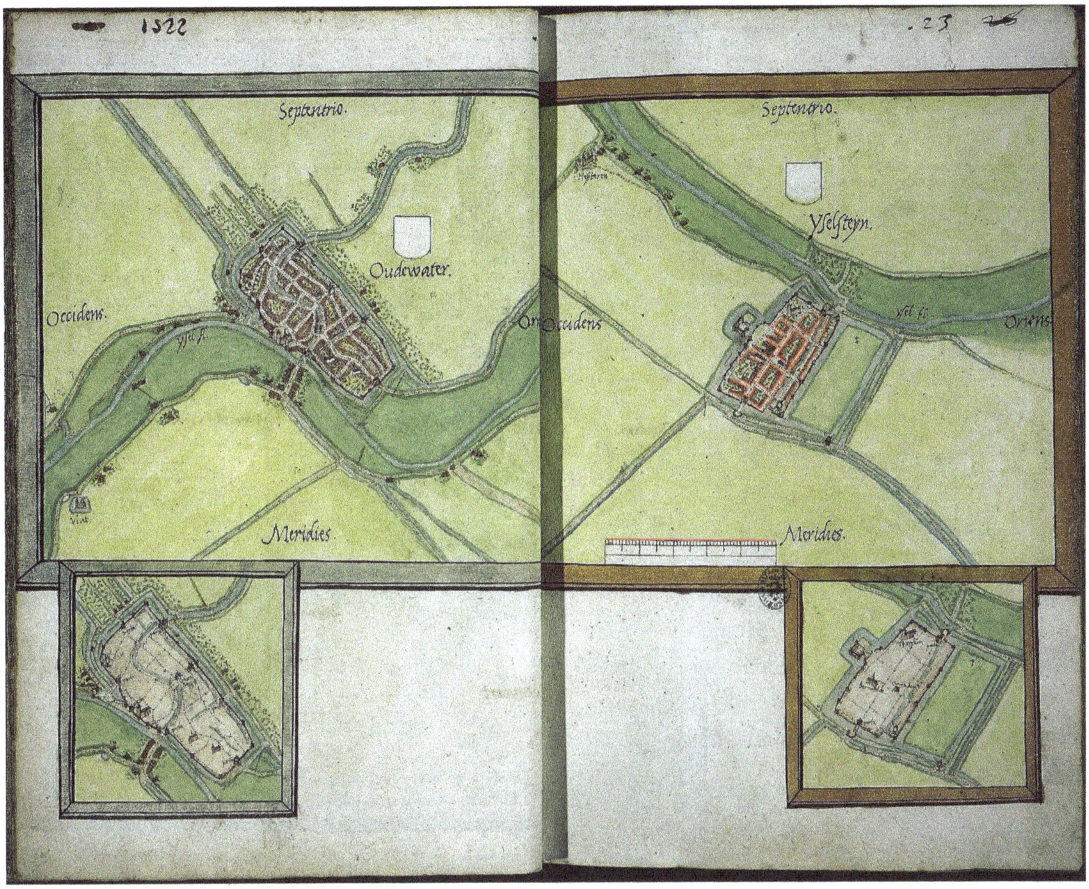
1522
Septentrio.
Oudewater.
Occidens.
Meridies.
Septentrio.
Yselsteyn.
Occidens
Oriens
Meridies.

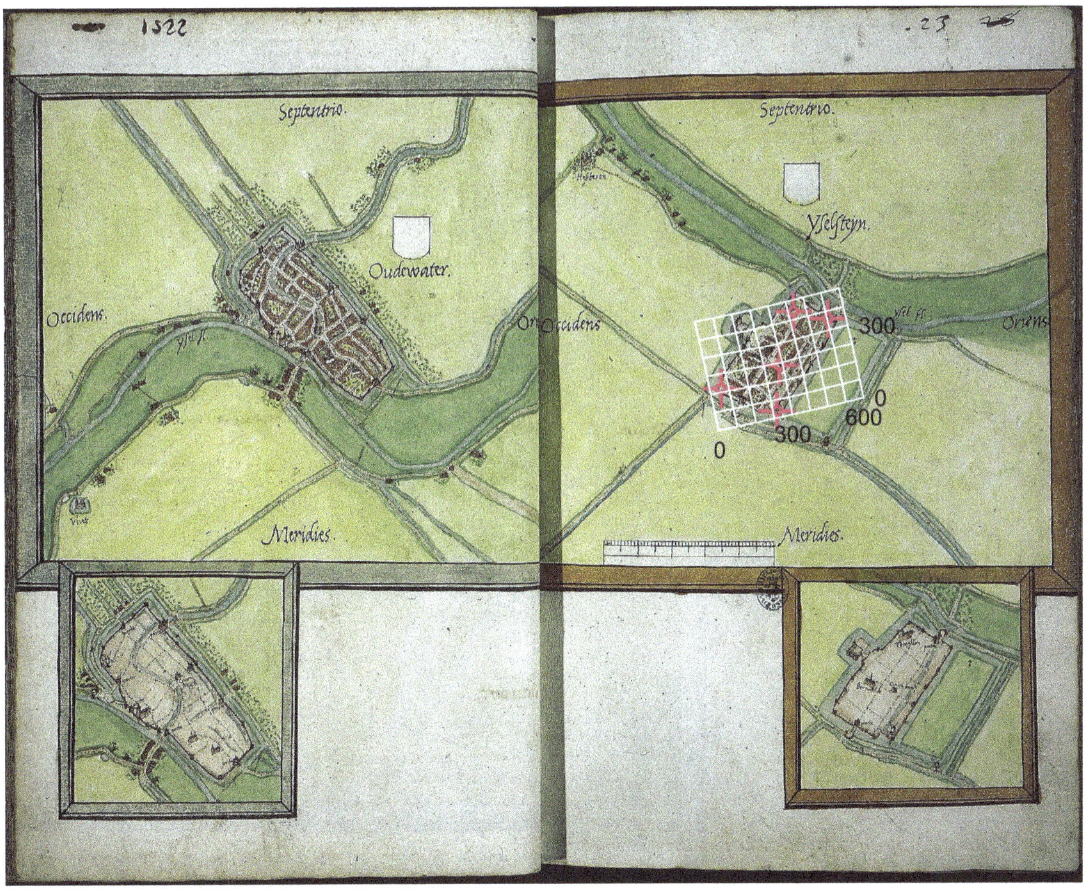
1522
Septentrio.
Oudewater.
Occidens.
Meridies.
Septentrio.
Yselsteyn.
Occidens
Oriens
300
0
600
0
300
Meridies.

116. Kampen

Tables 3.116 (measured and analysed).
Figures 3.116 (unmeasured, measured and analysed).

Measured

Pixels	Scale stick original map (pixels)	Houses	Street sides	Houses per street side	Inhabitants (RF 4.0)	Inhabitants (RF 5.5)	Inhabitants (RF 7.0)
9067	1004	1592	246	6.47	6369	8757	11,145

Analysed

May 23, 2023 11:19:27 PM	
Computation with 5 linked points | (*italics* points are less reliable):	1. Koornmarktpoort 2. Broederpoort 3. Cellebroederspoort 4. Bovenkerk 5. Buitenkerk
Reference points (Reliability 0–5)	5
Transformation with 3 parameters	
Scale	1:8050
Rotation	8 °
Mean position error	17m

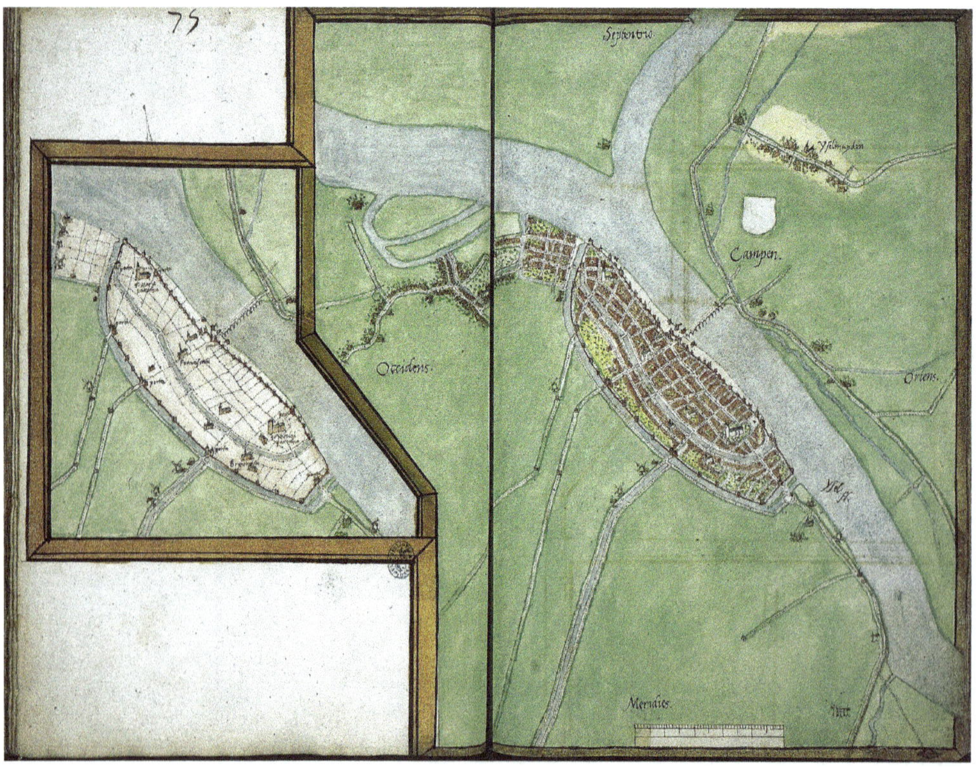

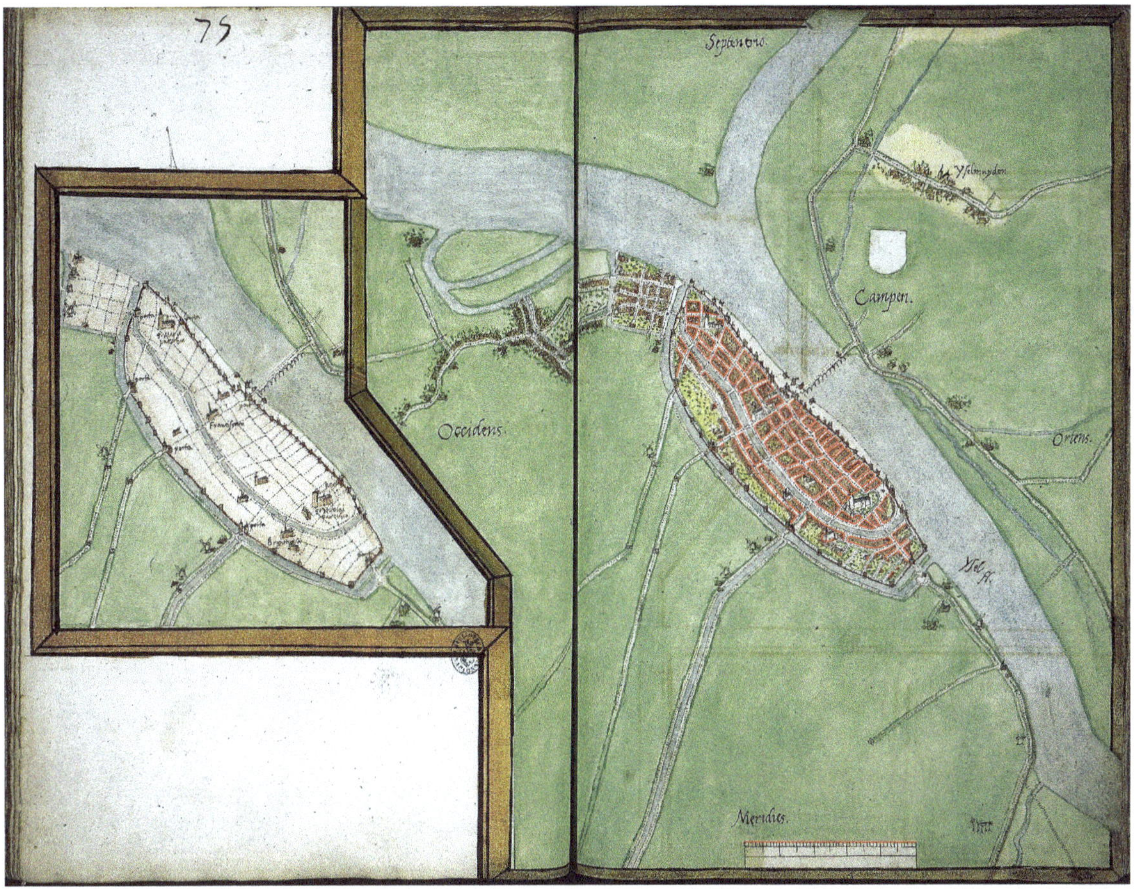
75
Septentrio.
Campen.
Occidens.
Oriens.
Meridies.

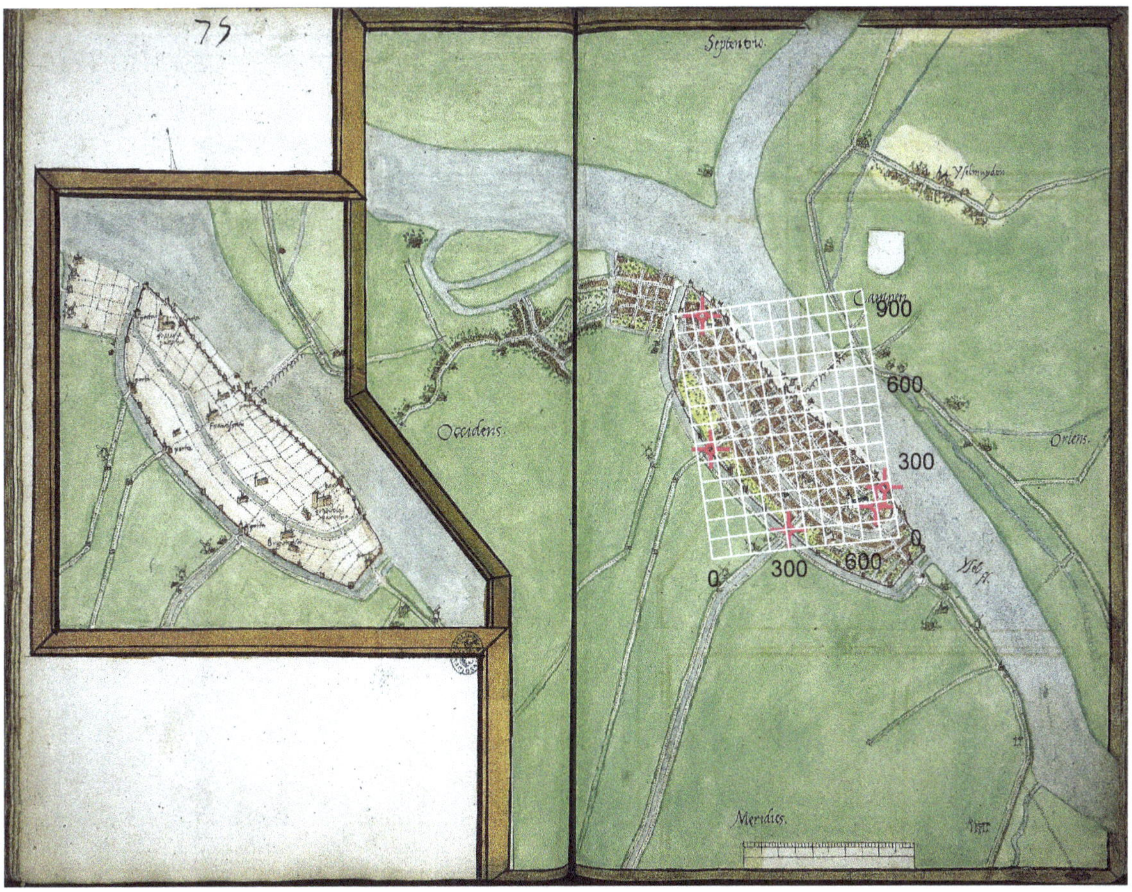
75
Septentrio.
900
600
300
0
0
300
600
Occidens.
Oriens.
Meridies.

117. Kortrijk

Tables 3.117 (measured and analysed).
Figures 3.117 (unmeasured, measured and analysed).

Measured							
Pixels	Scale stick original map (pixels)	Houses	Street sides	Houses per street side	Inhabitants (RF 4.0)	Inhabitants (RF 5.5)	Inhabitants (RF 7.0)
11,854	964	2082	269	7.74	8327	11,449	14,571

Analysed	
August 23, 2023 10:09:16 AM	
Computation with 5 linked points	1. Broeltorens
(*italics* points are less reliable):	2. Onze-Lieve-Vrouwkerk
	3. Sint-Maartenskerk
	4. Belfort
	5. Vlamingenstraat
Reference points (Reliability 0–5)	5
Transformation with 3 parameters	
Scale	1:7800
Rotation	13 °
Mean position error	9m

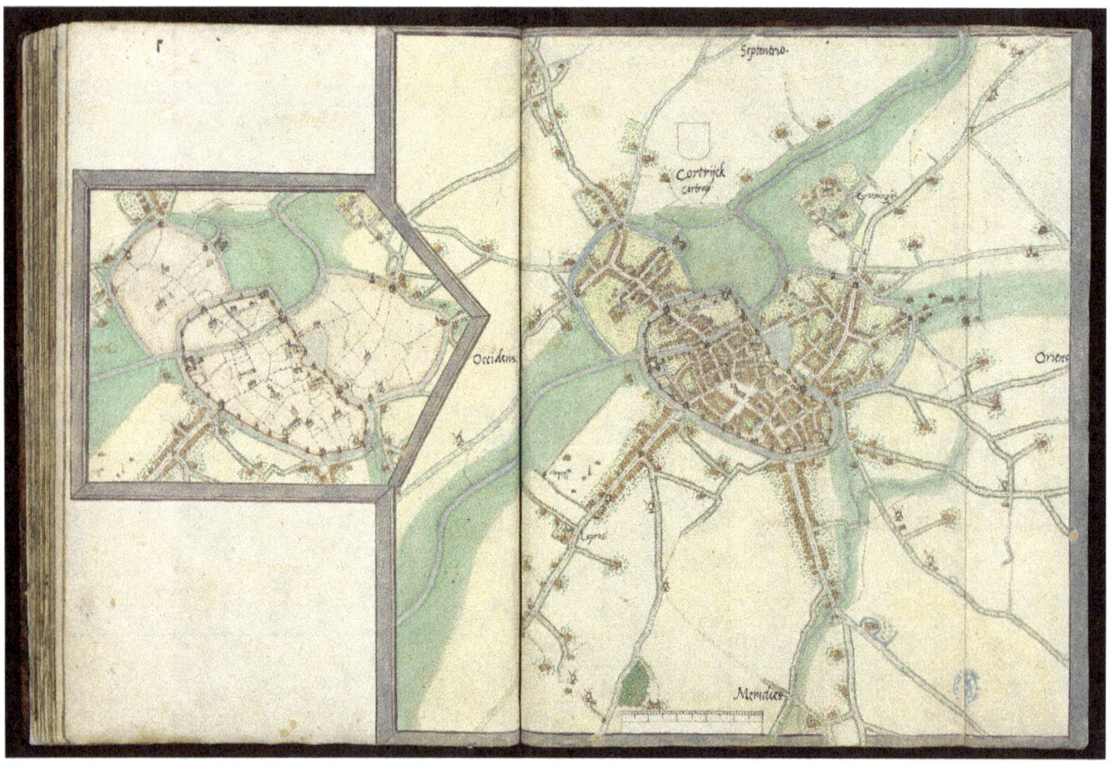

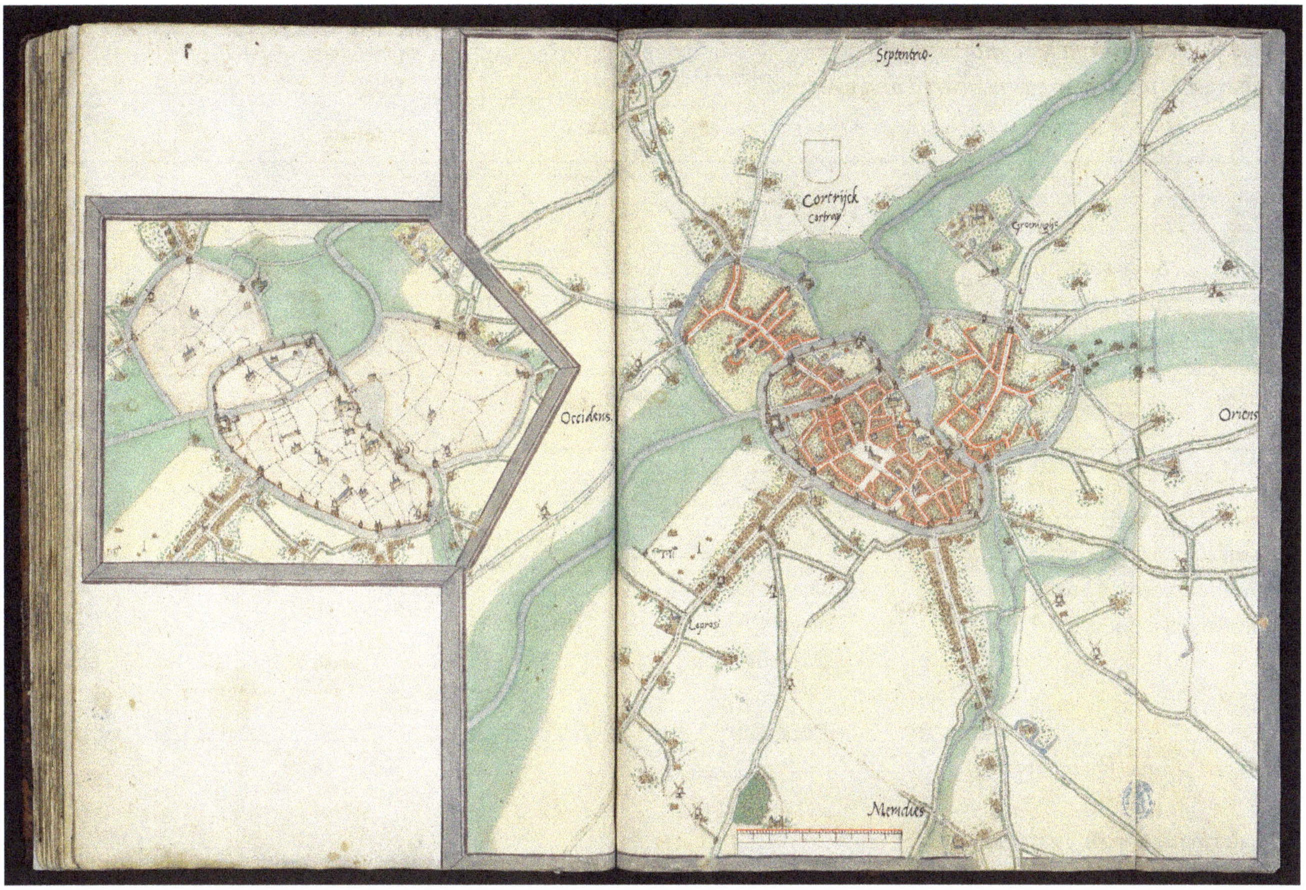
Septentrio
Cortrijck
Cortray
Occidens
Oriens
Meridies

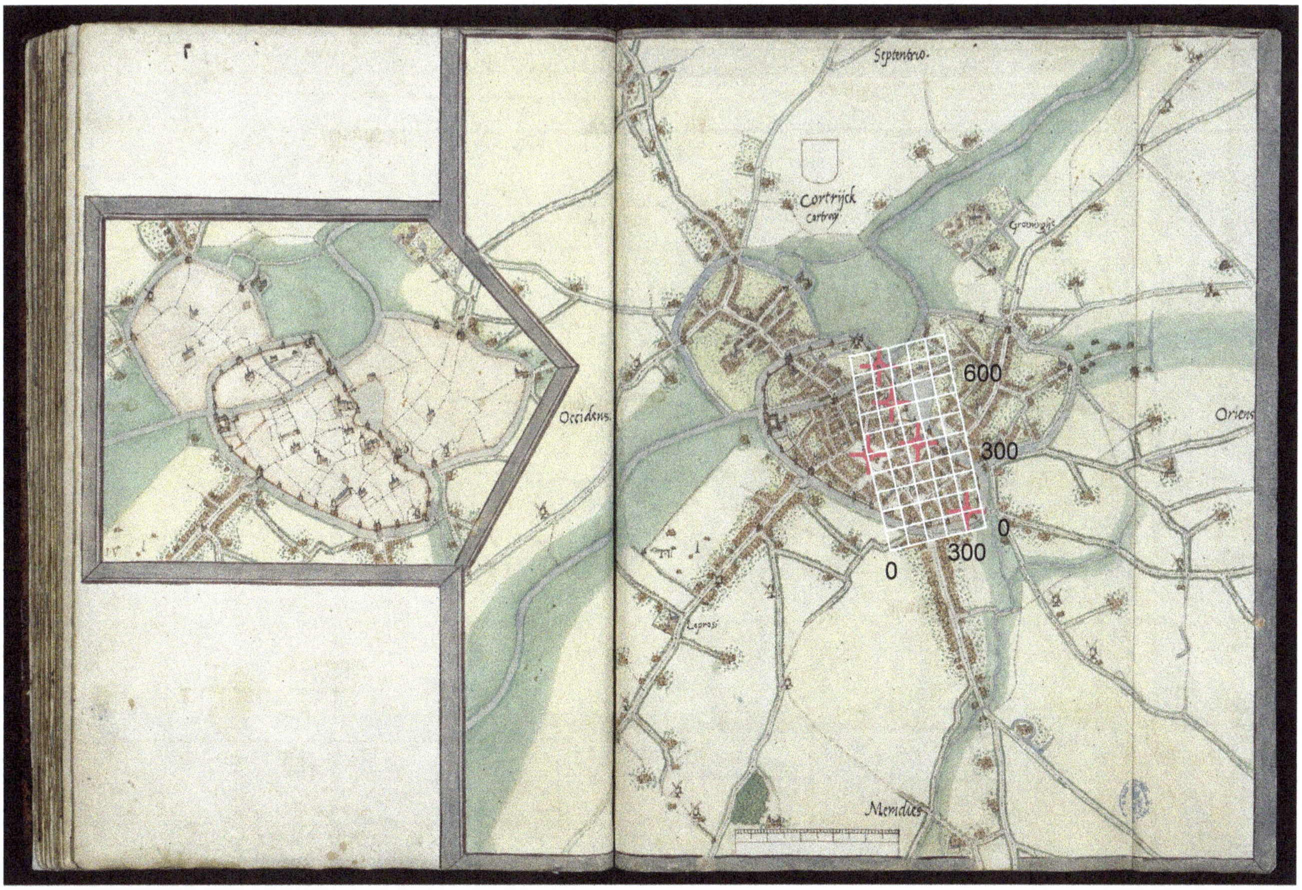
Septentrio
Cortrijck
Cortray
Occidens
Oriens
Meridies
600
300
0
0
300

118. Kuinre

Tables 3.118 (measured and analysed).
Figures 3.118 (unmeasured, measured and analysed).

Measured							
Pixels	Scale stick original map (pixels)	Houses	Street sides	Houses per street side	Inhabitants (RF 4.0)	Inhabitants (RF 5.5)	Inhabitants (RF 7.0)
1322	1004	232	27	8.60	967	1329	1691

Analysed	
May 23, 2023 11:56:19 PM	
Computation with 5 linked points (*italics* points are less reliable):	1. Sleep *2. Sleep* *3. De Linde* *4. De Linde* *5. De Linde*
Reference points (Reliability 0–5)	1
Transformation with 3 parameters	
Scale	1:6900
Rotation	3 °
Mean position error	62m

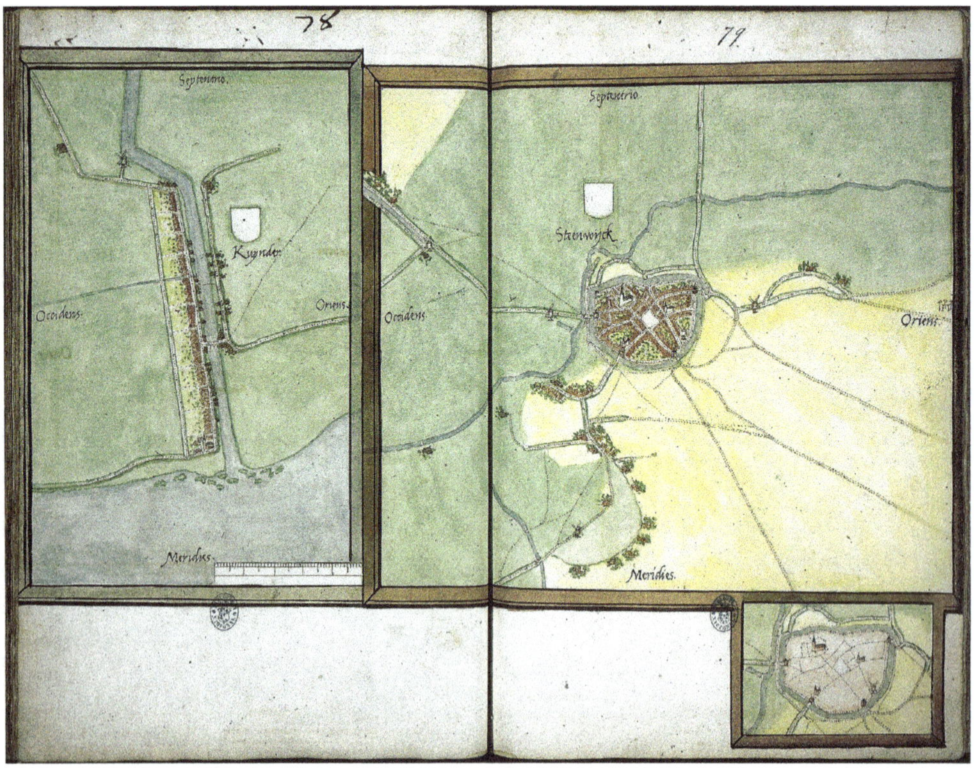

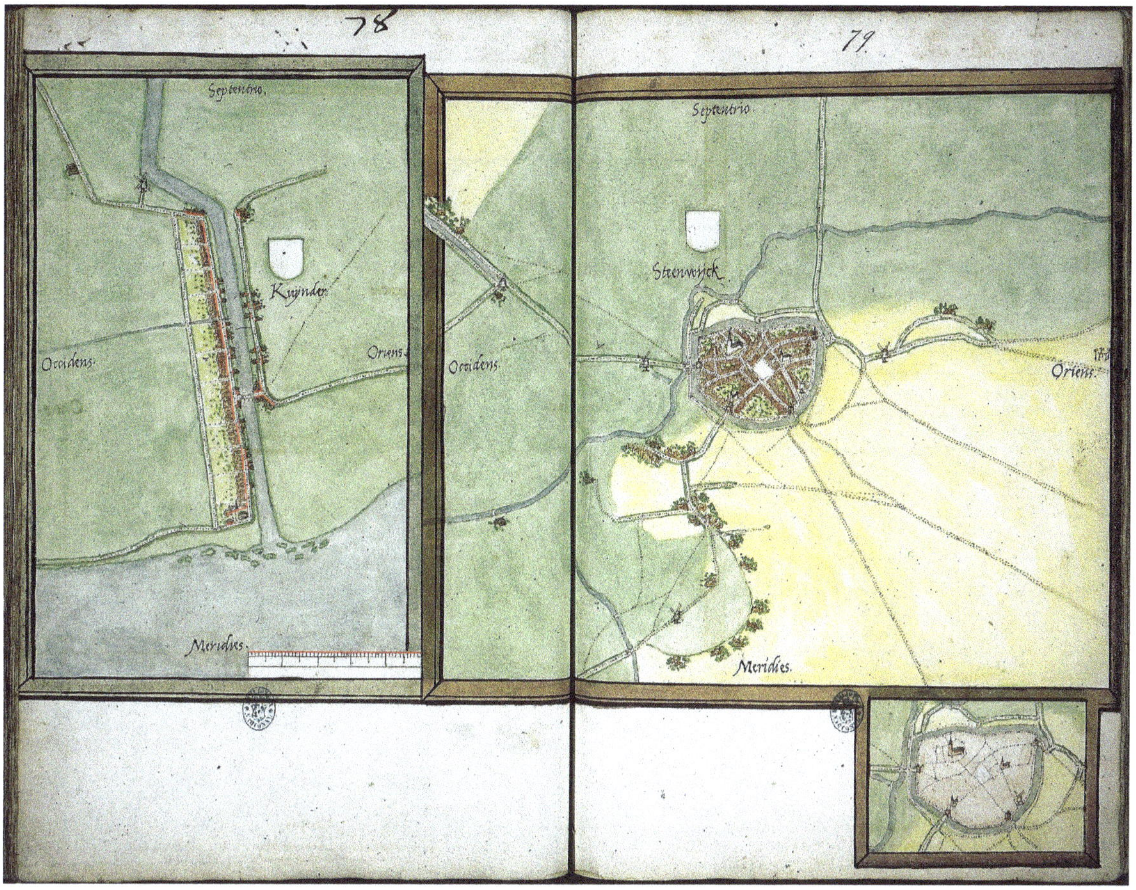
78
79
Septentrio.
Kuÿnder
Occidens.
Oriens.
Meridies.
Septentrio.
Steenwijck
Occidens.
Oriens.
Meridies.

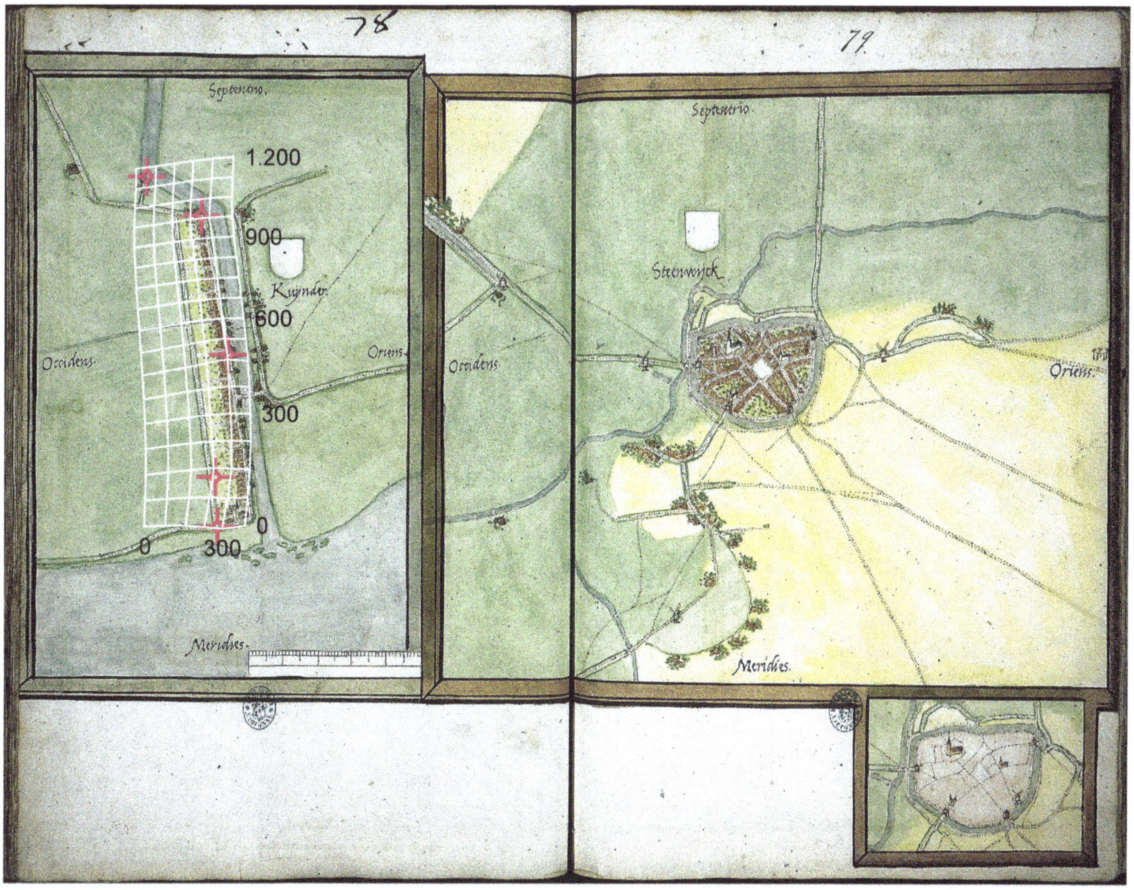
78
79
Septentrio.
1.200
900
600
300
0
0
300
Kuÿnder
Occidens.
Oriens.
Meridies.
Septentrio.
Steenwijck
Occidens.
Oriens.
Meridies.

119. La Bassée

Tables 3.119 (measured and analysed).
Figures 3.119 (unmeasured, measured and analysed).

Measured							
Pixels	Scale stick original map (pixels)	Houses	Street sides	Houses per street side	Inhabitants (RF 4.0)	Inhabitants (RF 5.5)	Inhabitants (RF 7.0)
2699	1027	474	46	10.30	1896	2607	3318

Analysed	
February 21, 2023 4:51:05 PM	
Computation with 5 linked points (*italics* points are less reliable):	1. Église Saint-Vaast 2. Rue du Gaz *3. Rue Lestarquit* 4. Place des Augustines 5. Rue Saint Aquart
Reference points (Reliability 0–5)	4
Transformation with 3 parameters	
Scale	1:8570
Rotation	7 °
Mean position error	24m

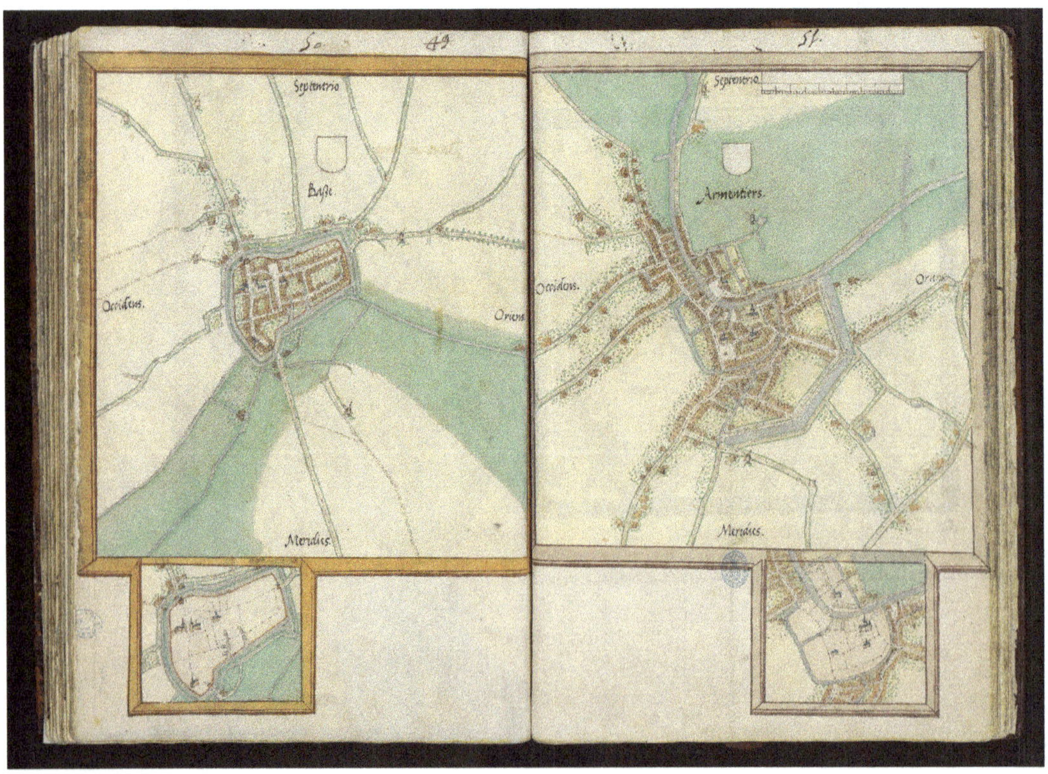

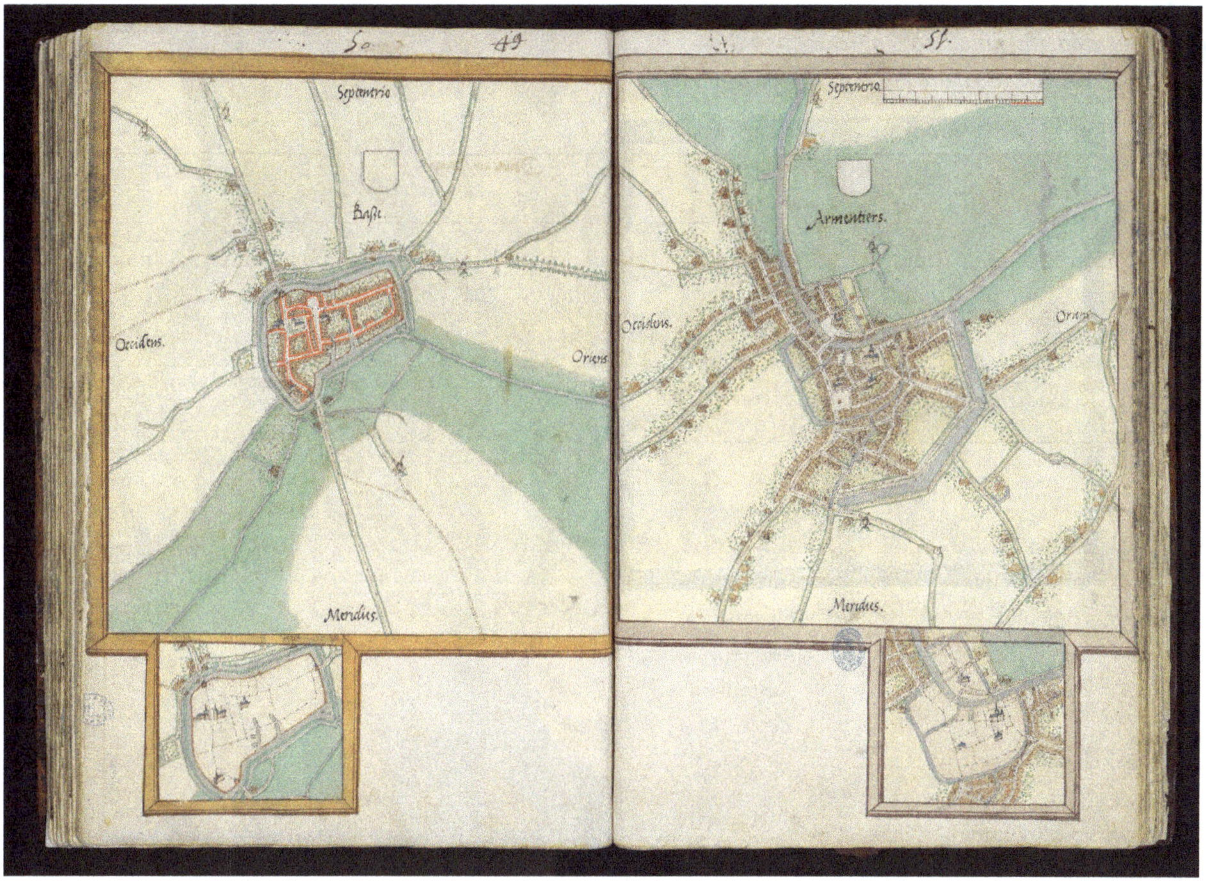
50
49
51
Septentrio
Basse
Occidens.
Oriens
Meridies.
Septentrio
Armentiers.
Occidens.
Meridies.

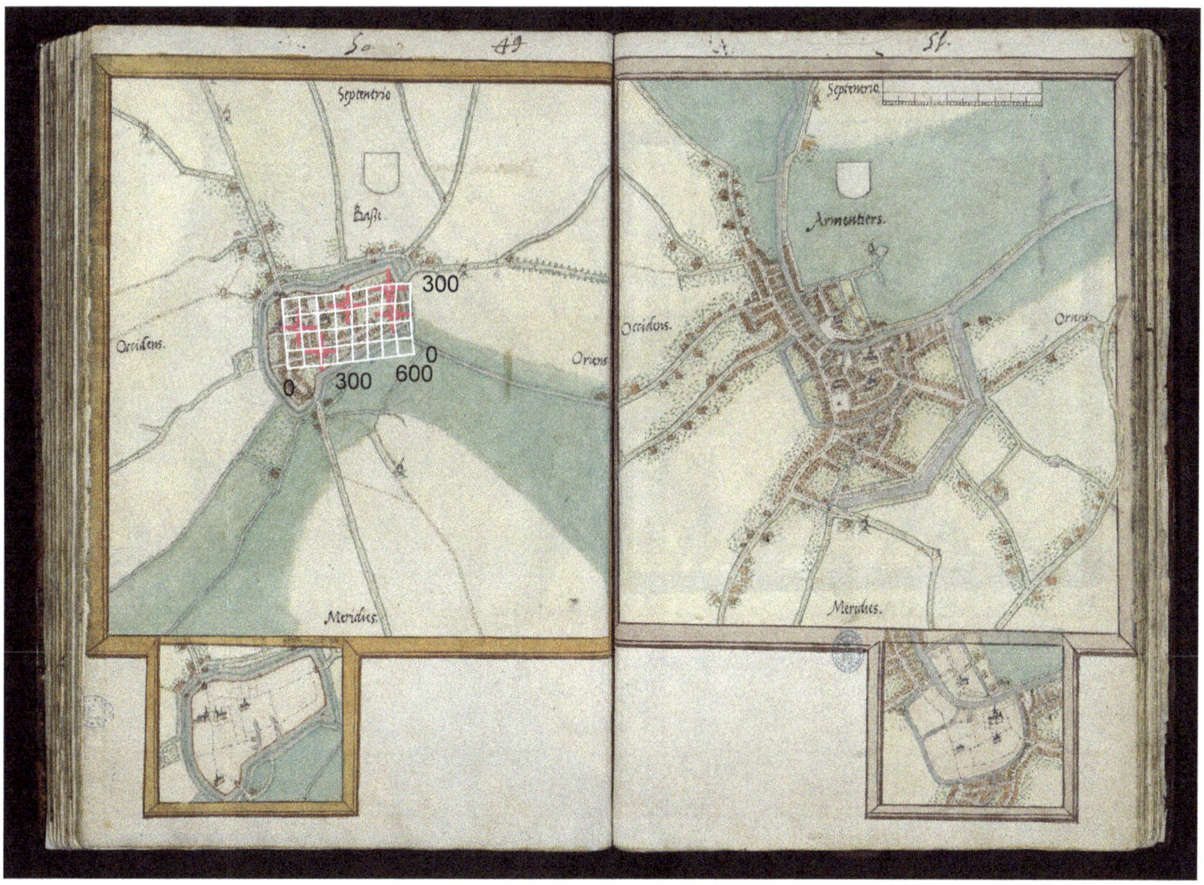
Septentrio
Basse
300
0
600
0
300
Occidens.
Oriens
Meridies.
Septentrio
Armentiers.
Occidens.
Meridies.

120. Landen
Tables 3.120 (measured and analysed).
Figures 3.120 (unmeasured, measured and analysed).

Measured							
Pixels	Scale stick original map (pixels)	Houses	Street sides	Houses per street side	Inhabitants (RF 4.0)	Inhabitants (RF 5.5)	Inhabitants (RF 7.0)
1396	missing	245	49	5.00	980	1348	1716

Analysed	
September 23, 2023 12:01:22 PM	
Computation with 5 linked points (*italics* points are less reliable):	1. Sint-Gertrudiskerk 2. Sinte-Gitter *3. Attenhovenstraat* 4. Kattenstraat 5. Zeyp
Reference points (Reliability 0–5)	4
Transformation with 3 parameters	
Scale	1:8710
Rotation	9 °
Mean position error	26m

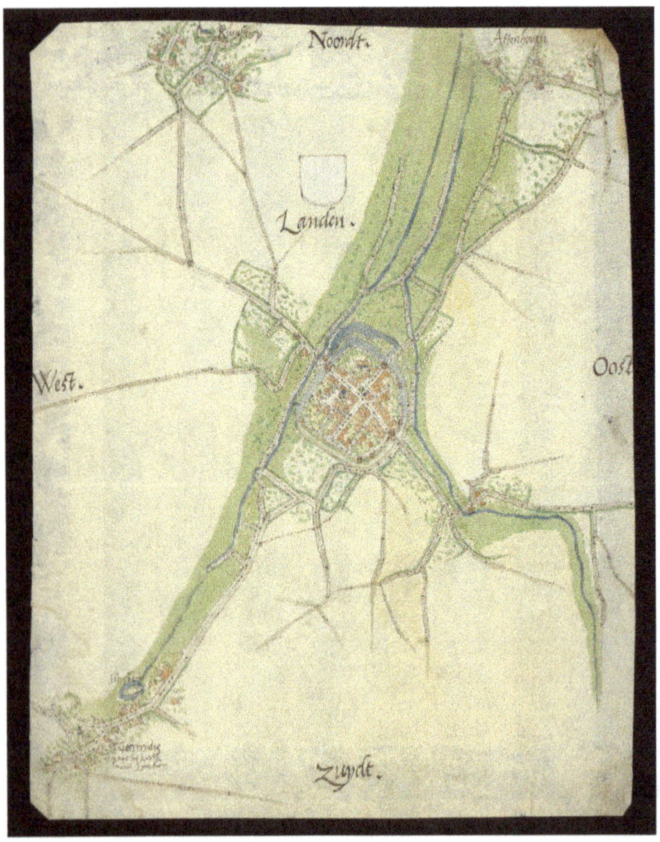

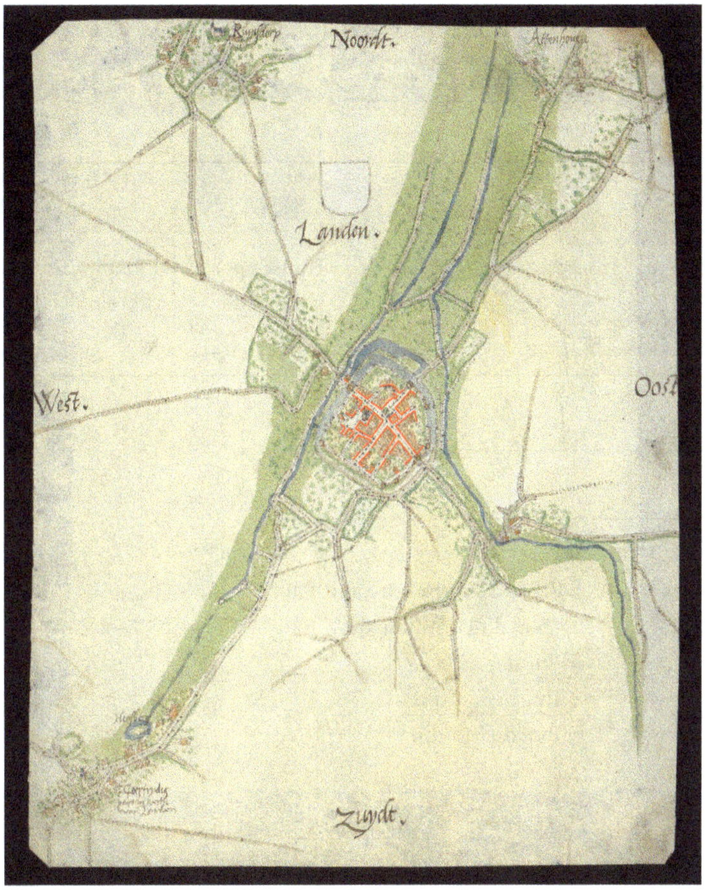
Noordt.
Landen.
West.
Oost
Zuydt.

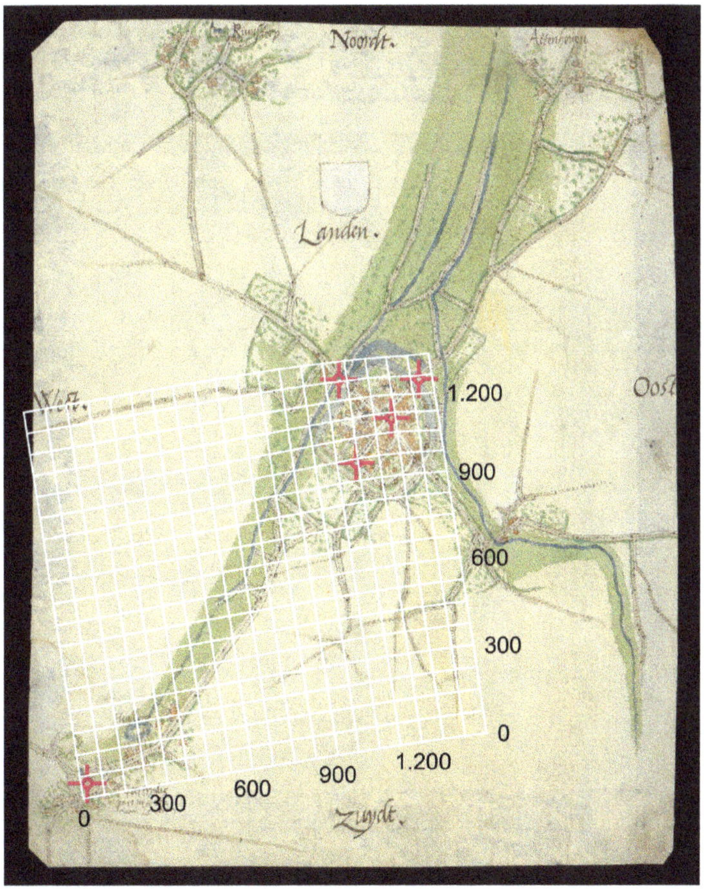
Noordt.
Landen.
West.
Oost
Zuydt.
1.200
900
600
300
0
0
300
600
900
1.200

121. Landrecies
Tables 3.121 (measured and analysed).
Figures 3.121 (unmeasured, measured and analysed).

Measured							
Pixels	Scale stick original map (pixels)	Houses	Street sides	Houses per street side	Inhabitants (RF 4.0)	Inhabitants (RF 5.5)	Inhabitants (RF 7.0)
1961	1015	344	75	4.59	1377	1894	2411

Analysed	
May 26, 2023 5:13:47 PM	
Computation with 5 linked points (*italics* points are less reliable):	1. Église Saint-Pierre-Saint-Paul 2. Quai du Pont Rouge *3. Rue des deux clefs* 4. Rue de la Cote 5. Rue du Moulin
Reference points (Reliability 0–5)	4
Transformation with 3 parameters	
Scale	1:8040
Rotation	23 °
Mean position error	25m

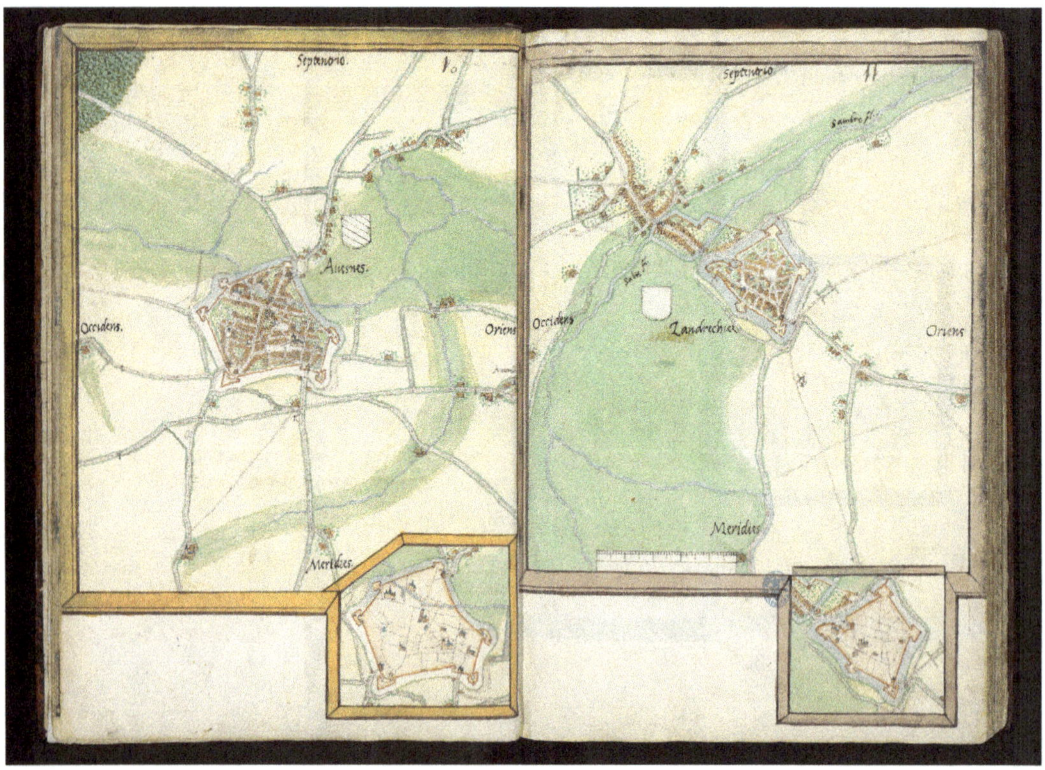

Septentrio
Auesnes
Occidens
Oriens
Meridies
Septentrio
Sambre fl.
Occidens
Oriens

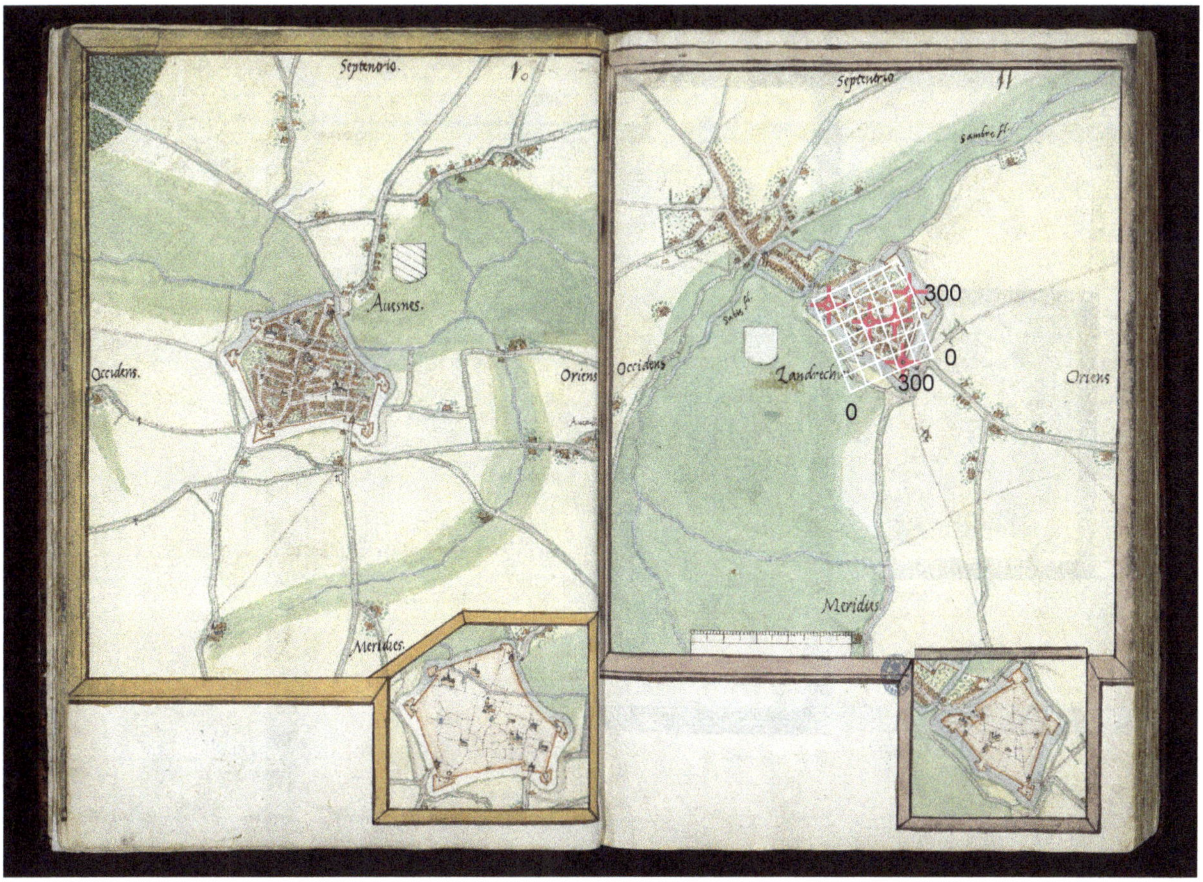
Septentrio
Auesnes
Occidens
Oriens
Meridies
Septentrio
Sambre fl.
Occidens
300
0
300
0
Oriens

122. Le Cateau-Cambrésis
Tables 3.122 (measured and analysed).
Figures 3.122 (unmeasured, measured and analysed).

Measured							
Pixels	Scale stick original map (pixels)	Houses	Street sides	Houses per street side	Inhabitants (RF 4.0)	Inhabitants (RF 5.5)	Inhabitants (RF 7.0)
3216	missing	565	75	7.53	2259	3106	3953

Analysed	
May 27, 2023 11:26:27 AM	
Computation with 5 linked points (*italics* points are less reliable):	1. Église Saint-Martin 2. La Selle *3. La Selle (sud)* 4. Rue du marché au chevaux 5. Rue des Remparts
Reference points (Reliability 0–5)	4
Transformation with 3 parameters	
Scale	1:8520
Rotation	5 °
Mean position error	23m

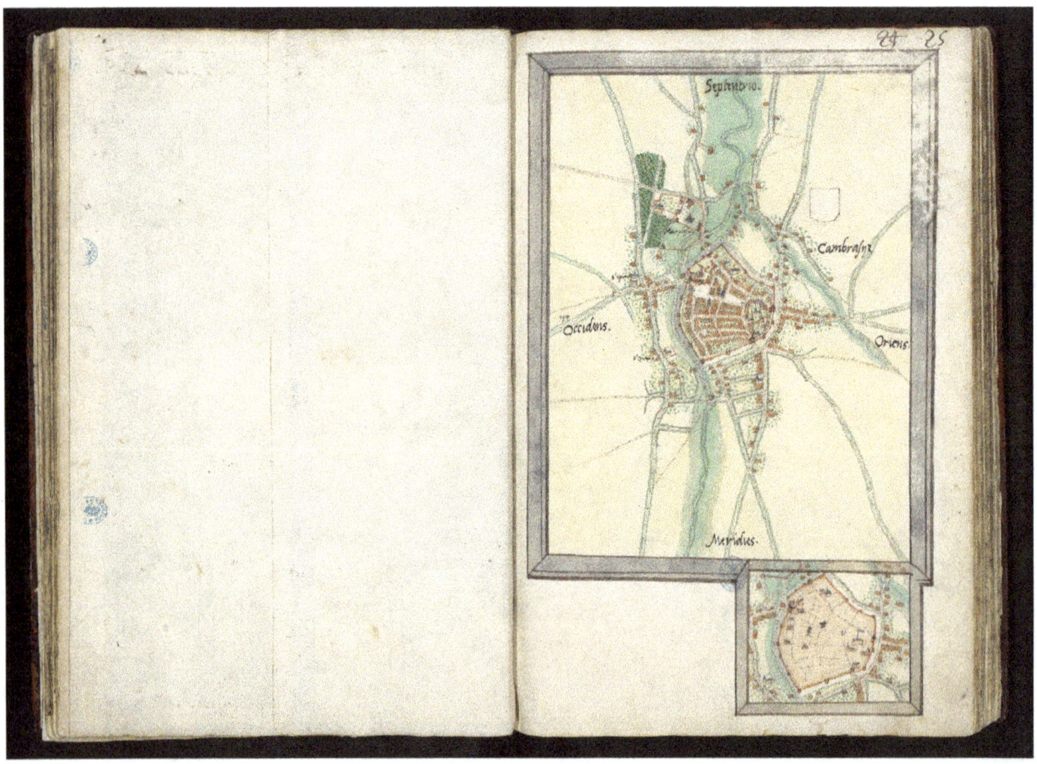

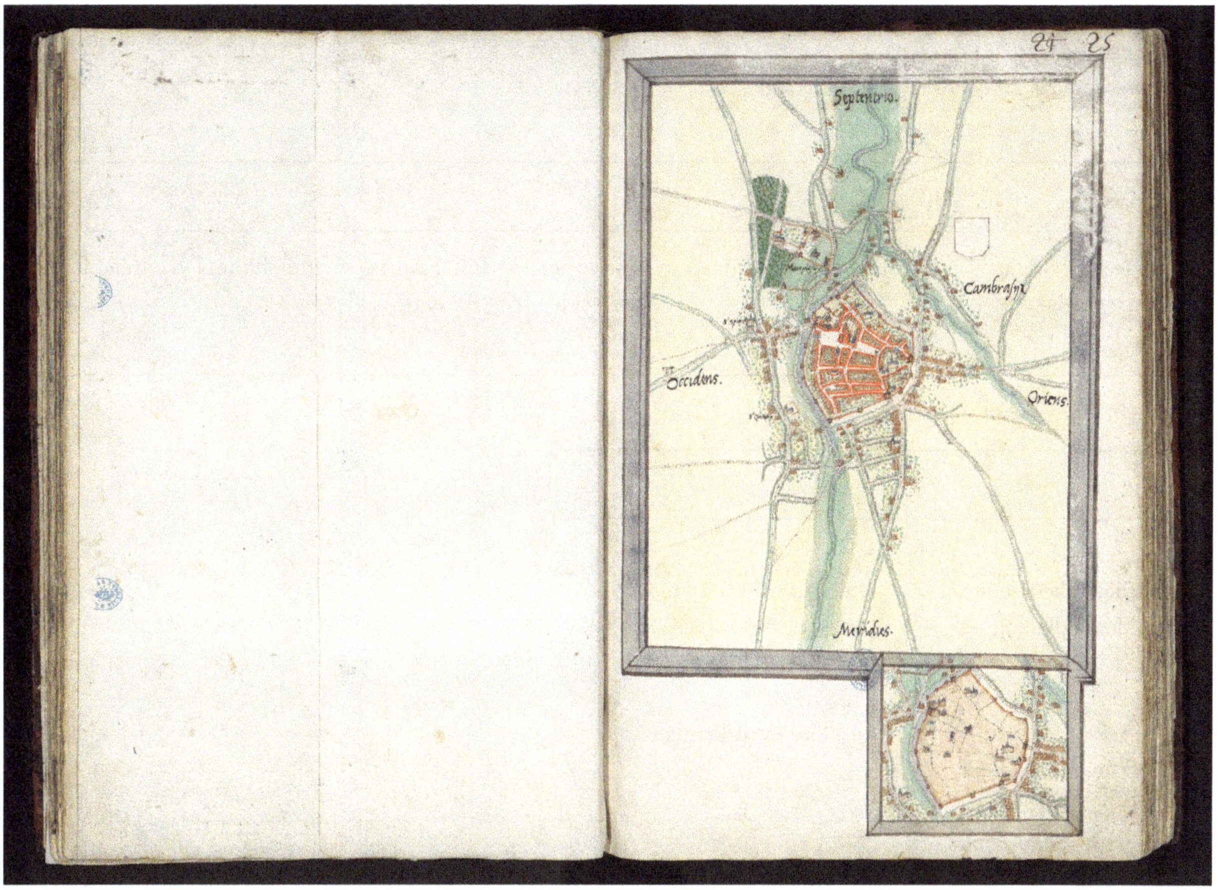
Septentrio.
Occidens.
Oriens.
Meridies.

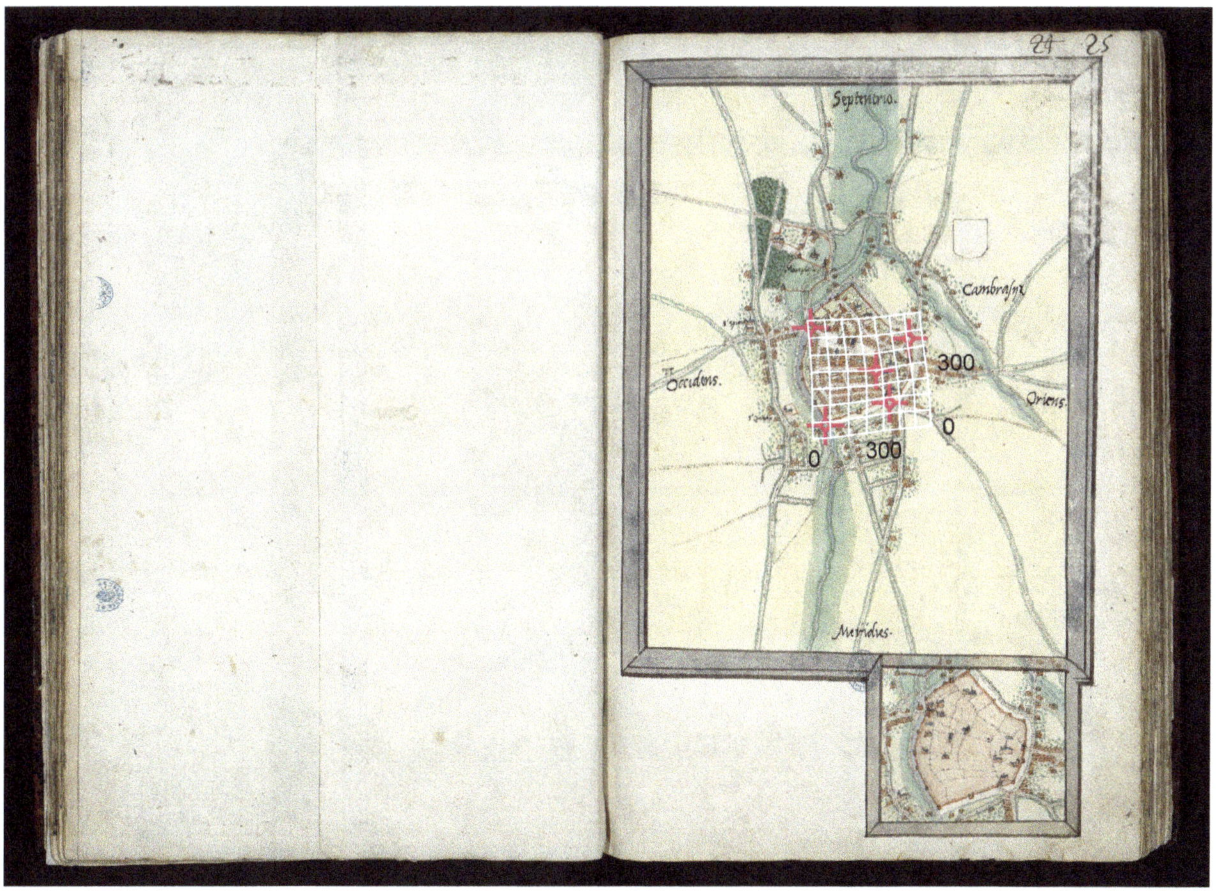
Septentrio.
Occidens.
Oriens.
Meridies.
300
0
0
300

123. Le Quesnoy

Tables 3.123 (measured and analysed).
Figures 3.123 (unmeasured, measured and analysed).

Measured							
Pixels	Scale stick original map (pixels)	Houses	Street sides	Houses per street side	Inhabitants (RF 4.0)	Inhabitants (RF 5.5)	Inhabitants (RF 7.0)
5174	1022	909	115	7.90	3634	4997	6360

Analysed	
April 18, 2023 9:47:22 AM	
Computation with 5 linked points (*italics* points are less reliable):	1. Église Notre Dame 2. Rues des Écoles 3. Rue Désiré Tanis 4. D114 5. Rue Saint-François
Reference points (Reliability 0–5)	5
Transformation with 3 parameters	
Scale	1:7910
Rotation	12 °
Mean position error	19m

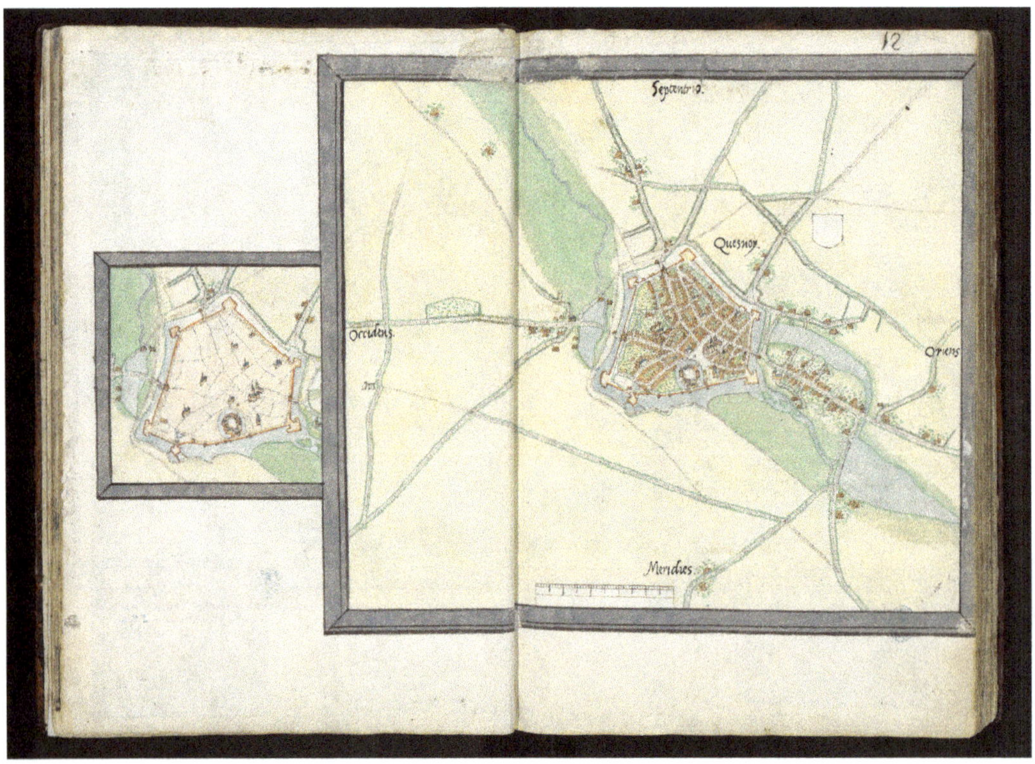

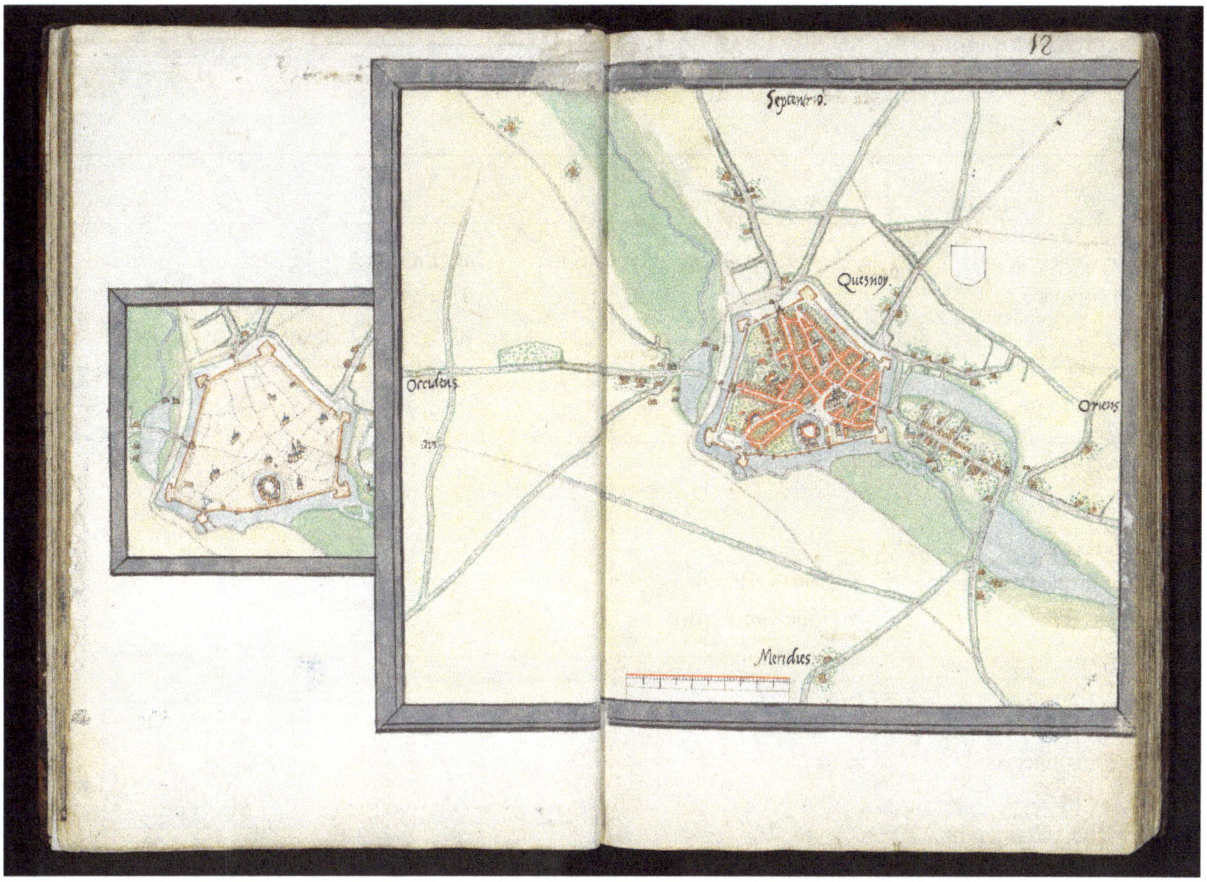
12
Septentrio
Quesnoy
Occidens
Oriens
Meridies

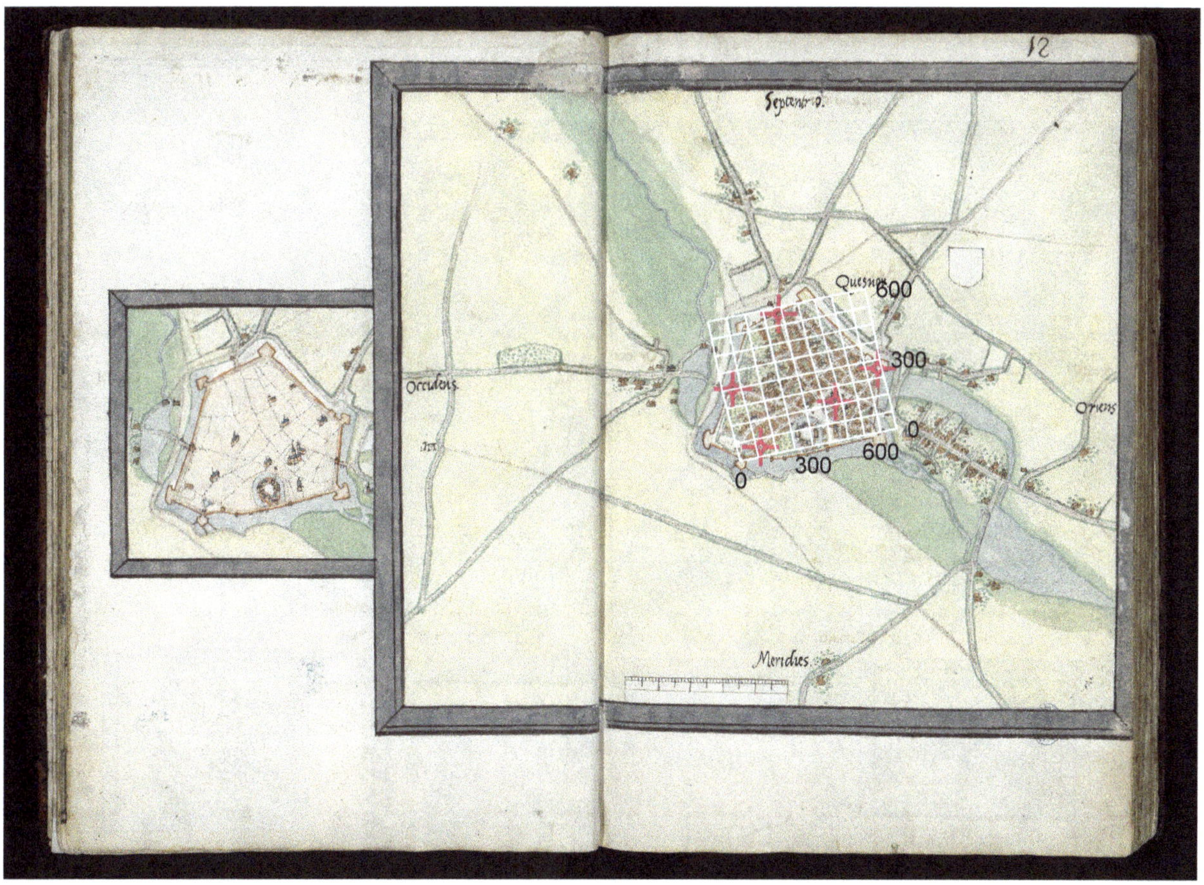
12
Septentrio
Quesnoy
600
300
0
Occidens
Oriens
0
300
600
Meridies

124. Leerdam

Tables 3.124 (measured and analysed).
Figures 3.124 (unmeasured, measured and analysed).

Measured							
Pixels	Scale stick original map (pixels)	Houses	Street sides	Houses per street side	Inhabitants (RF 4.0)	Inhabitants (RF 5.5)	Inhabitants (RF 7.0)
1839	1007	323	45	7.18	1292	1776	2260

Analysed	
June 5, 2023 10:44:20 AM	
Computation with 5 linked points	1. Grote Kerk
(*italics* points are less reliable):	2. Hofje van Aerden
	3. Noordwal
	4. Zuidwal
	5. Westwal
Reference points (Reliability 0–5)	5
Transformation with 3 parameters	
Scale	1:8300
Rotation	10 °
Mean position error	15m

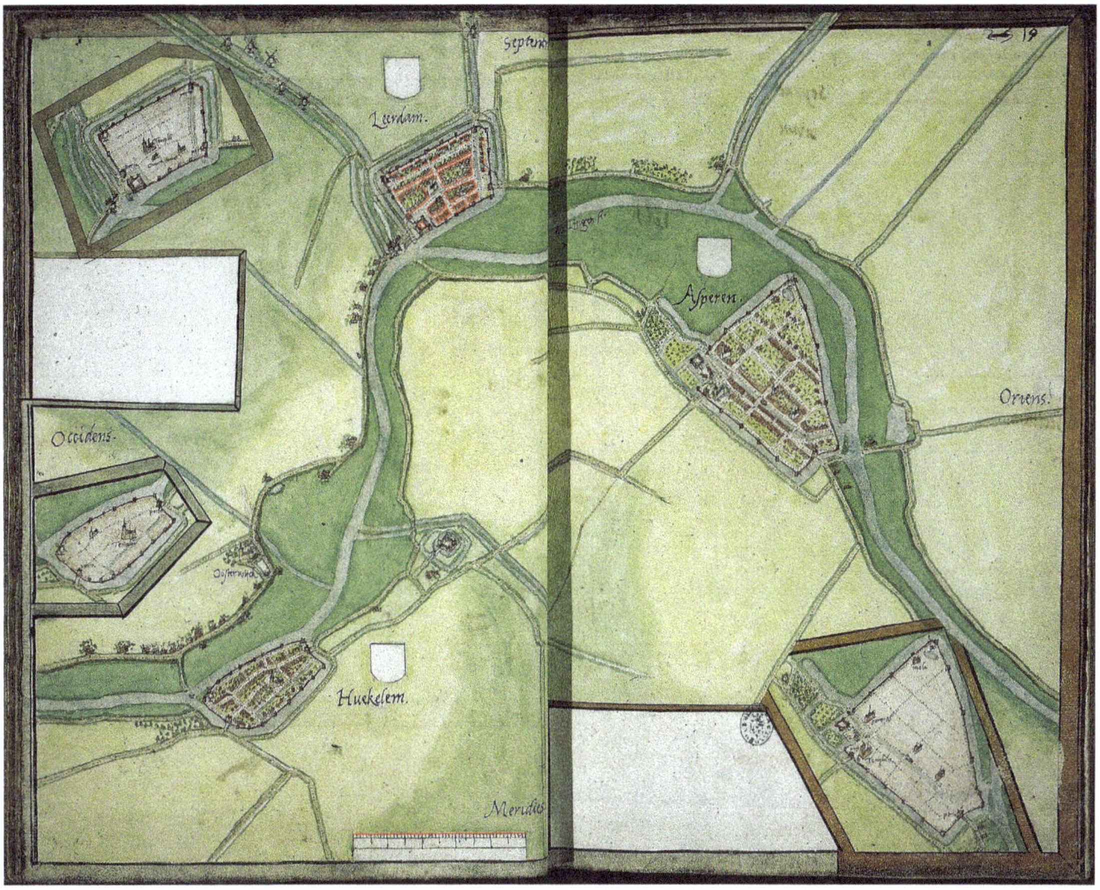
Leerdam.
Asperen.
Occidens.
Oriens.
Huekelem.
Meridies

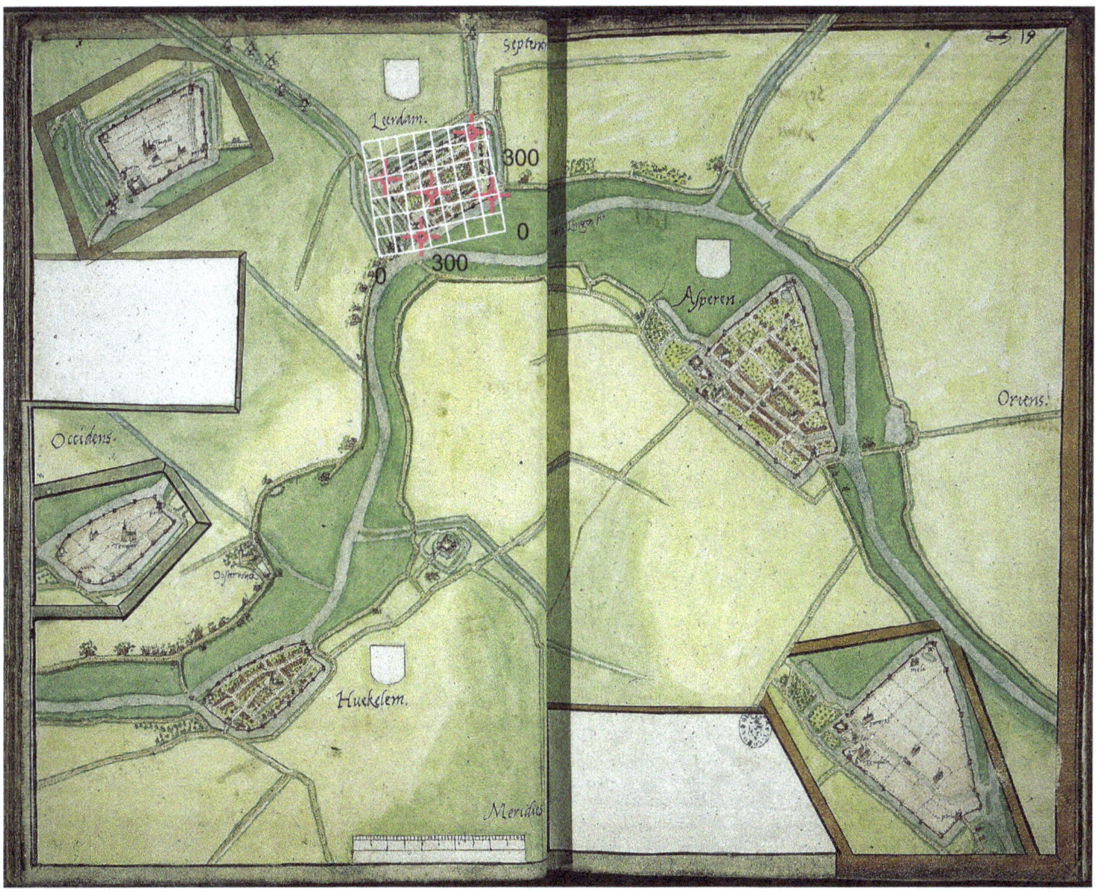
Leerdam.
300
0
0
300
Asperen.
Occidens.
Oriens.
Huekelem.
Meridies

125. Leeuwarden

Tables 3.125 (measured and analysed).
Figures 3.125 (unmeasured, measured and analysed).

Measured							
Pixels	Scale stick original map (pixels)	Houses	Street sides	Houses per street side	Inhabitants (RF 4.0)	Inhabitants (RF 5.5)	Inhabitants (RF 7.0)
8556	999	1502	248	6.06	6009	8263	10,517

Analysed	
June 5, 2023 10:22:18 PM	
Computation with 5 linked points (*italics* points are less reliable):	1. Grote Kerk 2. Nieuwestad 3. Voorstraat 4. Ruiterskwartier 5. *De Brol*
Reference points (Reliability 0–5)	4
Transformation with 3 parameters	
Scale	1:8740
Rotation	13 °
Mean position error	26m

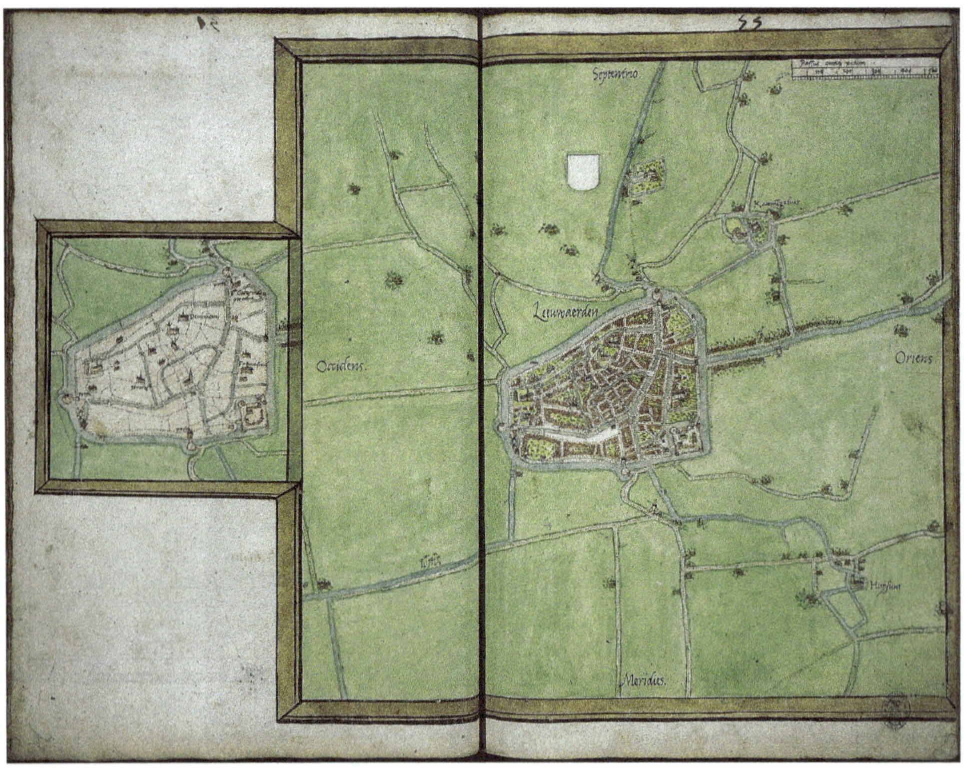

Septentrio.
Leeuwaerden
Occidens.
Oriens.
Meridies.
Huysum

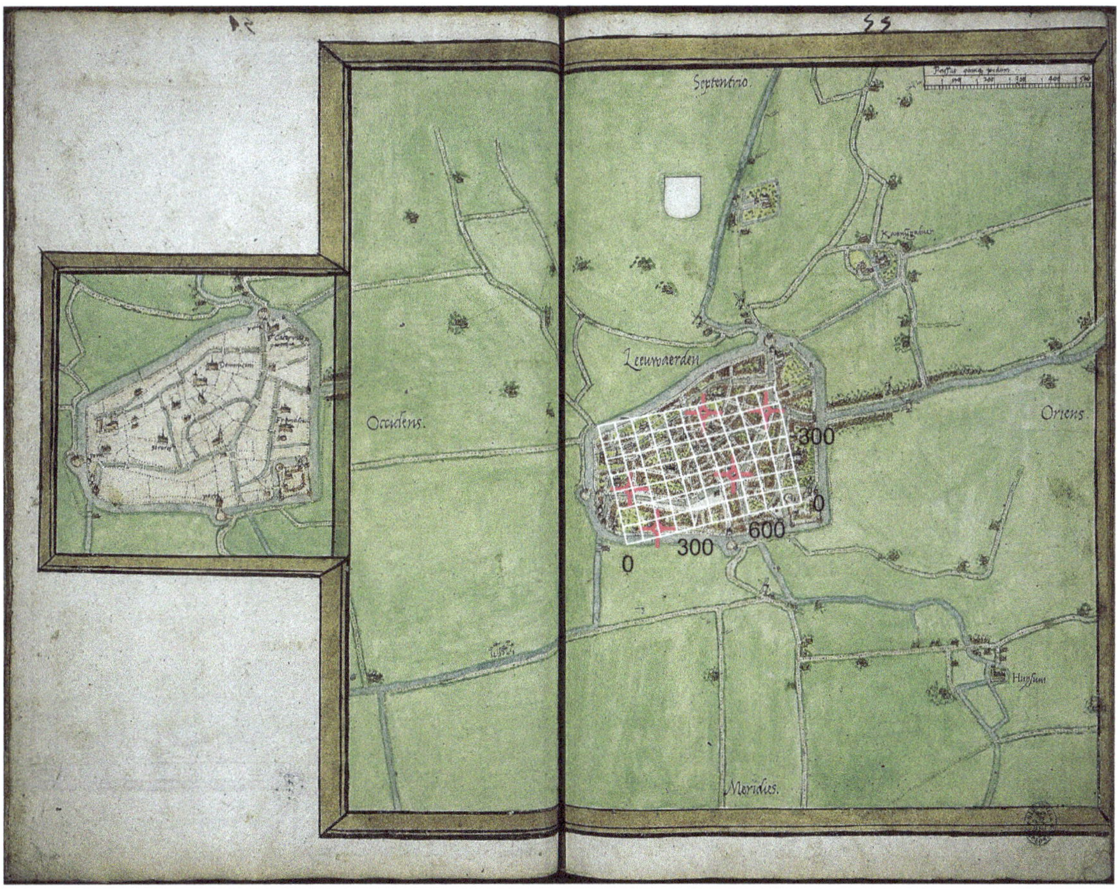
Septentrio.
Leeuwaerden
Occidens.
Oriens.
Meridies.
Huysum
300
0
600
300
0

126. Leiden

Tables 3.126 (measured and analysed).
Figures 3.126 (unmeasured, measured and analysed).

Measured

Pixels	Scale stick original map (pixels)	Houses	Street sides	Houses per street side	Inhabitants (RF 4.0)	Inhabitants (RF 5.5)	Inhabitants (RF 7.0)
14,313	996	2513	346	7.26	10,053	13,823	17,593

Analysed

January 23, 2023 7:44:31 PM	
Computation with 5 linked points (*italics* points are less reliable):	1. Pieterskerkhof 2. Burcht 3. Morriaanvest 4. Steenschuur 5. Oude Vest
Reference points (Reliability 0–5)	5
Transformation with 3 parameters	
Scale	1:8080
Rotation	8 °
Mean position error	18m

Septentrio.
Leyden
Renus
Renus
Oriens.
Meridies.
© Biblioteca Nacional de España

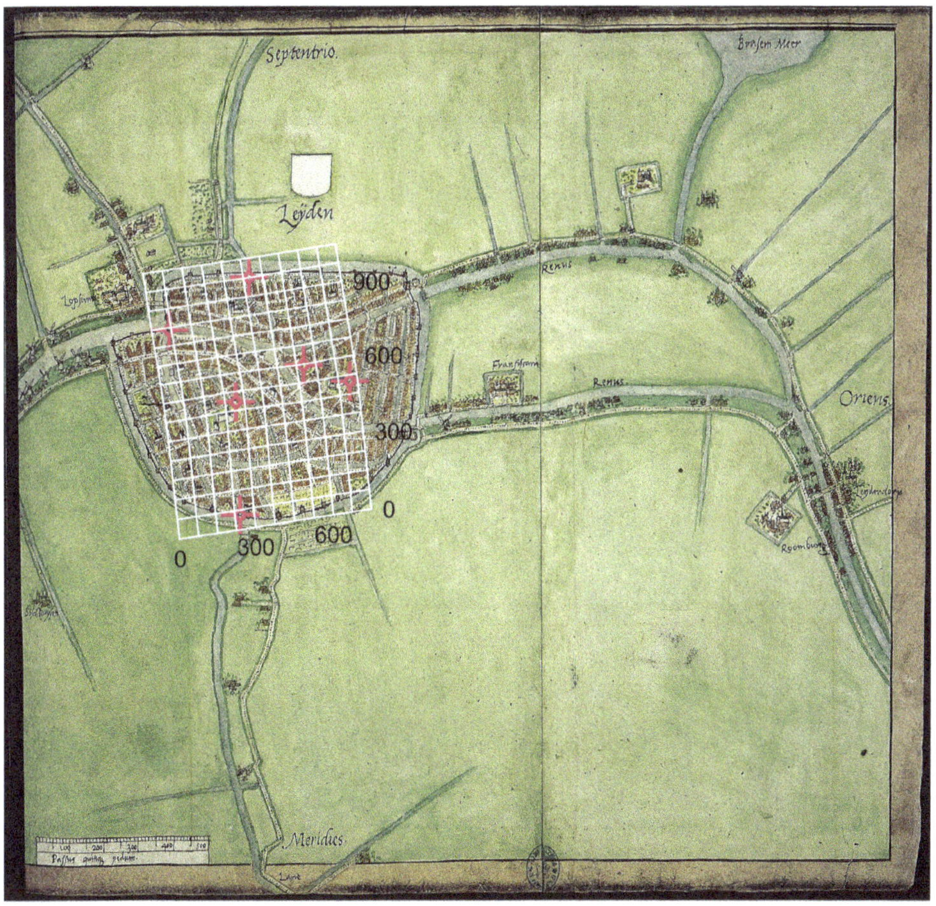
Septentrio.
Leyden
Renus
Renus
Oriens.
Meridies.
900
600
300
0
0
300
600

127. Lens
Tables 3.127 (measured and analysed).
Figures 3.127 (unmeasured, measured and analysed).

Measured							
Pixels	Scale stick original map (pixels)	Houses	Street sides	Houses per street side	Inhabitants (RF 4.0)	Inhabitants (RF 5.5)	Inhabitants (RF 7.0)
2726	1023	479	75	6.38	1915	2633	3351

Analysed	
June 6, 2023 9:06:49 AM	
Computation with 5 linked points (*italics* points are less reliable):	1. Église Saint-Léger 2. *Rue du Stade* 3. *Rue de Paris* 4. *Rue René Lanoy* 5. *Rue des Jardins*
Reference points (Reliability 0–5)	1
Transformation with 3 parameters	
Scale	1:8870
Rotation	9 °
Mean position error	40m

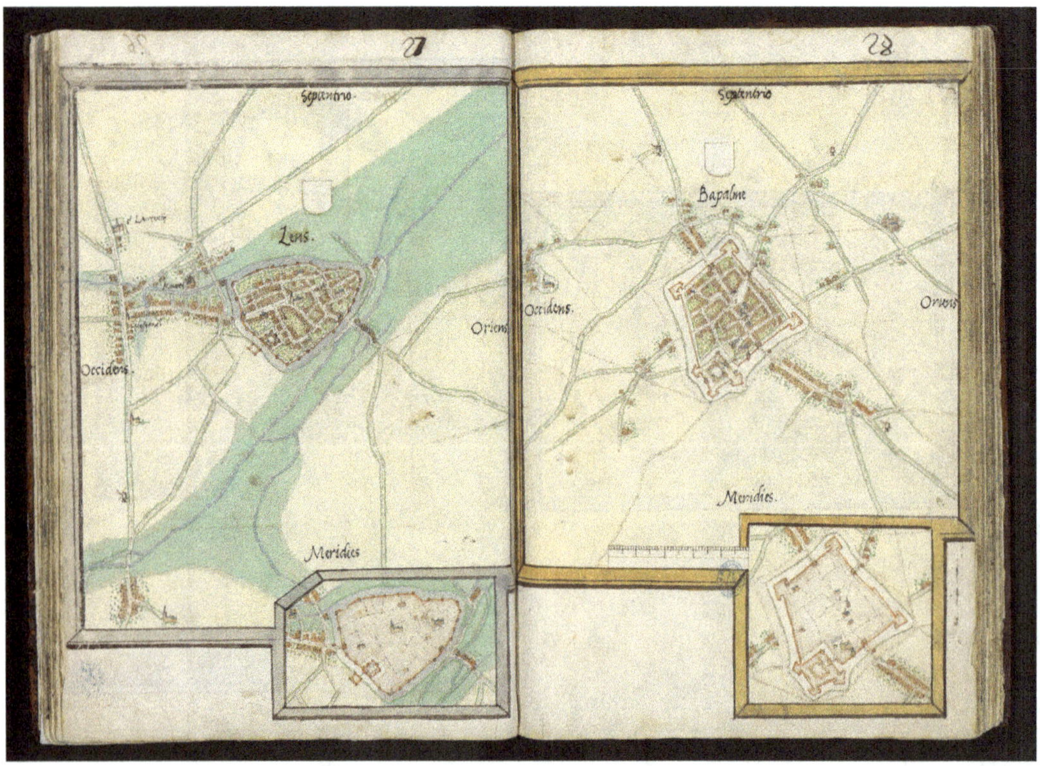

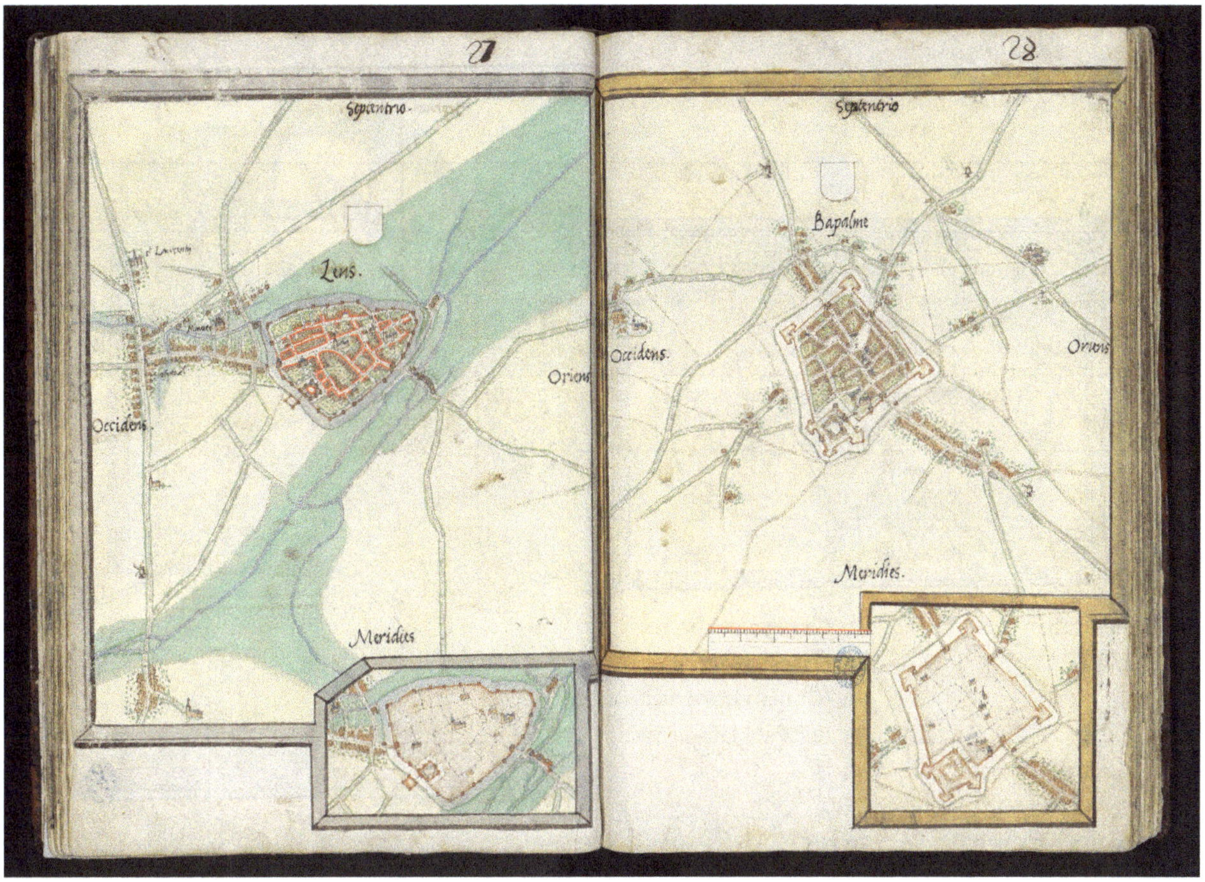
27
28
Septentrio
Septentrio
Lens.
Bapalme
Occidens.
Oriens
Occidens.
Oriens
Meridies
Meridies.

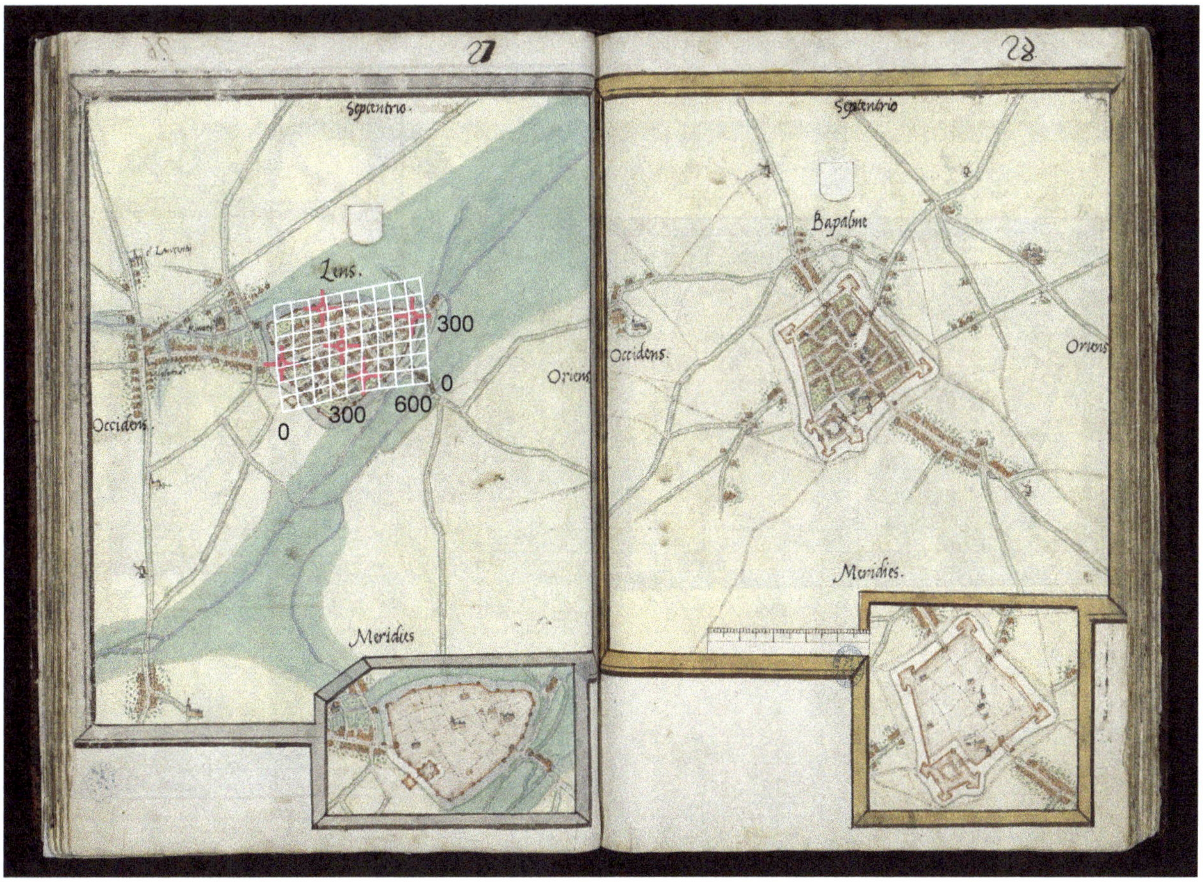
27
28
Septentrio
Septentrio
Lens.
Bapalme
300
0
0
300
600
Occidens.
Oriens
Occidens.
Oriens
Meridies
Meridies.

128. Le Roeulx
Tables 3.128 (measured and analysed).
Figures 3.128 (unmeasured, measured and analysed).

Measured							
Pixels	Scale stick original map (pixels)	Houses	Street sides	Houses per street side	Inhabitants (RF 4.0)	Inhabitants (RF 5.5)	Inhabitants (RF 7.0)
2456	1038	431	77	5.60	1725	2372	3019

Analysed	
April 18, 2023 9:47:22 AM	
Computation with 5 linked points (*italics* points are less reliable):	1. Église de Saint-Nicolas 2. Chateau des princes … *3. Rue du Cargies* 4. Rempart des Arbalestriers 5. Rue Paul Janson
Reference points (Reliability 0–5)	4
Transformation with 3 parameters	
Scale	1:8110
Rotation	9 °
Mean position error	21m

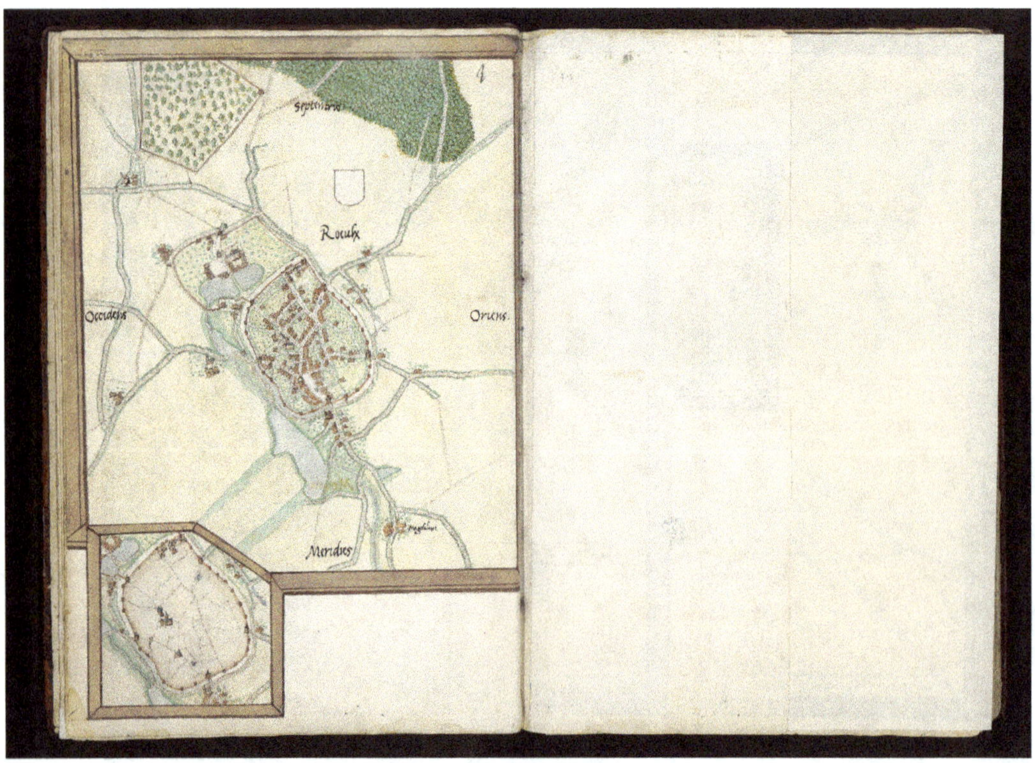

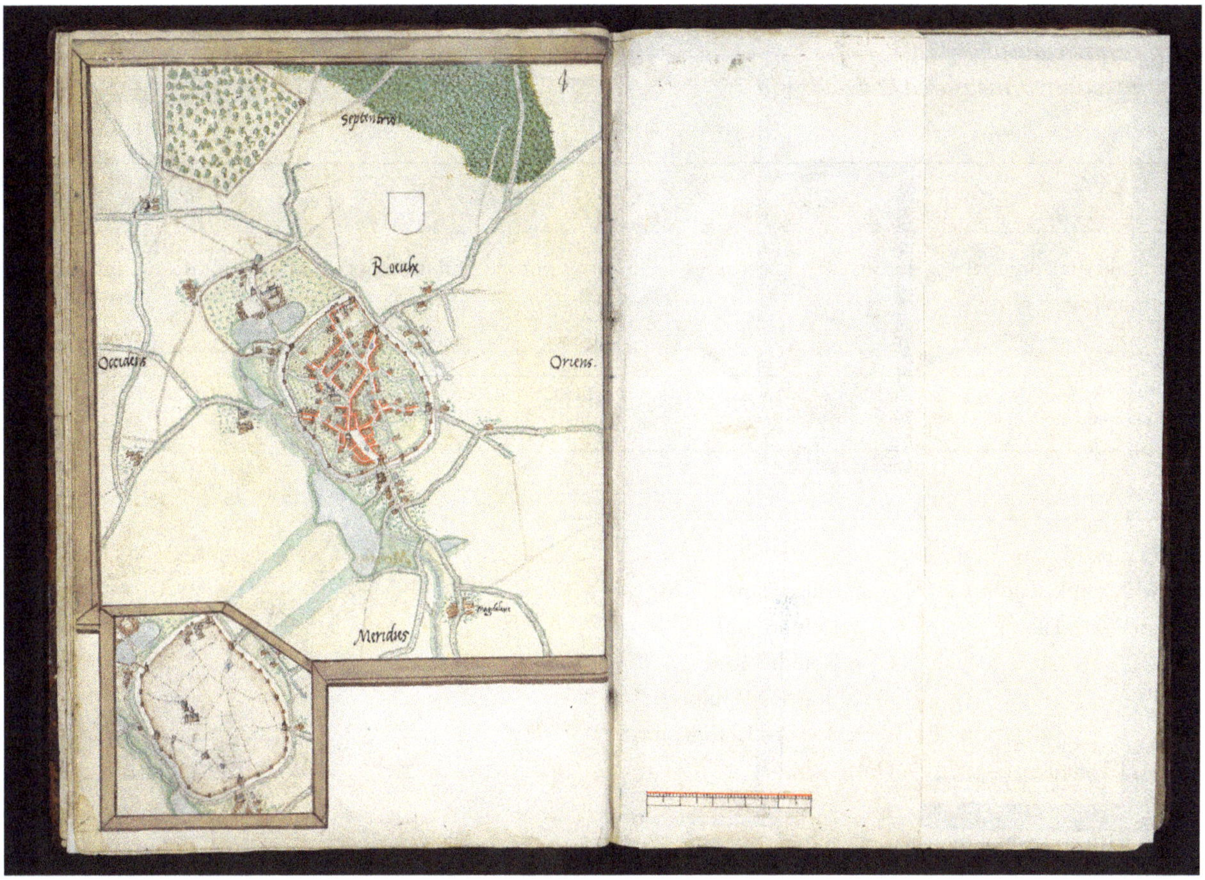
Septentrio
Roeulx
Occidens
Oriens
Meridies

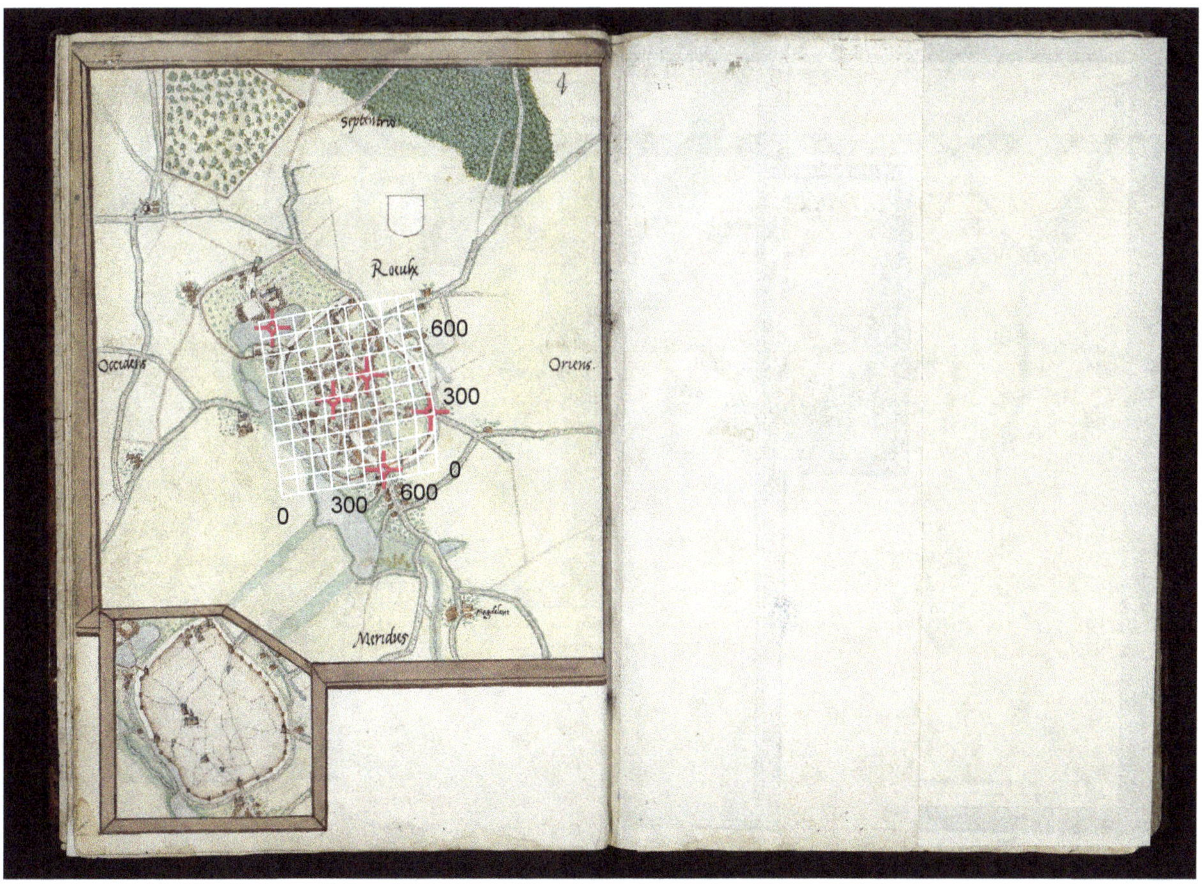
Septentrio
Roeulx
Occidens
Oriens
Meridies
600
300
0
0
300
600

129. Lessines
Tables 3.129 (measured and analysed).
Figures 3.129 (unmeasured, measured and analysed).

Measured							
Pixels	Scale stick original map (pixels)	Houses	Street sides	Houses per street side	Inhabitants (RF 4.0)	Inhabitants (RF 5.5)	Inhabitants (RF 7.0)
3508	1030	616	46	13.39	2464	3388	4312

Analysed	
June 7, 2023 10:24:11 AM	
Computation with 5 linked points (*italics* points are less reliable):	1. Église Saint-Pierre 2. Deindre sud 3. Deindre nord 4. Rue des Moulins 5. Rue de l'Herboristerie
Reference points (Reliability 0–5)	5
Transformation with 3 parameters	
Scale	1:8510
Rotation	8 °
Mean position error	14m

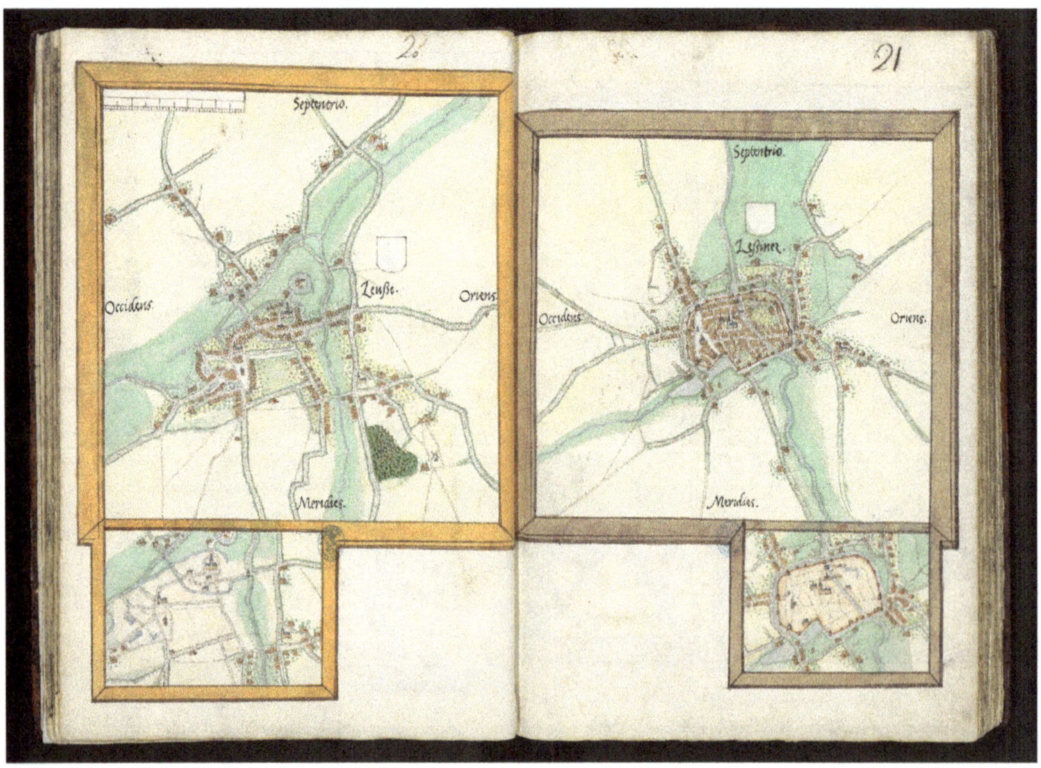

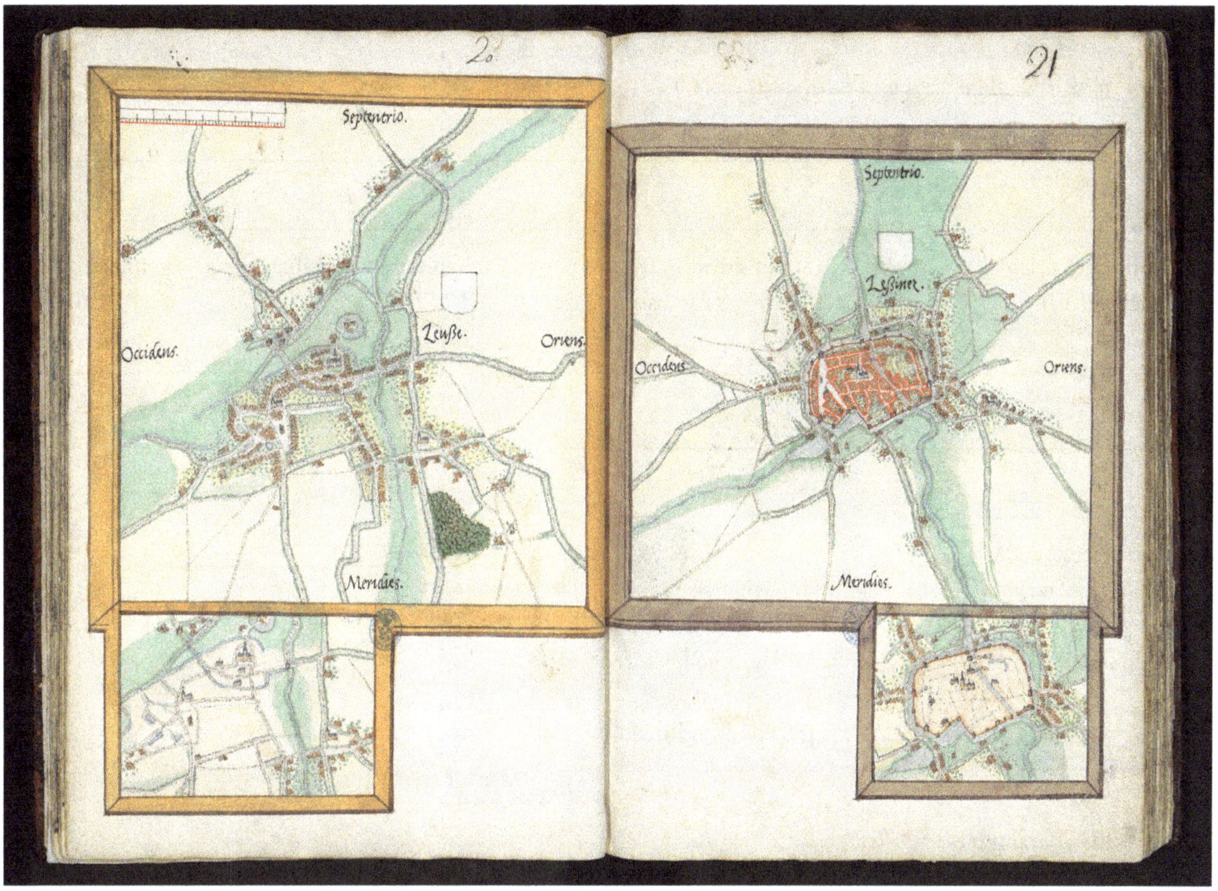
20
21
Septentrio.
Occidens.
Oriens.
Meridies.
Septentrio.
Occidens.
Oriens.
Meridies.

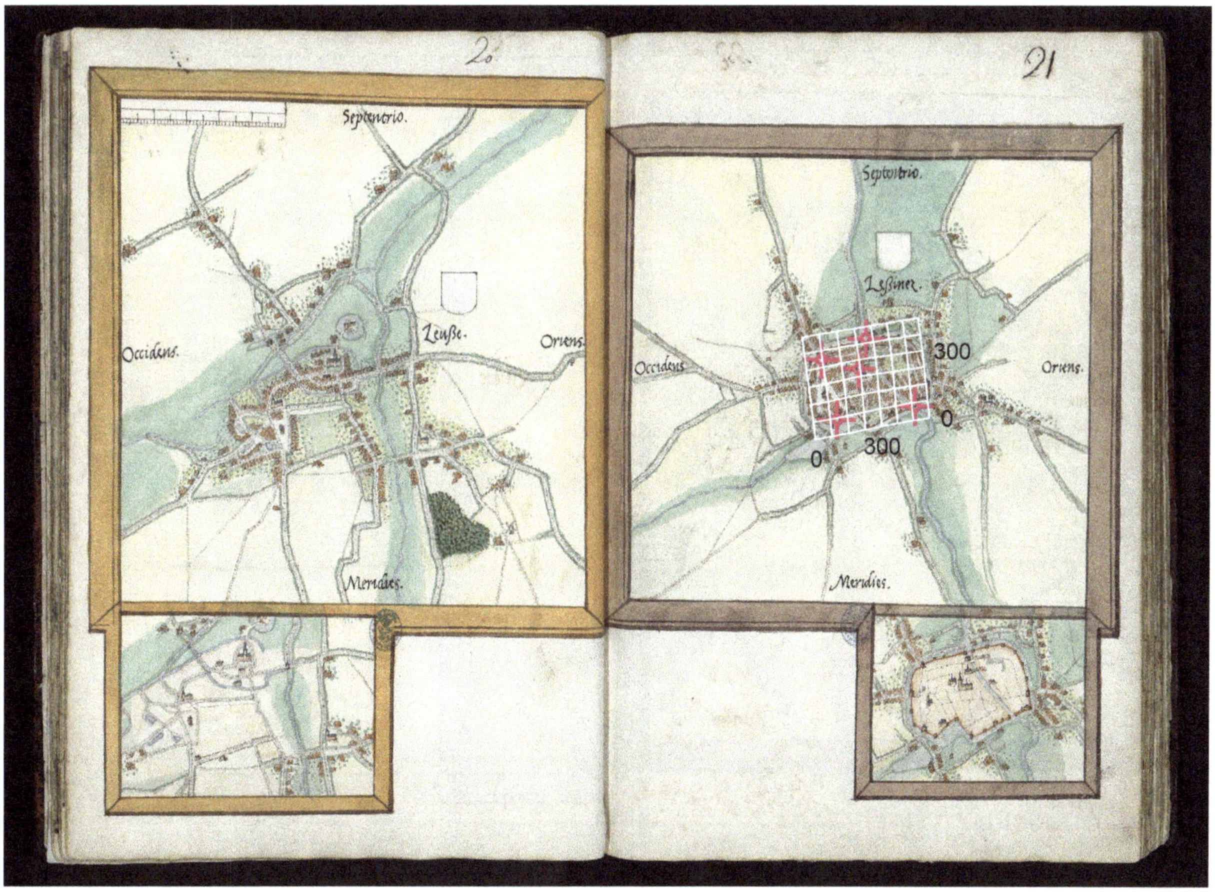
20
21
Septentrio.
Occidens.
Oriens.
Meridies.
Septentrio.
Occidens.
Oriens.
Meridies.
300
0
300
0

130. Leuven
Tables 3.130 (measured and analysed).
Figures 3.130 (unmeasured, measured and analysed).

Measured							
Pixels	Scale stick original map (pixels)	Houses	Street sides	Houses per street side	Inhabitants (RF 4.0)	Inhabitants (RF 5.5)	Inhabitants (RF 7.0)
25,367	missing	4454	456	9.77	17,817	24,499	31,181

Analysed	
September 24, 2023 12:02:03 PM	
Computation with 5 linked points (*italics* points are less reliable):	1. Sint-Pieterskerk 2. Sint-Anthoniuskapel 3. Minderbroederstoren 4. Tiensestraat 5. Deel oude vestingmuur
Reference points (Reliability 0–5)	5
Transformation with 3 parameters	
Scale	1:9190
Rotation	9 °
Mean position error	18m

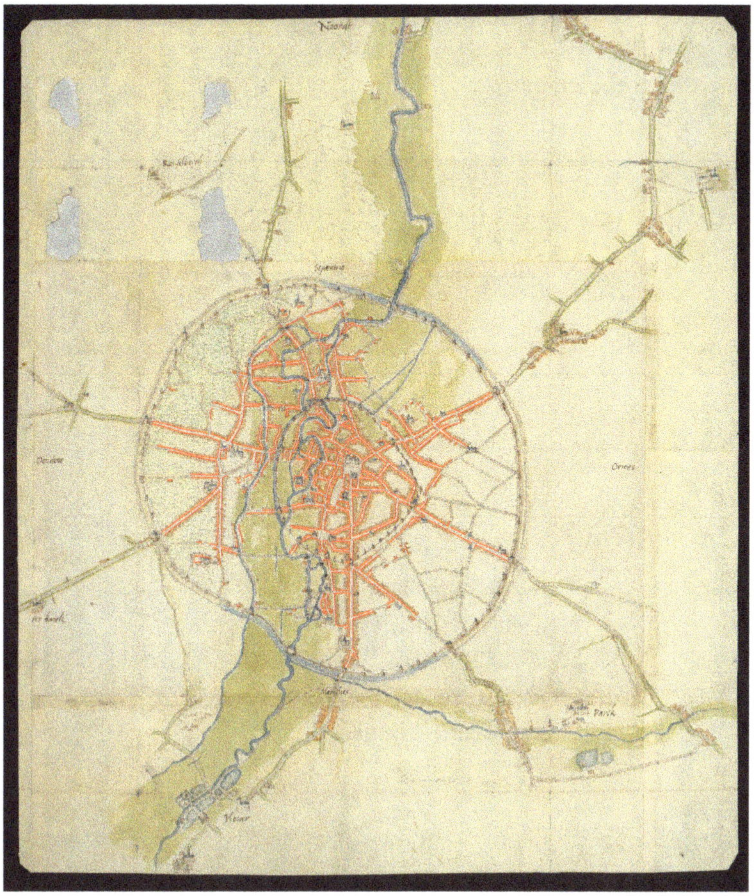

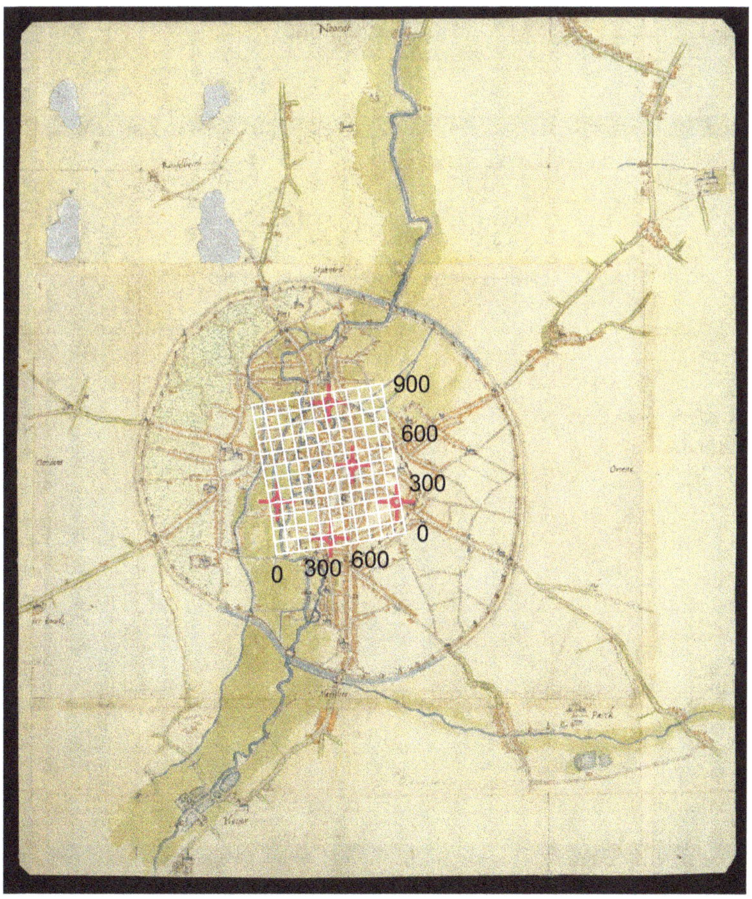
900
600
300
0
0
300
600

131. Leuze-en-Hainaut
Tables 3.131 (measured and analysed).
Figures 3.131 (unmeasured, measured and analysed).

Measured							
Pixels	Scale stick original map (pixels)	Houses	Street sides	Houses per street side	Inhabitants (RF 4.0)	Inhabitants (RF 5.5)	Inhabitants (RF 7.0)
2188	1030	384	46	8.35	1537	2113	2689

Analysed	
June 7, 2023 9:39:14 AM	
Computation with 5 linked points (*italics* points are less reliable):	1. Église de Saint Pierre 2. Rue du Pont Siant Martin 3. Rue de Condé 4. Rue du Rempart 5. Rue du Général Leman
Reference points (Reliability 0–5)	5
Transformation with 3 parameters	
Scale	1:8200
Rotation	11 °
Mean position error	11m

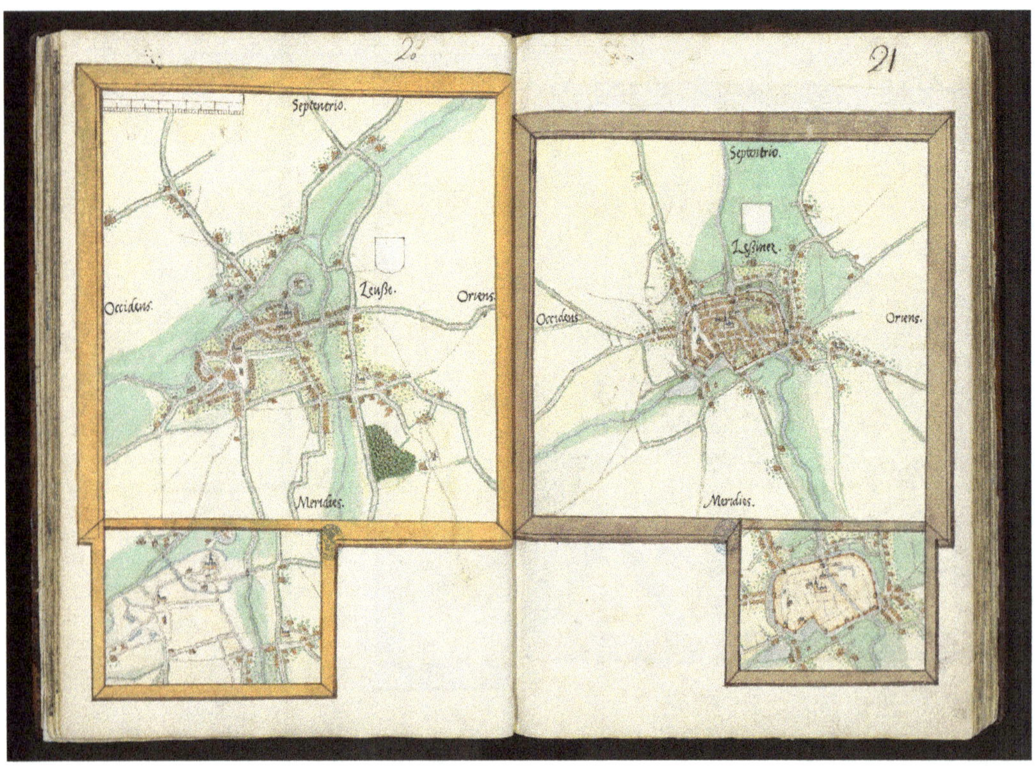

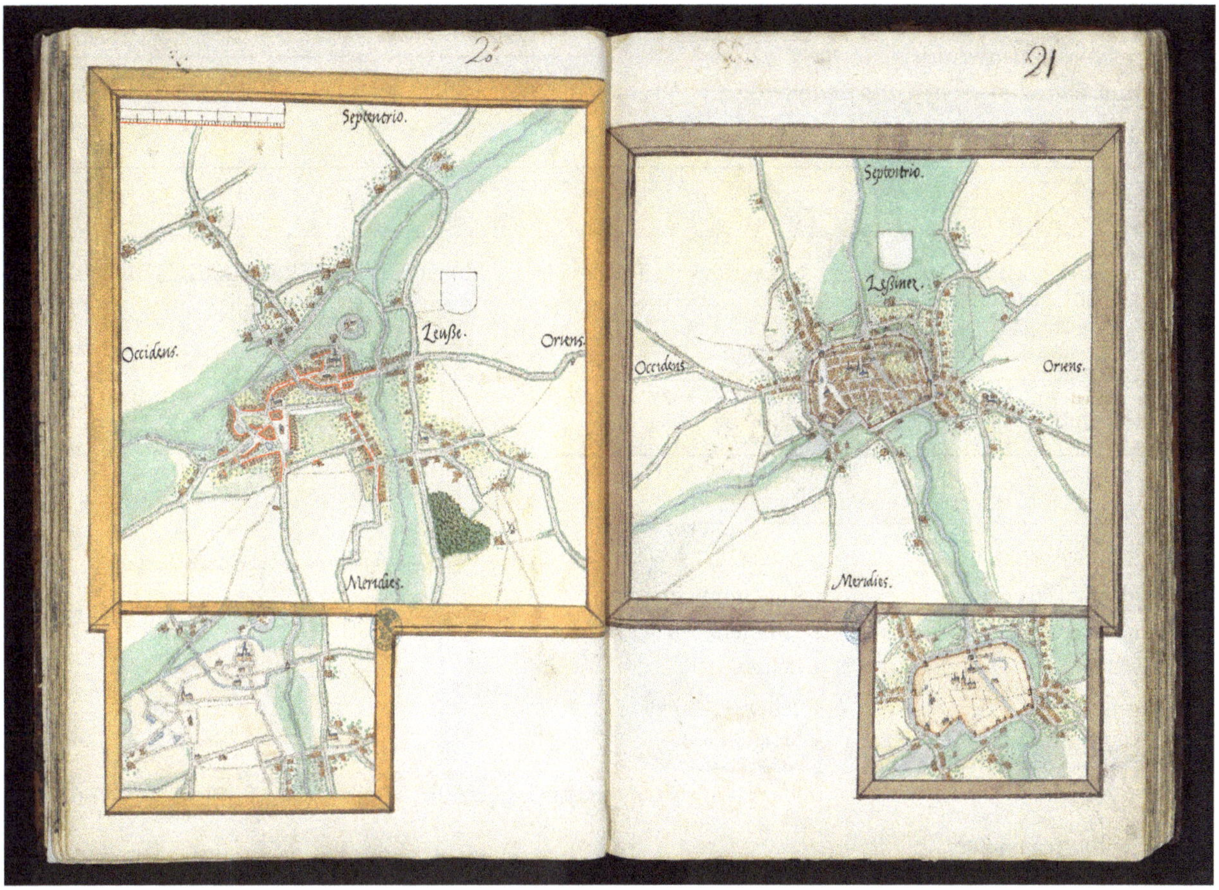
Septentrio.
Occidens.
Oriens.
Meridies.
Leuße.
Septentrio.
Occidens.
Oriens.
Meridies.

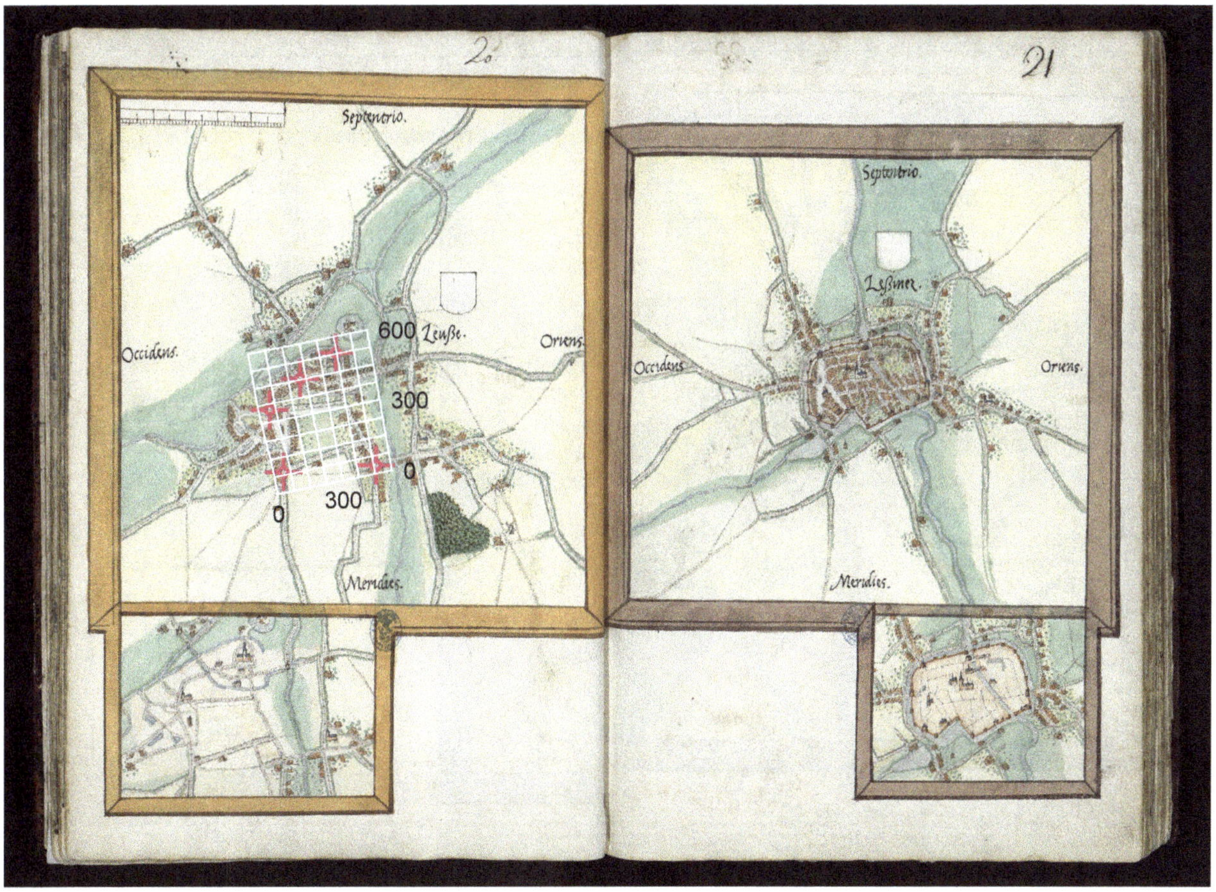
Septentrio.
Occidens.
Oriens.
Meridies.
Leuße.
600
300
0
0
300
Septentrio.
Occidens.
Oriens.
Meridies.

132. Lier

Tables 3.132 (measured and analysed).
Figures 3.132 (unmeasured, measured and analysed).

Measured							
Pixels	Scale stick original map (pixels)	Houses	Street sides	Houses per street side	Inhabitants (RF 4.0)	Inhabitants (RF 5.5)	Inhabitants (RF 7.0)
9730	missing	1709	214	7.98	6834	9397	11,960

Analysed	
September 25, 2023 3:34:53 PM	
Computation with 5 linked points (*italics* points are less reliable):	1. Sint-Gummaruskerk 2. Zimmertoren 3. Belfort 4. Wijngaardstraat 5. Mosdijk
Reference points (Reliability 0–5)	5
Transformation with 3 parameters	
Scale	1:8890
Rotation	19 °
Mean position error	17m

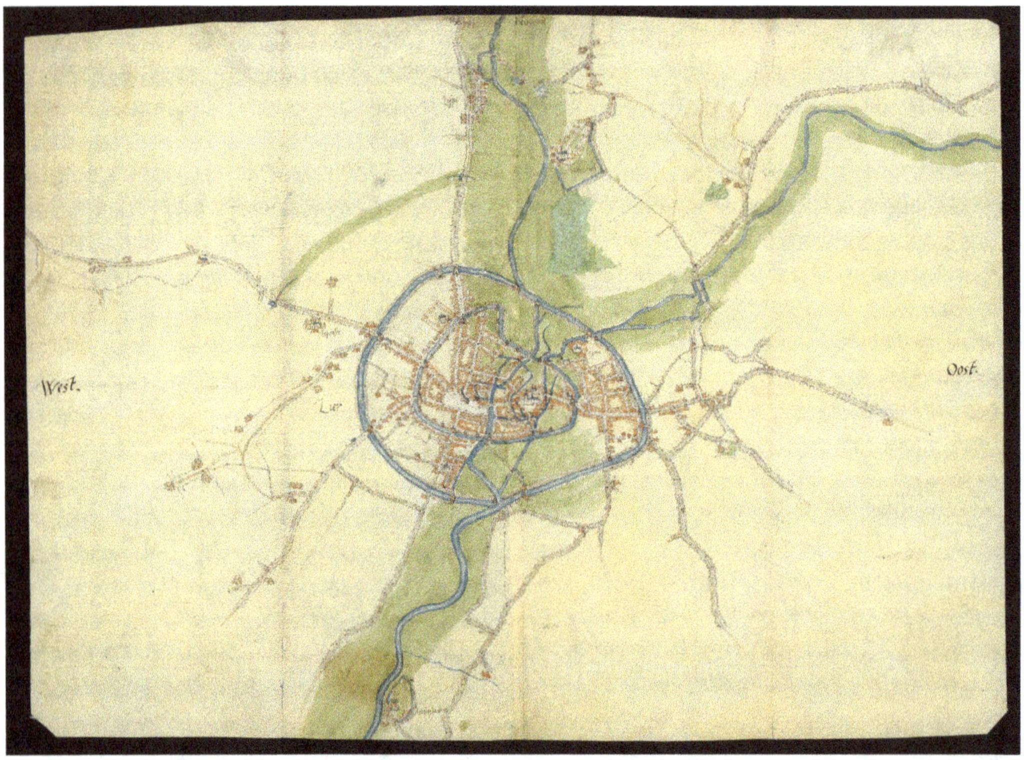

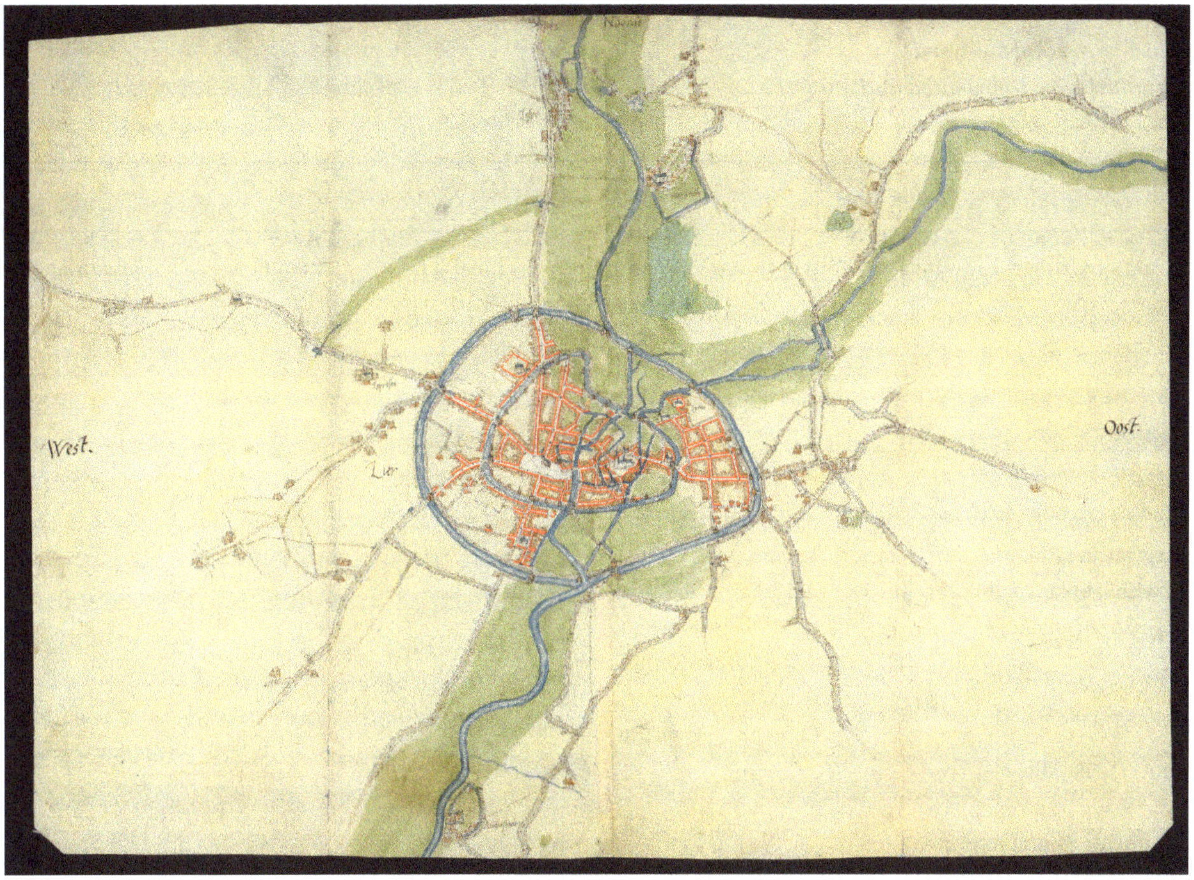
West.
Oost.

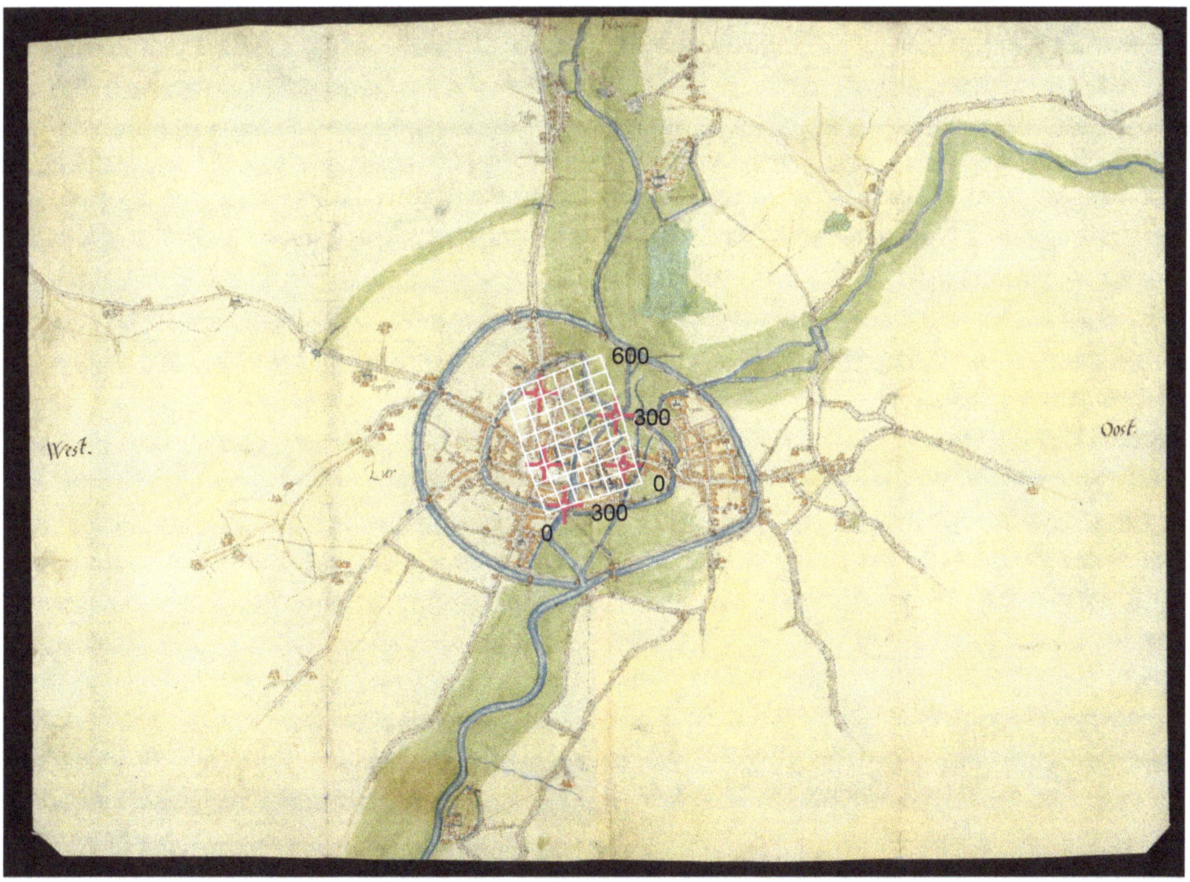
600
300
0
300
0
West.
Oost.

133. Lille
Tables 3.133 (measured and analysed).
Figures 3.133 (unmeasured, measured and analysed).

Measured							
Pixels	Scale stick original map (pixels)	Houses	Street sides	Houses per street side	Inhabitants (RF 4.0)	Inhabitants (RF 5.5)	Inhabitants (RF 7.0)
18,861	912	3312	480	6.90	13,135	18.216	22,985

Analysed	
October 10, 2023 5:25:12 PM	
Computation with 5 linked points (*italics* points are less reliable):	1. Église Saint Maurice 2. Église Saint-Sauveur 3. Église Sainte-Cathérine 4. Palais Rihour 5. Château de Courtrai
Reference points (Reliability 0–5)	5
Transformation with 3 parameters	
Scale	1:9070
Rotation	11 °
Mean position error	12m

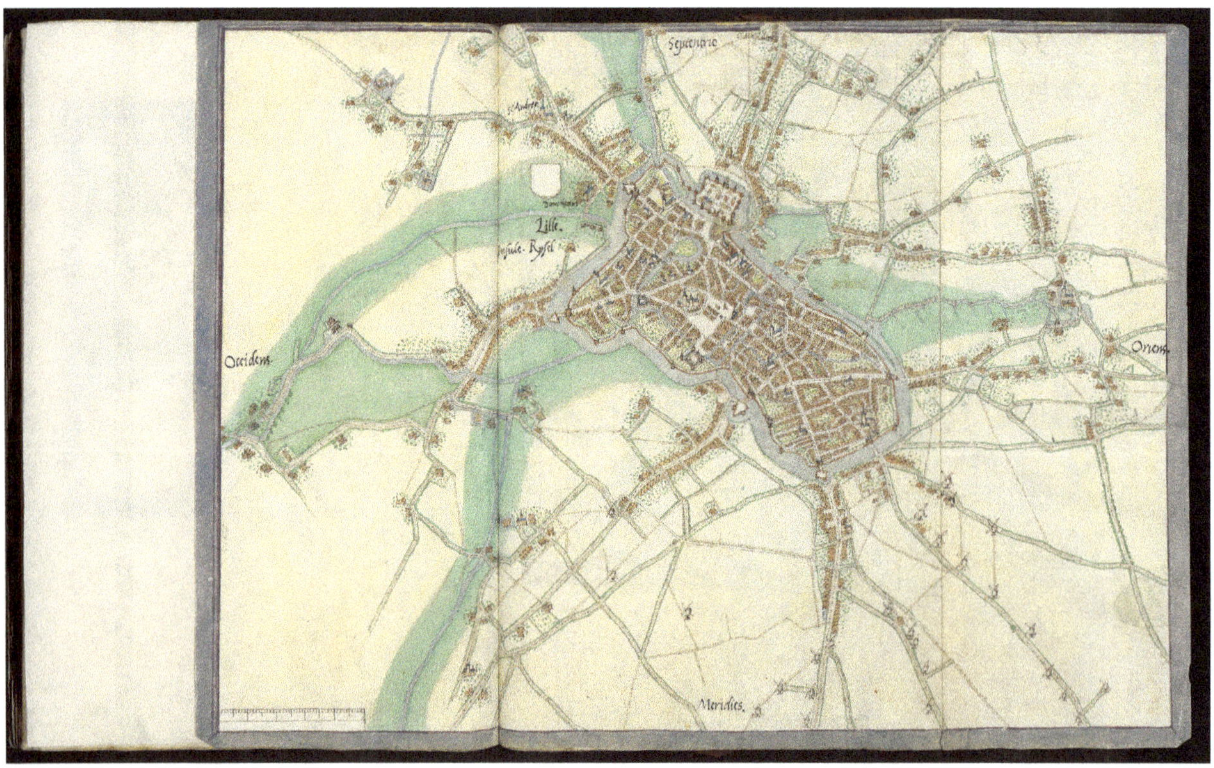

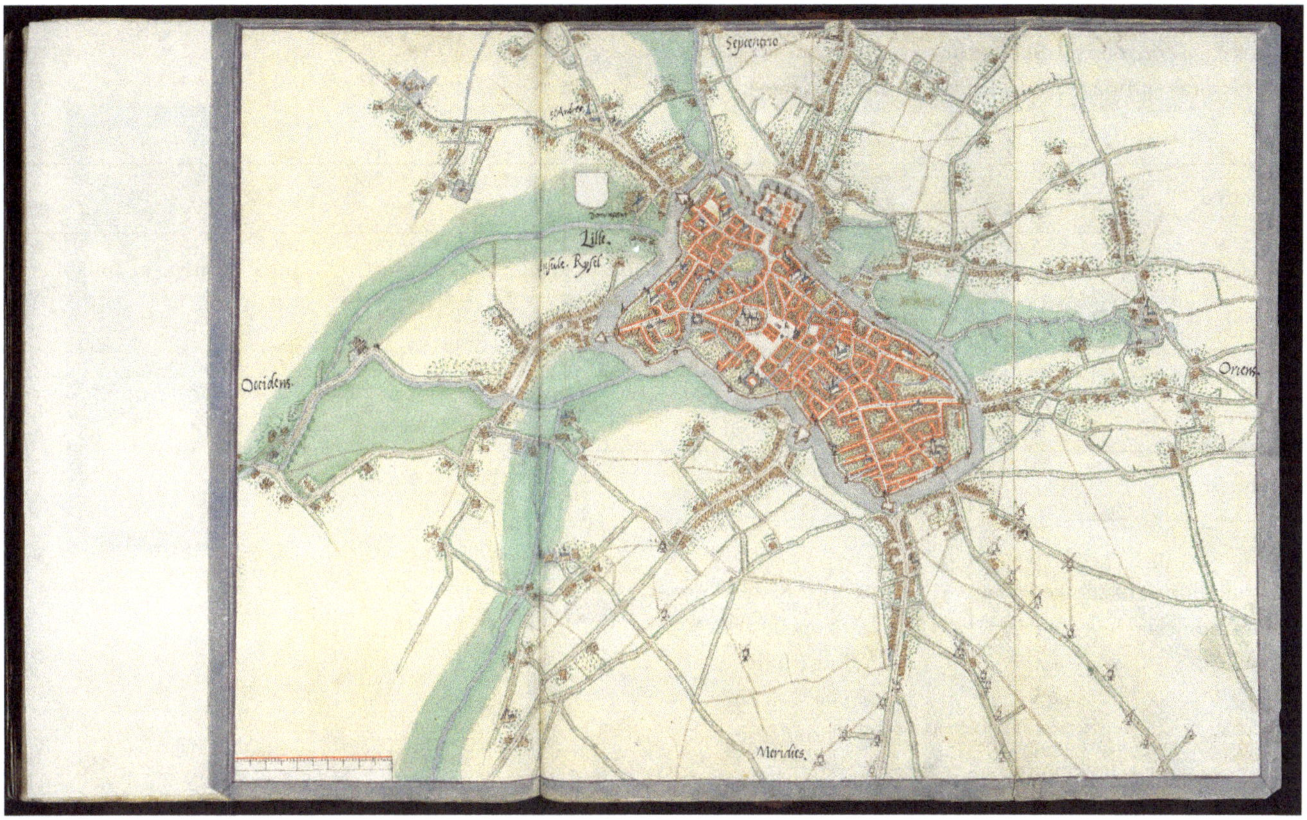
Septentrio
Lille.
Occidens.
Oriens.
Meridies.

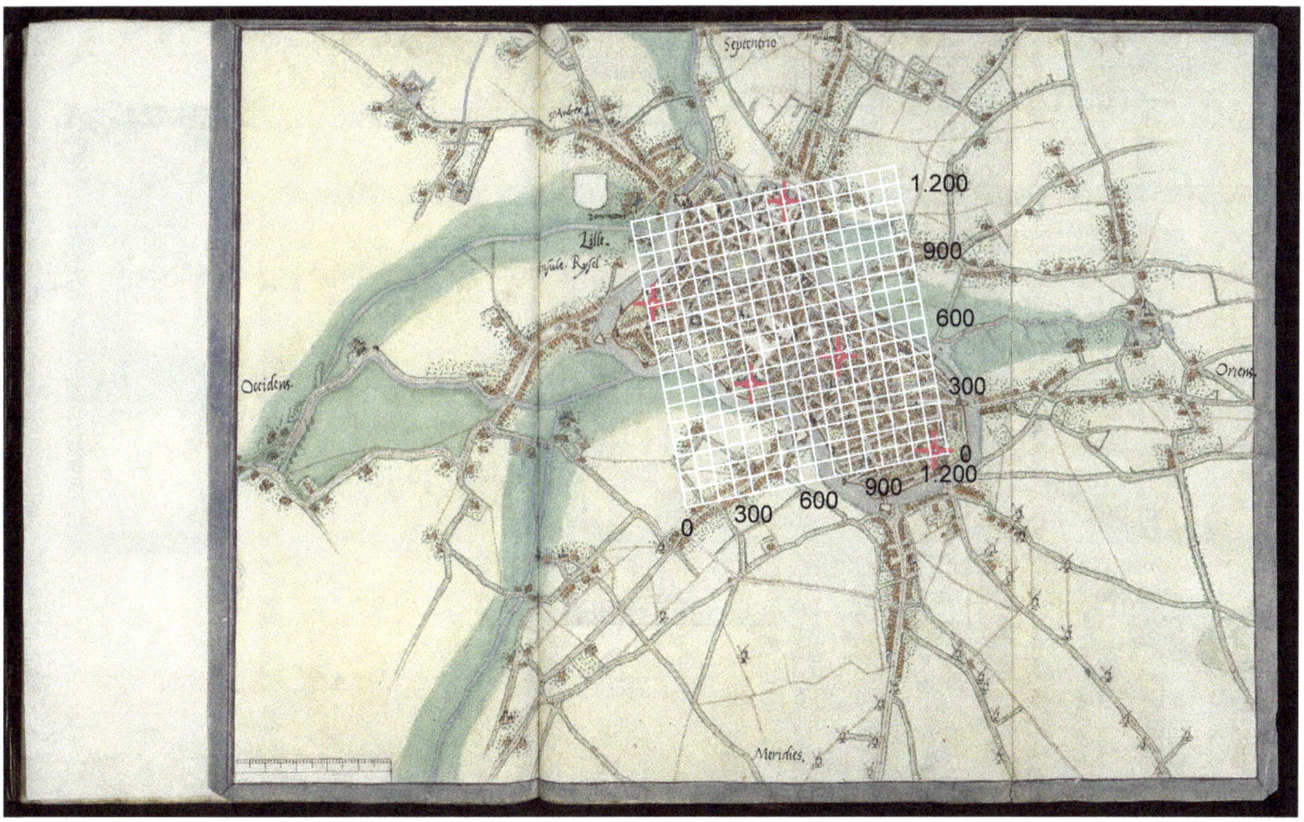
Septentrio
Lille.
Occidens.
Oriens.
Meridies.
1.200
900
600
300
0
0
300
600
900
1.200

134. Lillers

Tables 3.134 (measured and analysed).
Figures 3.134 (unmeasured, measured and analysed).

Measured							
Pixels	Scale stick original map (pixels)	Houses	Street sides	Houses per street side	Inhabitants (RF 4.0)	Inhabitants (RF 5.5)	Inhabitants (RF 7.0)
861	1015	151	31	4.88	605	832	1059

Analysed	
June 9, 2023 5:20:30 PM	
Computation with 5 linked points (*italics* points are less reliable):	1. Église Saint-Omer 2. Chapelle de la Miséricorde 3. Rue Fanien *4. Rue des Chats* *5. Rue d'Église*
Reference points (Reliability 0–5)	3
Transformation with 3 parameters	
Scale	1:9450
Rotation	13 °
Mean position error	19m

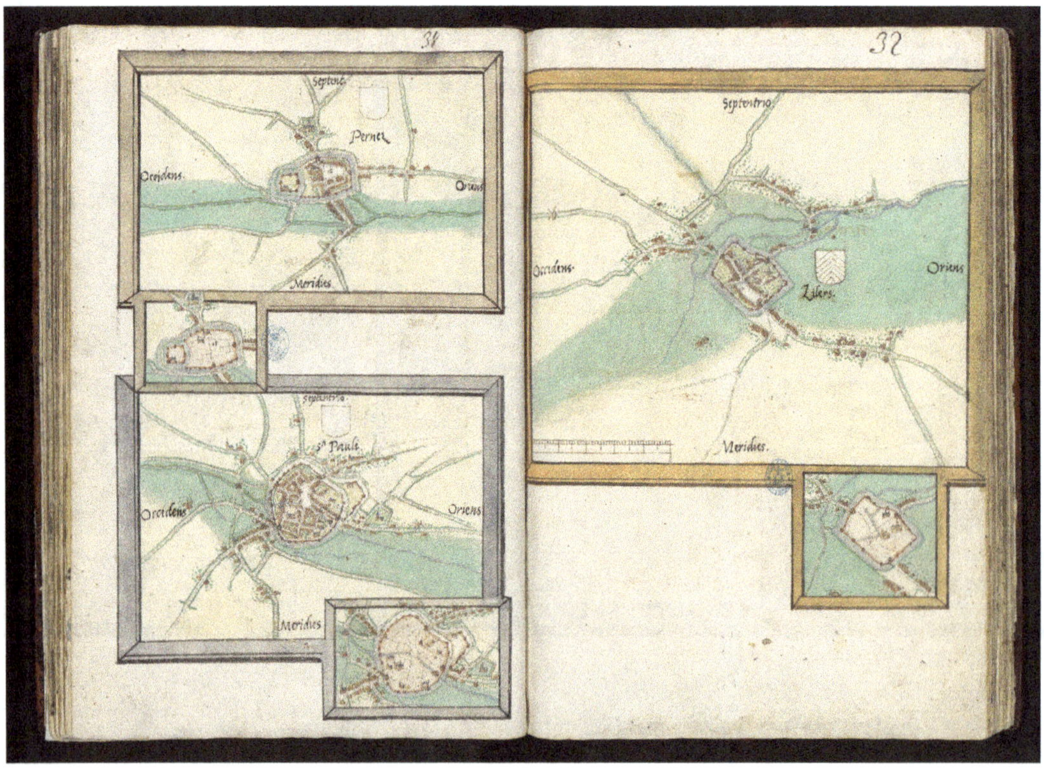

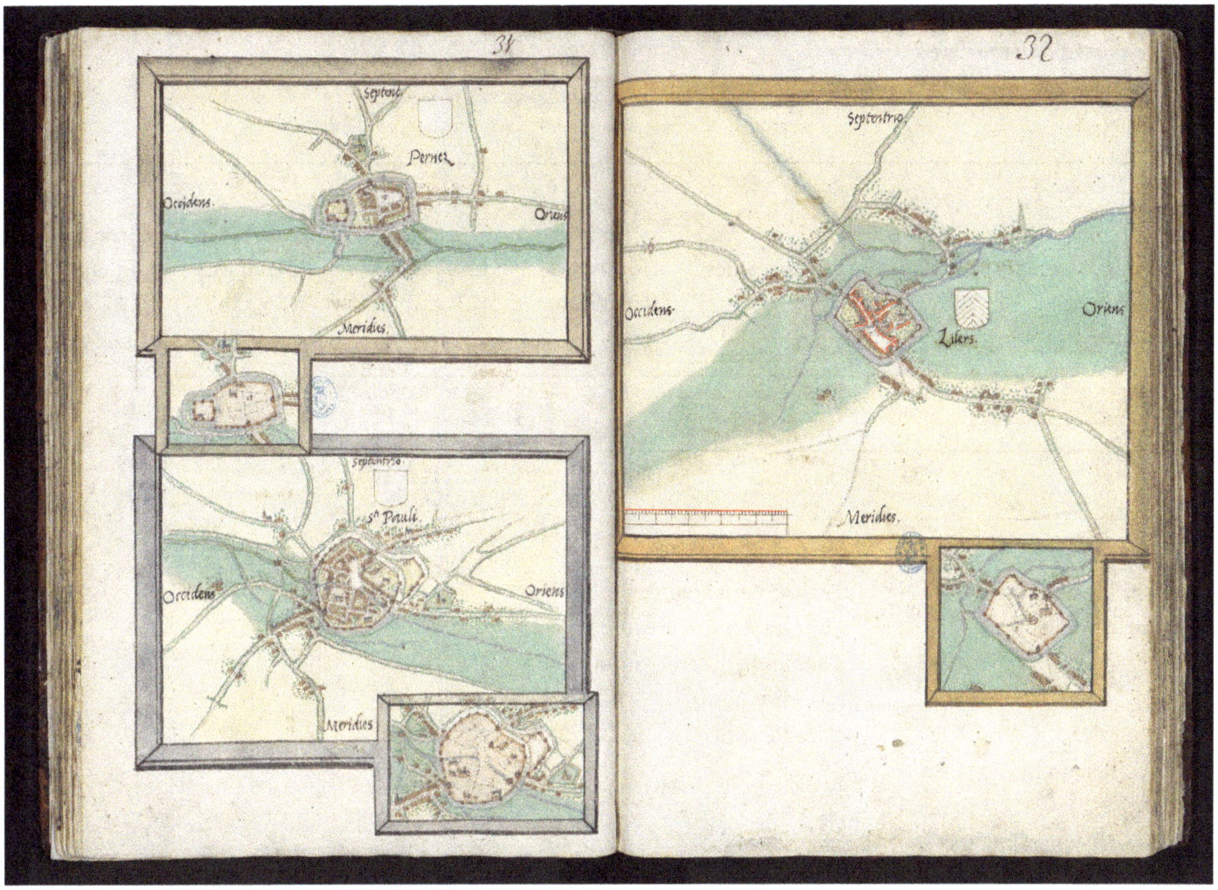
31
32
Septent.
Pernez
Occidens
Oriens
Meridies
Septentrio
St Pauli
Occidens
Oriens
Meridies
Septentrio
Occidens
Lilers
Oriens
Meridies

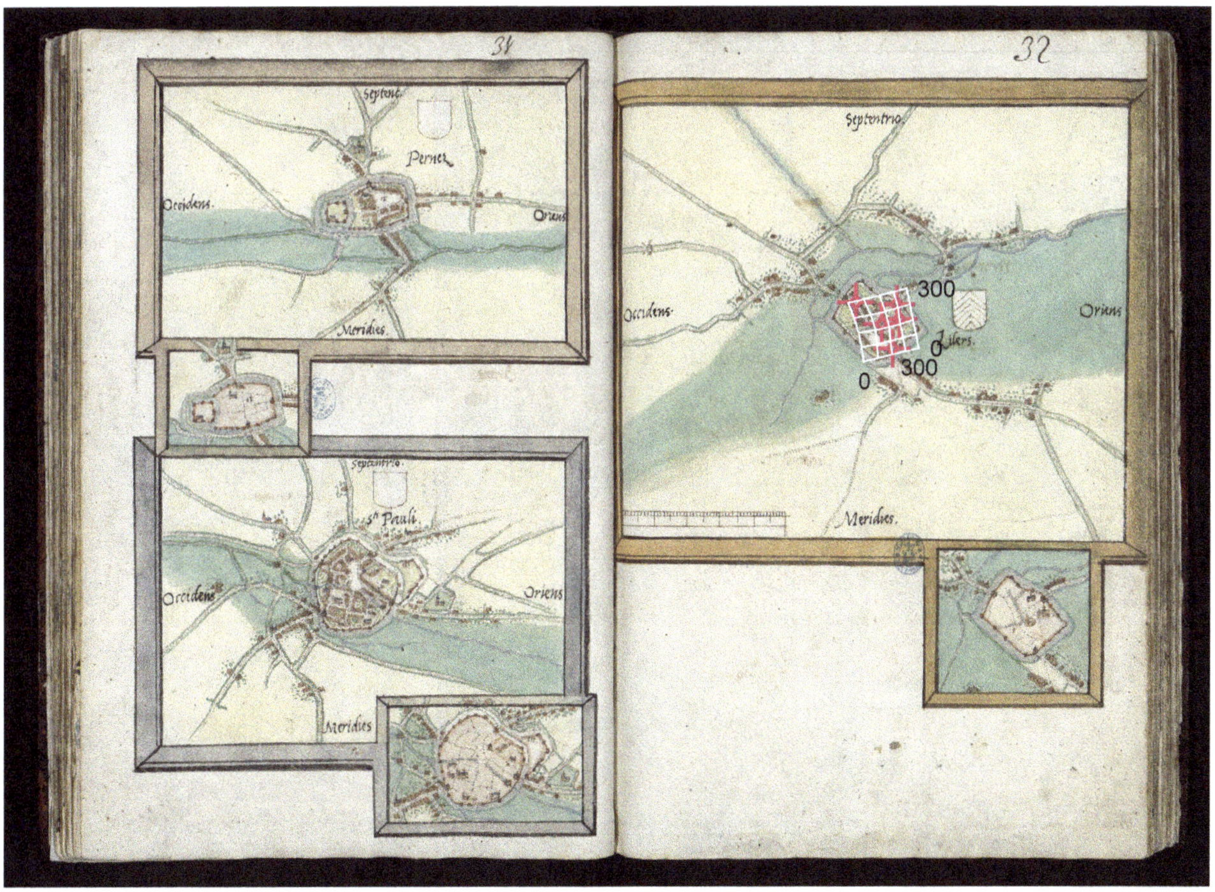
31
32
Septent.
Pernez
Occidens
Oriens
Meridies
Septentrio
St Pauli
Occidens
Oriens
Meridies
Septentrio
Occidens
300
Lilers
0
300
0
Oriens
Meridies

135. Limbourg

Tables 3.135 (measured and analysed).
Figures 3.135 (unmeasured, measured and analysed).

Measured							
Pixels	Scale stick original map (pixels)	Houses	Street sides	Houses per street side	Inhabitants (RF 4.0)	Inhabitants (RF 5.5)	Inhabitants (RF 7.0)
1906	missing	335	66	5.07	2423	3332	4241

Analysed	
October 15, 2023 6:56:57 PM	
Computation with 5 linked points (*italics* points are less reliable):	1. Église Saint-Georges 2. Sur les Remparts (nord) 3. Sur les Remparts (ouest) 4. Vesdre (nord) 5. Vesdre (sud)
Reference points (Reliability 0–5)	5
Transformation with 3 parameters	
Scale	1:7460
Rotation	25 °
Mean position error	74m

Noordt.
Daelhem
West.
Oost.
Limbourg.
Gulken
Zuijdt.

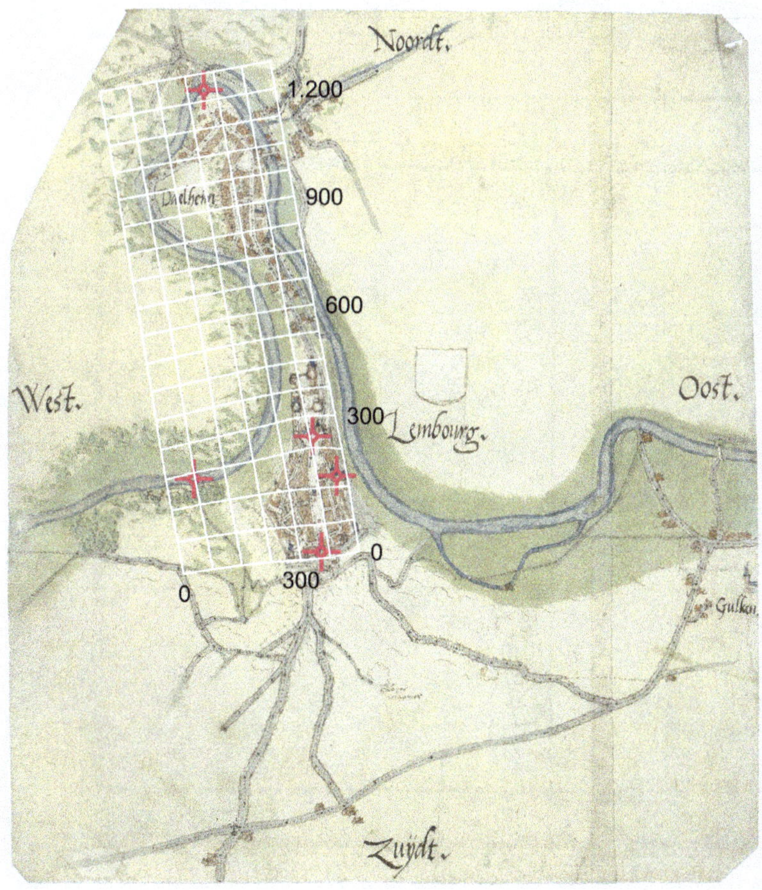
Noordt.
1.200
900
600
300
0
0
300
Daelhem
West.
Oost.
Limbourg.
Gulken
Zuijdt.

136. Lingen

Tables 3.136 (measured and analysed).
Figures 3.136 (unmeasured, measured and analysed).

Measured							
Pixels	Scale stick original map (pixels)	Houses	Street sides	Houses per street side	Inhabitants (RF 4.0)	Inhabitants (RF 5.5)	Inhabitants (RF 7.0)
1201	1007	211	33	6.39	844	1160	1476

Analysed	
June 10, 2023 10:23:23 AM	
Computation with 5 linked points (*italics* points are less reliable):	1. Evangelisch-reformierte Kirche 2. Am Wall Nord 3. Grosse Strasse 4. Neue Strasse 5. Pulverturm
Reference points (Reliability 0–5)	5
Transformation with 3 parameters	
Scale	1:7960
Rotation	9 °
Mean position error	19m

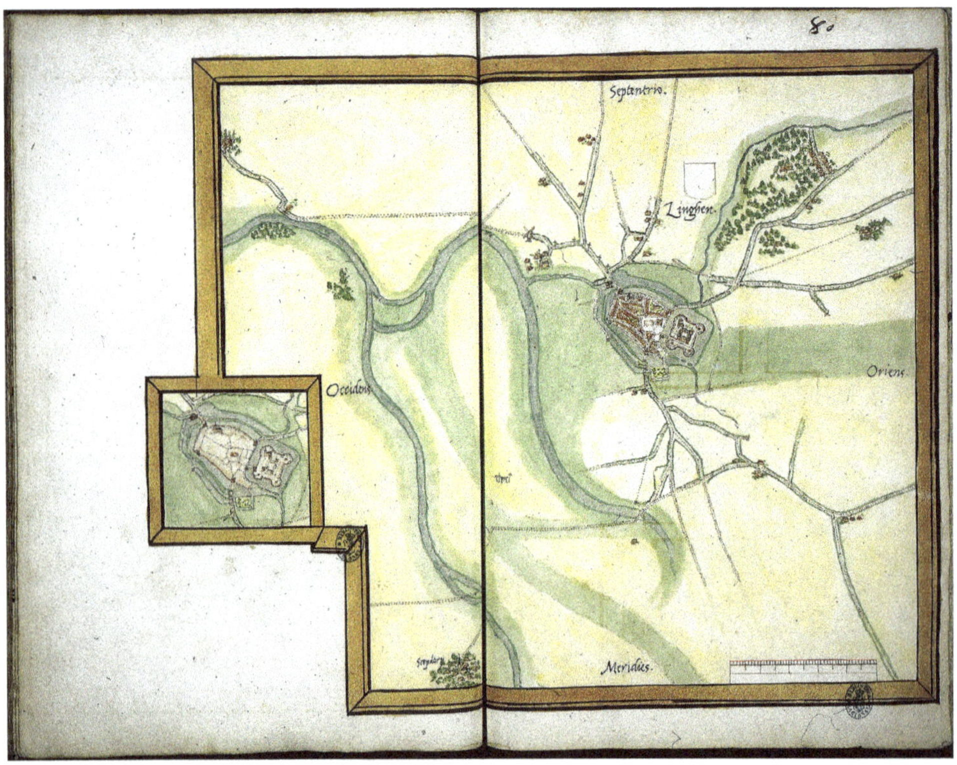

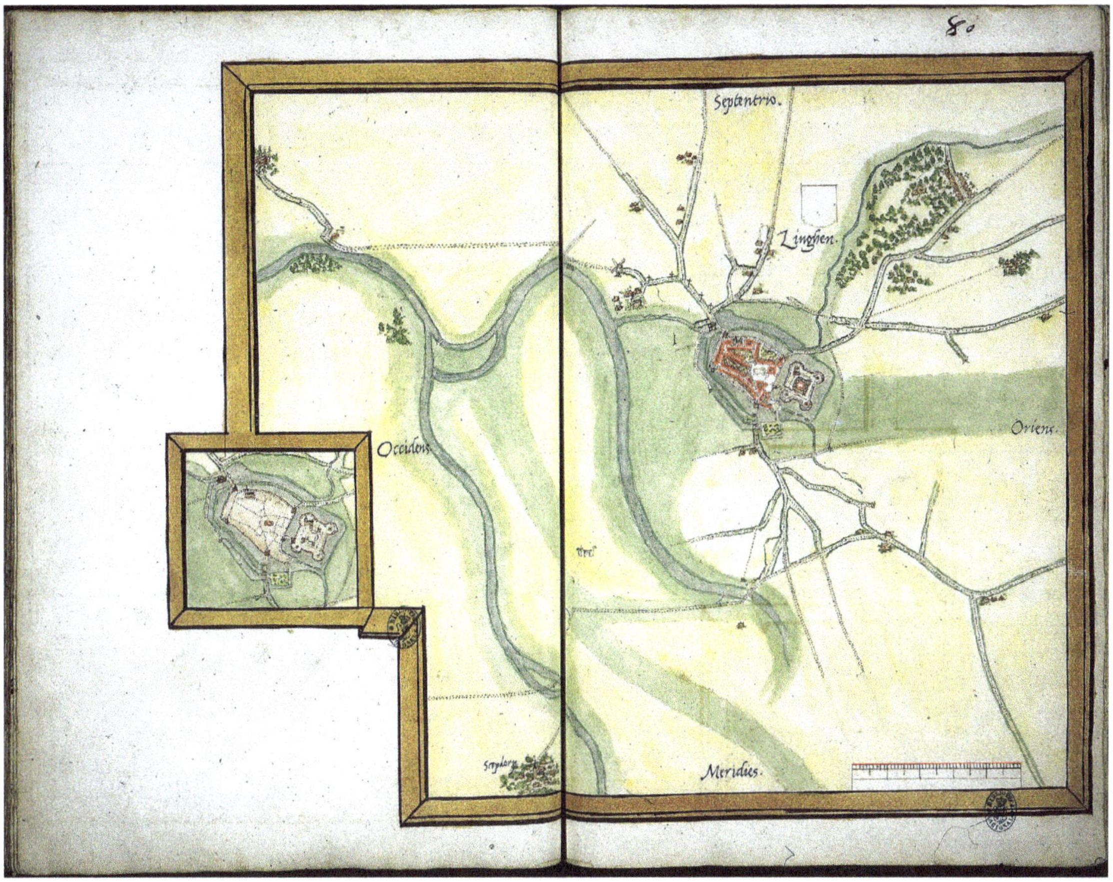
Septentrio.
Linghen.
Occidens
Oriens.
Meridies.

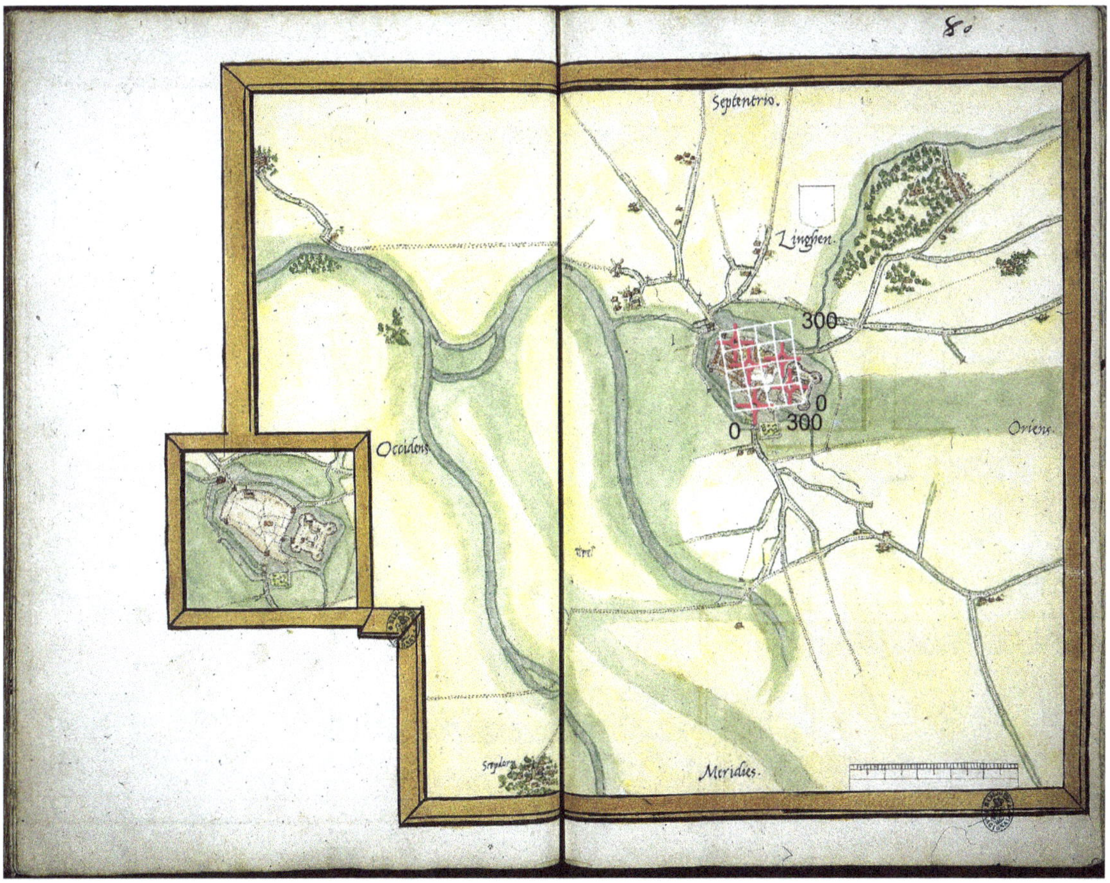
Septentrio.
Linghen.
Occidens
Oriens.
Meridies.
300
0
300
0

137. Lochem

Tables 3.137 (measured and analysed).
Figures 3.137 (unmeasured, measured and analysed).

Measured

Pixels	Scale stick original map (pixels)	Houses	Street sides	Houses per street side	Inhabitants (RF 4.0)	Inhabitants (RF 5.5)	Inhabitants (RF 7.0)
1972	1021	346	52	6.66	1385	1905	2425

Analysed

June 12, 2023 10:20:00 AM	
Computation with 5 linked points (*italics* points are less reliable):	1. Sint Gudula Kerk 2. Poststeeg 3. Walderstraat 4. Oosterwal 5. Westerwal
Reference points (Reliability 0–5)	5
Transformation with 3 parameters	
Scale	1:7890
Rotation	15 °
Mean position error	11m

99 bis
Septentrio.
Lochum.
Occidens.
Meridies.
Septentrio.
Groll.
Occidens.
Meridies..

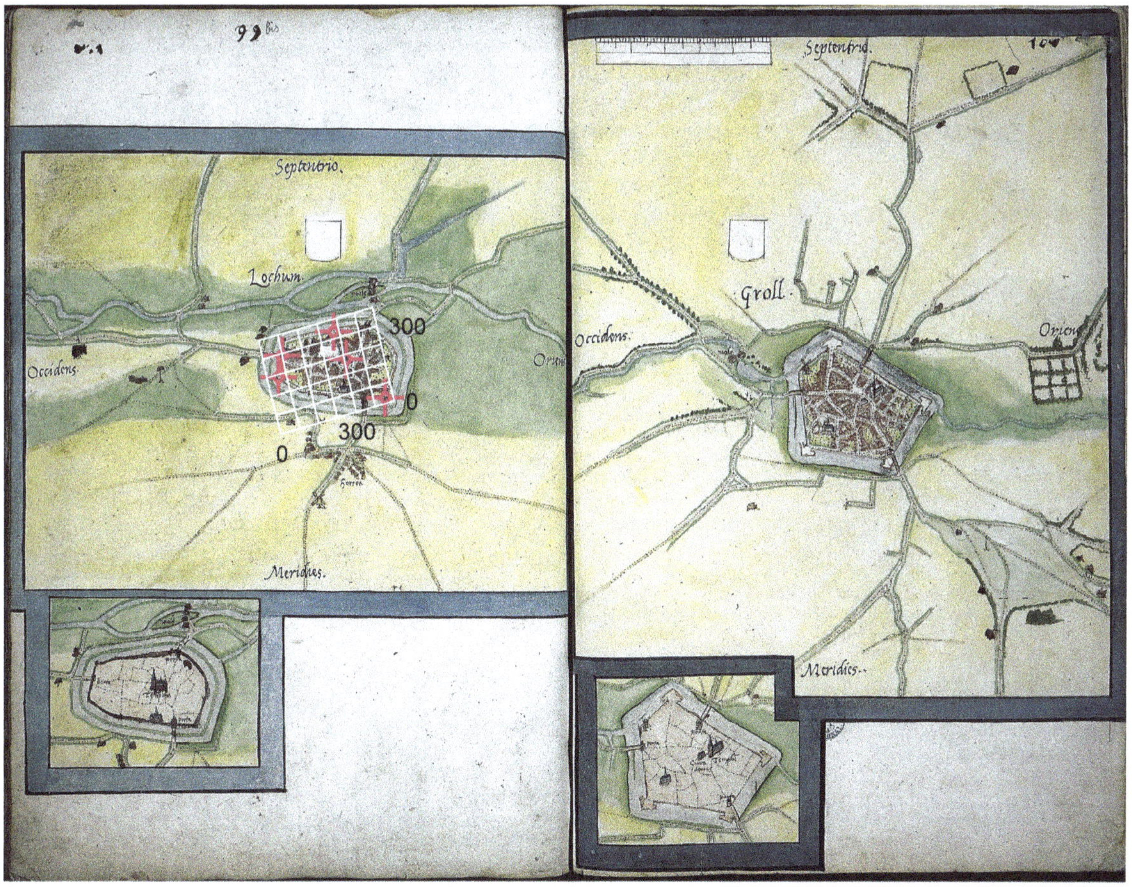
99 bis
Septentrio.
Lochum.
300
0
300
0
Occidens.
Meridies.
Septentrio.
Groll.
Occidens.
Meridies..

138. Luxembourg

Tables 3.138 (measured and analysed).
Figures 3.138 (unmeasured, measured and analysed).

Measured							
Pixels	Scale stick original map (pixels)	Houses	Street sides	Houses per street side	Inhabitants (RF 4.0)	Inhabitants (RF 5.5)	Inhabitants (RF 7.0)
6362	missing	1117	156	7.16	4468	6144	7820

Analysed	
September 25, 2023 11:41:02 PM	
Computation with 5 linked points (*italics* points are less reliable):	1. Saint-Michel 2. Biisserbreck 3. Grand Rue 4. Boulevard Royal 5. Rue des Capucines
Reference points (Reliability 0–5)	5
Transformation with 3 parameters	
Scale	1:7400
Rotation	11 °
Mean position error	14m

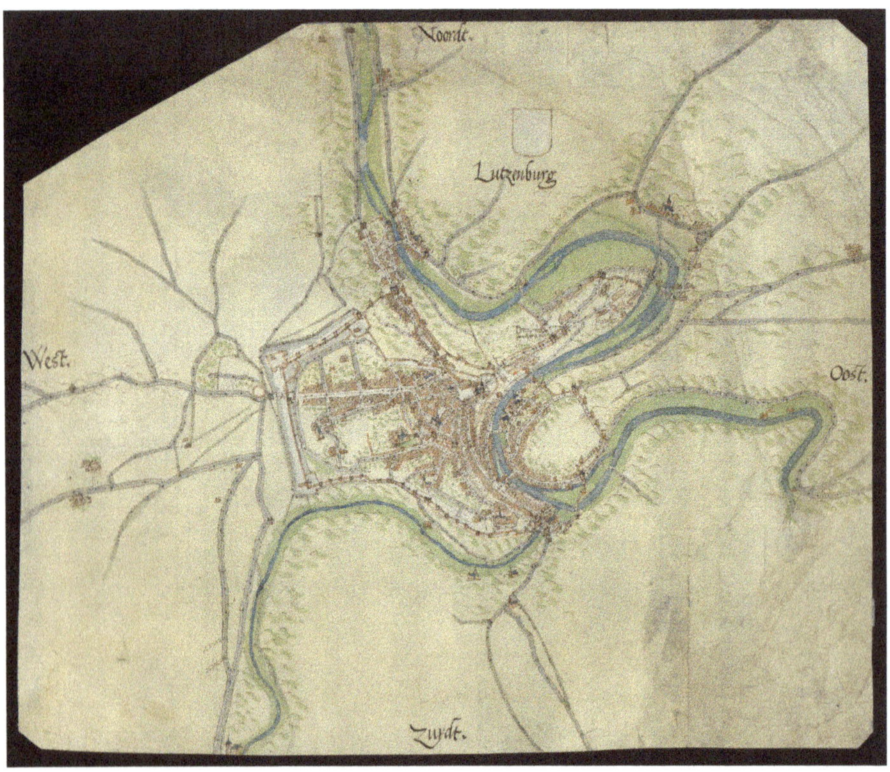

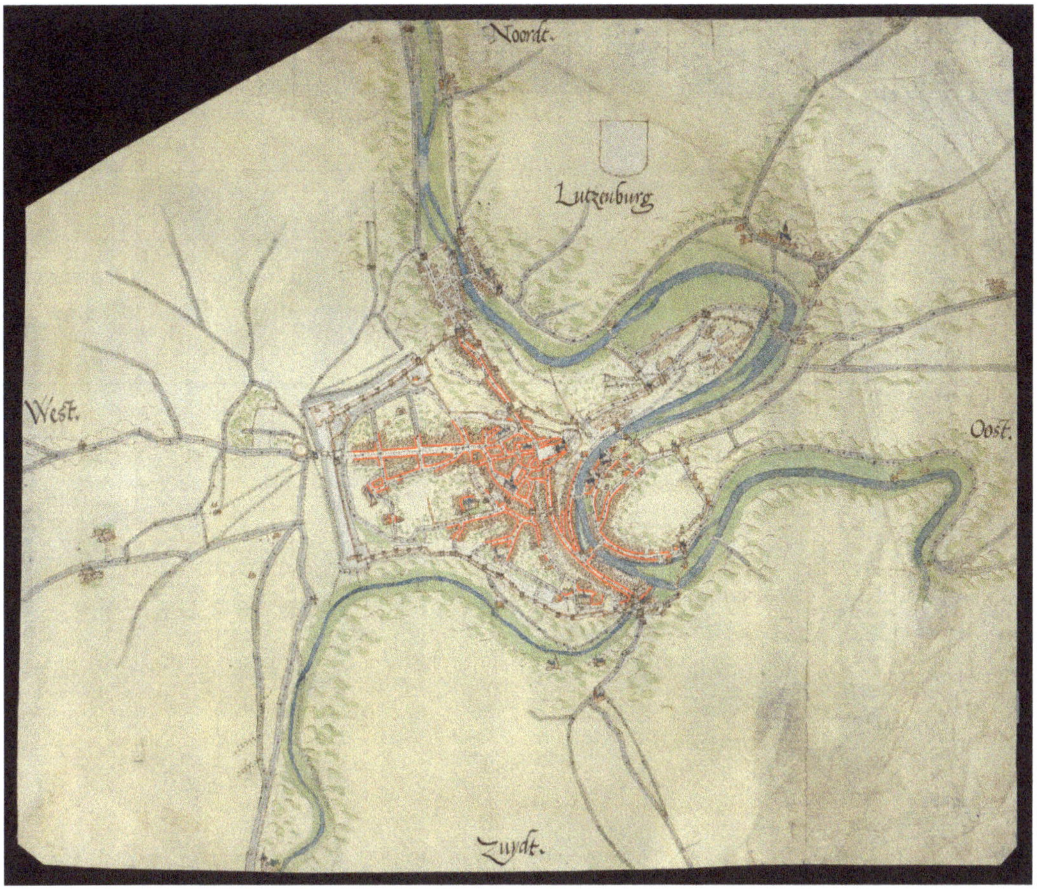
Noordt.
Lutzenburg
West.
Oost.
Zuydt.

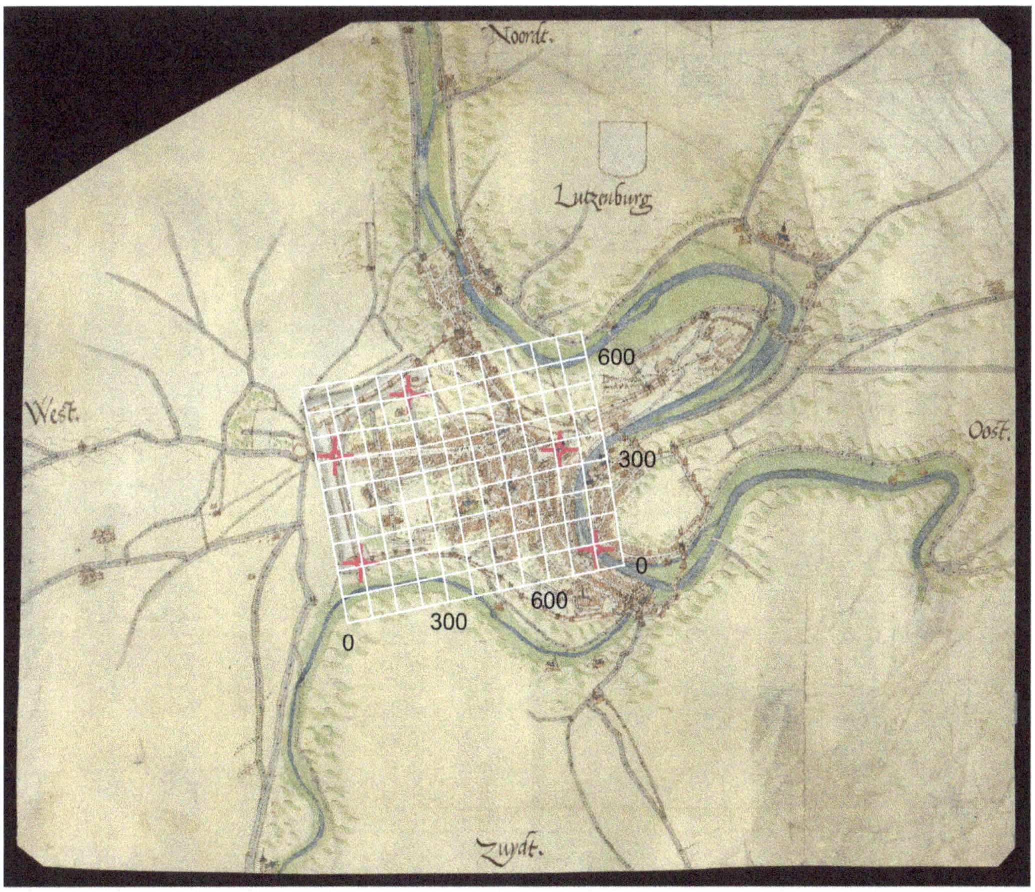
Noordt.
Lutzenburg
West.
Oost.
Zuydt.
600
300
0
0
300
600

139. Marche-en-Famenne

Tables 3.139 (measured and analysed).
Figures 3.139 (unmeasured, measured and analysed).

Measured							
Pixels	Scale stick original map (pixels)	Houses	Street sides	Houses per street side	Inhabitants (RF 4.0)	Inhabitants (RF 5.5)	Inhabitants (RF 7.0)
2806	missing	493	94	5.24	1971	2710	3449

Analysed	
September 26, 2023 9:31:27 AM	
Computation with 5 linked points (*italics* points are less reliable):	1. Église Saint-Remacie 2. Rempart du Jésuites 3. Boulevard du Midi 4. Place Tucrée 5. Rue Porte Basse
Reference points (Reliability 0–5)	5
Transformation with 3 parameters	
Scale	1:7.050
Rotation	9 °
Mean position error	18m

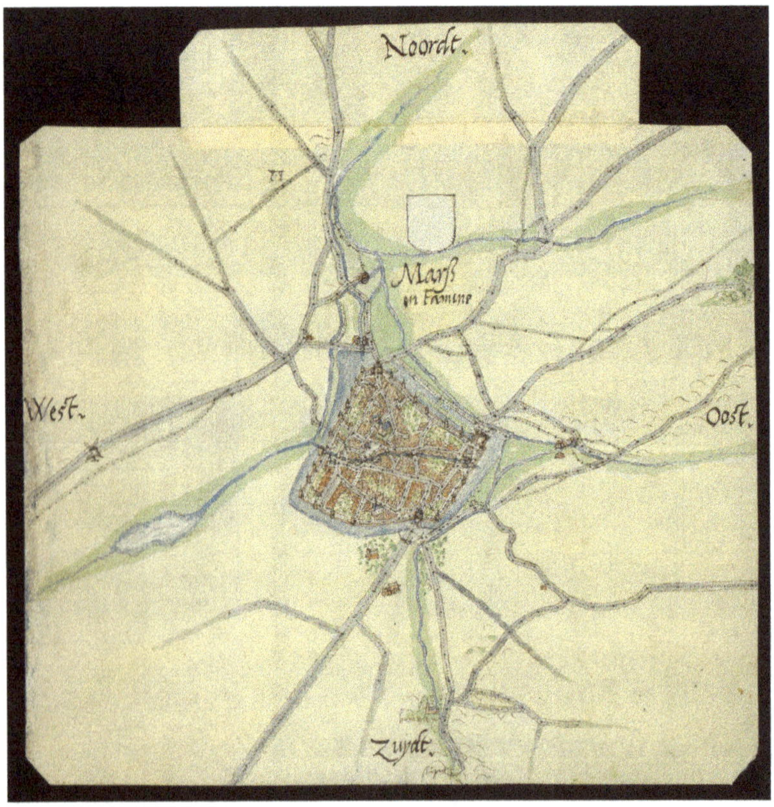

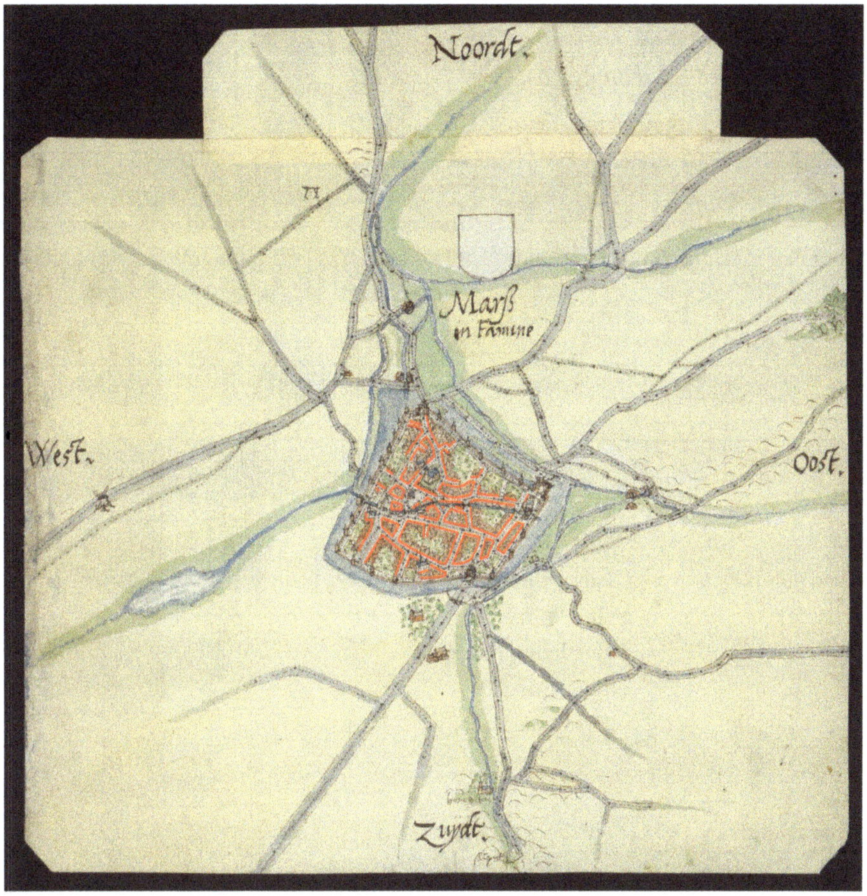
Noordt.
Marß
in Famine
West.
Oost.
Zuydt.

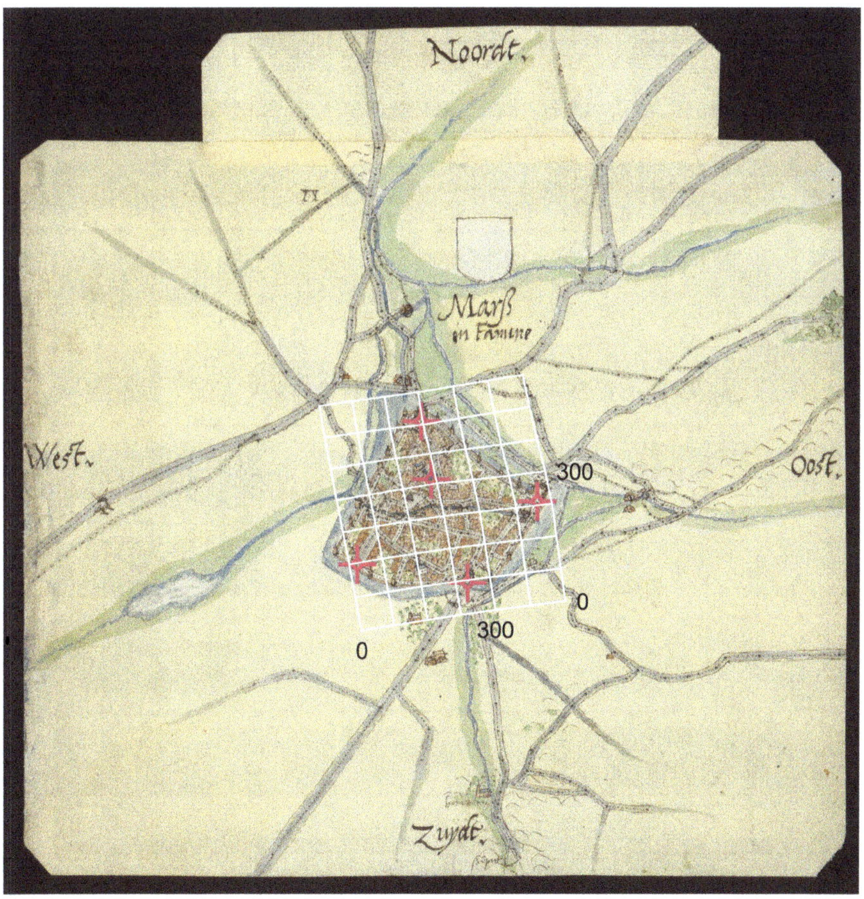
Noordt.
Marß
in Famine
West.
Oost.
Zuydt.
300
0
300
0

140. Mariembourg
Tables 3.140 (measured and analysed).
Figures 3.140 (unmeasured, measured and analysed).

Measured							
Pixels	Scale stick original map (pixels)	Houses	Street sides	Houses per street side	Inhabitants (RF 4.0)	Inhabitants (RF 5.5)	Inhabitants (RF 7.0)
1303	missing	229	46	4.97	915	1258	1601

Analysed	
April 18, 2023 9:47:22 AM	
Computation with 5 linked points (*italics* points are less reliable):	1. Église Saint-Madeleine 2. Rue Dauphine 3. Rue Reine Astrid 4. Rue d'Arschot 5. Rue Saint-Louis
Reference points (Reliability 0–5)	5
Transformation with 3 parameters	
Scale	1:7990
Rotation	12 °
Mean position error	10m

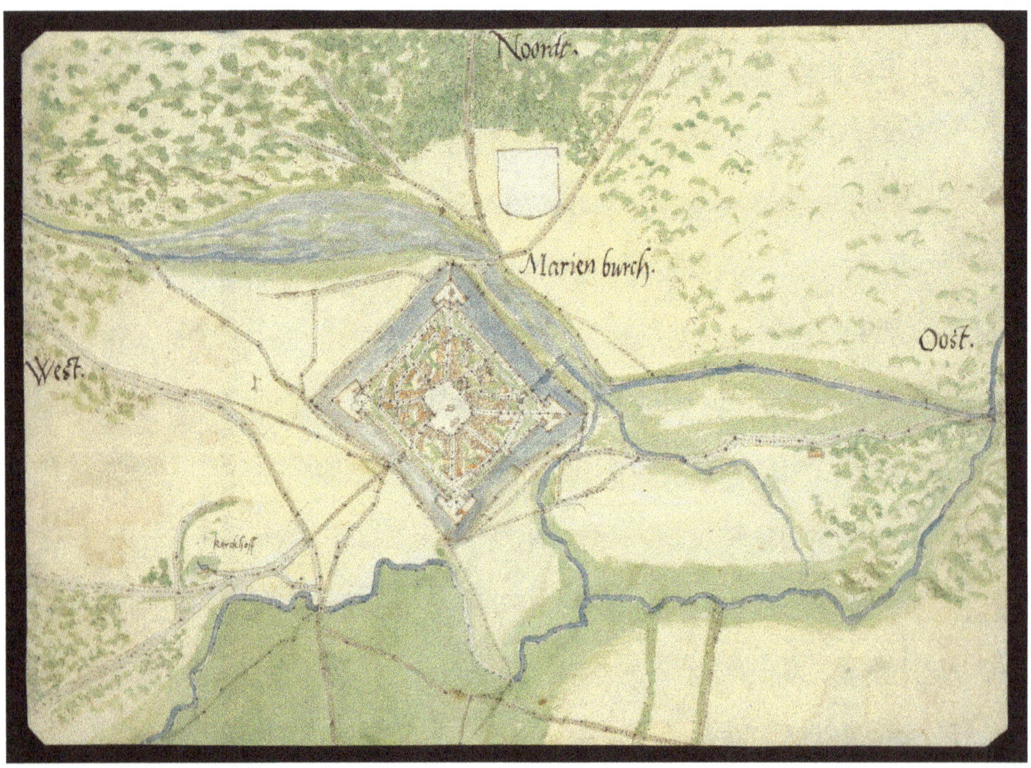

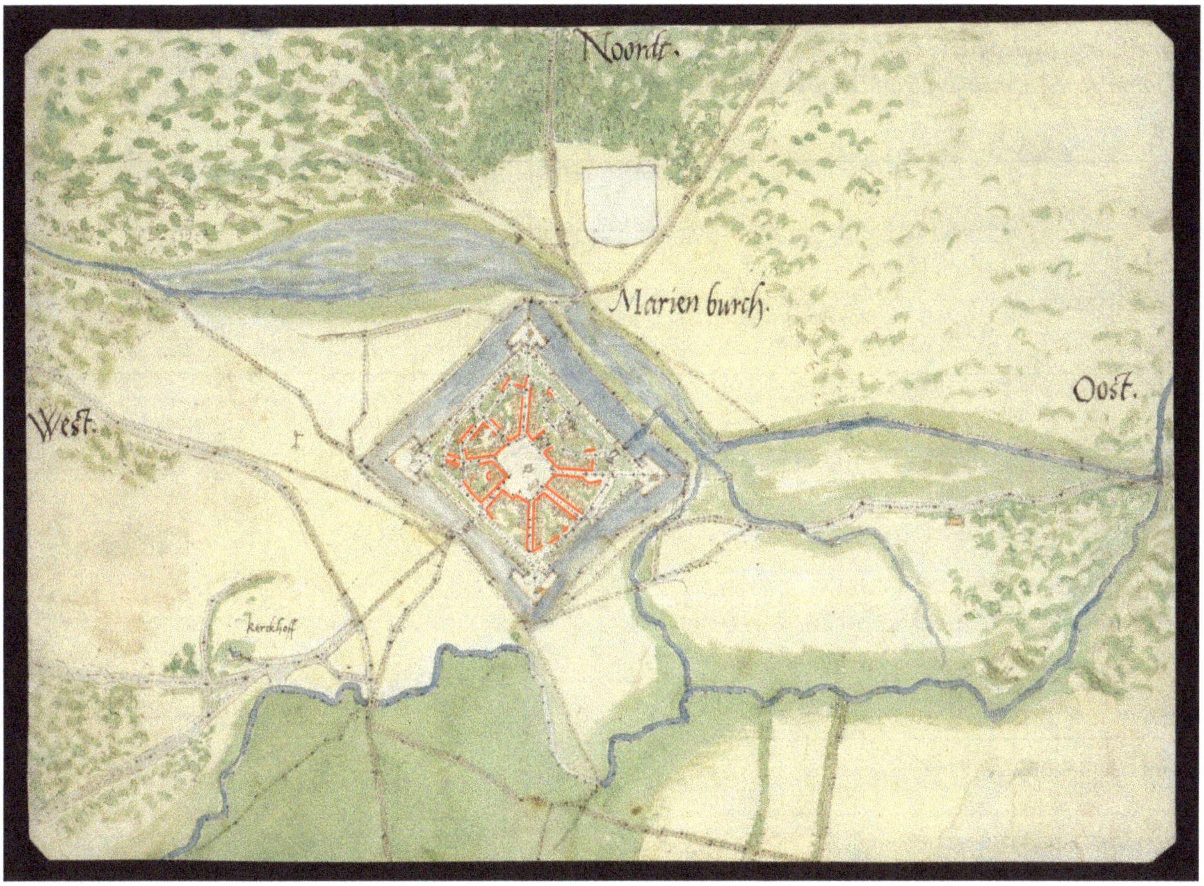
Noordt.
Marien burch.
West.
Oost.
Kerckhoff

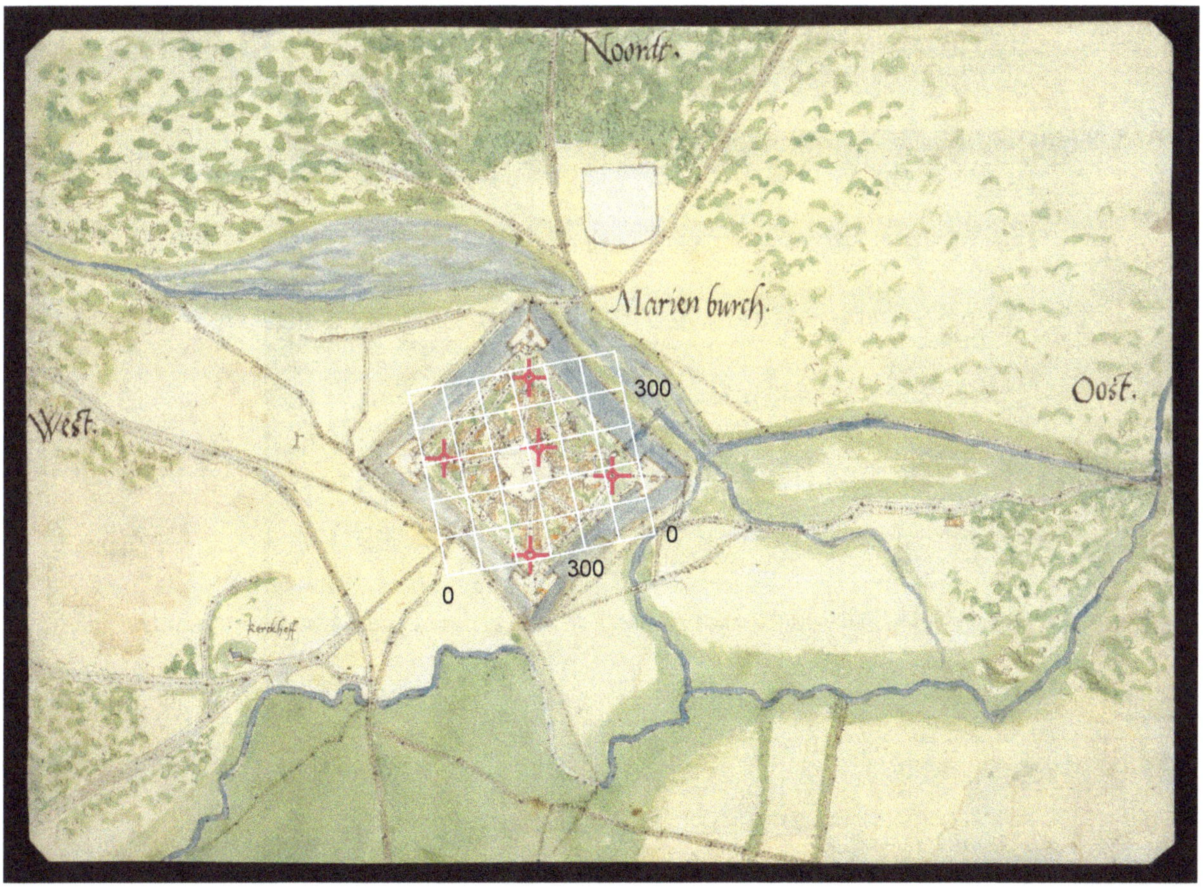
Noordt.
Marien burch.
West.
Oost.
Kerckhoff
300
0
300
0

141. Maubeuge
Tables 3.141 (measured and analysed).
Figures 3.141 (unmeasured, measured and analysed).

Measured							
Pixels	Scale stick original map (pixels)	Houses	Street sides	Houses per street side	Inhabitants (RF 4.0)	Inhabitants (RF 5.5)	Inhabitants (RF 7.0)
6890	1031	1210	169	7.16	4839	6654	8469

Analysed	
October 15, 2023 7:43:21 PM	
Computation with 5 linked points (*italics* points are less reliable):	*1. Avenue de France* *2. La Sambre (centre)* *3. Rue de Fleurus* *4. La Sambre (ouest)* *5. Rue Vauban*
Reference points (Reliability 0–5)	0
Transformation with 3 parameters	
Scale	1:7330
Rotation	13 °
Mean position error	88m

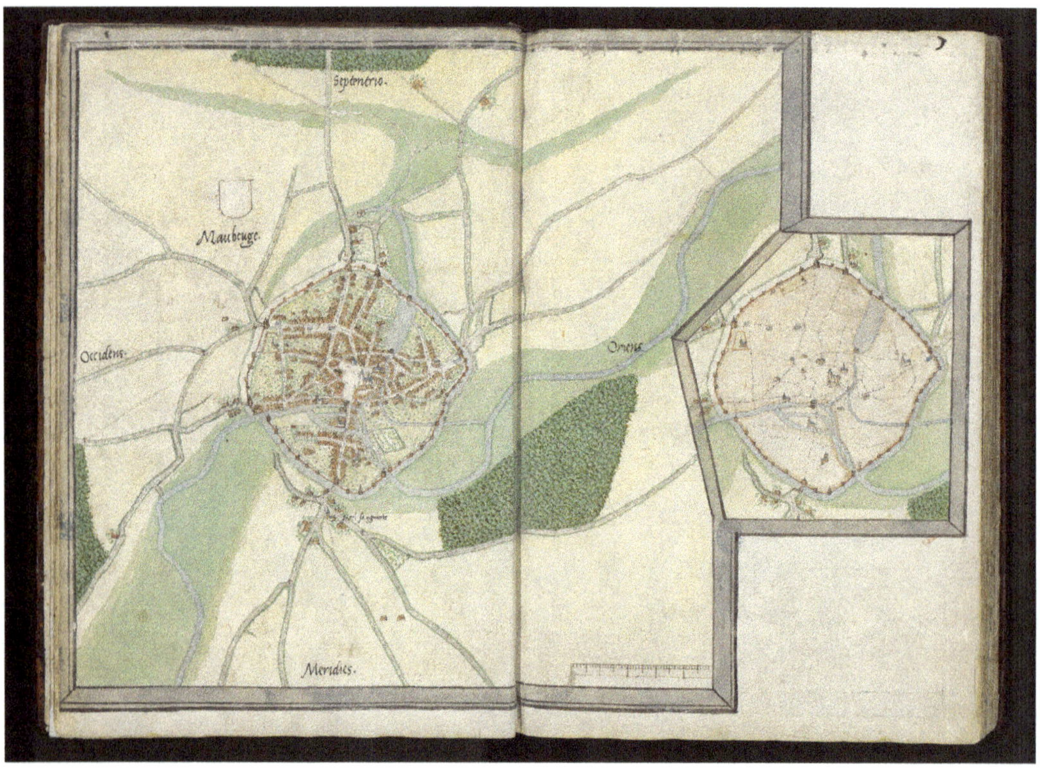

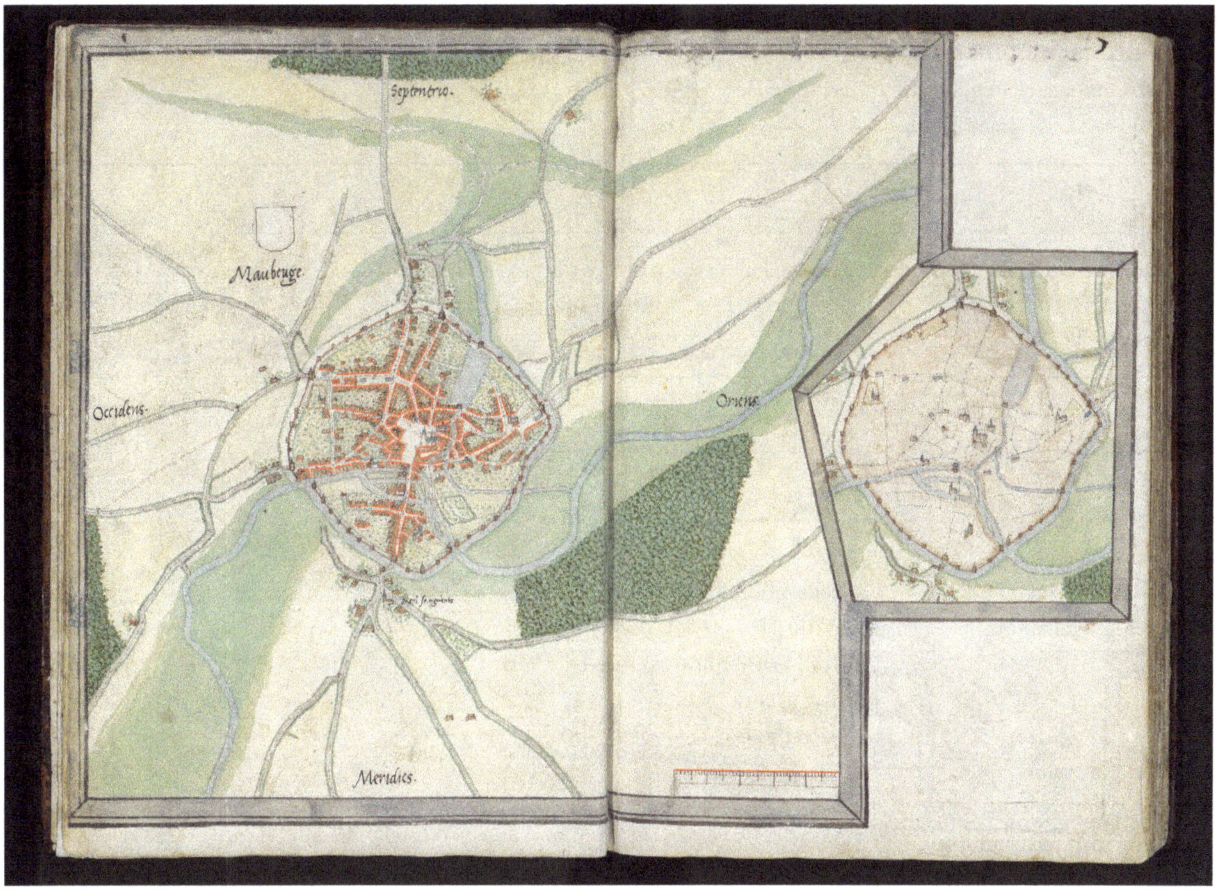
Septentrio.
Maubeuge
Occidens.
Oriens
Meridies.

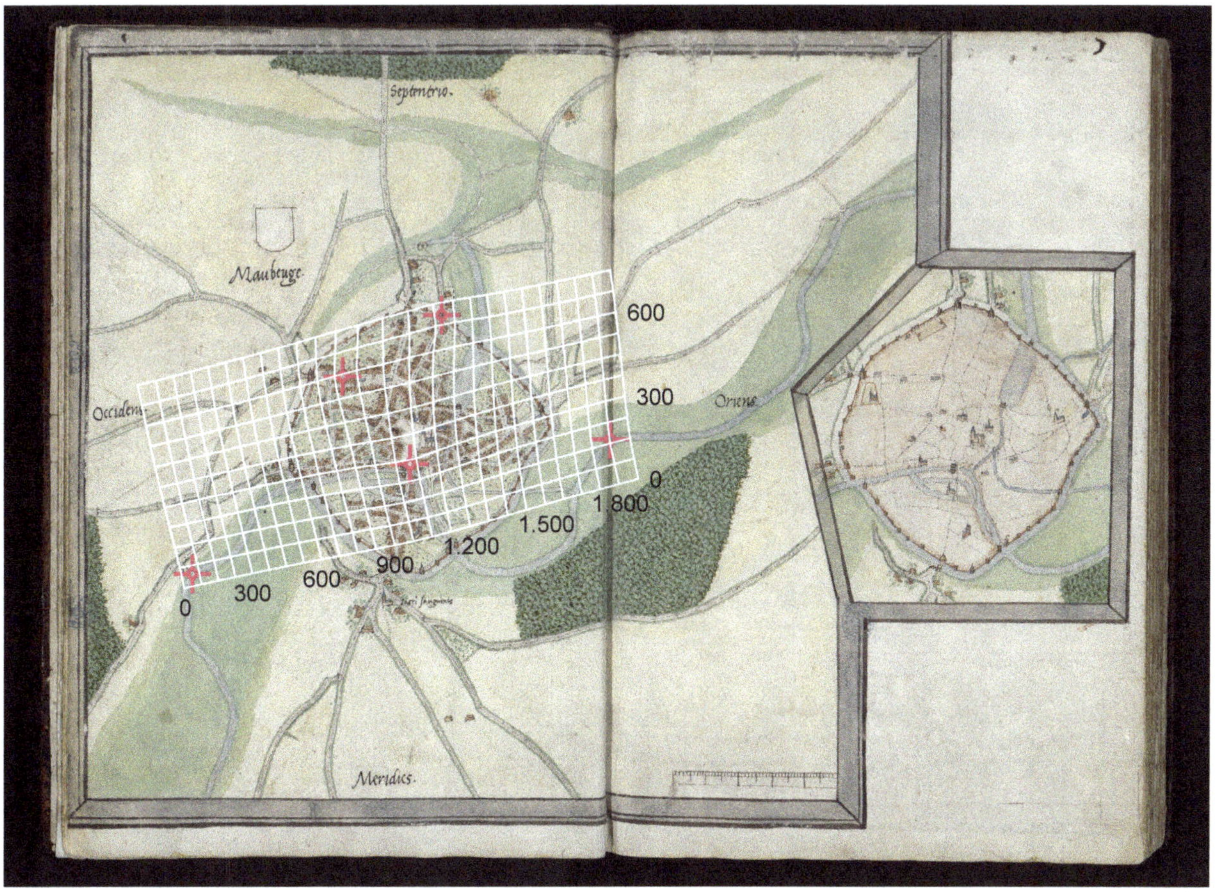
Septentrio.
Maubeuge
Occidens.
Oriens
Meridies.
600
300
0
0
300
600
900
1.200
1.500
1.800

142. Mechelen

Tables 3.142 (measured and analysed).
Figures 3.142 (unmeasured, measured and analysed).

Measured

Pixels	Scale stick original map (pixels)	Houses	Street sides	Houses per street side	Inhabitants (RF 4.0)	Inhabitants (RF 5.5)	Inhabitants (RF 7.0)
24,783	1316	4352	535	8.13	17,407	23,935	30,463

Analysed

September 5, 2023 10:49:00 AM	
Computation with 5 linked points (*italics* points are less reliable):	1. Sint-Romboutskathedraal 2. Grootbrug 3. Onze-Lieve-Vrouw-over-de-Dijlekerk 4. Sint-Janskerk 5. Keizerstraat
Reference points (Reliability 0–5)	5
Transformation with 3 parameters	
Scale	1:8430
Rotation	13 °
Mean position error	18m

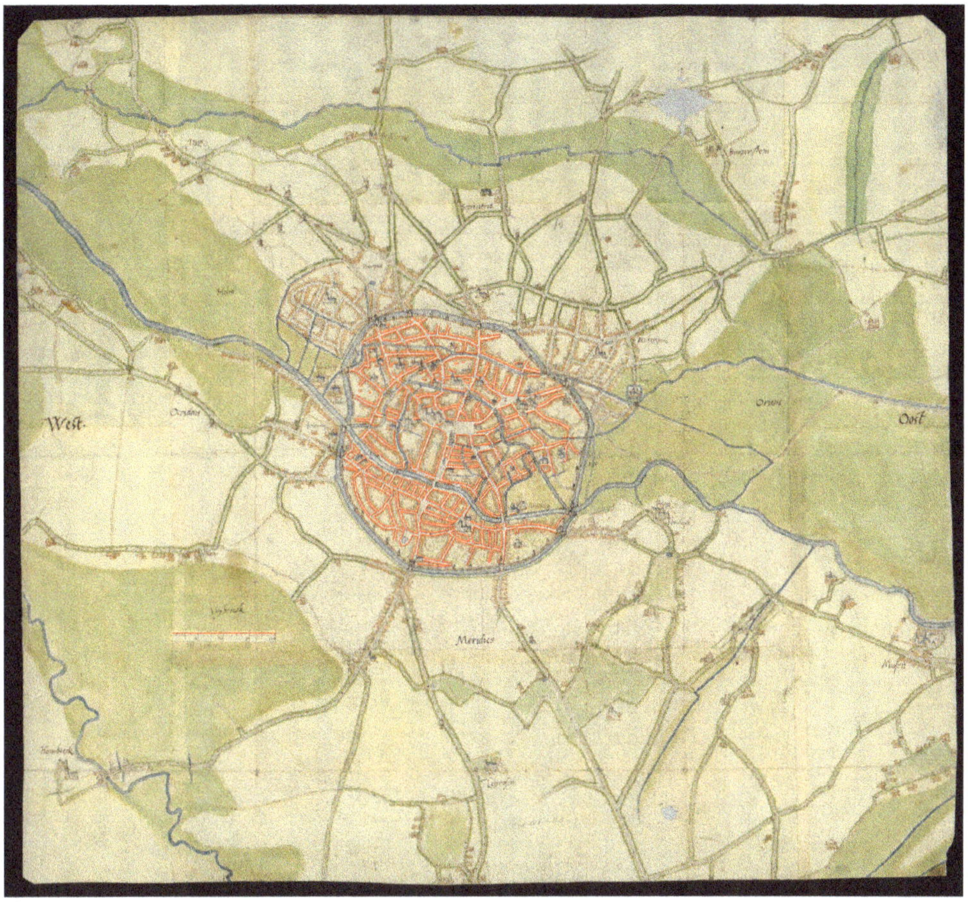
West
Oost

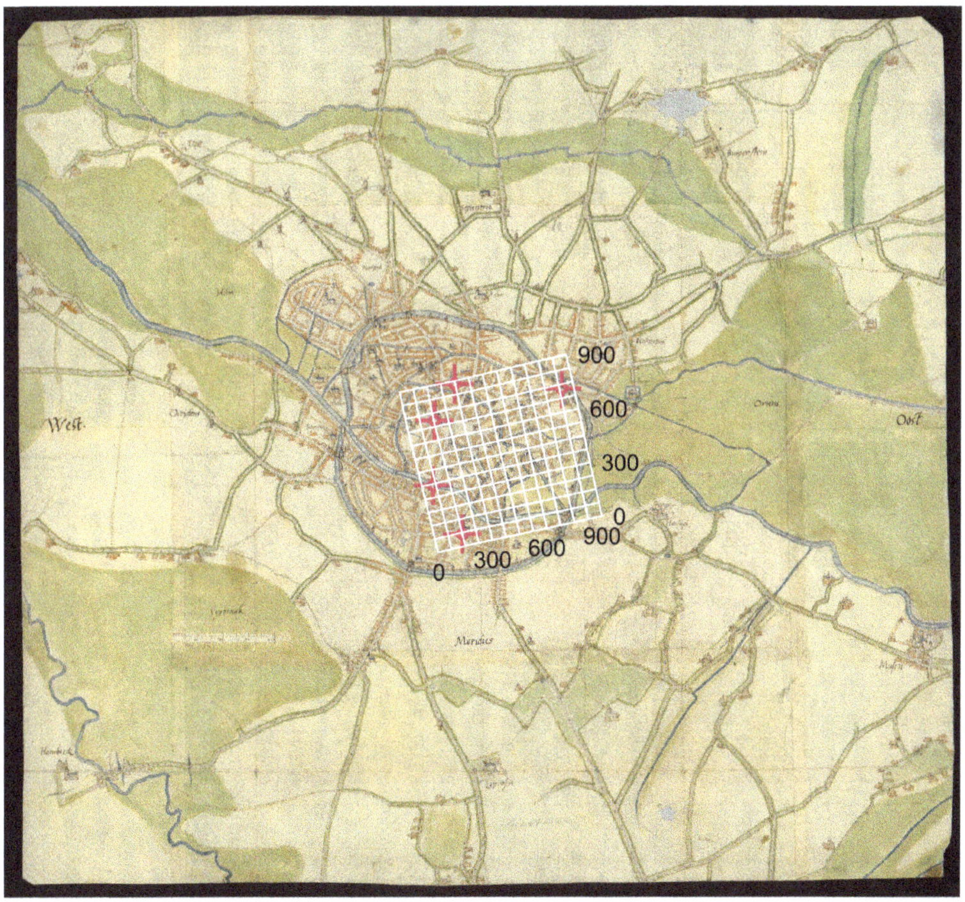
West
Oost
900
600
300
0
900
0
300
600

143. Medemblik

Tables 3.143 (measured and analysed).
Figures 3.143 (unmeasured, measured and analysed).

Measured							
Pixels	Scale stick original map (pixels)	Houses	Street sides	Houses per street side	Inhabitants (RF 4.0)	Inhabitants (RF 5.5)	Inhabitants (RF 7.0)
3084	1011	542	75	7.22	2167	2979	3791

Analysed	
June 13, 2023 5:46:49 PM	
Computation with 5 linked points (*italics* points are less reliable):	1. Bonifaciuskerk 2. Kasteel Radboud 3. *Oude Haven* 4. Achterom 5. *Ridderstraat*
Reference points (Reliability 0–5)	3
Transformation with 3 parameters	
Scale	1:8610
Rotation	11 °
Mean position error	36m

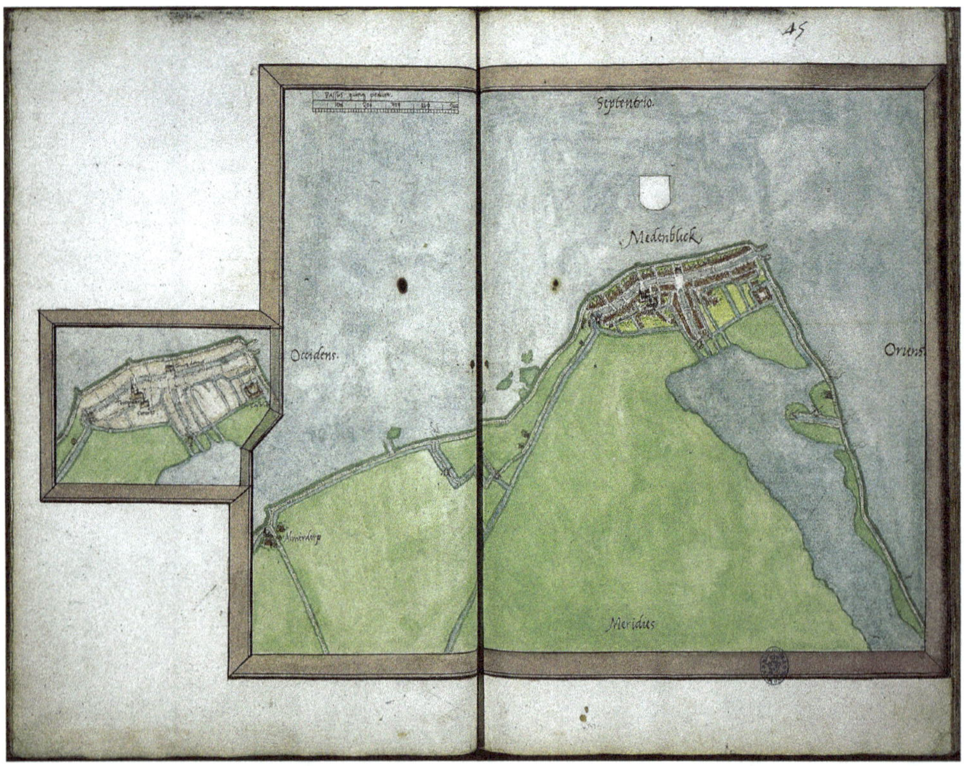

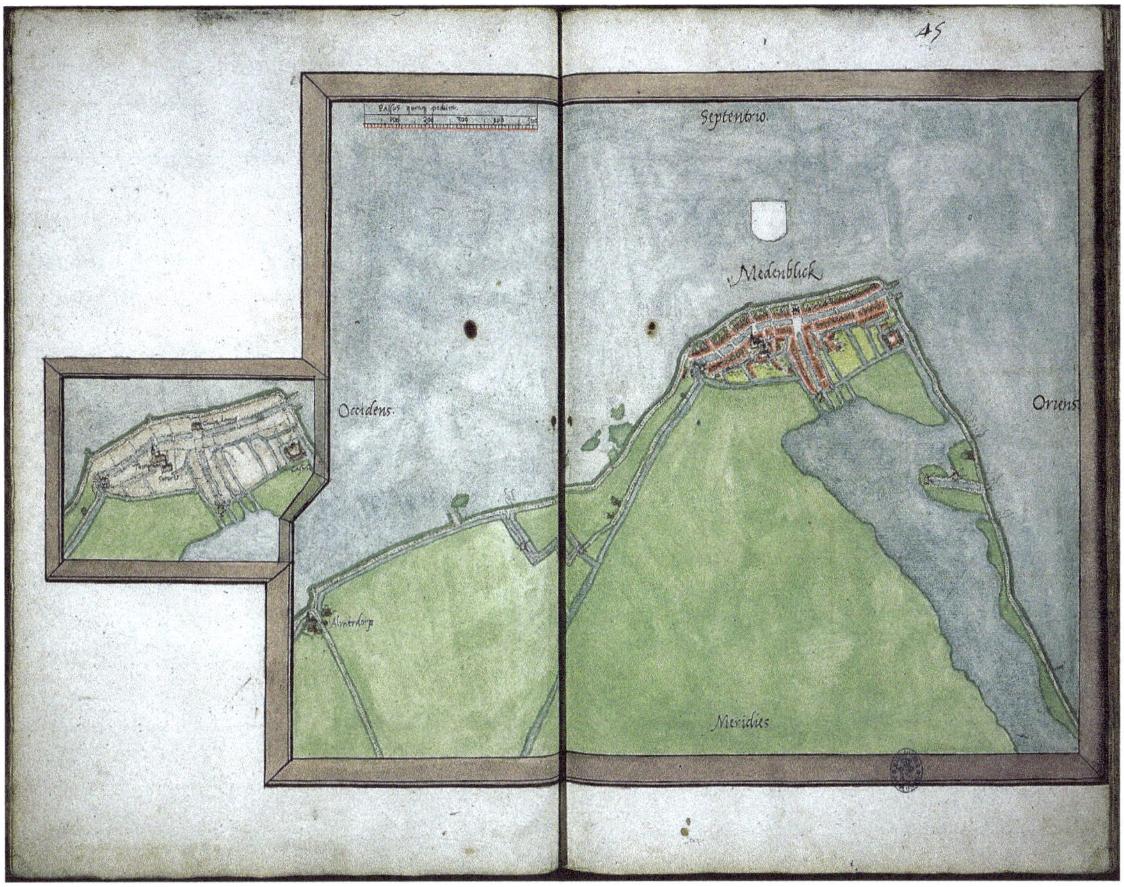
Septentrio
Medenblick
Occidens
Oriens
Meridies

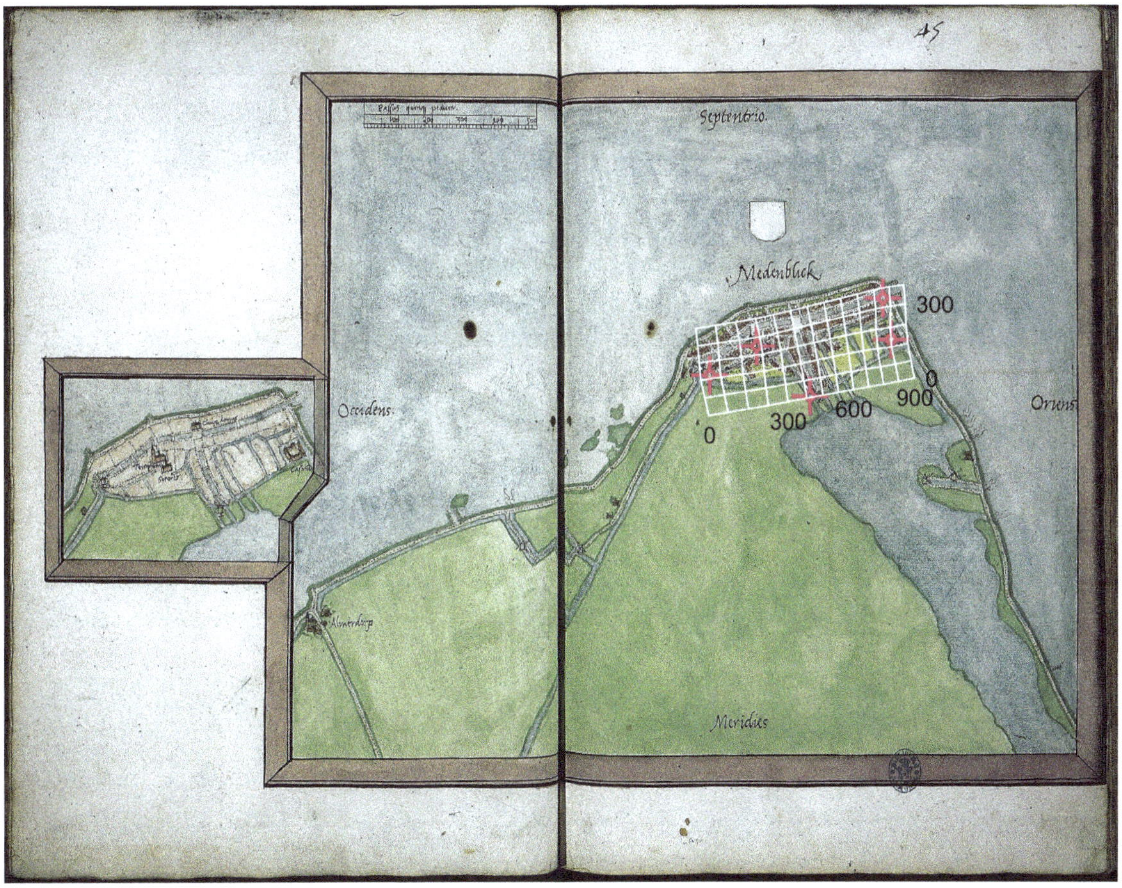
Septentrio
Medenblick
300
0
900
600
300
0
Occidens
Oriens
Meridies

144. Megen

Tables 3.144 (measured and analysed).
Figures 3.144 (unmeasured, measured and analysed).

Measured							
Pixels	Scale stick original map (pixels)	Houses	Street sides	Houses per street side	Inhabitants (RF 4.0)	Inhabitants (RF 5.5)	Inhabitants (RF 7.0)
1266	missing	222	38	5.85	889	1223	1334

Analysed	
September 26, 2023 10:32:37 AM	
Computation with 5 linked points	1. Gevangentoren
(*italics* points are less reliable):	2. Koolmarkt
	3. Walstraat
	4. Maasdijk
	5. Torenstraat
Reference points (Reliability 0–5)	5
Transformation with 3 parameters	
Scale	1:7300
Rotation	13 °
Mean position error	9m

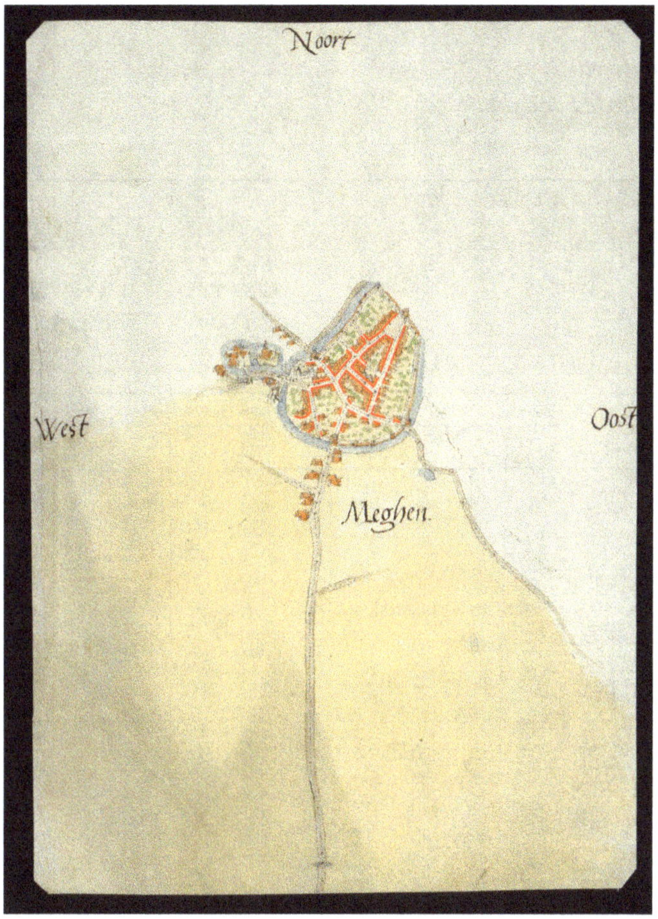
Noort
West
Oost
Meghen.

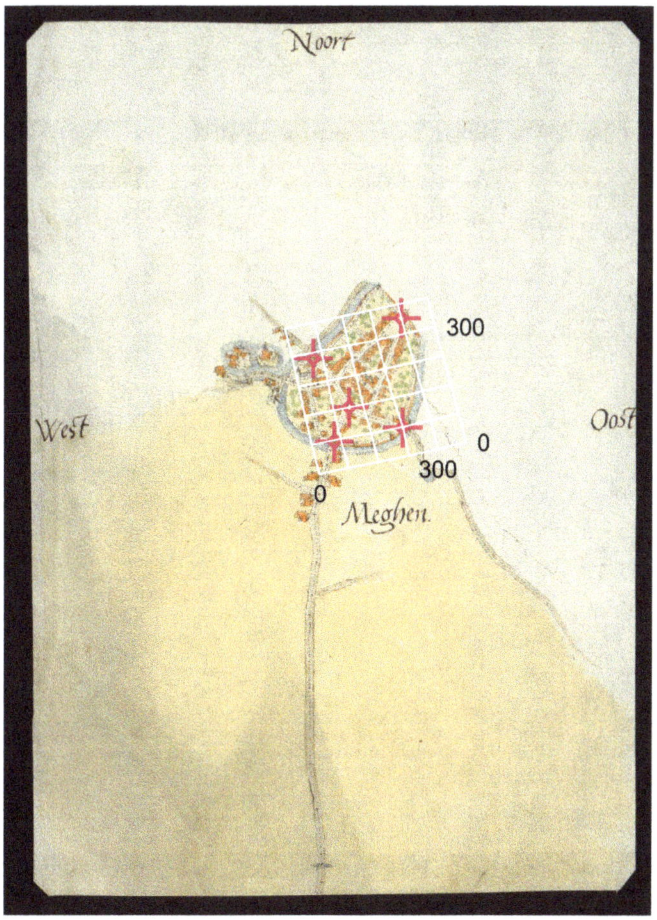
Noort
West
Oost
Meghen.
300
0
300
0

145. Menen

Tables 3.145 (measured and analysed).
Figures 3.145 (unmeasured, measured and analysed).

Measured							
Pixels	Scale stick original map (pixels)	Houses	Street sides	Houses per street side	Inhabitants (RF 4.0)	Inhabitants (RF 5.5)	Inhabitants (RF 7.0)
6800	1027	1194	113	10.57	4776	6567	8358

Analysed	
May 10, 2023 4:02:28 PM	
Computation with 5 linked points (*italics* points are less reliable):	1. Sint-Vedastuskerk *2. Belfort* 3. Rijselsestraat *4. Waalvest* 5. Bruggestraat
Reference points (Reliability 0–5)	3
Transformation with 3 parameters	
Scale	1:8680
Rotation	7 °
Mean position error	63m

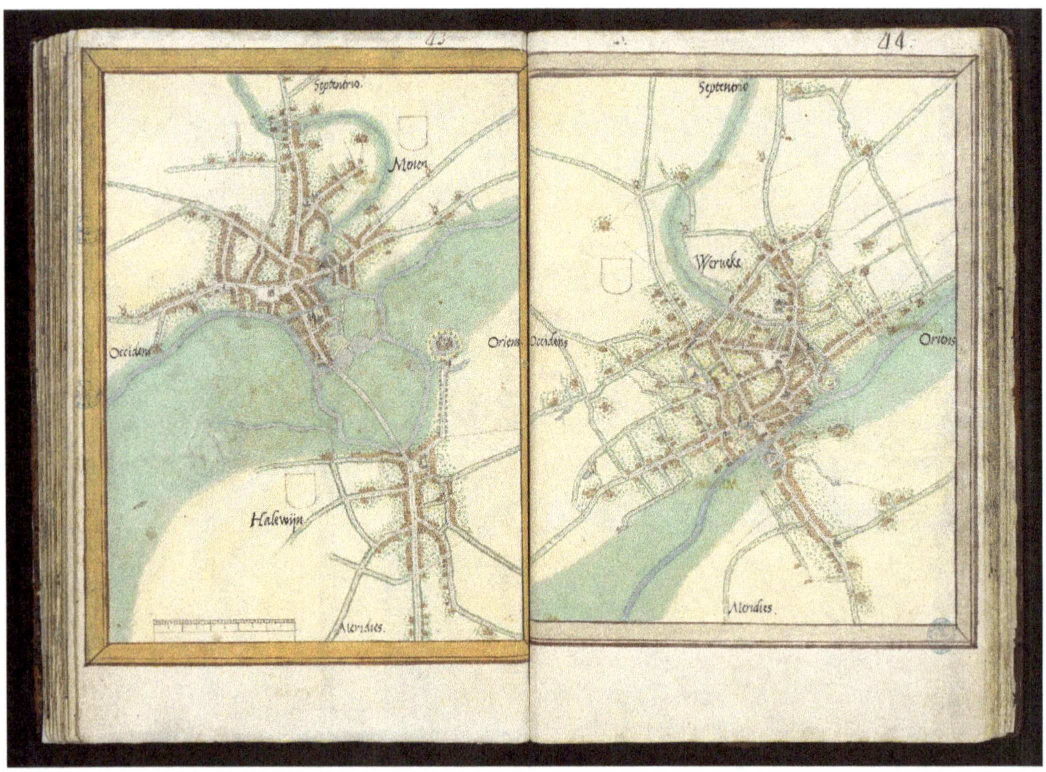

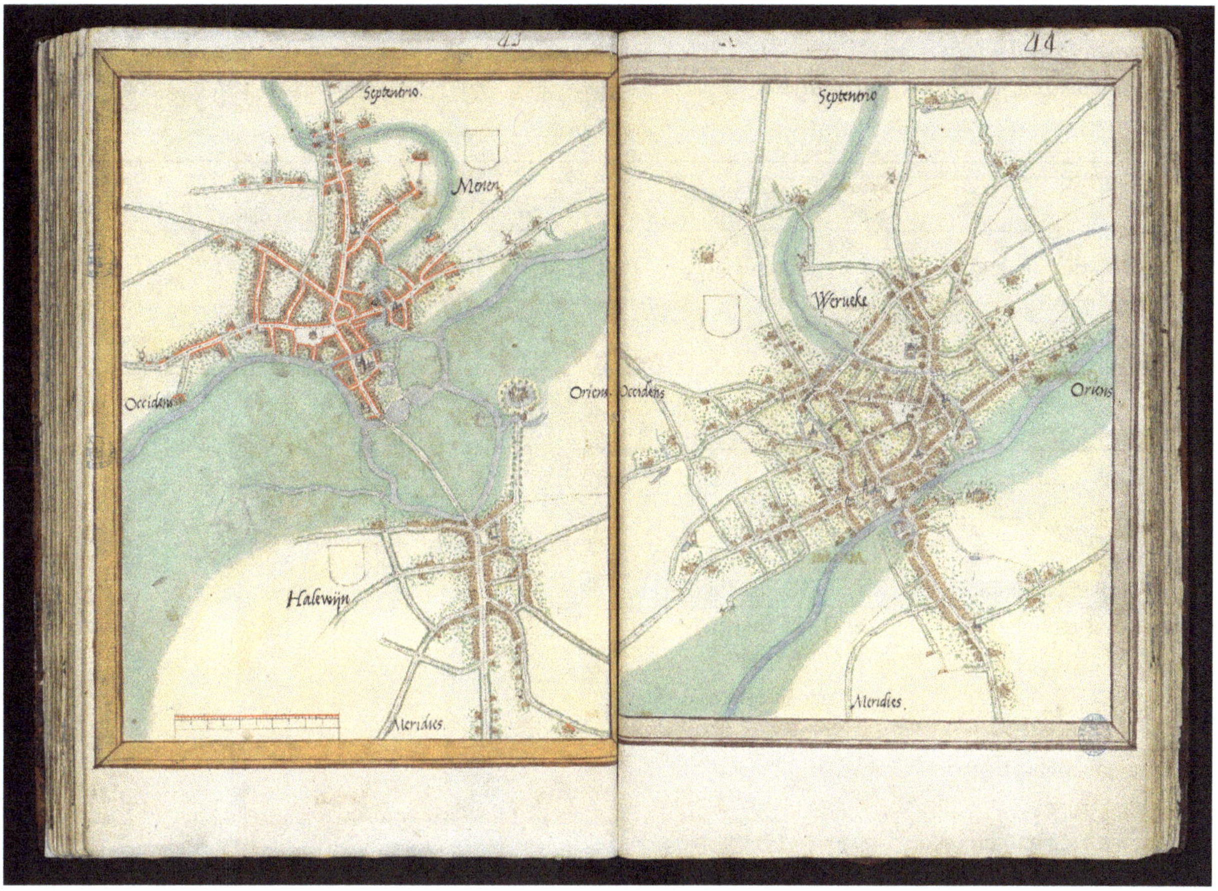
44
Septentrio
Menen
Occidens
Oriens
Halewijn
Meridies
Septentrio
Occidens
Oriens
Meridies

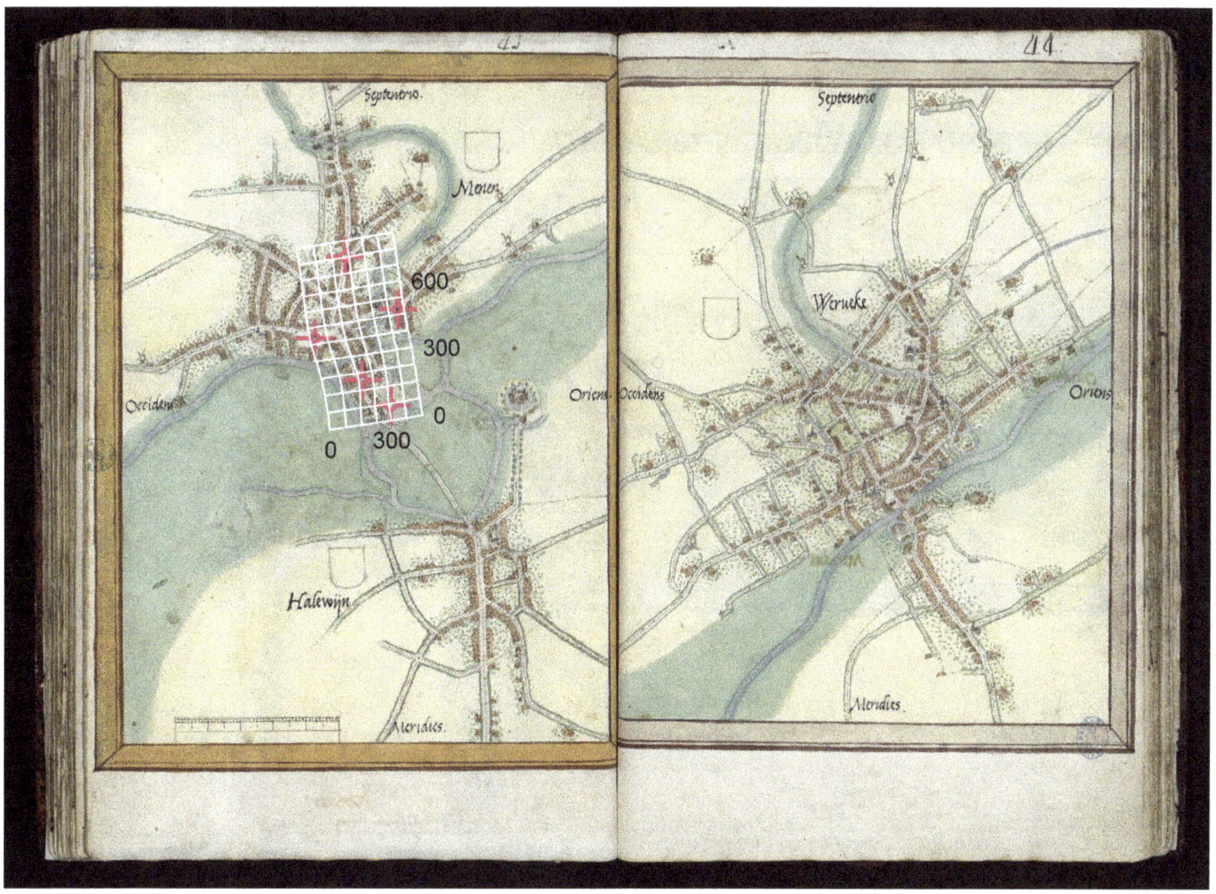
44
Septentrio
Menen
600
300
0
0
300
Occidens
Oriens
Halewijn
Meridies
Septentrio
Occidens
Oriens
Meridies

146. Mesen
Tables 3.146 (measured and analysed).
Figures 3.146 (unmeasured, measured and analysed).

Measured							
Pixels	Scale stick original map (pixels)	Houses	Street sides	Houses per street side	Inhabitants (RF 4.0)	Inhabitants (RF 5.5)	Inhabitants (RF 7.0)
3206	1027	563	80	7.04	2252	3096	3940

Analysed	
June 13, 2023 6:52:27 PM	
Computation with 5 linked points (*italics* points are less reliable):	1. Sint-Niklaaskerk 2. Markt oost *3. Markt west* 4. Armentierstraat 5. Slijpstraat
Reference points (Reliability 0–5)	4
Transformation with 3 parameters	
Scale	1:7240
Rotation	8 °
Mean position error	20m

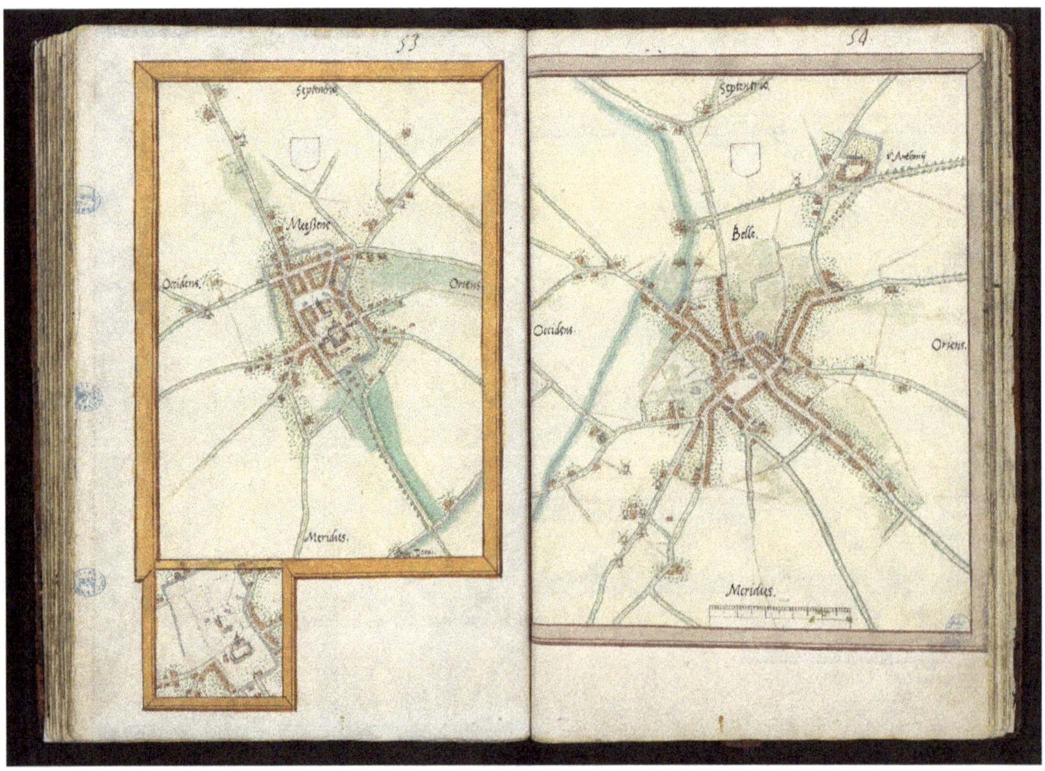

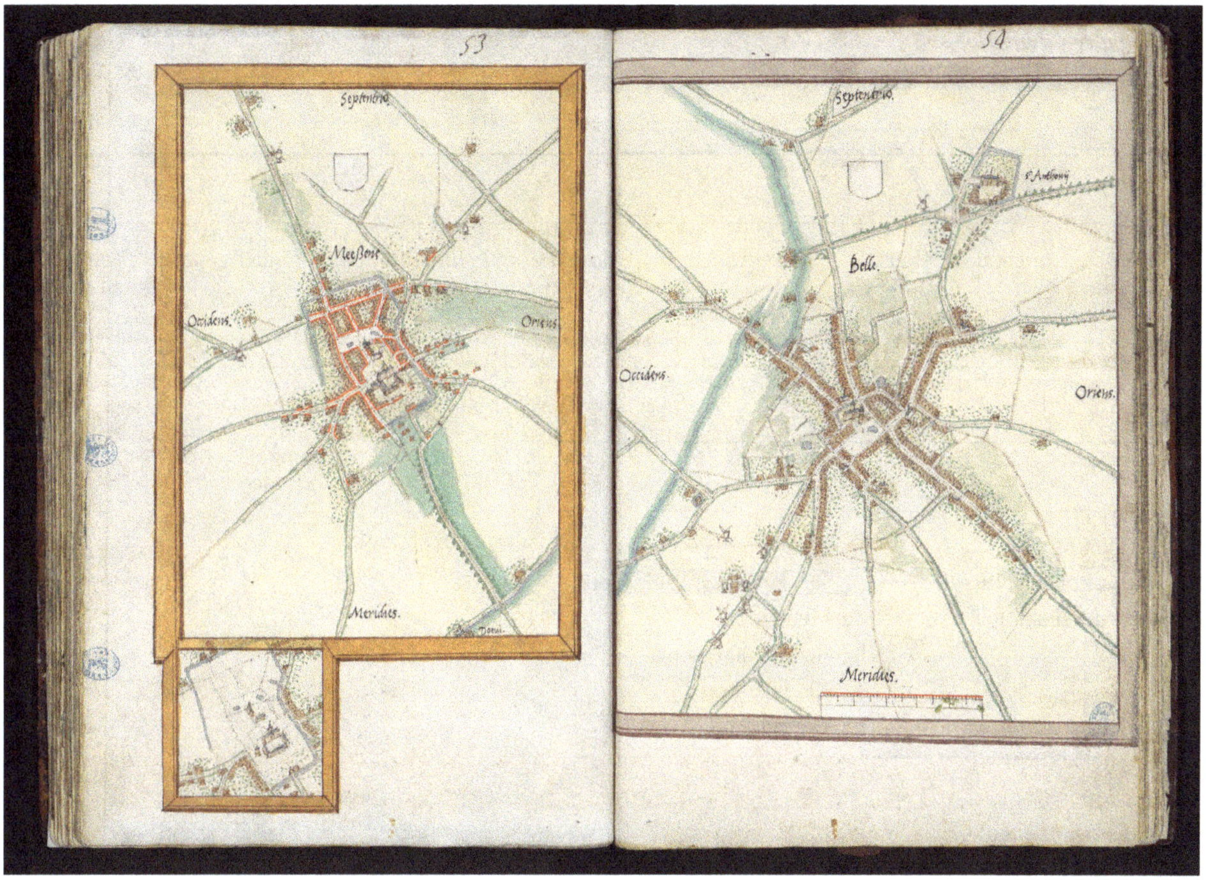
53
54
Septentrio
Occidens
Oriens
Meridies
Belle

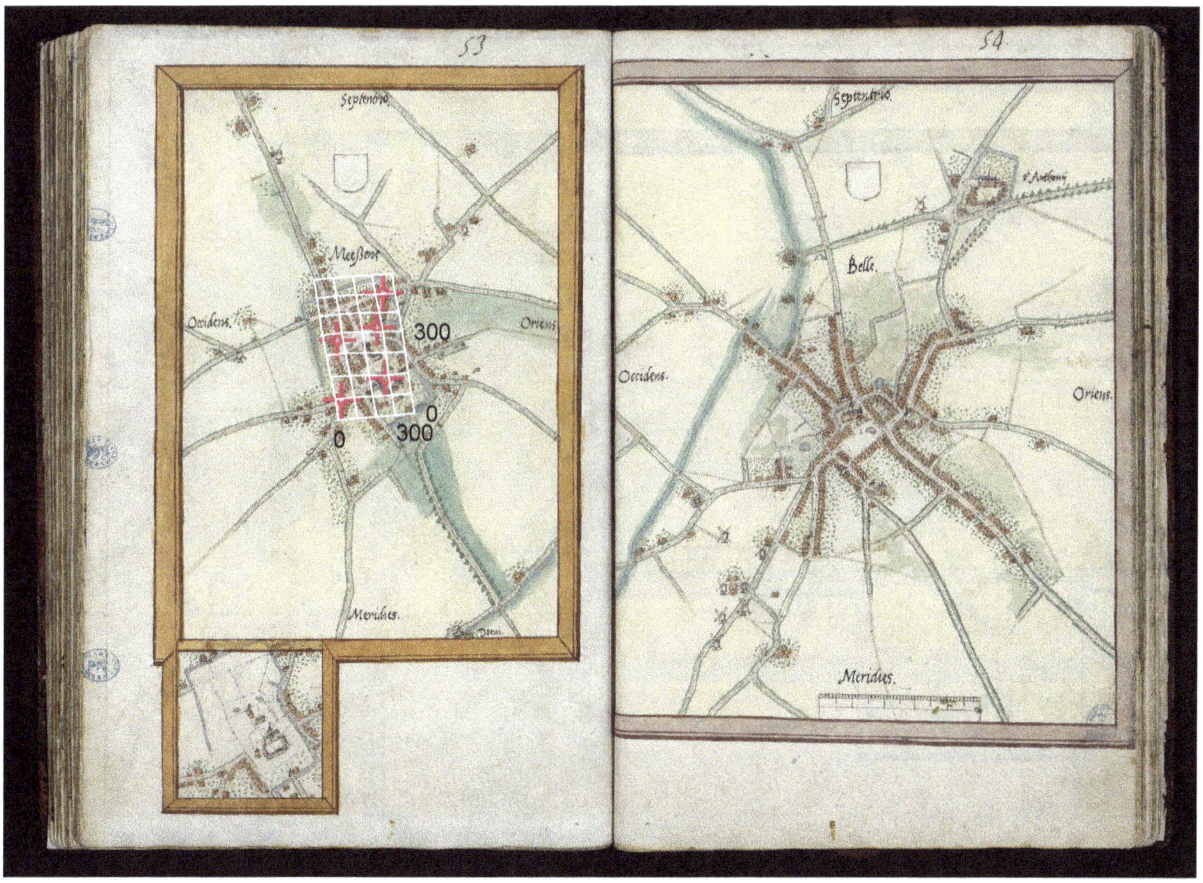
53
54
Septentrio
Occidens
Oriens
300
0
0
300
Meridies
Belle

147. Middelburg (B)
Tables 3.147 (measured and analysed).
Figures 3.147 (unmeasured, measured and analysed).

Measured							
Pixels	Scale stick original map (pixels)	Houses	Street sides	Houses per street side	Inhabitants (RF 4.0)	Inhabitants (RF 5.5)	Inhabitants (RF 7.0)
1306	1021	229	35	6.55	917	1261	1605

Analysed	
June 15, 2023 6:03:06 PM	
Computation with 5 linked points (*italics* points are less reliable):	1. Sint-Petrus en Pauluskerk 2. Meulekreek 3. Rentestraat 4. Kloosterstraat 5. Hoornstraat
Reference points (Reliability 1–5)	5
Transformation with 3 parameters	
Scale	1:7240
Rotation	13 °
Mean position error	16m

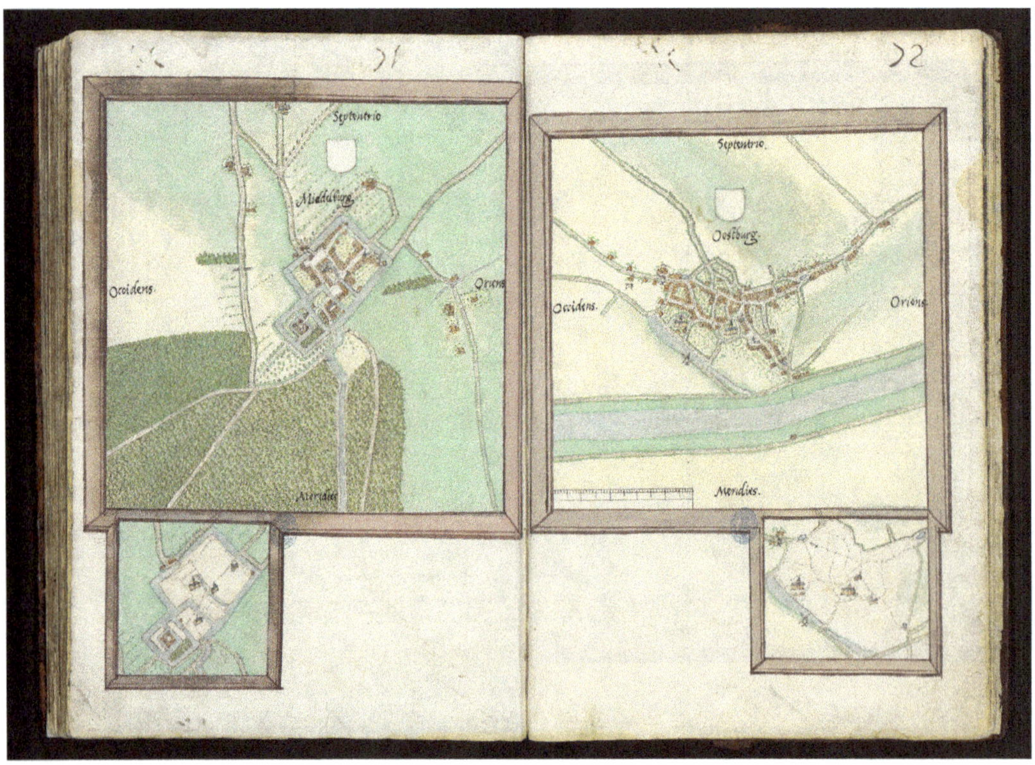

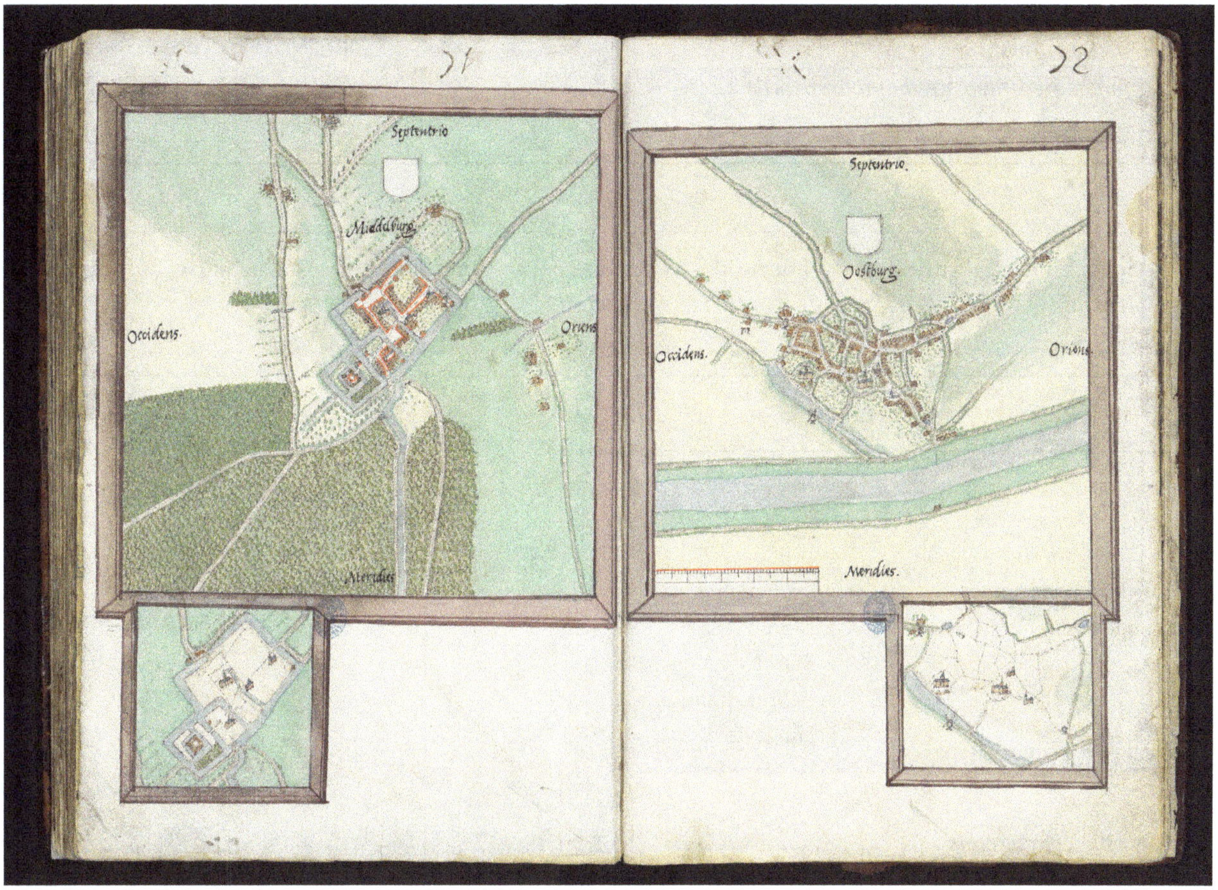
Septentrio
Middelburg
Occidens.
Oriens
Meridies
Septentrio.
Oostburg.
Occidens.
Oriens
Meridies.

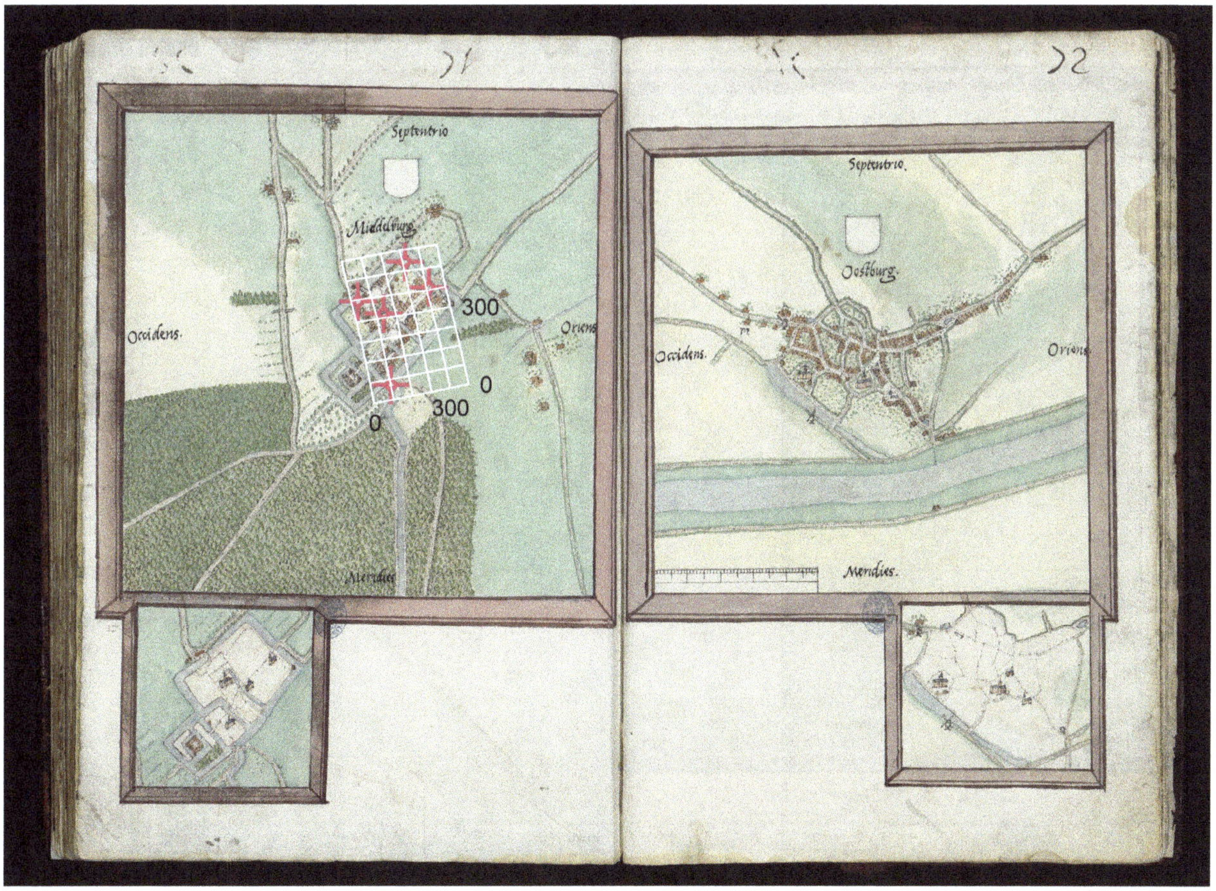
Septentrio
Middelburg
300
0
0
300
Occidens.
Oriens
Meridies
Septentrio.
Oostburg.
Occidens.
Oriens
Meridies.

148. Middelburg (Nl)

Tables 3.148 (measured and analysed).
Figures 3.148 (unmeasured, measured and analysed).

Measured							
Pixels	Scale stick original map (pixels)	Houses	Street sides	Houses per street side	Inhabitants (RF 4.0)	Inhabitants (RF 5.5)	Inhabitants (RF 7.0)
7328	1008	1287	124	10.38	5147	7077	9007

Analysed	
January 23, 2023 9:53:32 AM	
Computation with 5 linked points (*italics* points are less reliable):	1. Lambrechtstraat 2. Dam 3. Markt 4. Bellinkstraat 5. Langeviele
Reference points (Reliability 0–5)	5
Transformation with 3 parameters	
Scale	1:8830
Rotation	16 °
Mean position error	5m

Septentrio.
Middelburg.
Occidens.
Meridies.
© Biblioteca Nacional de España

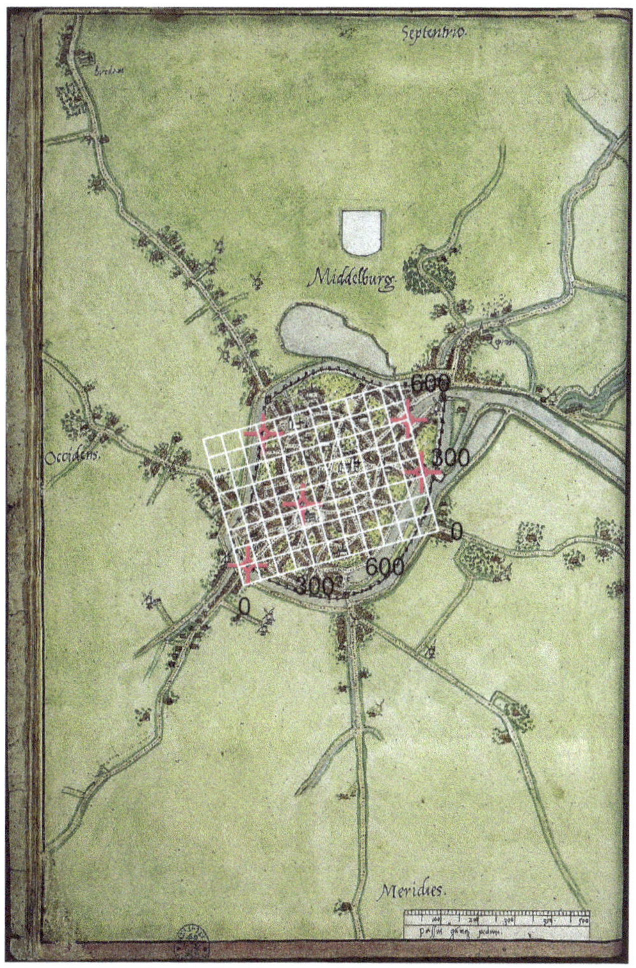
Septentrio.
Middelburg.
Occidens.
600
300
0
0
300
600
Meridies.

149. Monnickendam
Tables 3.149 (measured and analysed).
Figures 3.149 (unmeasured, measured and analysed).

Measured							
Pixels	Scale stick original map (pixels)	Houses	Street sides	Houses per street side	Inhabitants (RF 4.0)	Inhabitants (RF 5.5)	Inhabitants (RF 7.0)
3323	989	584	95	6.14	2334	3209	4084

Analysed	
June 15, 2023 11:02:38 PM	
Computation with 5 linked points (*italics* points are less reliable):	1. HH Nicolaas en Anthoniuskerk 2. Noordeinde 3. Speeltoren 4. Zuideinde 5. De Haven
Reference points (Reliability 0–5)	5
Transformation with 3 parameters	
Scale	1:8390
Rotation	10 °
Mean position error	11m

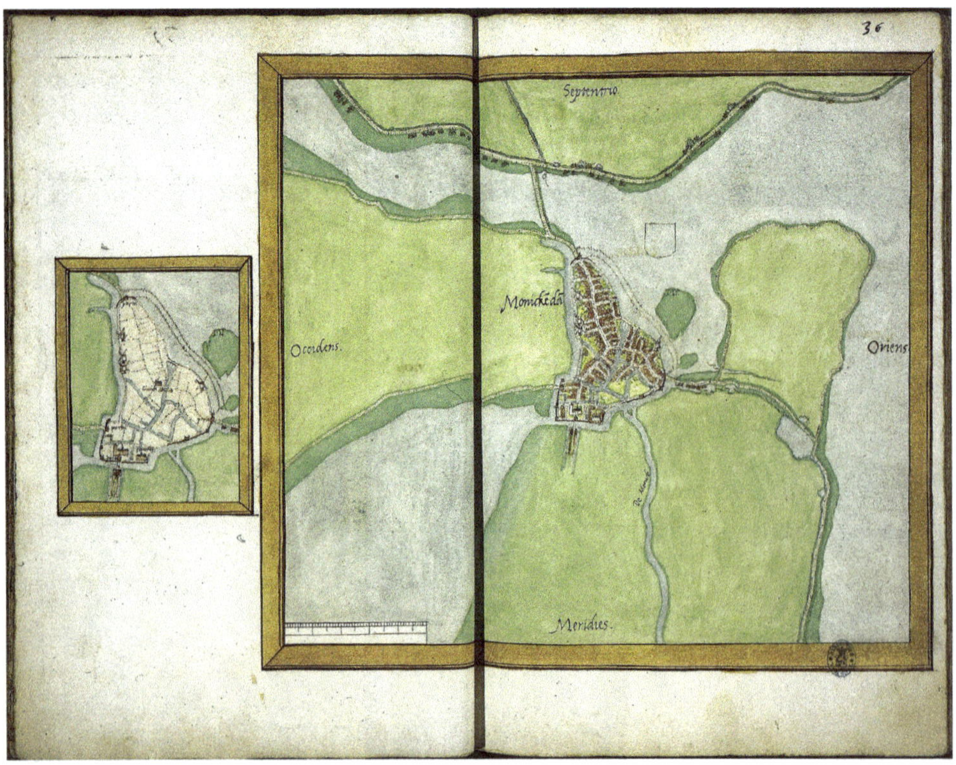

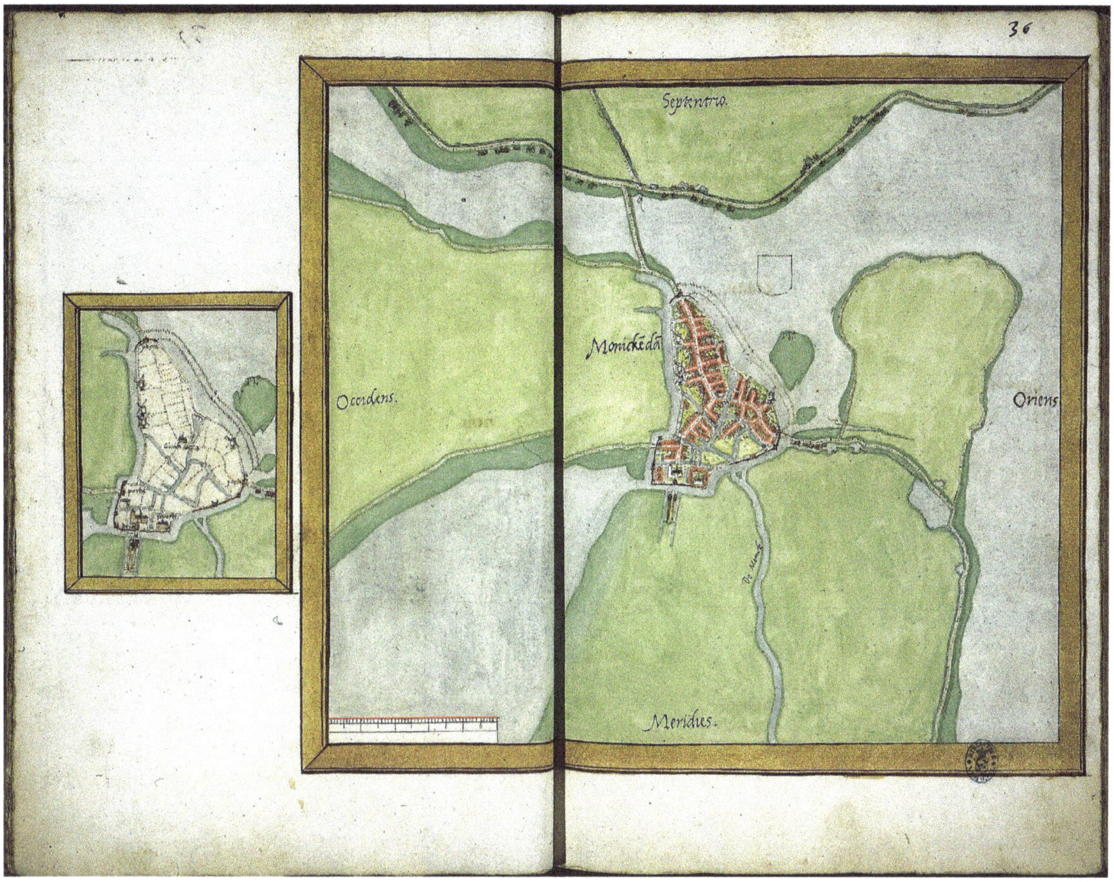
Septentrio.
Monickēdā
Occidens.
Oriens
Meridies.

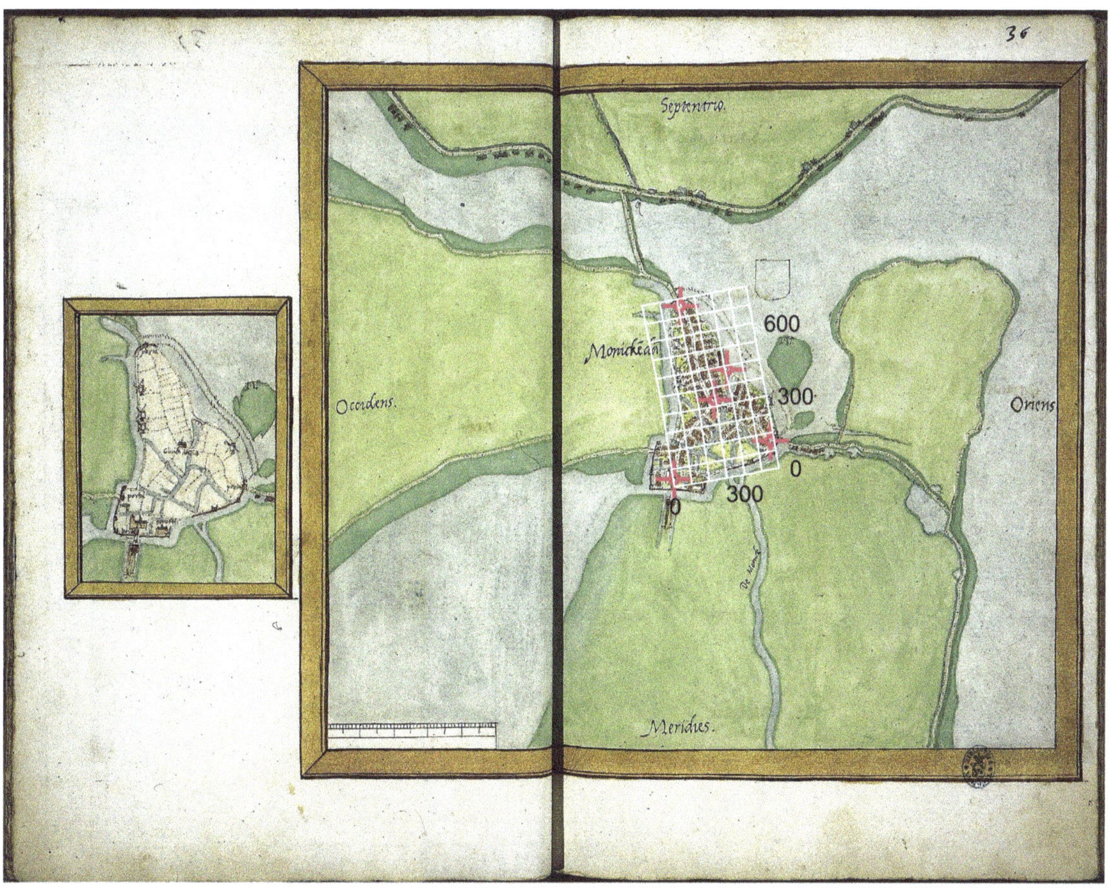
Septentrio.
Monickēdā
Occidens.
Oriens
Meridies.
600
300
0
300
0

150. Mons
Tables 3.150 (measured and analysed).
Figures 3.150 (unmeasured, measured and analysed).

Measured

Pixels	Scale stick original map (pixels)	Houses	Street sides	Houses per street side	Inhabitants (RF 4.0)	Inhabitants (RF 5.5)	Inhabitants (RF 7.0)
17,129	881	3008	304	9.89	12,031	16,543	21,055

Analysed

June 16, 2023 11:27:01 AM	
Computation with 5 linked points (*italics* points are less reliable):	1. Sint-Waltrudiskerk 2. Belfort *3. Rue d'Havré* 4. Rue de Nimy *5. Rue André Masquelier*
Reference points (Reliability 0–5)	3
Transformation with 3 parameters	
Scale	1:8590
Rotation	11°
Mean position error	37m

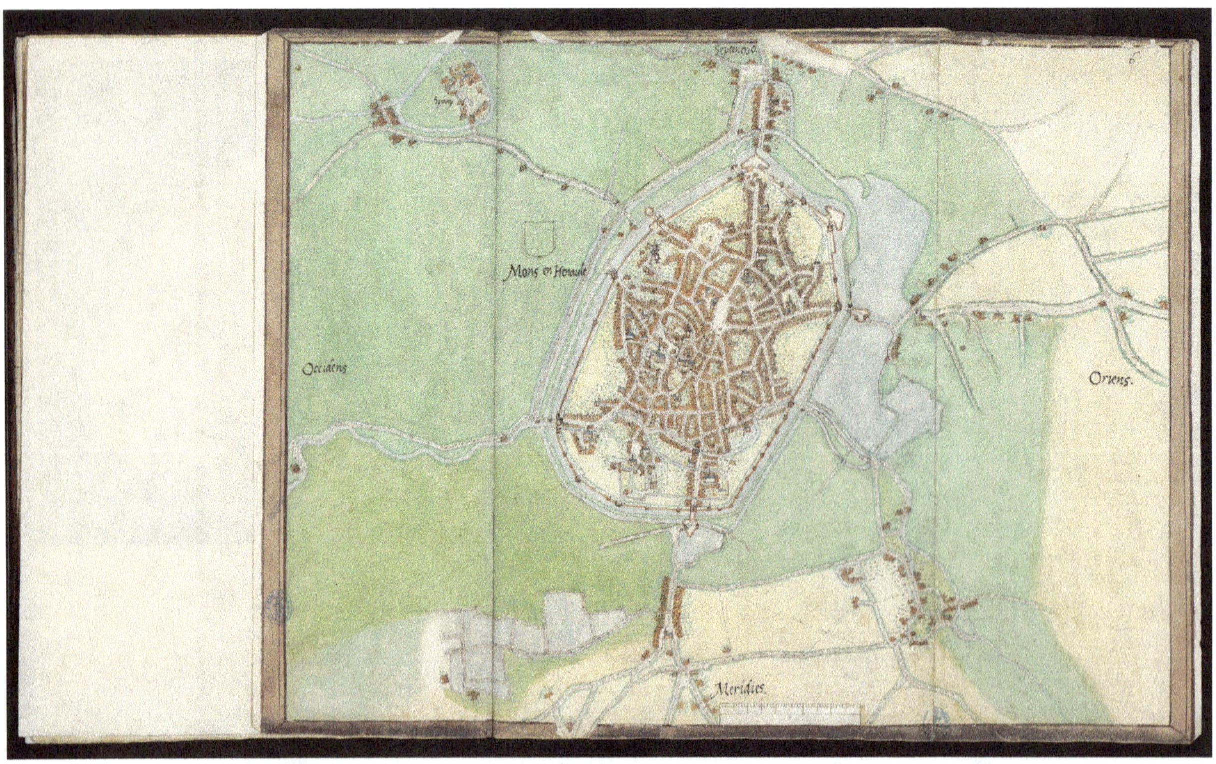

Occidens
Oriens.
Meridies.

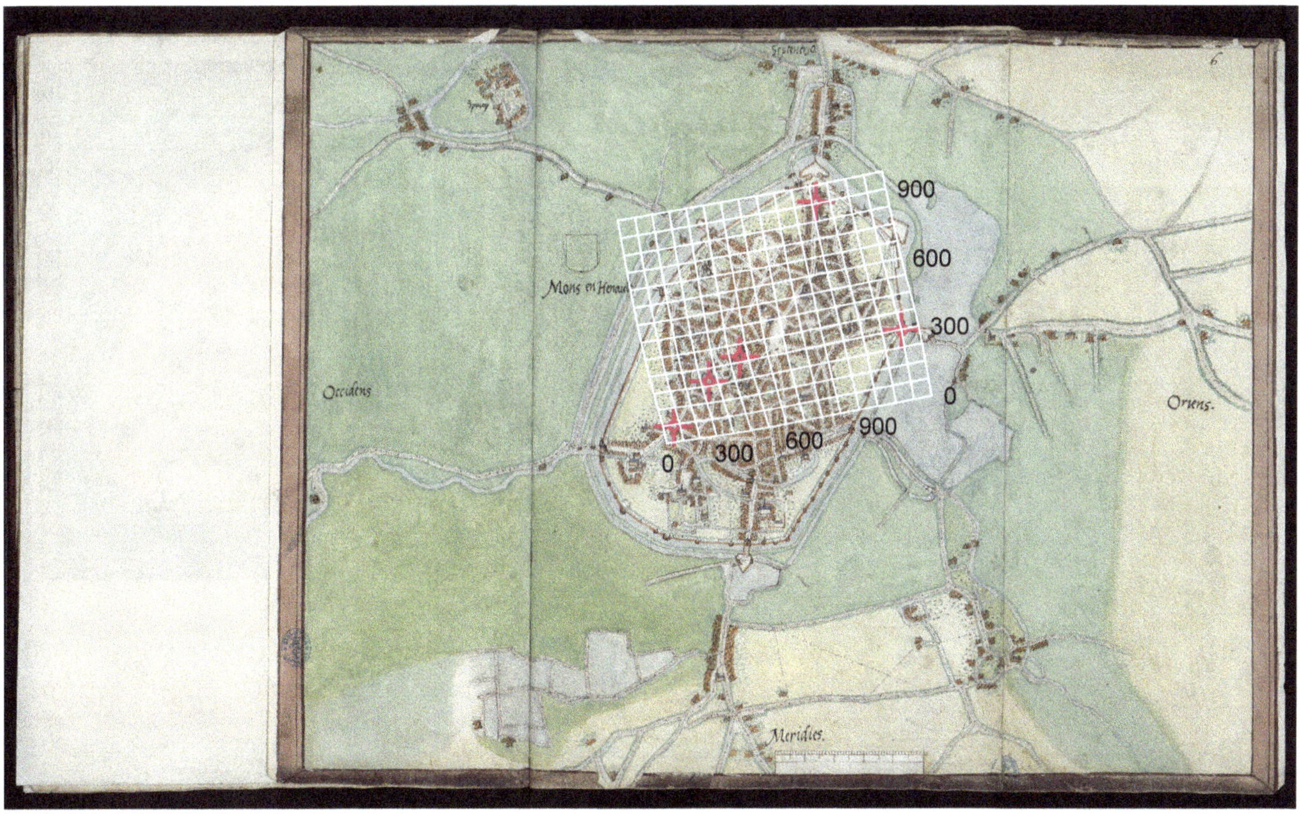
Occidens
Oriens.
Meridies.
900
600
300
0
0
300
600
900

151. Montfoort

Tables 3.151 (measured and analysed).
Figures 3.151 (unmeasured, measured and analysed).

Measured							
Pixels	Scale stick original map (pixels)	Houses	Street sides	Houses per street side	Inhabitants (RF 4.0)	Inhabitants (RF 5.5)	Inhabitants (RF 7.0)
2244	996	394	61	6.46	1576	2167	2758

Analysed	
June 18, 2023 9:47:34 PM	
Computation with 5 linked points (*italics* points are less reliable):	1. Grote of Sint-Janskerk 2. Kasteel Montfoort 3. IJsselpoort 4. Molenstraat 5. Achterbaan
Reference points (Reliability 0–5)	5
Transformation with 3 parameters	
Scale	1:7580
Rotation	16 °
Mean position error	13m

61
Septentrio
Montfort
Occidens
Oriens
Meridies

61
Septentrio
Montfort
300
Occidens
Oriens
0
300
0
Meridies

152. Montfort

Tables 3.152 (measured and analysed).
Figures 3.152 (unmeasured, measured and analysed).

Measured							
Pixels	Scale stick original map (pixels)	Houses	Street sides	Houses per street side	Inhabitants (RF 4.0)	Inhabitants (RF 5.5)	Inhabitants (RF 7.0)
1476	1008	259	32	8.10	1037	1426	1815

Analysed	
April 3, 2023 2:08:35 PM	
Computation with 5 linked points (*italics* points are less reliable):	1. Slot Montfort 2. Catharinakerk 3. Vlootbeek 4. Markt 5. Dijkstraat
Reference points (Reliability 0–5)	5
Transformation with 3 parameters	
Scale	1:8700
Rotation	15 °
Mean position error	12m

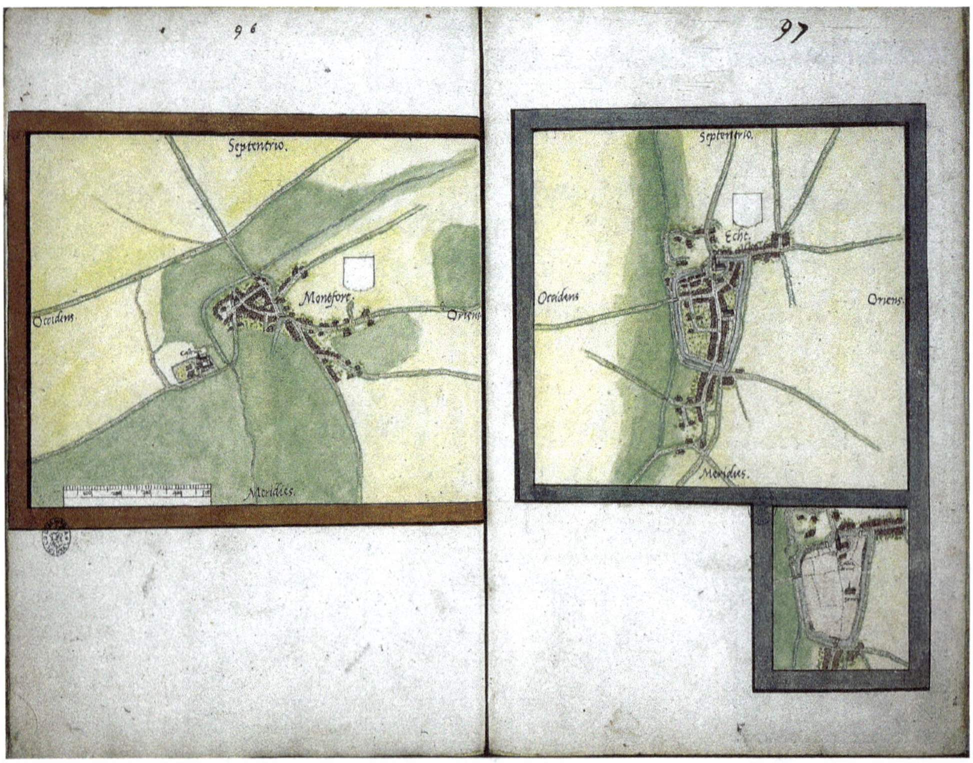

96
97
Septentrio.
Occidens.
Montfort.
Oriens
Meridies.
Septentrio.
Echt.
Occidens
Oriens.
Meridies.

96
97
Septentrio.
Occidens.
Montfort.
Oriens
300
Meridies.
Septentrio.
Echt.
Occidens
Oriens.
Meridies.

153. Montmédy

Tables 3.153 (measured and analysed).
Figures 3.153 (unmeasured, measured and analysed).

Measured							
Pixels	Scale stick original map (pixels)	Houses	Street sides	Houses per street side	Inhabitants (RF 4.0)	Inhabitants (RF 5.5)	Inhabitants (RF 7.0)
1383	missing	243	56	4.34	972	1336	1700

Analysed	
September 26, 2023 2:38:11 PM	
Computation with 5 linked points (*italics* points are less reliable):	1. Eglise Saint-Martin 2. Rue de l'Arsenal 3. Rue Traversièire 4. Rue de la Fausse Porte 5. Place de l'Infirmerie
Reference points (Reliability 0–5)	5
Transformation with 3 parameters	
Scale	1:7300
Rotation	15 °
Mean position error	7m

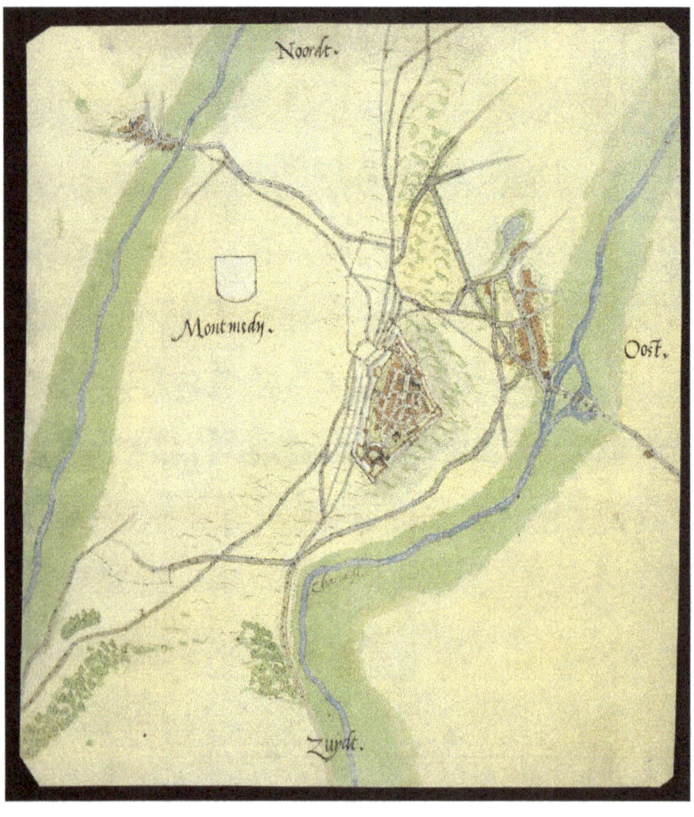

Noordt.
Montmedy.
Oost.
Zuydt.

Noordt.
Montmedy.
Oost.
300
0
300
0
Zuydt.

154. Muiden

Tables 3.154 (measured and analysed).
Figures 3.154 (unmeasured, measured and analysed).

Measured							
Pixels	Scale stick original map (pixels)	Houses	Street sides	Houses per street side	Inhabitants (RF 4.0)	Inhabitants (RF 5.5)	Inhabitants (RF 7.0)
1061	1010	186	33	5.65	745	1025	1305

Analysed	
June 19, 2023 9:21:18 AM	
Computation with 5 linked points (*italics* points are less reliable):	1. Muiderslot 2. Grote of Sint-Nicolaaskerk 3. Hellingstraat 4. Heerengracht 5. Heerengracht
Reference points (Reliability 0–5)	5
Transformation with 3 parameters	
Scale	1:8780
Rotation	4 °
Mean position error	17m

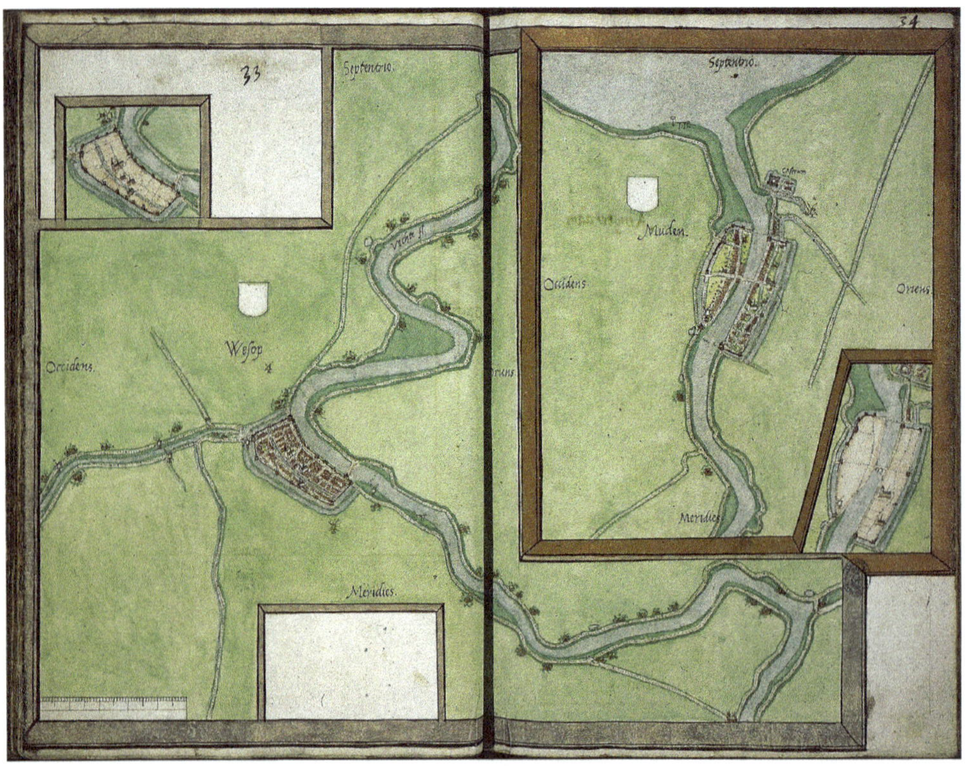

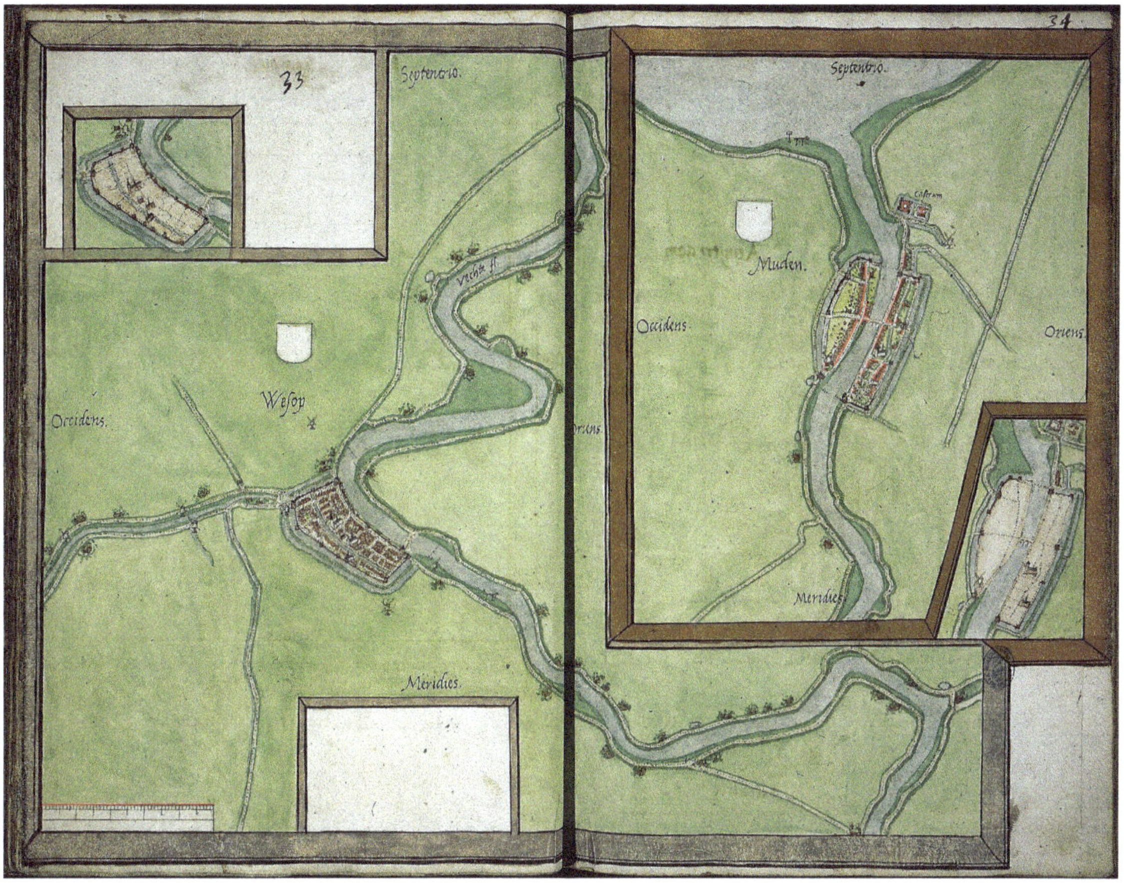
33
Septentrio.
Occidens.
Wesop
Vechte fl
Oriens.
Meridies.
34
Septentrio.
Castrum
Muden.
Occidens.
Oriens.
Meridies.

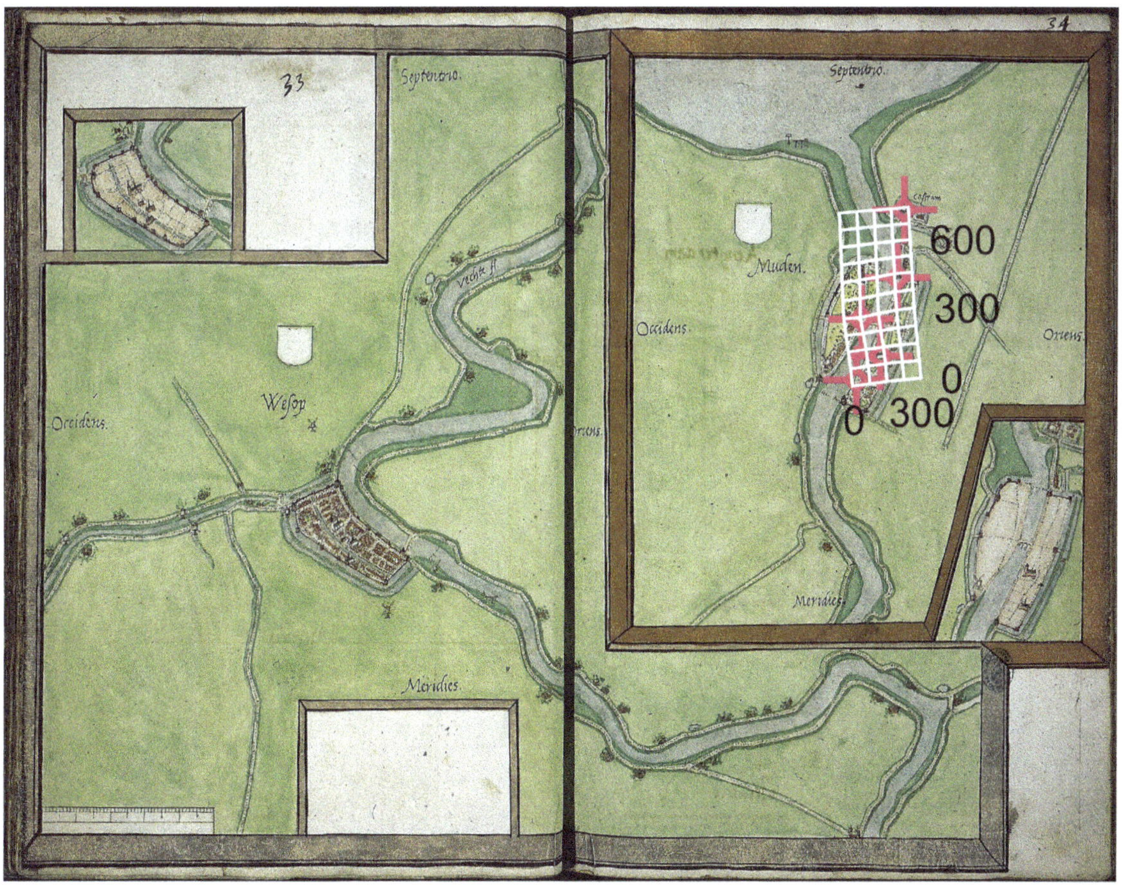
33
Septentrio.
Occidens.
Wesop
Vechte fl
Oriens.
Meridies.
34
Septentrio.
Castrum
Muden.
600
300
0
0 300
Occidens.
Oriens.
Meridies.

155. Naarden

Tables 3.155 (measured and analysed).
Figures 3.155 (unmeasured, measured and analysed).

Measured							
Pixels	Scale stick original map (pixels)	Houses	Street sides	Houses per street side	Inhabitants (RF 4.0)	Inhabitants (RF 5.5)	Inhabitants (RF 7.0)
4006	missing	703	89	7.90	2814	3869	4924

Analysed	
June 19, 2023 10:56:17 AM	
Computation with 5 linked points (*italics* points are less reliable):	1. Sint-Vituskerk 2. Nieuwe Haven 3. Oostwalstraat 4. Westwalstraat 5. Sint Vitusstraat
Reference points (Reliability 0–5)	5
Transformation with 3 parameters	
Scale	1:8070
Rotation	10 °
Mean position error	17m

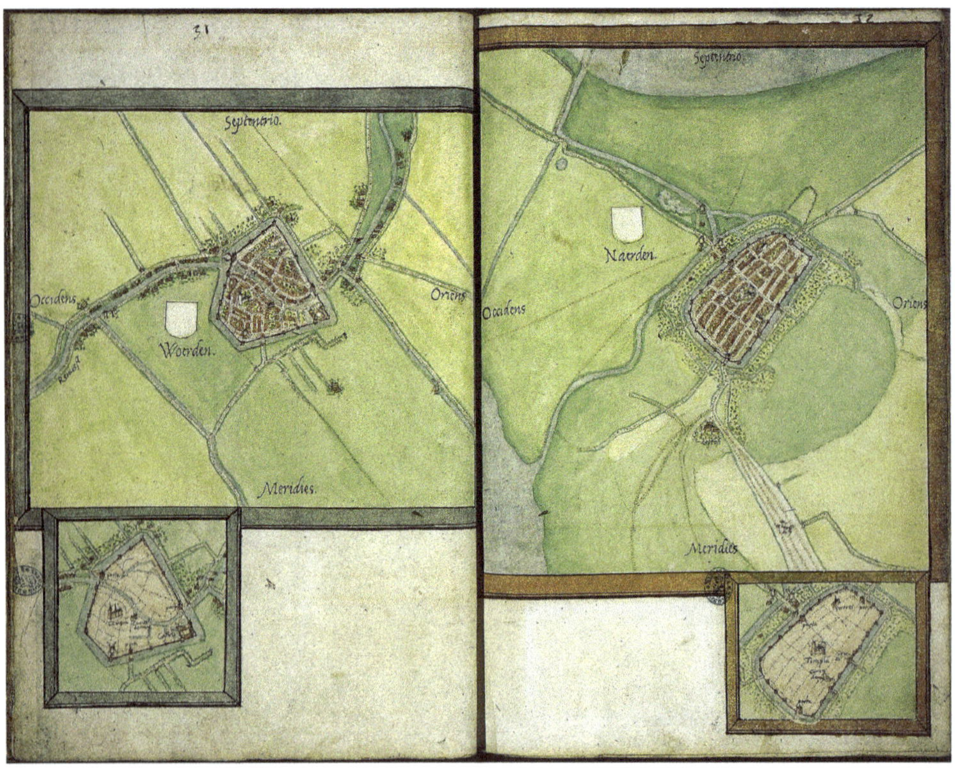

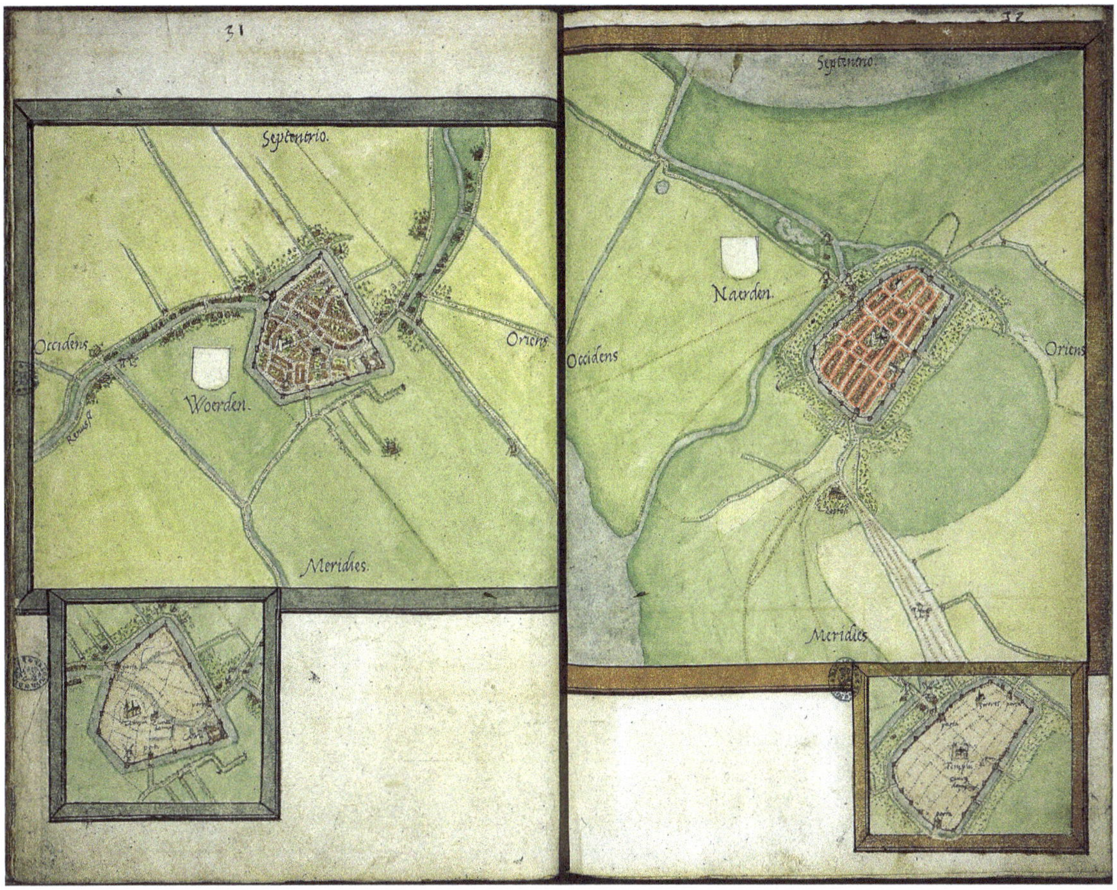
31
Septentrio.
Occidens
Oriens
Woerden.
Meridies.
Septentrio.
Naerden.
Occidens
Oriens
Meridies.
Templu

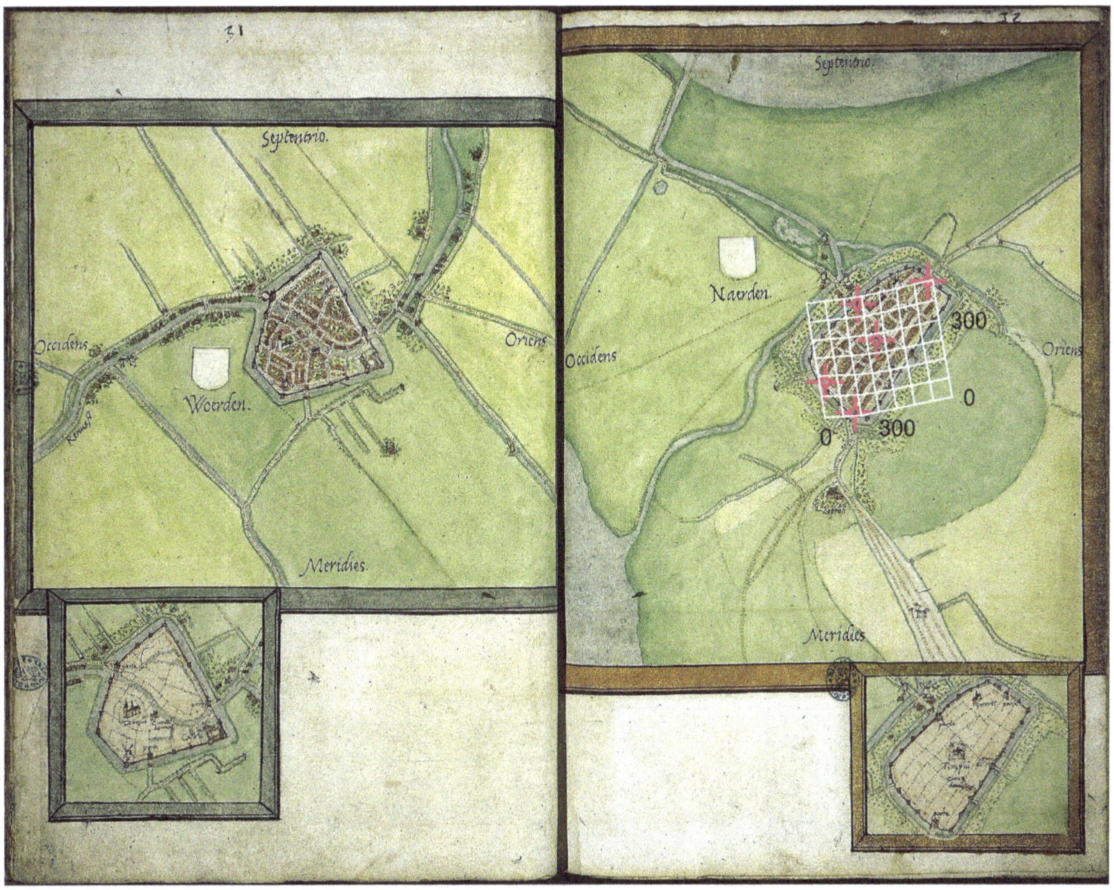
31
Septentrio.
Occidens
Oriens
Woerden.
Meridies.
Septentrio.
Naerden.
Occidens
Oriens
300
0
0
300
Meridies.
Templu

156. Nieuwpoort (B)

Tables 3.156 (measured and analysed).
Figures 3.156 (unmeasured, measured and analysed).

Measured

Pixels	Scale stick original map (pixels)	Houses	Street sides	Houses per street side	Inhabitants (RF 4.0)	Inhabitants (RF 5.5)	Inhabitants (RF 7.0)
6336	1027	1113	124	8.97	5563	6119	6676

Analysed

June 21, 2023 9:57:13 AM	
Computation with 5 linked points	1. Onze-Lieve-Vrouwekerk
(*italics* points are less reliable):	2. Kaai
	3. Langestraat
	4. Langestraat
	5. Kerkstraat
Reference points (Reliability 1–5)	5
Transformation with 3 parameters	
Scale	1:7780
Rotation	9 °
Mean position error	12m

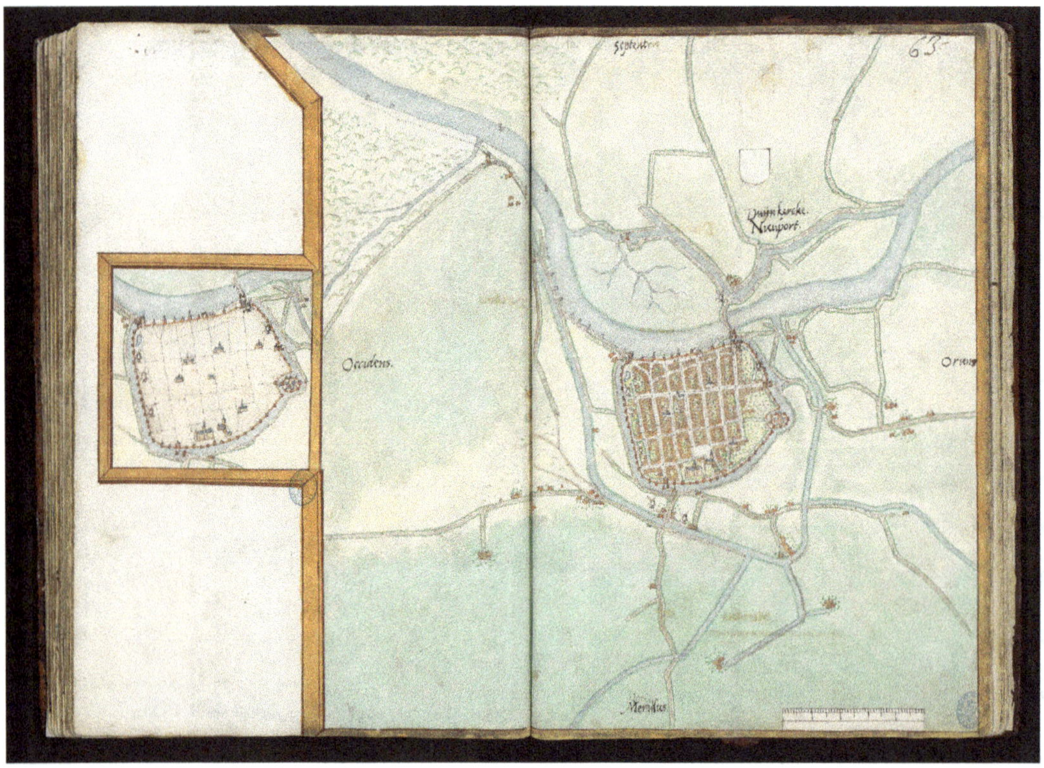

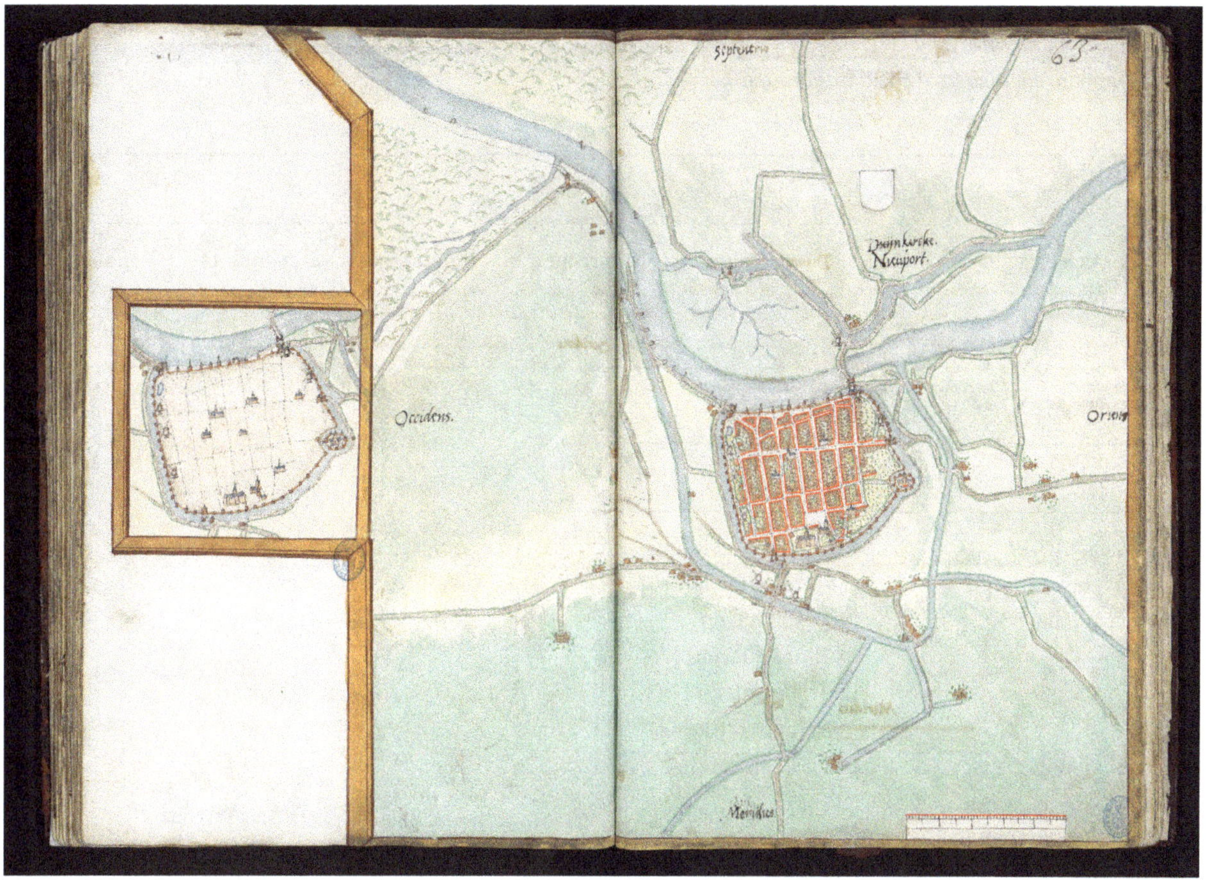
Occidens.
Nieuport.

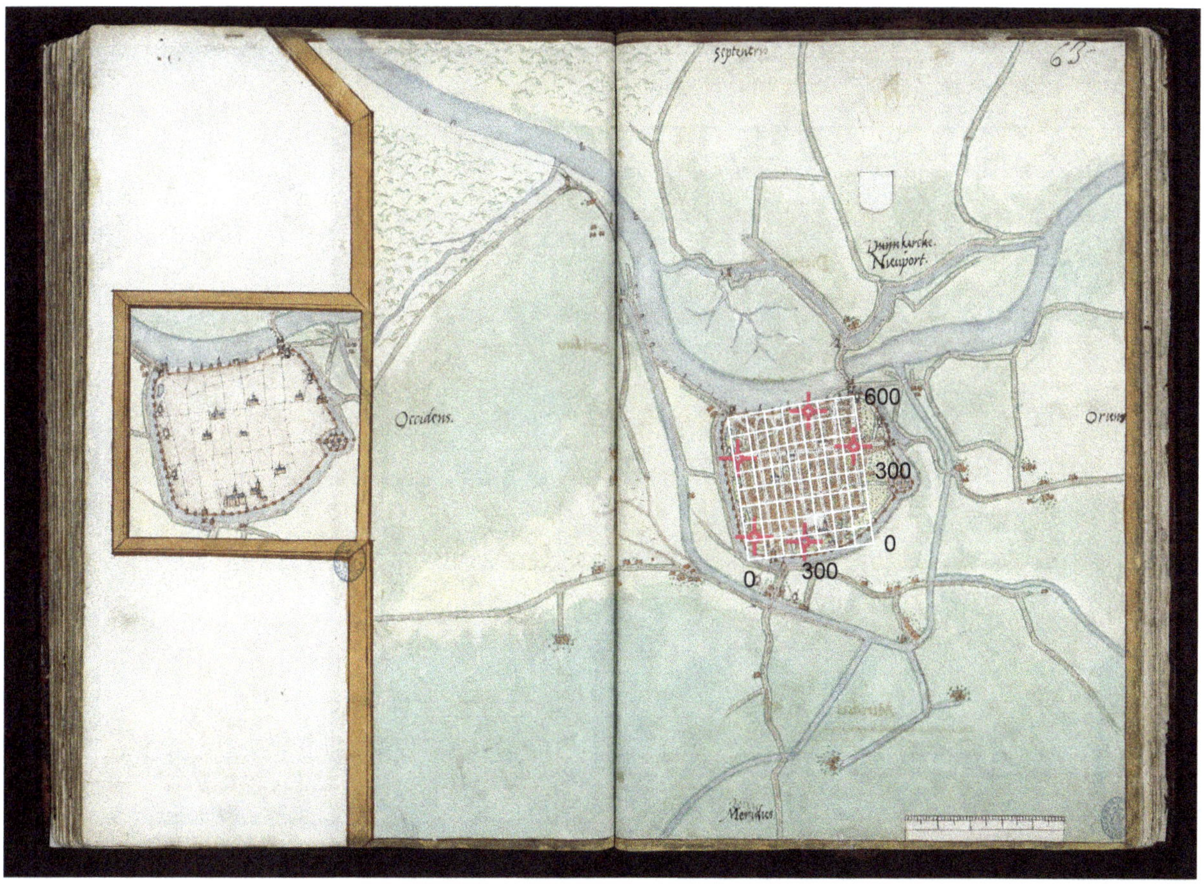
Occidens.
Nieuport.
600
300
0
0
300

157. Nieuwpoort (Nl)

Tables 3.157 (measured and analysed).
Figures 3.157 (unmeasured, measured and analysed).

Measured							
Pixels	Scale stick original map (pixels)	Houses	Street sides	Houses per street side	Inhabitants (RF 4.0)	Inhabitants (RF 5.5)	Inhabitants (RF 7.0)
1119	993	196	25	7.86	786	1081	1376

Analysed	
June 19, 2023 10:49:23 PM	
Computation with 5 linked points (*italics* points are less reliable):	1. Hervormde Kerk 2. Buitenhaven 3. Binnenhaven 4. Hoogstraat 5. Schipperweg
Reference points (Reliability 0–5)	5
Transformation with 3 parameters	
Scale	1:7510
Rotation	12 °
Mean position error	16m

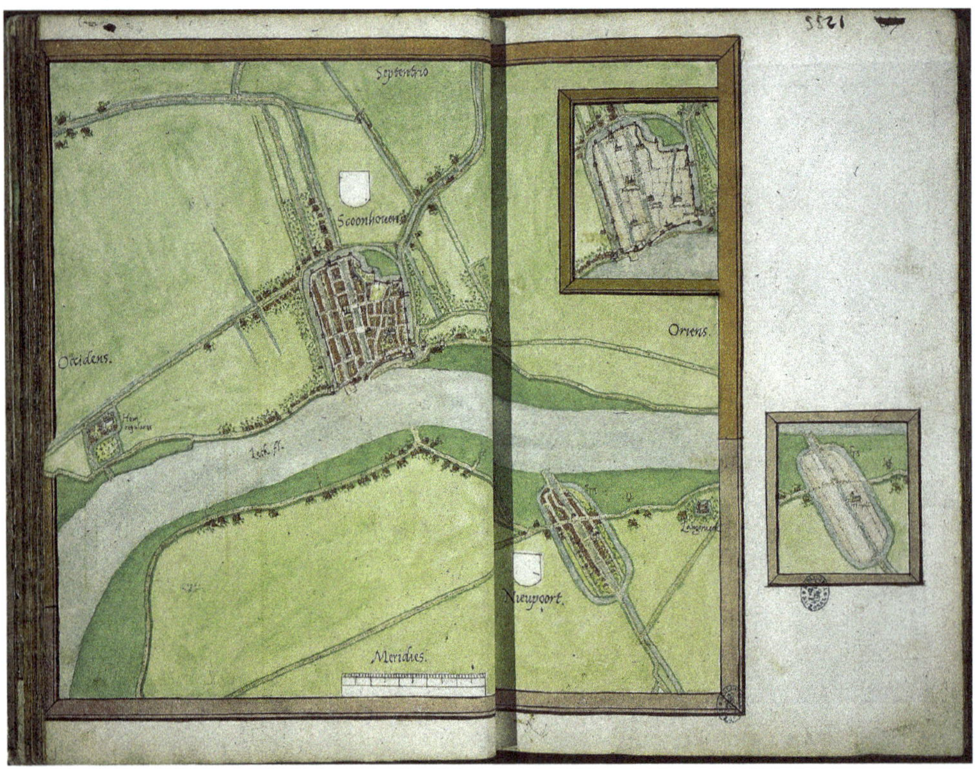

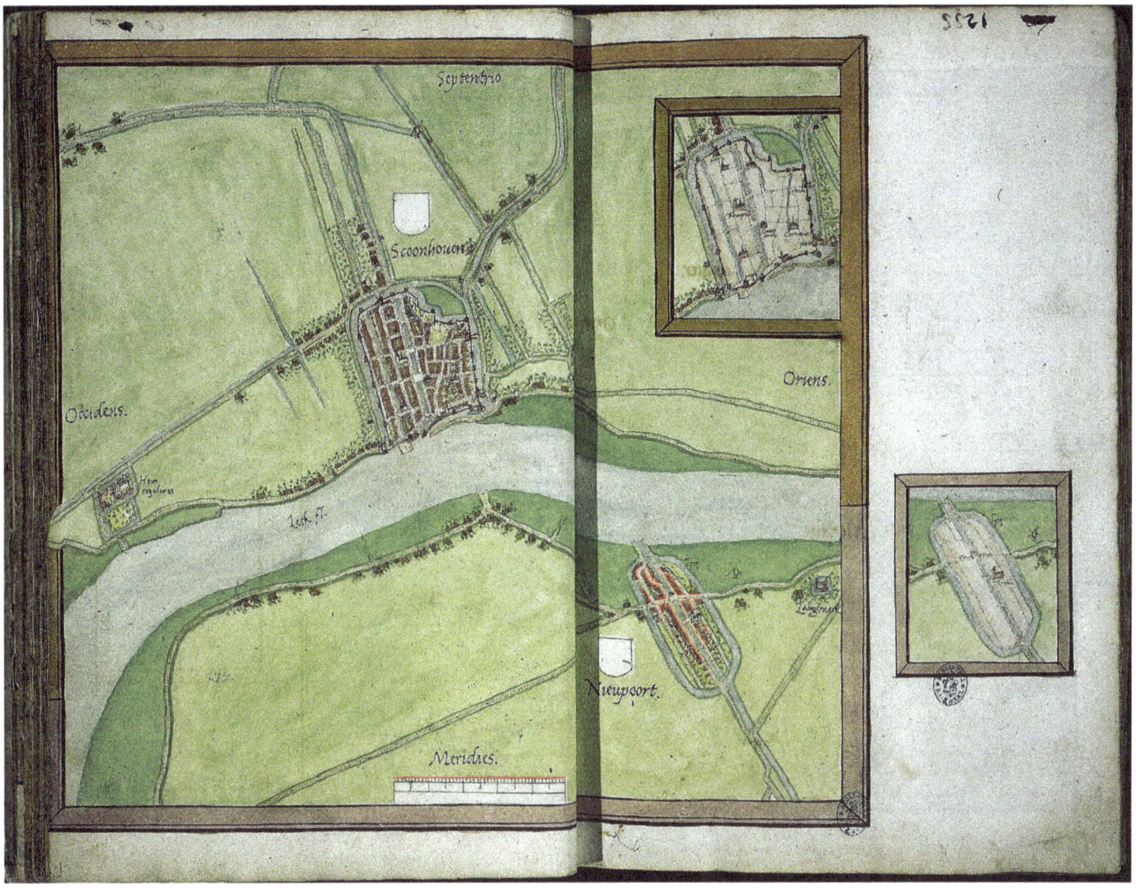
Septentrio
Scoonhouen
Oriens.
Occidens.
Lek fl.
Nieupoort.
Meridies.

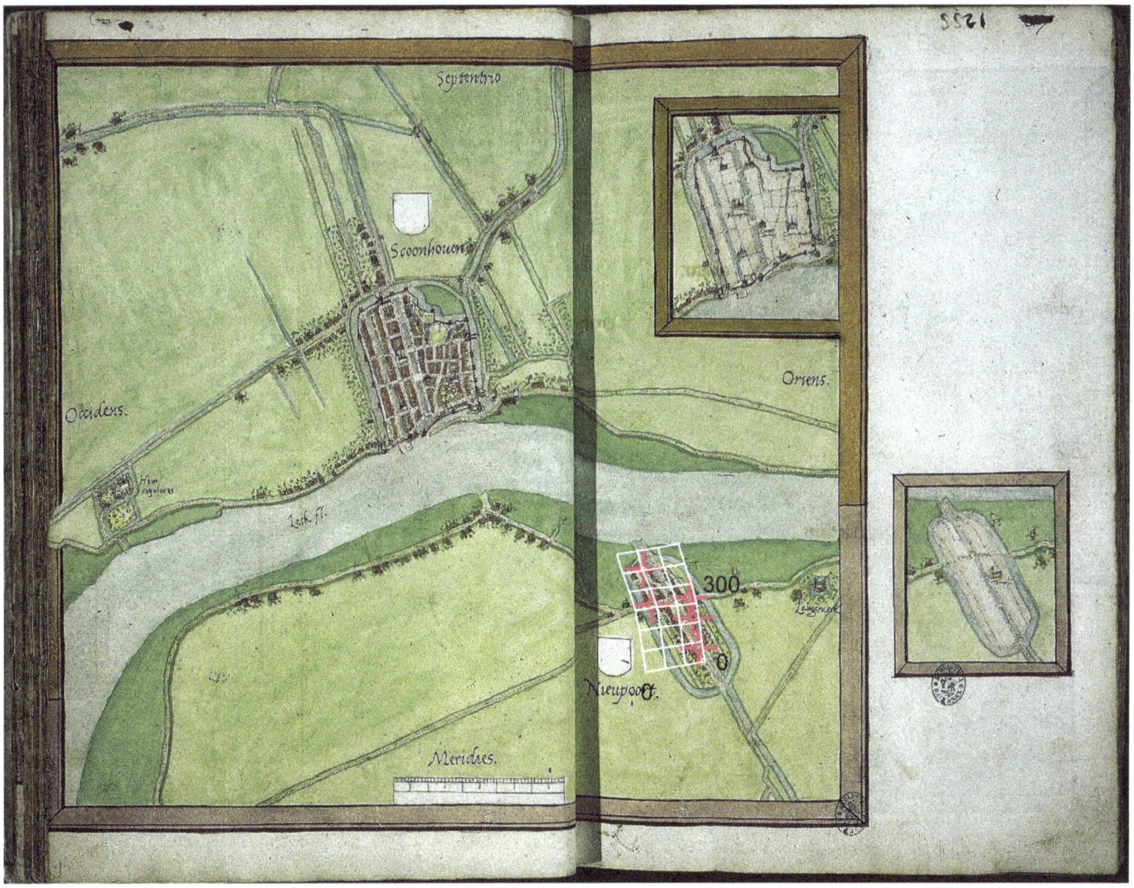
Septentrio
Scoonhouen
Oriens.
Occidens.
Lek fl.
300
0
Meridies.

158. Nijmegen

Tables 3.158 (measured and analysed).
Figures 3.158 (unmeasured, measured and analysed).

Measured							
Pixels	Scale stick original map (pixels)	Houses	Street sides	Houses per street side	Inhabitants (RF 4.0)	Inhabitants (RF 5.5)	Inhabitants (RF 7.0)
10,930	1003	1919	271	7.08	7677	10,556	13,435

Analysed	
October 15, 2023 2:21:01 PM	
Computation with 5 linked points (*italics* points are less reliable):	1. Sint-Stevenskerk 2. Houtstraat 3. Sint-Petrus Canisius 4. Jacobskapel 5. Hofberg
Reference points (Reliability 0–5)	5
Transformation with 3 parameters	
Scale	1:8490
Rotation	12 °
Mean position error	13m

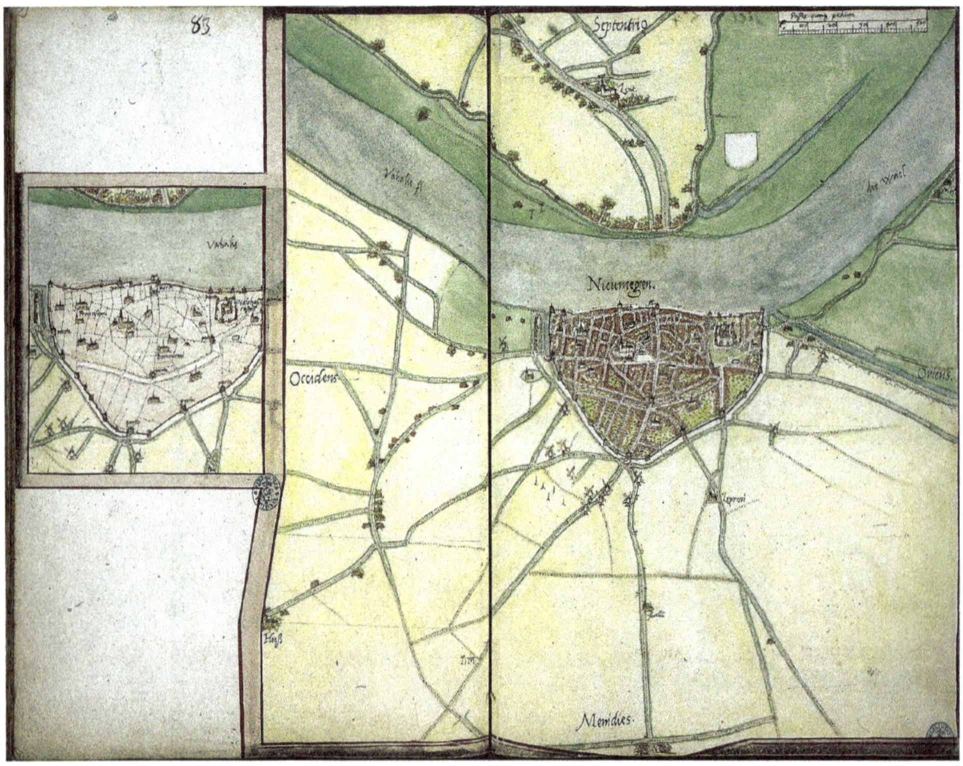

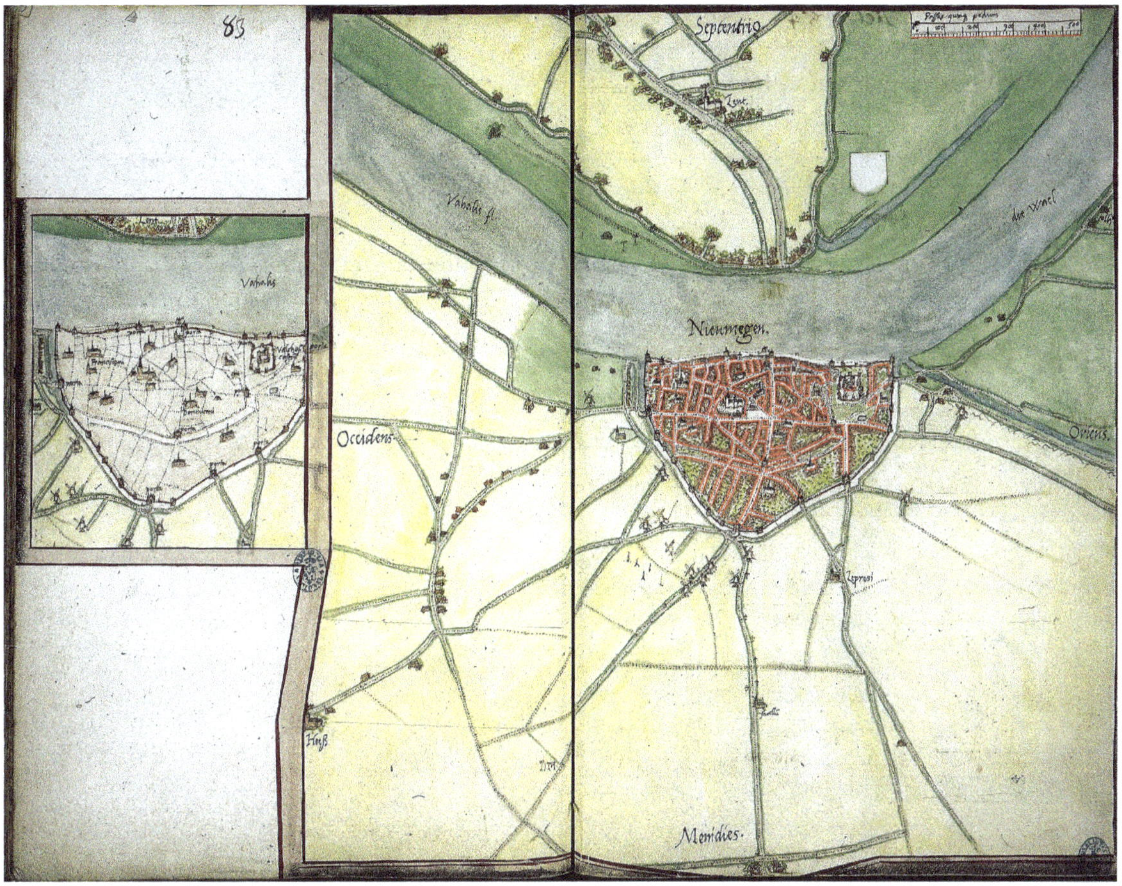
83
Septentrio
Nieumegen.
Occidens.
Oriens.
Meridies.

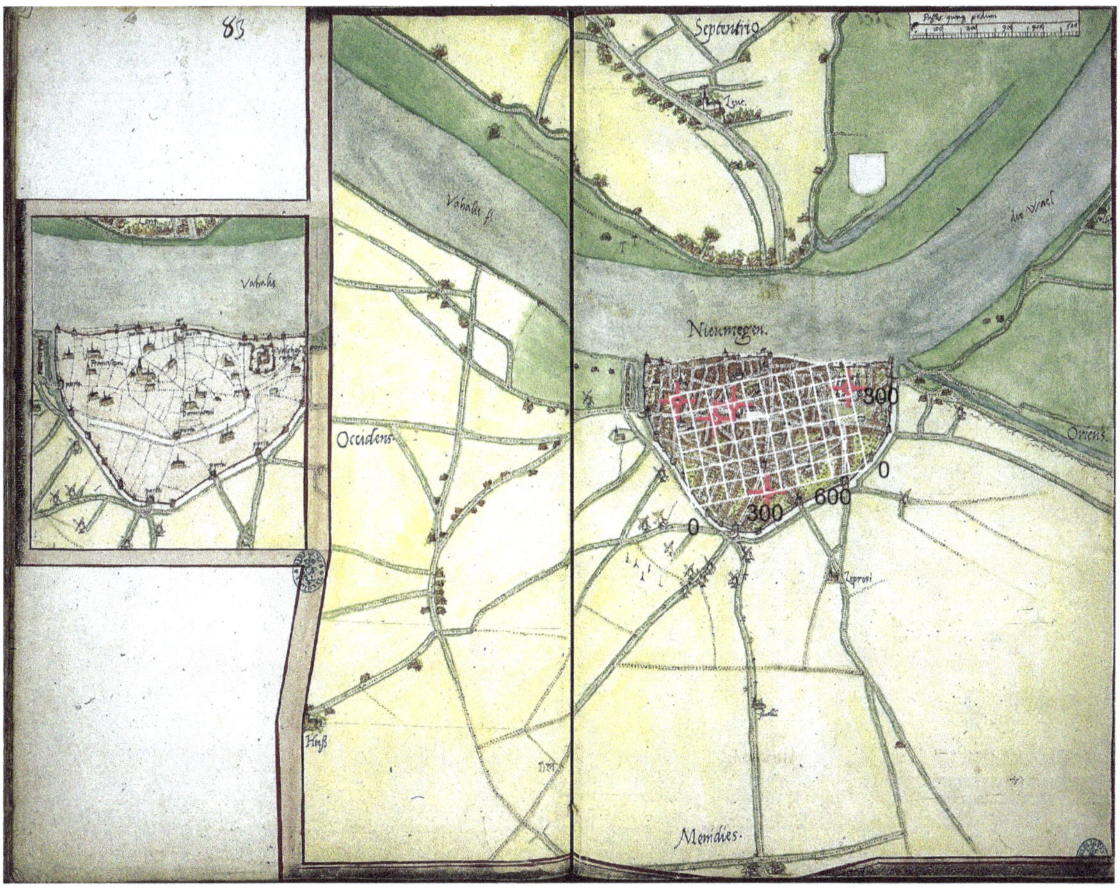
83
Septentrio
Nieumegen.
Occidens.
Oriens.
Meridies.
0
300
600
300
0

159. Ninove

Tables 3.159 (measured and analysed).
Figures 3.159 (unmeasured, measured and analysed).

Measured

Pixels	Scale stick original map (pixels)	Houses	Street sides	Houses per street side	Inhabitants (RF 4.0)	Inhabitants (RF 5.5)	Inhabitants (RF 7.0)
3066	1028	538	76	7.08	2153	2961	3769

Analysed

June 22, 2023 6:38:19 PM	
Computation with 5 linked points (*italics* points are less reliable):	1. Onze-Lieve-Vrouw Hemelvaartkerk 2. Koepoort 3. *Lavendelstraat* 4. Oude Kaai 5. *Graanmarkt*
Reference points (Reliability 0–5)	3
Transformation with 3 parameters	
Scale	1:8350
Rotation	9 °
Mean position error	36m

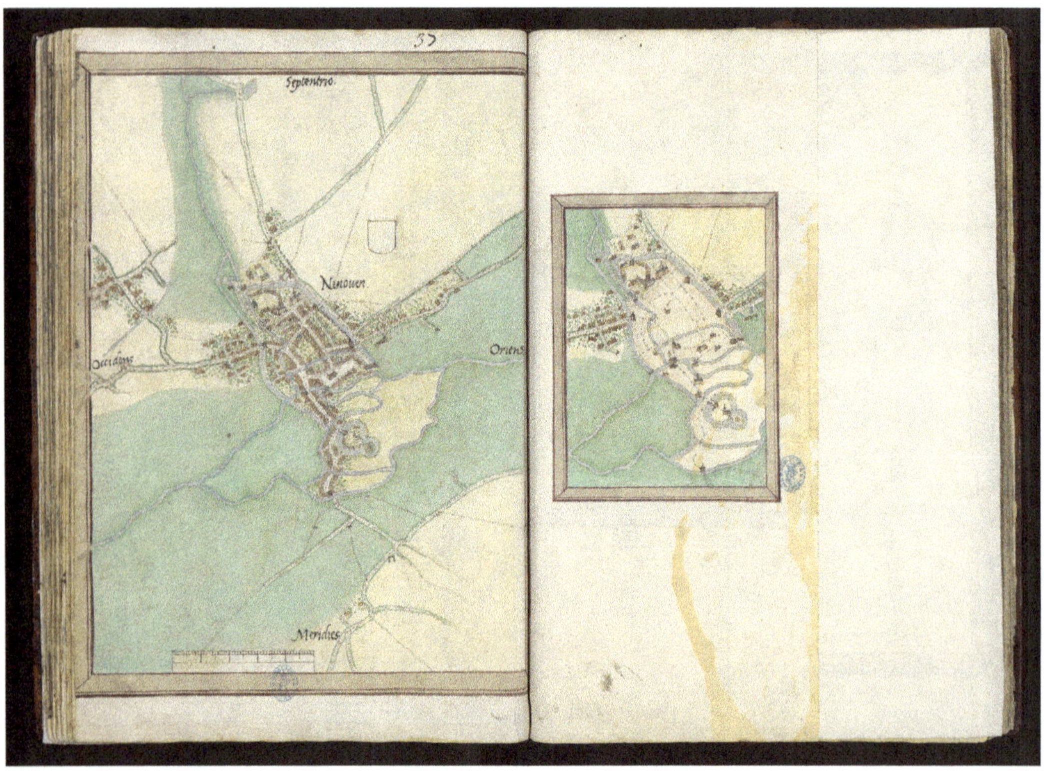

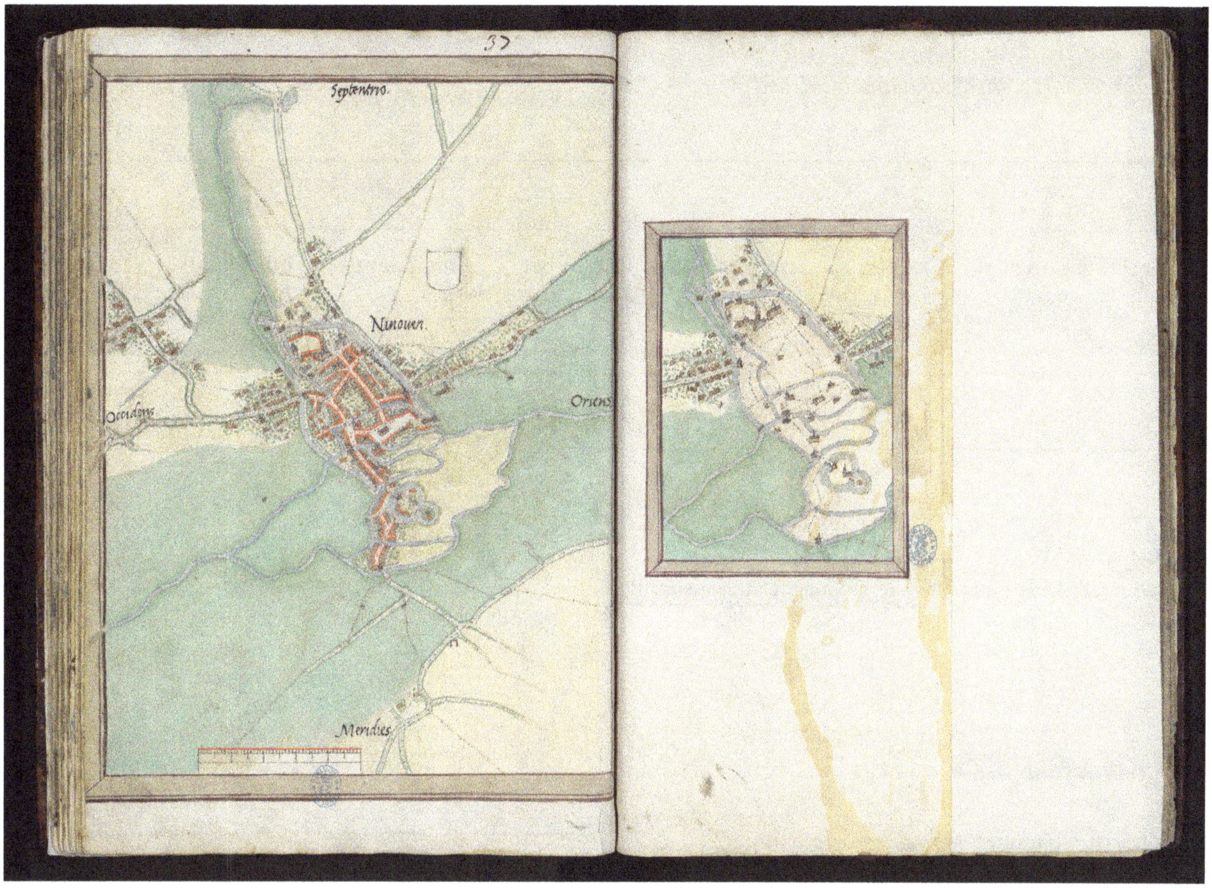
37
Septentrio.
Ninouen.
Occidens
Oriens
Meridies

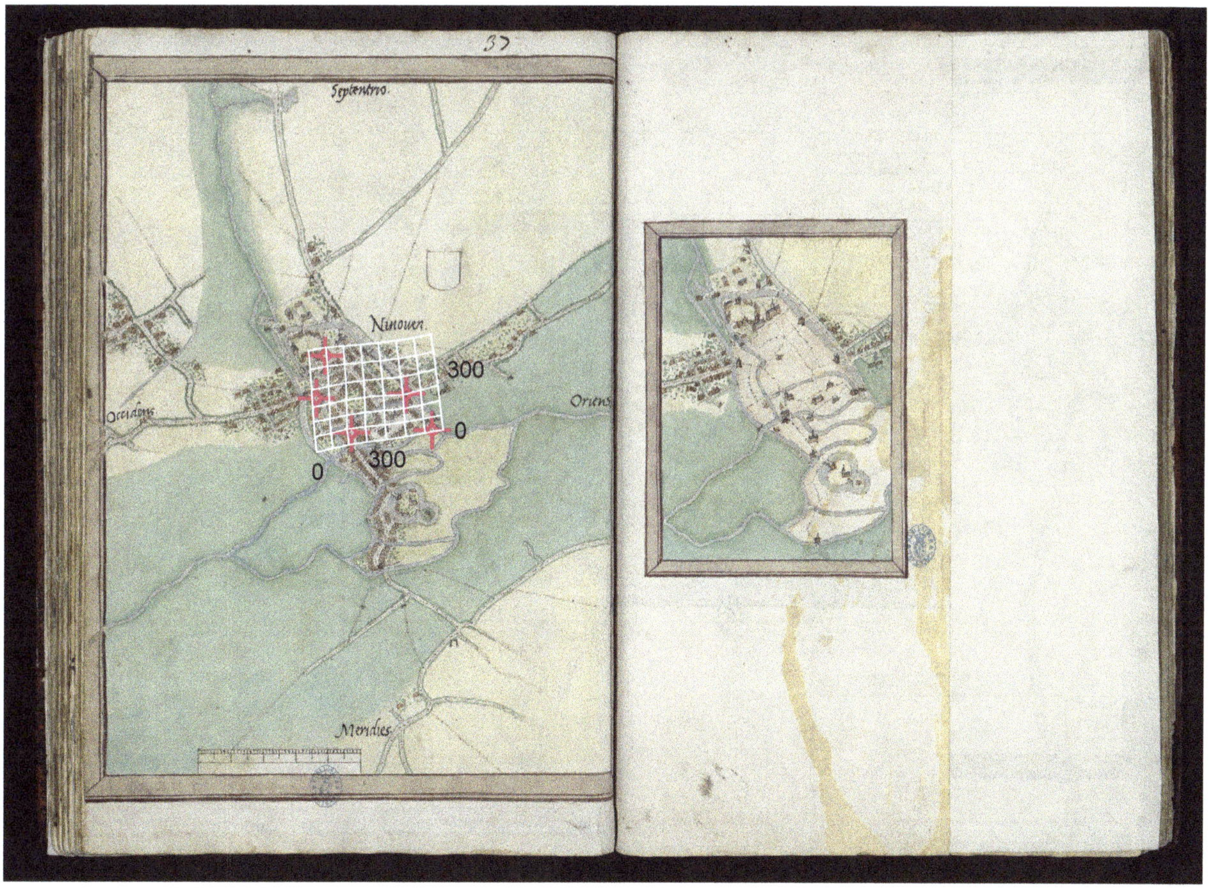
37
Septentrio.
Ninouen.
300
0
0
300
Occidens
Oriens
Meridies

160. Oldenzaal

Tables 3.160 (measured and analysed).
Figures 3.160 (unmeasured, measured and analysed).

Measured							
Pixels	Scale stick original map (pixels)	Houses	Street sides	Houses per street side	Inhabitants (RF 4.0)	Inhabitants (RF 5.5)	Inhabitants (RF 7.0)
4171	941	732	111	6.60	2929	4.028	5127

Analysed	
June 24, 2023 11:49:00 AM	
Computation with 5 linked points (*italics* points are less reliable):	1. Sint-Plechelmusbasiliek 2. Oostwal 3. Steenstraat 4. Vestingstraat 5. Marktstraat
Reference points (Reliability 0–5)	5
Transformation with 3 parameters	
Scale	1:8090
Rotation	12 °
Mean position error	11m

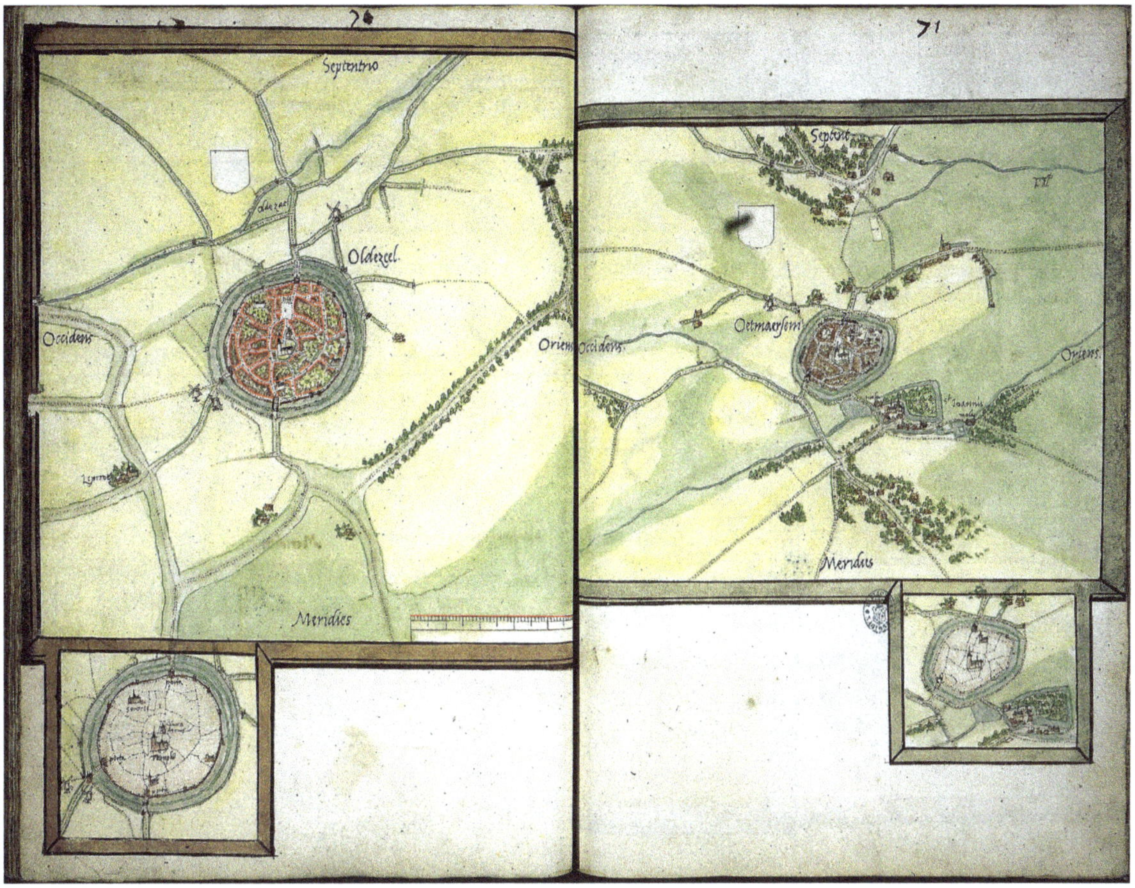

Septentrio
Oldezel.
Occidens
Meridies
Ootmaarsem
Occidens
Oriens
Meridus

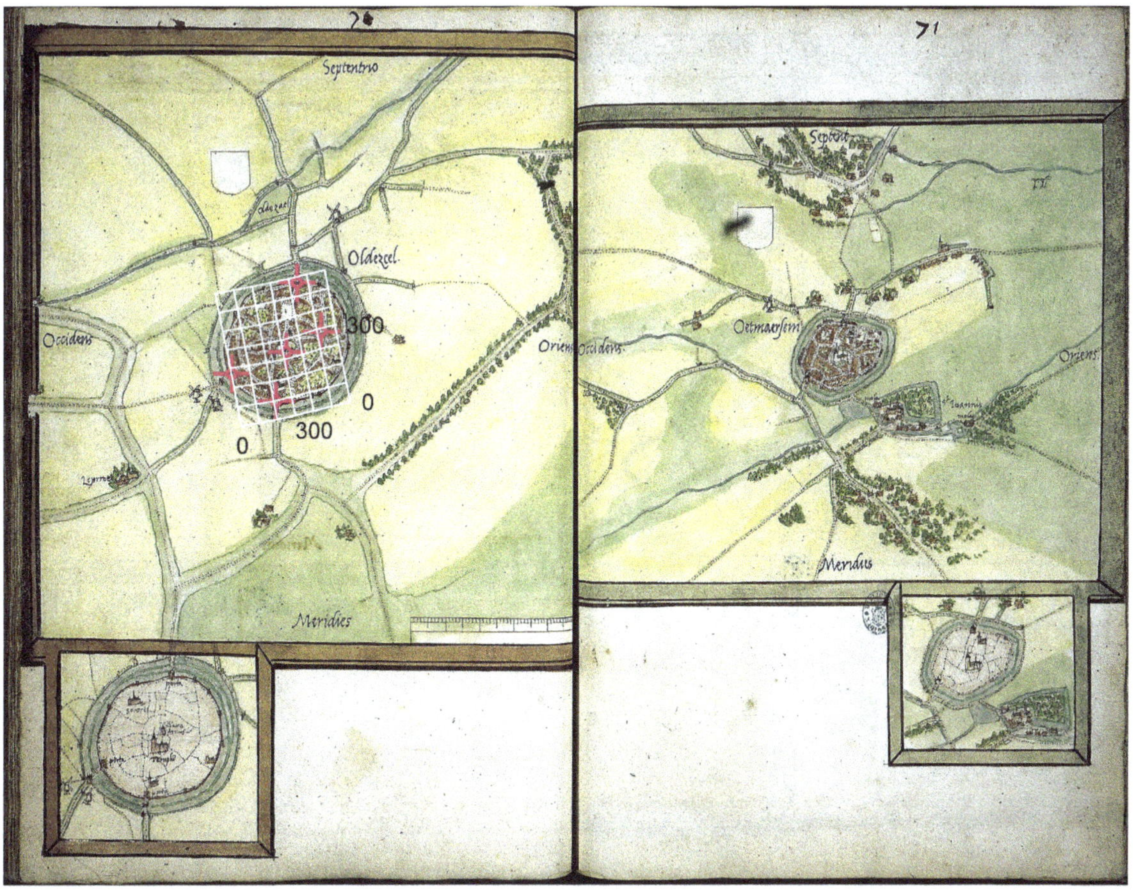

Septentrio
Oldezel.
300
0
300
0
Occidens
Meridies
Ootmaarsem
Occidens
Oriens
Meridus

161. Oostburg

Tables 3.161 (measured and analysed).
Figures 3.161 (unmeasured, measured and analysed).

Measured							
Pixels	Scale stick original map (pixels)	Houses	Street sides	Houses per street side	Inhabitants (RF 4.0)	Inhabitants (RF 5.5)	Inhabitants (RF 7.0)
2389	1021	420	56	7.49	1678	2307	2936

Analysed	
June 25, 2023 6:28:26 PM	
Computation with 5 linked points (*italics* points are less reliable):	1. Zuidwal 2. Walstraat *3. Markt* 4. Noordwal 5. Torenweidelaan
Reference points (Reliability 0–5)	4
Transformation with 3 parameters	
Scale	1:7660
Rotation	3 °
Mean position error	21m

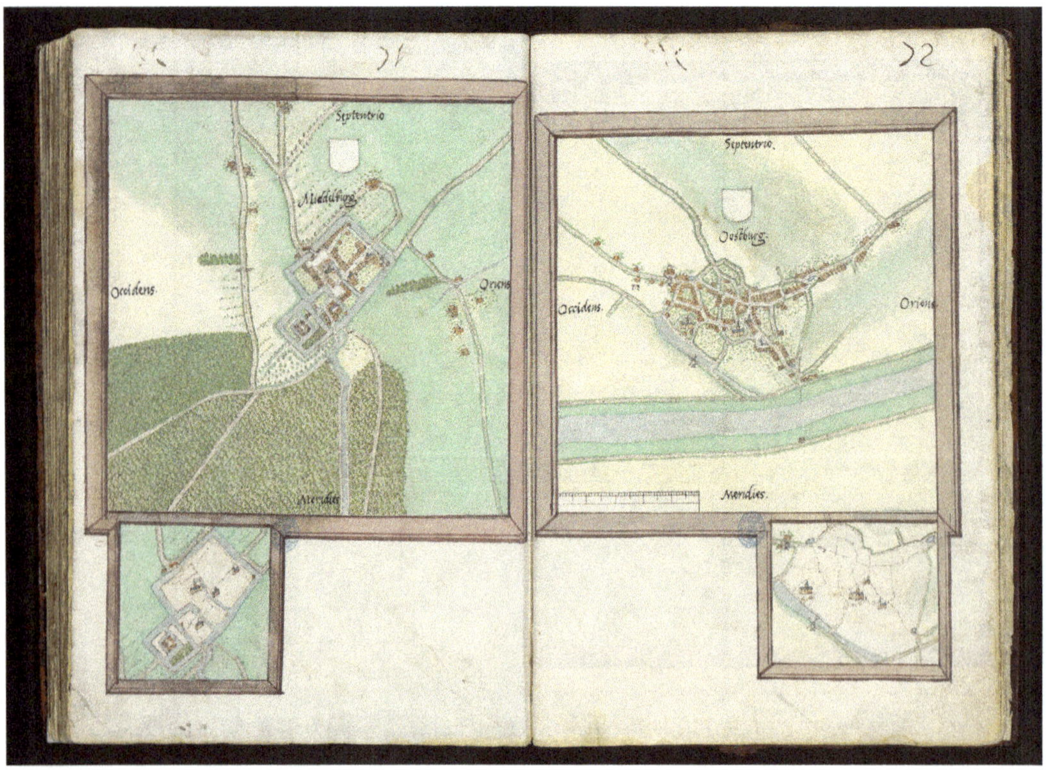

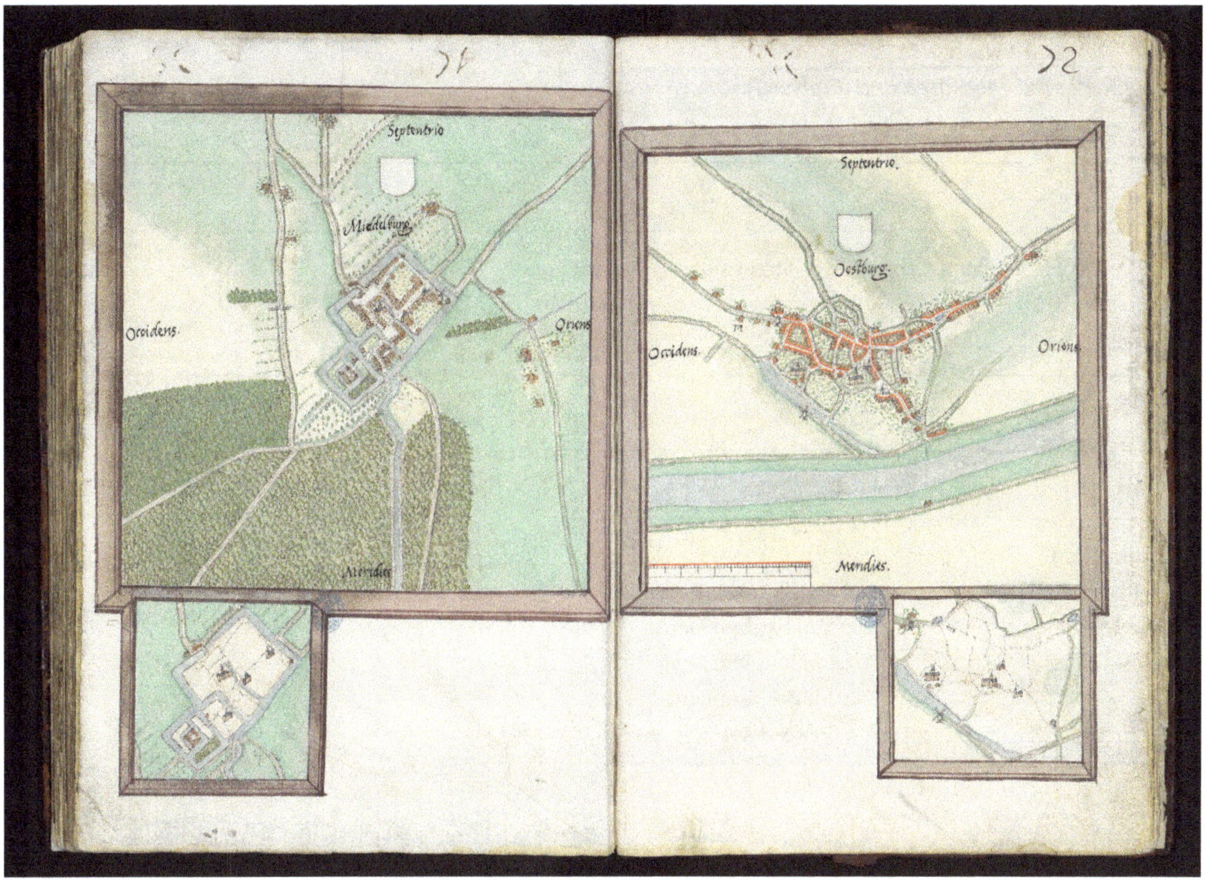
Septentrio
Middelburg
Occidens
Oriens
Septentrio
Oostburg
Occidens
Oriens
Meridies

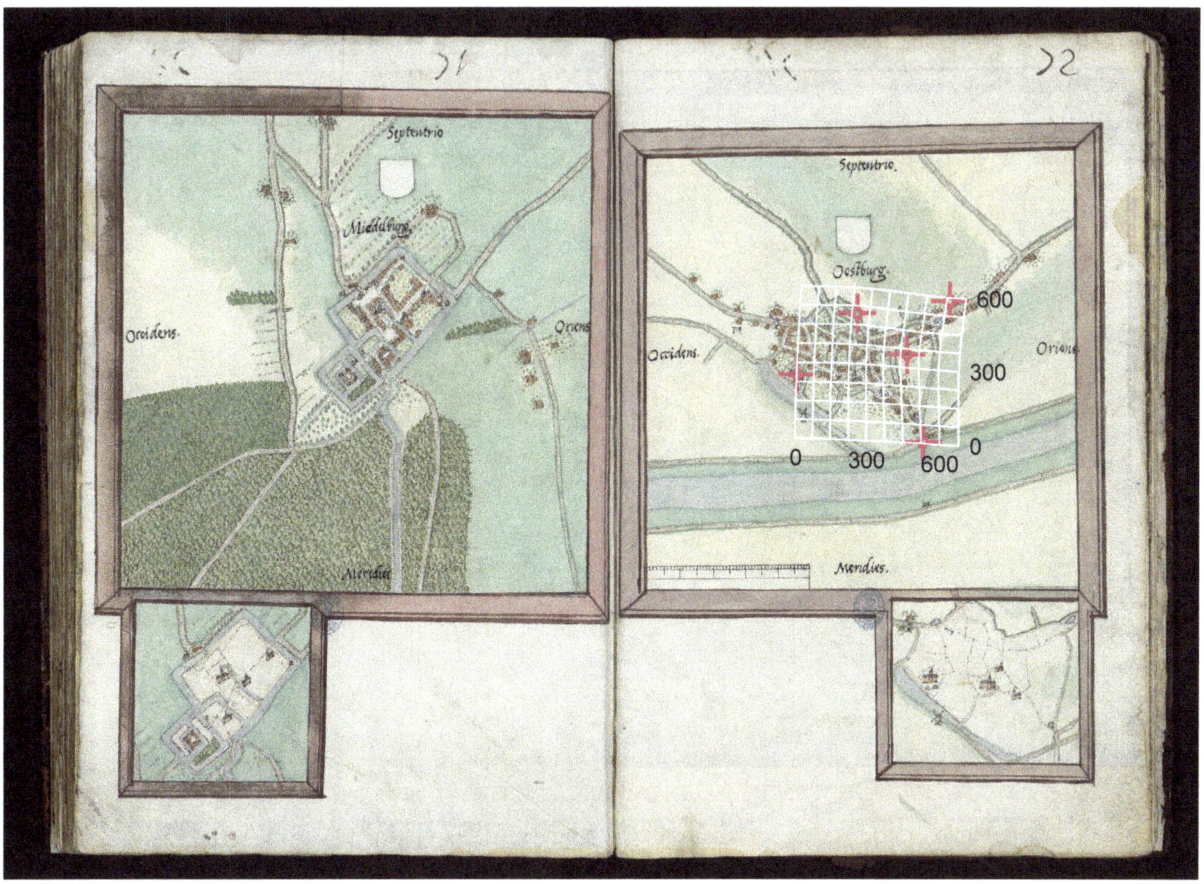
Septentrio
Middelburg
Occidens
Oriens
Septentrio
Oostburg
Occidens
600
300
0
0
300
600
Meridies

162. Oostende

Tables 3.162 (measured and analysed).
Figures 3.162 (unmeasured, measured and analysed).

Measured							
Pixels	Scale stick original map (pixels)	Houses	Street sides	Houses per street side	Inhabitants (RF 4.0)	Inhabitants (RF 5.5)	Inhabitants (RF 7.0)
5933	1023	1042	136	7.66	4167	5730	7293

Analysed	
June 29, 2023 11:01:52 AM	
Computation with 5 linked points (*italics* points are less reliable):	1. Sint-Petrus en -Pauluskerk 2. Van Iseghemlaan 3. Wapenplein 4. Marie-Joséplein 5. Visserskaai
Reference points (Reliability 0–5)	5
Transformation with 3 parameters	
Scale	1:8100
Rotation	8 °
Mean position error	11m

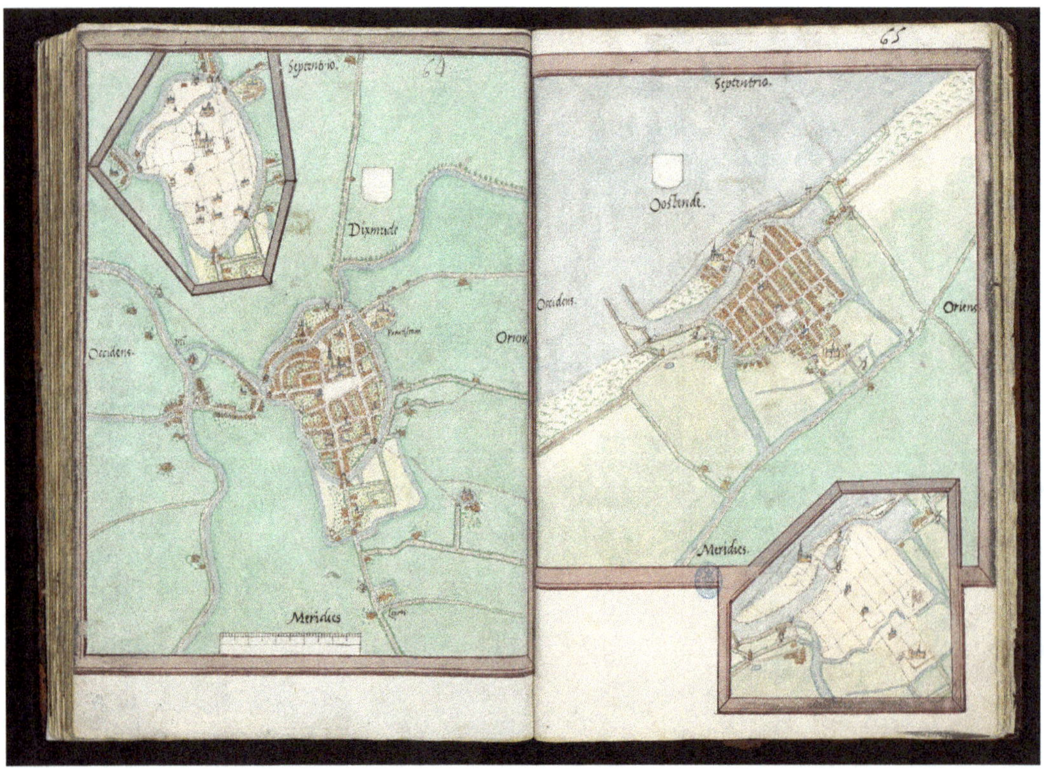

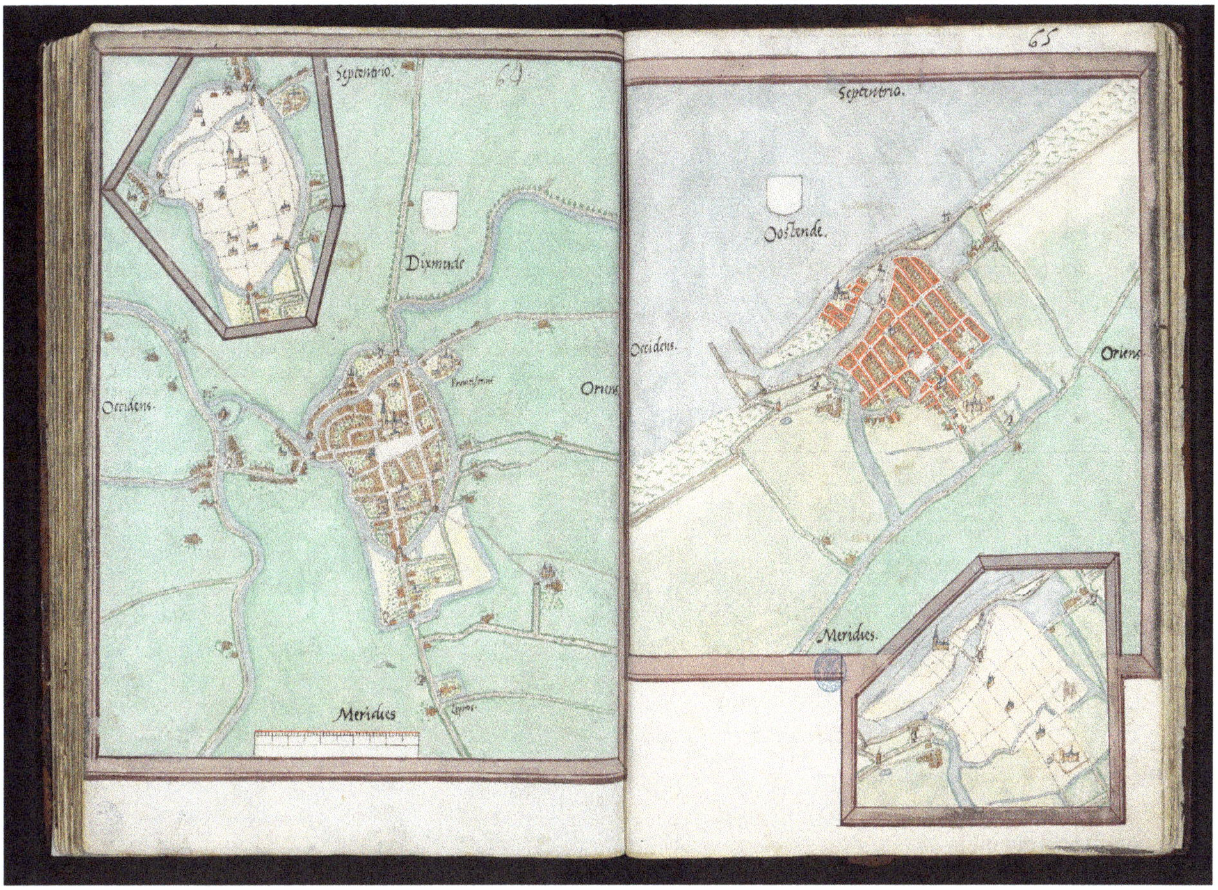
Septentrio.
64
Dixmude
Oriens
Occidens.
Meridies
65
Septentrio.
Oostende.
Occidens.
Oriens
Meridies.

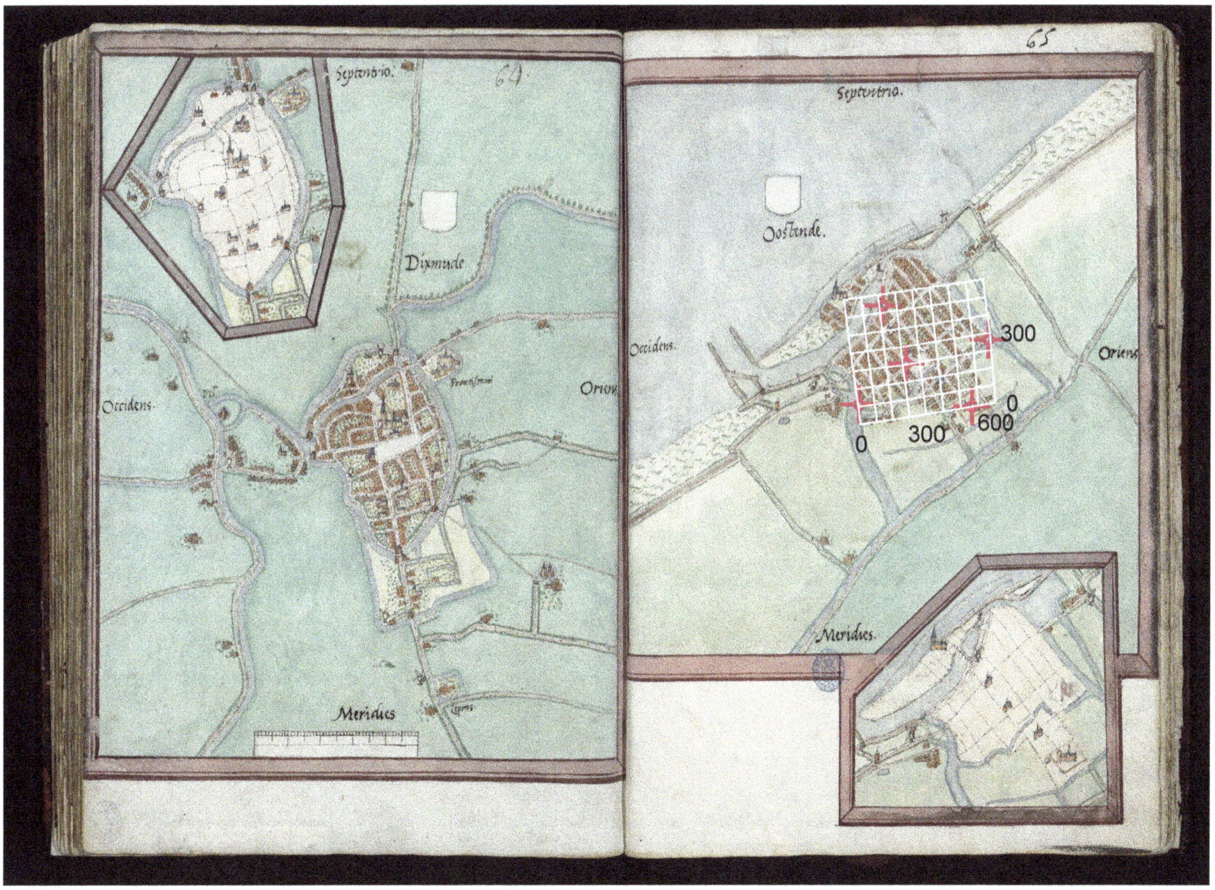
Septentrio.
64
Dixmude
Oriens
Occidens.
Meridies
65
Septentrio.
Oostende.
300
Occidens.
Oriens
0
0
300
600
Meridies.

163. Ootmarsum

Tables 3.163 (measured and analysed).
Figures 3.163 (unmeasured, measured and analysed).

Measured							
Pixels	Scale stick original map (pixels)	Houses	Street sides	Houses per street side	Inhabitants (RF 4.0)	Inhabitants (RF 5.5)	Inhabitants (RF 7.0)
2190	941	385	60	6.41	1538	2115	2692

Analysed	
June 25, 2023 4:42:13 PM	
Computation with 5 linked points (*italics* points are less reliable):	1. HH. Simon en Judaskerk 2. Bergstraat 3. Walstraat 4. Gasthuisstraat 5. Giltstraat
Reference points (Reliability 0–5)	5
Transformation with 3 parameters	
Scale	1:7510
Rotation	12 °
Mean position error	13m

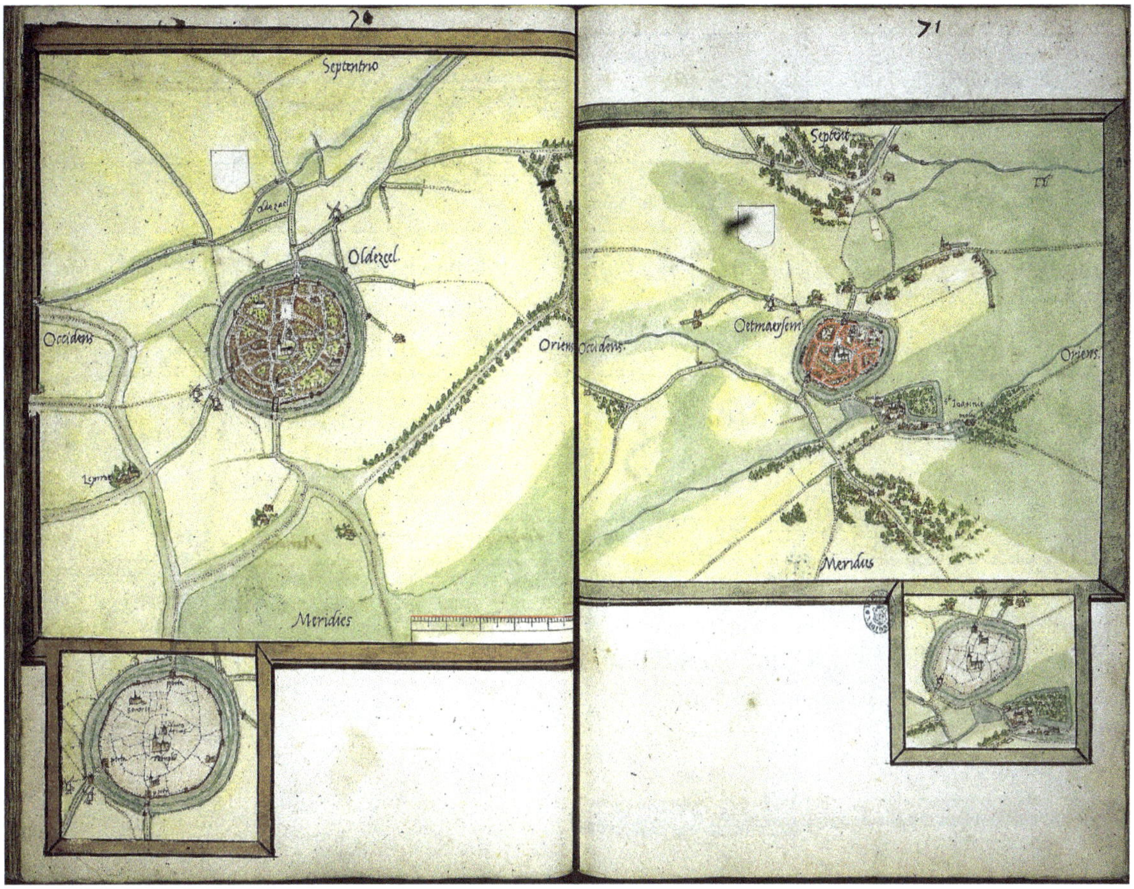
Septentrio
Oldezeel.
Occidens
Oriens
Meridies
Septent.
Ottmaersem
Occidens.
Oriens.
Meridus

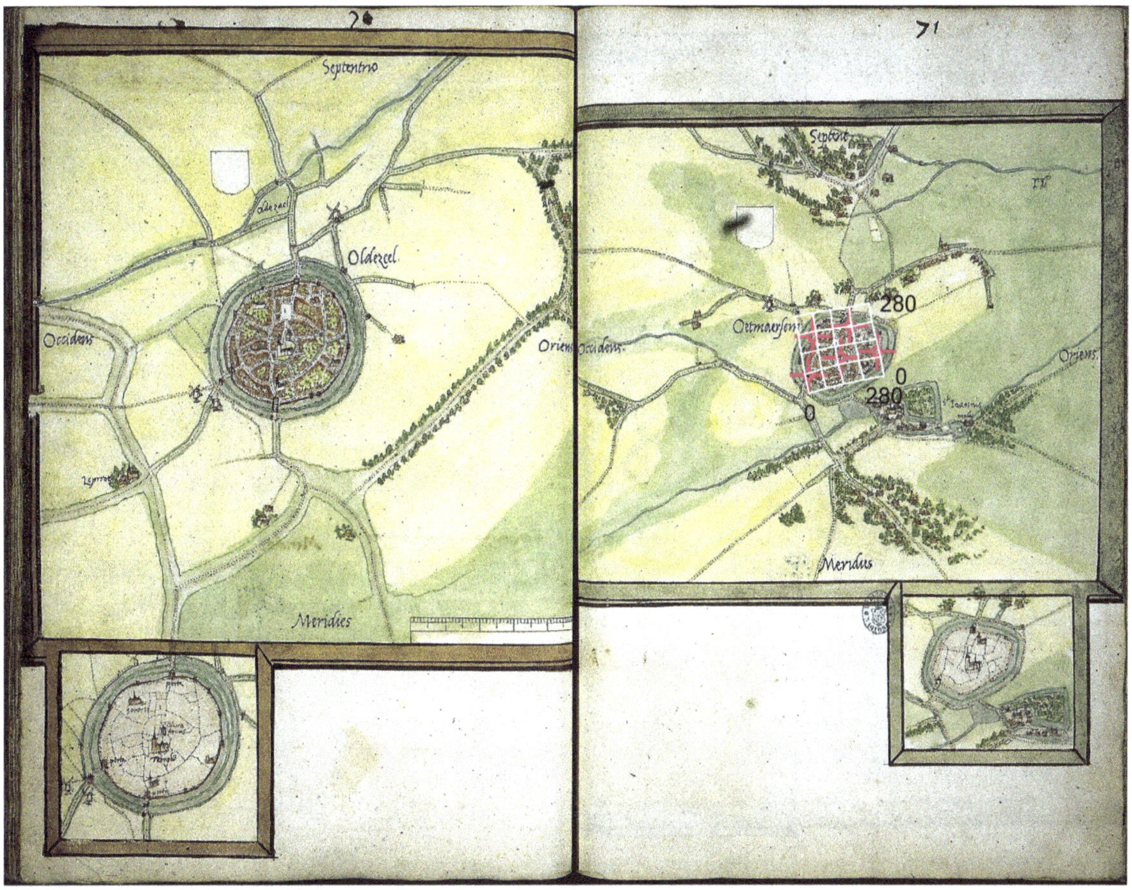
Septentrio
Oldezeel.
Occidens
Oriens
Meridies
Septent.
Ottmaersem
Occidens.
Oriens.
280
0
280
0
Meridus

164. Orchies

Tables 3.164 (measured and analysed).
Figures 3.164 (unmeasured, measured and analysed).

Measured							
Pixels	Scale stick original map (pixels)	Houses	Street sides	Houses per street side	Inhabitants (RF 4.0)	Inhabitants (RF 5.5)	Inhabitants (RF 7.0)
1747	1021	307	46	6.67	1227	1687	2147

Analysed	
June 29, 2023 11:35:39 AM	
Computation with 5 linked points (*italics* points are less reliable):	1. Église Notre-Dame 2. Rue de la Paterne 3. Rue Jules Roche 4. Rue Jules Ferry 5. Rue de l'Église
Reference points (Reliability 0–5)	5
Transformation with 3 parameters	
Scale	1:7410
Rotation	15 °
Mean position error	16m

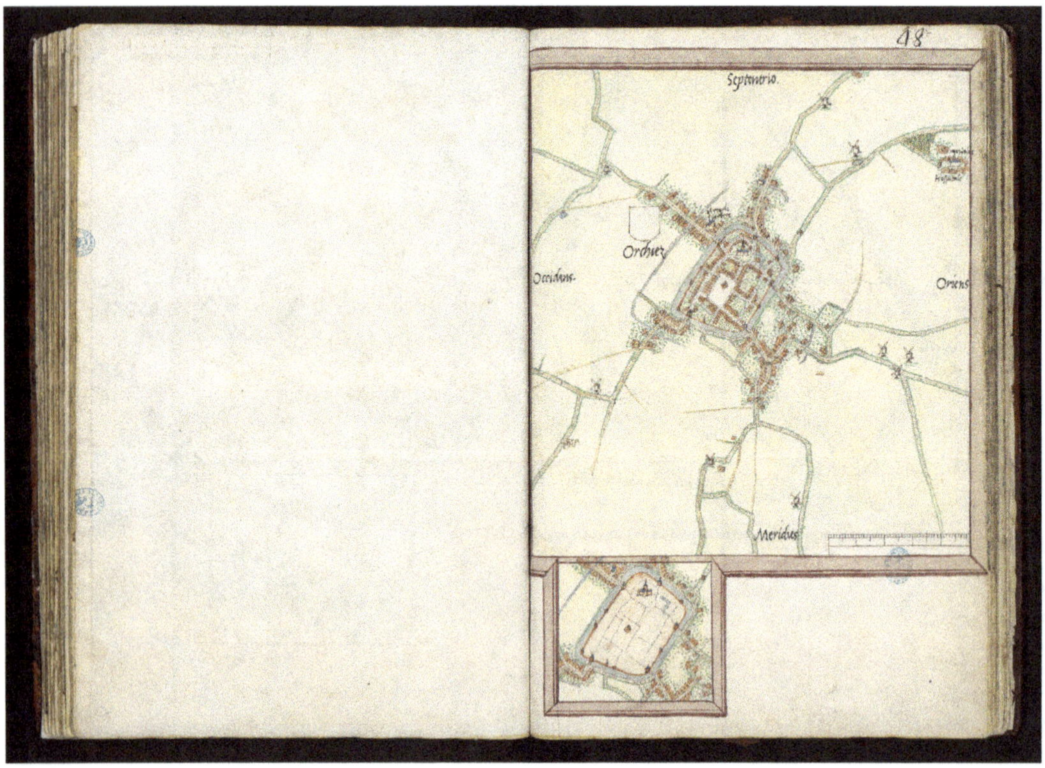

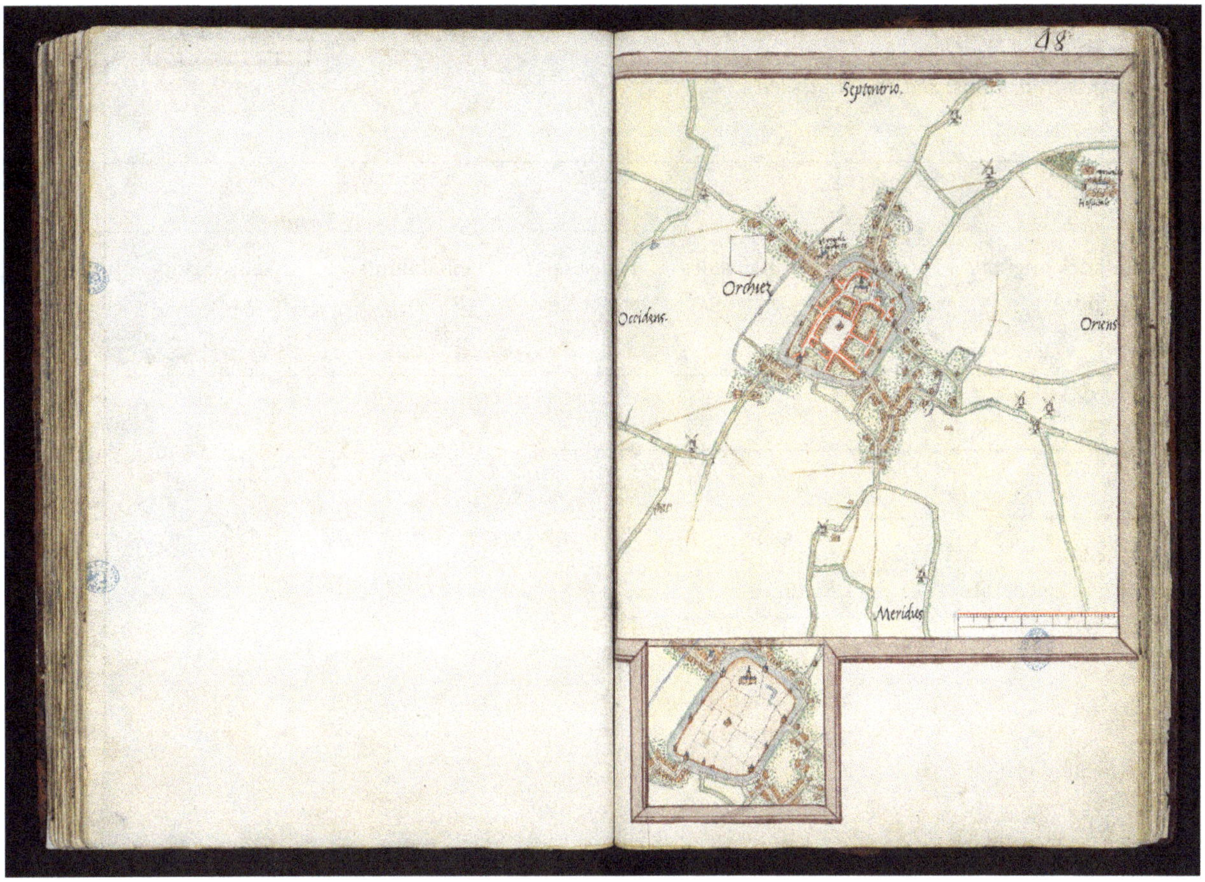
48
Septentrio.
Orchiez
Occidens.
Oriens
Meridies

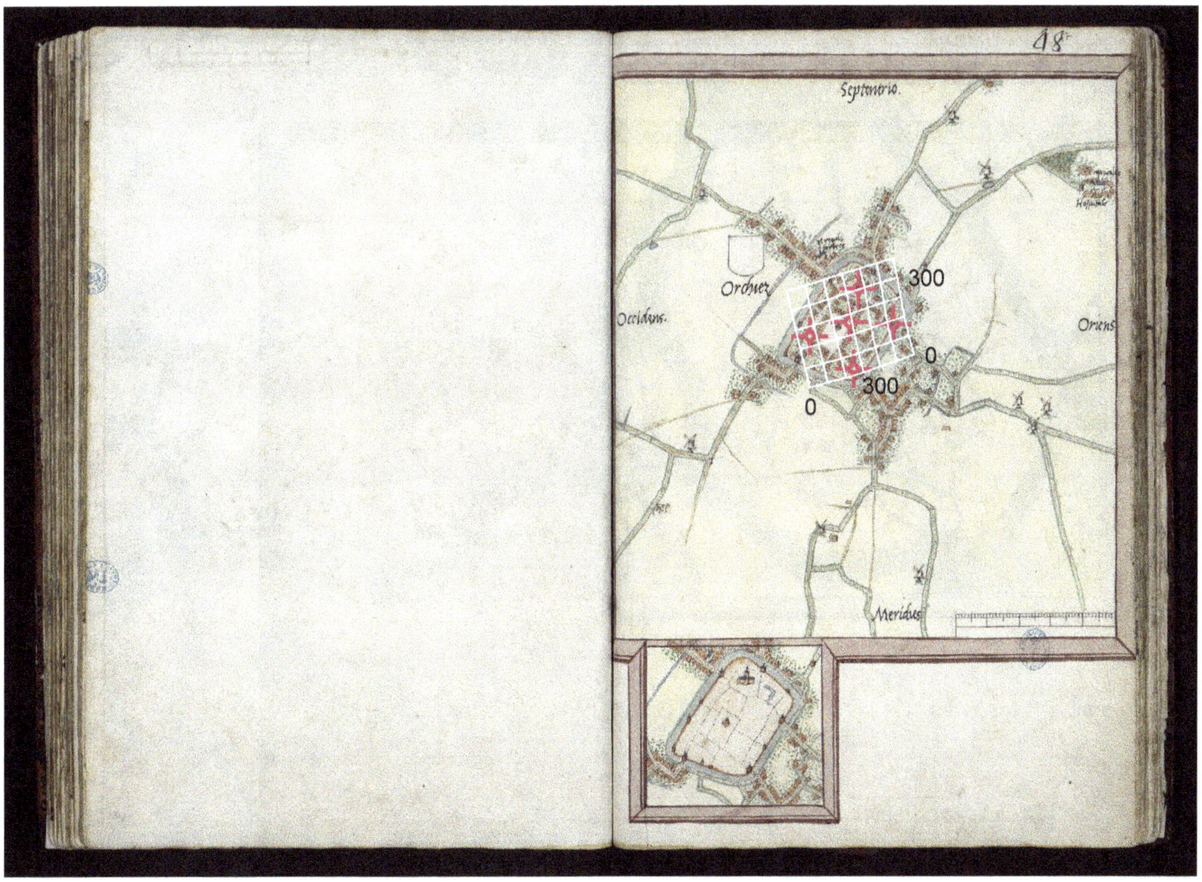
48
Septentrio.
Orchiez
Occidens.
Oriens
300
0
300
0
Meridies

165. Oudenaarde

Tables 3.165 (measured and analysed).
Figures 3.165 (unmeasured, measured and analysed).

Measured							
Pixels	Scale stick original map (pixels)	Houses	Street sides	Houses per street side	Inhabitants (RF 4.0)	Inhabitants (RF 5.5)	Inhabitants (RF 7.0)
5960	979	1047	128	8.18	4186	5756	7326

Analysed	
June 30, 2023 2:25:54 PM	
Computation with 5 linked points (*italics* points are less reliable):	1. Stadhuis 2. *Sint-Walburgakerk* 3. Onze-Lieve-Vrouwekerk 4. Begijnhof 5. Kattestraat
Reference points (Reliability 0–5)	4
Transformation with 3 parameters	
Scale	1:7670
Rotation	10 °
Mean position error	23m

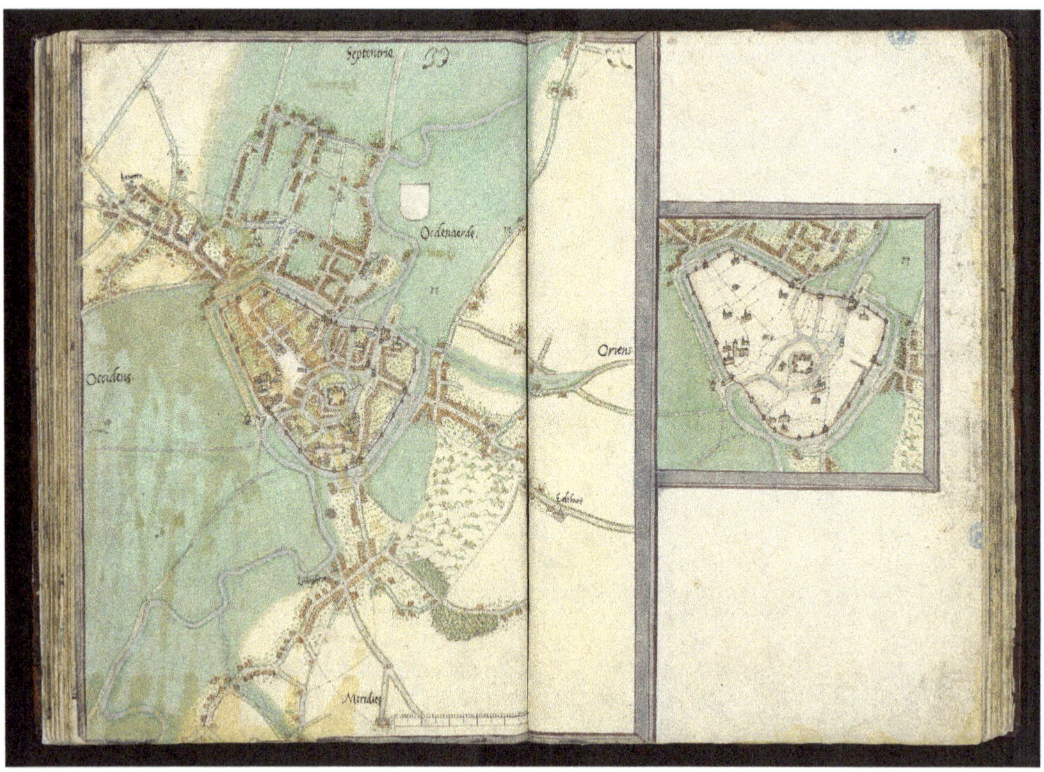

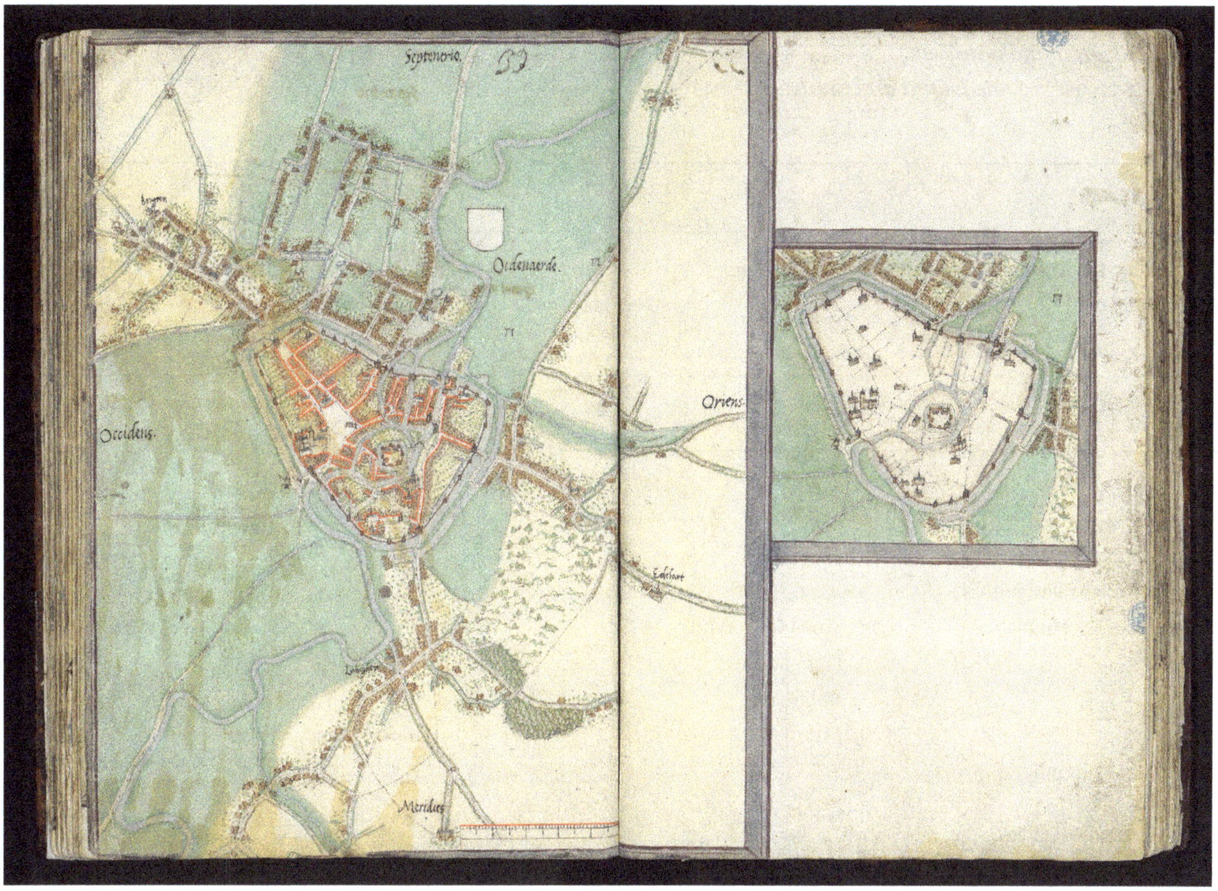
Septentrio.
Oudenaarde.
Occidens.
Oriens.
Meridies.

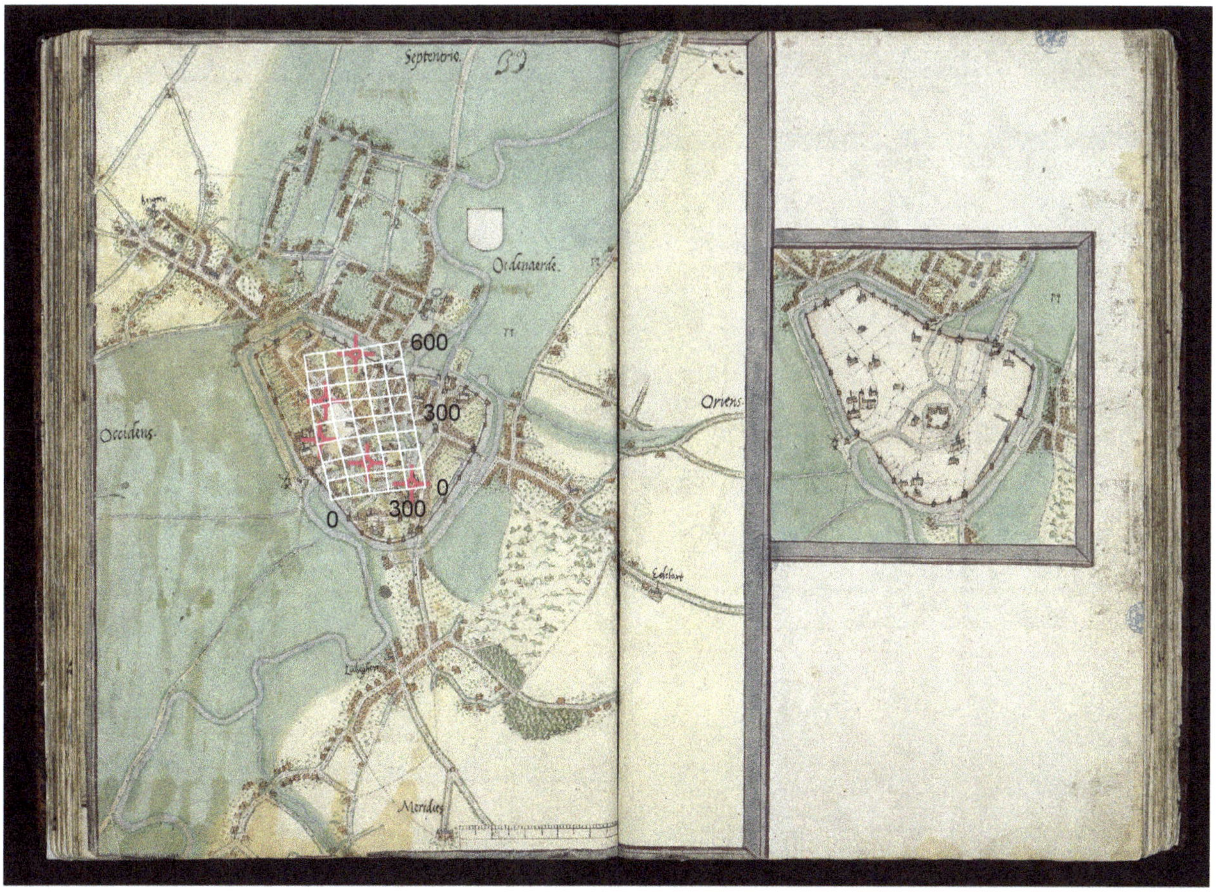
Septentrio.
Oudenaarde.
600
300
0
300
0
Occidens.
Oriens.
Meridies.

166. Oudenburg
Tables 3.166 (measured and analysed).
Figures 3.166 (unmeasured, measured and analysed).

Measured							
Pixels	Scale stick original map (pixels)	Houses	Street sides	Houses per street side	Inhabitants (RF 4.0)	Inhabitants (RF 5.5)	Inhabitants (RF 7.0)
2835	1021	498	55	9.05	1991	2738	3485

Analysed	
June 30, 2023 3:09:15 PM	
Computation with 5 linked points (*italics* points are less reliable):	1. Onze-Lieve-Vrouwekerk 2. Sint-Pietersabdij 3. Hoogstraat 4. Bekestraat 5. Mariastraat
Reference points (Reliability 0–5)	5
Transformation with 3 parameters	
Scale	1:8010
Rotation	5 °
Mean position error	8m

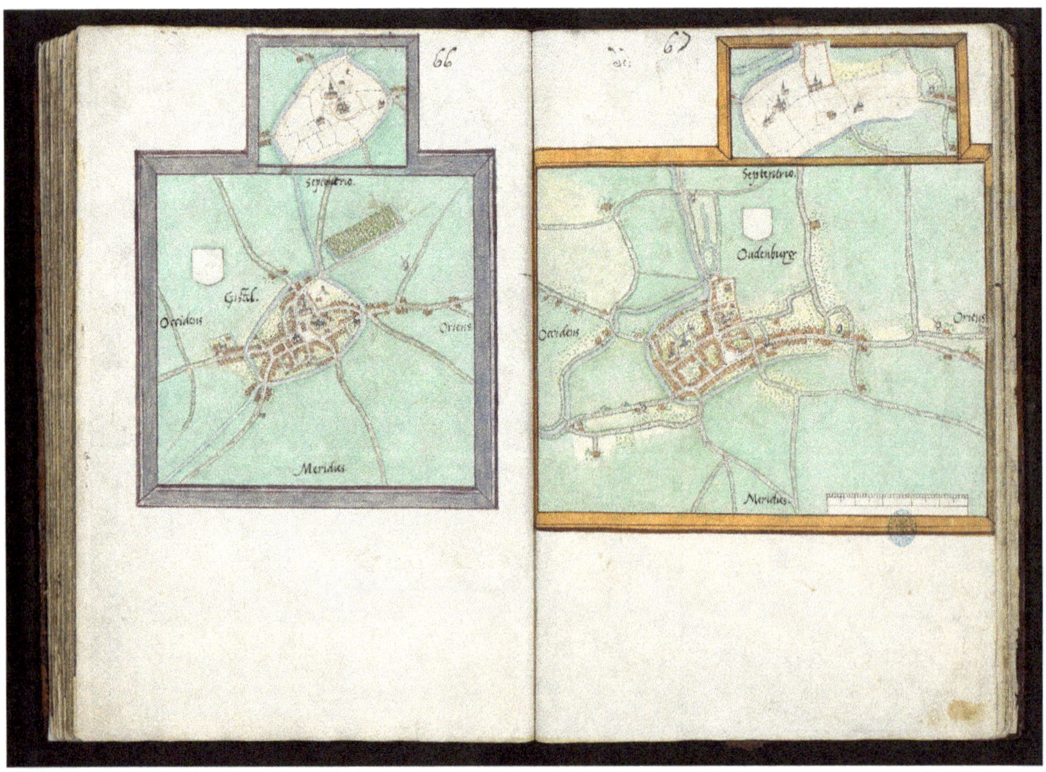

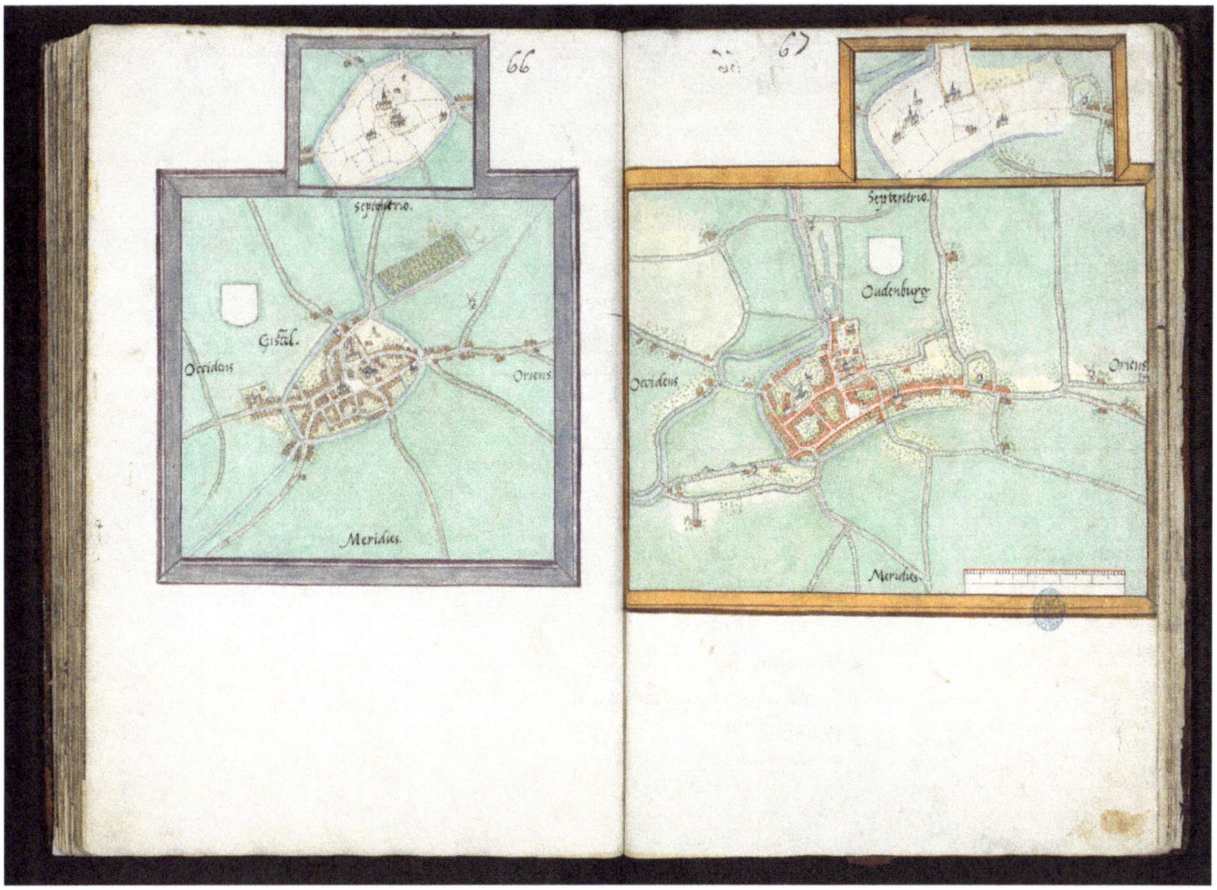
66
67
Septentrio.
Gistel.
Occidens
Oriens
Meridies.
Septentrio.
Oudenburg
Occidens
Oriens
Meridies.

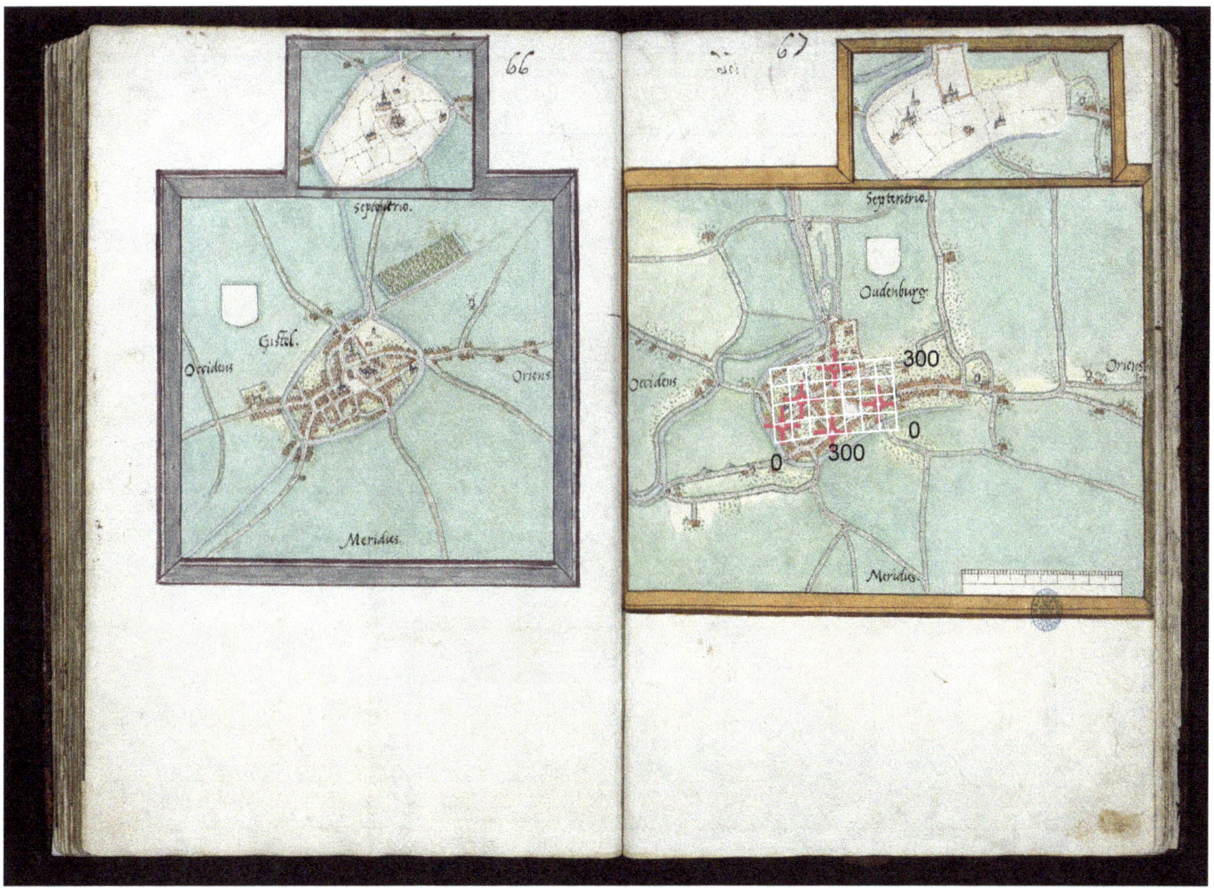
66
67
Septentrio.
Gistel.
Occidens
Oriens
Meridies.
Septentrio.
Oudenburg
300
Occidens
Oriens
0
0
300
Meridies.

167. Oudewater

Tables 3.167 (measured and analysed).
Figures 3.167 (unmeasured, measured and analysed).

Measured							
Pixels	Scale stick original map (pixels)	Houses	Street sides	Houses per street side	Inhabitants (RF 4.0)	Inhabitants (RF 5.5)	Inhabitants (RF 7.0)
3492	995	613	75	8.18	2453	3373	4293

Analysed	
October 15, 2023 11:16:04 PM	
Computation with 5 linked points (*italics* points are less reliable):	*1. Lange Lindschoten* *2. Kromme Haven* *3. Heksenwaag* 4. Grote of Sint-Michaëlskerk *5. Waardsedijk*
Reference points (Reliability 0–5)	1
Transformation with 3 parameters	
Scale	1:7690
Rotation	15 °
Mean position error	44m

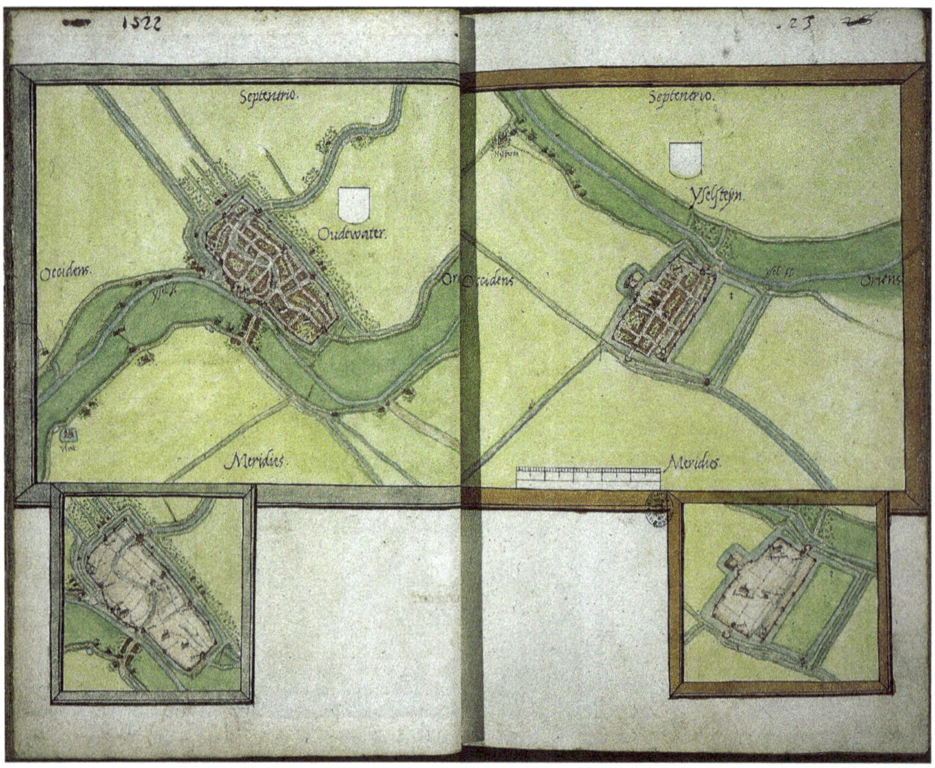

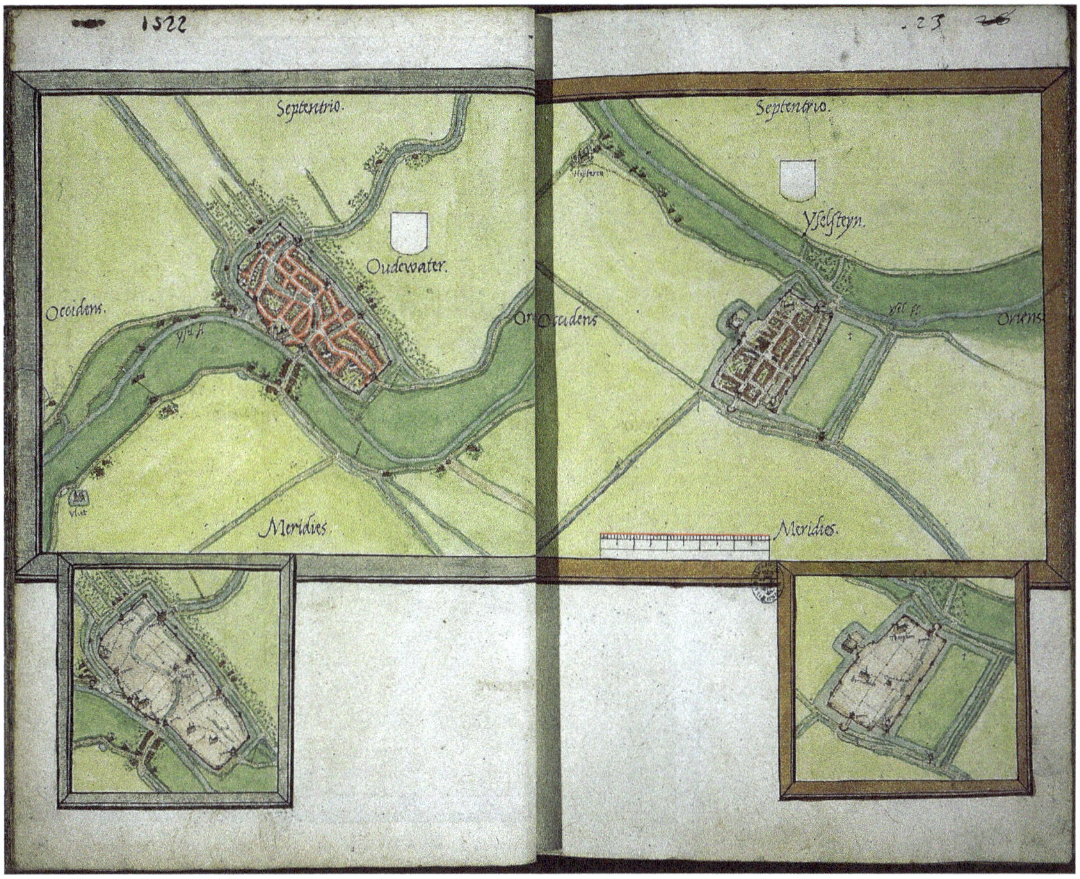
Septentrio.
Oudewater.
Occidens.
Meridies.
Septentrio.
Yselsteyn.
Occidens
Oriens
Meridies.

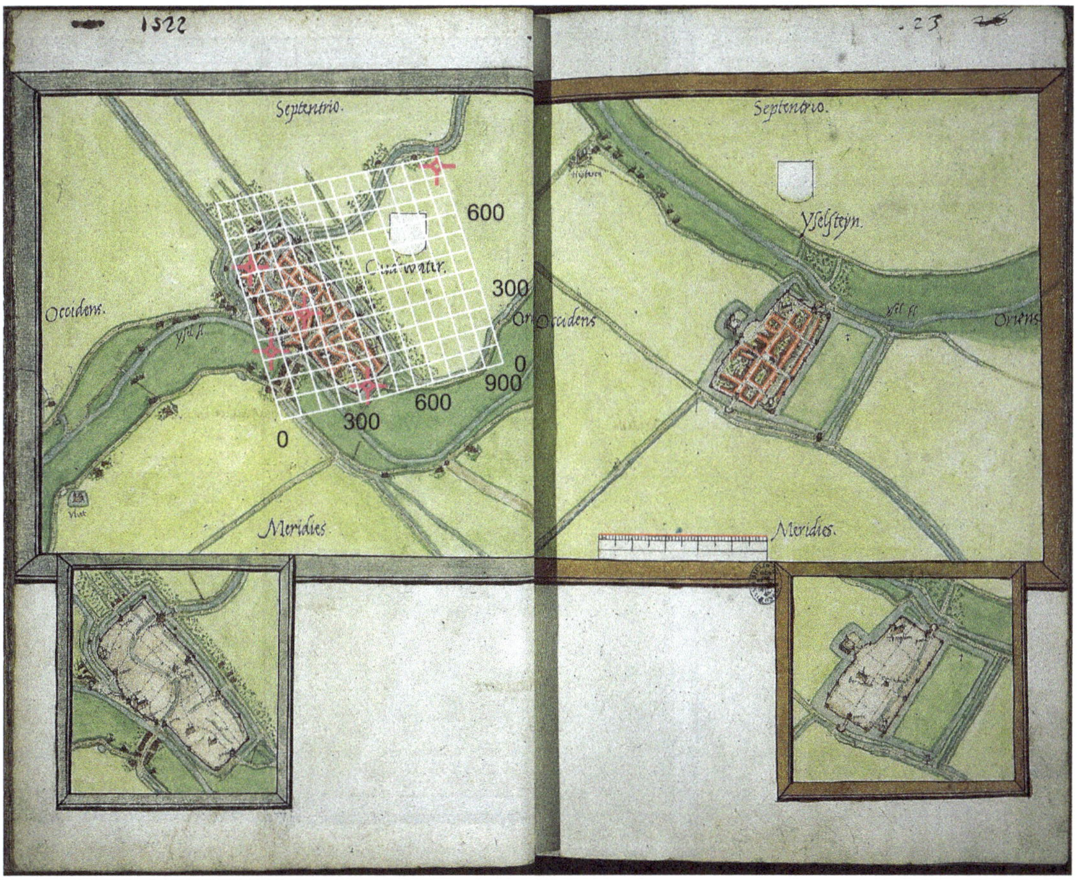
Septentrio.
600
300
0
900
600
300
0
Occidens.
Meridies.
Septentrio.
Yselsteyn.
Occidens
Oriens
Meridies.

168. Pernes

Tables 3.168 (measured and analysed).
Figures 3.168 (unmeasured, measured and analysed).

Measured							
Pixels	Scale stick original map (pixels)	Houses	Street sides	Houses per street side	Inhabitants (RF 4.0)	Inhabitants (RF 5.5)	Inhabitants (RF 7.0)
689	1015	121	23	5.26	484	665	846

Analysed	
July 1, 2023 6:49:37 PM	
Computation with 5 linked points (*italics* points are less reliable):	1. Église Saint-Pierre 2. Rue du Chateau 3. Place du Récolets 4. Rue de l'Église 5. Chateau
Reference points (Reliability 0–5)	5
Transformation with 3 parameters	
Scale	1:7120
Rotation	8 °
Mean position error	18m

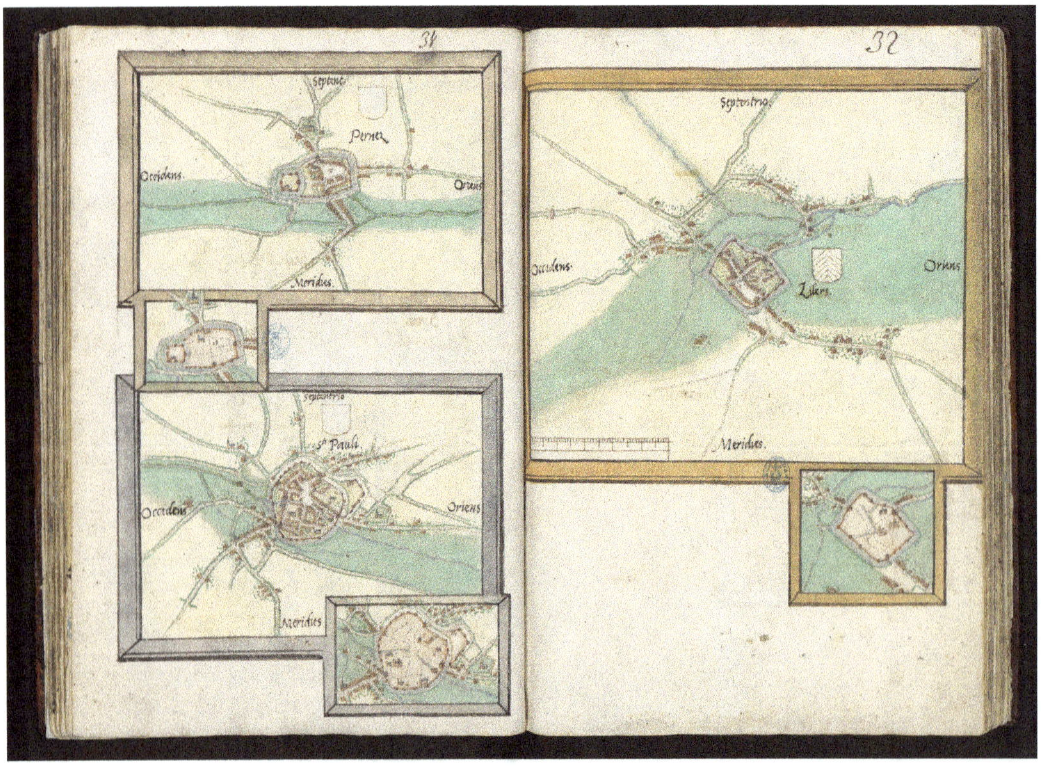

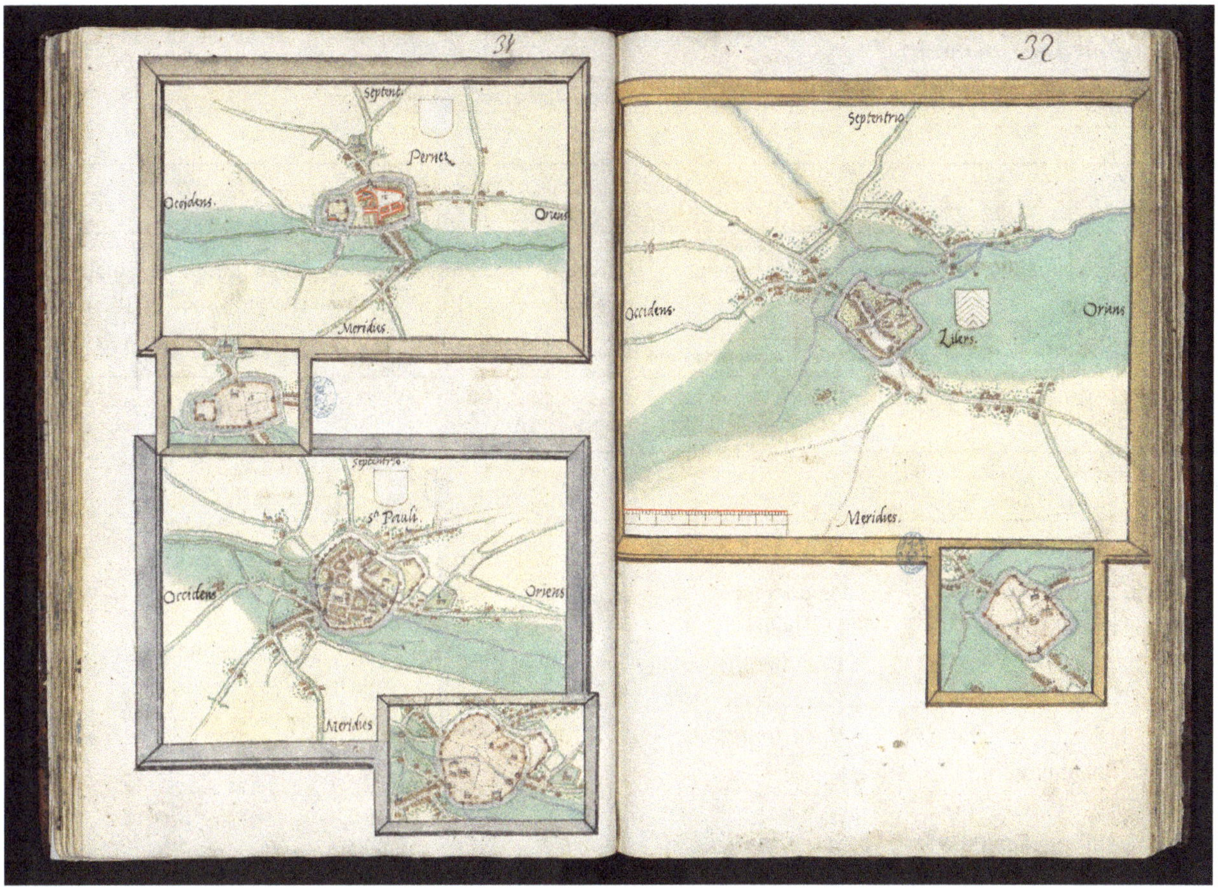
31
32
Septent.
Pernez
Occidens.
Oriens
Meridies.
Septentrio
St Pauli
Occidens
Oriens
Meridies
Septentrio
Occidens.
Liliers.
Oriens
Meridies.

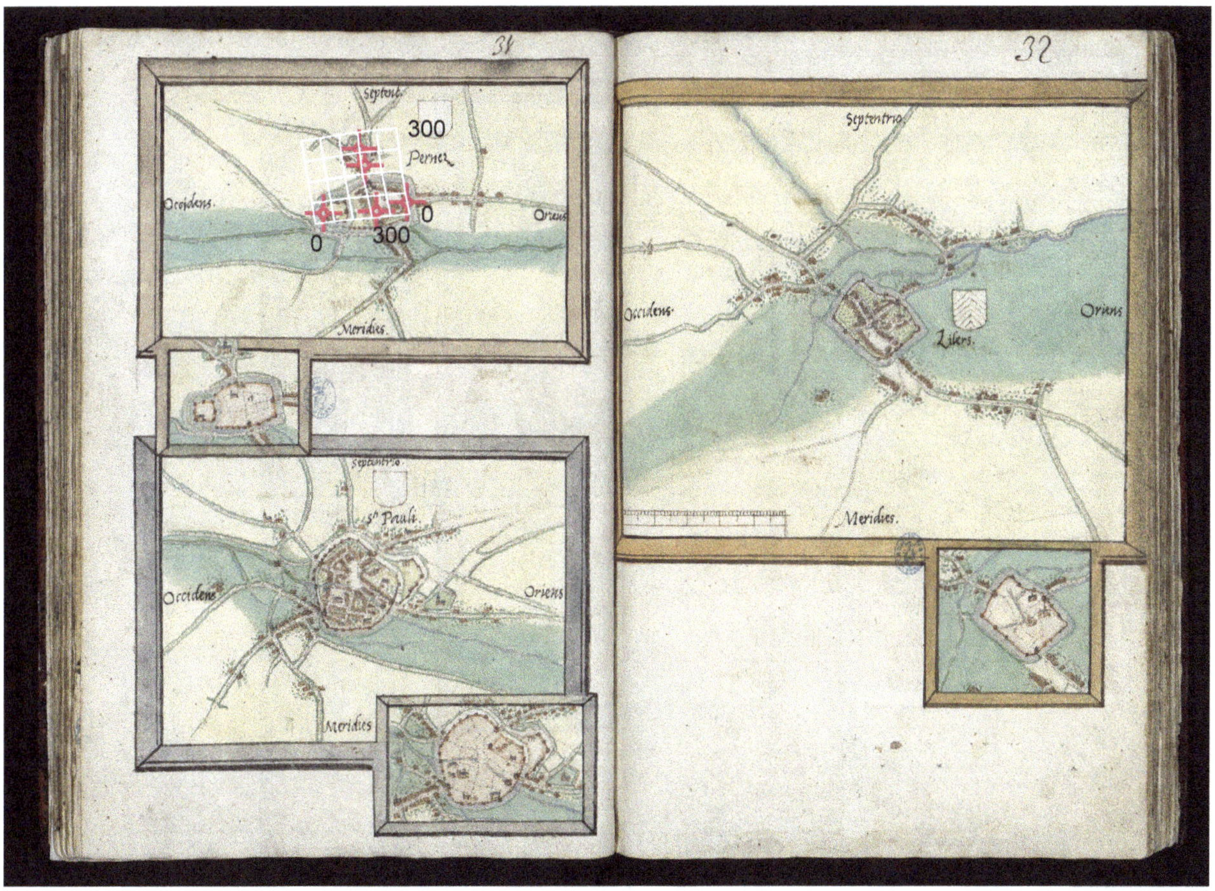
31
32
Septent.
300
Pernez
Occidens.
0
Oriens
0
300
Meridies.
Septentrio
St Pauli
Occidens
Oriens
Meridies
Septentrio
Occidens.
Liliers.
Oriens
Meridies.

169. Poperinge

Tables 3.169 (measured and analysed).
Figures 3.169 (unmeasured, measured and analysed).

Measured							
Pixels	Scale stick original map (pixels)	Houses	Street sides	Houses per street side	Inhabitants (RF 4.0)	Inhabitants (RF 5.5)	Inhabitants (RF 7.0)
4198	1021	737	87	8.47	2948	4054	5160

Analysed	
July 5, 2023 10:42:28 AM	
Computation with 5 linked points (*italics* points are less reliable):	1. Peperstraat 2. De Halle 3. Benedictijnenstraat 4. Casselstraat *5. Hondstraat*
Reference points (Reliability 0–5)	4
Transformation with 3 parameters	
Scale	1:8150
Rotation	11 °
Mean position error	27m

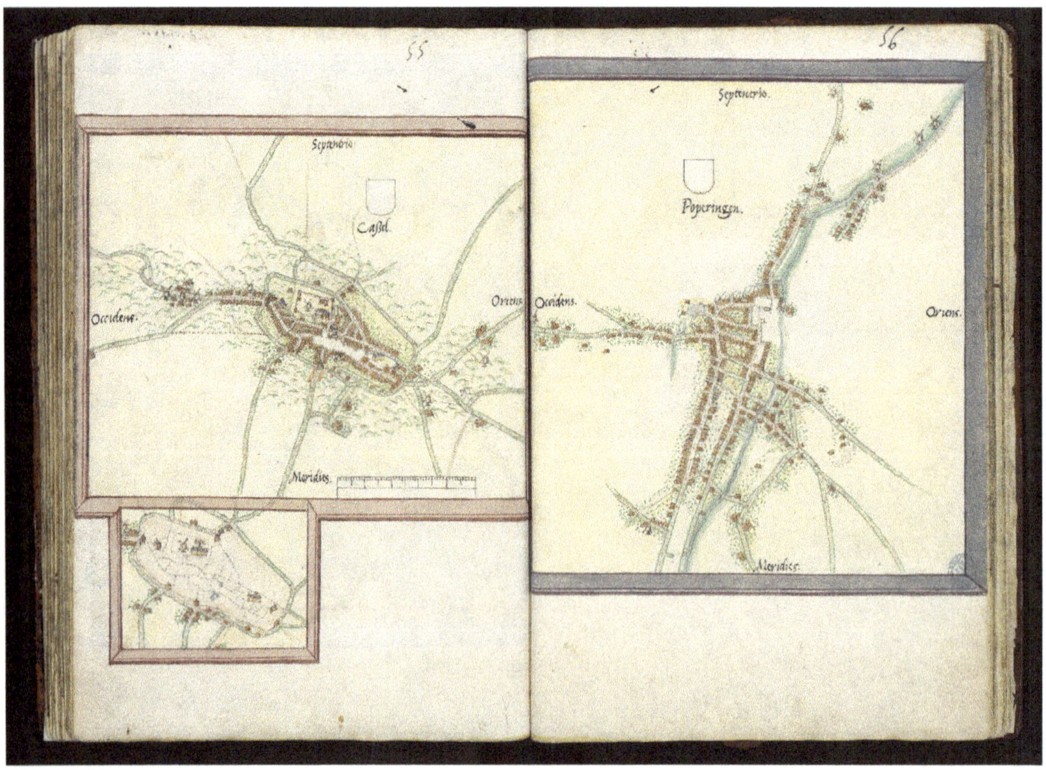

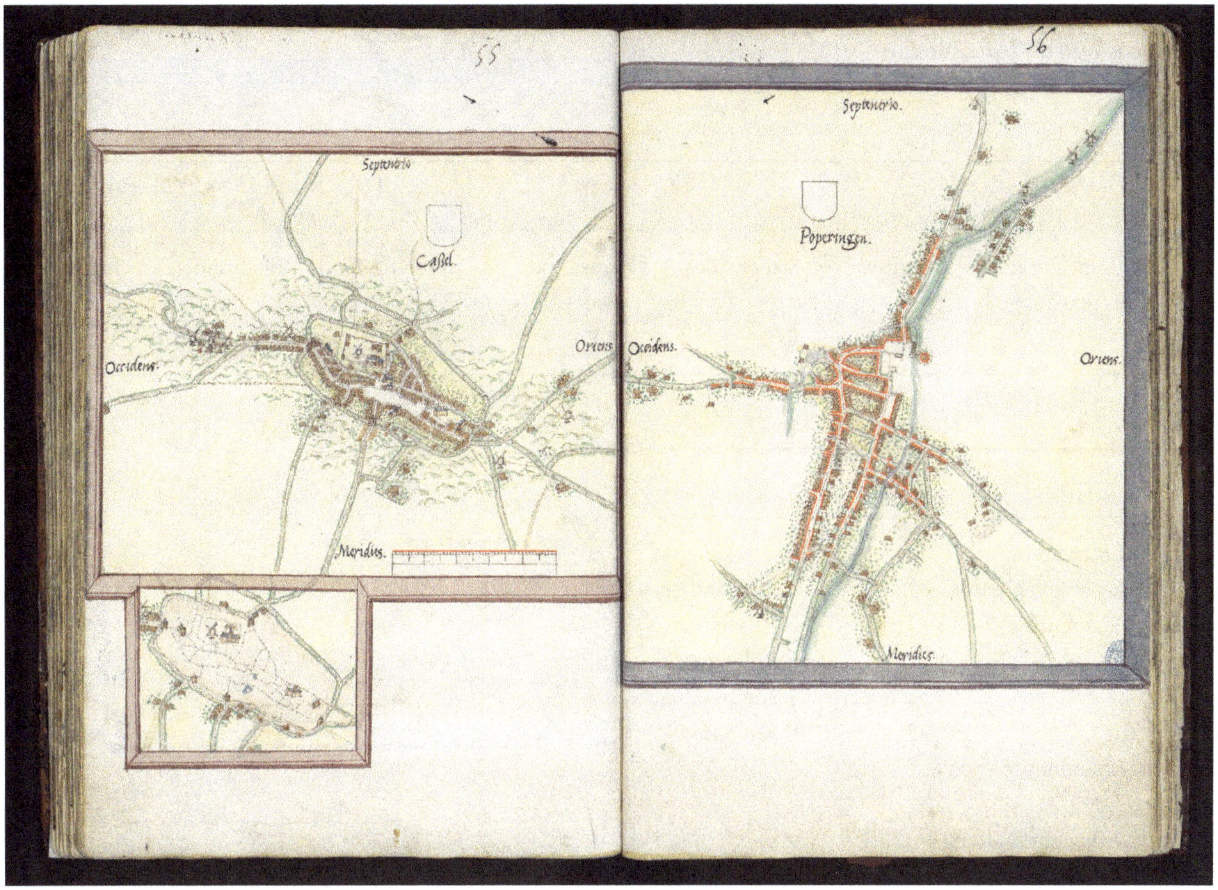
55
56
Septentrio
Cassel.
Occidens.
Oriens.
Meridies.
Septentrio.
Poperingen.
Occidens.
Oriens.
Meridies.

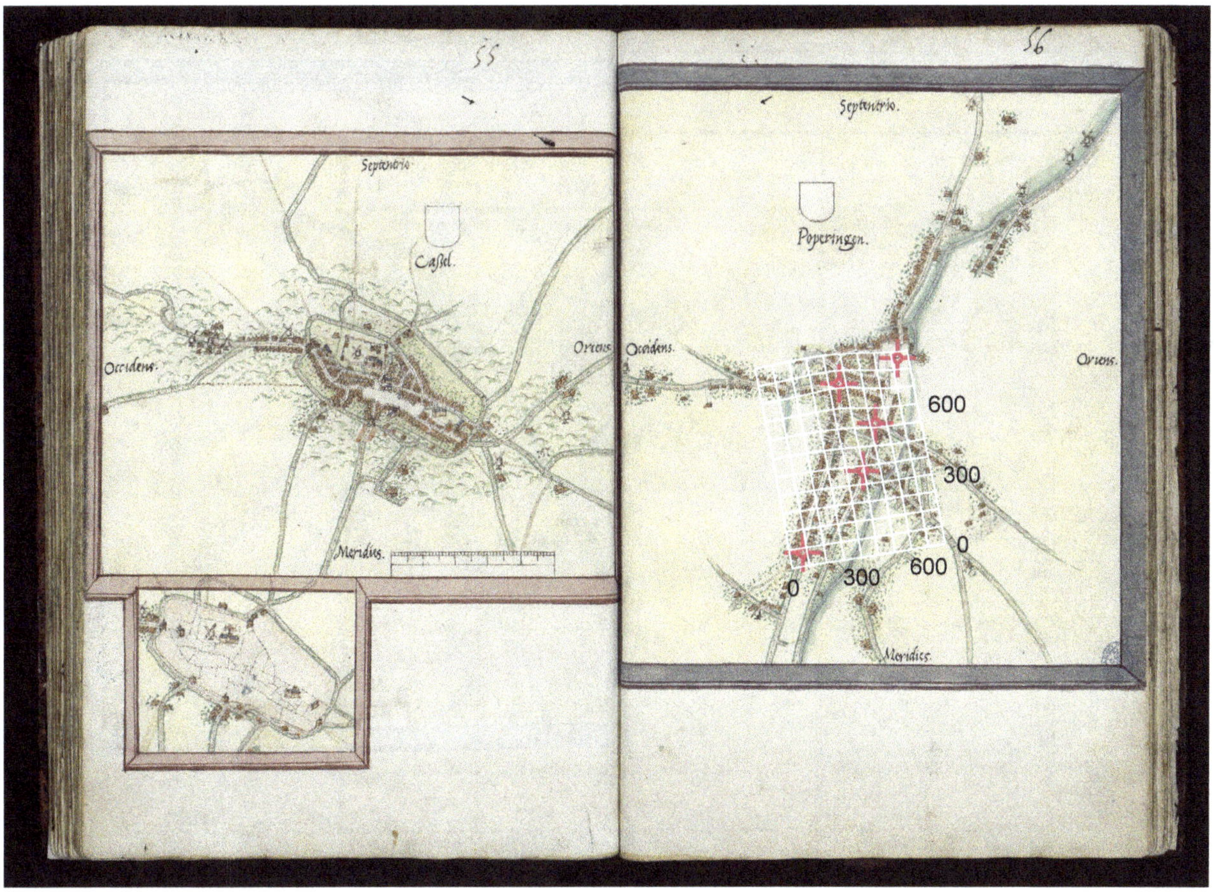
55
56
Septentrio
Cassel.
Occidens.
Oriens.
Meridies.
Septentrio.
Poperingen.
Occidens.
Oriens.
600
300
0
0
300
600
Meridies.

170. Purmerend

Tables 3.170 (measured and analysed).
Figures 3.170 (unmeasured, measured and analysed).

Measured							
Pixels	Scale stick original map (pixels)	Houses	Street sides	Houses per street side	Inhabitants (RF 4.0)	Inhabitants (RF 5.5)	Inhabitants (RF 7.0)
1897	997	333	61	5.46	1332	1832	2332

Analysed	
October 12, 2023 10:54:50 AM	
Computation with 5 linked points (*italics* points are less reliable):	1. Nicolaaskerk 2. Peperstraat 3. Kolkstraat 4. Nieuwsstraat 5. Koemarkt
Reference points (Reliability 0–5)	5
Transformation with 3 parameters	
Scale	1:9190
Rotation	11 °
Mean position error	7m

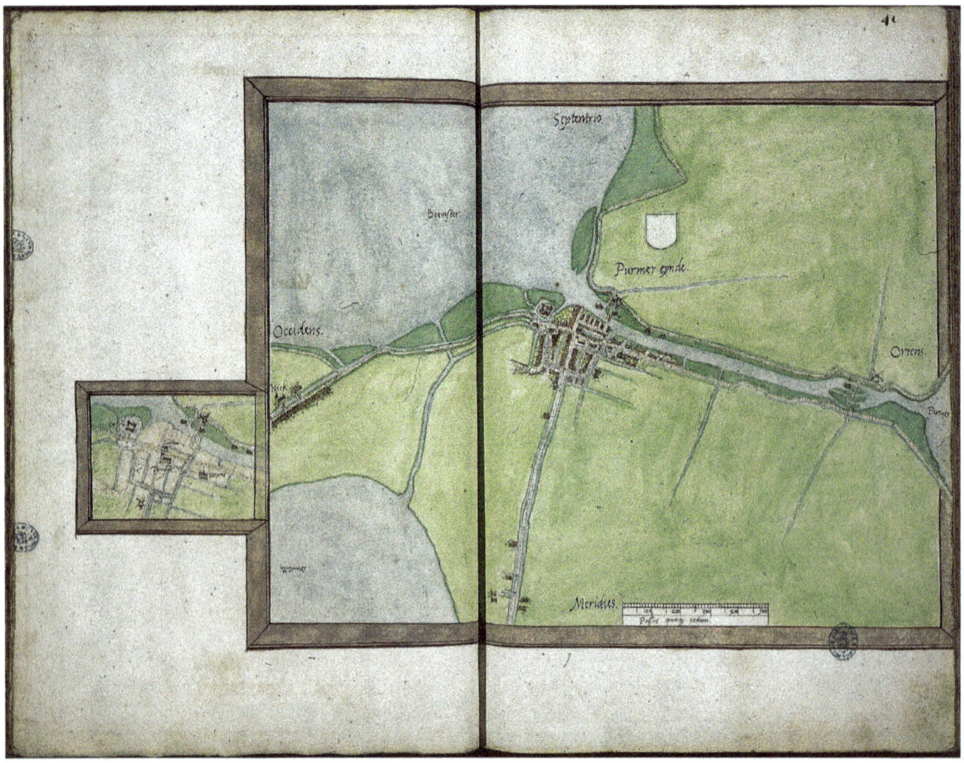

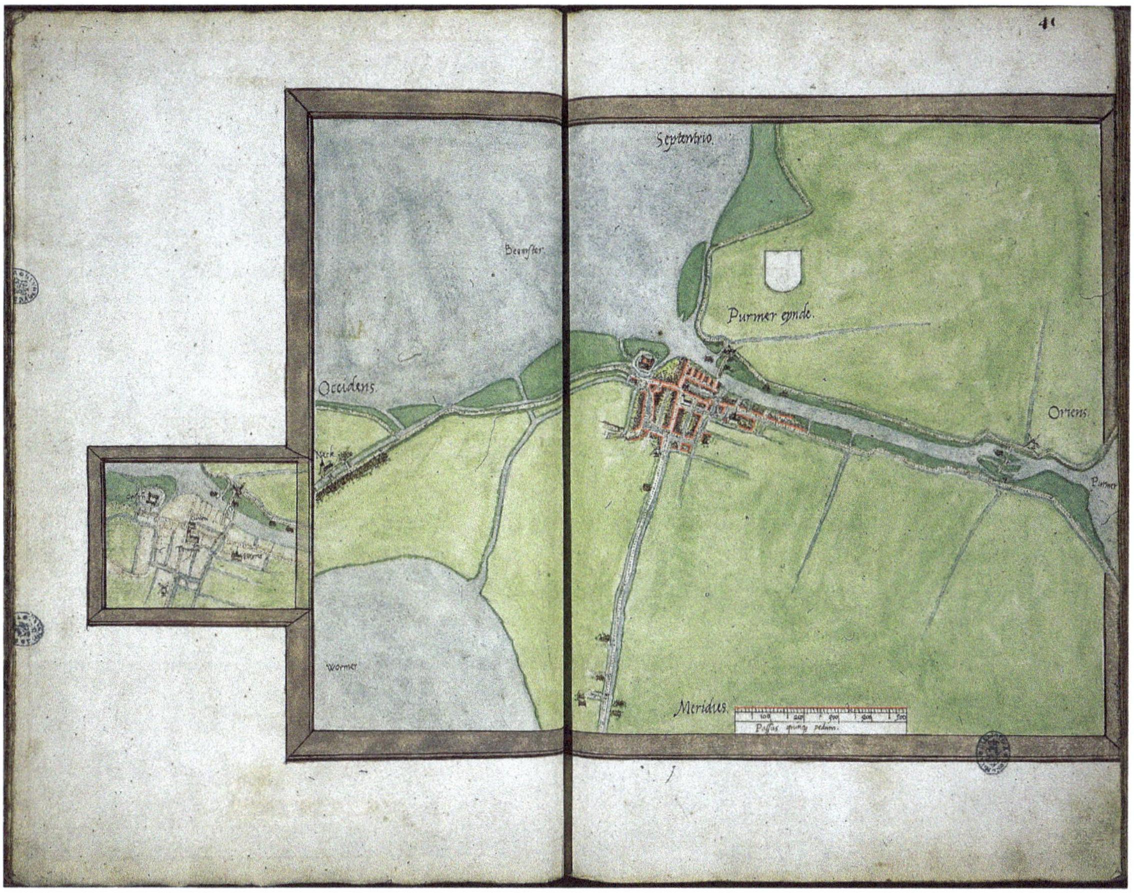
Septentrio.
Purmer eynde.
Occidens.
Oriens.
Meridies.

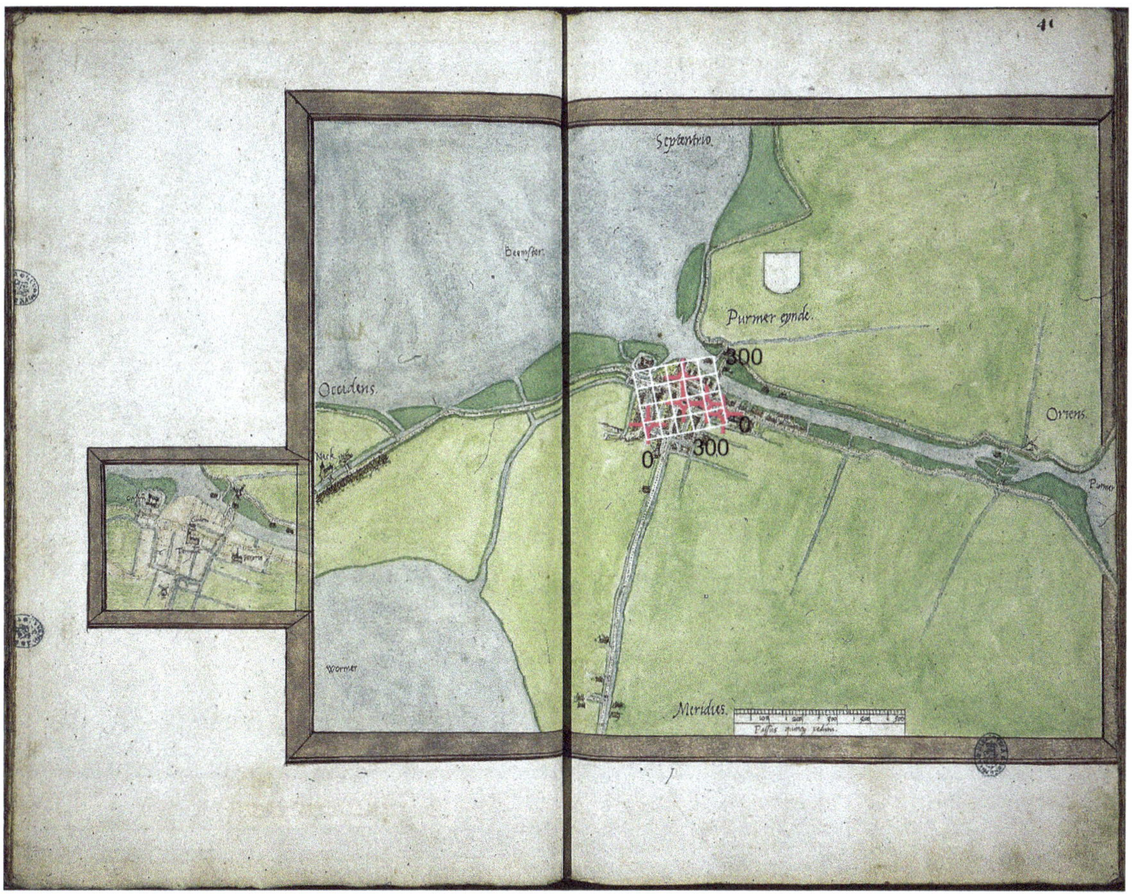
Septentrio.
Purmer eynde.
300
Occidens.
0
0
300
Oriens.
Meridies.

171. Reimerswaal

Table 3.171 (measured and analysed).
Figures 3.171 (unmeasured and measured).

Measured							
Pixels	Scale stick original map (pixels)	Houses	Street sides	Houses per street side	Inhabitants (RF 4.0)	Inhabitants (RF 5.5)	Inhabitants (RF 7.0)
2617	1010	460	65	7.07	1838	2527	3216

The historic city of Reimerswaal, also known as Rommerswael or Reymerswael, was located on South Beveland on the south bank of the Oosterschelde, more than four kilometers southwest of the city of Tholen. During the late Middle Ages it was the third largest city in Zeeland after Middelburg and Zierikzee. Reimerswaal had to deal with continuous flooding from the end of the 16th century. The last inhabitants left in the middle of the 17th century. Since the beginning of the 18th century, Reimerswaal has been a completely drowned city.

Septentrio.
Occidens
Meridies

172. Rhenen
Tables 3.172 (measured and analysed).
Figures 3.172 (unmeasured, measured and analysed).

Measured							
Pixels	Scale stick original map (pixels)	Houses	Street sides	Houses per street side	Inhabitants (RF 4.0)	Inhabitants (RF 5.5)	Inhabitants (RF 7.0)
2667	1002	468	76	6.16	1873	2576	3279

Analysed	
July 9, 2023 11:05:59 PM	
Computation with 5 linked points (*italics* points are less reliable):	1. Markt 2. Keldermanspad 3. Kloosterstraat 4. Molenstraat 5. Koningstraat
Reference points (Reliability 0–5)	5
Transformation with 3 parameters	
Scale	1:8430
Rotation	15 °
Mean position error	15m

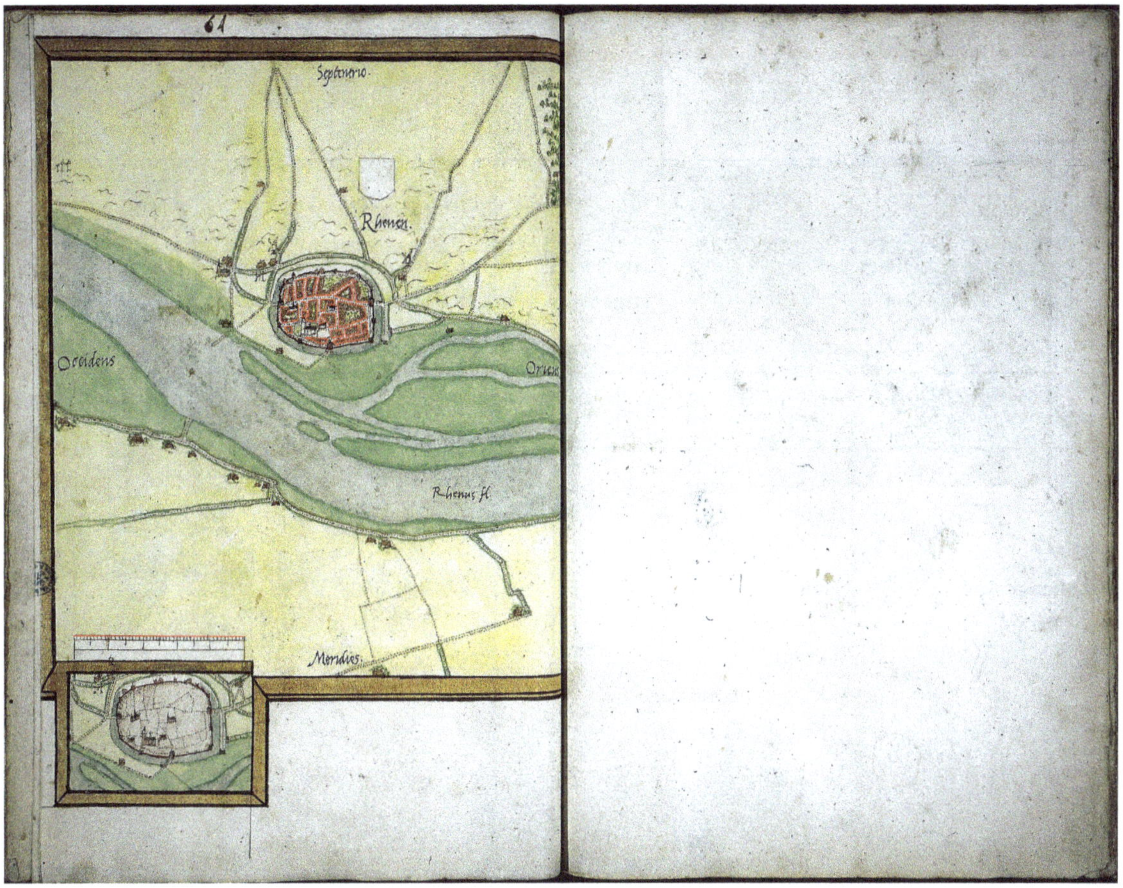
Septentrio.
Rhenen.
Occidens
Rhenus fl.
Meridies.

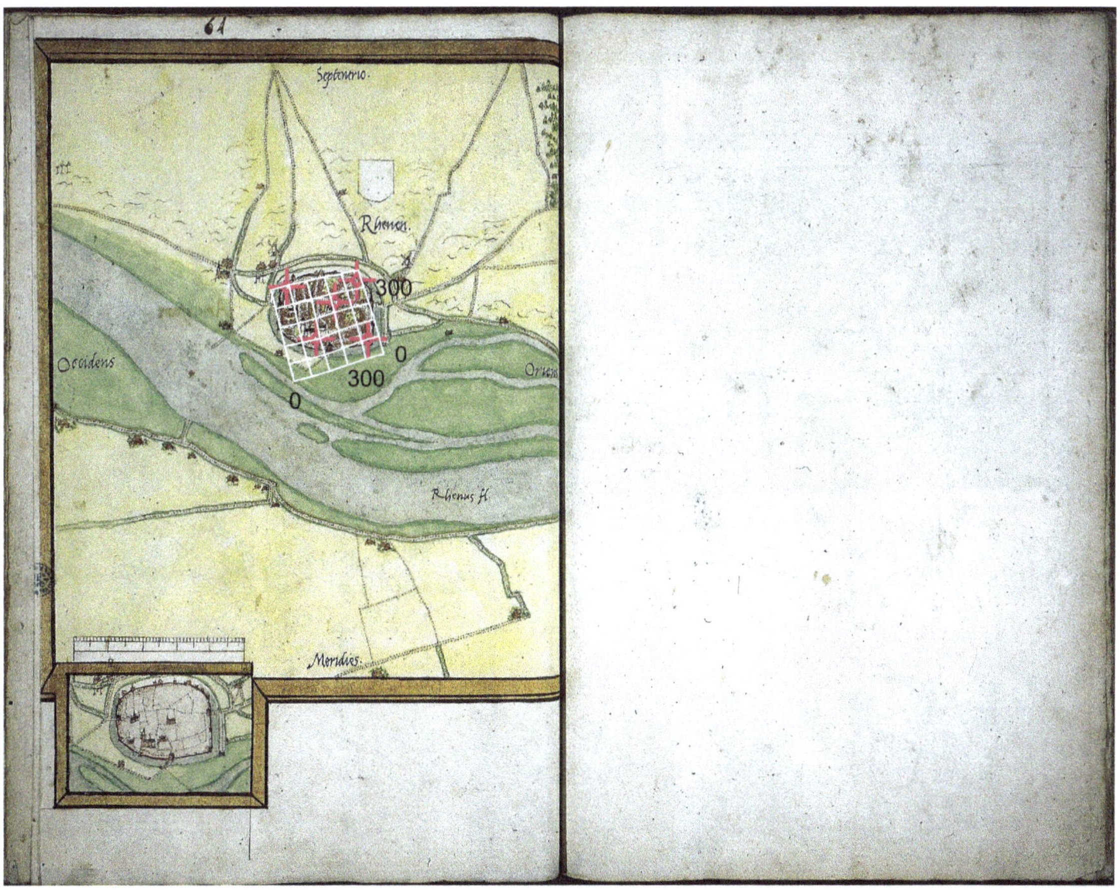
Septentrio.
Rhenen.
300
0
300
0
Occidens
Rhenus fl.
Meridies.

173. Roermond

Tables 3.173 (measured and analysed).
Figures 3.173 (unmeasured, measured and analysed).

Measured							
Pixels	Scale stick original map (pixels)	Houses	Street sides	Houses per street side	Inhabitants (RF 4.0)	Inhabitants (RF 5.5)	Inhabitants (RF 7.0)
8339	missing	1464	161	9.10	5857	8054	10,251

Analysed	
July 11, 2023 10:26:36 AM	
Computation with 5 linked points (*italics* points are less reliable):	1. Roer 2. Munsterkerk 3. Minderbroederskerk 4. Kartuizerklooster 5. Kruisherenstraat
Reference points (Reliability 0–5)	5
Transformation with 3 parameters	
Scale	1:8670
Rotation	14 °
Mean position error	14m

95
Septentrio.
Mosa fl.
Ruermond
Occidens.
Oriens.

95
Septentrio.
Mosa fl.
Ruermond
Occidens.
Oriens.
600
300
0
300
600
0

174. Rotterdam

Tables 3.174 (measured and analysed).
Figures 3.174 (unmeasured, measured and analysed).

Measured							
Pixels	Scale stick original map (pixels)	Houses	Street sides	Houses per street side	Inhabitants (RF 4.0)	Inhabitants (RF 5.5)	Inhabitants (RF 7.0)
7291	1016	1280	227	5.64	5121	7042	8963

Analysed	
July 12, 2023 10:40:28 AM	
Computation with 5 linked points (*italics* points are less reliable):	1. Sint-Laurenskerk 2. *Boompjeskade* 3. Blaak 4. Oostplein 5. *Binnenrotte*
Reference points (Reliability 0–5)	3
Transformation with 3 parameters	
Scale	1:8620
Rotation	11 °
Mean position error	32m

Rotterdam.
Occidens.
Meridies.

Rotterdam.
Occidens.
Meridies.
600
300
0
900
600
300
0

175. Saint-Ghislain
Tables 3.175 (measured and analysed).
Figures 3.175 (unmeasured, measured and analysed).

Measured							
Pixels	Scale stick original map (pixels)	Houses	Street sides	Houses per street side	Inhabitants (RF 4.0)	Inhabitants (RF 5.5)	Inhabitants (RF 7.0)
1421	1029	250	40	6.24	998	1372	1746

Analysed	
October 11, 2023 9:33:02 AM	
Computation with 5 linked points (*italics* points are less reliable):	1. Église Saint-Géry 2. Grande Place 3. Rue d'Ath *4. Rue du Moulin (ouest)* *5. Rue du Moulin (est)*
Reference points (Reliability 1–5)	3
Transformation with 3 parameters	
Scale	1:8.130
Rotation	9 °
Mean Position Error	24m

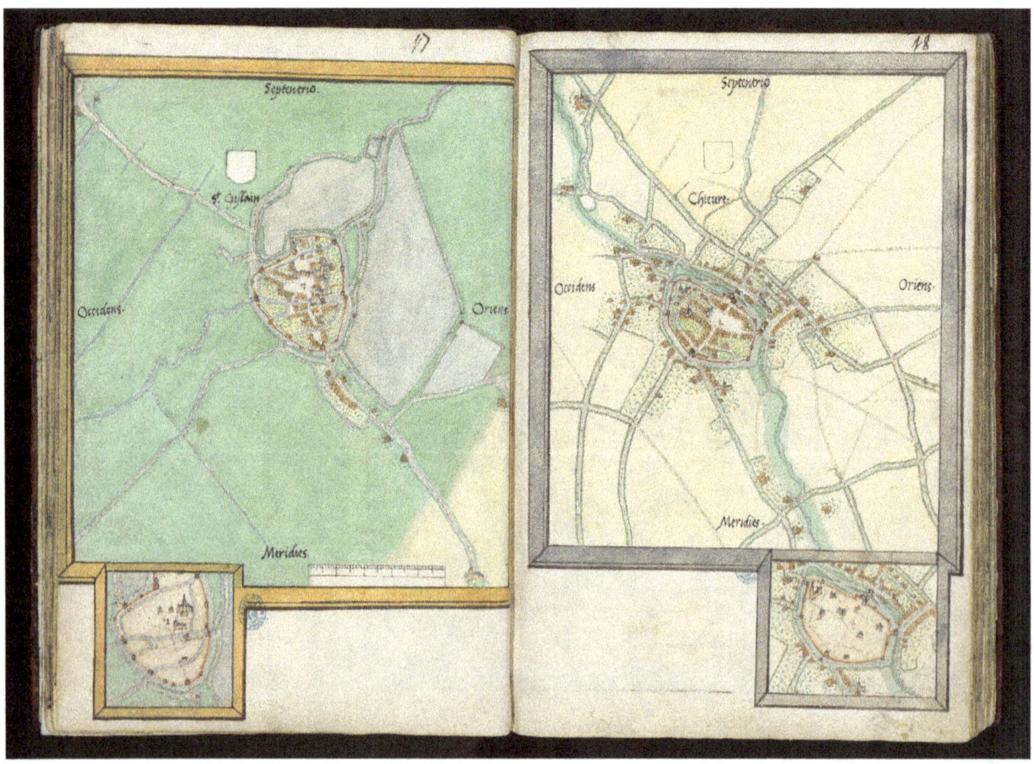

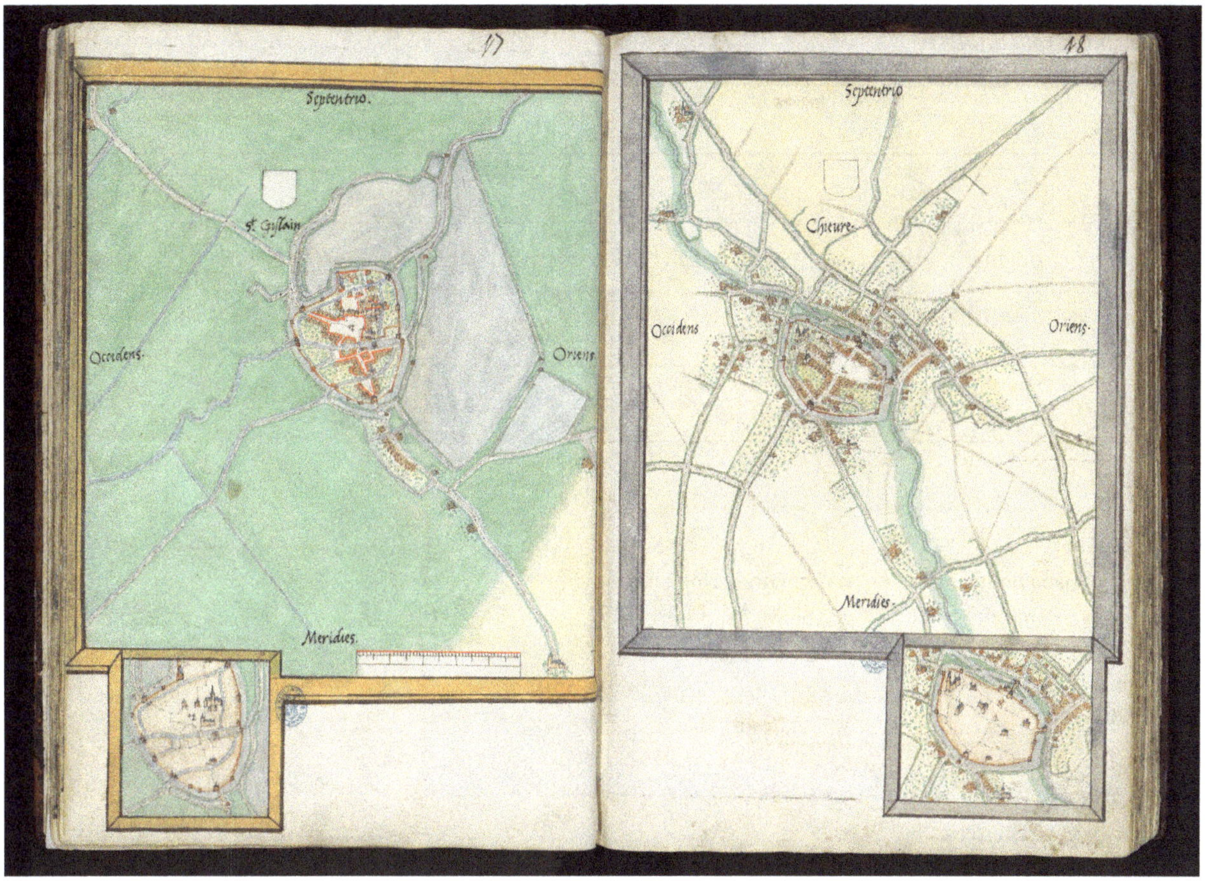
47
48
Septentrio.
St. Gyslain
Occidens.
Oriens
Meridies.
Septentrio
Chieure.
Occidens
Oriens.
Meridies.

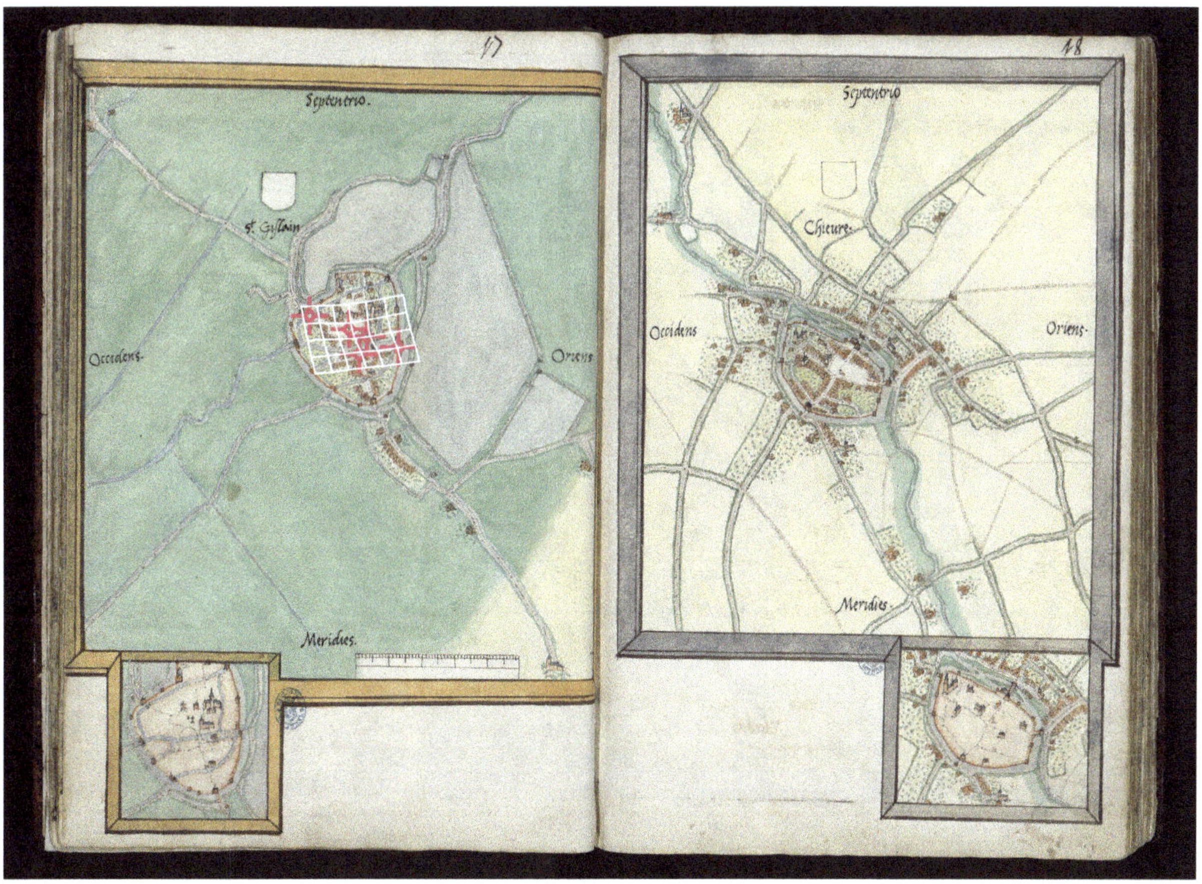
47
48
Septentrio.
St. Gyslain
Occidens.
Oriens
Meridies.
Septentrio
Chieure.
Occidens
Oriens.
Meridies.

176. Saint-Omer

Tables 3.176 (measured and analysed).
Figures 3.176 (unmeasured, measured and analysed).

Measured

Pixels	Scale stick original map (pixels)	Houses	Street sides	Houses per street side	Inhabitants (RF 4.0)	Inhabitants (RF 5.5)	Inhabitants (RF 7.0)
17,216	884	3023	392	7.71	12,092	16,627	21,162

Analysed

July 13, 2023 9:21:08 AM	
Computation with 5 linked points (*italics* points are less reliable):	1. Cathédrale Notre-Dame 2. Abbaye Saint-Bertin 3. Église Saint-Denis 4. Église Saint-Sépulcre 5. Rue Courteville
Reference points (Reliability 0–5)	5
Transformation with 3 parameters	
Scale	1:9200
Rotation	7 °
Mean position error	15m

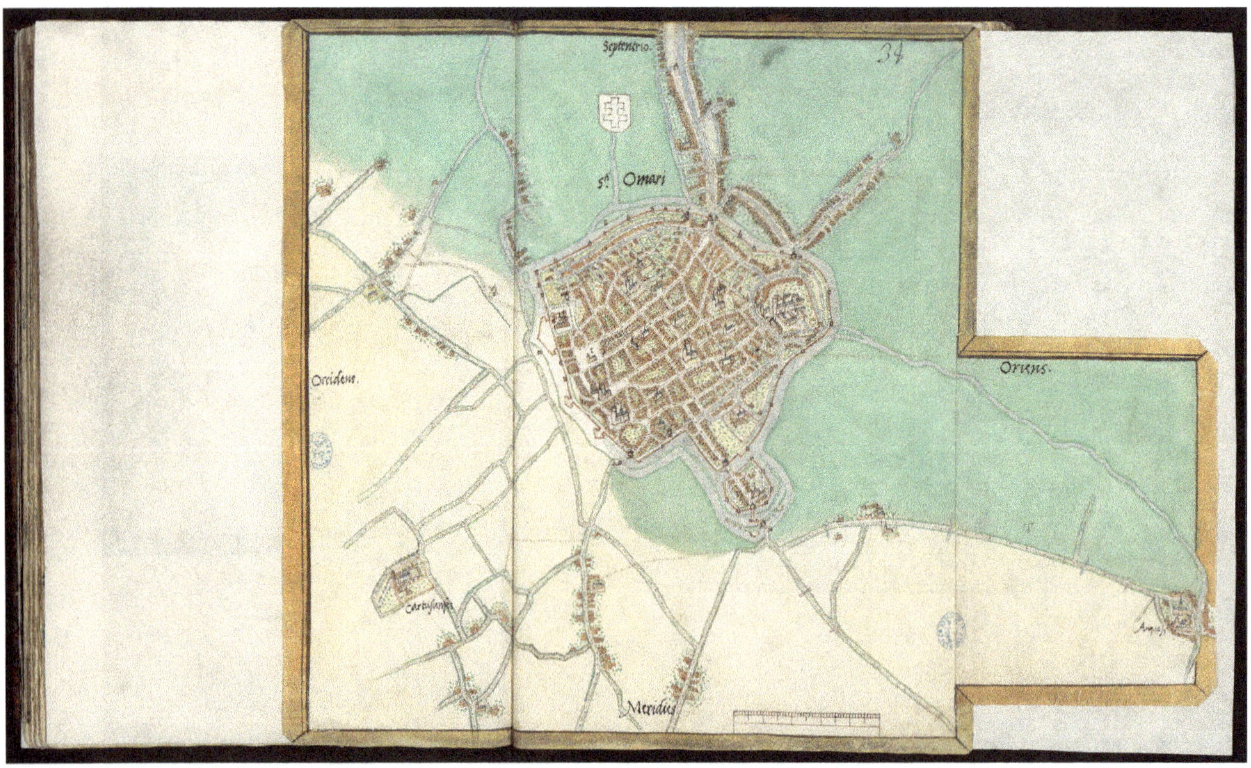

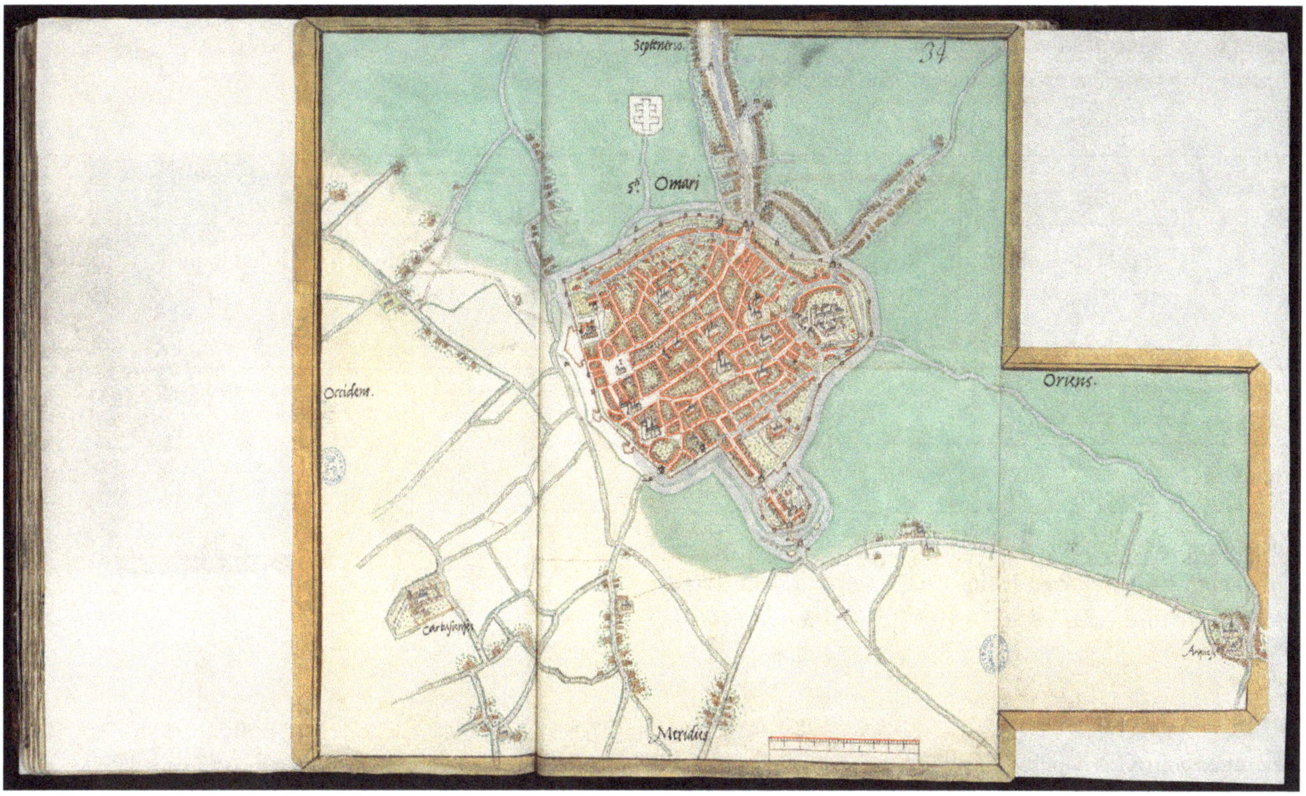
Septentrio
St Omari
Occidens
Oriens
Meridies

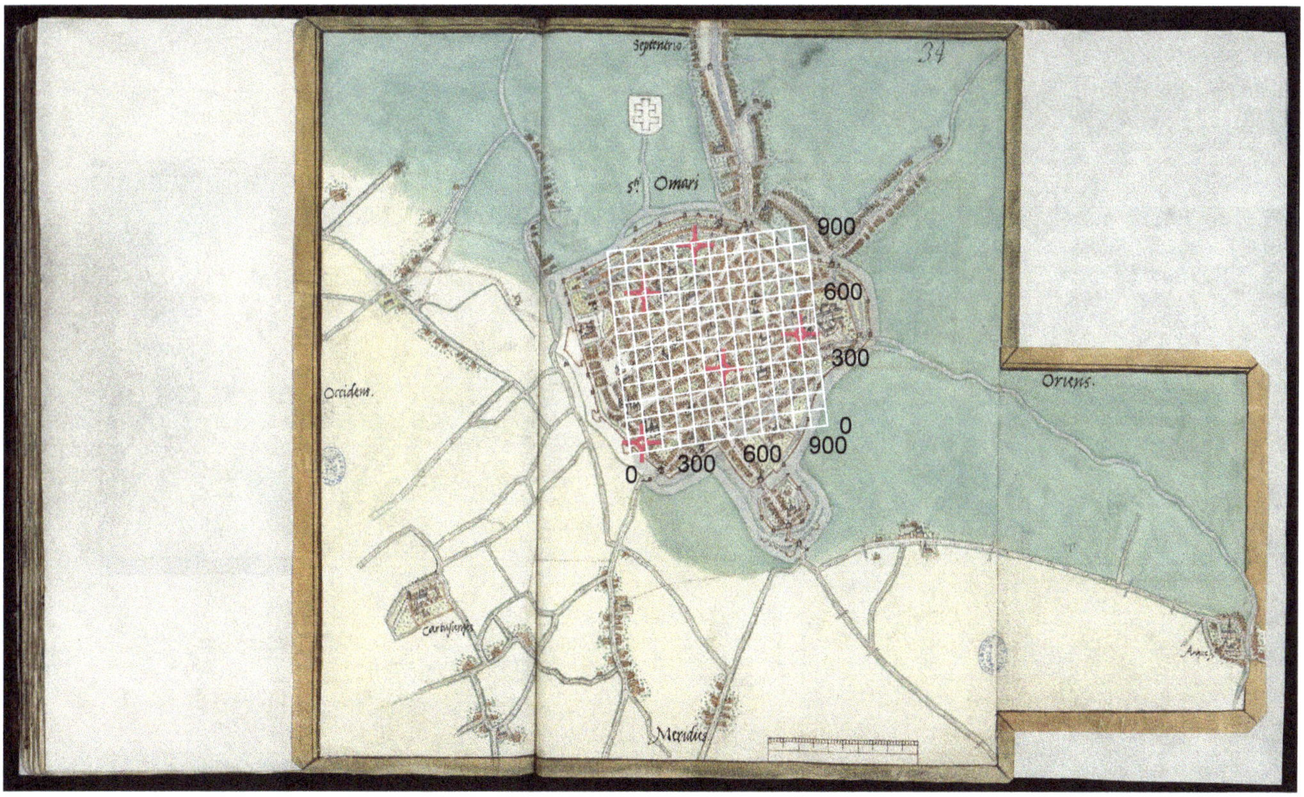
Septentrio
St Omari
900
600
300
0
900
600
300
0
Occidens
Oriens
Meridies

177. Saint-Pol-sur-Ternoise

Tables 3.177 (measured and analysed).
Figures 3.177 (unmeasured, measured and analysed).

Measured

Pixels	Scale stick original map (pixels)	Houses	Street sides	Houses per street side	Inhabitants (RF 4.0)	Inhabitants (RF 5.5)	Inhabitants (RF 7.0)
1848	1015	325	54	6.01	1298	1785	2272

Analysed

July 3, 2023 9:57:47 AM	
Computation with 5 linked points (*italics* points are less reliable):	*1. La Ternoise (est)* *2. Rue de Béthune* *3. Campanile* *4. La Ternoise Ouest* *5. Rue de Frévent*
Reference points (Reliability 0–5)	0
Transformation with 3 parameters	
Scale	1:8360
Rotation	3°
Mean position error	54m

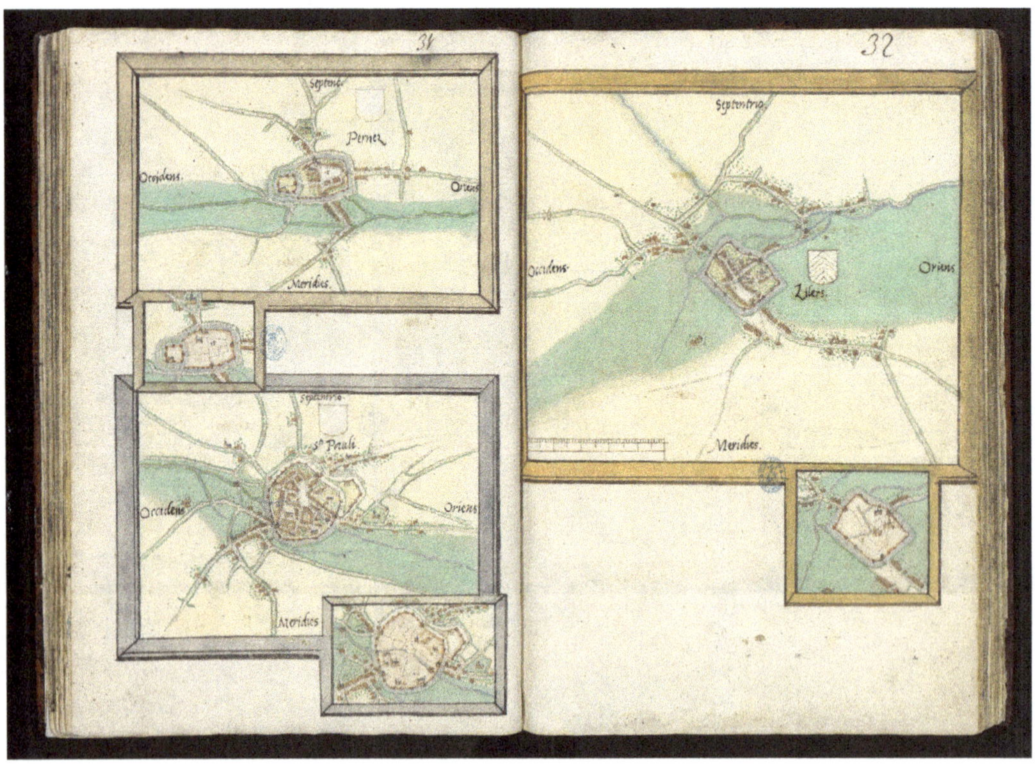

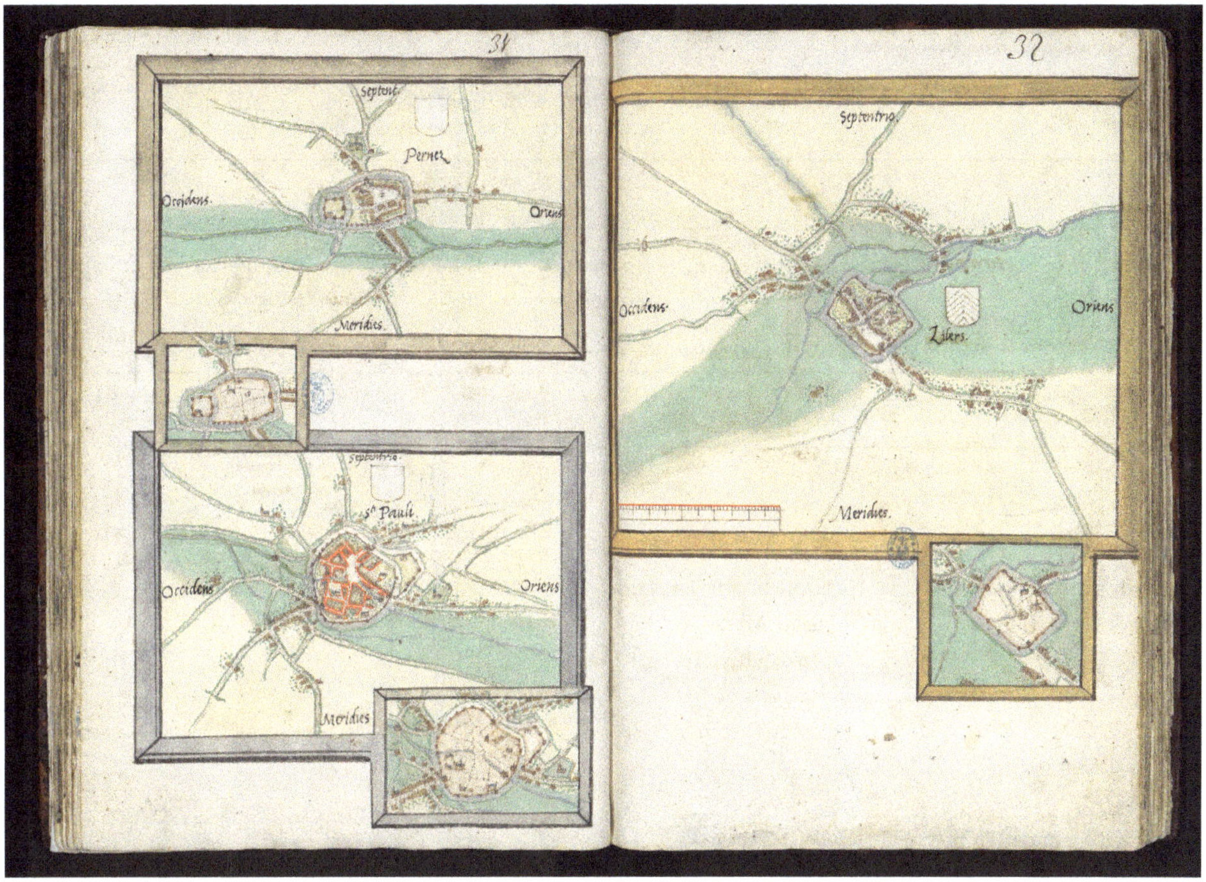
31
32
Septent.
Pernez
Occidens.
Oriens
Meridies.
Septentrio
St Pauli
Occidens
Oriens
Meridies
Septentrio
Occidens.
Lilers.
Oriens
Meridies.

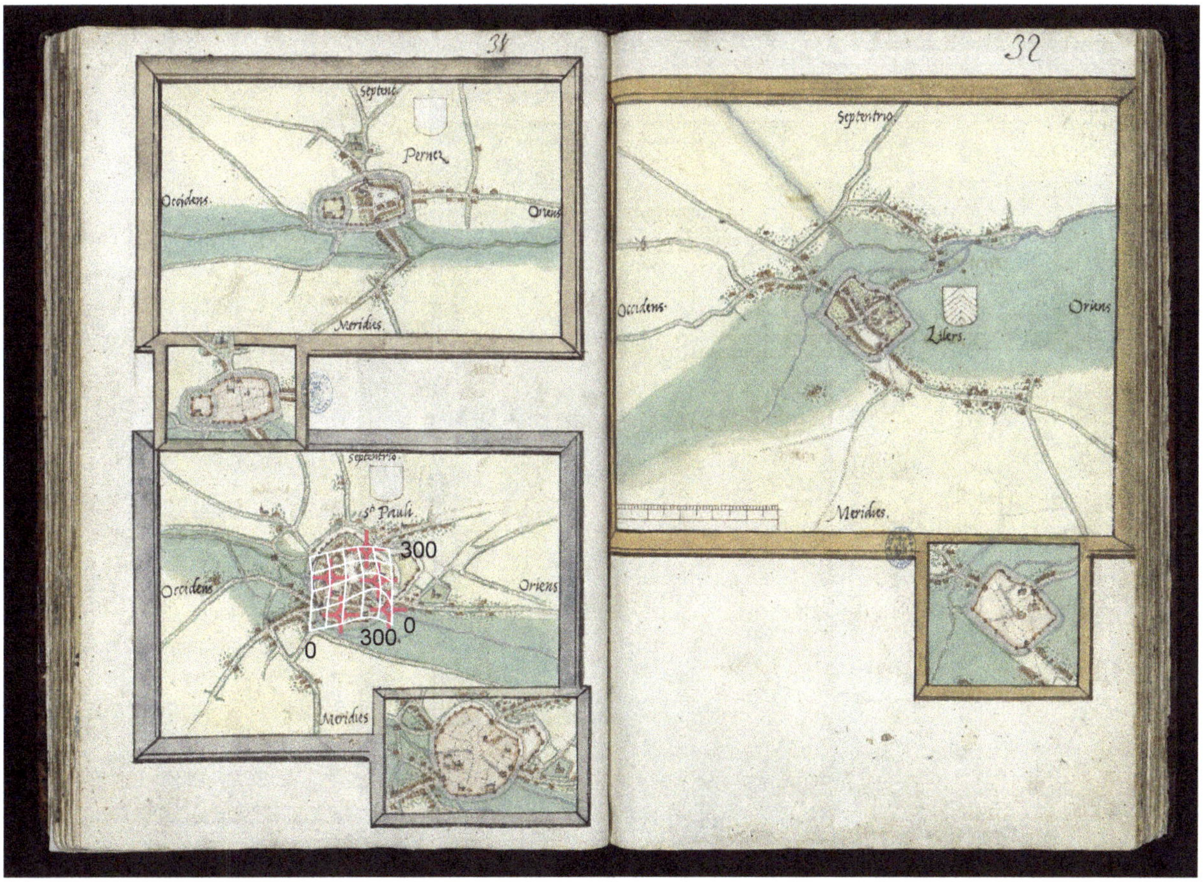
31
32
Septent.
Pernez
Occidens.
Oriens
Meridies.
Septentrio
St Pauli
300
Occidens
Oriens
0
300
0
Meridies
Septentrio
Occidens.
Lilers.
Oriens
Meridies.

178. Schiedam

Tables 3.178 (measured and analysed).
Figures 3.178 (unmeasured, measured and analysed).

Measured							
Pixels	Scale stick original map (pixels)	Houses	Street sides	Houses per street side	Inhabitants (RF 4.0)	Inhabitants (RF 5.5)	Inhabitants (RF 7.0)
3817	994	670	127	5.28	2681	3686	4691

Analysed	
July 13, 2023 10:02:41 AM	
Computation with 5 linked points (*italics* points are less reliable):	1. Grote of Sint-Janskerk 2. Grote Markt 3. Huis te Riviere 4. *Hoogstraat* 5. Lange Haven
Reference points (Reliability 0–5)	4
Transformation with 3 parameters	
Scale	1:9410
Rotation	16 °
Mean position error	30m

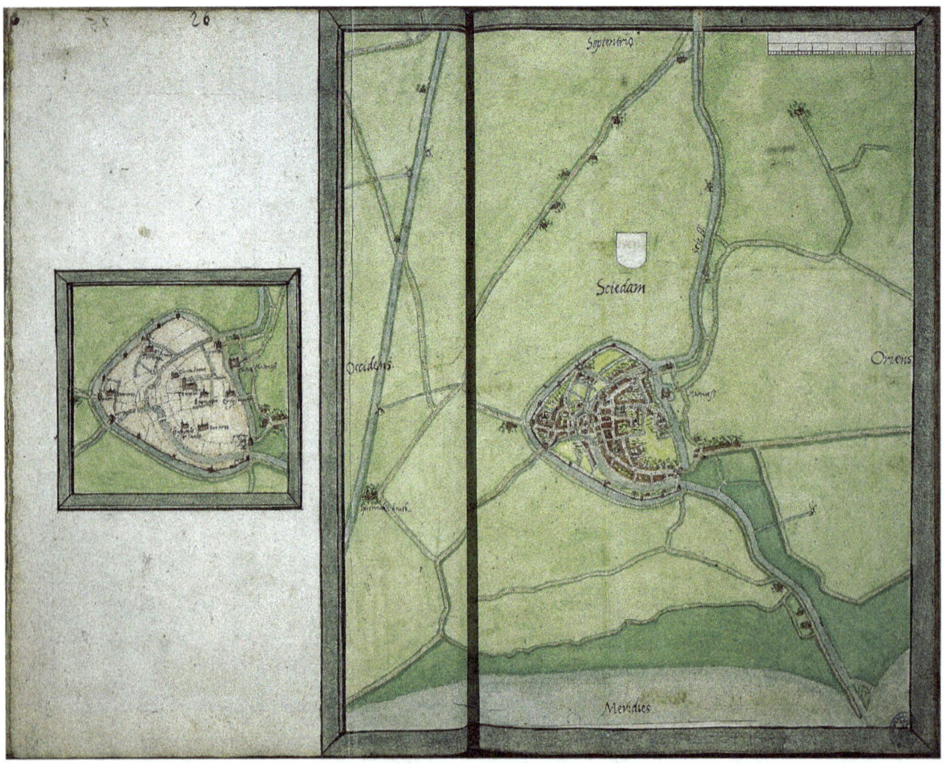

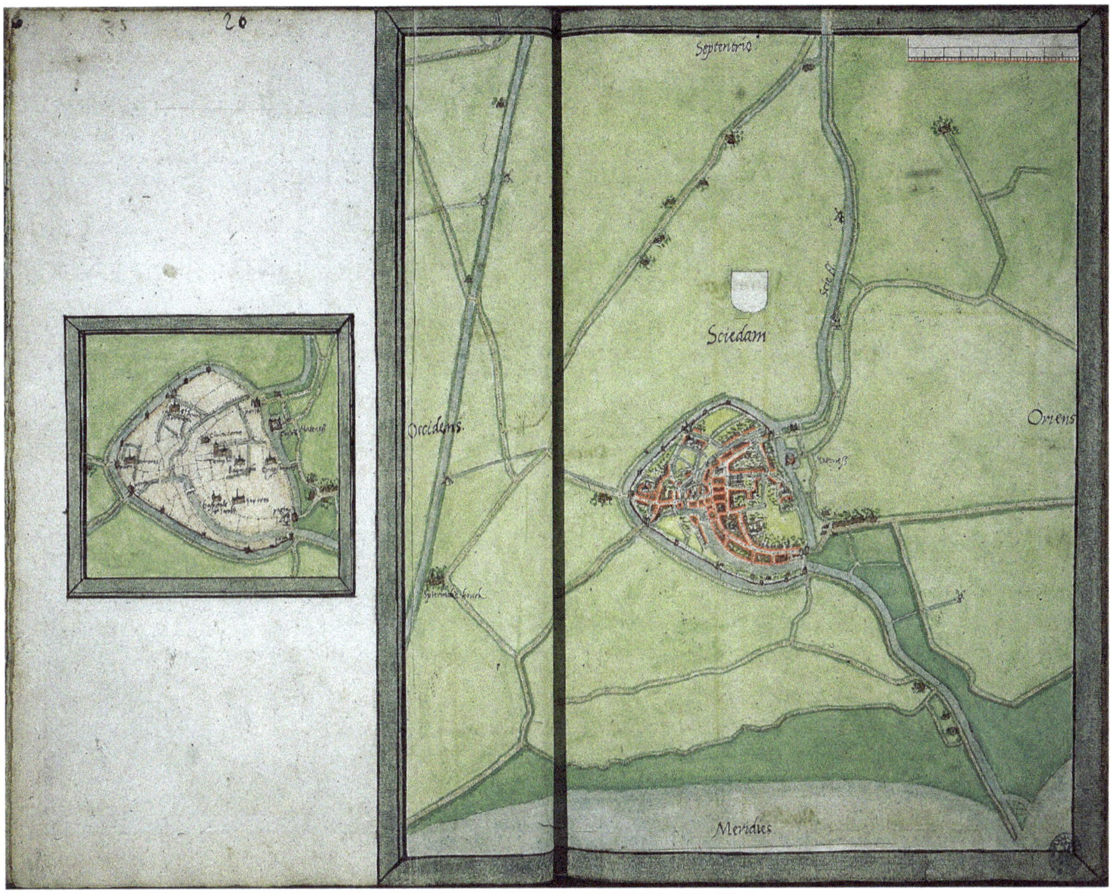
Septentrio
Sciedam
Occidens
Oriens
Meridies

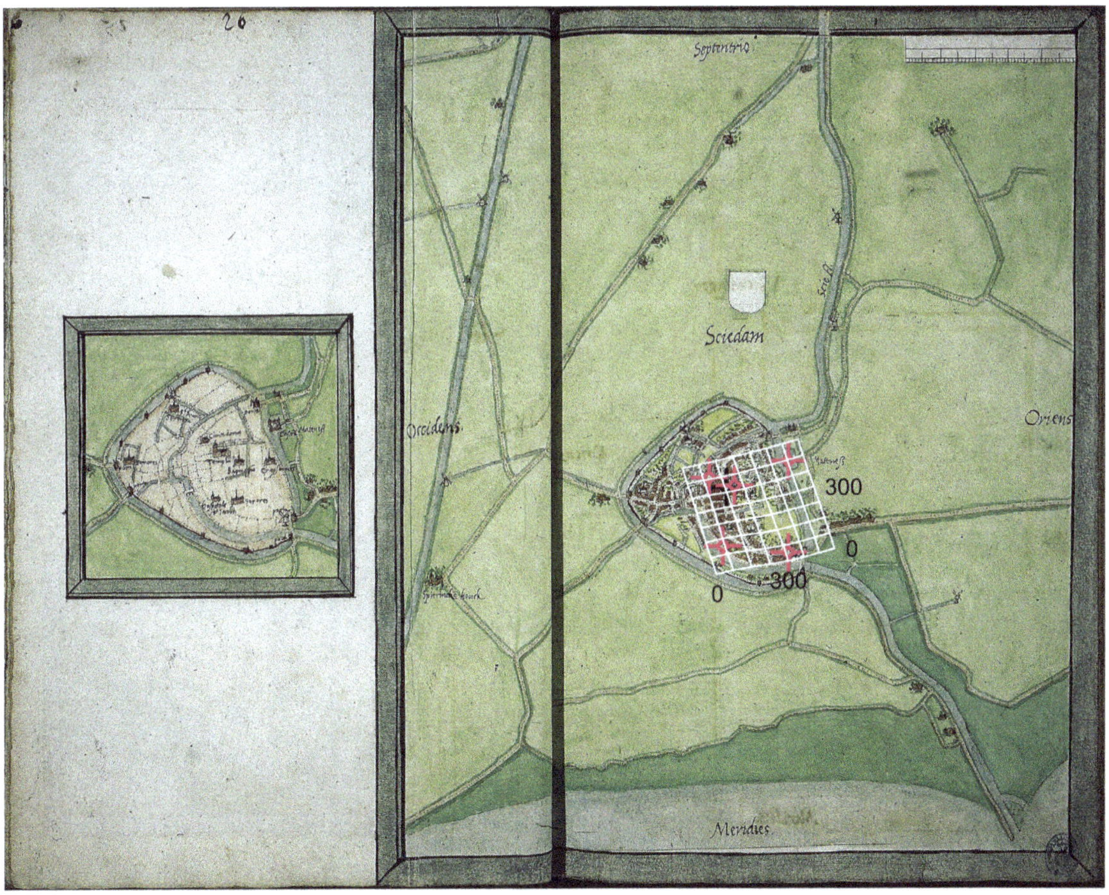
Septentrio
Sciedam
Occidens
Oriens
300
0
0
300
Meridies

179. Schoonhoven

Tables 3.179 (measured and analysed).
Figures 3.179 (unmeasured, measured and analysed).

Measured

Pixels	Scale stick original map (pixels)	Houses	Street sides	Houses per street side	Inhabitants (RF 4.0)	Inhabitants (RF 5.5)	Inhabitants (RF 7.0)
4305	993	756	213	3.55	3024	4158	5292

Analysed

July 15, 2023 4:04:48 PM	
Computation with 5 linked points (*italics* points are less reliable):	1. Bartholomeüskerk 2. Stadhuis 3. Lekdijk-West 4. Koestraat 5. Wal
Reference points (Reliability 0–5)	5
Transformation with 3 parameters	
Scale	1:8360
Rotation	12 °
Mean position error	18m

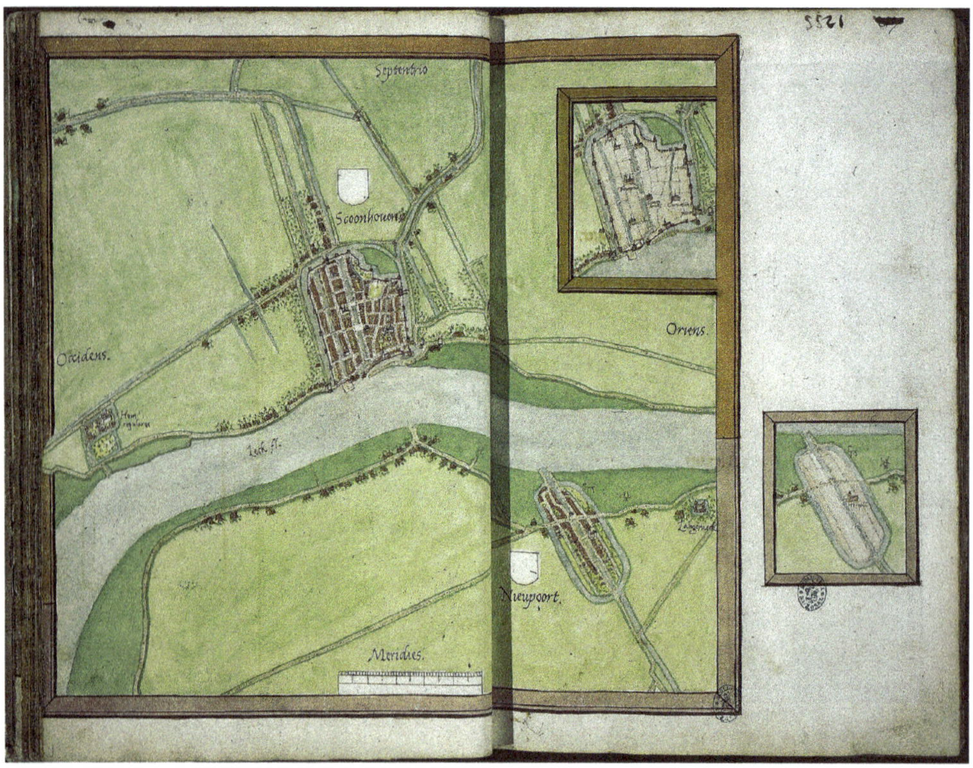

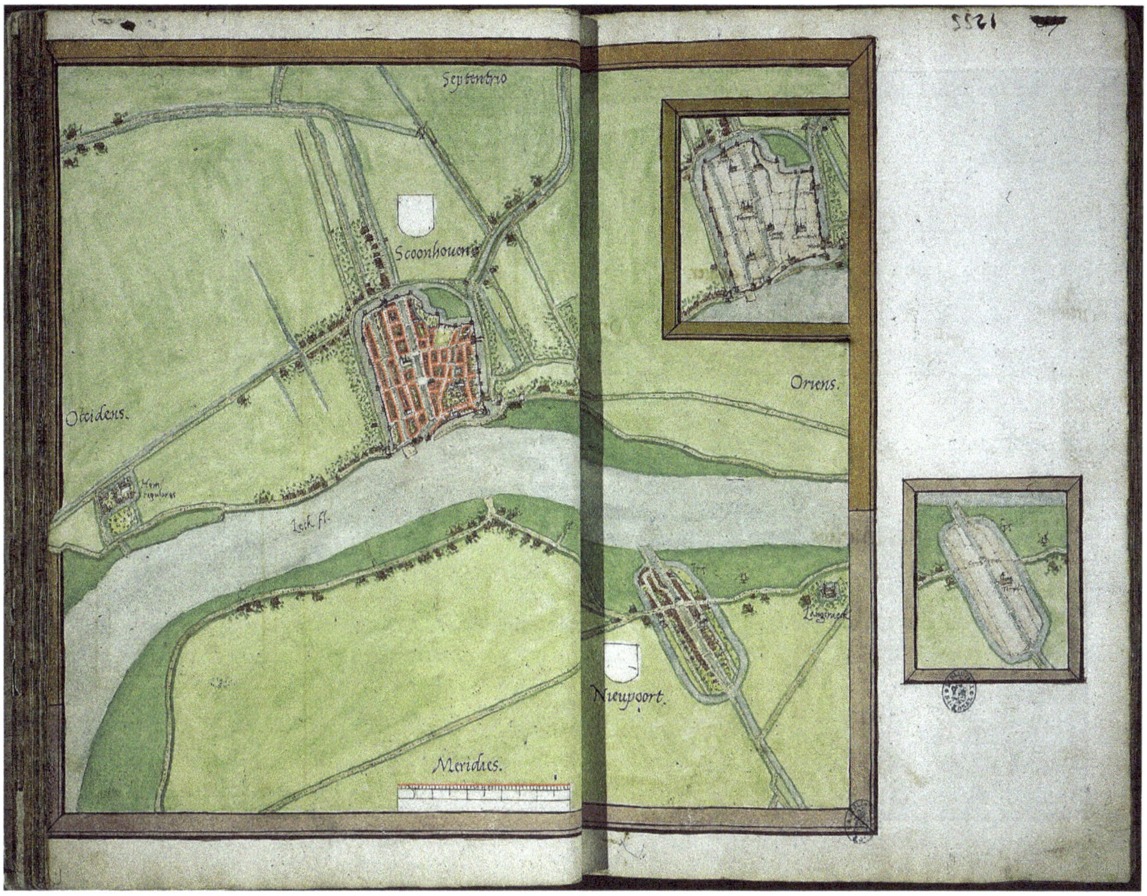
Septentrio
Scoonhouen
Occidens.
Oriens.
Leck fl.
Nieupoort.
Meridies.

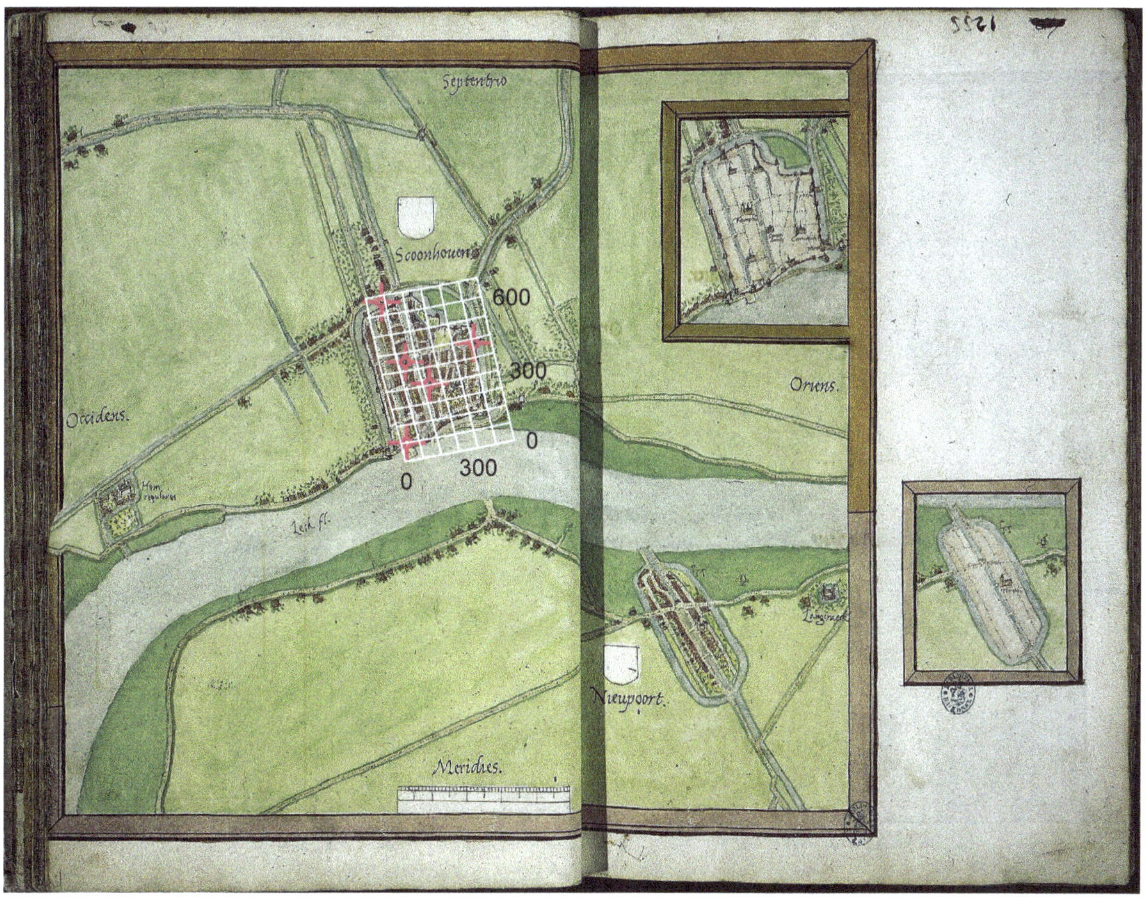
Septentrio
Scoonhouen
600
300
0
0
300
Occidens.
Oriens.
Leck fl.
Nieupoort.
Meridies.

180. Sint-Anna-ter-Muiden

Tables 3.180 (measured and analysed).
Figures 3.180 (unmeasured, measured and analysed).

Measured							
Pixels	Scale stick original map (pixels)	Houses	Street sides	Houses per street side	Inhabitants (RF 4.0)	Inhabitants (RF 5.5)	Inhabitants (RF 7.0)
328	missing	58	2	28.80	231	317	403

Analysed	
July 15, 2023 4:44:53 PM	
Computation with 5 linked points (*italics* points are less reliable):	1. Hervormde Kerk Used were the analyse and the reference points of Sluis.
Reference points (Reliability 0–5)	1
Transformation with 3 parameters	
Scale	1:8180
Rotation	12 °
Mean position error	27m

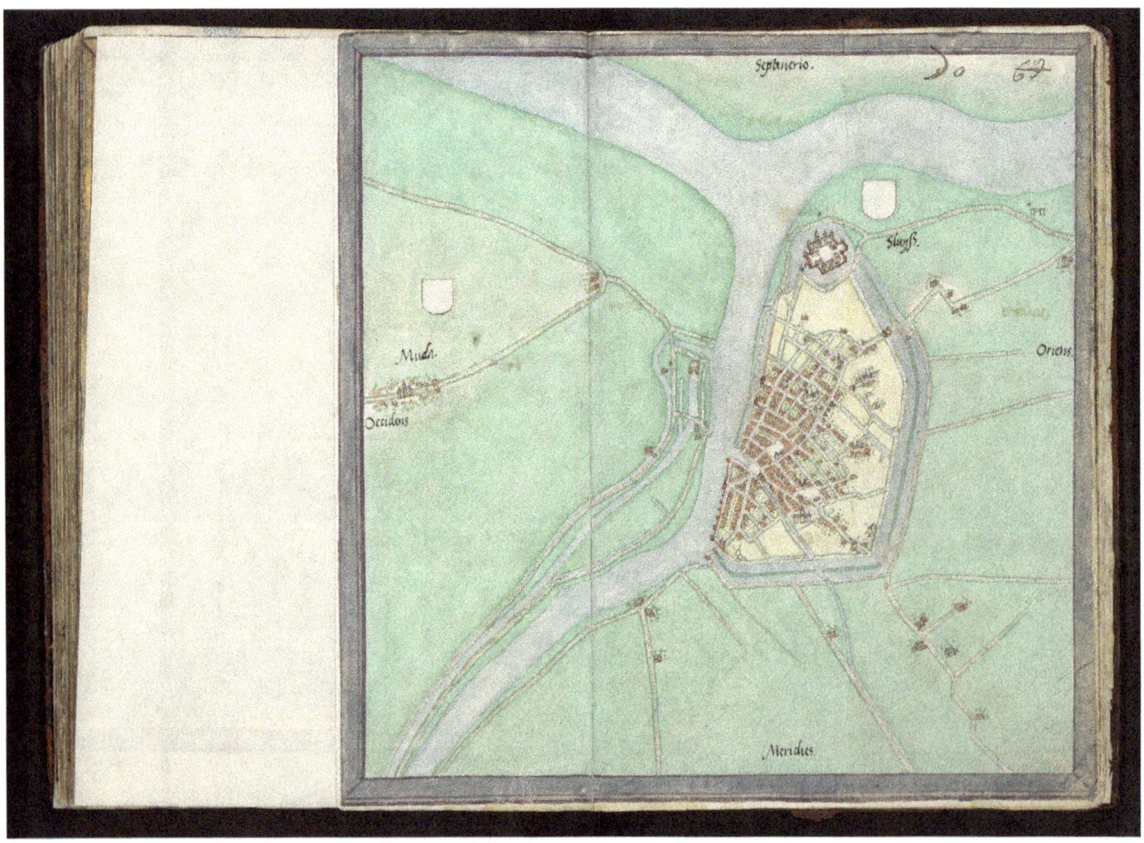

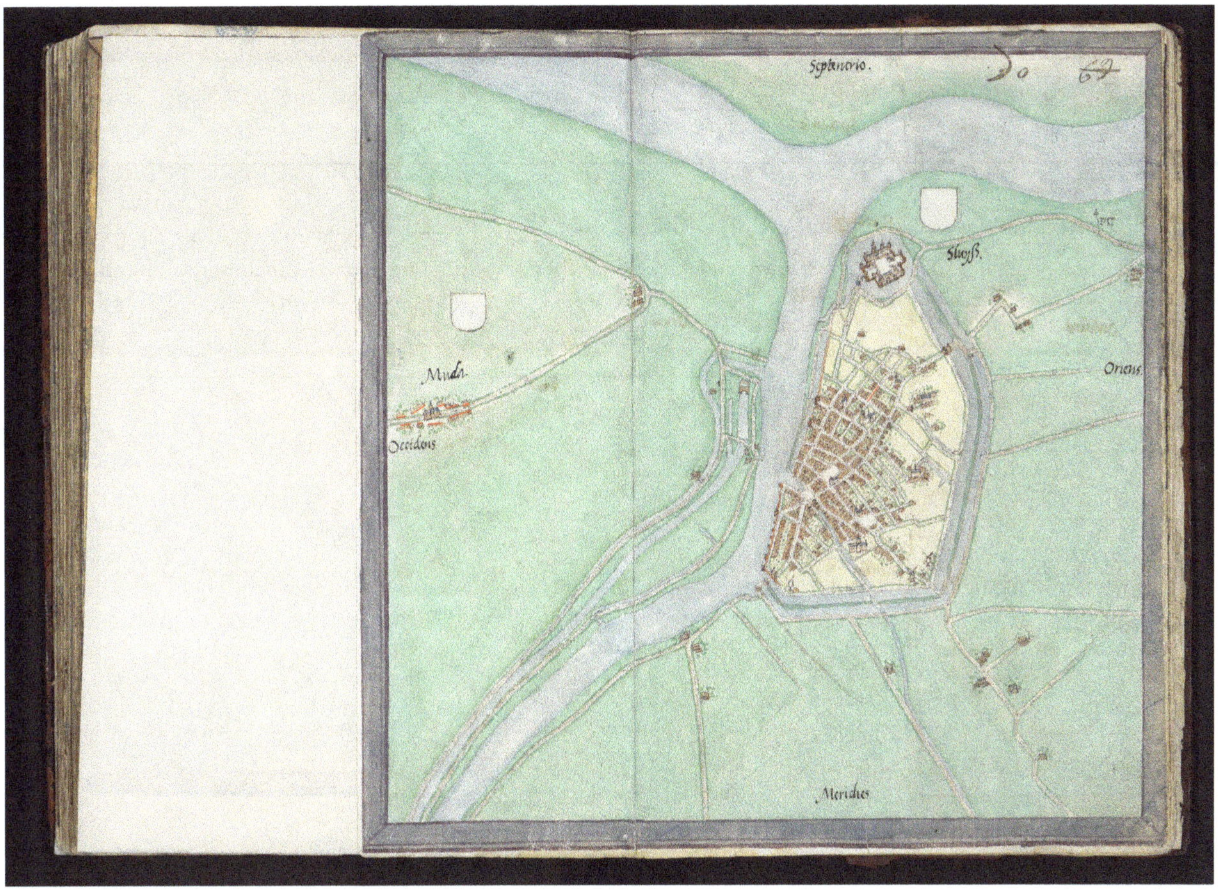
Septentrio.
Sluyß.
Muda.
Occidens
Oriens.
Meridies

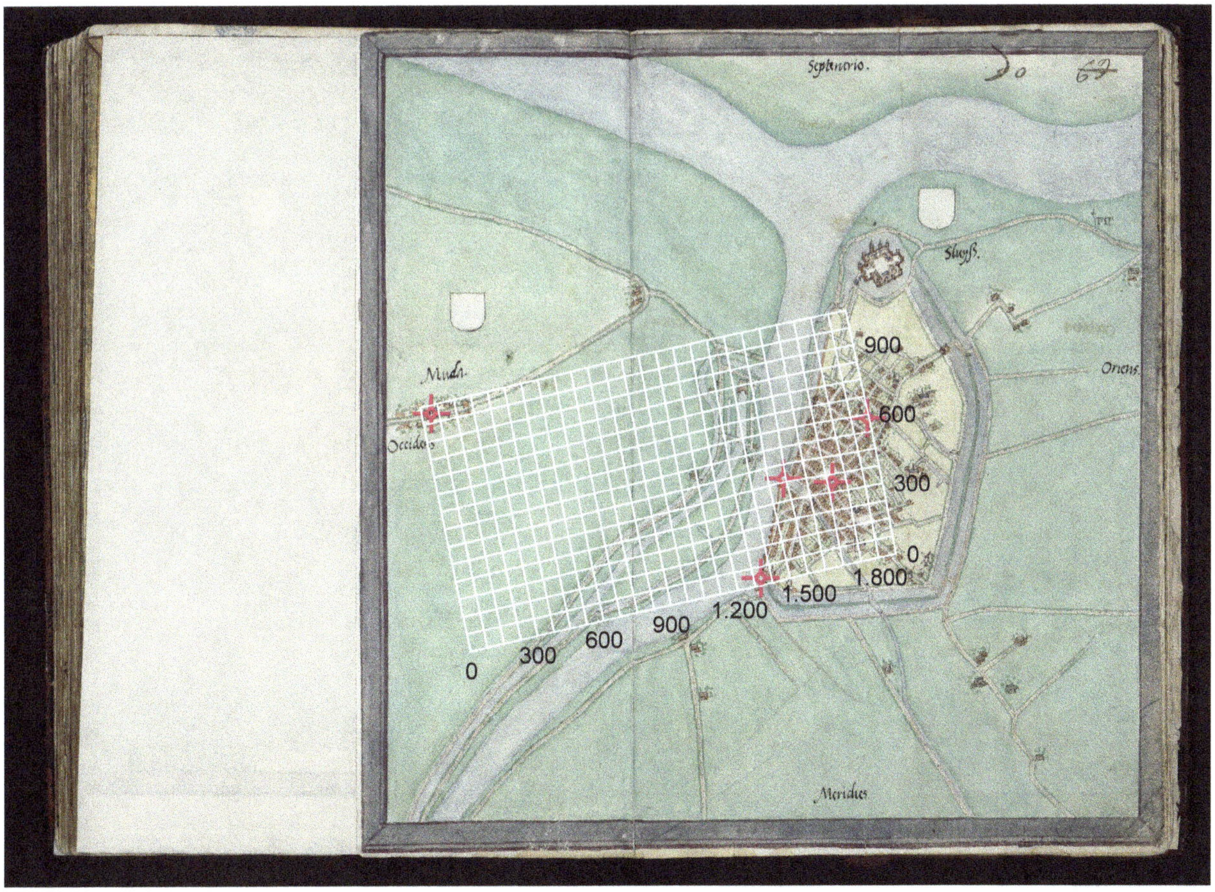
Septentrio.
Sluyß.
Muda.
Oriens.
Meridies
900
600
300
0
1.800
1.500
1.200
900
600
300
0

181. Sint- Maartensdijk

Tables 3.181 (measured and analysed).
Figures 3.181 (unmeasured, measured and analysed).

Measured							
Pixels	Scale stick original map (pixels)	Houses	Street sides	Houses per street side	Inhabitants (RF 4.0)	Inhabitants (RF 5.5)	Inhabitants (RF 7.0)
1142	1001	201	33	6.08	802	1103	1404

Analysed	
July 17, 2023 10:13:49 AM	
Computation with 5 linked points (*italics* points are less reliable):	1. Sint-Maartenskerk 2. Markt 3. Westvest 4. Oostvest 5. Keethil
Reference points (Reliability 0–5)	5
Transformation with 3 parameters	
Scale	1:8130
Rotation	10 °
Mean position error	11m

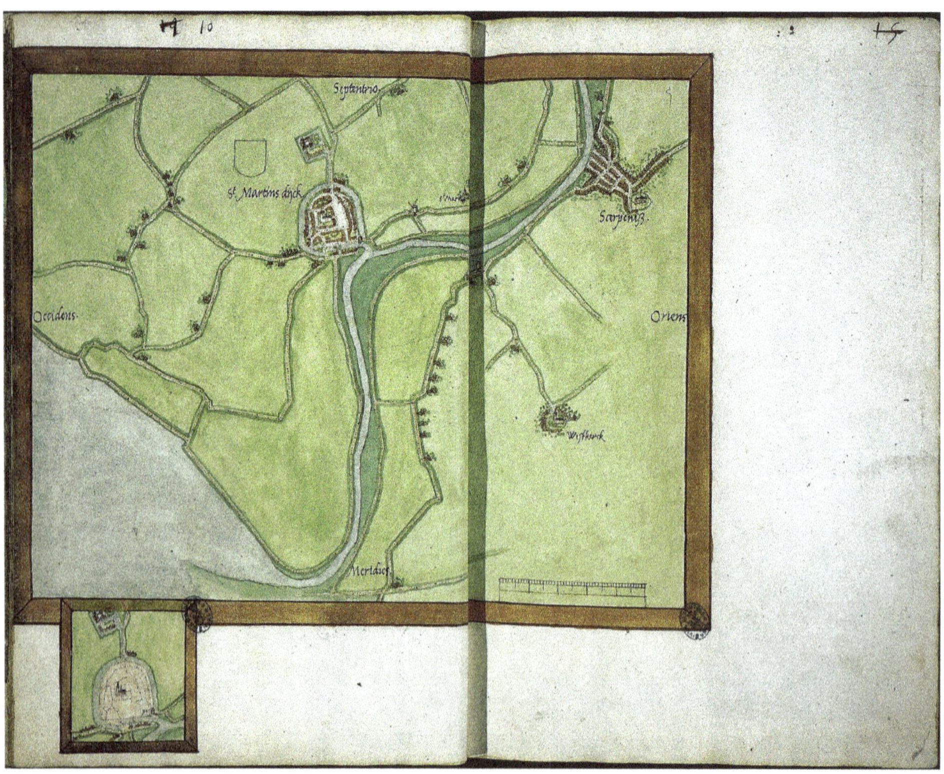

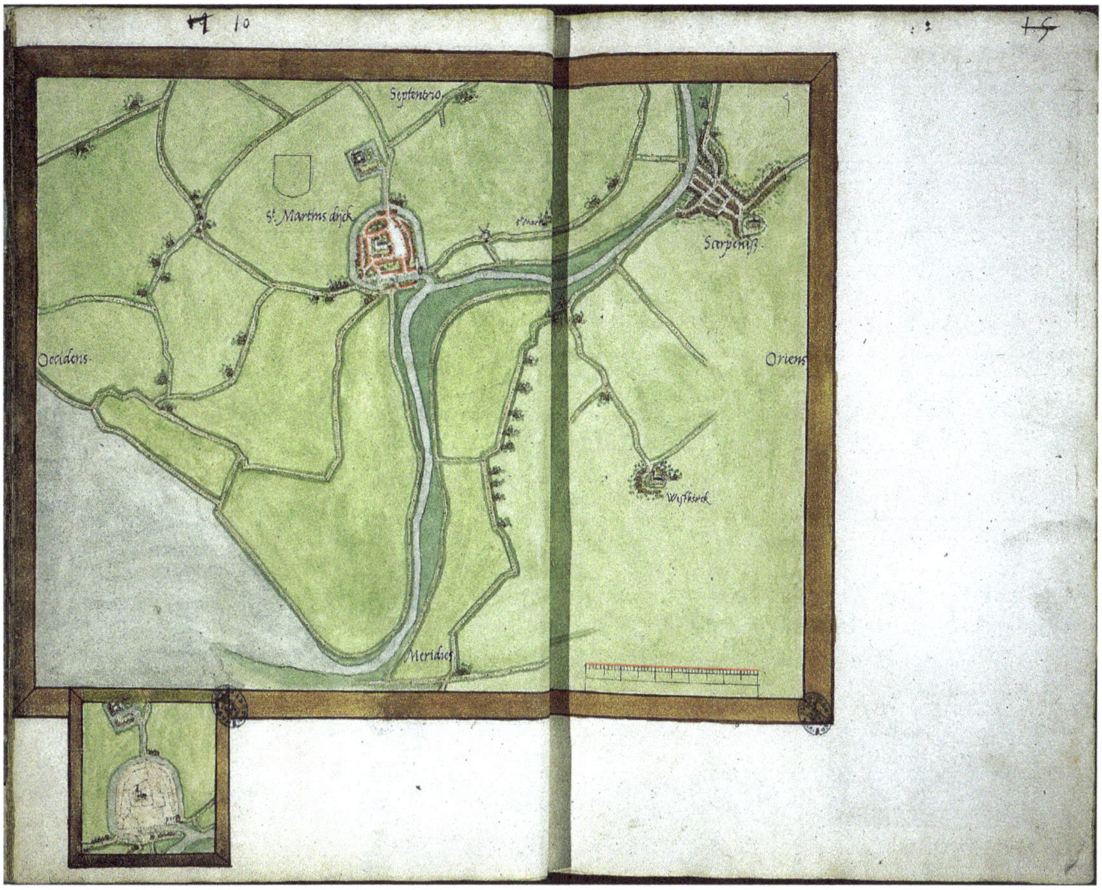
Septentrio
St. Martins dijck
Occidens
Oriens
Meridies

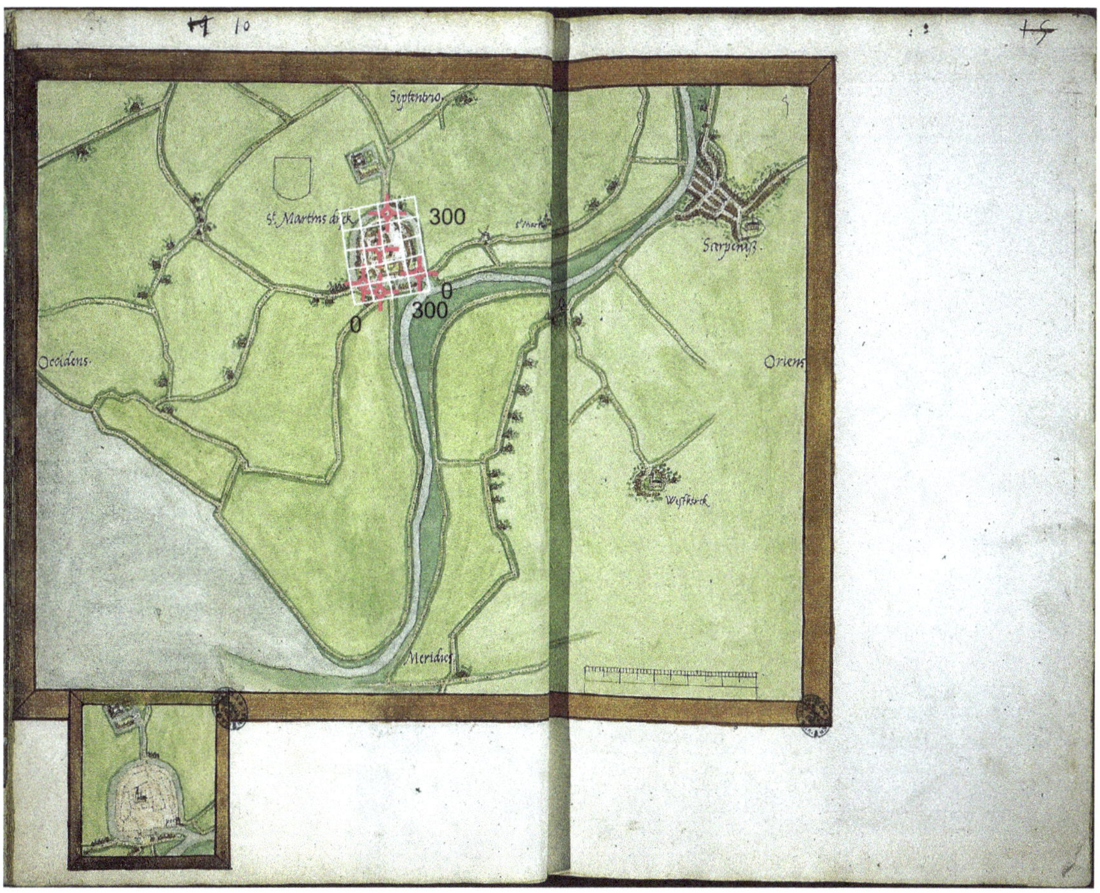
Septentrio
St. Martins dijck
300
0
0
300
Occidens
Oriens
Meridies

182. Sloten

Tables 3.182 (measured and analysed).
Figures 3.182 (unmeasured, measured and analysed).

Measured

Pixels	Scale stick original map (pixels)	Houses	Street sides	Houses per street side	Inhabitants (RF 4.0)	Inhabitants (RF 5.5)	Inhabitants (RF 7.0)
1232	989	216	50	4.33	865	1190	1515

Analysed

July 17, 2023 10:58:16 AM	
Computation with 5 linked points (*italics* points are less reliable):	1. Dubbelstraat 2. Bolwerk Zuidzijde 3. Bolwerk Noordzijde 4. Koestraat 5. Achterom
Reference points (Reliability 0–5)	5
Transformation with 3 parameters	
Scale	1:8380
Rotation	4 °
Mean position error	7m

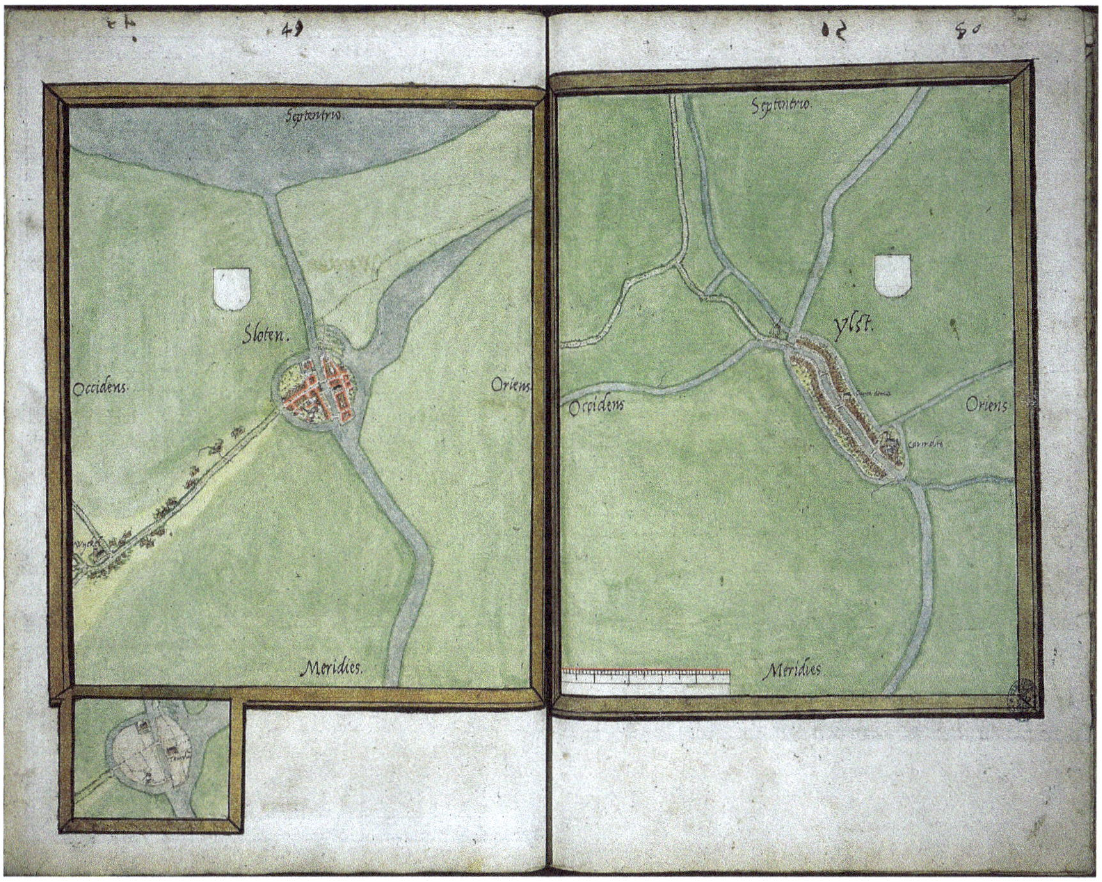
Septentrio.
Sloten.
Occidens.
Oriens.
Meridies.
Septentrio.
Ylst.
Occidens
Oriens
Meridies

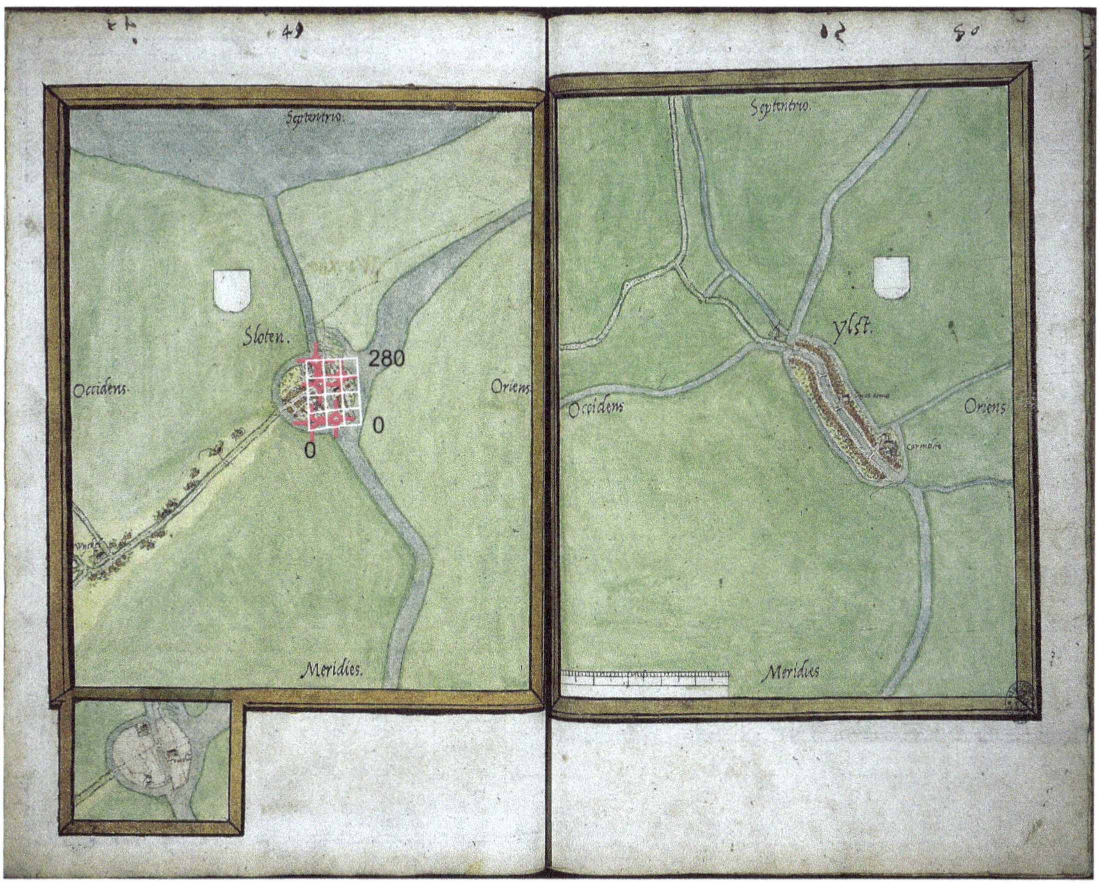
Septentrio.
Sloten.
280
Occidens.
0
0
Oriens.
Meridies.
Septentrio.
Ylst.
Occidens
Oriens
Meridies

183. Sluis

Tables 3.183 (measured and analysed).
Figures 3.183 (unmeasured, measured and analysed).

Measured							
Pixels	Scale stick original map (pixels)	Houses	Street sides	Houses per street side	Inhabitants (RF 4.0)	Inhabitants (RF 5.5)	Inhabitants (RF 7.0)
5560	missing	976	161	6.06	3905	5370	6835

Analysed	
July 15, 2023 4:44:53 PM	
Computation with 5 linked points (*italics* points are less reliable):	1. Belfort 2. Sint-Janskerk *3. Kaai* 4. Oude Kerkstraat 5. Sint-Donaes
Reference points (Reliability 0–5)	4
Transformation with 3 parameters	
Scale	1:8180
Rotation	12 °
Mean position error	27m

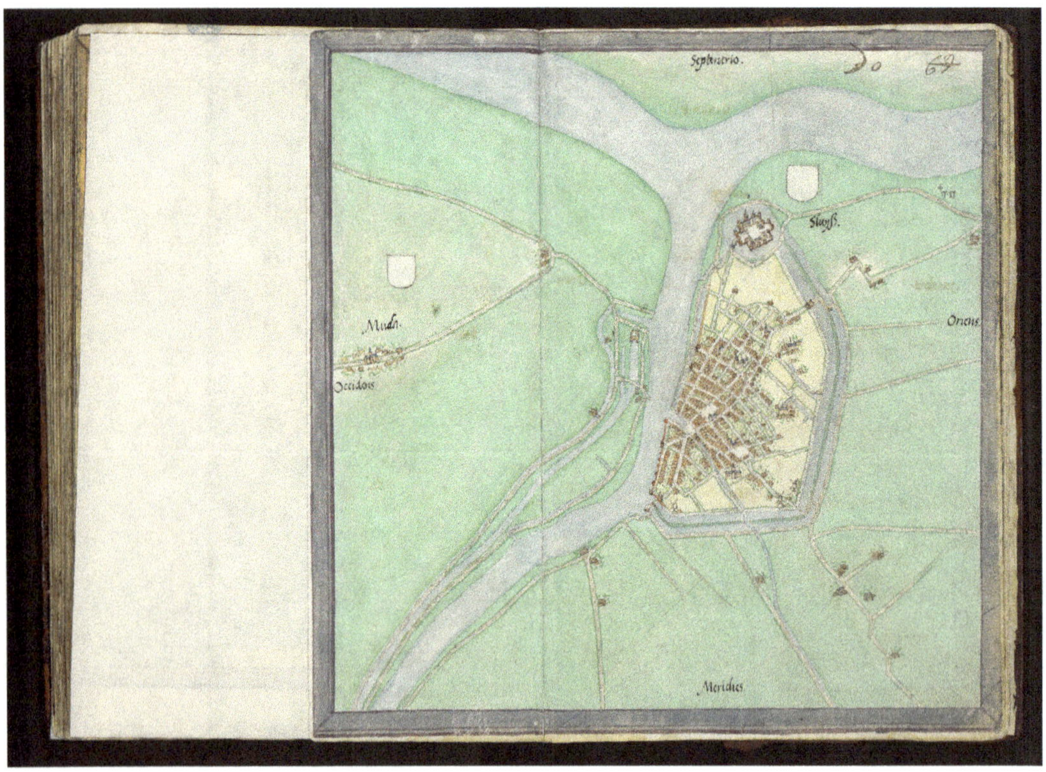

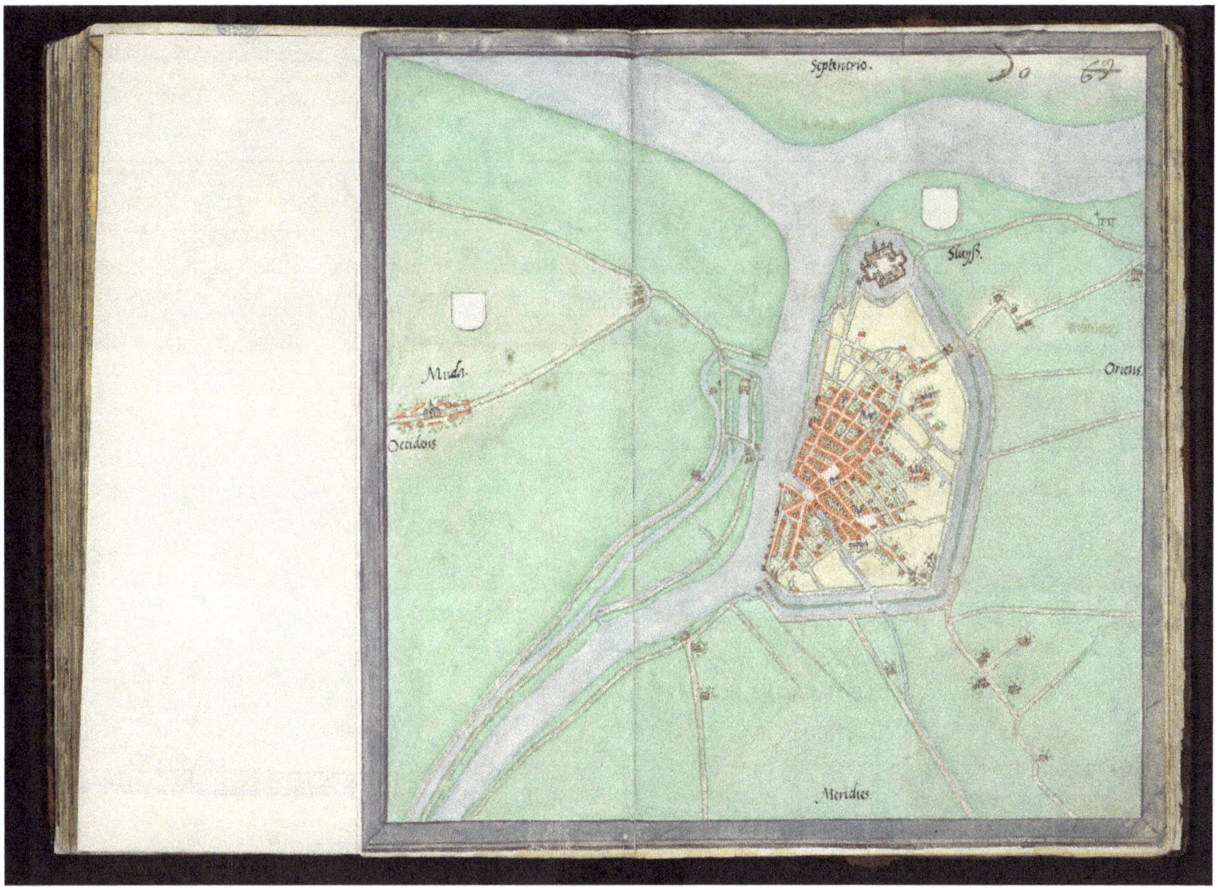
Septentrio.
Sluyß.
Muda.
Occidens
Oriens
Meridies

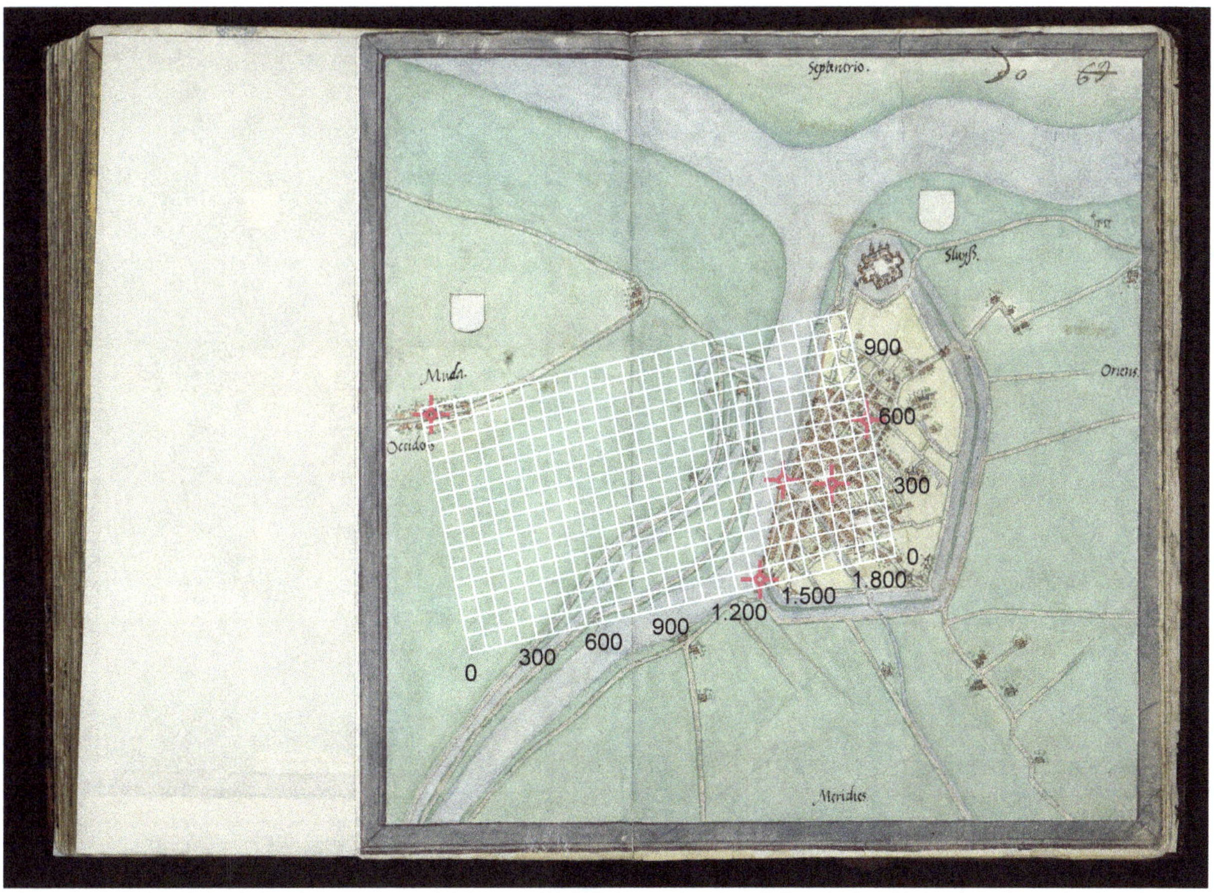
Septentrio.
Sluyß.
Muda.
Occidens
Oriens
Meridies
900
600
300
0
1.800
1.500
1.200
900
600
300
0

184. Sneek

Tables 3.184 (measured and analysed).
Figures 3.184 (unmeasured, measured and analysed).

Measured

Pixels	Scale stick original map (pixels)	Houses	Street sides	Houses per street side	Inhabitants (RF 4.0)	Inhabitants (RF 5.5)	Inhabitants (RF 7.0)
5163	1016	907	135	6.72	3626	4986	6346

Analysed

July 17, 2023 2:59:33 PM	
Computation with 5 linked points (*italics* points are less reliable):	1. Martinikerk 2. Leeuwenburg 3. Waterpoortsgracht 4. Jousterkade 5. Kerkgracht
Reference points (Reliability 0–5)	5
Transformation with 3 parameters	
Scale	1:7750
Rotation	10 °
Mean position error	7m

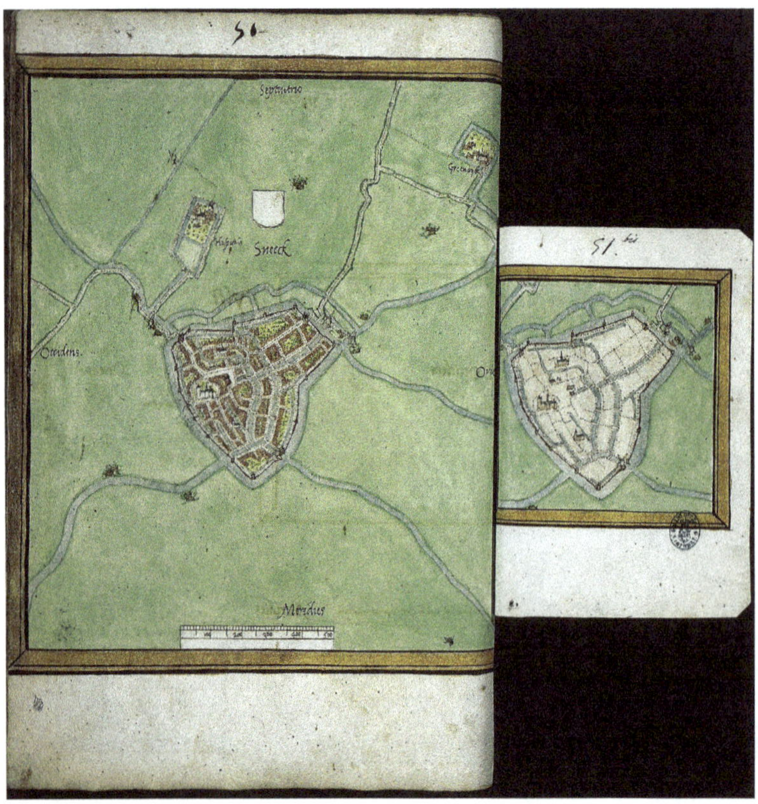

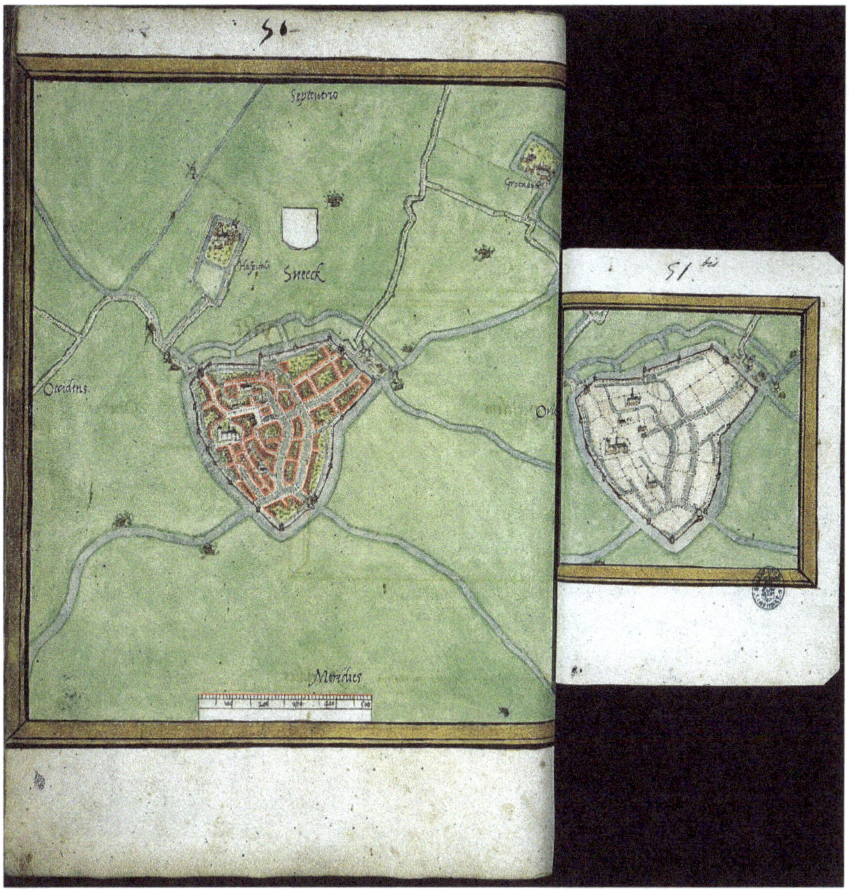
Septentrio
Sneeck
Meridies

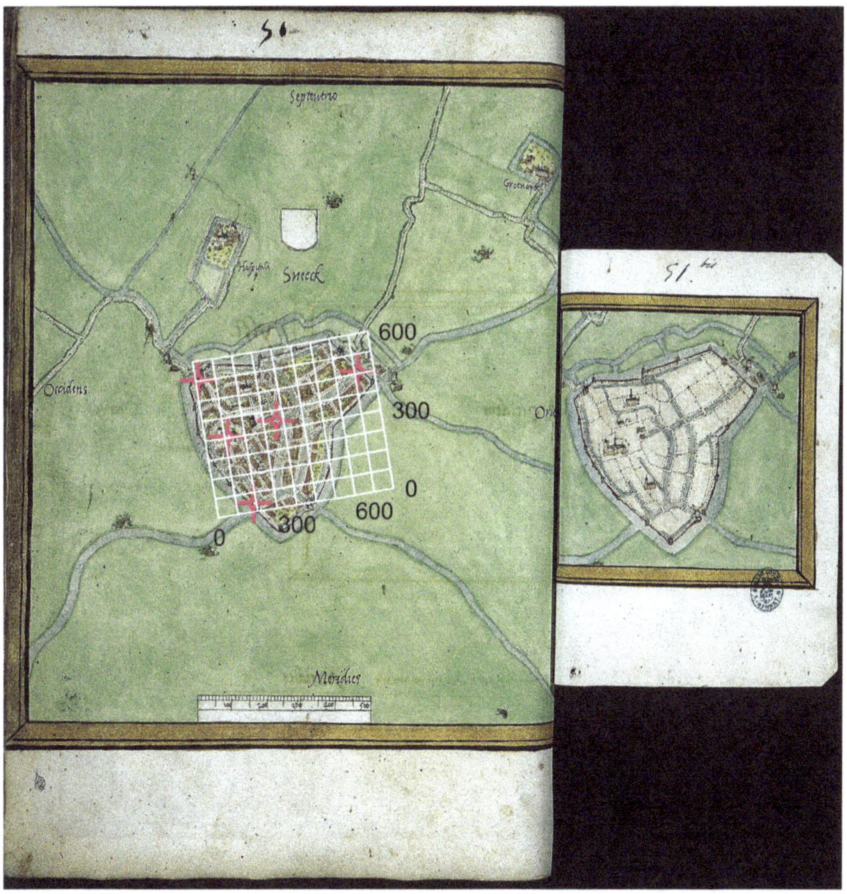
Septentrio
Sneeck
600
300
0
0
300
600
Meridies

185. Soignies

Tables 3.185 (measured and analysed).
Figures 3.185 (unmeasured, measured and analysed).

Measured							
Pixels	Scale stick original map (pixels)	Houses	Street sides	Houses per street side	Inhabitants (RF 4.0)	Inhabitants (RF 5.5)	Inhabitants (RF 7.0)
2628	1034	461	75	6.15	1846	2538	3230

Analysed	
July 21, 2023 9:54:08 AM	
Computation with 5 linked points (*italics* points are less reliable):	1. Collégiale Saint-Vincent 2. Rempart Legros 3. Rempart du vieux Cimetière 4. Rue Neuve 5. Rue Léon Hachez
Reference points (Reliability 0–5)	5
Transformation with 3 parameters	
Scale	1:7730
Rotation	6 °
Mean position error	8m

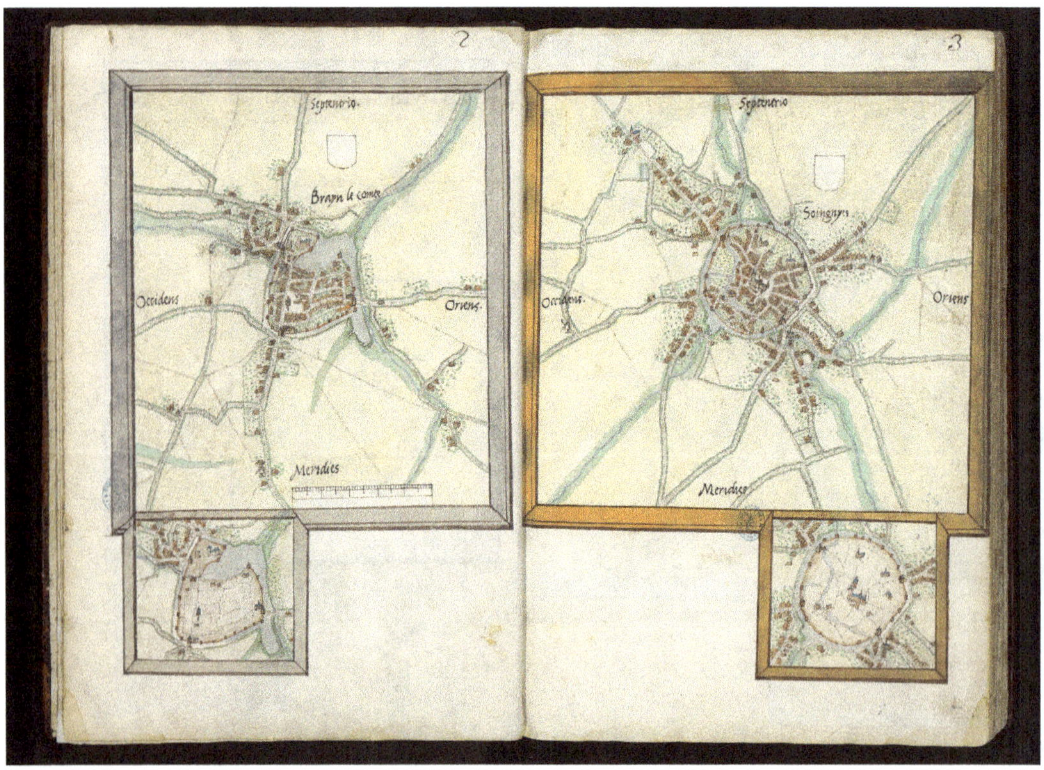

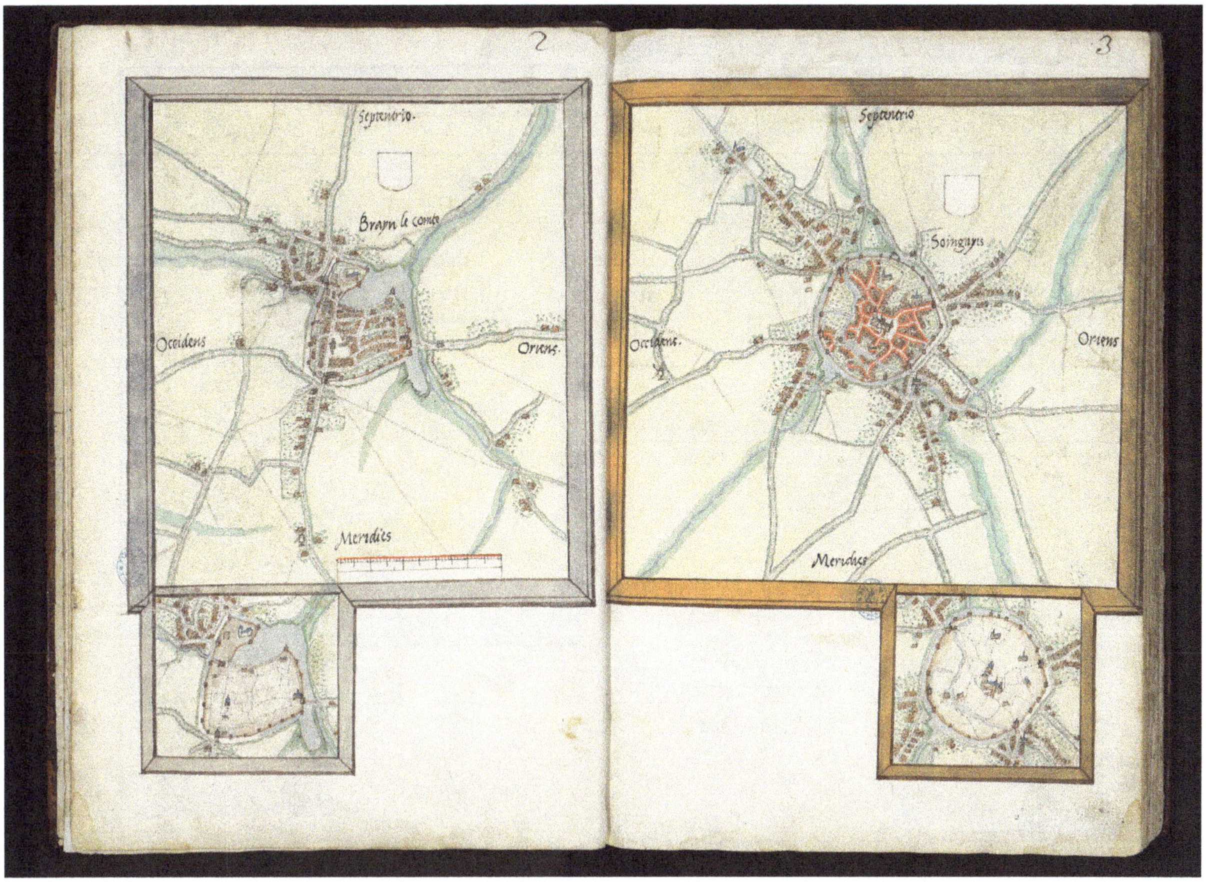
Septentrio.
Brayn le comte
Occidens
Oriens.
Meridies
Septentrio
Soingnys
Occidens.
Oriens
Meridies

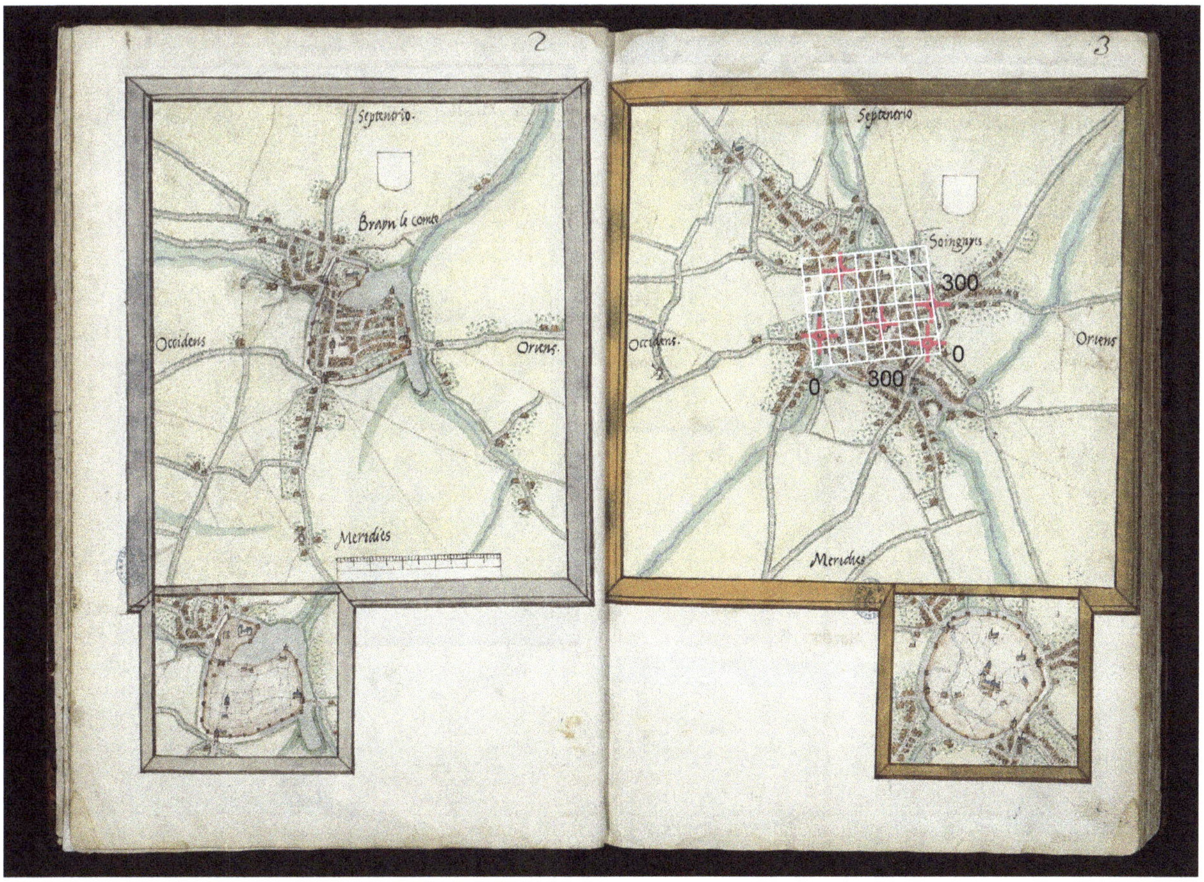
Septentrio.
Brayn le comte
Occidens
Oriens.
Meridies
Septentrio
Soingnys
300
0
0
300
Occidens.
Oriens
Meridies

186. Stavoren

Tables 3.186 (measured and analysed).
Figures 3.186 (unmeasured, measured and analysed).

Measured							
Pixels	Scale stick original map (pixels)	Houses	Street sides	Houses per street side	Inhabitants (RF 4.0)	Inhabitants (RF 5.5)	Inhabitants (RF 7.0)
1540	1000	270	41	6.60	1081	1487	1893

Analysed	
July 21, 2023 10:42:58 AM	
Computation with 5 linked points (*italics* points are less reliable):	1. Dwinger 2. gracht 3. Achterdelft 4. Stadsfenne 5. Voorstraat
Reference points (Reliability 0–5)	5
Transformation with 3 parameters	
Scale	1:8450
Rotation	9 °
Mean position error	17m

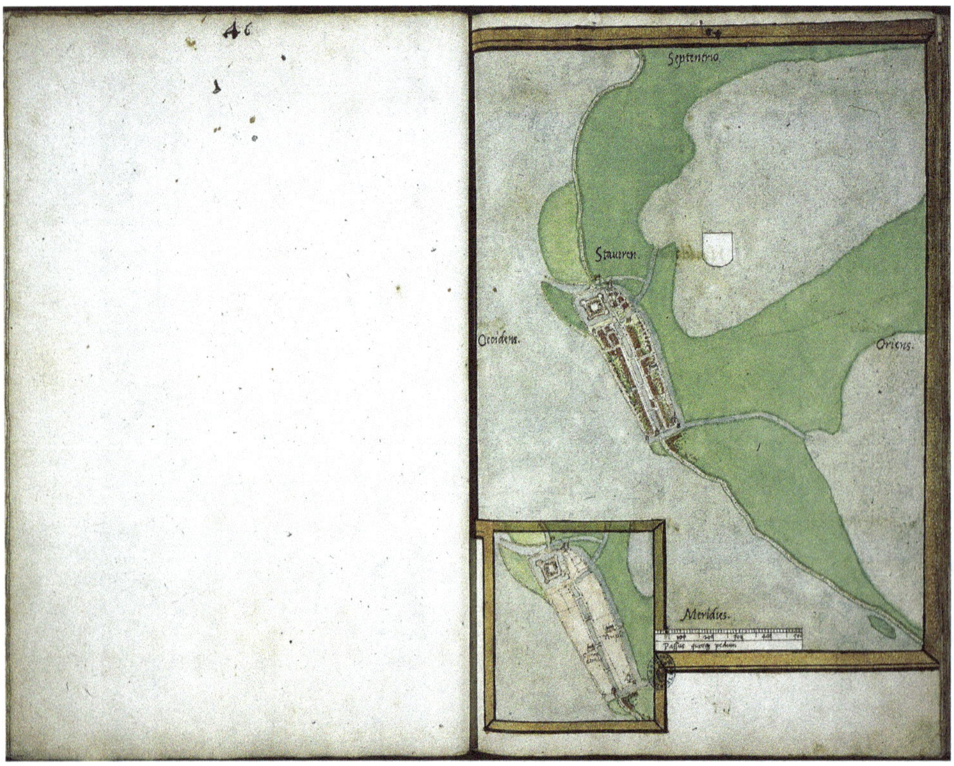

Septentrio.
Stauren.
Occidens.
Oriens.
Meridies.

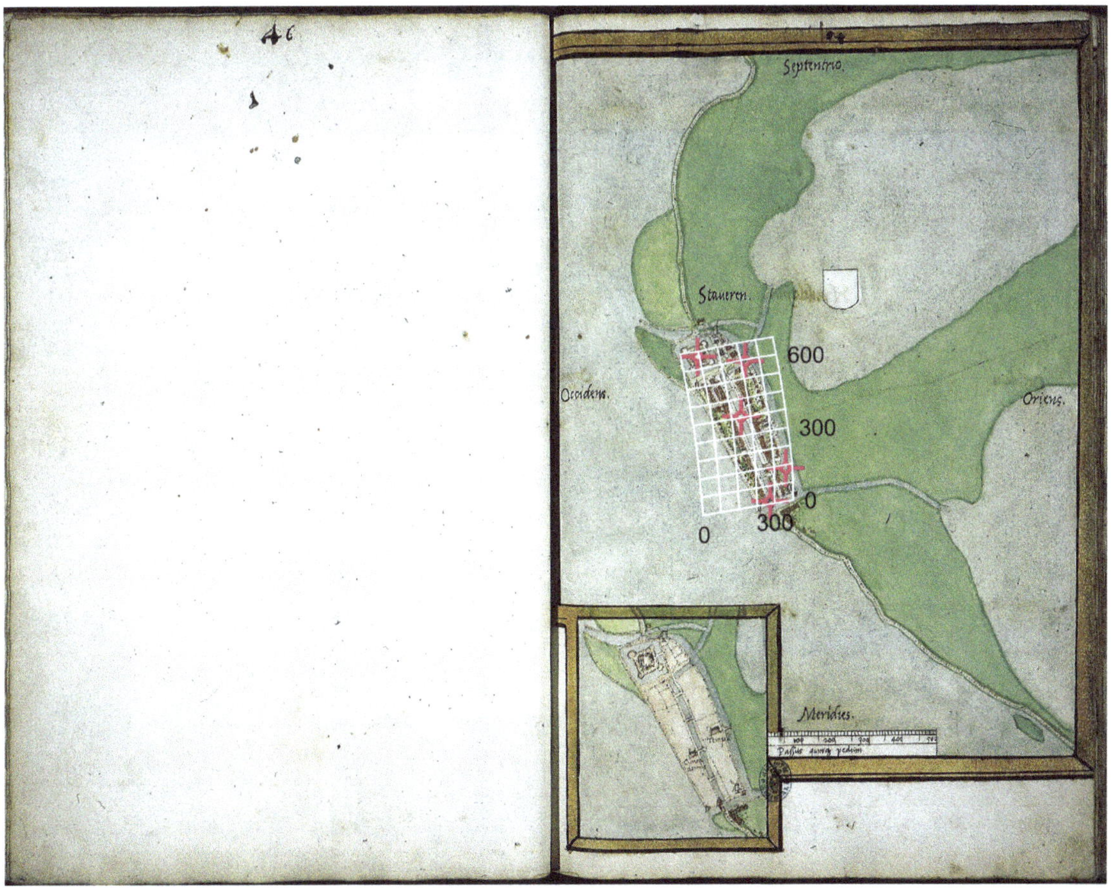
Septentrio.
Stauren.
Occidens.
Oriens.
Meridies.
600
300
0
0
300

187. Steenwijk

Tables 3.187 (measured and analysed).
Figures 3.187 (unmeasured, measured and analysed).

Measured							
Pixels	Scale stick original map (pixels)	Houses	Street sides	Houses per street side	Inhabitants (RF 4.0)	Inhabitants (RF 5.5)	Inhabitants (RF 7.0)
2632	1004	462	60	7.70	1849	2542	3235

Analysed	
July 24, 2023 11:03:06 AM	
Computation with 5 linked points (*italics* points are less reliable):	1. Sint-Clemenskerk 2. Markt 3. Onnastraat 4. Kornputsingel 5. Scholestraat
Reference points (Reliability 0–5)	5
Transformation with 3 parameters	
Scale	1:7750
Rotation	12 °
Mean position error	11m

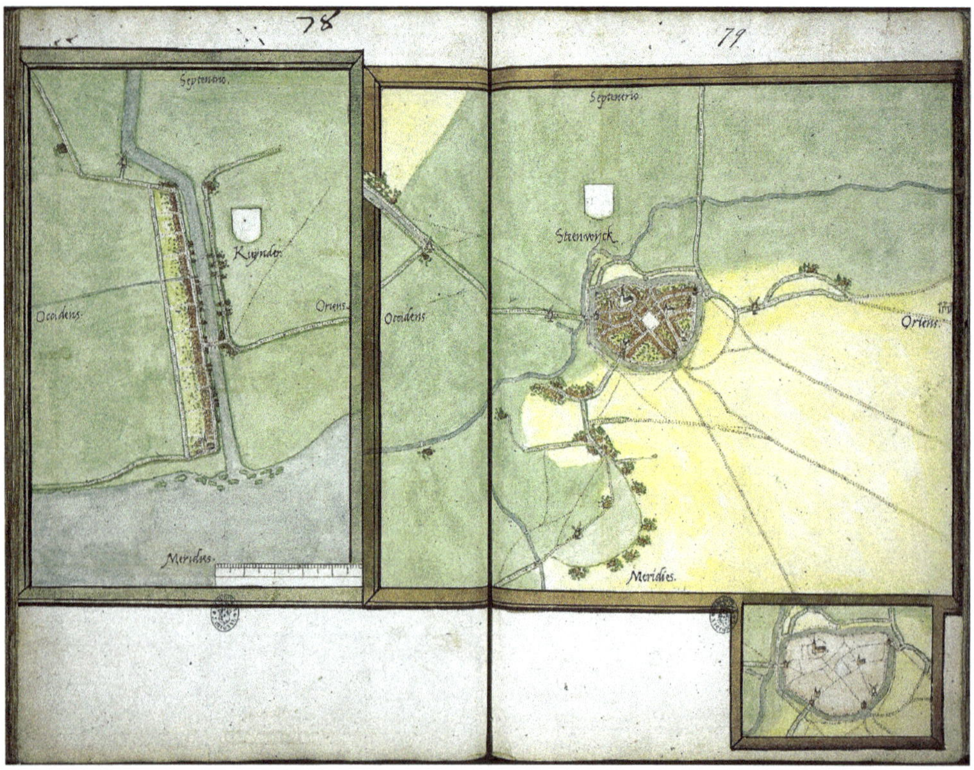

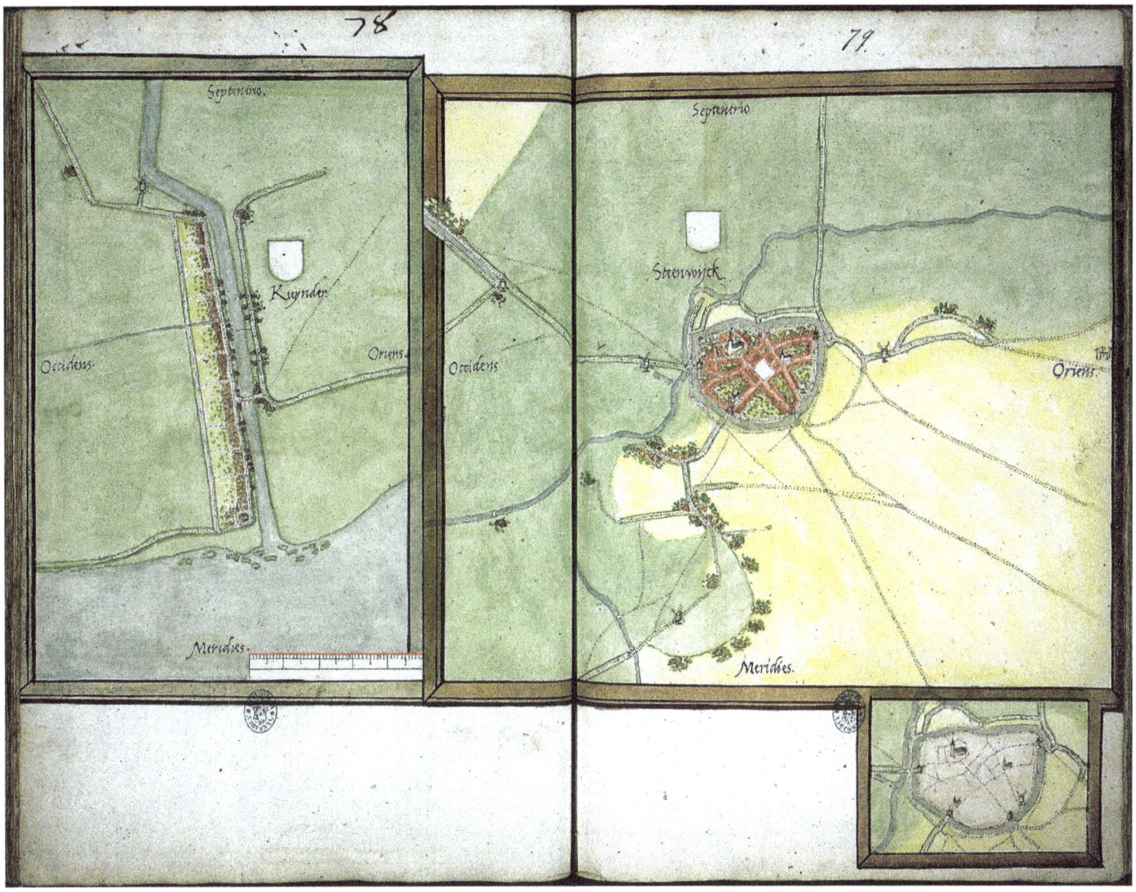
78
79
Septentrio.
Septentrio.
Kuynder.
Steenwijck.
Occidens.
Oriens.
Occidens.
Oriens.
Meridies.
Meridies.

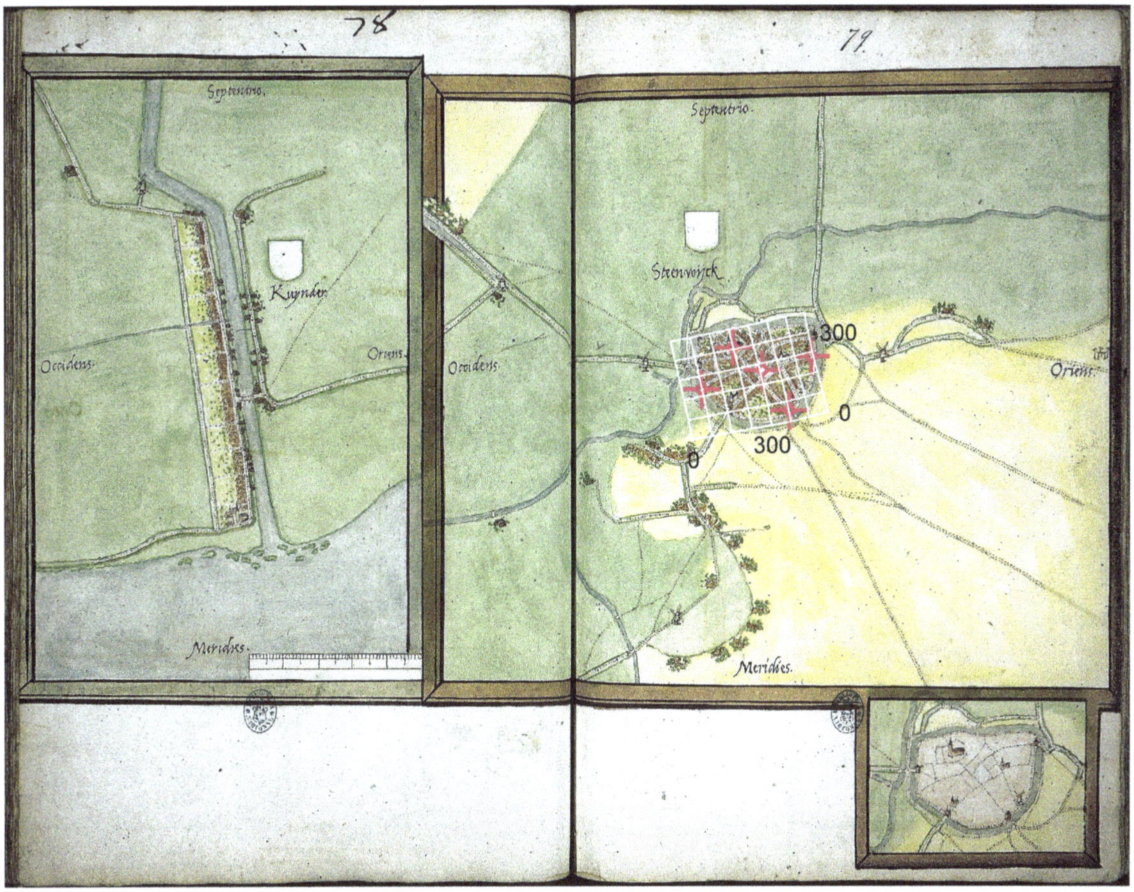
78
79
Septentrio.
Septentrio.
Kuynder.
Steenwijck.
Occidens.
Oriens.
Occidens.
Oriens.
300
0
300
0
Meridies.
Meridies.

188. Straelen

Tables 3.188 (measured and analysed).
Figures 3.188 (unmeasured, measured and analysed).

Measured							
Pixels	Scale stick original map (pixels)	Houses	Street sides	Houses per street side	Inhabitants (RF 4.0)	Inhabitants (RF 5.5)	Inhabitants (RF 7.0)
1357	1007	238	39	6.11	953	1311	1669

Analysed	
July 24, 2023 11:38:36 AM	
Computation with 5 linked points (*italics* points are less reliable):	1. Pfarrkirche St. Peter und Paul 2. Nordwall 3. Südwall 4. Westwall 5. Ostwall
Reference points (Reliability 0–5)	5
Transformation with 3 parameters	
Scale	1:9050
Rotation	11 °
Mean position error	9m

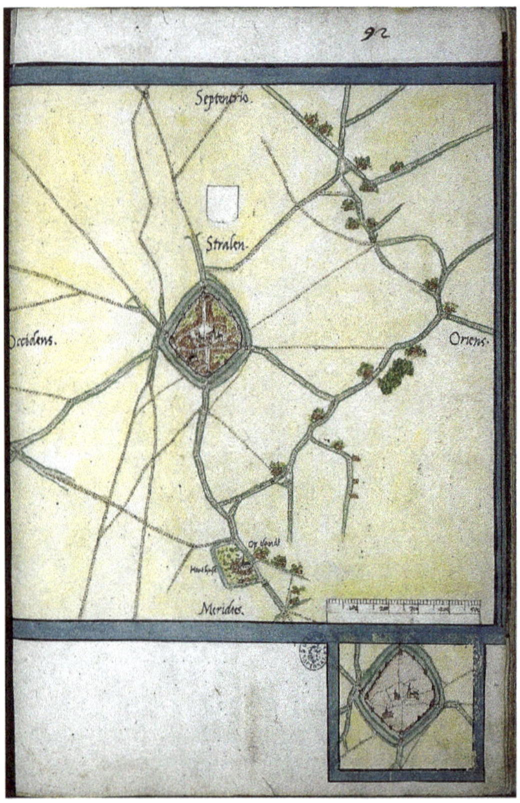

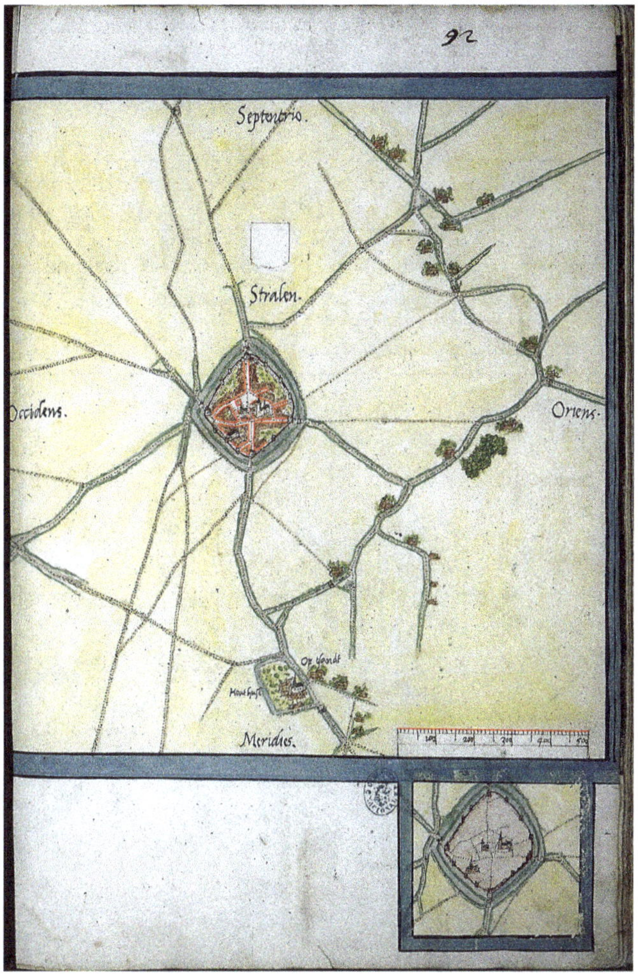
92
Septentrio.
Stralen.
Occidens.
Oriens.
Meridies.

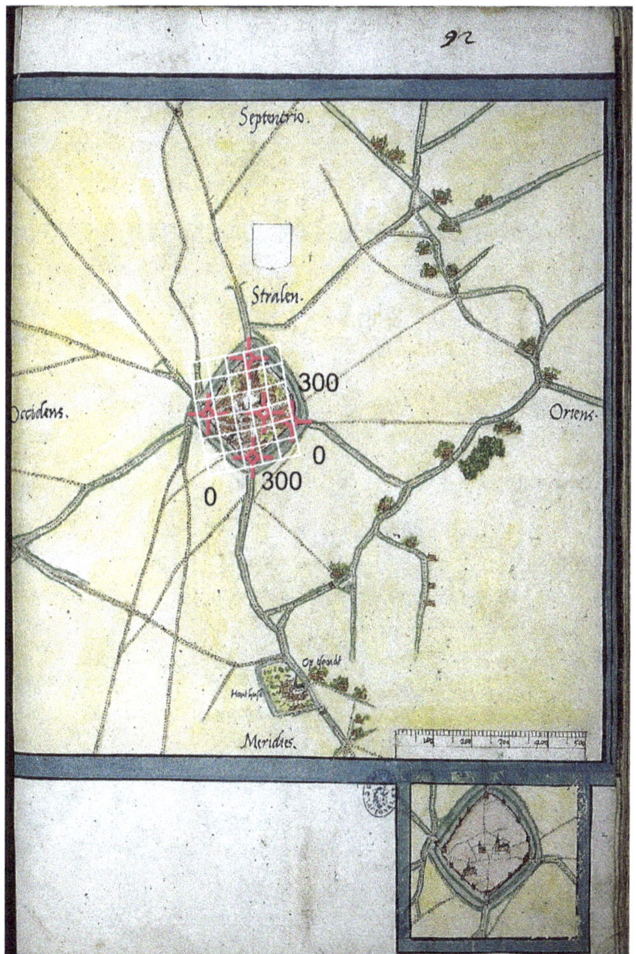
92
Septentrio.
Stralen.
300
Occidens.
Oriens.
0
0
300
Meridies.

189. Thérouanne

Tables 3.189 (measured and analysed).
Figures 3.189 (unmeasured and analysed).

Measured							
Pixels	Scale stick original map (pixels)	Houses	Street sides	Houses per street side	Inhabitants (RF 4.0)	Inhabitants (RF 5.5)	Inhabitants (RF 7.0)
0	1035	0	0	0	0	0	0

Analysed	
February 11, 2023 6:06:57 PM	
Computation with 5 linked points (*italics* points are less reliable):	1. Grande Rue 2. Remparts nord-ouest 3. Remparts nord-est 4. Remparts ouest *5. Remparts sud-ouest*
Reference points (Reliability 0–5)	4
Transformation with 3 parameters	
Scale	1:8170
Rotation	14 °
Mean position error	23m

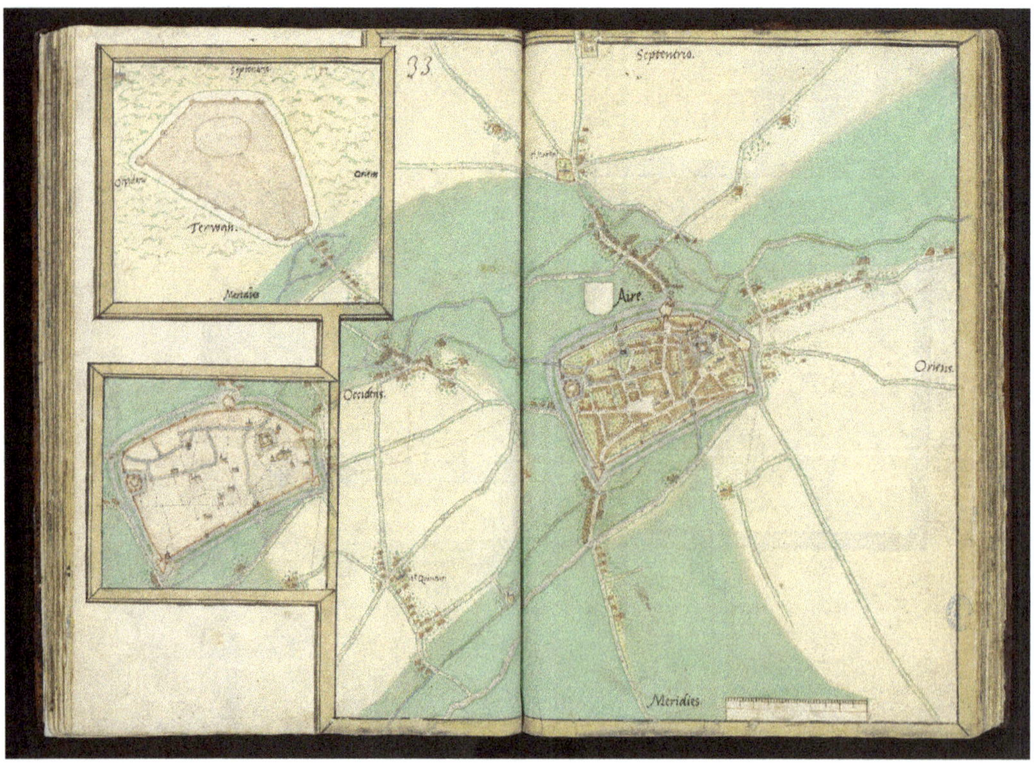

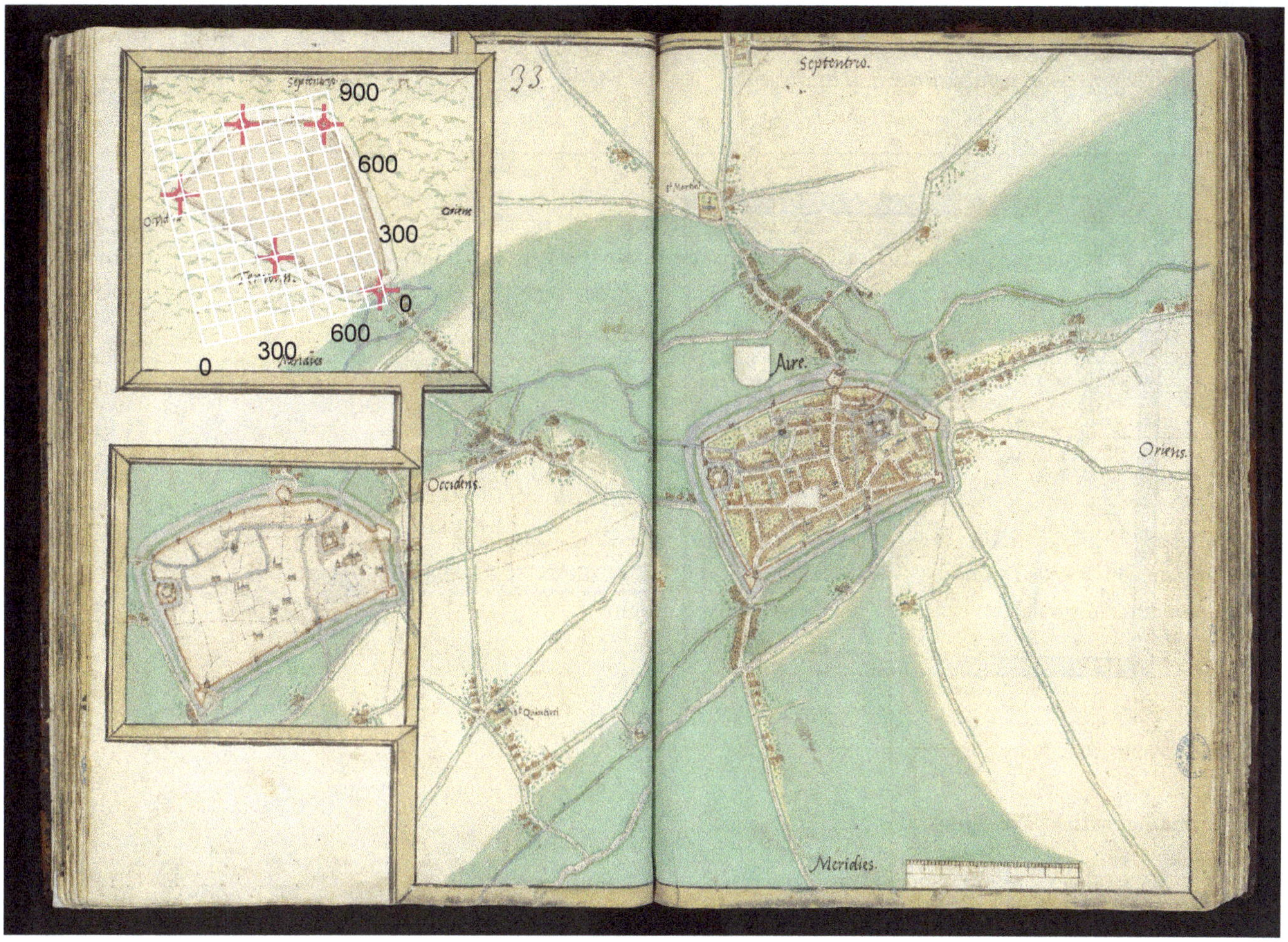

Until the mid-sixteenth century, the town of Thérouanne was a French enclave in the Netherlands. In 1553, the city was destroyed by order of Charles V. When Jacob Van Deventer drew the nearby town of Aire-sur-la-Lys, he drew a side map of Thérouanne. Only the ramparts can be seen on that map. Street side measurements are therefore not possible. However, the contours of the old town can be seen on the contemporary.

190. Thionville

Tables 3.190 (measured and analysed).
Figures 3.190 (unmeasured, measured and analysed).

Measured							
Pixels	Scale stick original map (pixels)	Houses	Street sides	Houses per street side	Inhabitants (RF 4.0)	Inhabitants (RF 5.5)	Inhabitants (RF 7.0)
3737	missing	656	133	4.93	2625	3609	4593

Analysed	
September 28, 2023 10:39:01 AM	
Computation with 5 linked points (*italics* points are less reliable):	1. Tour aux Puces 2. Église Saint-Maximin 3. Bastion de Saint-François 4. Parvis des Capucins 5. Place de la Veille Porte
Reference points (Reliability 0–5)	5
Transformation with 3 parameters	
Scale	1:9090
Rotation	15 °
Mean position error	12m

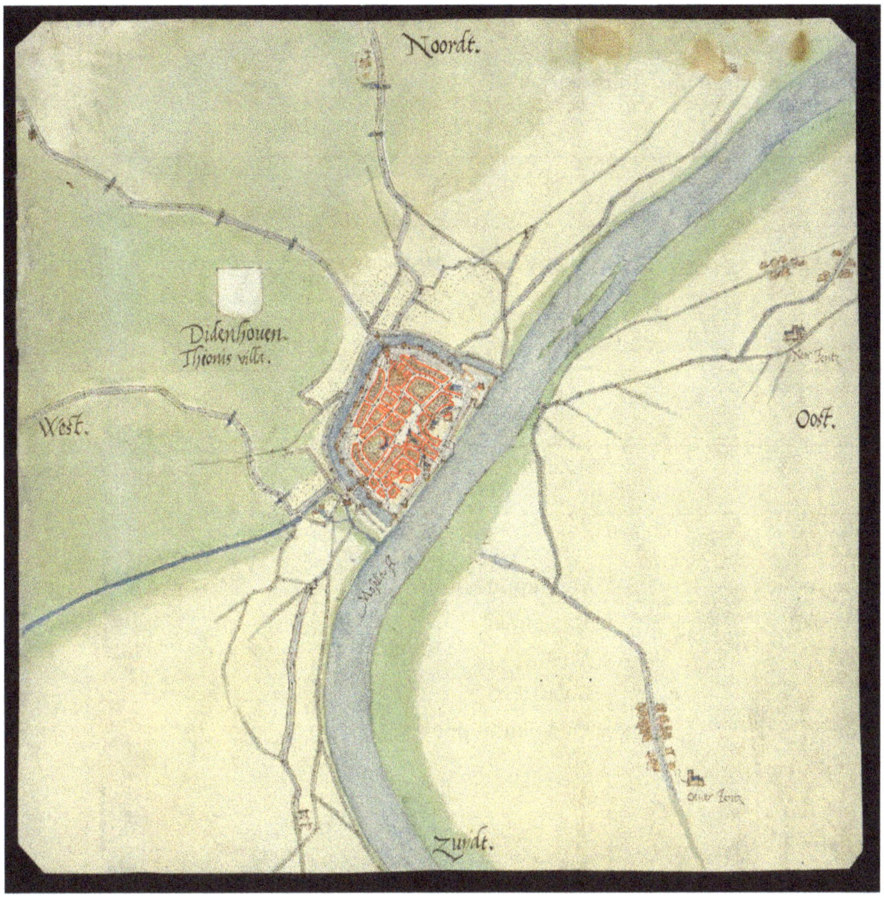
Noordt.
Didenhouen.
Thionis villa.
West.
Oost.
Zuydt.

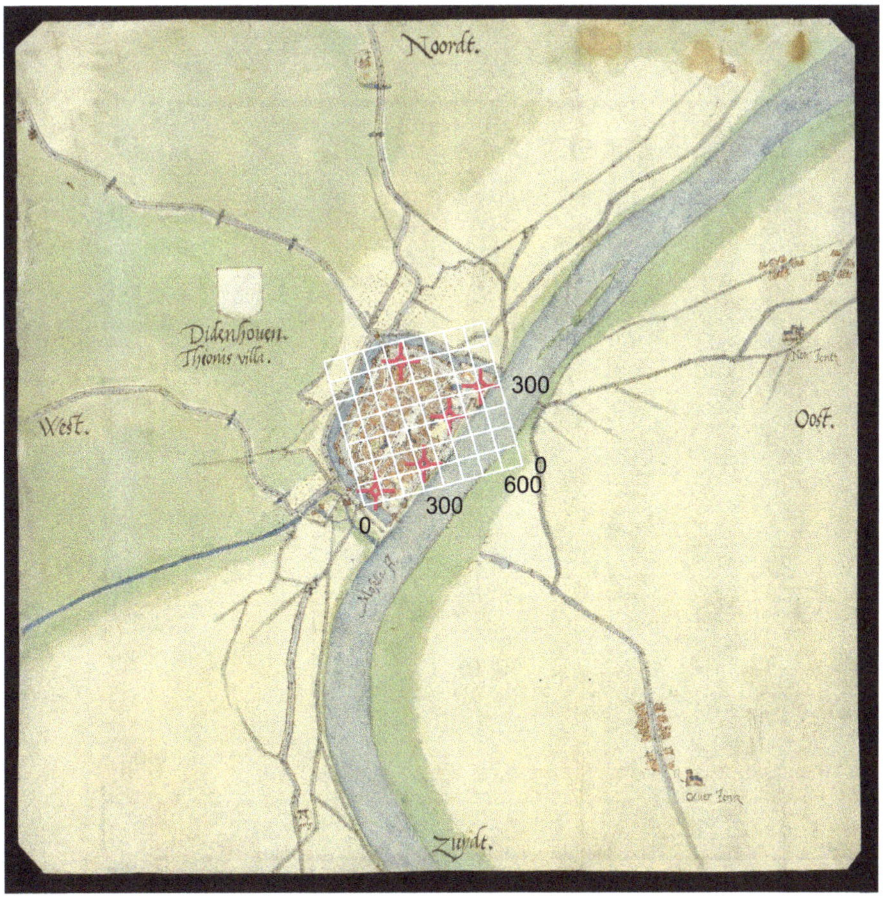
Noordt.
Didenhouen.
Thionis villa.
West.
Oost.
Zuydt.
300
0
600
300
0

191. Tholen

Tables 3.191 (measured and analysed).
Figures 3.191 (unmeasured, measured and analysed).

Measured							
Pixels	Scale stick original map (pixels)	Houses	Street sides	Houses per street side	Inhabitants (RF 4.0)	Inhabitants (RF 5.5)	Inhabitants (RF 7.0)
2045	1013	359	47	7.64	1436	1975	2514

Analysed	
January 23, 2023 10:49:28 AM	
Computation with 5 linked points (*italics* points are less reliable):	1. Minneputterstraat 2. Brugstraat *3. Haven* *4. Bolwerk* 5. Oudelandsepoort
Reference points (Reliability 0–5)	**3**
Transformation with 3 parameters	
Scale	1:8450
Rotation	8 °
Mean position error	23m

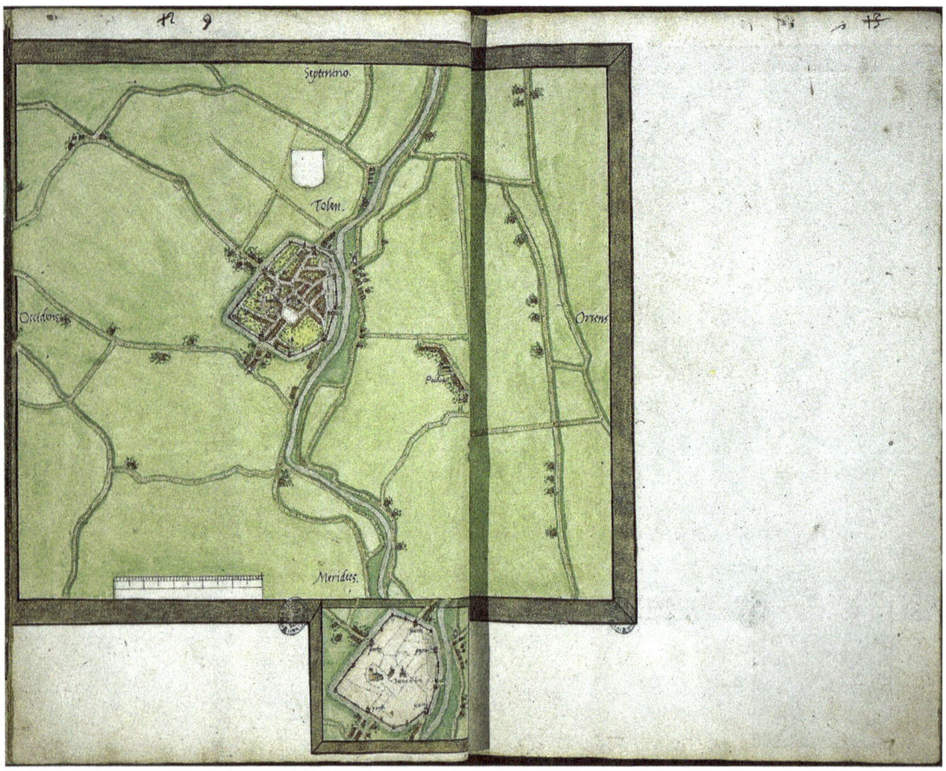

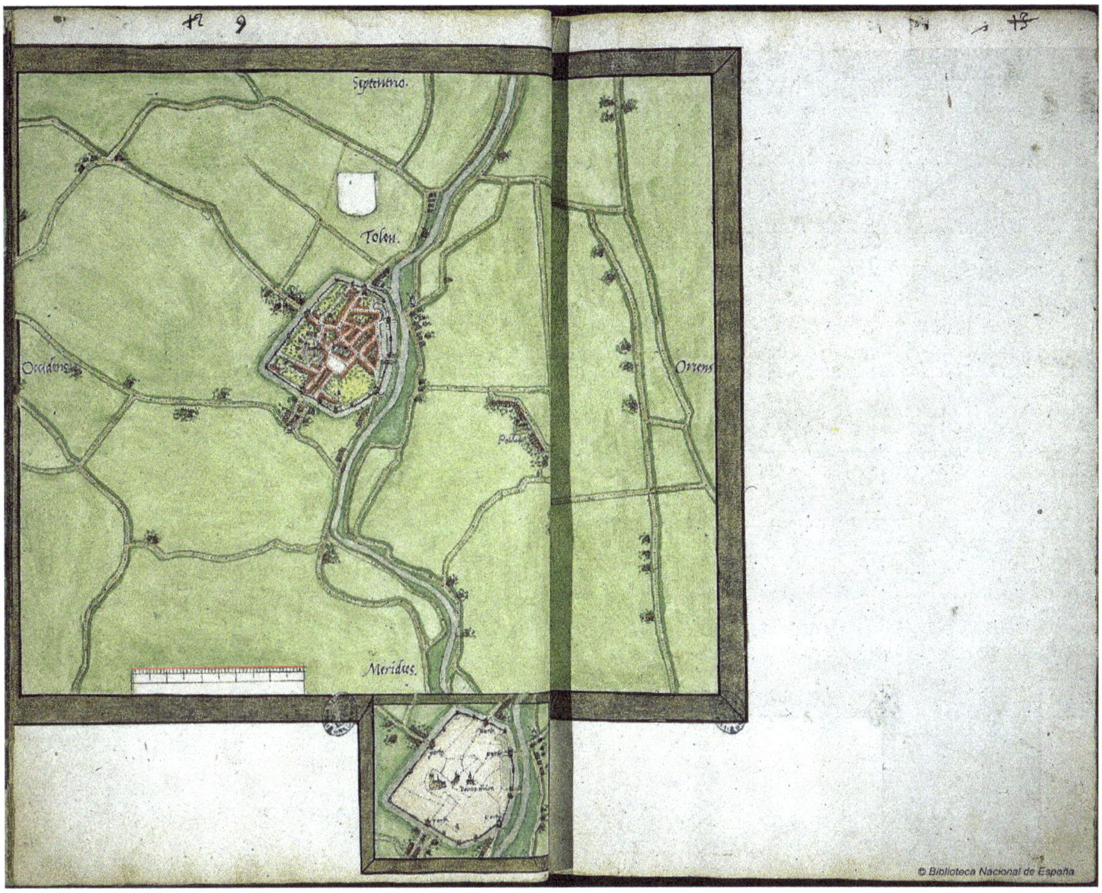
Septentrio.
Tolen.
Occidens
Oriens
Meridies.
© Biblioteca Nacional de España

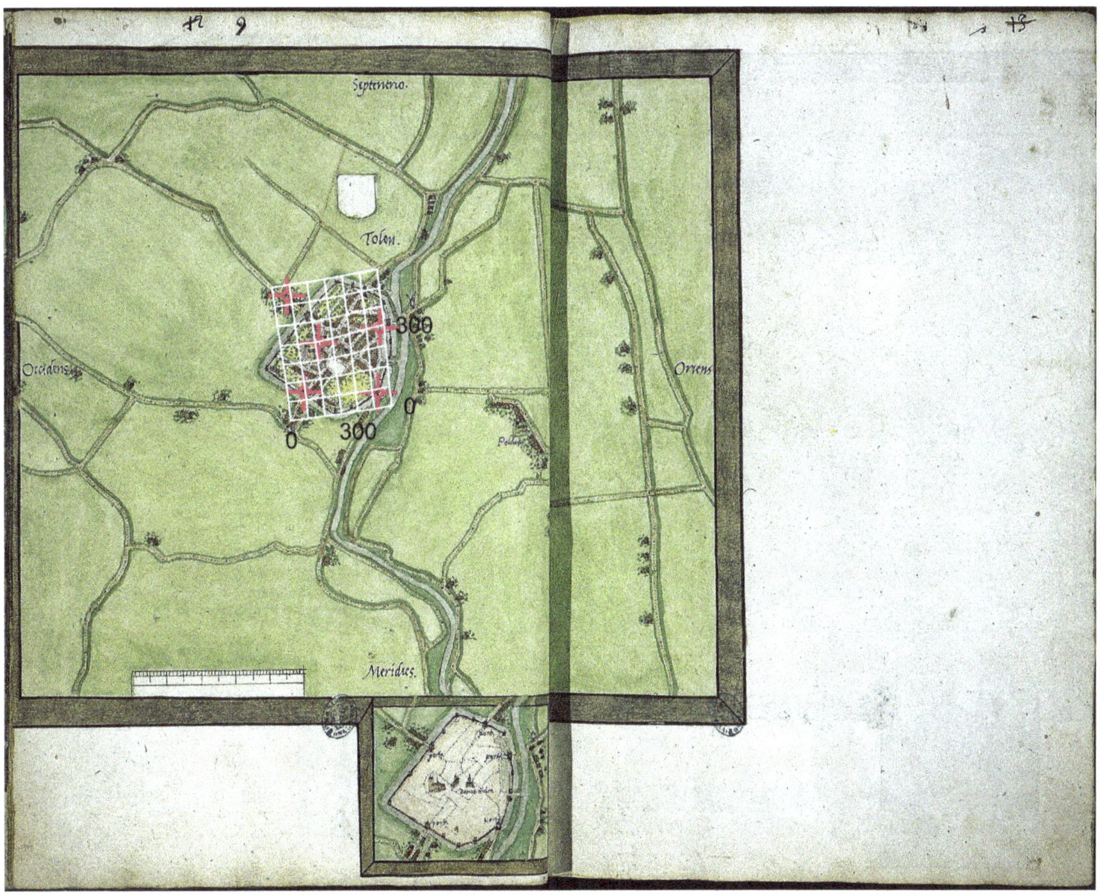
Septentrio.
Tolen.
Occidens
Oriens
Meridies.
300
0
0
300

192. Tiel

Tables 3.192 (measured and analysed).
Figures 3.192 (unmeasured, measured and analysed).

Measured							
Pixels	Scale stick original map (pixels)	Houses	Street sides	Houses per street side	Inhabitants (RF 4.0)	Inhabitants (RF 5.5)	Inhabitants (RF 7.0)
3912	999	687	99	6.94	2748	3778	4808

Analysed	
July 26, 2023 1:00:39 PM	
Computation with 5 linked points (*italics* points are less reliable):	1. Sint-Maartenskerk 2. Ambtmanshuis 3. Sint Caeciliakapel 4. Oliemolenwal 5. Molenhoek
Reference points (Reliability 0–5)	5
Transformation with 3 parameters	
Scale	1:8940
Rotation	13 °
Mean position error	13m

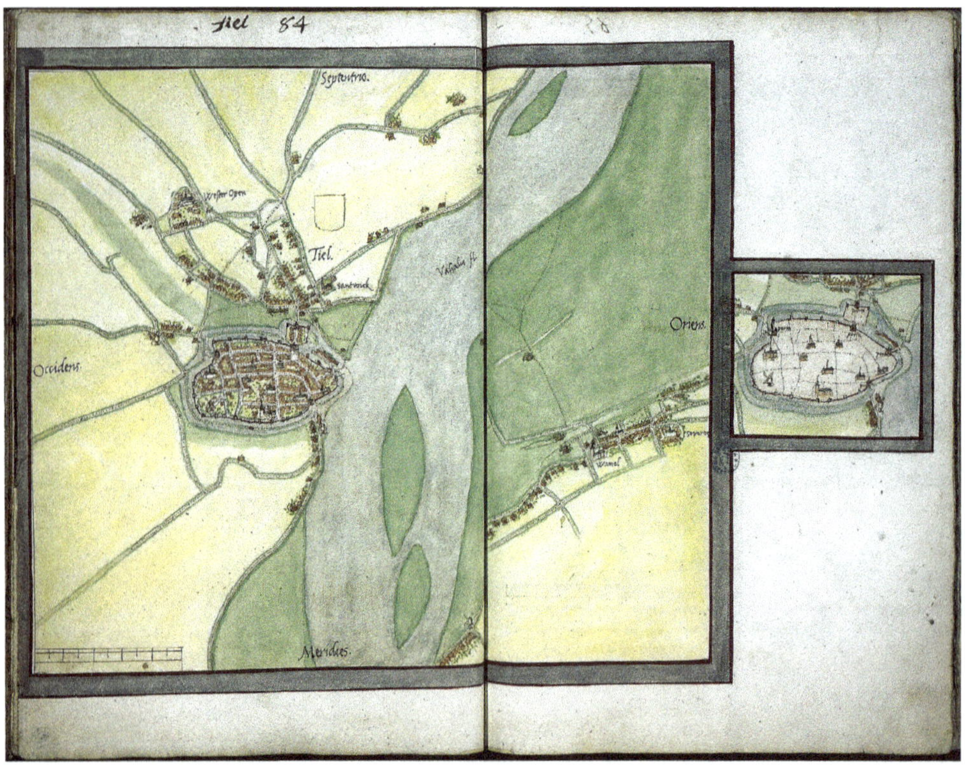

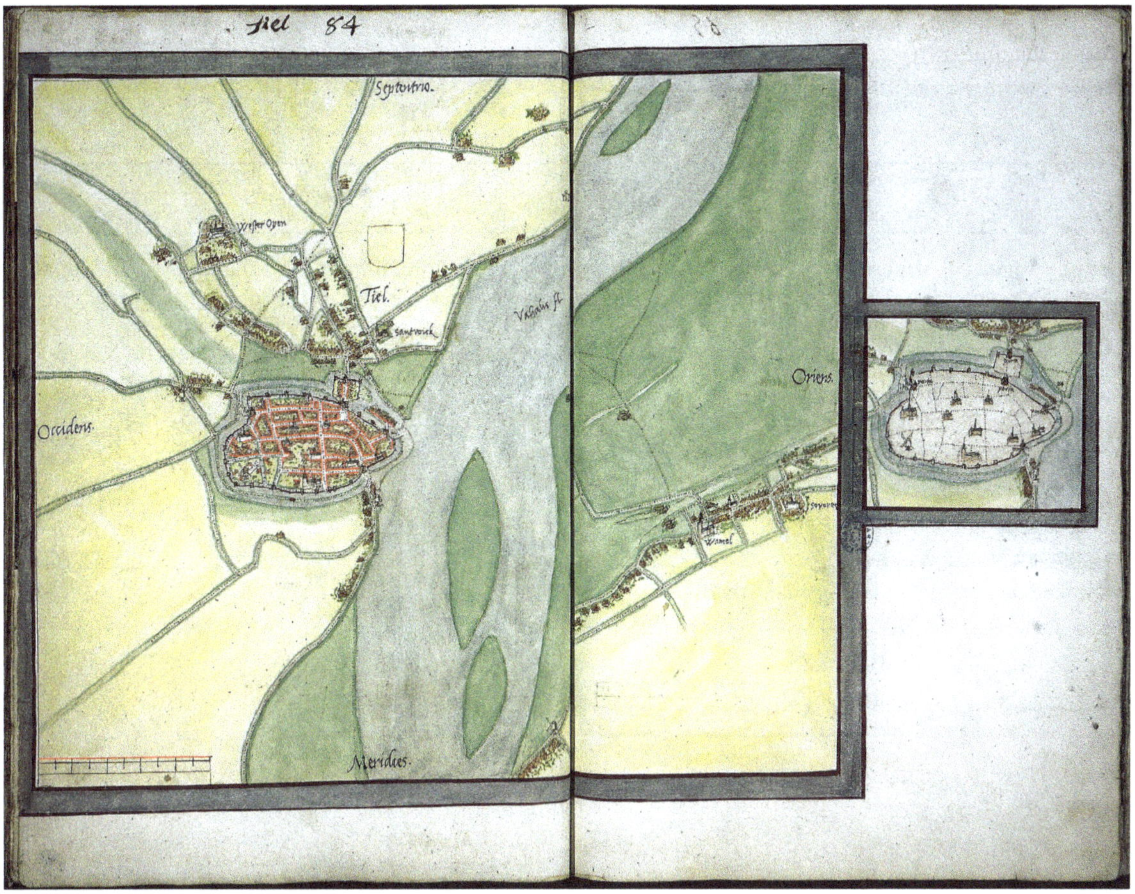
Tiel 84
Septentrio.
Tiel.
Occidens.
Oriens.
Meridies.

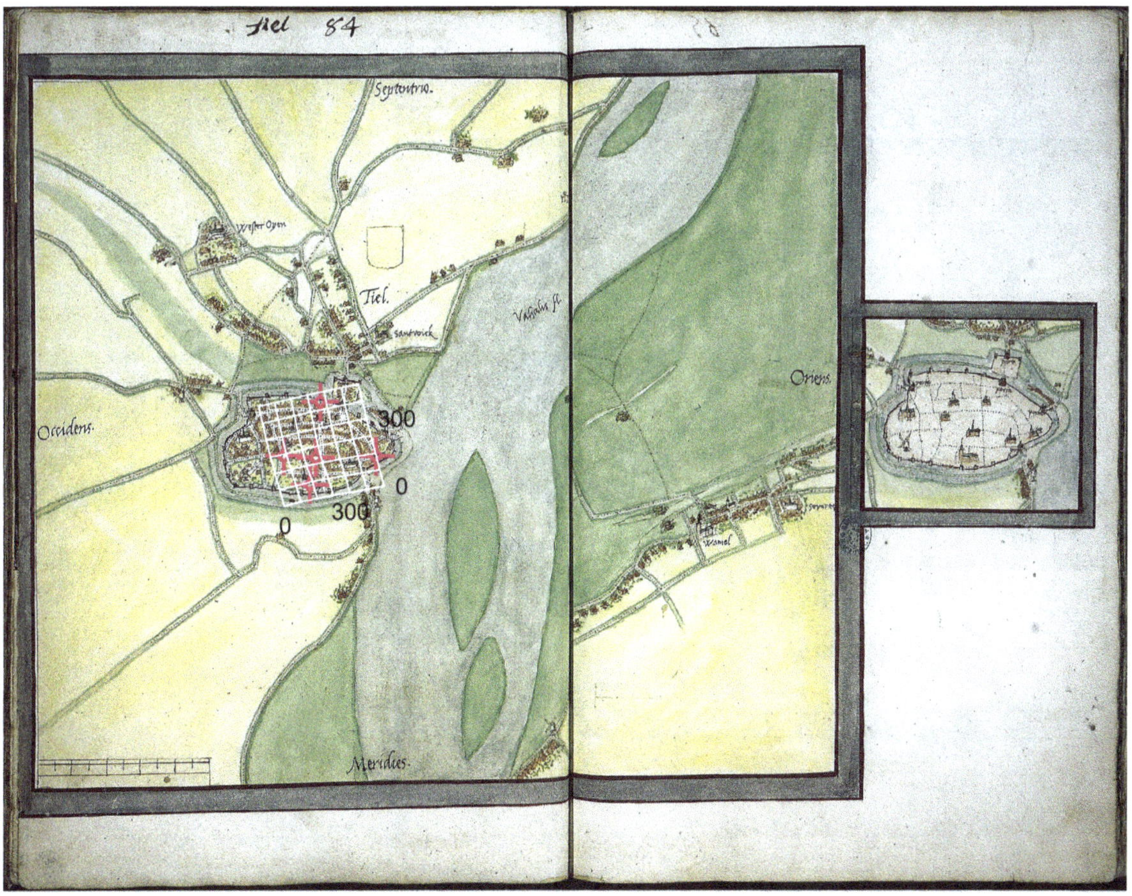
Tiel 84
Septentrio.
Tiel.
Occidens.
Oriens.
300
0
0
300
Meridies.

193. Tienen

Tables 3.193 (measured and analysed).
Figures 3.193 (unmeasured, measured and analysed).

Measured

Pixels	Scale stick original map (pixels)	Houses	Street sides	Houses per street side	Inhabitants (RF 4.0)	Inhabitants (RF 5.5)	Inhabitants (RF 7.0)
11,597	missing	2036	256	7.95	8145	11,200	14,255

Analysed

September 28, 2023 12:15:04 PM	
Computation with 5 linked points (*italics* points are less reliable):	1. Sint-Germanuskerk 2. Onze-Lieve-Vrouw-ten-Poelkerk 3. Paterskerk 4. Reizigersstraat 5. Grote Gete
Reference points (Reliability 0–5)	5
Transformation with 3 parameters	
Scale	1:8510
Rotation	13 °
Mean position error	12 m

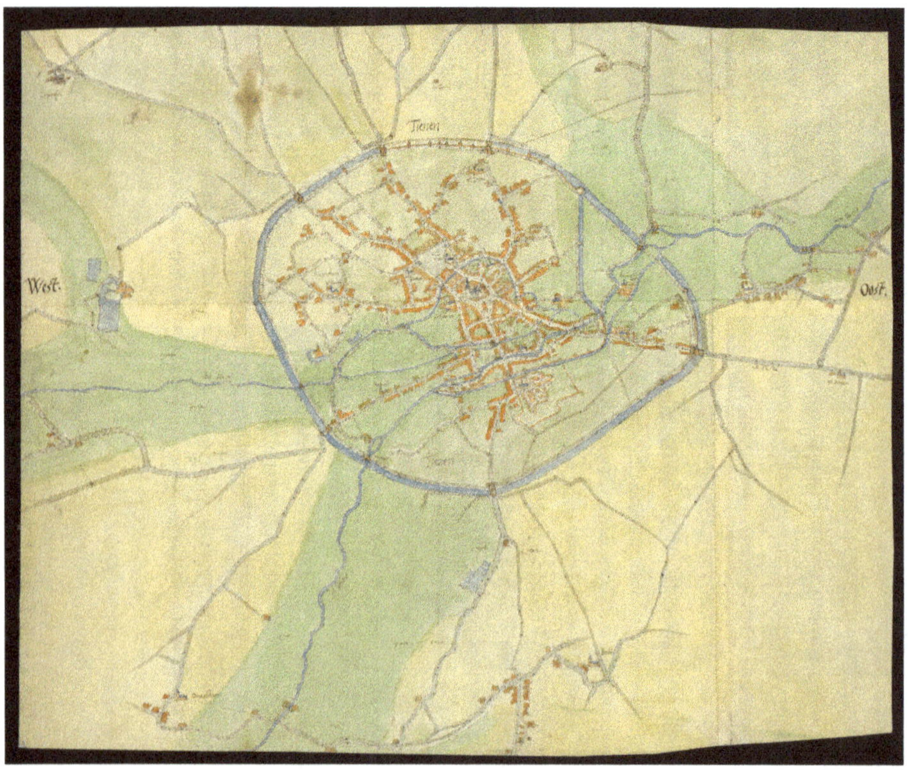

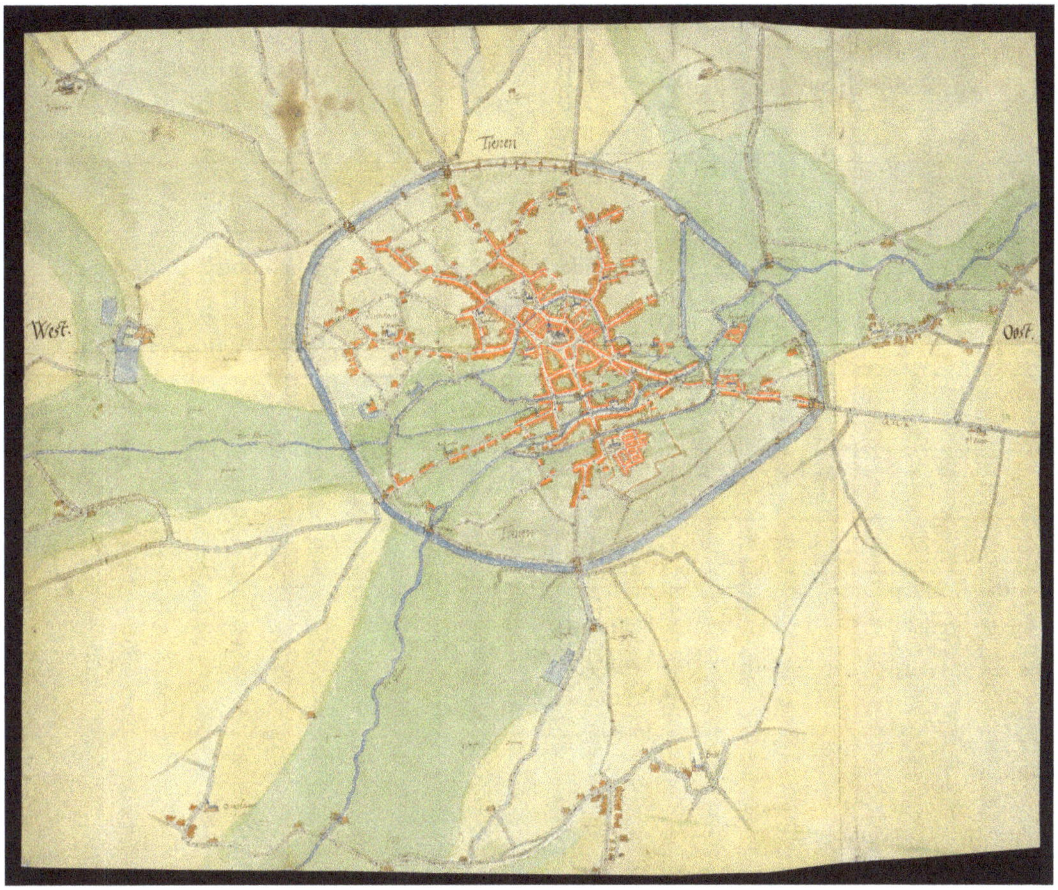
Tienen
West.
Oost.
Tienen

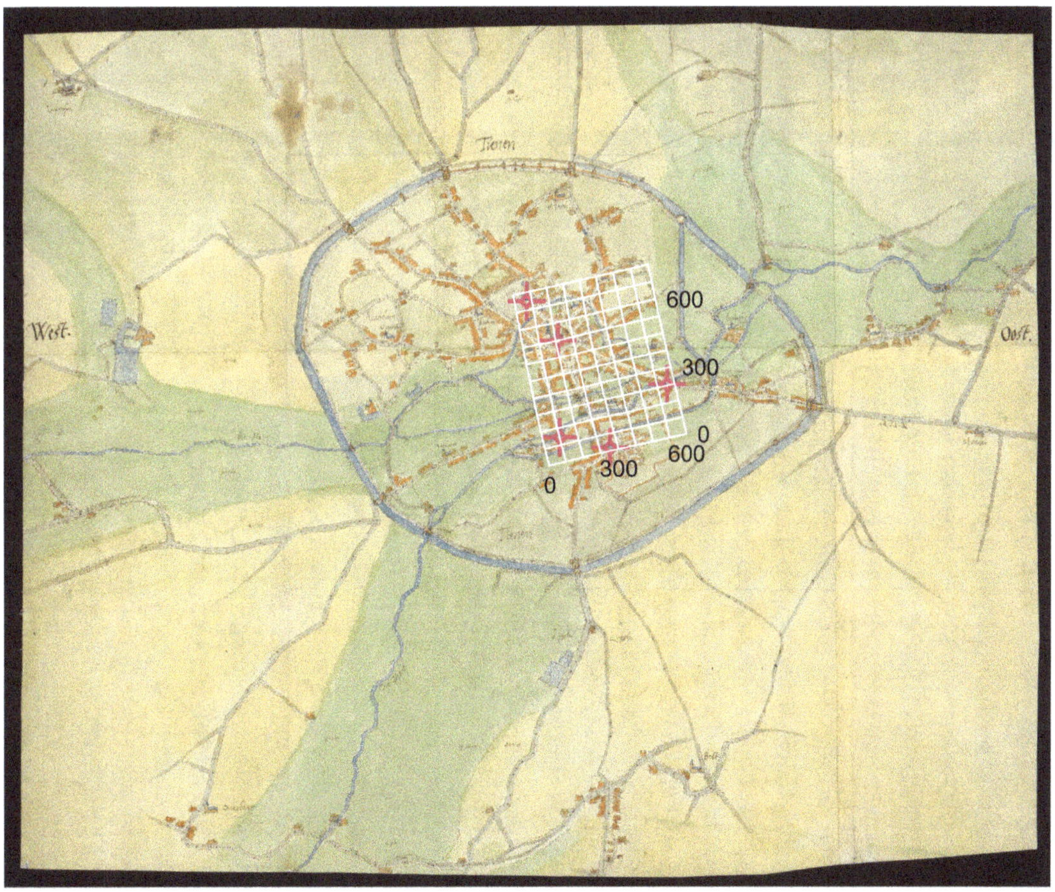
Tienen
West.
Oost.
600
300
0
600
0
300
Tienen

194. Tournai

Tables 3.194 (measured and analysed).
Figures 3.194 (unmeasured, measured and analysed).

Measured							
Pixels	Scale stick original map (pixels)	Houses	Street sides	Houses per street side	Inhabitants (RF 4.0)	Inhabitants (RF 5.5)	Inhabitants (RF 7.0)
26,844	861	4714	558	8.45	18,855	25,926	32,997

Analysed	
August 1, 2023 8:03:16 PM	
Computation with 5 linked points (*italics* points are less reliable):	1. Pont de Fer 2. Sint-Piatuskerk 3. Sint-Kwintenskerk 4. Sint-Brixiuskerk 5. Rue du Ballon
Reference points (Reliability 1–5)	5
Transformation with 3 parameters	
Scale	1:8760
Rotation	12 °
Mean position error	17m

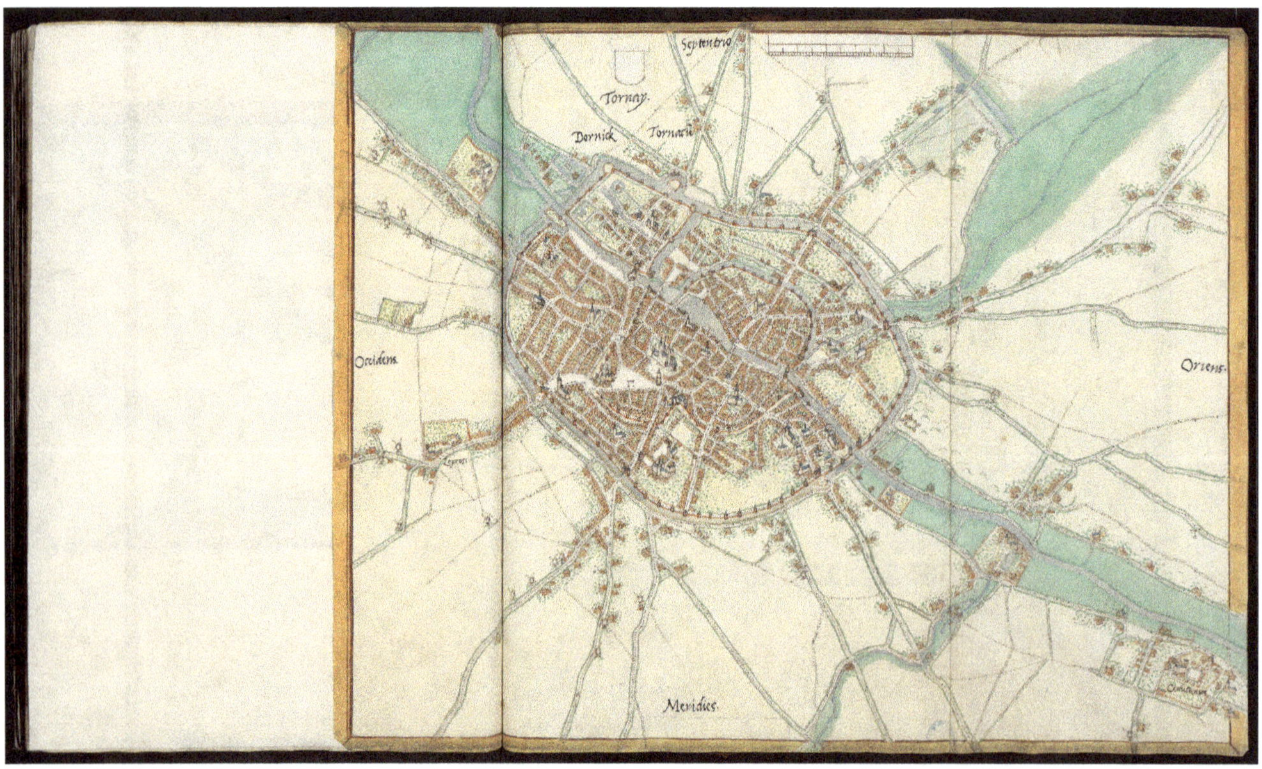

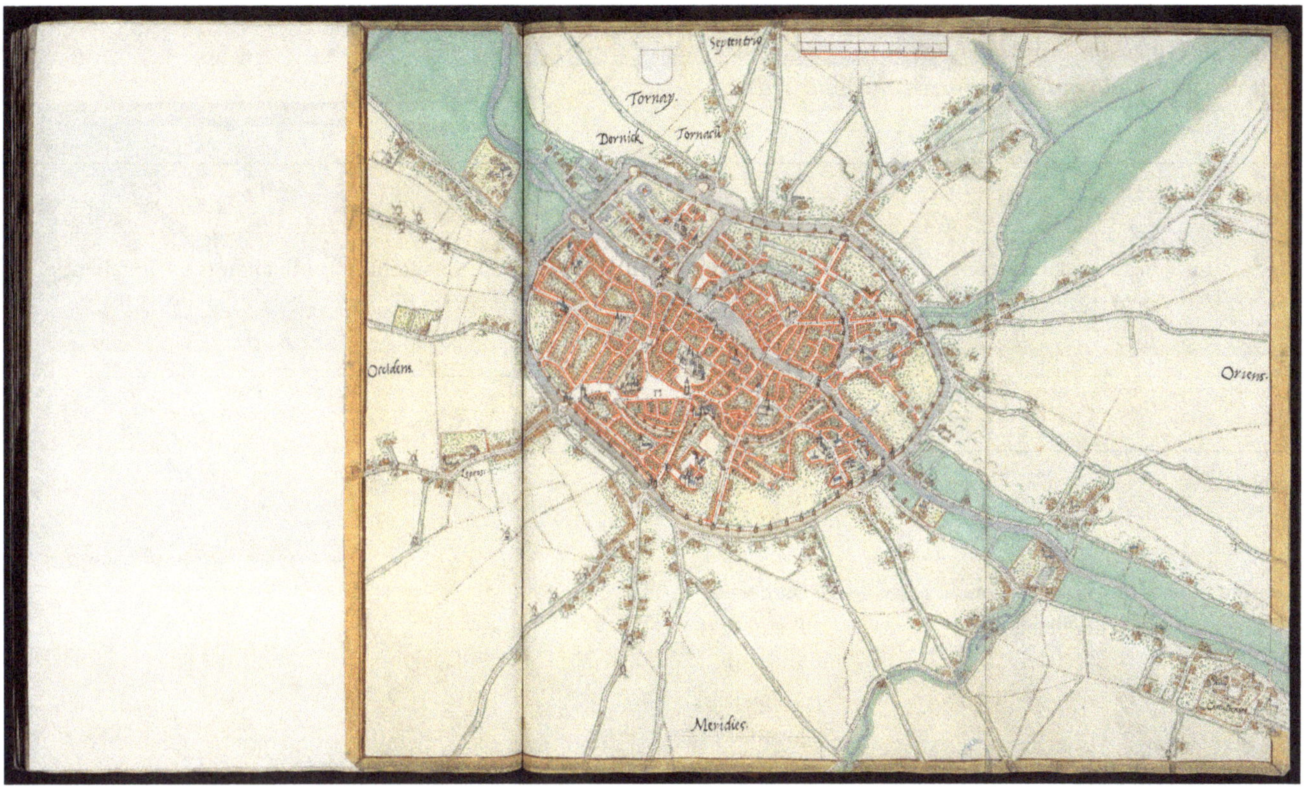
Septentrio
Tornay.
Dernick
Tornacu
Occidens
Oriens.
Meridies.

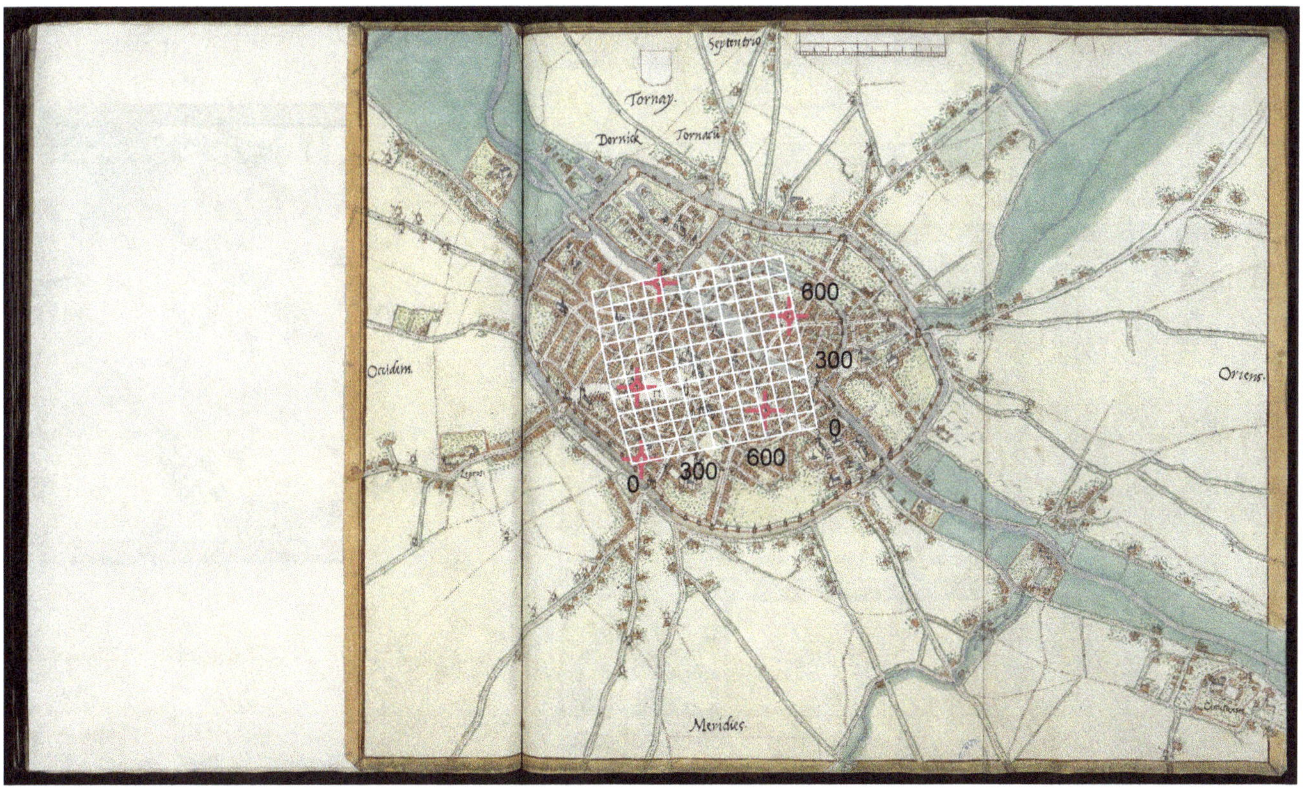
Septentrio
Tornay.
Dernick
Tornacu
Occidens
Oriens.
Meridies.
600
300
0
0
300
600

195. Utrecht

Tables 3.195 (measured and analysed).
Figures 3.195 (unmeasured, measured and analysed).

Measured							
Pixels	Scale stick original map (pixels)	Houses	Street sides	Houses per street side	Inhabitants (RF 4.0)	Inhabitants (RF 5.5)	Inhabitants (RF 7.0)
22,616	1010	3971	517	7.68	15,886	21,843	27,800

Analysed	
August 3, 2023 10:22:15 AM	
Computation with 5 linked points (*italics* points are less reliable):	1. Domkerk 2. Pieterskerk 3. Janskerk 4. Nicolaikerk 5. Jacobikerk
Reference points (Reliability 1–5)	5
Transformation with 3 parameters	
Scale	1:8410
Rotation	14 °
Mean position error	19m

Septentrio
Traiectum Inferius
Vtrecht
Meridies

Septentrio
1200
900
600
300
0
600
300
0
Traiectum Inferius
Vtrecht
Meridies

196. Valenciennes

Tables 3.196 (measured and analysed).
Figures 3.196 (unmeasured, measured and analysed).

Measured

Pixels	Scale stick original map (pixels)	Houses	Street sides	Houses per street side	Inhabitants (RF 4.0)	Inhabitants (RF 5.5)	Inhabitants (RF 7.0)
20,517	897	3603	573	6.29	14,411	19,815	25,219

Analysed

August 4, 2023 9:23:01 AM	
Computation with 5 linked points (*italics* points are less reliable):	1. La Rhonelle 2. *Vieil Escaut (nord)* 3. *Vieil Escaut (sud)* 4. Place d'Armes (nord) 5. Place d'Armes (sud)
Reference points (Reliability 0–5)	3
Transformation with 3 parameters	
Scale	1:9680
Rotation	10 °
Mean position error	28m

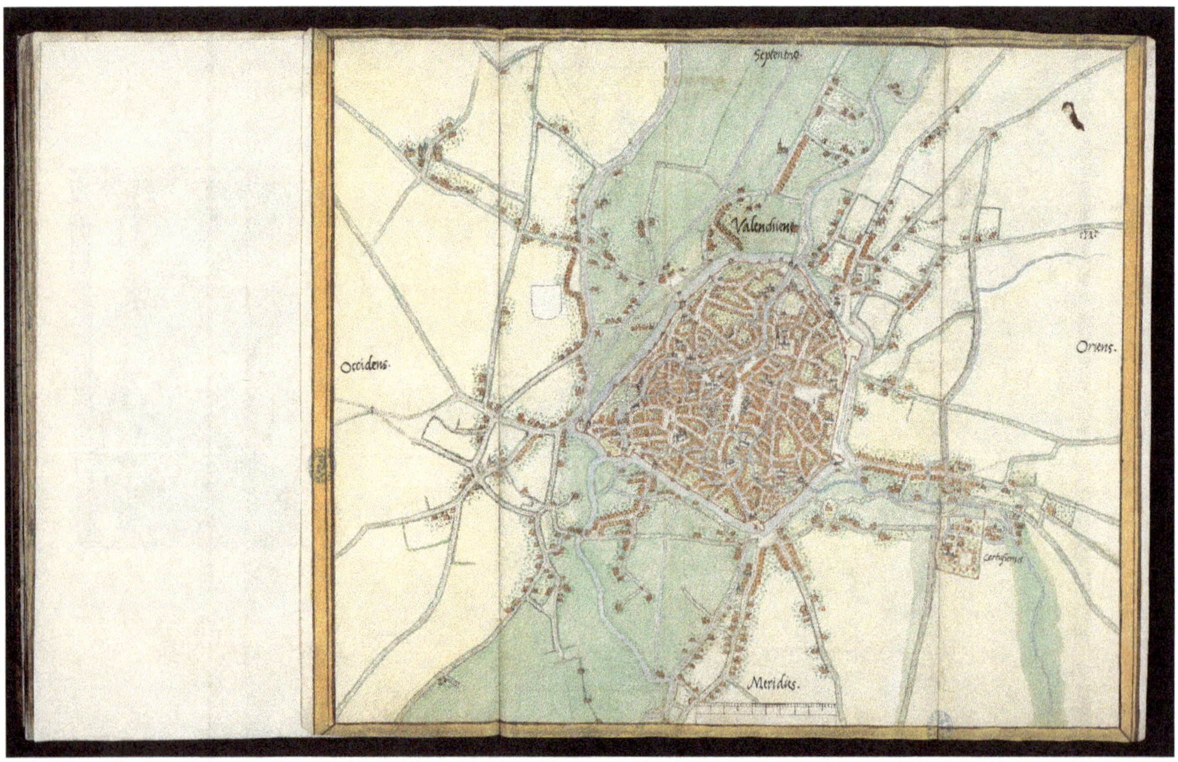

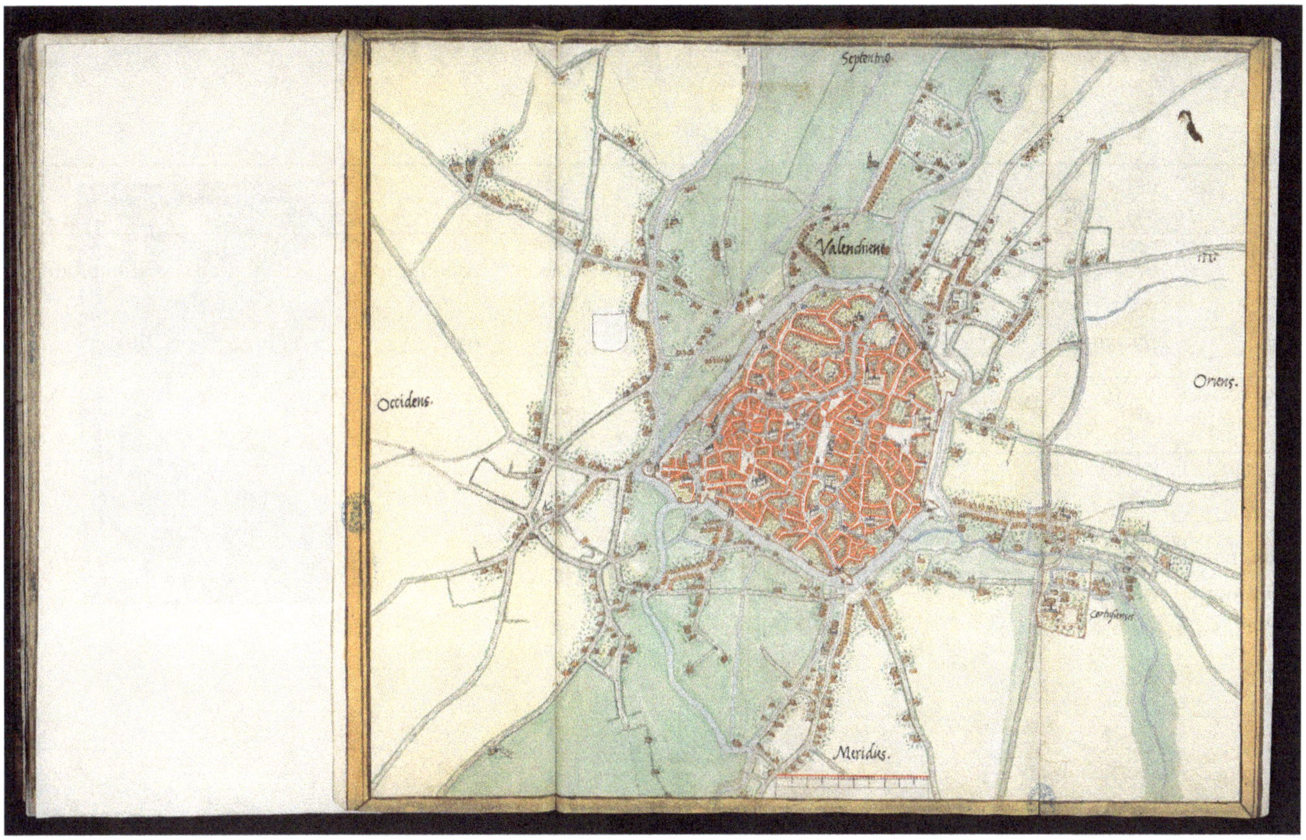
Septentrio.
Occidens.
Oriens.
Meridies.

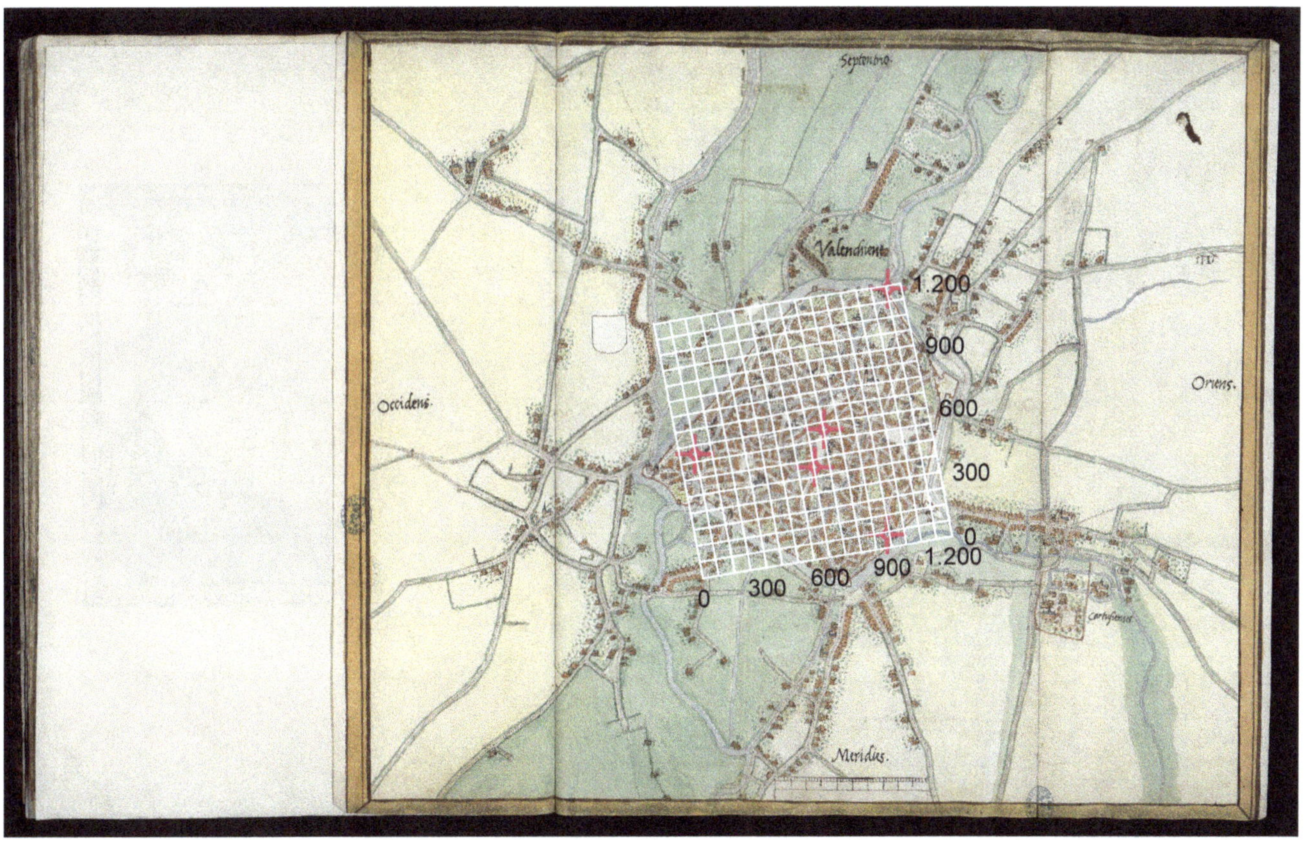
Septentrio.
Occidens.
Oriens.
Meridies.
1.200
900
600
300
0
0
300
600
900
1.200

197. Valkenburg

Tables 3.197 (measured and analysed).
Figures 3.197 (unmeasured, measured and analysed).

Measured							
Pixels	Scale stick original map (pixels)	Houses	Street sides	Houses per street side	Inhabitants (RF 4.0)	Inhabitants (RF 5.5)	Inhabitants (RF 7.0)
1336	missing	235	52	4.51	938	1290	1642

Analysed	
September 28, 2023 2:18:07 PM	
Computation with 5 linked points (*italics* points are less reliable):	1. HH. Nicolaas en Barbarakerk 2. Kasteel Valkenburg 3. Geulpoort 4. Grendelpoort 5. Walramplein
Reference points (Reliability 0–5)	5
Transformation with 3 parameters	
Scale	1:7670
Rotation	9 °
Mean position error	6m

Noordt.
Valkenburg.
West.
Oost.
Zuydt.

Noordt.
Valkenburg.
West.
Oost.
300
0
300
0
Zuydt.

198. Veere
Tables 3.198 (measured and analysed).
Figures 3.198 (unmeasured, measured and analysed).

Measured							
Pixels	Scale stick original map (pixels)	Houses	Street sides	Houses per street side	Inhabitants (RF 4.0)	Inhabitants (RF 5.5)	Inhabitants (RF 7.0)
4229	725	743	112	6.63	2970	4084	5198

Analysed	
January 23, 2023 2:11:22:46 AM	
Computation with 5 linked points (*italics* points are less reliable):	1. Stadhuis 2. Kaai 3. Oudestraat 4. Bastion 5. Baljuwstraat
Reference points (Reliability 0–5)	5
Transformation with 3 parameters	
Scale	1:8070
Rotation	14 °
Mean position error	12m

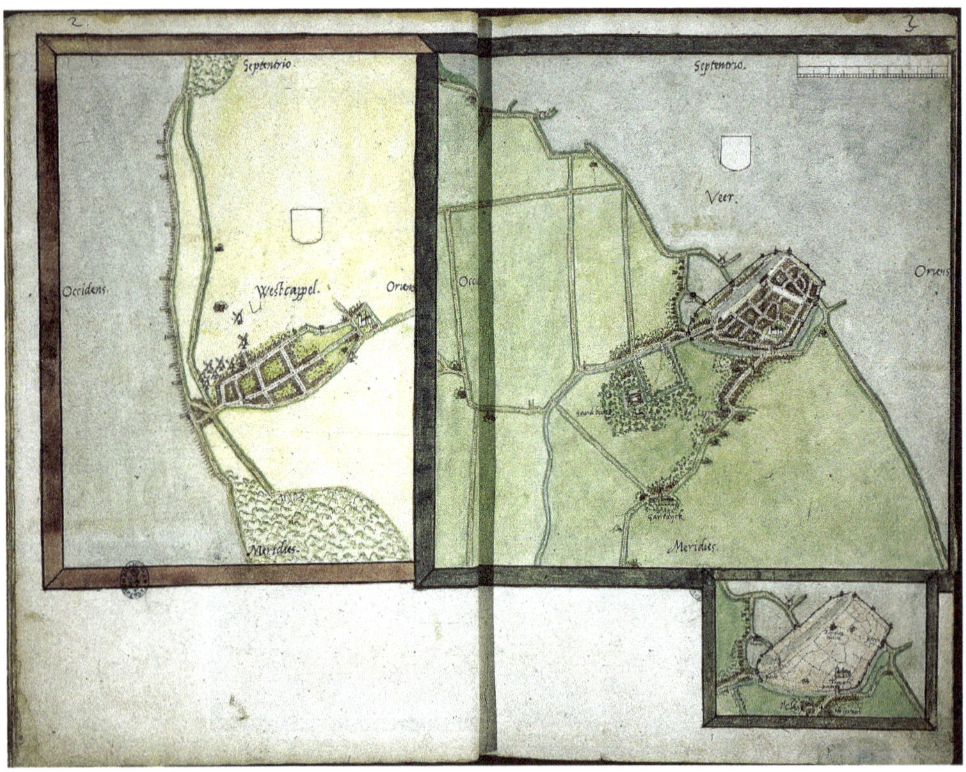

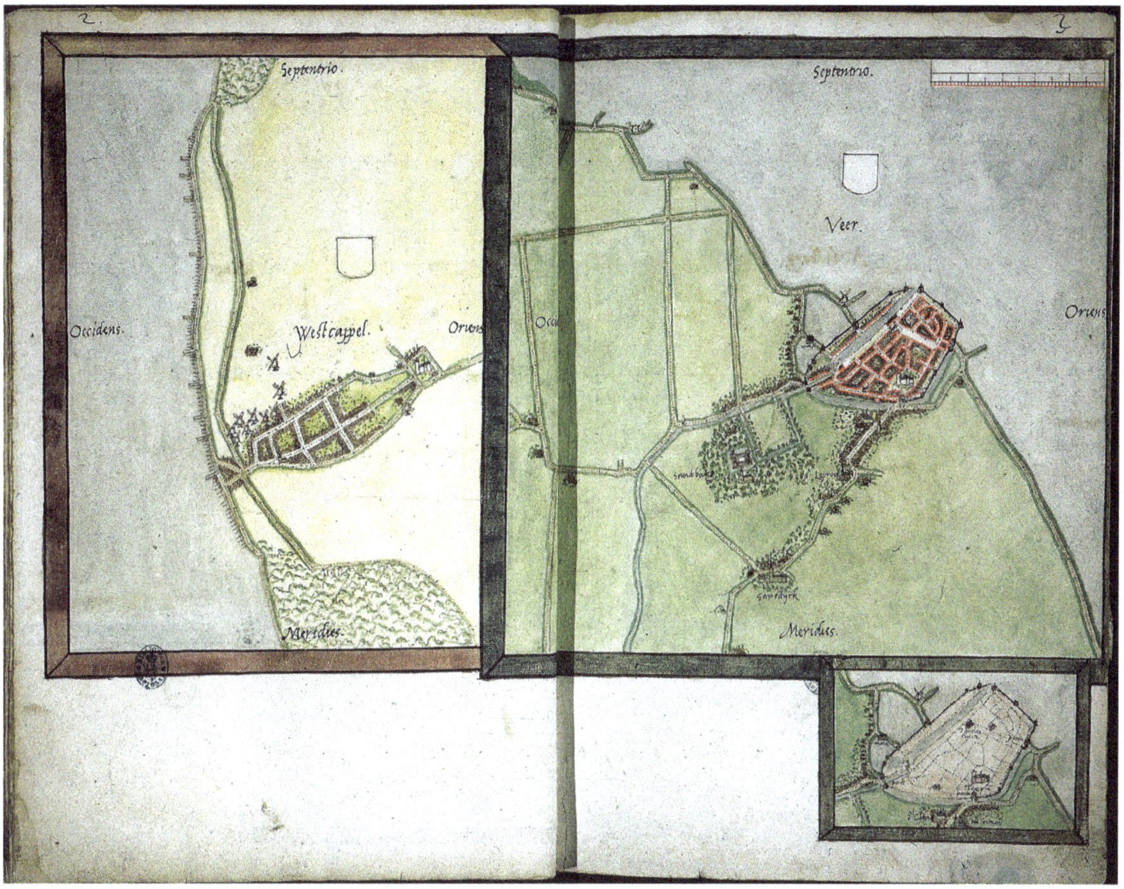
Septentrio.
Westcappel.
Occidens.
Oriens
Meridies.
Septentrio.
Veer.
Oriens
Meridies.

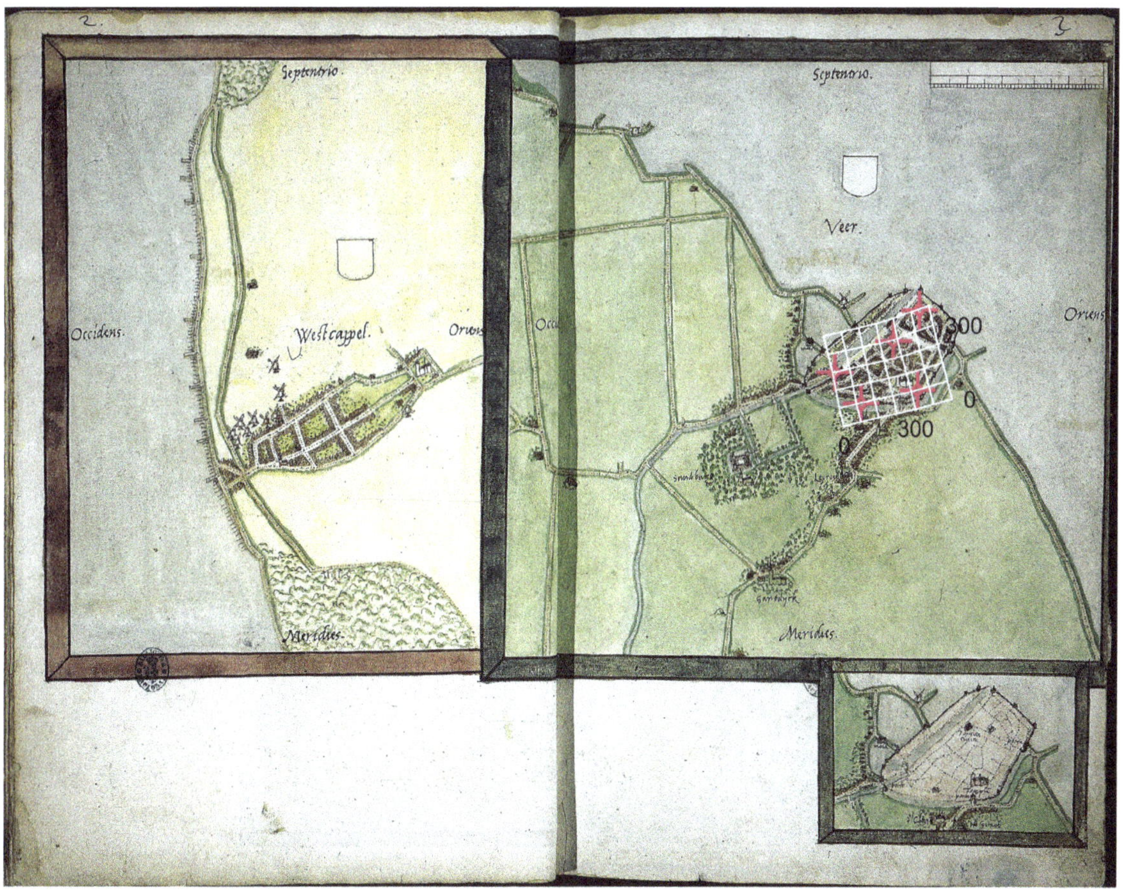
Septentrio.
Westcappel.
Occidens.
Oriens
Meridies.
Septentrio.
Veer.
Oriens
300
0
300
0
Meridies.

199. Venlo
Tables 3.199 (measured and analysed).
Figures 3.199 (unmeasured, measured and analysed).

Measured							
Pixels	Scale stick original map (pixels)	Houses	Street sides	Houses per street side	Inhabitants (RF 4.0)	Inhabitants (RF 5.5)	Inhabitants (RF 7.0)
4084	missing	717	96	7.47	2868	3944	5020

Analysed	
August 5, 2023 11:30:16 AM	
Computation with 5 linked points (*italics* points are less reliable):	1. Sint-Martinusbasiliek 2. Sint-Joriskerk 3. Maas *4. Maas (midden)* 5. Nieuwstraat
Reference points (Reliability 0–5)	4
Transformation with 3 parameters	
Scale	1:8700
Rotation	13 °
Mean position error	21m

23
Septentrio.
Venloo
Occidens
Oriens
Meridies.

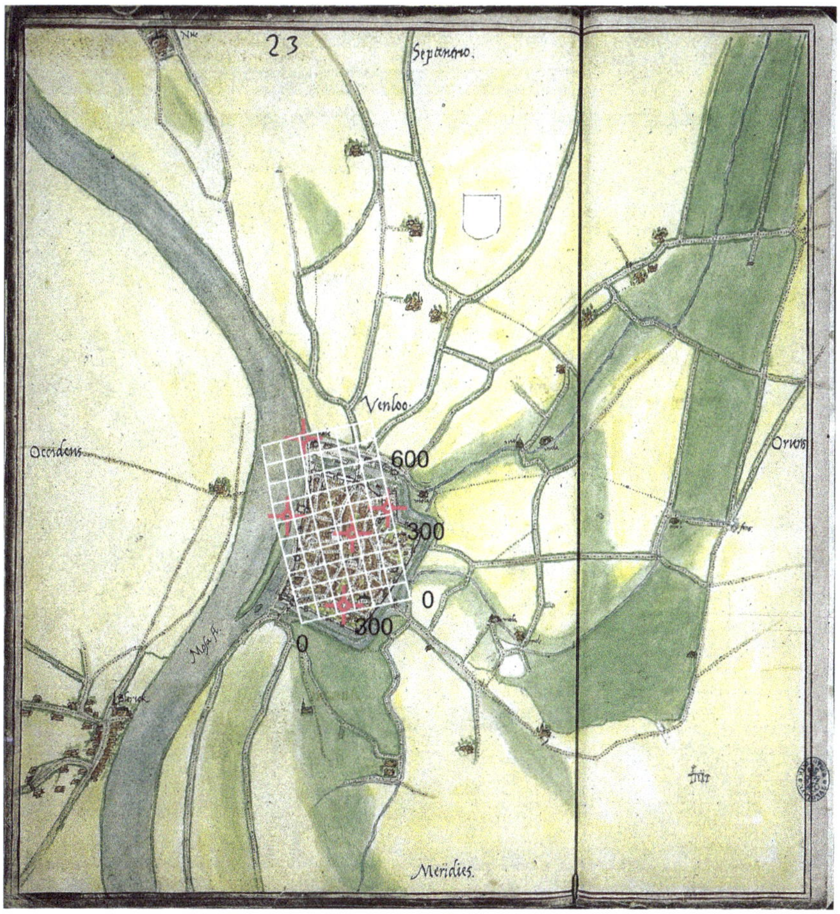
23
Septentrio.
Venloo
Occidens
Oriens
600
300
0
300
0
Meridies.

200. Veurne
Tables 3.200 (measured and analysed).
Figures 3.200 (unmeasured, measured and analysed).

Measured							
Pixels	Scale stick original map (pixels)	Houses	Street sides	Houses per street side	Inhabitants (RF 4.0)	Inhabitants (RF 5.5)	Inhabitants (RF 7.0)
4932	1026	866	120	7.22	3464	4763	6062

Analysed	
August 6, 2023 10:22:10 AM	
Computation with 5 linked points (*italics* points are less reliable):	1. Sint-Walburgakerk 2. Sint-Niklaaskerk 3. Zuidstraat 4. Noordstraat 5. Oude Beestenmarkt
Reference points (Reliability 0–5)	5
Transformation with 3 parameters	
Scale	1:7200
Rotation	11 °
Mean position error	14m

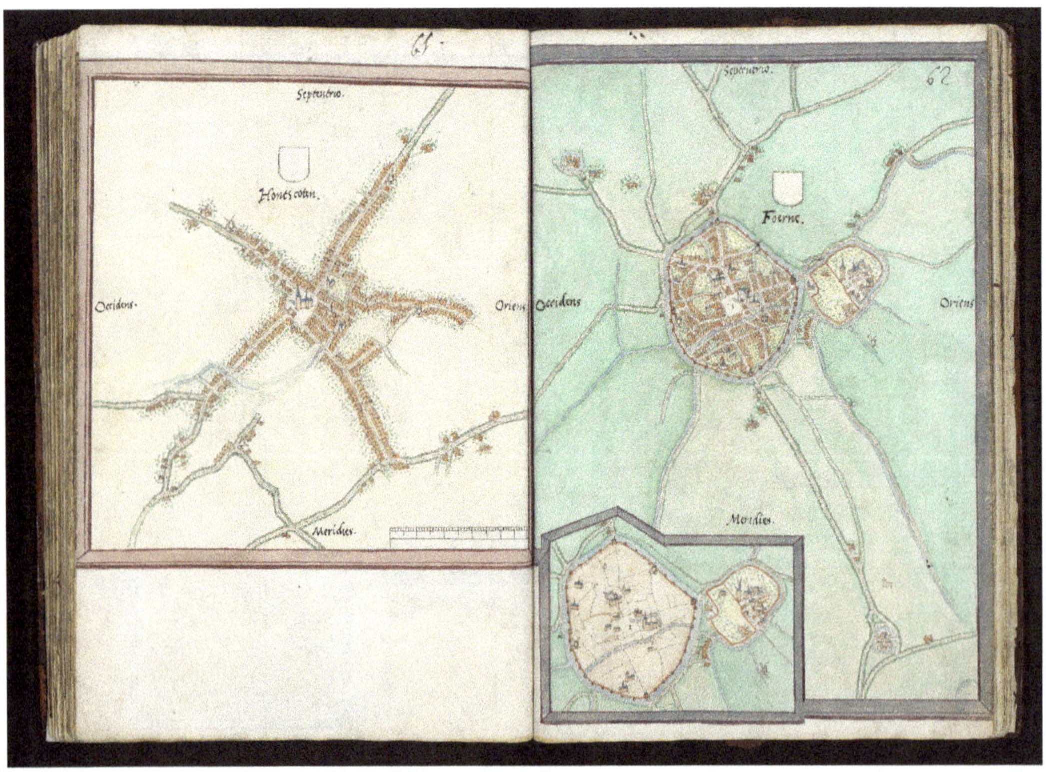

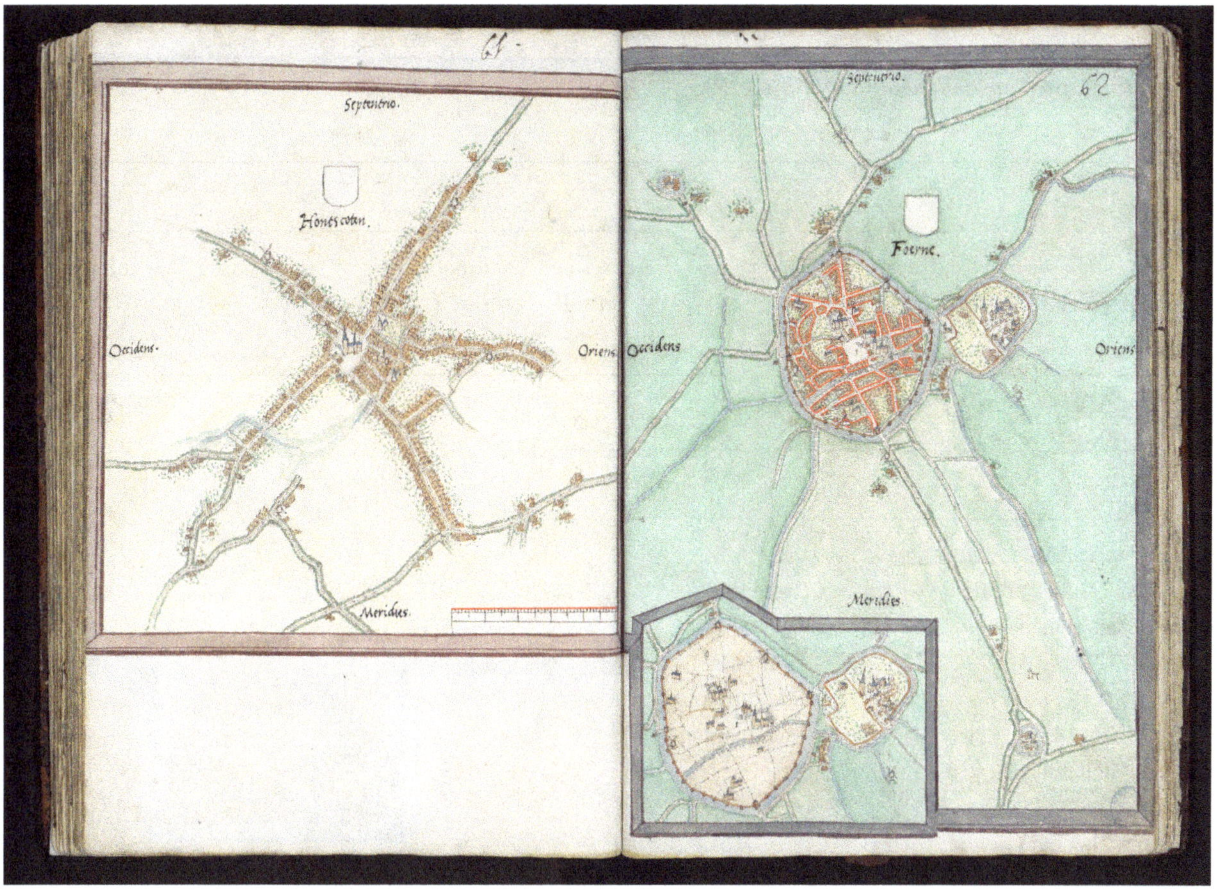
Septentrio.
Hones coten.
Occidens.
Oriens
Meridies.
Septentrio.
62
Fourne.
Occidens
Oriens
Meridies.

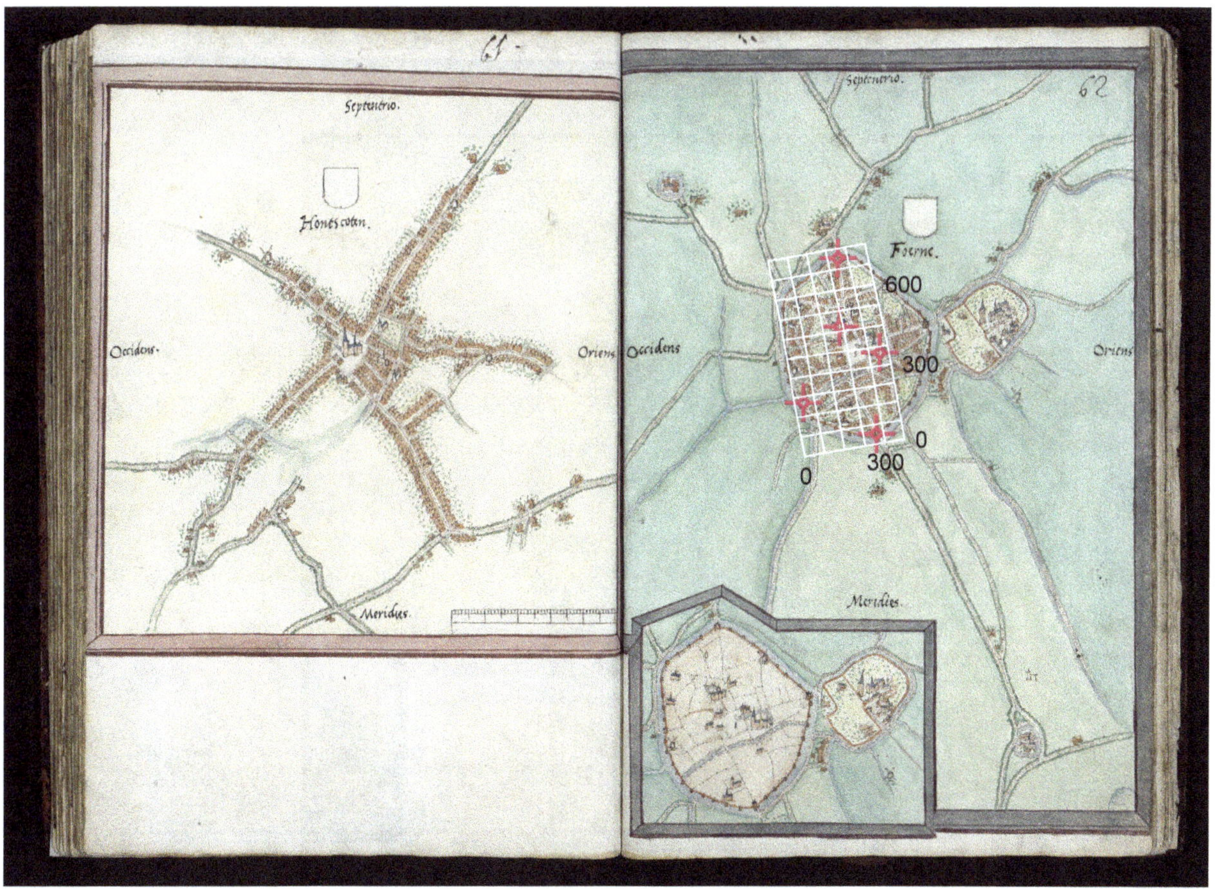
Septentrio.
Hones coten.
Occidens.
Oriens
Meridies.
Septentrio.
62
Fourne.
600
300
0
300
0
Occidens
Oriens
Meridies.

201. Vianen

Tables 3.201 (measured and analysed).
Figures 3.201 (unmeasured, measured and analysed)

Measured							
Pixels	Scale stick original map (pixels)	Houses	Street sides	Houses per street side	Inhabitants (RF 4.0)	Inhabitants (RF 5.5)	Inhabitants (RF 7.0)
4.034	1009	708	79	8.97	2833	3896	4959

Analysed	
August 6, 2023 11:05:37 AM	
Computation with 5 linked points (*italics* points are less reliable):	1. Grote Kerk 2. Lekpoort 3. Hofpoort 4. Stadhuisstraat 5. Molenstraat
Reference points (Reliability 0–5)	5
Transformation with 3 parameters	
Scale	1:7280
Rotation	9 °
Mean position error	12m

Septant
Lek fl.
Vianen
Occidens.
Meridies.

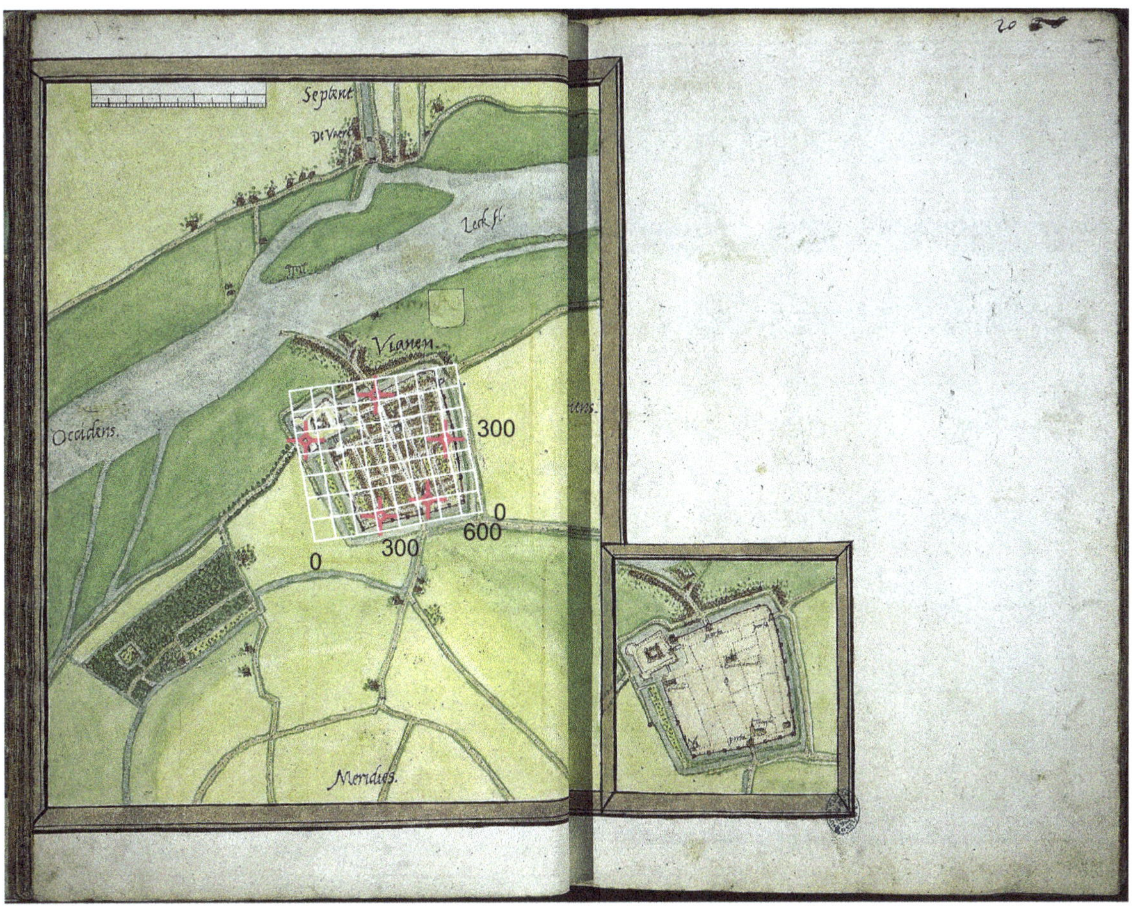
Septant
Lek fl.
Vianen
Occidens.
300
0
600
0
300
Meridies.

202. Vieil-Hesdin

Figure 3.202

The original Hesdin (Vieil-Hesdin) lies 6 kilometres north of Hesdin. The two maps that have been surveyed and analysed are of present-day Hesdin (see also pp. 274 and 275). As a result of the feud between France and the Habsburg Netherlands, this city was partially destroyed in 1553 by its lord Charles V after he latter had decided for strategic reasons to build a new citadel six kilometers to the west. From this the later Hesdin would arise (see page 274, 275: Hesdin). Vieil-Hesdin fell into disrepair, but still exists as a village.

30
Septentrio.
Hesdyn fort.
Occidens.
Oriens.
Meridies.
Hesdyn
Meridies.

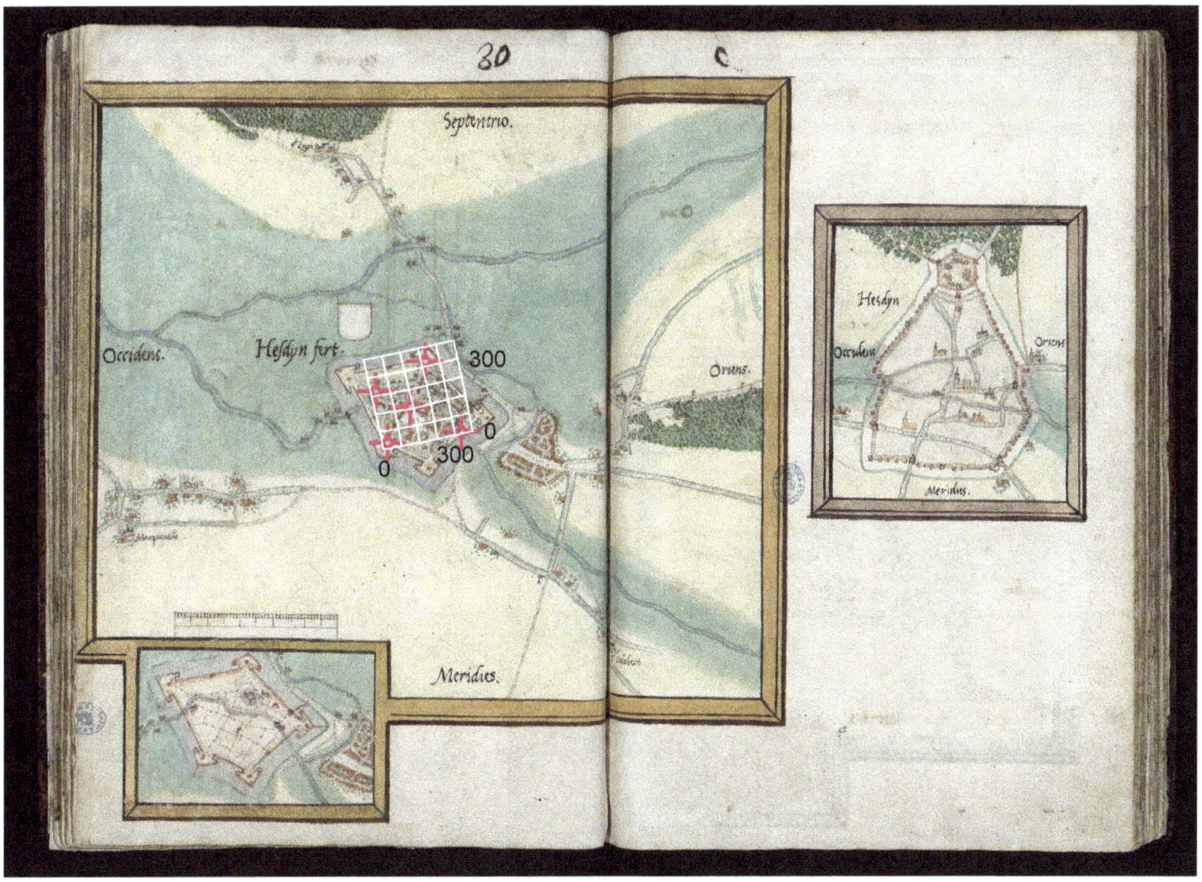
30
Septentrio.
Hesdyn fort.
Occidens.
Oriens.
300
0
0
300
Meridies.
Hesdyn
Meridies.

203. Vilvoorde

Tables 3.203 (measured and analysed).
Figures 3.203 (unmeasured, measured and analysed)

Measured

Pixels	Scale stick original map (pixels)	Houses	Street sides	Houses per street side	Inhabitants (RF 4.0)	Inhabitants (RF 5.5)	Inhabitants (RF 7.0)
5677	missing	997	168	5.93	3988	5483	6978

Analysed

September 29, 2023 9:35:07 AM	
Computation with 5 linked points (*italics* points are less reliable):	1. Onze-Lieve-Vrouw van Goede Hoopkerk 2. Basiliek van OLV- van-Troost 3. Spuymolenstraat 4. Toekomststraat 5. Stationlei
Reference points (Reliability 0–5)	5
Transformation with 3 parameters	
Scale	1:8040
Rotation	10 °
Mean position error	11m

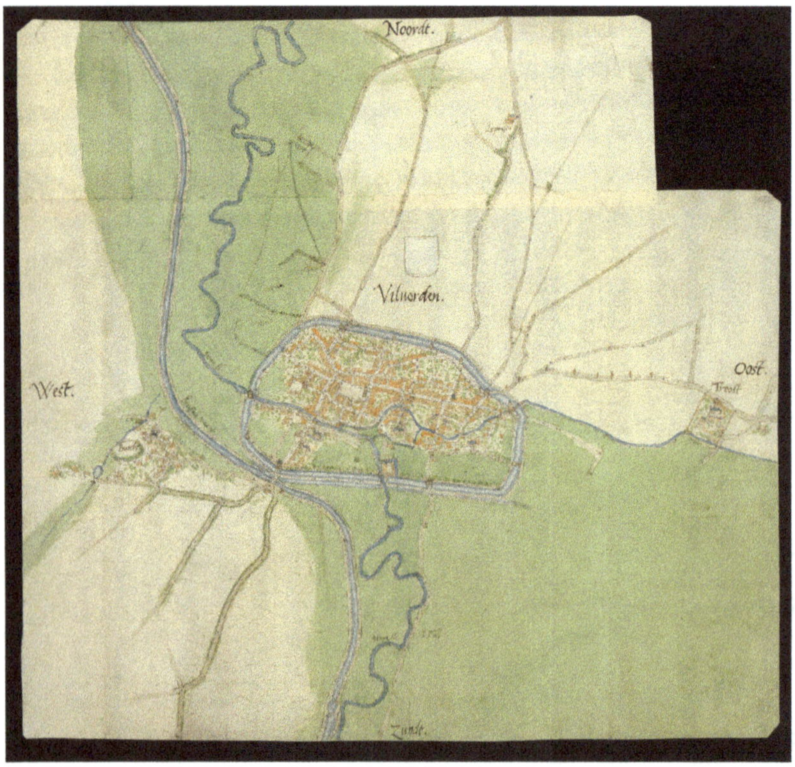

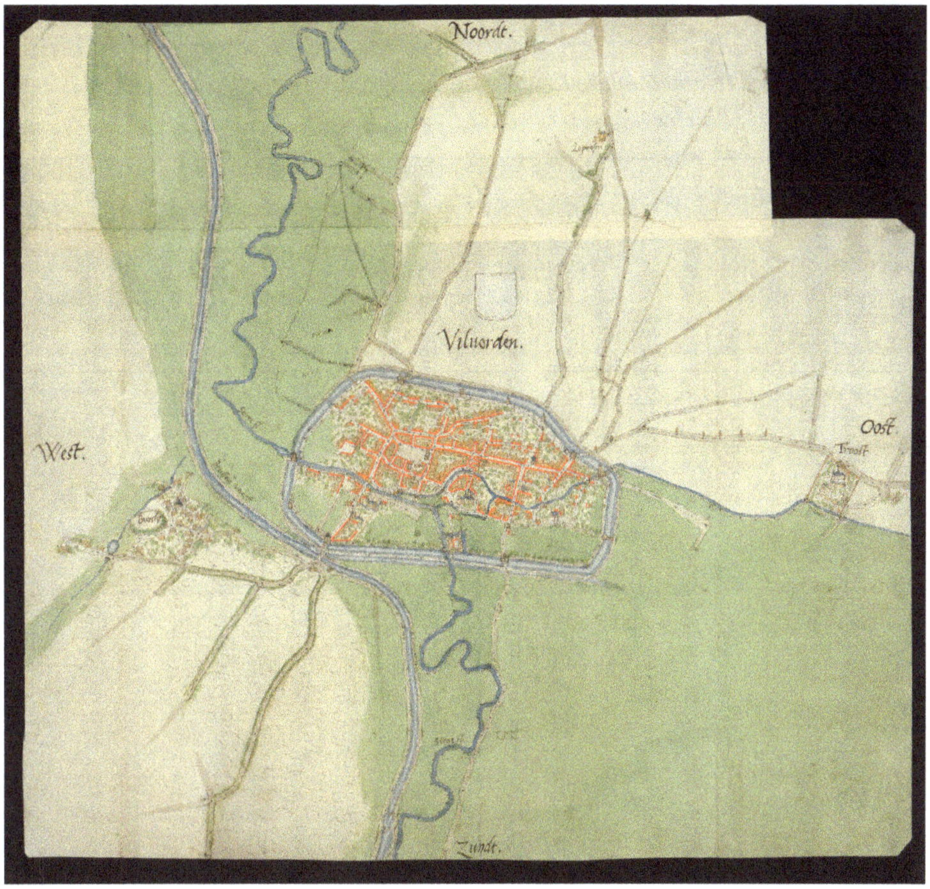
Noordt.
Vilvorden.
West.
Oost.
Zuydt.

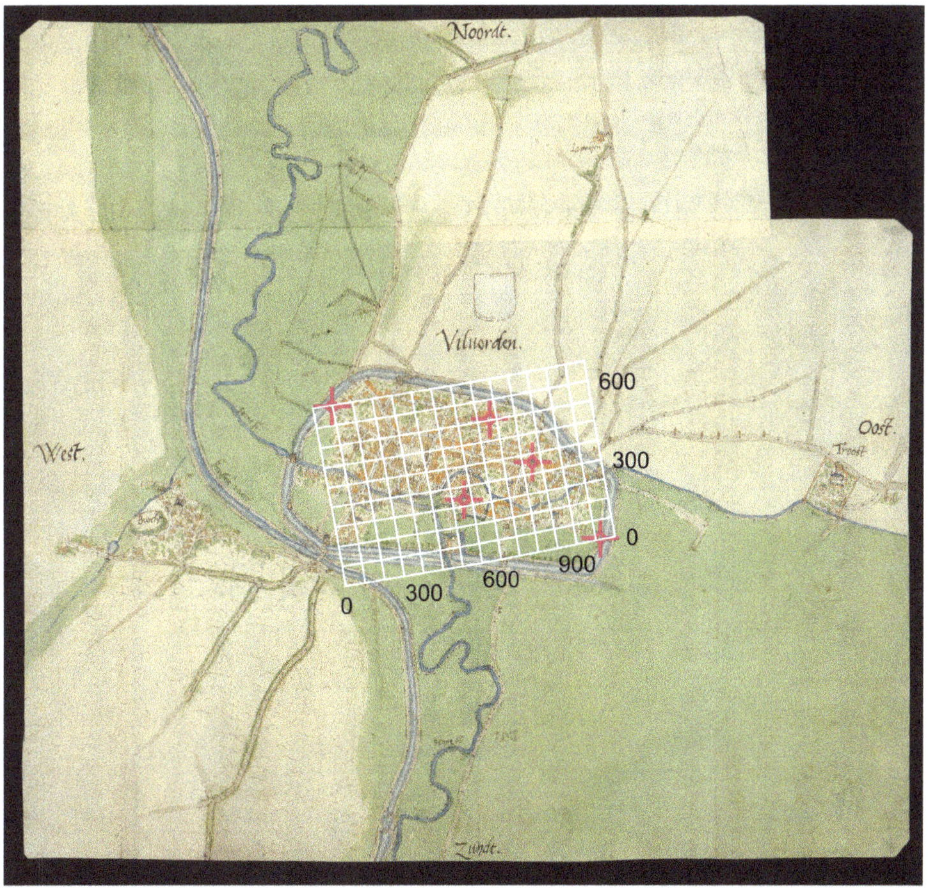
Noordt.
Vilvorden.
West.
Oost.
Zuydt.
600
300
0
0
300
600
900

204. Vlaardingen

Tables 3.204 (measured and analysed).
Figures 3.204 (unmeasured, measured and analysed).

Measured							
Pixels	Scale stick original map (pixels)	Houses	Street sides	Houses per street side	Inhabitants (RF 4.0)	Inhabitants (RF 5.5)	Inhabitants (RF 7.0)
2043	999	359	46	7.80	1435	1973	2511

Analysed	
October 15, 2023 10:26:19 PM	
Computation with 5 linked points (*italics* points are less reliable):	1. Grote Kerk 2. Kortedijk 3. Gedempte Biersloot 4. Hoogstraat 5. Maassluisedijk
Reference points (Reliability 0–5)	5
Transformation with 3 parameters	
Scale	1:8680
Rotation	14 °
Mean position error	5m

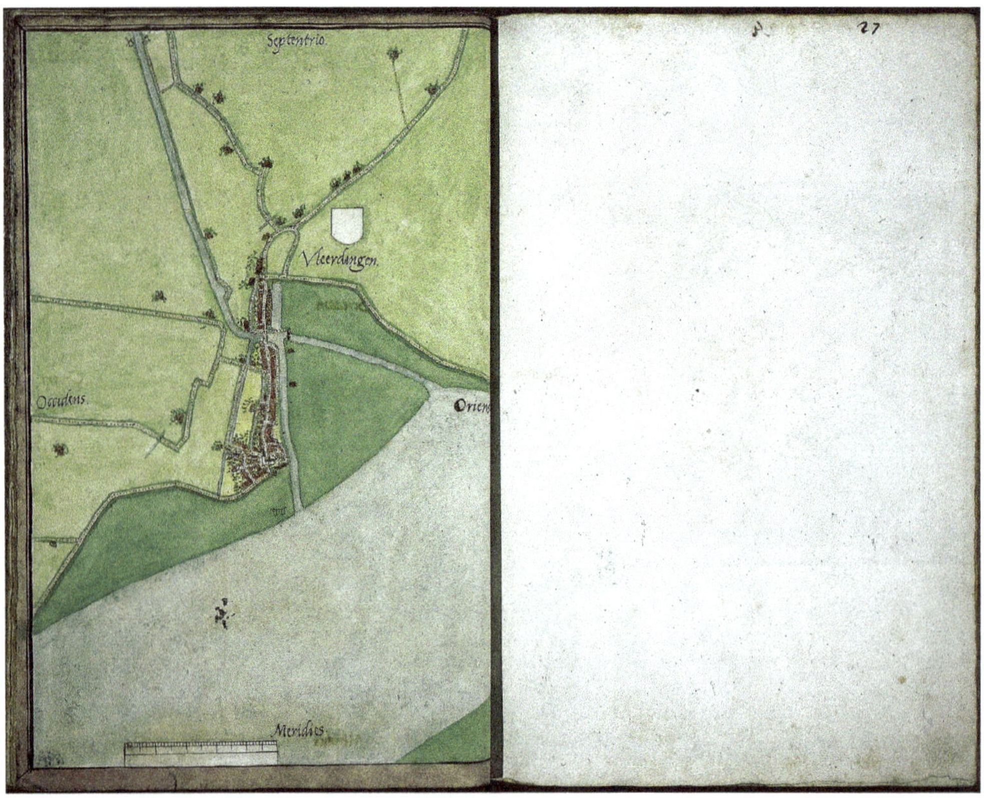

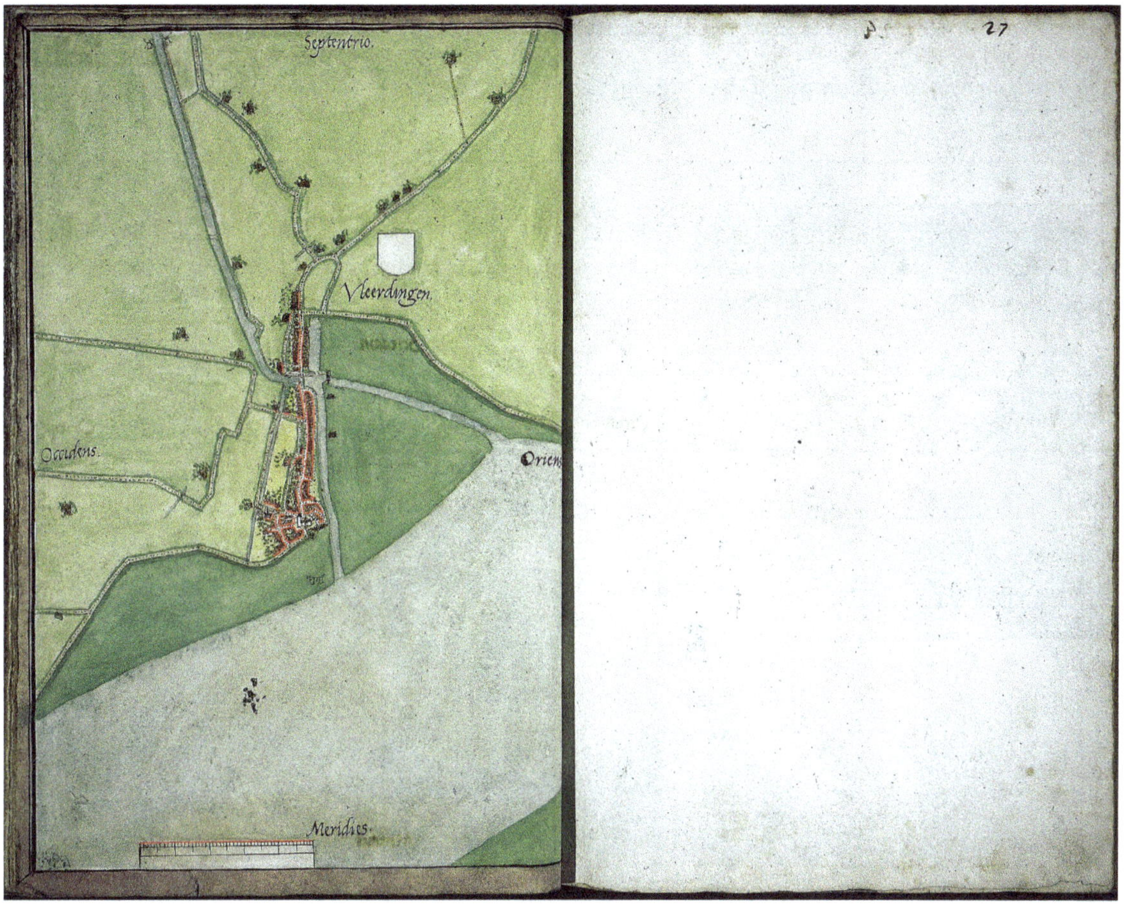
Septentrio.
Vleerdingen.
Occidens.
Meridies.
27

Septentrio.
Vleerdingen.
Occidens.
600
300
0
0
300
Meridies.
27

205. Vlissingen

Tables 3.205 (measured and analysed).
Figures 3.205 (unmeasured, measured and analysed).

Measured							
Pixels	Scale stick original map (pixels)	Houses	Street sides	Houses per street side	Inhabitants (RF 4.0)	Inhabitants (RF 5.5)	Inhabitants (RF 7.0)
4245	1025	745	97	7.68	2982	4100	5218

Analysed	
January 31, 2024 9:50:47 PM	
Computation with 5 linked points (*italics* points are less reliable):	1. Leugenaarshoofd 2. Keizersbolwerk 3. Nieuwstraat 4. Kerkstraat 5. Spuistraat
Reference points (Reliability 1–5)	5
Transformation with 3 parameters	
Scale	1:8030
Rotation	12 °
Mean Position Error	17m

Septentrio
Vlißinge
Occidens
Meridies

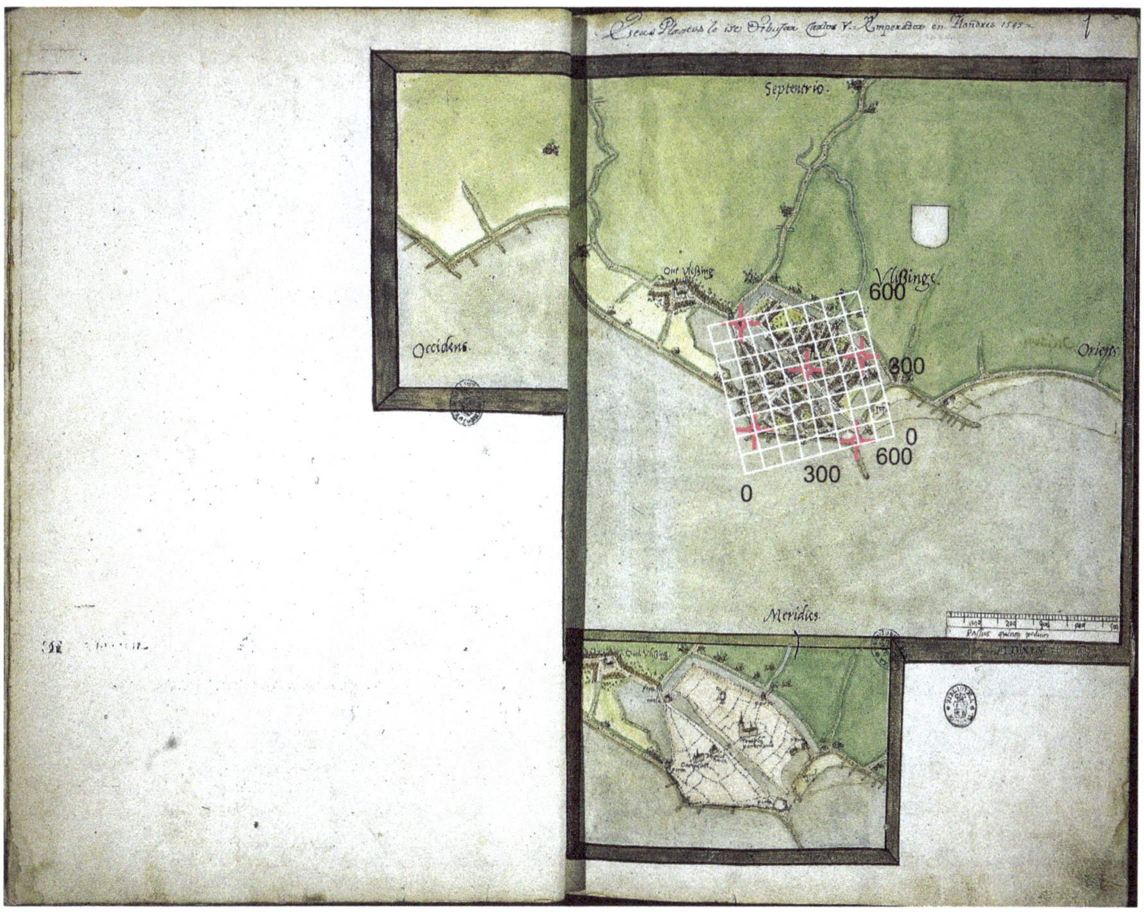
Septentrio
Vlißinge
Occidens
Meridies
600
300
0
0
300
600

206. Vollenhove

Tables 3.206 (measured and analysed).
Figures 3.206 (unmeasured, measured and analysed).

Measured							
Pixels	Scale stick original map (pixels)	Houses	Street sides	Houses per street side	Inhabitants (RF 4.0)	Inhabitants (RF 5.5)	Inhabitants (RF 7.0)
2880	993	506	56	9.03	2023	2782	3541

Analysed	
August 10, 2023 11:13:42 AM	
Computation with 5 linked points (*italics* points are less reliable):	1. Grote of Sint-Nicolaaskerk 2. Kleine of Lieve-Vrouwekerk 3. kasteel Toutenburg 4. Havezate Oldruitenborgh 5. Molenberg
Reference points (Reliability 0–5)	5
Transformation with 3 parameters	
Scale	1:8150
Rotation	11 °
Mean position error	16m

76
Septentrio.
Swarte water.
Occidens.
Genemuyden
Oriens
Meridies.
Septentrio.
77
Vollenhove.
Oriens.
Meridies.

76
Septentrio.
Swarte water.
Occidens.
Genemuyden
Oriens
Meridies.
Septentrio.
77
Vollenhove.
300
0
0
300
600
Oriens.
Meridies.

207. Wachtendonk

Tables 3.207 (measured and analysed).
Figures 3.207 (unmeasured, measured and analysed).

Measured

Pixels	Scale stick original map (pixels)	Houses	Street sides	Houses per street side	Inhabitants (RF 4.0)	Inhabitants (RF 5.5)	Inhabitants (RF 7.0)
1369	951	240	38	6.33	961	1322	1683

Analysed

August 10, 2023 11:38:26 AM	
Computation with 5 linked points (*italics* points are less reliable):	1. Sankt Michael 2. Burg Wachtendonk 3. Wall 4. Mühlenstrasse 5. Neustrasse
Reference points (Reliability 0–5)	5
Transformation with 3 parameters	
Scale	1:8260
Rotation	13 °
Mean position error	7m

Septentrio.
Wachtendonck.
Occidens
Oriens
Meridies

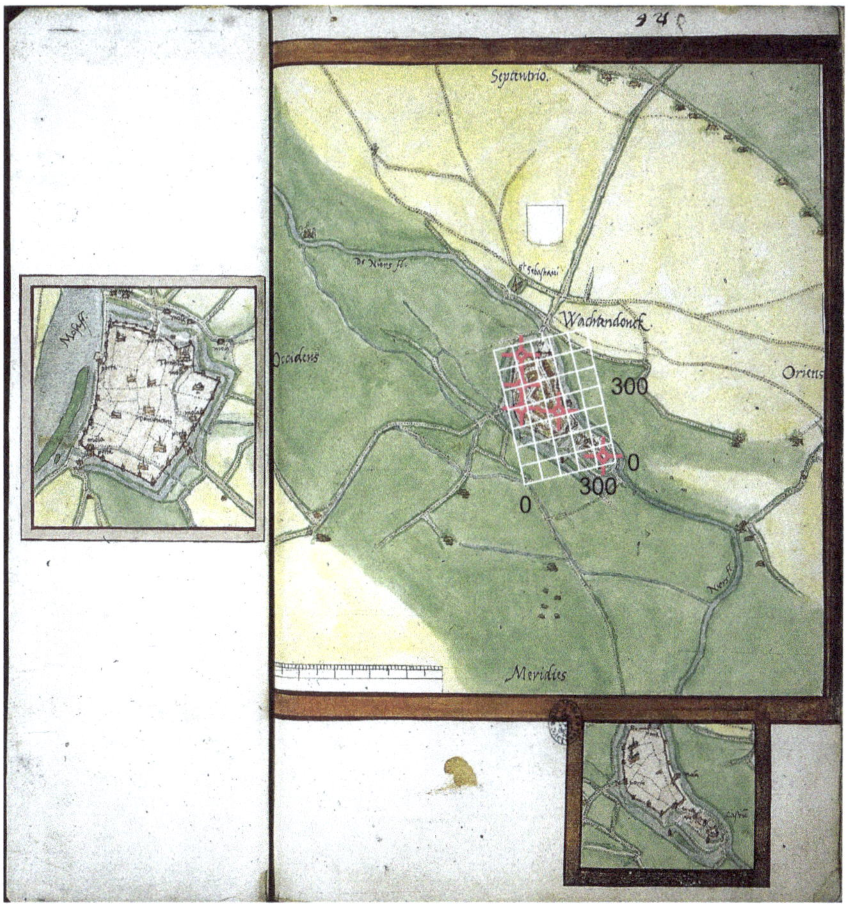
Septentrio.
Wachtendonck.
Occidens
Oriens
300
0
300
0
Meridies

208. Wageningen

Tables 3.208 (measured and analysed).
Figures 3.208 (unmeasured, measured and analysed).

Measured

Pixels	Scale stick original map (pixels)	Houses	Street sides	Houses per street side	Inhabitants (RF 4.0)	Inhabitants (RF 5.5)	Inhabitants (RF 7.0)
2133	missing	375	59	6.35	1498	2060	2622

Analysed

August 10, 2023 11:56:43 AM	
Computation with 5 linked points (*italics* points are less reliable):	1. Grote Kerk 2. Nudepoort 3. Bowlespark 4. Walstraat 5. Molenstraat
Reference points (Reliability 0–5)	5
Transformation with 3 parameters	
Scale	1:9480
Rotation	20 °
Mean position error	16m

86
Septentrio.
Wageningen.
Occidens.
Oriens.
Meridies.

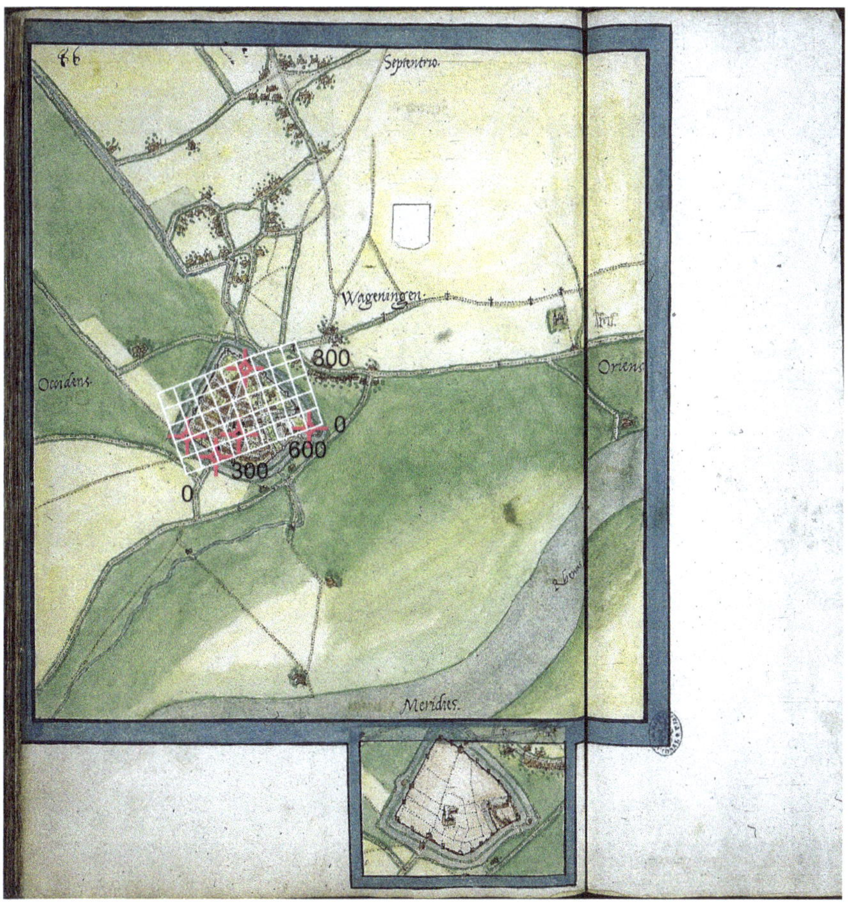

86
Septentrio.
Wageningen.
Occidens.
Oriens.
Meridies.
300
0
600
300
0

209. Warneton

Tables 3.209 (measured and analysed).
Figures 3.209 (unmeasured, measured and analysed).

Measured							
Pixels	Scale stick original map (pixels)	Houses	Street sides	Houses per street side	Inhabitants (RF 4.0)	Inhabitants (RF 5.5)	Inhabitants (RF 7.0)
2322	1015	408	61	6.68	1631	2243	2855

Analysed	
March 16, 2023 10:34:45 AM	
Computation with 5 linked points (*italics* points are less reliable):	1. Église Saints-Pierre-et-Paul 2. Quai Verboeckhoven 3. Chemin des Brilles 4. Rue d'Ypres 5. Rue de Lille
Reference points (Reliability 0–5)	5
Transformation with 3 parameters	
Scale	1:7470
Rotation	8 °
Mean position error	22m

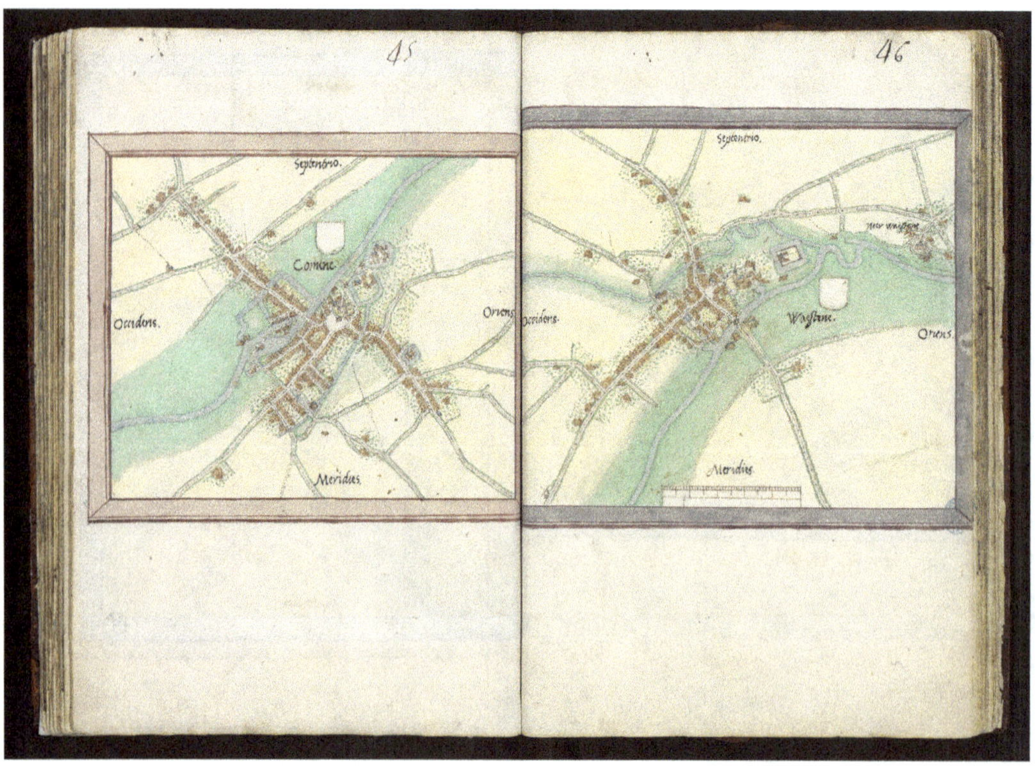

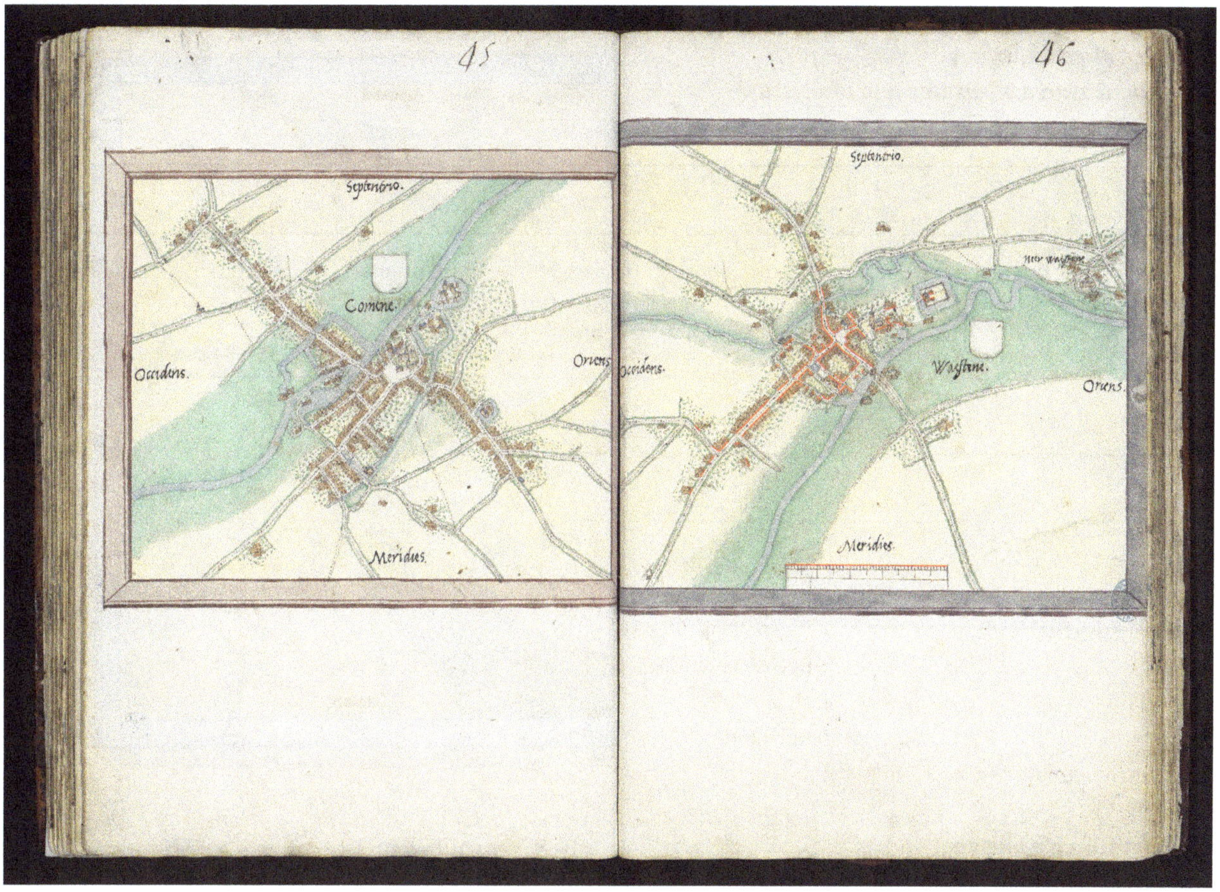
45
46
Septentrio
Occidens
Oriens
Meridies

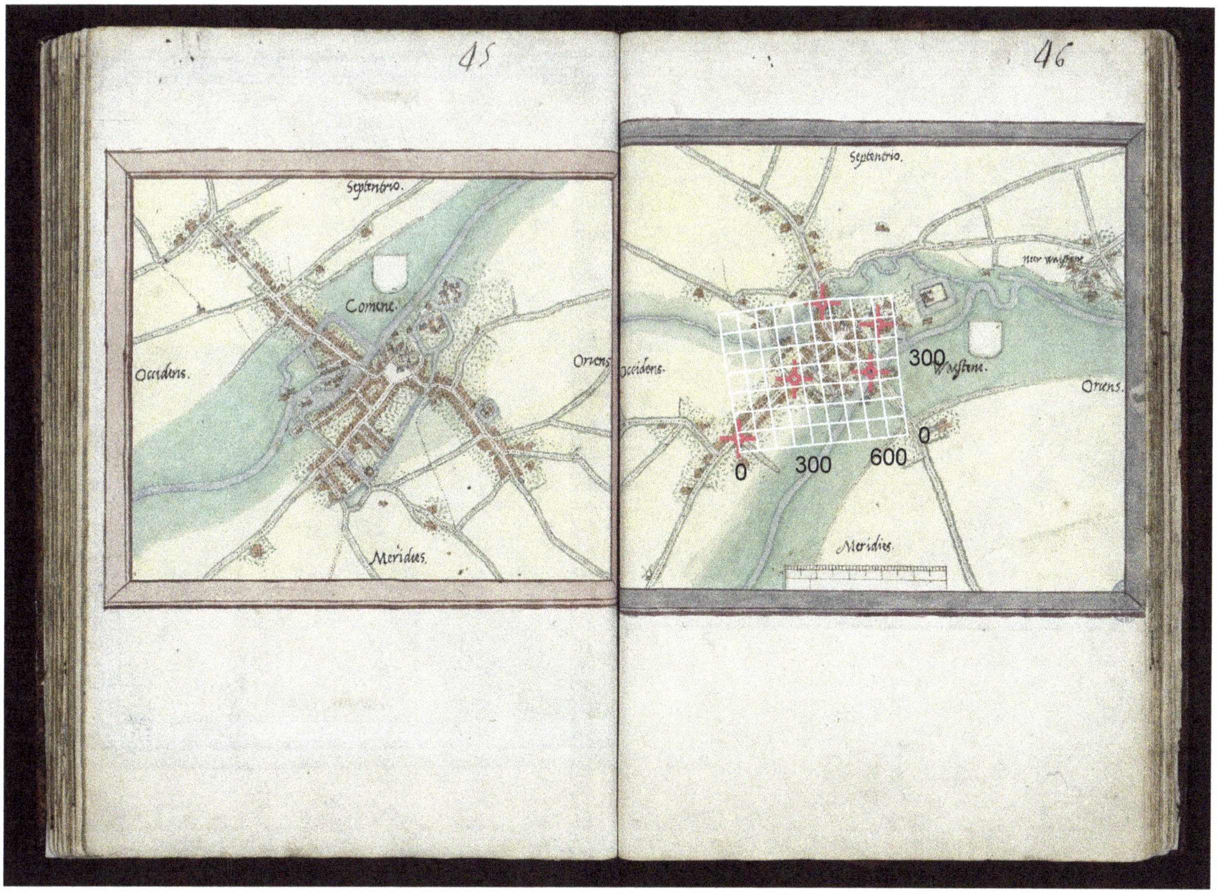
45
46
Septentrio
Occidens
Oriens
300
0
300
600
0
Meridies

210. Wavre
Tables 3.210 (measured and analysed).
Figures 3.210 (unmeasured, measured and analysed).

Measured							
Pixels	Scale stick original map (pixels)	Houses	Street sides	Houses per street side	Inhabitants (RF 4.0)	Inhabitants (RF 5.5)	Inhabitants (RF 7.0)
2560	missing	450	87	5.17	1798	2472	3146

Analysed	
September 29, 2023 10:31:51 AM	
Computation with 5 linked points (*italics* points are less reliable):	1. Sint-Jan-de-Doperkerk 2. Rue de Bruxelles 3. Rue de la Limité 4. Dyle 5. Rue haute
Reference points (Reliability 0–5)	5
Transformation with 3 parameters	
Scale	1:7600
Rotation	6 °
Mean position error	5m

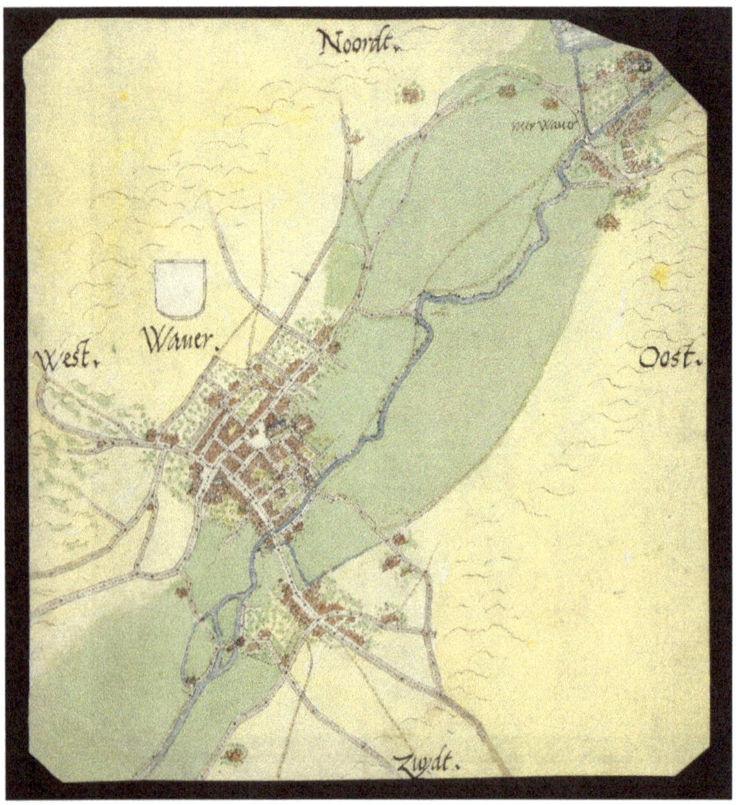

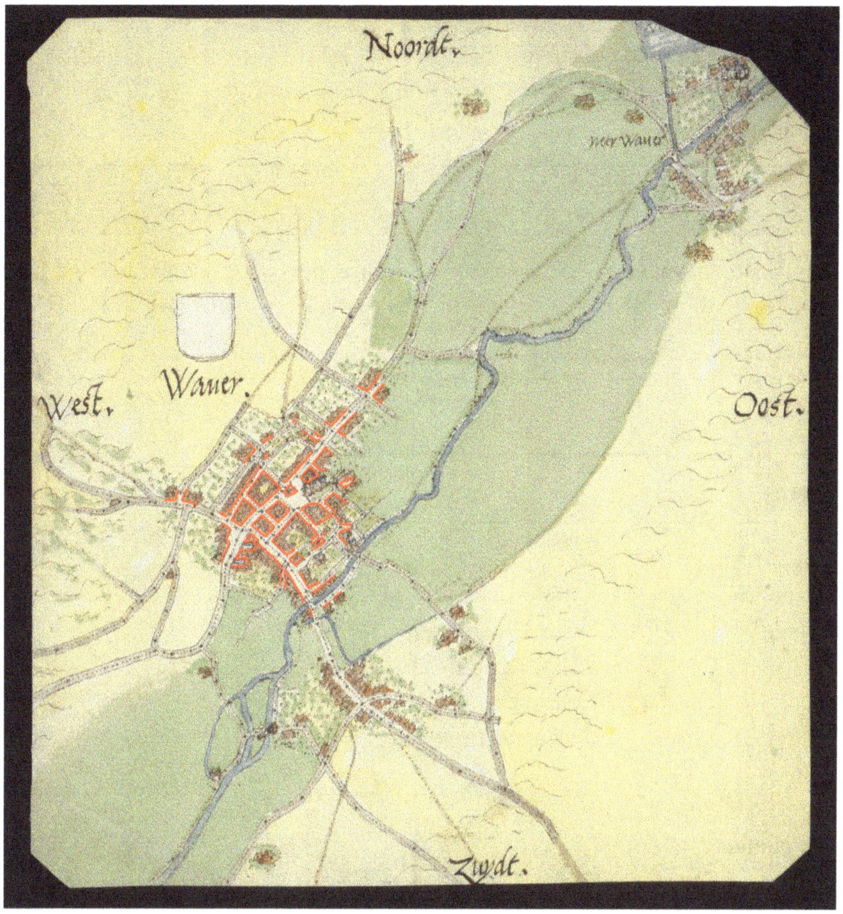
Noordt.
West.
Wauer.
Oost.
Zuydt.

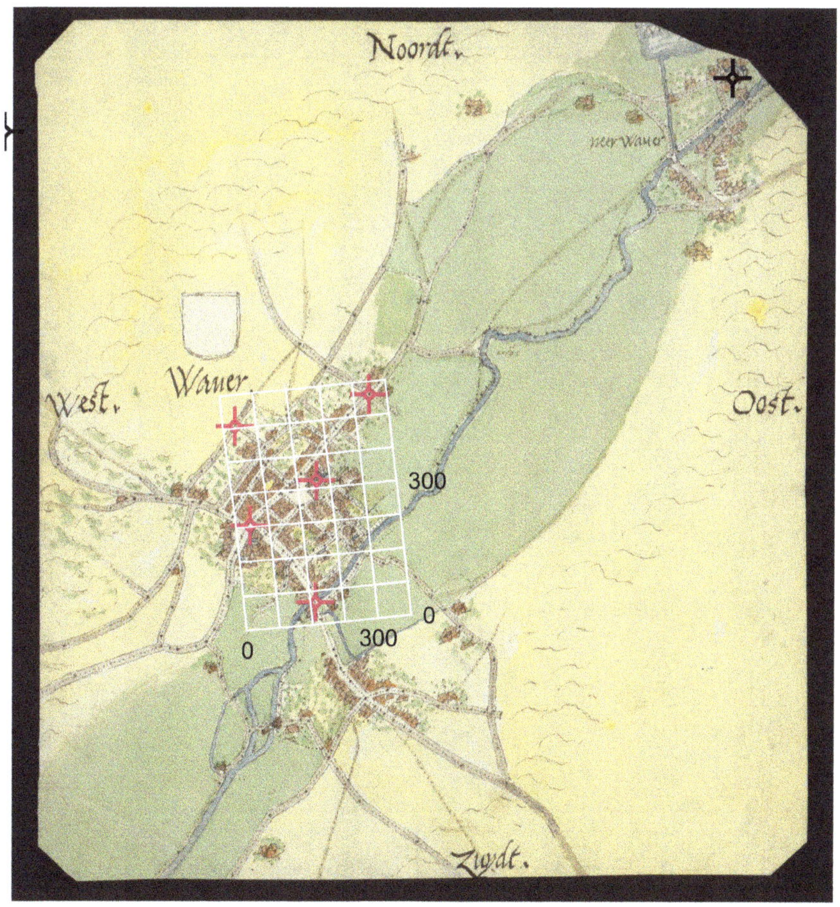
Noordt.
West.
Wauer.
Oost.
300
0
300
0
Zuydt.

211. Weert
Tables 3.211 (measured and analysed).
Figures 3.211 (unmeasured, measured and analysed).

Measured							
Pixels	Scale stick original map (pixels)	Houses	Street sides	Houses per street side	Inhabitants (RF 4.0)	Inhabitants (RF 5.5)	Inhabitants (RF 7.0)
4512	missing	792	94	8.43	3169	4358	5.547

Analysed	
September 29, 2023 11:04:34 AM	
Computation with 5 linked points (*italics* points are less reliable):	1. Sint-Martinuskerk 2. Kasteel Nijenborgh 3. Beekstraat 4. Langstraat 5. Molenpoort
Reference points (Reliability 0–5)	5
Transformation with 3 parameters	
Scale	1:8200
Rotation	10 °
Mean position error	16m

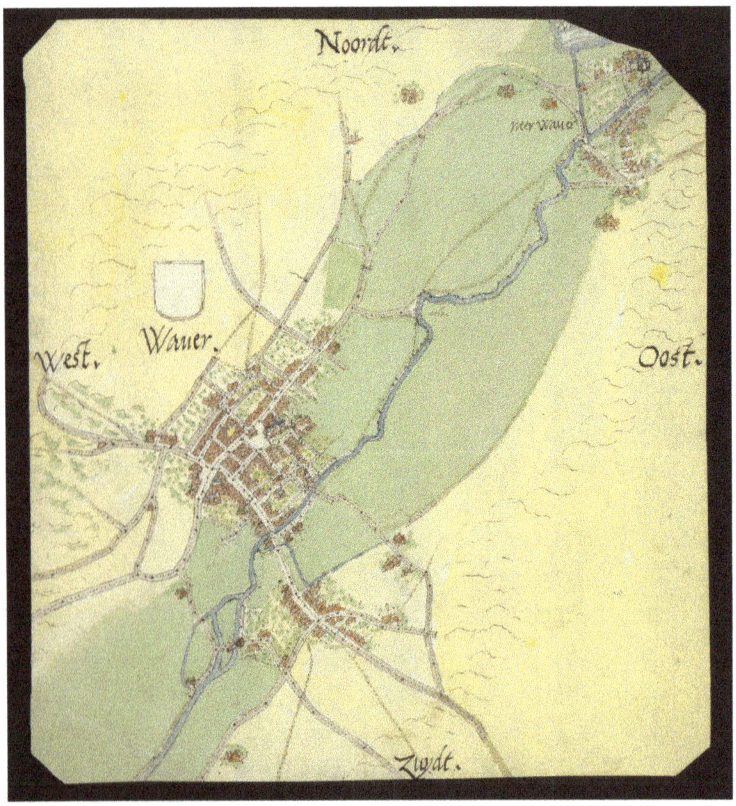

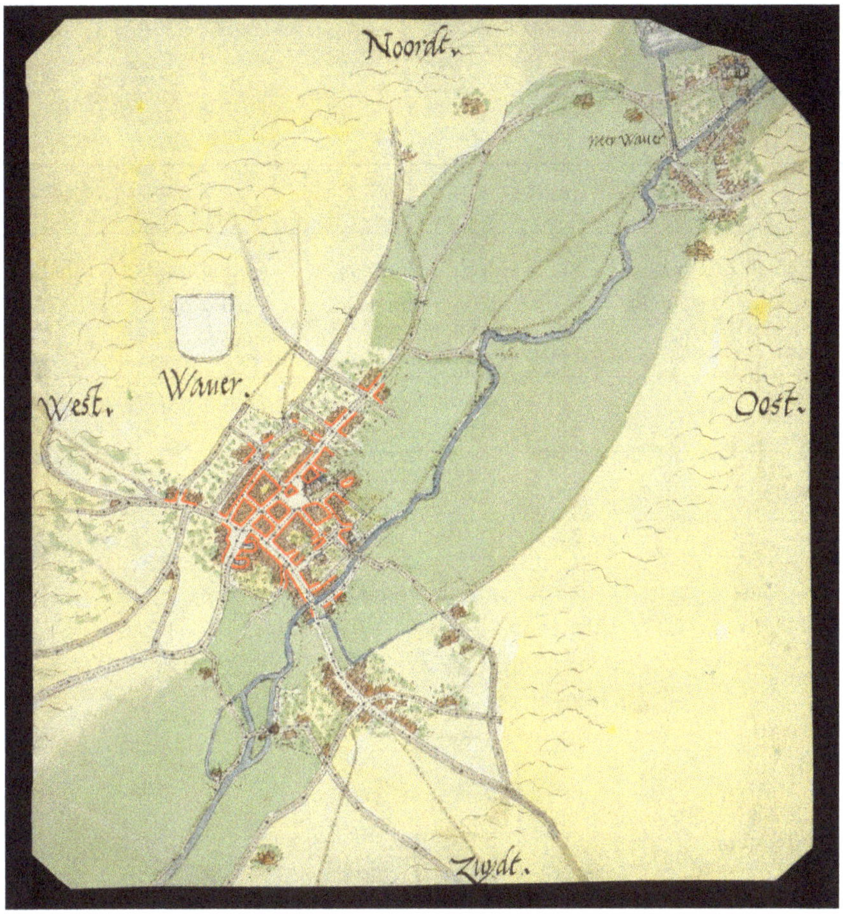
Noordt.
West.
Wauer.
Oost.
Zuydt.

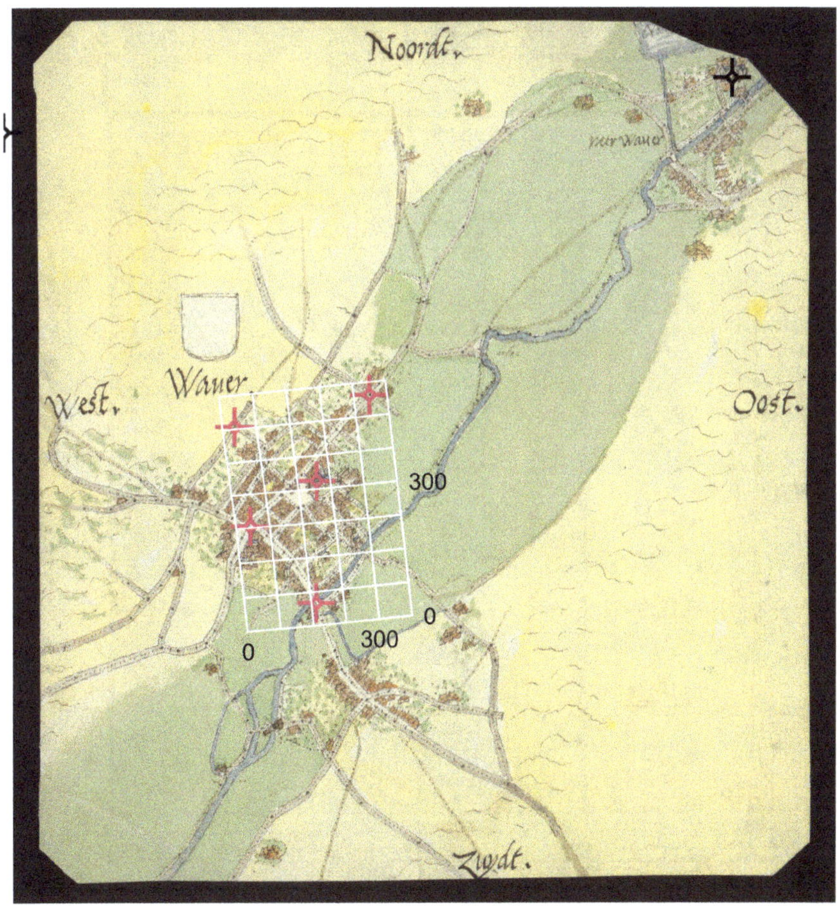
Noordt.
West.
Wauer.
Oost.
Zuydt.
300
0
300
0

212. Weesp
Tables 3.212 (measured and analysed).
Figures 3.212 (unmeasured, measured and analysed).

Measured							
Pixels	Scale stick original map (pixels)	Houses	Street sides	Houses per street side	Inhabitants (RF 4.0)	Inhabitants (RF 5.5)	Inhabitants (RF 7.0)
2003	1010	352	73	4,82	1407	1934	2462

Analysed	
June 19, 2023 9:36:33 AM	
Computation with 5 linked points (*italics* points are less reliable):	1. Grote of Sint Laurenskerk 2. Achterom 3. Oude Gracht 4. Hoogstraat 5. Achteromstraat
Reference points (Reliability 0–5)	5
Transformation with 3 parameters	
Scale	1:8260
Rotation	7 °
Mean position error	11m

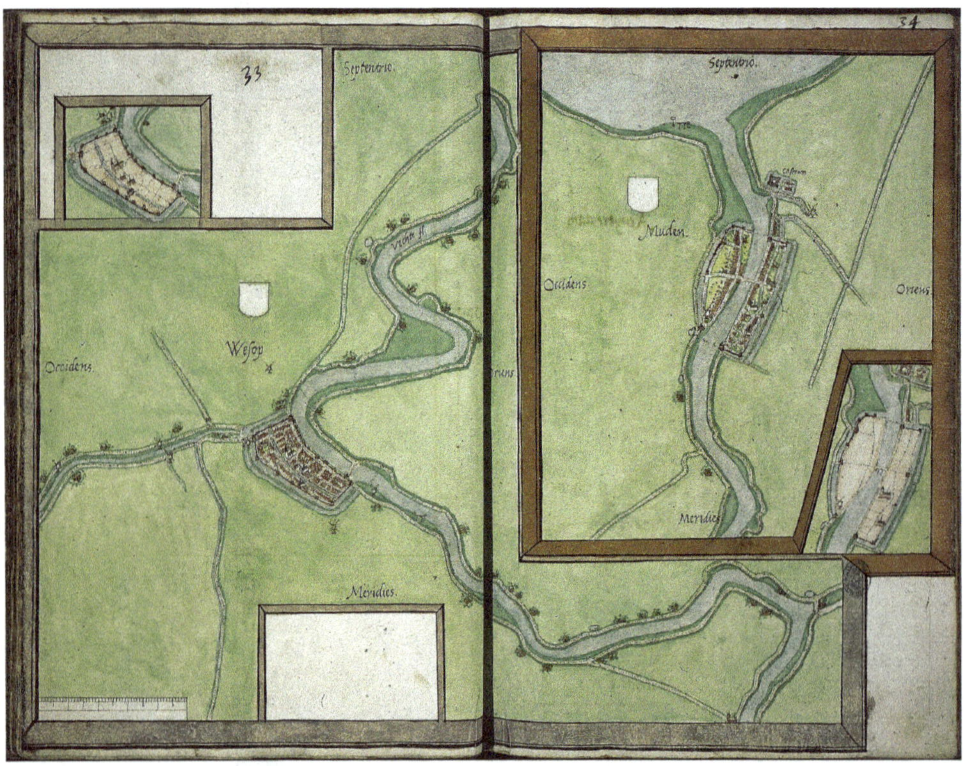

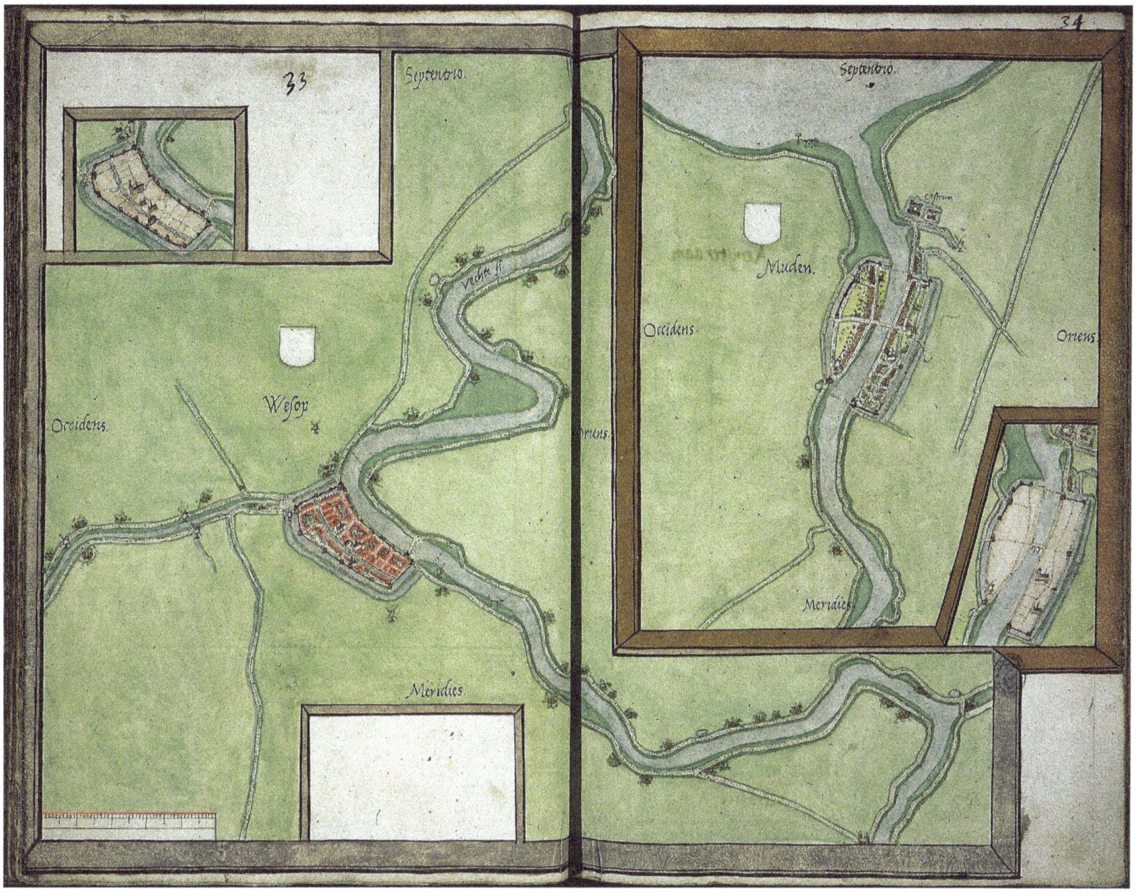
33
34
Septentrio.
Wesop
Occidens.
Oriens.
Meridies.
Muden.

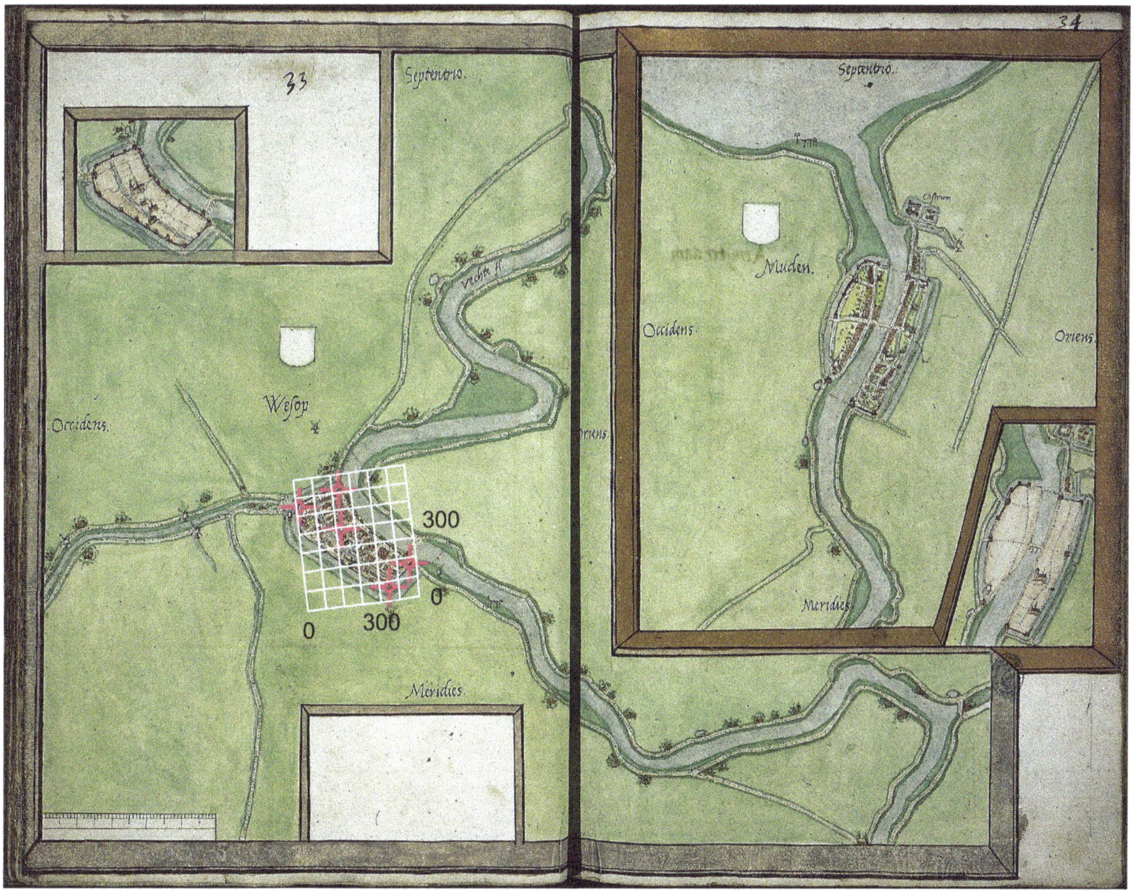
33
34
Septentrio.
Wesop
Occidens.
Oriens.
Meridies.
Muden.
300
0
0
300

213. Wervik

Tables 3.213 (measured and analysed).
Figures 3.213 (unmeasured, measured and analysed).

Measured

Pixels	Scale stick original map (pixels)	Houses	Street sides	Houses per street side	Inhabitants (RF 4.0)	Inhabitants (RF 5.5)	Inhabitants (RF 7.0)
7990	1027	1403	205	6.84	7.015	7717	8418

Analysed

August 10, 2023 12:27:00 PM	
Computation with 5 linked points (*italics* points are less reliable):	1. Sint-Medarduskerk 2. Grauwzustersklooster 3. Kruisekemolen *4. Lege Kruise* 5. Molenstraat
Reference points (Reliability 0–5)	4
Transformation with 3 parameters	
Scale	1:7460
Rotation	5 °
Mean position error	23m

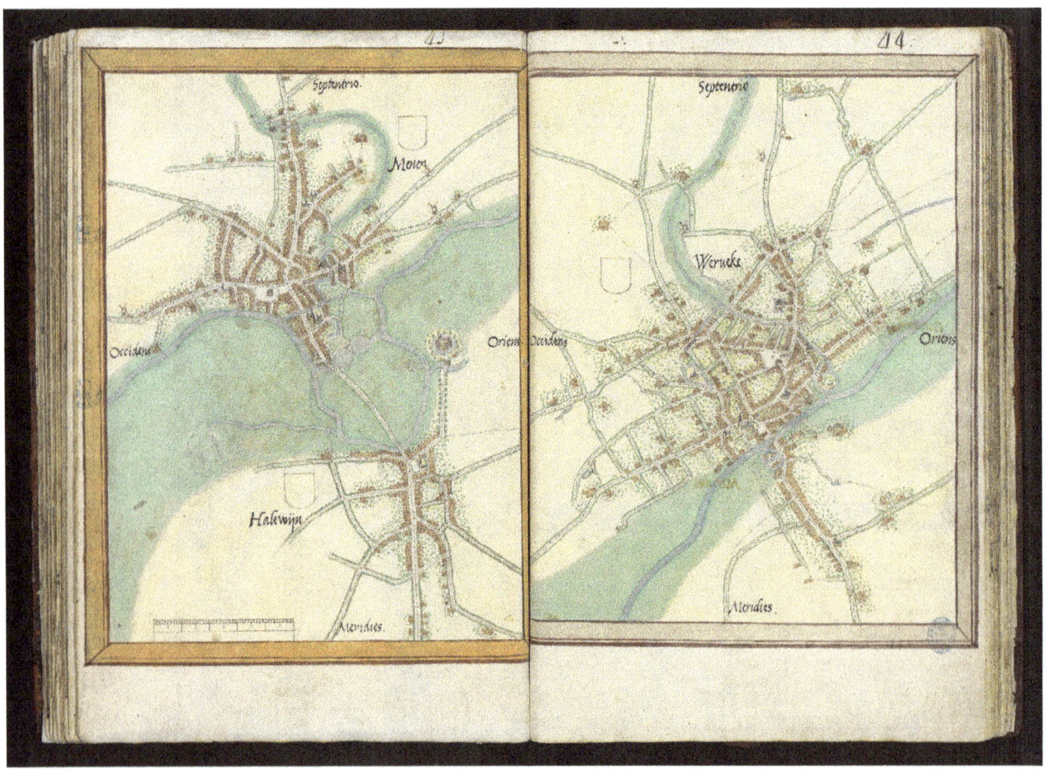

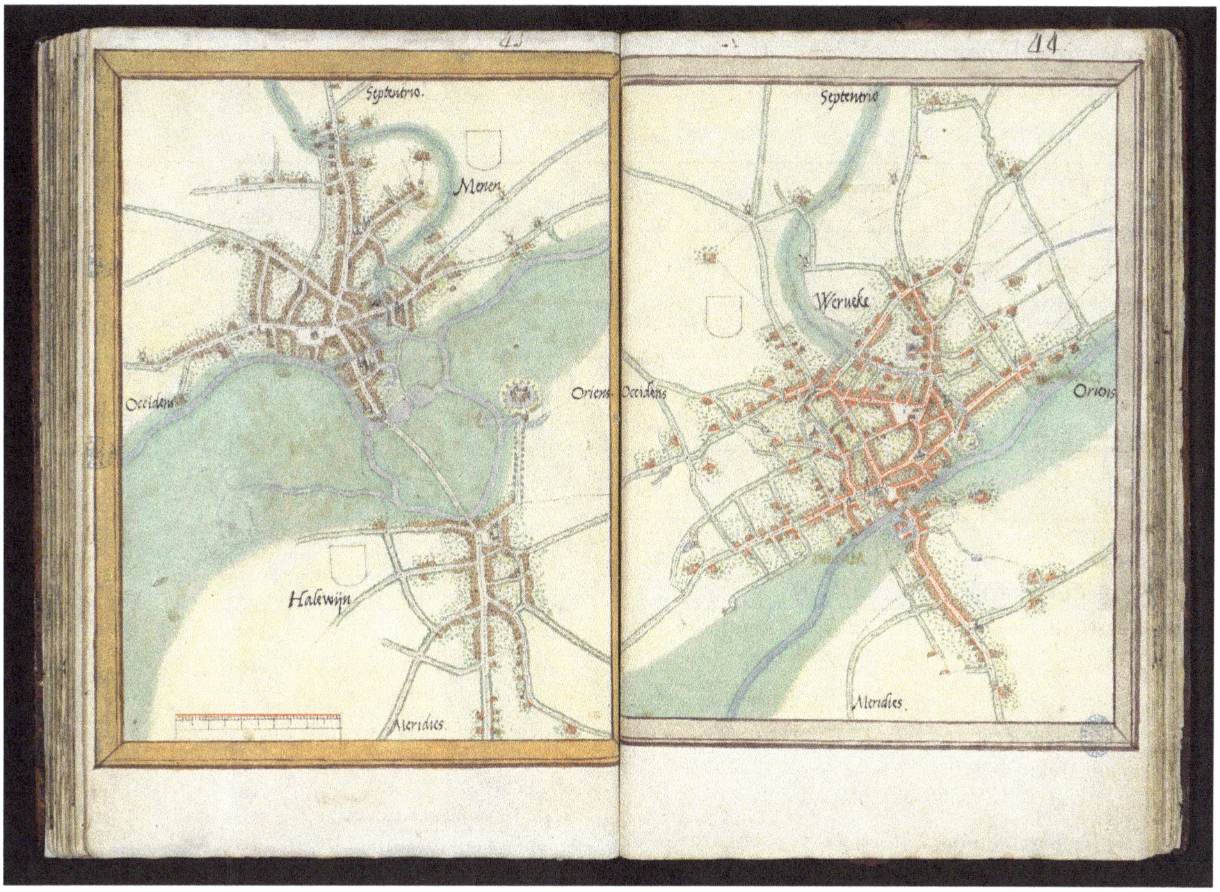
Septentrio
Menen
Occidens
Oriens
Halewijn
Meridies
Septentrio
Werueke
Occidens
Oriens
Meridies

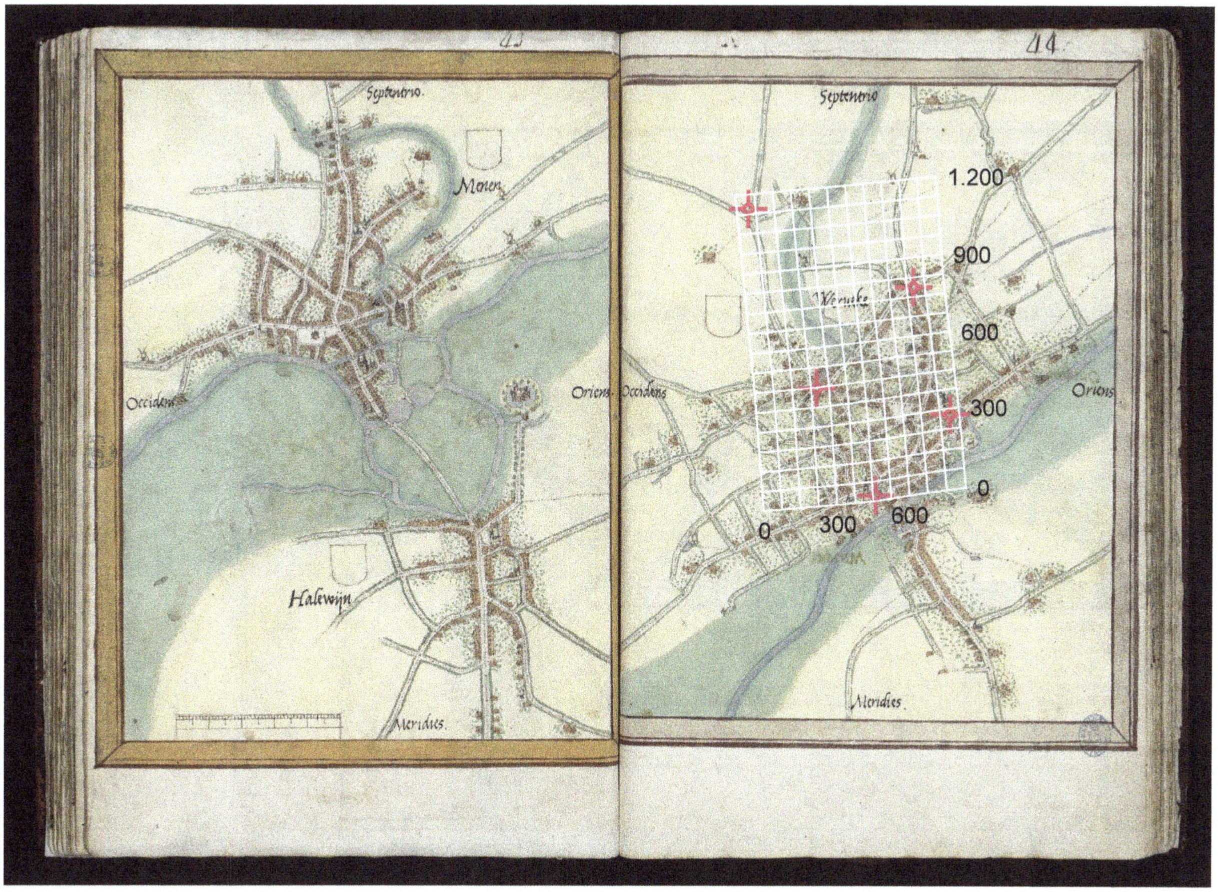
Septentrio
Menen
Occidens
Oriens
Halewijn
Meridies
Septentrio
1.200
900
Werueke
600
Occidens
Oriens
300
0
0
300
600
Meridies

214. Westkapelle
Tables 3.214 (measured and analysed).
Figures 3.214 (unmeasured, measured and analysed).

Measured							
Pixels	Scale stick original map (pixels)	Houses	Street sides	Houses per street side	Inhabitants (RF 4.0)	Inhabitants (RF 5.5)	Inhabitants (RF 7.0)
2663	725	468	40	11.69	1871	2572	3273

Analysed	
March 22, 2023 5:05:57 PM	
Computation with 5 linked points (*italics* points are less reliable):	1. Markt 2. *Noordstraat* 3. Zeedijk 4. Papestraat 5. Vuurtoren
Reference points (Reliability 0–5)	4
Transformation with 3 parameters	
Scale	1:9100
Rotation	12 °
Mean position error	16m

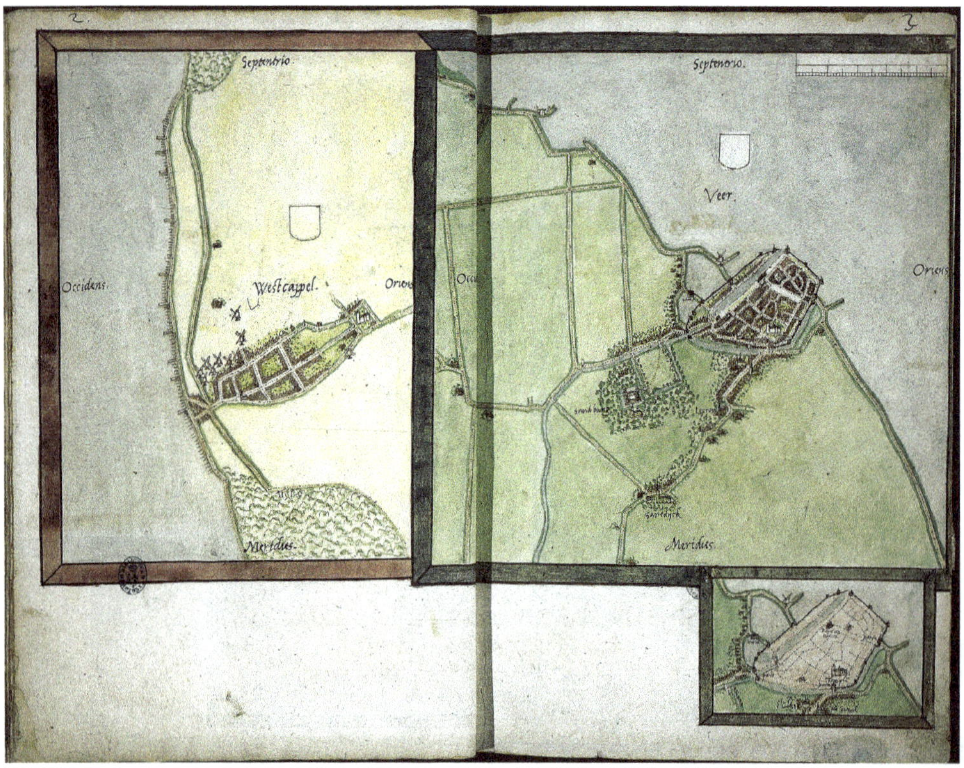

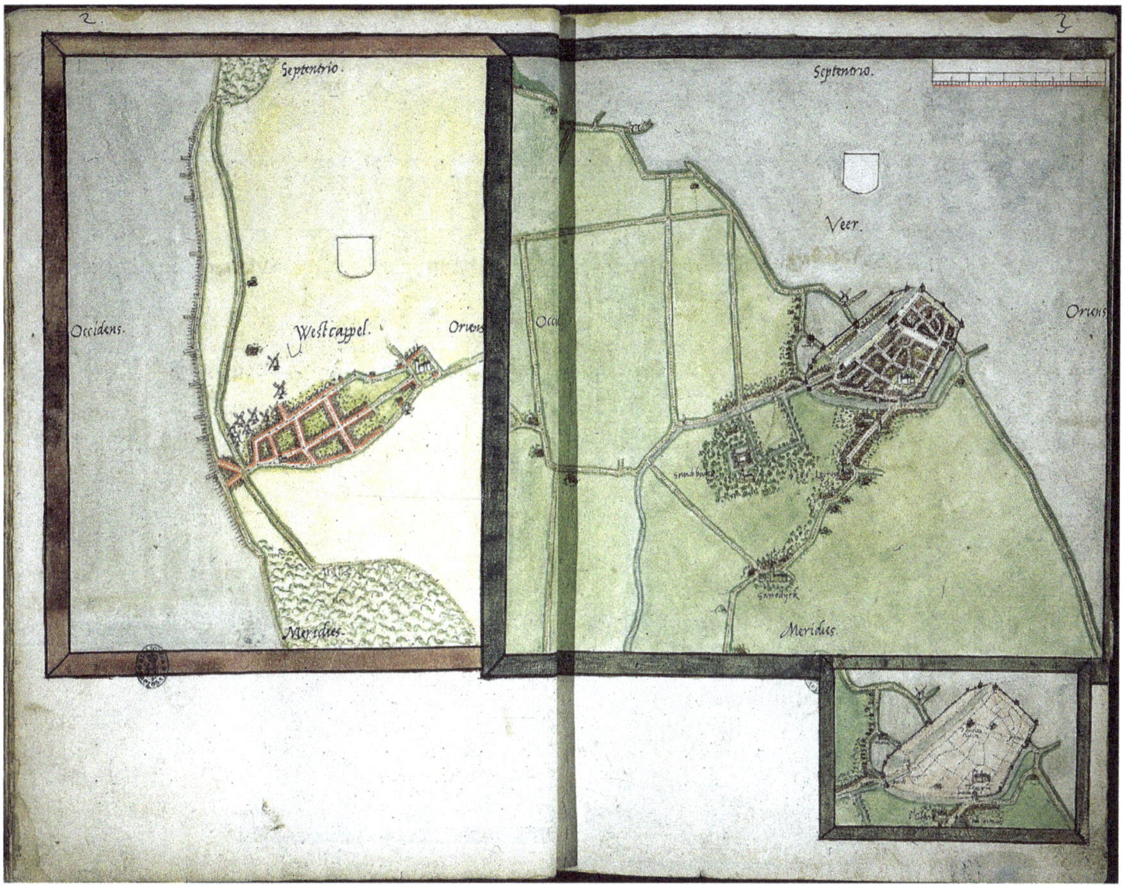
Septentrio.
Occidens.
Westcappel.
Oriens
Meridies.
Septentrio.
Veer.
Oriens
Meridies.

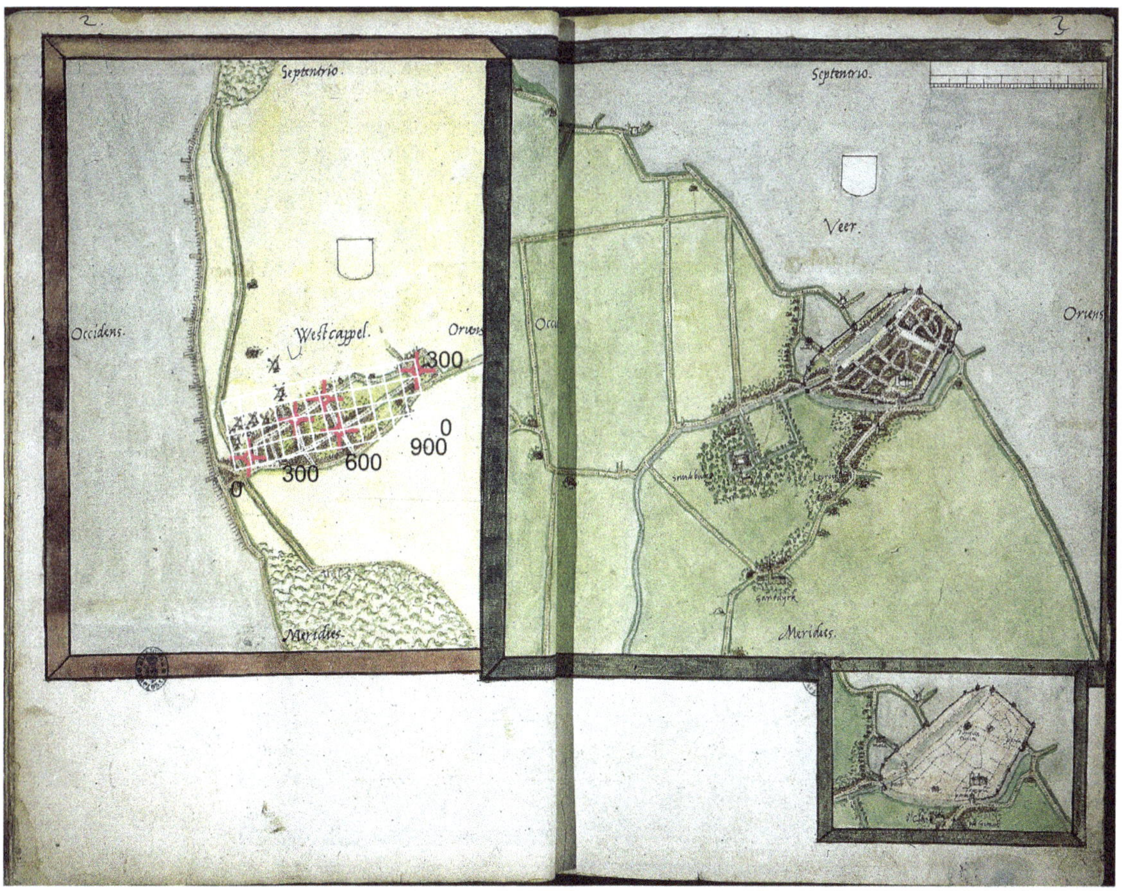
Septentrio.
Occidens.
Westcappel.
Oriens
300
0
900
0
300
600
Meridies.
Septentrio.
Veer.
Oriens
Meridies.

215. Wijk bij Duurstede
Tables 3.215 (measured and analysed).
Figures 3.215 (unmeasured, measured and analysed).

Measured							
Pixels	Scale stick original map (pixels)	Houses	Street sides	Houses per street side	Inhabitants (RF 4.0)	Inhabitants (RF 5.5)	Inhabitants (RF 7.0)
3451	996	606	68	8.91	2424	3333	4242

Analysed	
August 14, 2023 9:58:22 AM	
Computation with 5 linked points (*italics* points are less reliable):	1. Kasteel Duurstede 2. Sint Jan de Doperkerk 3. St-Janskerk 4. Dijkstraat 5. Langs de Wal
Reference points (Reliability 0–5)	5
Transformation with 3 parameters	
Scale	1:8.310
Rotation	13 °
Mean position error	17m

61
Septentrio
Montfort
Occidens
Oriens
Meridies
Septentrio
Occidens
Oriens
Meridies

61
Septentrio
Montfort
Occidens
Oriens
Meridies
Septentrio
Oriens
Meridies
300
0
0
300

216. Woerden

Tables 3.216 (measured and analysed).
Figures 3.216 (unmeasured, measured and analysed).

Measured							
Pixels	Scale stick original map (pixels)	Houses	Street sides	Houses per street side	Inhabitants (RF 4.0)	Inhabitants (RF 5.5)	Inhabitants (RF 7.0)
3043	missing	534	84	6.36	2137	2939	3741

Analysed	
August 15, 2023 9:56:01 AM	
Computation with 5 linked points (*italics* points are less reliable):	1. Petruskerk 2. Kasteel van Woerden 3. Hogewalstraat 4. Voorstraat 5. Molenstraat
Reference points (Reliability 0–5)	5
Transformation with 3 parameters	
Scale	1:7760
Rotation	11 °
Mean position error	13m

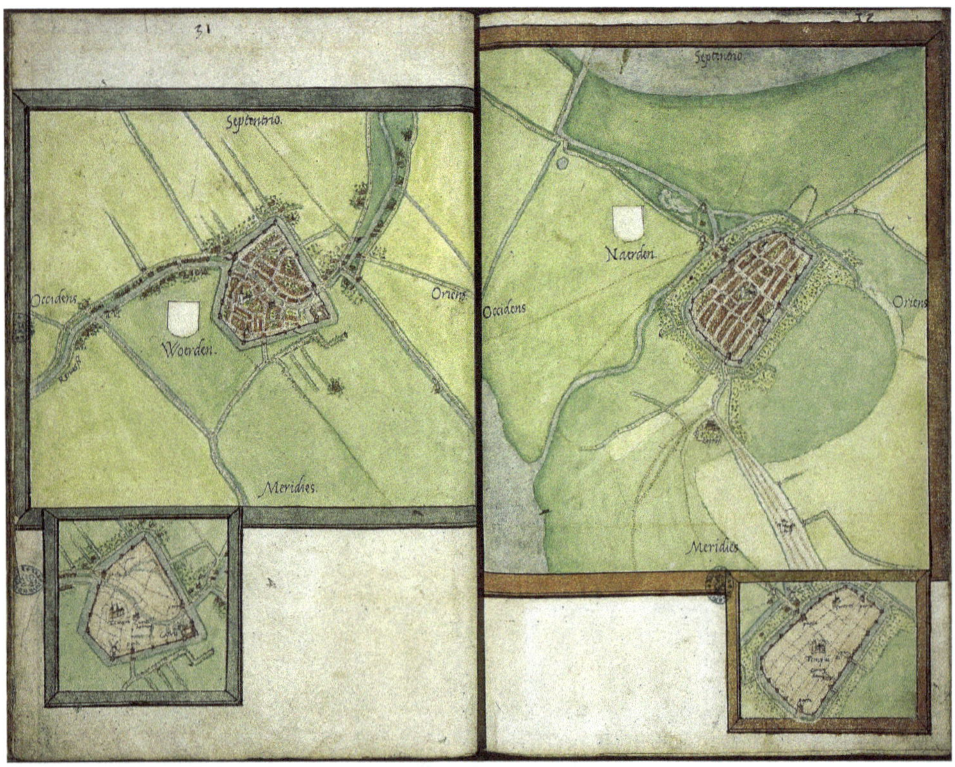

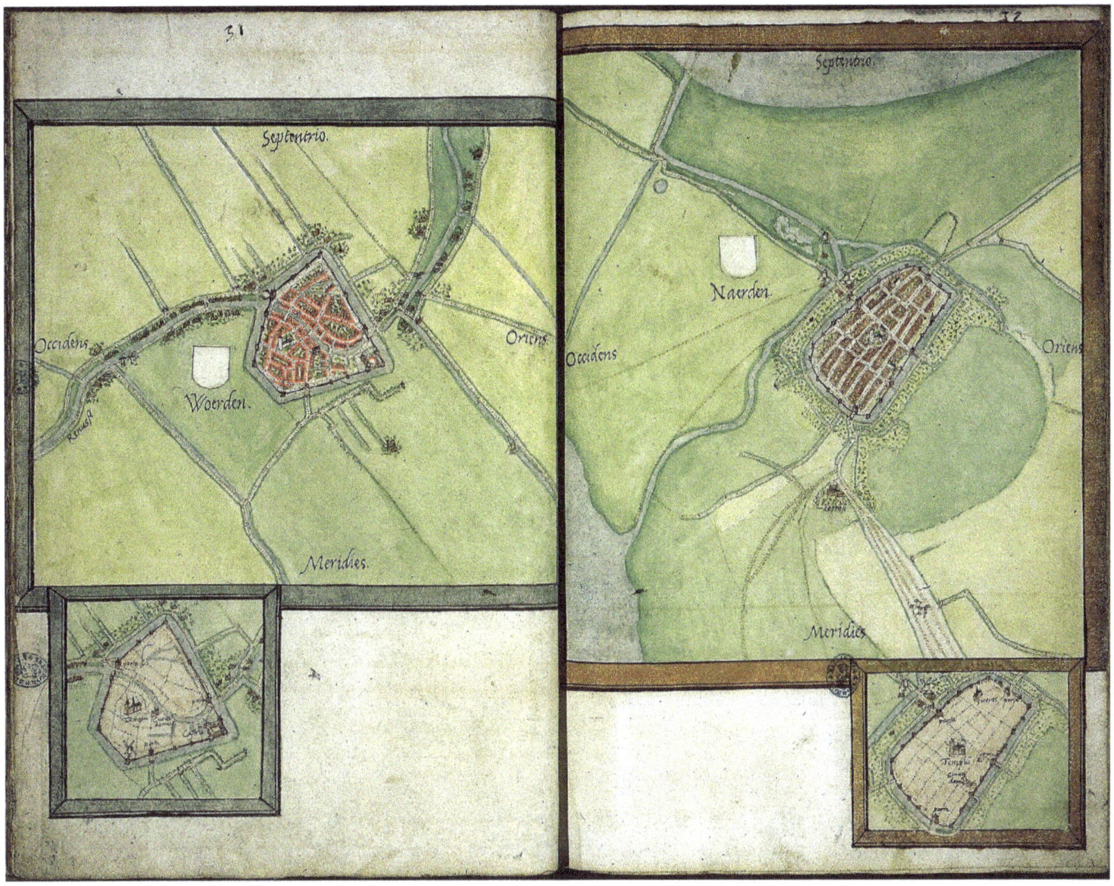
Septentrio.
Occidens
Oriens
Woerden.
Meridies.
Septentrio.
Naerden.
Occidens
Oriens
Meridies

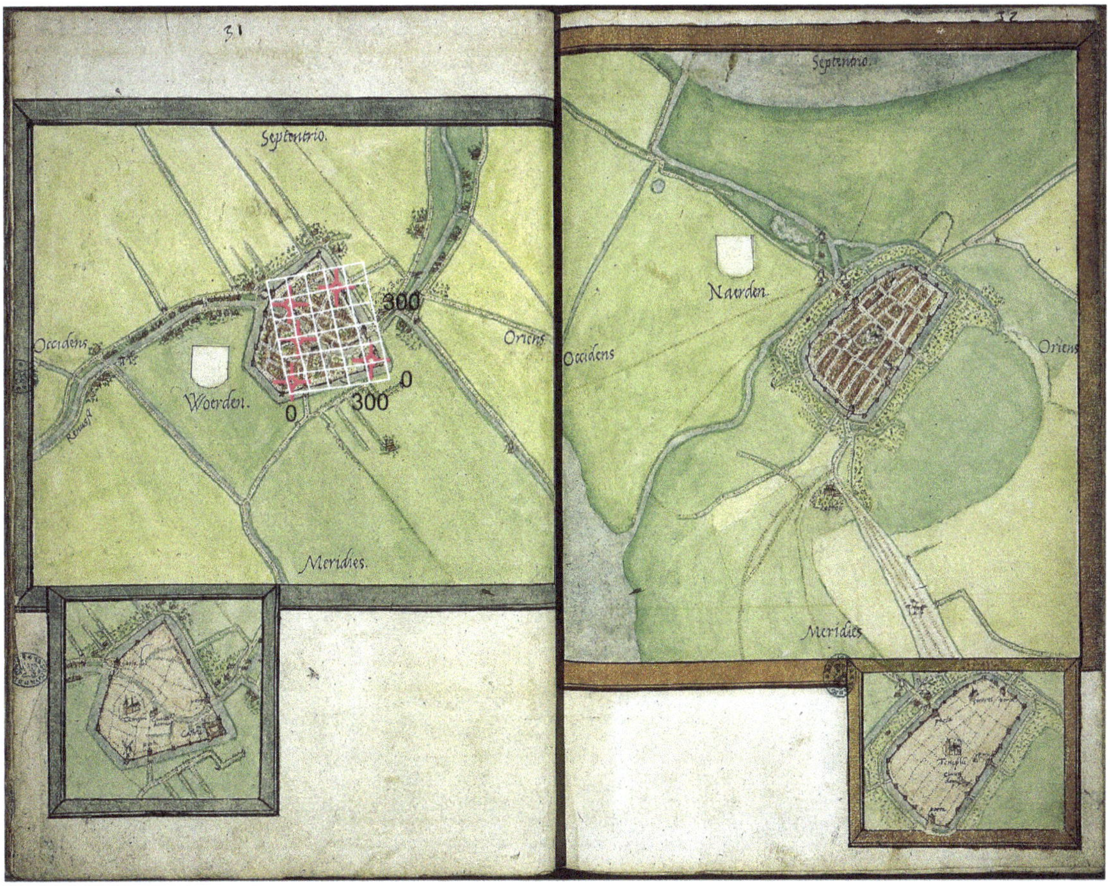
Septentrio.
300
Occidens
Oriens
Woerden.
0
300
0
Meridies.
Septentrio.
Naerden.
Occidens
Oriens
Meridies

217. Workum

Tables 3.217 (measured and analysed).
Figures 3.217 (unmeasured, measured and analysed).

Measured							
Pixels	Scale stick original map (pixels)	Houses	Street sides	Houses per street side	Inhabitants (RF 4.0)	Inhabitants (RF 5.5)	Inhabitants (RF 7.0)
2401	996	422	43	9.81	1687	2319	2951

Analysed	
May 22, 2023 3:34:24 PM	
Computation with 5 linked points (*italics* points are less reliable):	1. Grote of Sint-Gertrudiskerk 2. Djippe Dolte 3. Lynbaen 4. *Noard* 5. Trekwei
Reference points (Reliability 0–5)	4
Transformation with 3 parameters	
Scale	1:8340
Rotation	13 °
Mean position error	26m

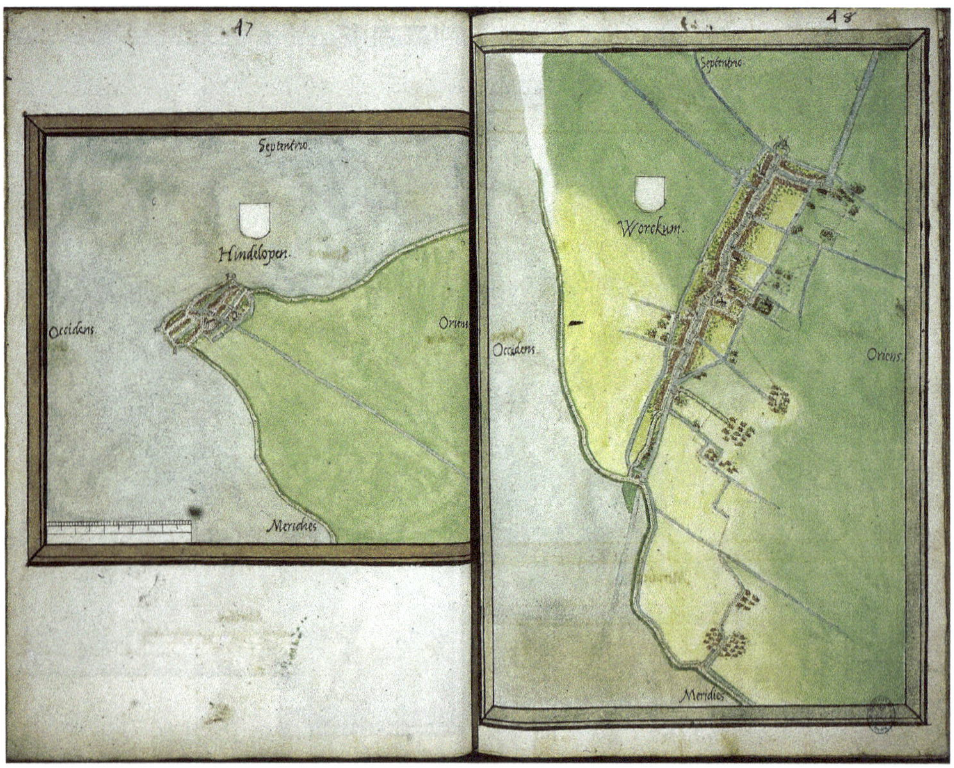

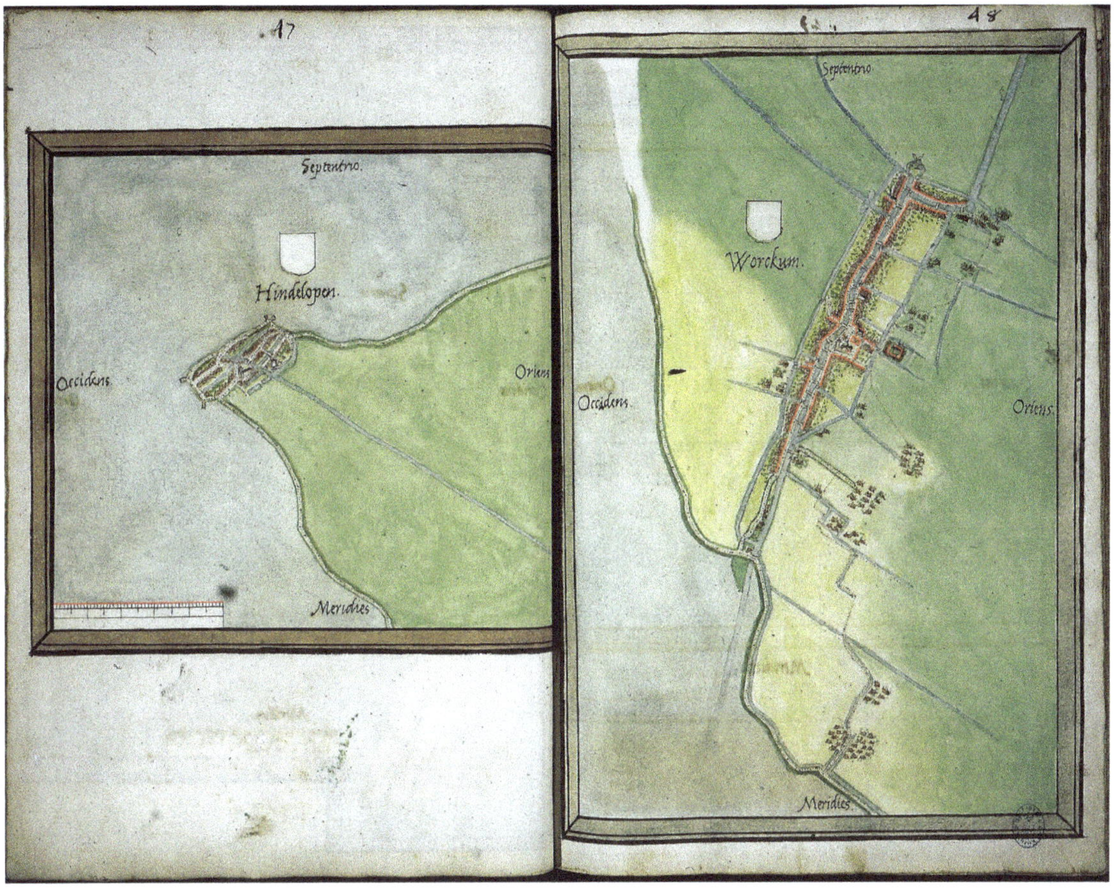
47
48
Septentrio.
Hindelopen.
Occidens.
Oriens.
Meridies
Septentrio
Worckum.
Occidens.
Oriens.
Meridies.

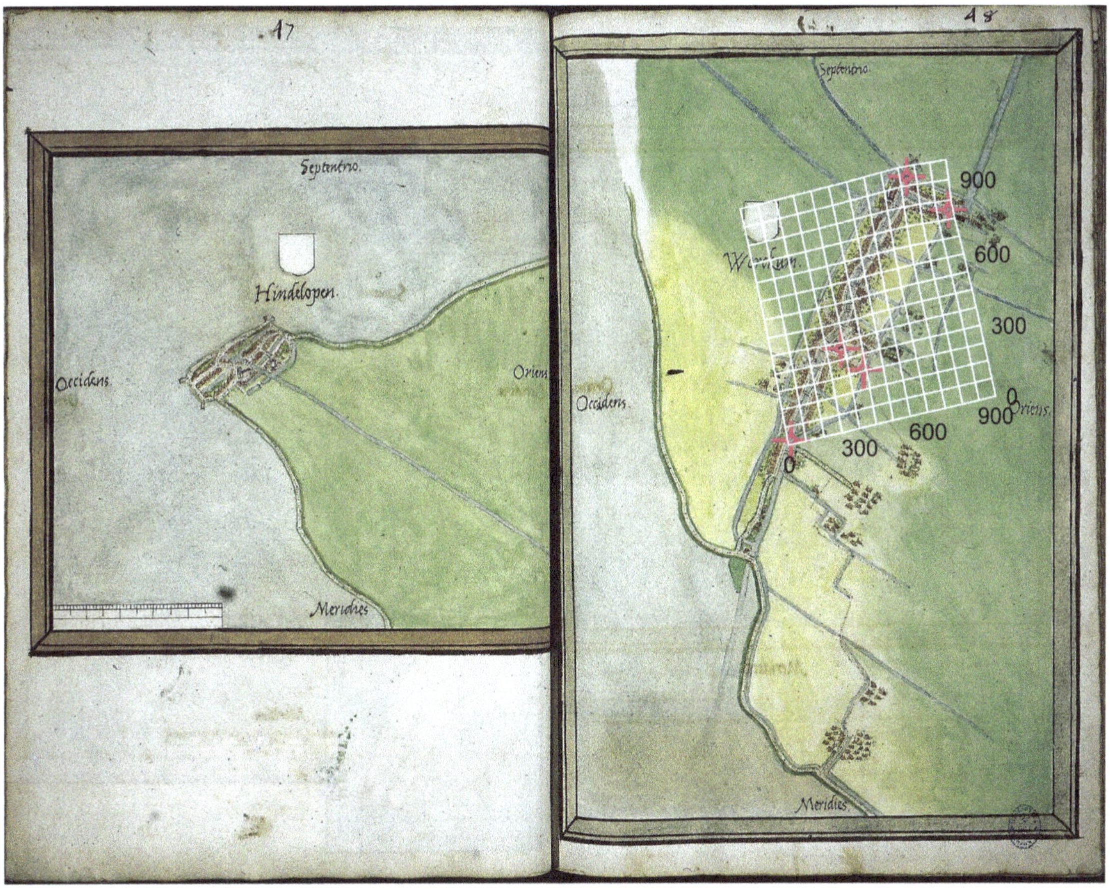
47
48
Septentrio.
Hindelopen.
Occidens.
Oriens.
Meridies
Septentrio
Occidens.
Oriens.
Meridies.
900
600
300
0
0
300
600
900

218. Woudrichem

Tables 3.218 (measured and analysed).
Figures 3.218 (unmeasured, measured and analysed).

Measured

Pixels	Scale stick original map (pixels)	Houses	Street sides	Houses per street side	Inhabitants (RF 4.0)	Inhabitants (RF 5.5)	Inhabitants (RF 7.0)
1215	998	213	33	6.47	853	1173	1493

Analysed

October 15, 2023 10:50:59 PM	
Computation with 5 linked points	1. Martinuskerk
(*italics* points are less reliable):	2. Slot Loevestein
	3. Molenstraat
	4. Bagijnestraat
	5. Rijkswal
Reference points (Reliability 0–5)	5
Transformation with 3 parameters	
Scale	1:7990
Rotation	14 °
Mean position error	14m

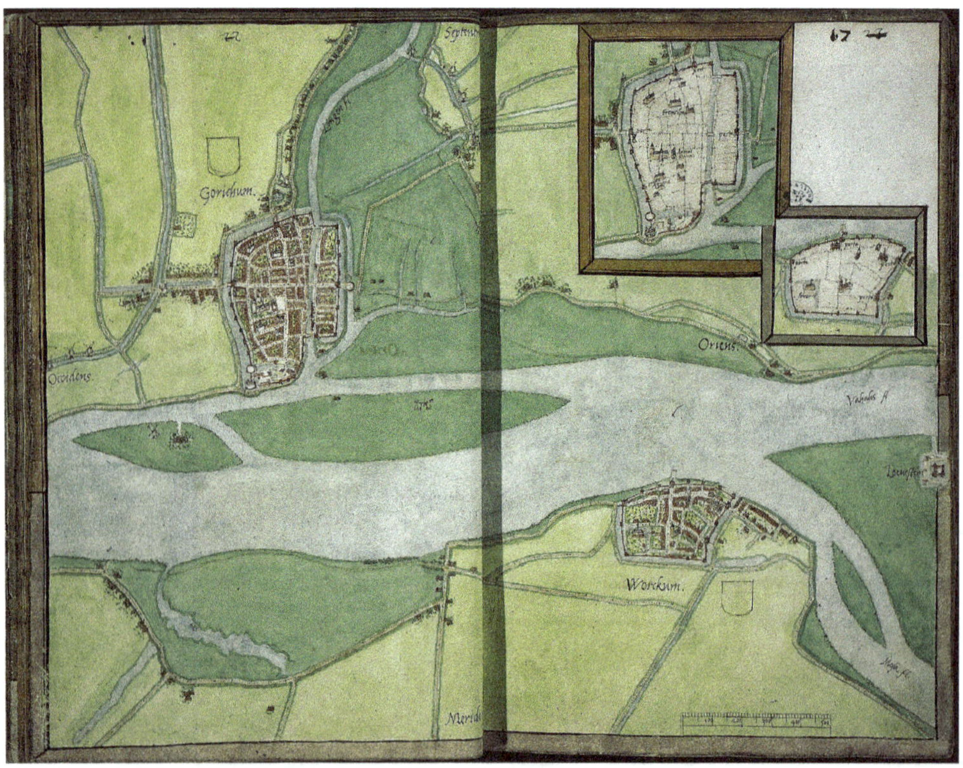

Gorichum.
Occidens.
Oriens.
Worckum.
Vahalis fl.

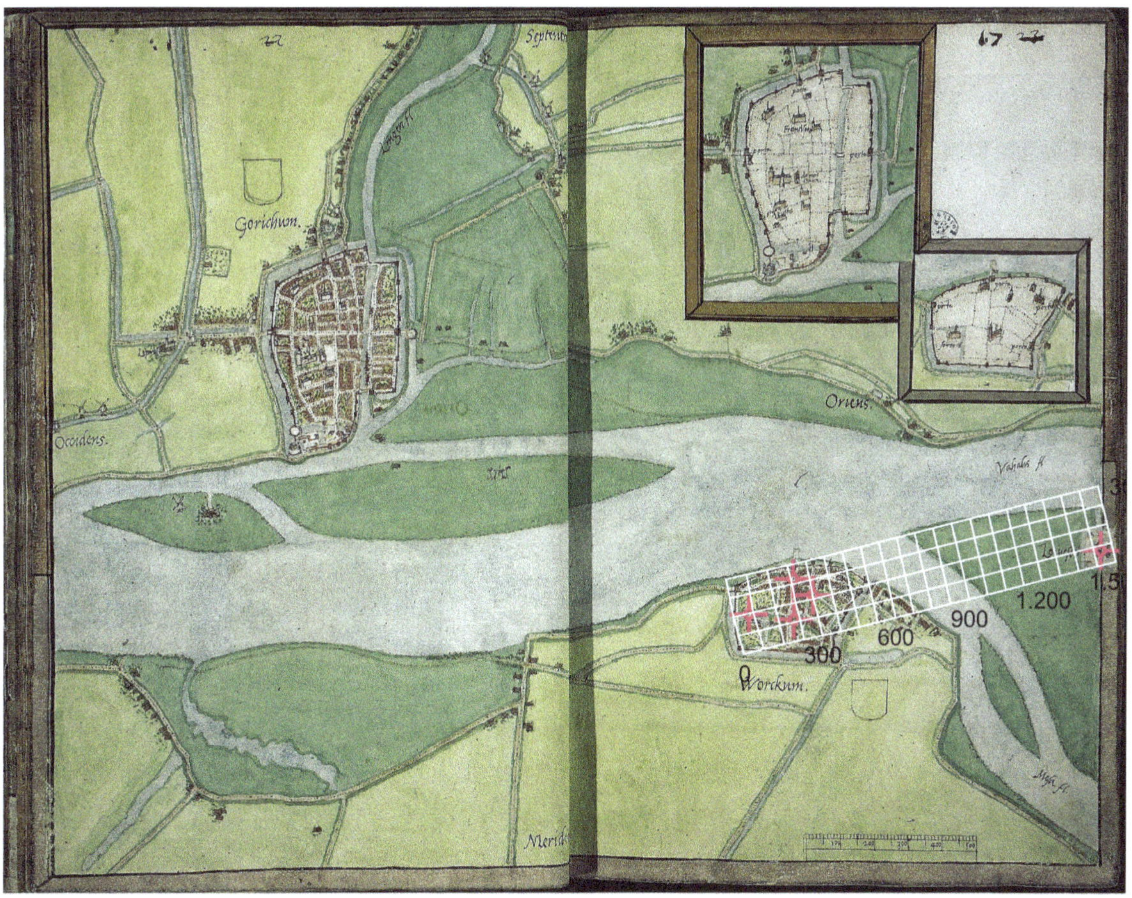
Gorichum.
Occidens.
Oriens.
Worckum.
Vahalis fl.
0
300
600
900
1.200

219. Yvois (Carignan)

Tables 3.219 (measured and analysed).
Figures 3.219 (unmeasured, measured and analysed).

Measured							
Pixels	Scale stick original map (pixels)	Houses	Street sides	Houses per street side	Inhabitants (RF 4.0)	Inhabitants (RF 5.5)	Inhabitants (RF 7.0)
2386	missing	419	84	4.99	1676	2304	2932

Analysed	
September 29, 2023 12:36:48 PM	
Computation with 5 linked points (*italics* points are less reliable):	1. Église Notre-Dame 2. Rue des Écoles (nord) 3. Promenade des Remparts 4. Promenade des Remparts (sud-est) 5. Rue des Écoles (ouest)
Reference points (Reliability 0–5)	5
Transformation with 3 parameters	
Scale	1:7320
Rotation	18 °
Mean position error	11m

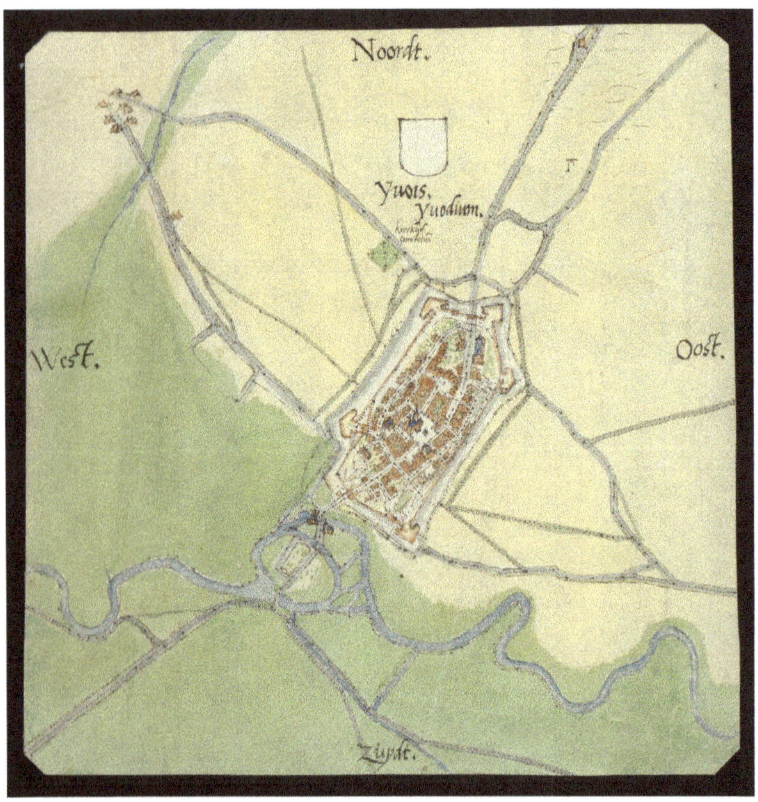

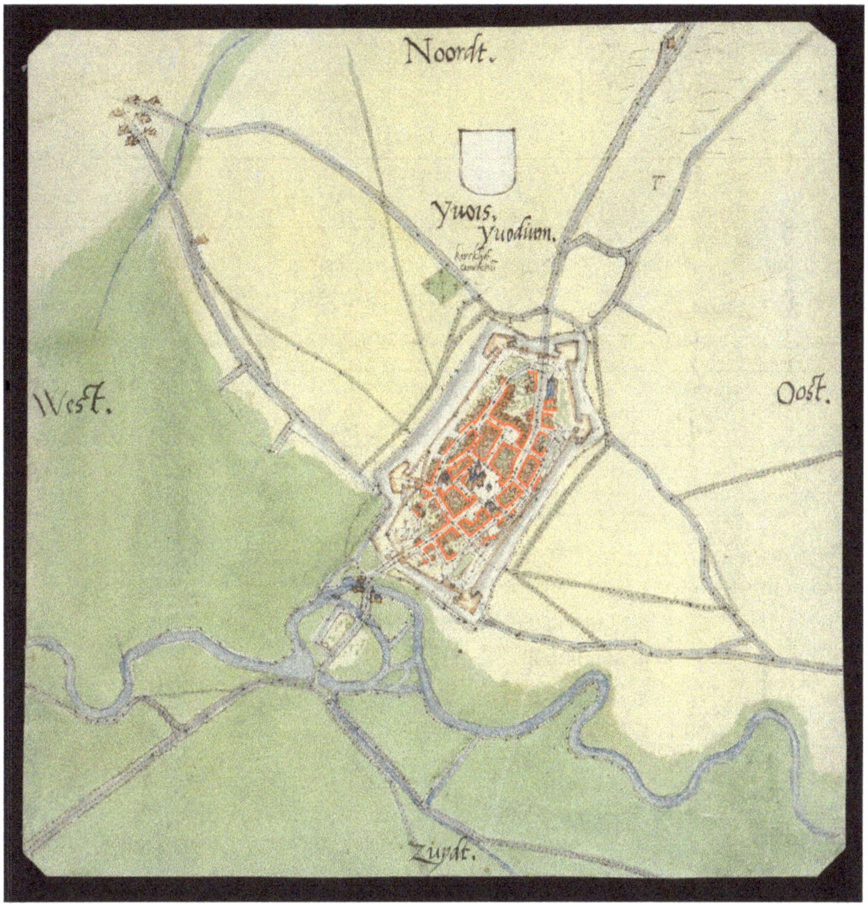
Noordt.
Yvois,
yuodium.
West.
Oost.
Zuydt.

Noordt.
Yvois,
yuodium.
West.
Oost.
300
0
300
0
Zuydt.

220. Zaltbommel

Tables 3.220 (measured and analysed).
Figures 3.220 (unmeasured, measured and analysed).

Measured

Pixels	Scale stick original map (pixels)	Houses	Street sides	Houses per street side	Inhabitants (RF 4.0)	Inhabitants (RF 5.5)	Inhabitants (RF 7.0)
5106	1012	897	114	7.87	3586	4931	6276

Analysed

August 19, 2023 6:43:16 PM	
Computation with 5 linked points (*italics* points are less reliable):	1. Grote of Sint-Maartenskerk 2. Stadskasteel Zaltbommel 3. Zandstraat 4. Waalkade 5. Heksenwal
Reference points (Reliability 0–5)	5
Transformation with 3 parameters	
Scale	1:8950
Rotation	12 °
Mean position error	13m

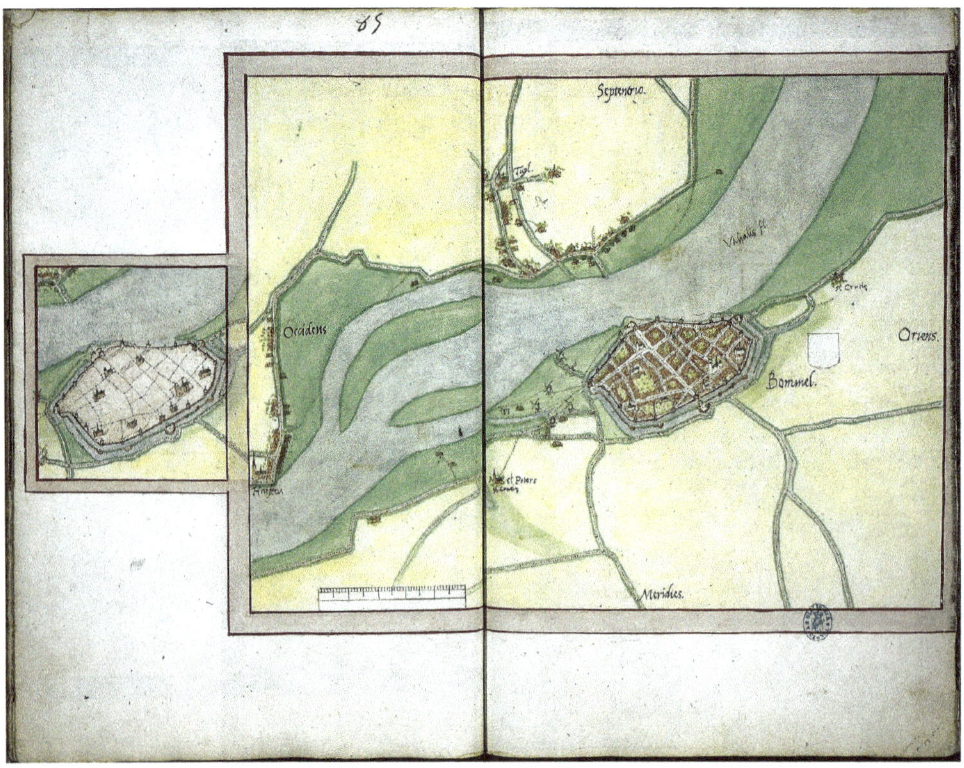

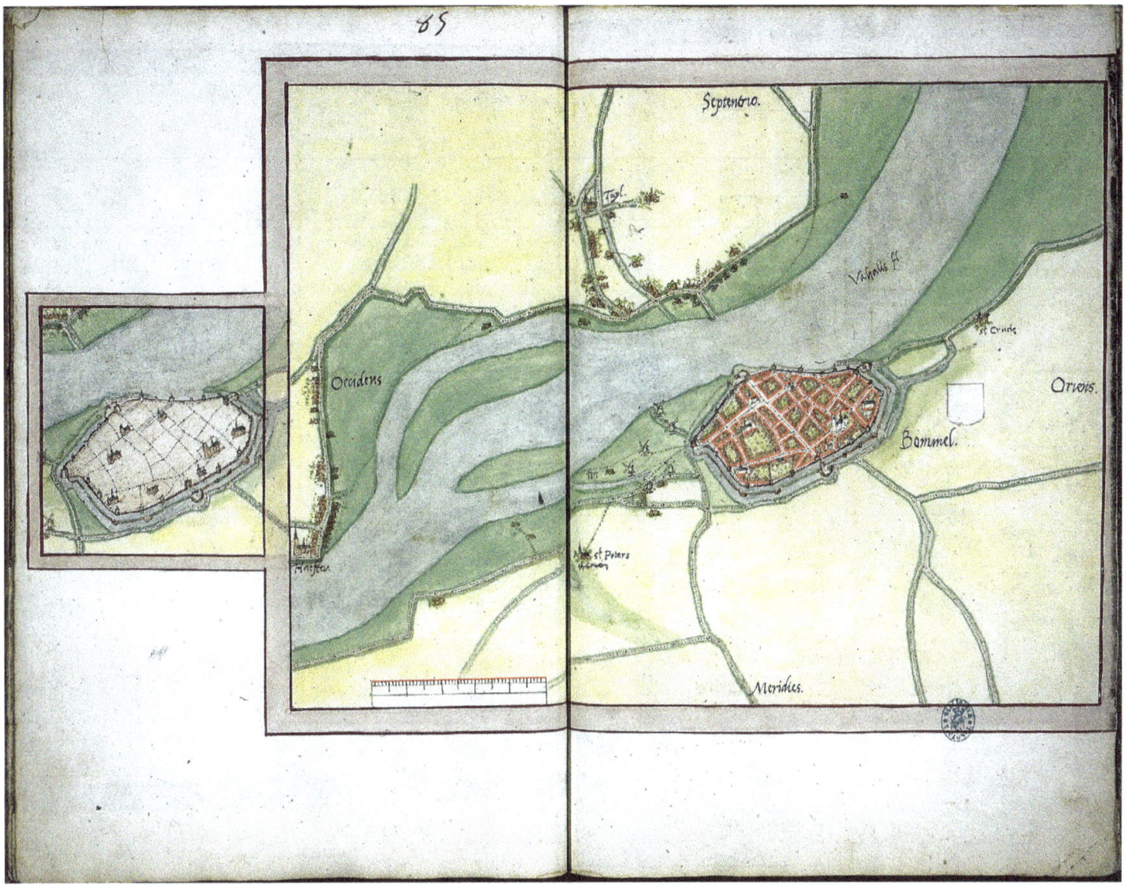
85
Septentrio.
Vahalis fl.
Occidens
Oriens.
Bommel.
Meridies.

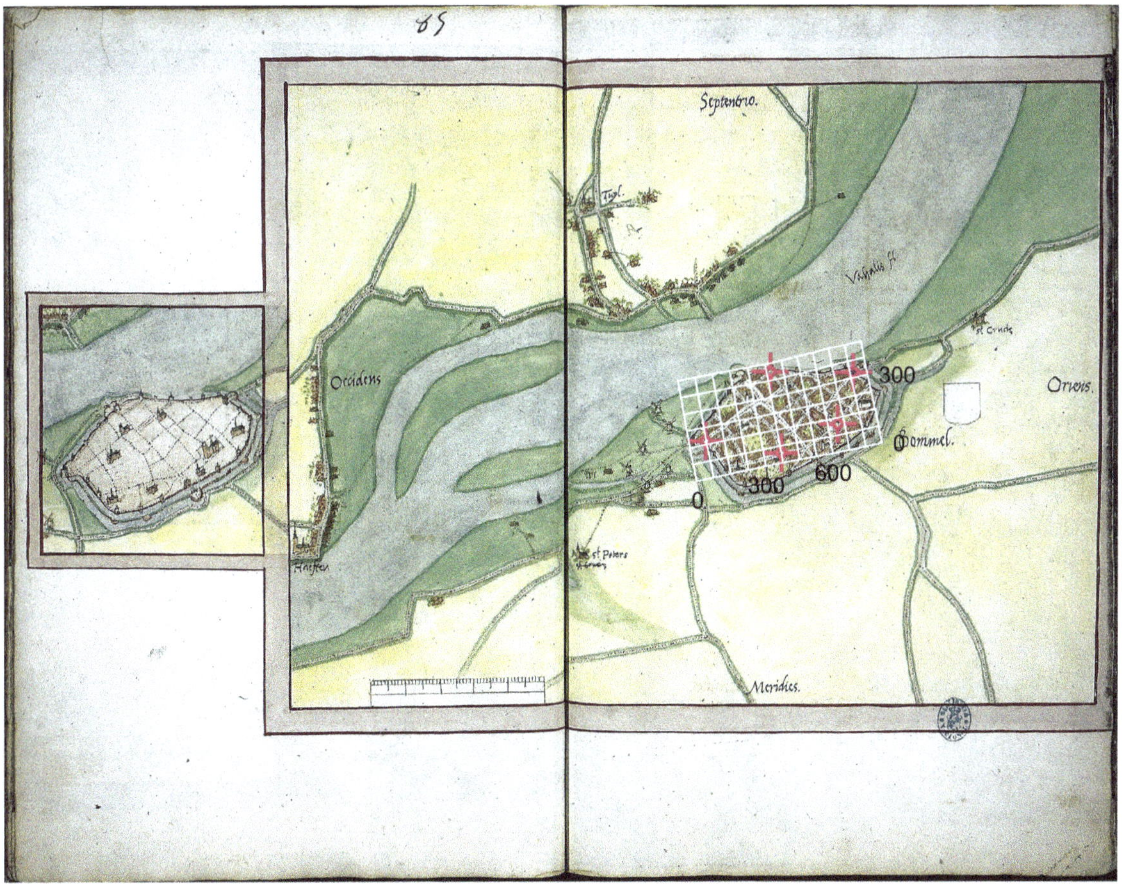
85
Septentrio.
Vahalis fl.
Occidens
Oriens.
300
0
300
600
Bommel.
Meridies.

221. Zevenbergen

Tables 3.221 (measured and analysed).
Figures 3.221 (unmeasured, measured and analysed).

Measured

Pixels	Scale stick original map (pixels)	Houses	Street sides	Houses per street side	Inhabitants (RF 4.0)	Inhabitants (RF 5.5)	Inhabitants (RF 7.0)
1468	999	258	19	13.57	1031	1418	1805

Analysed

August 19, 2023 10:42:44 PM	
Computation with 5 linked points (*italics* points are less reliable):	1. Catharinakerk 2. Roode Vaart *3. Zuidhaven* 4. Lage Wipstraat *5. Stationsstraat*
Reference points (Reliability 0–5)	**3**
Transformation with 3 parameters	
Scale	1:9070
Rotation	11 °
Mean position error	30m

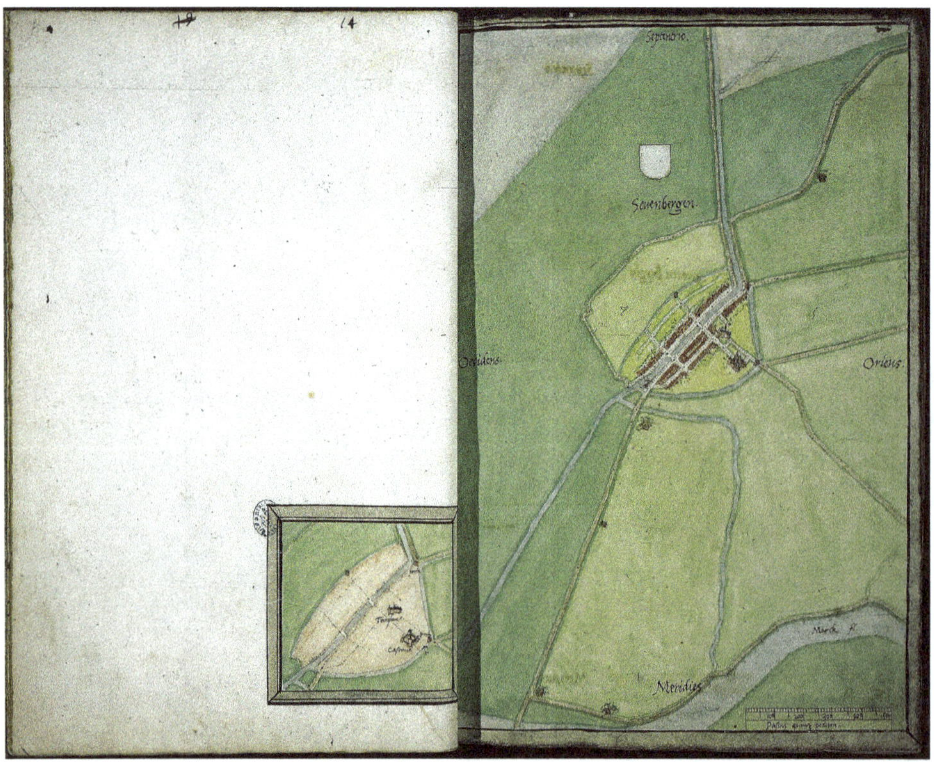

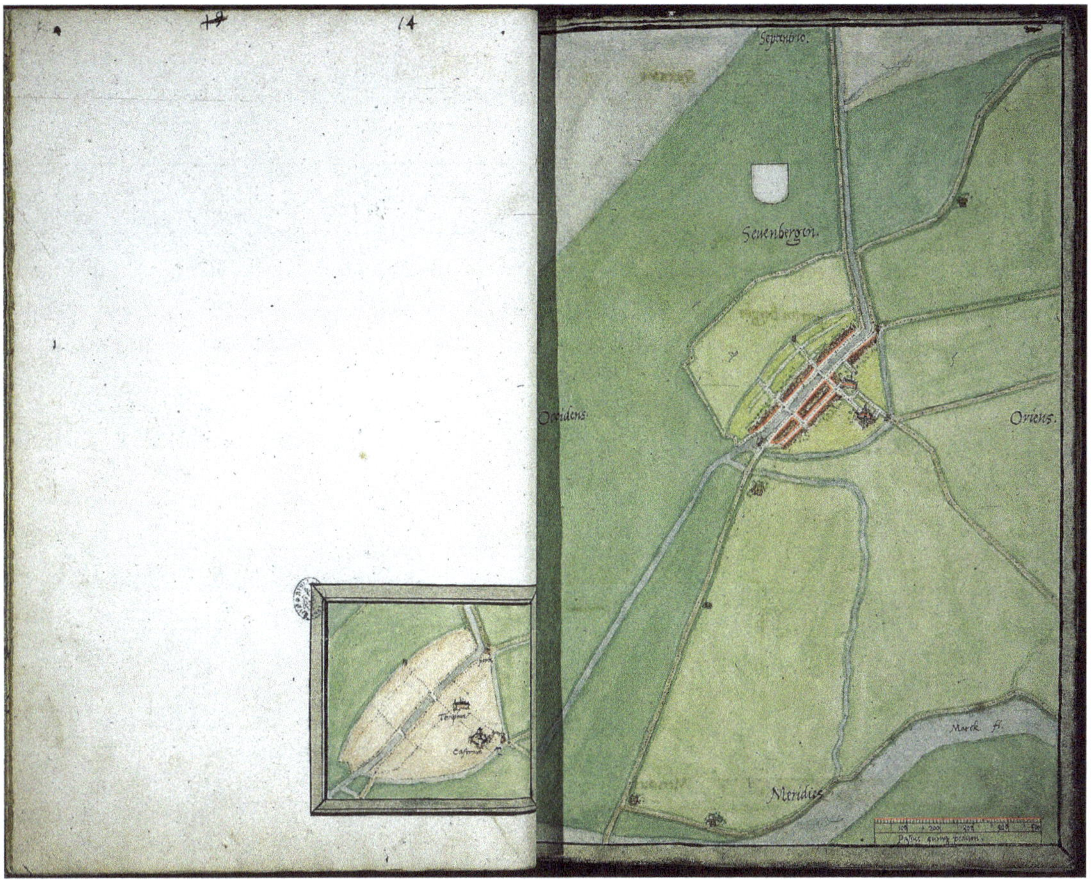
Septentrio.
Sevenbergen.
Occidens.
Oriens.
Marck fl.
Meridies.

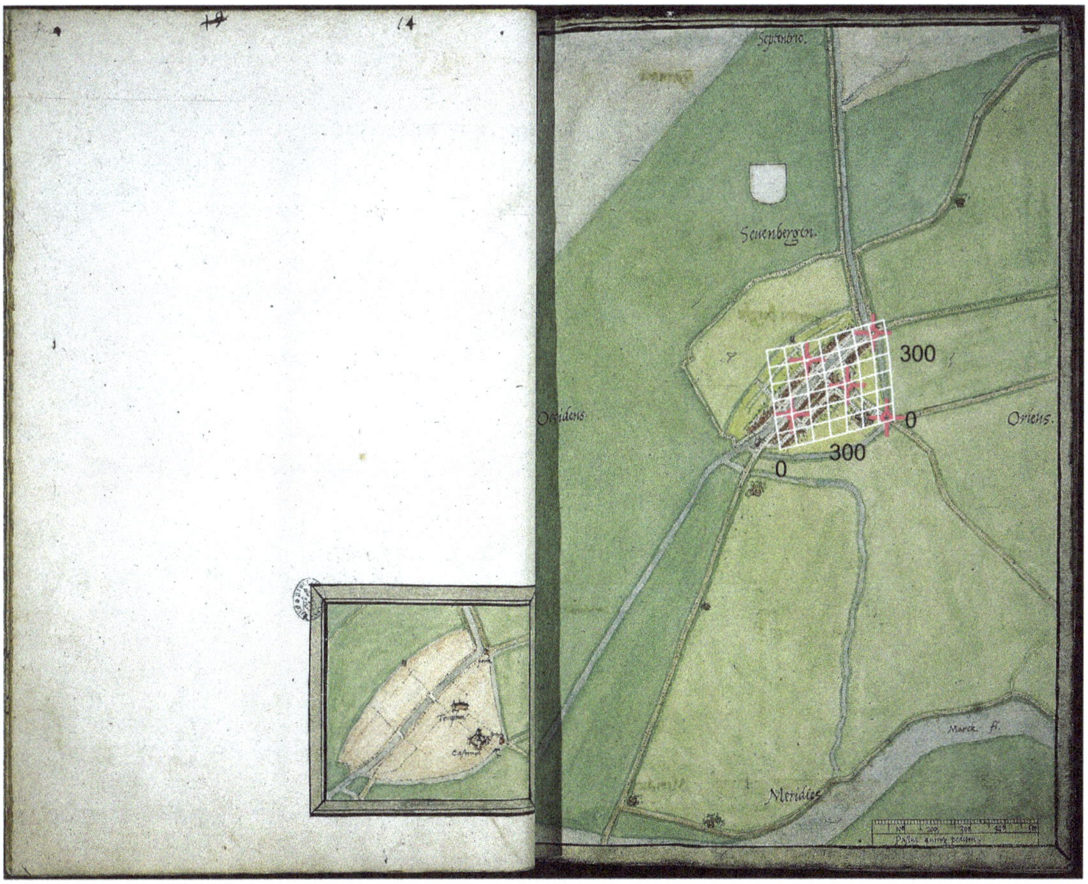
Septentrio.
Sevenbergen.
300
Occidens.
0
Oriens.
0
300
Marck fl.
Meridies.

222. Zichem

Tables 3.222 (measured and analysed).
Figures 3.222 (unmeasured, measured and analysed).

Measured							
Pixels	Scale stick original map (pixels)	Houses	Street sides	Houses per street side	Inhabitants (RF 4.0)	Inhabitants (RF 5.5)	Inhabitants (RF 7.0)
2598	missing	456	64	7.13	1825	2509	3193

Analysed	
September 29, 2023 1:45:50 PM	
Computation with 5 linked points (*italics* points are less reliable):	1. Sint-Eustachiuskerk 2. Oranjekasteel 3. Maagdentoren 4. Veststraat 5. Ter Elzen
Reference points (Reliability 0–5)	5
Transformation with 3 parameters	
Scale	1:8610
Rotation	12 °
Mean position error	13m

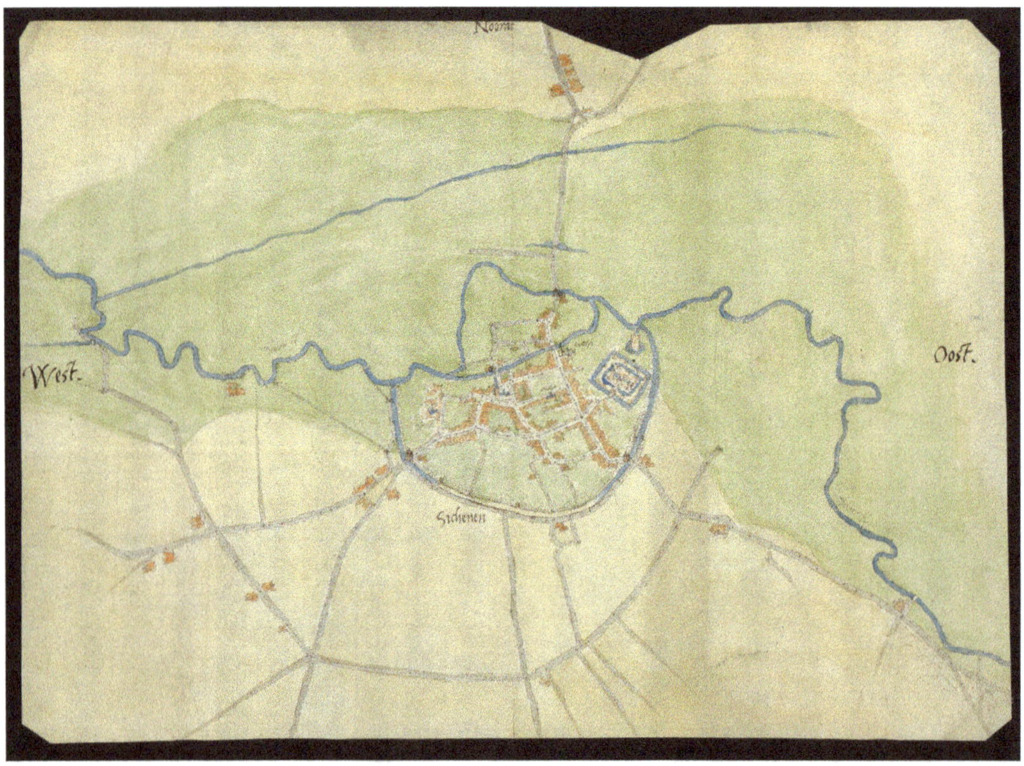

Noorn
West.
Oost.
Sichenen

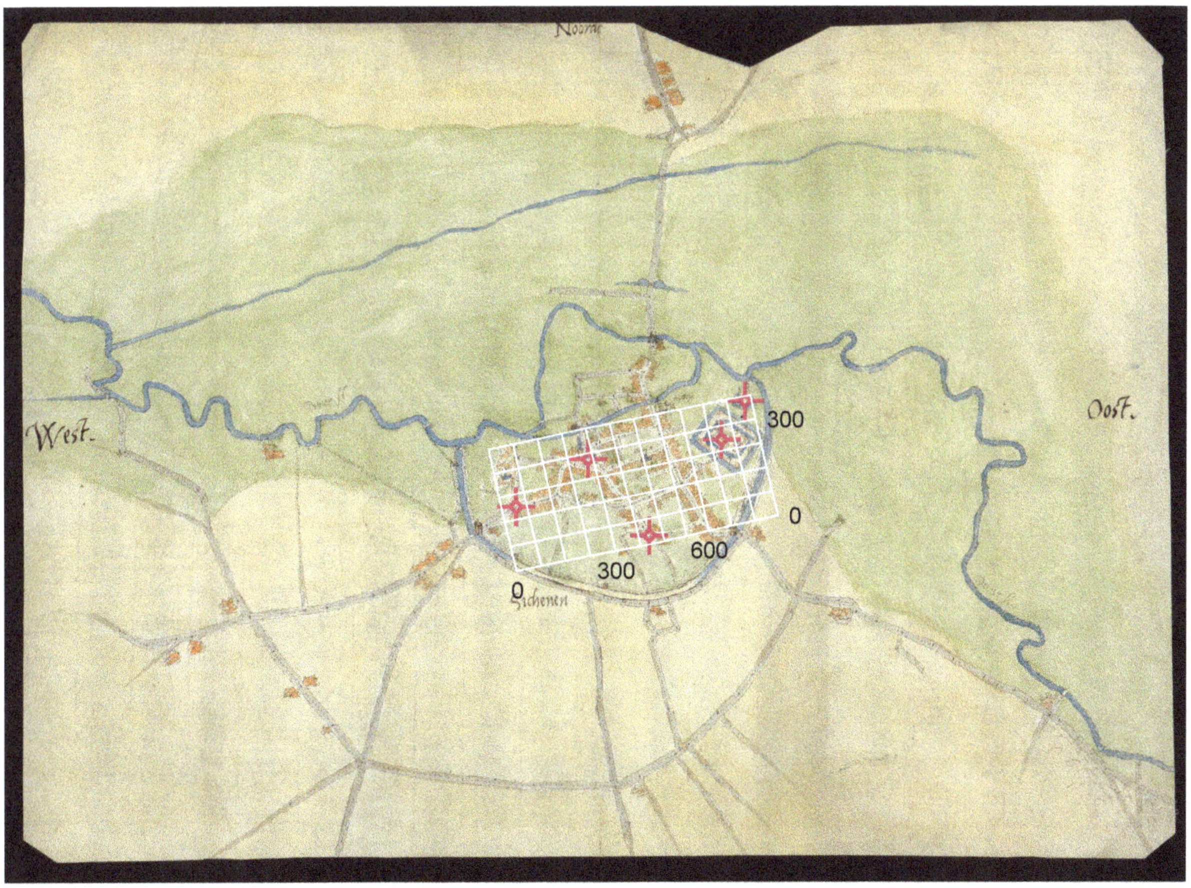
Noorn
West.
Oost.
300
0
600
300
0
Sichenen

223. Zierikzee

Tables 3.223 (measured and analysed).
Figures 3.223 (unmeasured, measured and analysed).

Measured

Pixels	Scale stick original map (pixels)	Houses	Street sides	Houses per street side	Inhabitants (RF 4.0)	Inhabitants (RF 5.5)	Inhabitants (RF 7.0)
7962	1006	1398	151	9.26	5.593	7690	9787

Analysed

January 23, 2023 11:57:57 AM	
Computation with 5 linked points (*italics* points are less reliable):	1. Kerkplein 2. Noorderhavenpoort 3. Nobelpoort 4. Hoofdpoortstraat 5. Zuidwellestraat
Reference points (Reliability 0–5)	5
Transformation with 3 parameters	
Scale	1:9040
Rotation	11 °
Mean position error	10m

Septentrio.
Zurczee
Occidens.
Oriens
Meridies

Septentrio.
Zurczee
600
300
0
0
300
600
900
Occidens.
Oriens
Meridies

224. Zoutleeuw
Tables 3.224 (measured and analysed).
Figures 3.224 (unmeasured, measured and analysed).

Measured							
Pixels	Scale stick original map (pixels)	Houses	Street sides	Houses per street side	Inhabitants (RF 4.0)	Inhabitants (RF 5.5)	Inhabitants (RF 7.0)
5364	missing	942	147	6.41	3768	5181	6594

Analysed	
September 29, 2023 7:05:04 PM	
Computation with 5 linked points (*italics* points are less reliable):	1. Schipstraat 2. Kapelstraat 3. Beekstraat 4. Predikherenstraat 5. Vleesstraat
Reference points (Reliability 0–5)	5
Transformation with 3 parameters	
Scale	1:6900
Rotation	14 °
Mean position error	6m

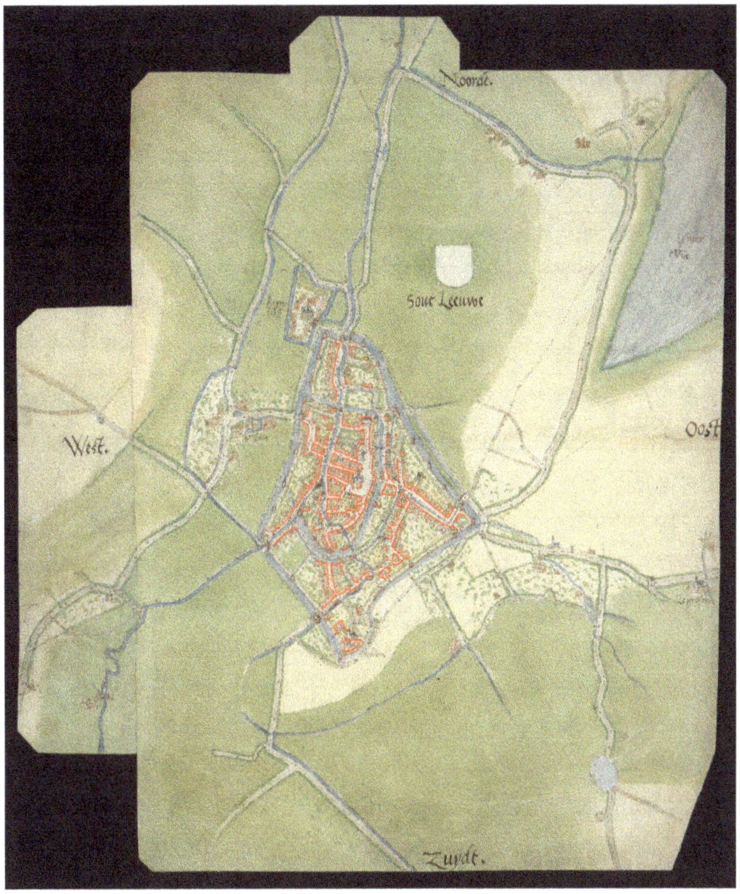
Noorde.
Sout Leeuwe
West.
Oost
Zuydt.

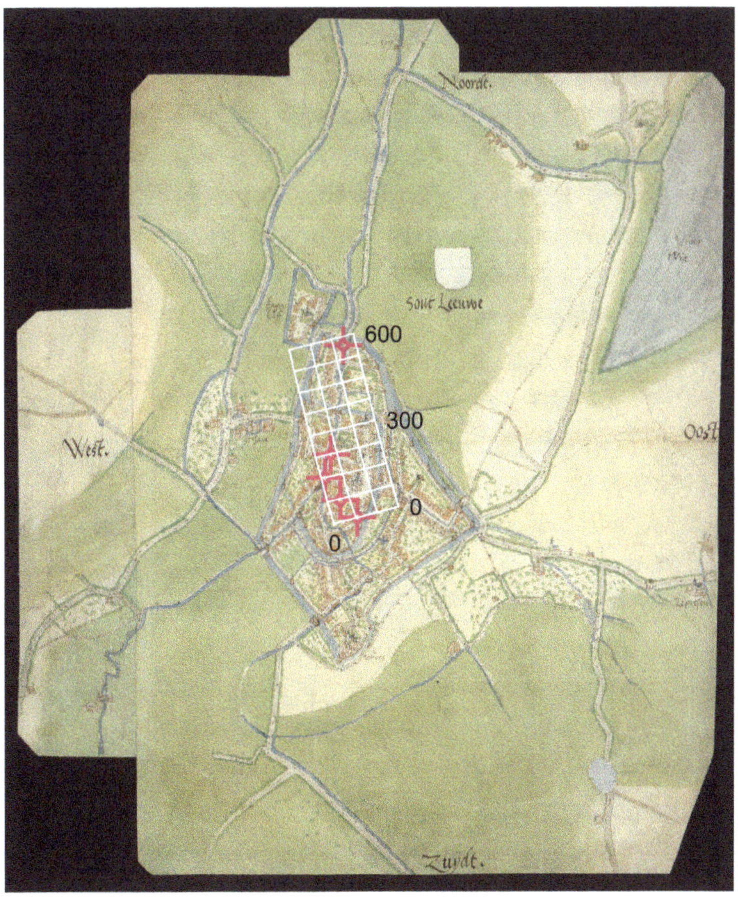
Noorde.
Sout Leeuwe
600
300
0
0
West.
Oost
Zuydt.

225. Zutphen

Tables 3.225 (measured and analysed).
Figures 3.225 (unmeasured, measured and analysed).

Measured							
Pixels	Scale stick original map (pixels)	Houses	Street sides	Houses per street side	Inhabitants (RF 4.0)	Inhabitants (RF 5.5)	Inhabitants (RF 7.0)
9111	1001	1600	205	7.80	6399	8799	11,199

Analysed	
August 22, 2023 1:52:45 PM	
Computation with 5 linked points	1. Walburgiskerk
(*italics* points are less reliable):	2. Broederenkerk
	3. Nieuwstadkerk
	4. Drogenapstoren
	5. Berkelpoort
Reference points (Reliability 0–5)	5
Transformation with 3 parameters	
Scale	1:7490
Rotation	11 °
Mean position error	10m

Septentrio.
Zutphen.
Occidens.
Oriens.
Meridies.
99.

Septentrio.
Zutphen.
Occidens.
Oriens.
Meridies.
99.
600
300
0
300
0

226. Zwolle

Tables 3.226 (measured and analysed).
Figures 3.226 (unmeasured, measured and analysed).

Measured

Pixels	Scale stick original map (pixels)	Houses	Street sides	Houses per street side	Inhabitants (RF 4.0)	Inhabitants (RF 5.5)	Inhabitants (RF 7.0)
7594	985	1334	238	5.60	5334	7334	9334

Analysed

August 23, 2023 8:47:40 AM	
Computation with 5 linked points (*italics* points are less reliable):	1. Grote Kerk 2. Peperbus 3. Diezerpoort 4, Sassenpoort 5. Buitenkant
Reference points (Reliability 0–5)	5
Transformation with 3 parameters	
Scale	1:7870
Rotation	9 °
Mean position error	7m

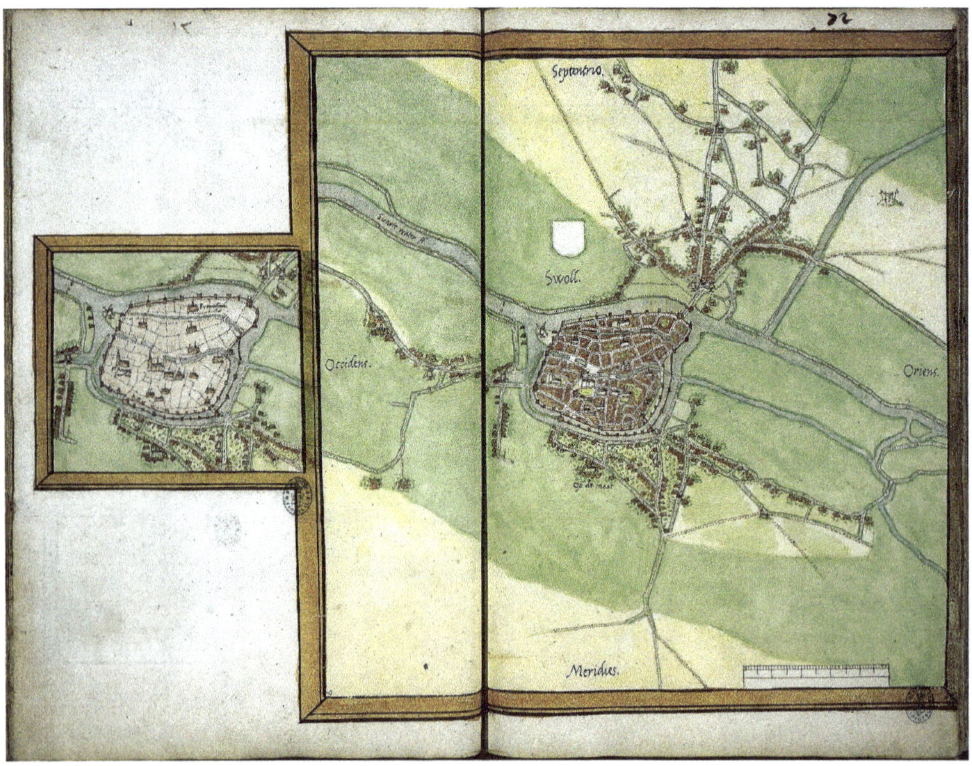

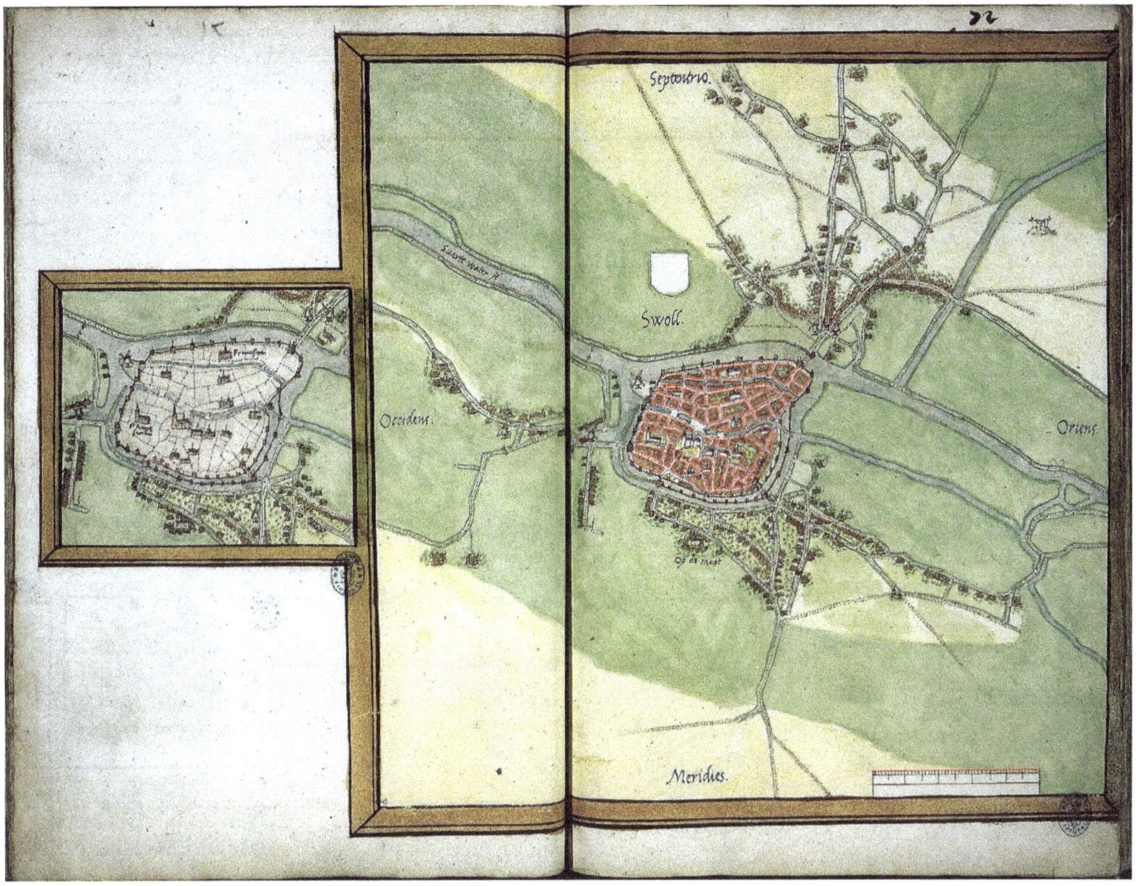
Septentrio.
Swoll.
Occidens.
Oriens
Meridies.

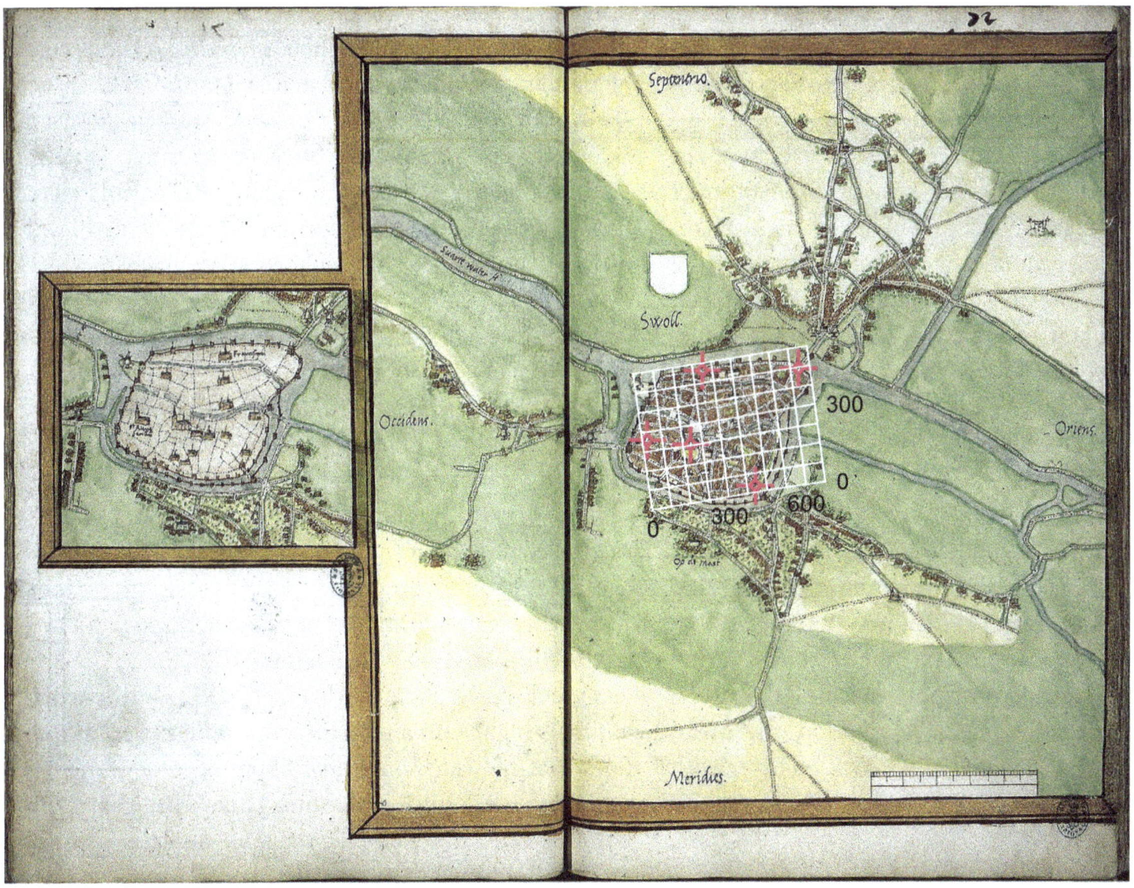
Septentrio.
Swoll.
Occidens.
Oriens
Meridies.
300
0
0
300
600

3.2 Comparisons

TABLE 3.227 The 226 cities: Dimensions (summary)

	Pixels	Scale stick in pixels (original maps)	Houses	Street sides	Houses per street side	Inhabitants (RF 5.5)
Total	1,204,457		211,543	28,904		1,165,020
Average	5401	1009	949	130	7	5224

TABLE 3.228 The 226 cities: Accuracy (summary)

	Scale 1: (± cm)	Scale deviation from Vlissingen en Leiden (± m)	Rotation (°)	Mean Position Error (± m)	Weighted Mean Position Error (meters per house)	Recognizable reference points (1–5)
Average	8147	5	11	20	0.0432	4
Correlation Coefficient (with the Weighted Mean Position Error)	0.01	0.01	0.18	0.7	–0.5	1.0

Table 3.227 and 3.228 show the totals and averages of all measurements, counts, calculations, and analyses: pixels, scale stick, houses, street sides, houses per street side, inhabitants, scale, scale deviation, rotation, mean position error absolute, mean position weighted, and reliability. The results can be divided into two groups.

The first group includes the totals and averages. For example, the total number of measured pixels of all street sides on all 226 maps is 1,204,457, with an average of 5401 per city. The average scale stick is 1009 pixels. In total, the 226 cities count 211,543 houses standing on 28,904 street sides, averaging 7 houses per street side. Using a reduction factor of 5.5 inhabitants per house, there were 1,165,020 people living in the 226 cities, averaging 5224 per city. Further details on Tables 3.2.02 to 3.2.05 will be discussed later.

The second group includes the analyses from MapAnalyst. The results are as follows: the average scale of all 226 maps is 1:8147. The deviation from the 'ideal' scales of the test cities Vlissingen and Leiden is 5 meter, at a standard deviation factor of 7 percent (not presented in the table), indicating considerable uniformity of the maps. The same applies to the degree of rotation, the Mean Position Error and the Weighted Mean Position Error, with averages of respectively 11 degrees and 20 (absolute) and 0.0432 (weighted) meters.

An important analysis is the correlation between the reliability analysis added in the context of this study (the number of reliable reference points) and the Weighted Mean Position Error (the global measure of the accuracy of all reference points calculated by MapAnalyst. This can score from –1 to +1. A score of 0 means no correlation at all. Scores of –1 or +1 indicate a perfect correlation. The correlation coefficient, as shown in Table 3.227, calculated over the 226 cities is –0.7 (absolute) and –0.5 (weighted), which means that there is a significant correlation where a higher deviation in meters corresponds with a lower rating of the quality of the reference points. This is shown in Graph 3.1.

The conclusion that the addition of the reliability analysis to MapAnalyst is valuable, is now justified. Based on this, the other analyses of using MapAnalyst can be assessed for reliability, where a score of 0 is low and a score of 5 is high. When a particular city shows a Weighted Mean Position Error that deviates strongly from the average, the reliability score can be used to assess the extent to which the deviation should be taken seriously and investigated further. A map with a low MPE is not necessarily less precise than a map with a high MPE. Only when reliability is high, with a score of 4 or 5, and the MPE is low, does that indicate something about accuracy. The three other analyses of using MapAnalysthave little correlation with the reliability analysis: the scale analyses score 0.01 and the Rotation analysis 0.18.

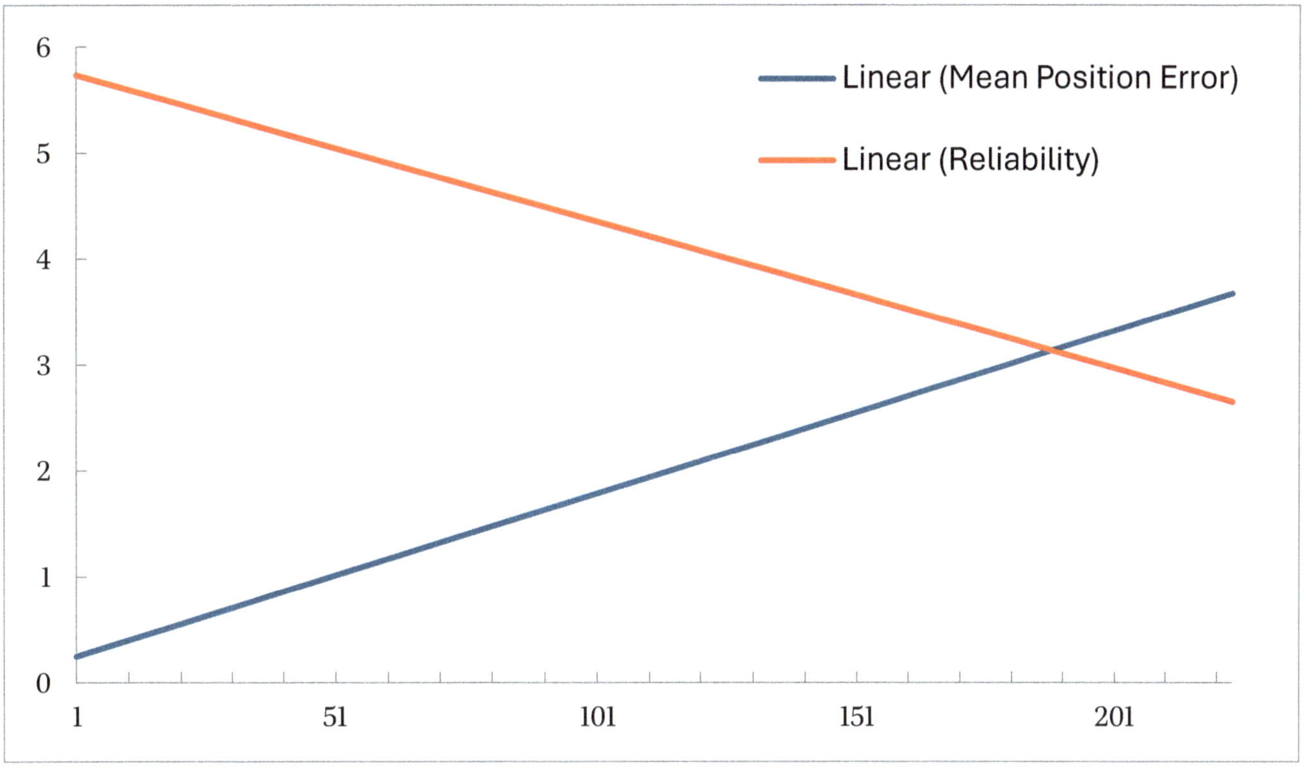

GRAPH 3.1 The 226 cities: Cities: Correlation (0.7, linear trend) between the MapAnalist analysis 'Mean Position Error' and the new added analysis 'Reliability'

TABLE 3.229 The 226 cities: Alphabetically

No.	Madrid or Brussel	City	Year or period
1	M	Aalst	1551–1559
2	M	Aardenburg	1560–1569
3	B	Aarschot	1550–1572
4	M	Aire-sur-la-Lys	1560–1570
5	M	Alkmaar	1560–1561
6	M	Almelo	1557–1558
7	M	Amersfoort	1560
8	M	Amsterdam	1560–1561
9	M	Appingedam	1568
10	B	Arlon	1554–1558
11	M	Armentières	1560–1563
12	M	Arnemuiden	1550 (about)
13	M	Arnhem	1550 (after)
14	M	Arras	1565 (in or after)
15	M	Asperen	1566 (before)
16	M	Ath	1550–1560
17	M	Avesnes-sur-Helpe	1556 (after)
18	M	Axel	1560–1571
19	M	Bailleul	1560–1572
20	M	Bapaume	1554–1570
21	B	Bastogne	(unknown)
22	M	Bavay	1560–1572

TABLE 3.229 The 226 cities: Alphabetically (*cont.*)

No.	Madrid or Brussel	City	Year or period
23	M	Beaumont	1548–1568
24	B	Bergen op Zoom	1542–1545
25	M	Bergues	1543–1558
26	M	Béthune	1553–1570
27	M	Beverwijk	1560 (about)
28	M	Binche	1550–1554
29	B	Bitburg	1558
30	M	Bolsward	1559–1562
31	M	Bouchain	1572 (in or after)
32	M	Bourbourg	1551 (in or after)
33	B	Bouvignes-sur-Meuse	1554–1566
34	M	Braine-le-Comte	1550–1560
35	M	Bredevoort	1561–1562
36	M	Brielle	1550–1558
37	M	Brouwershaven	unknown
38	M	Brugge	1566 (in or after)
39	B	Brussel	1554–1557
40	M	Buren	1557 (about)
41	M	Cambrai	1543–1559
42	M	Cassel	1569–1571
43	M	Chièvres	1560–1570
44	M	Chimay	1560–1565
45	B	Chiny	1550–1565
46	M	Coevorden	1552 (after)
47	M	Comines	1560
48	M	Conde-sur-l'Escaut	1562
49	M	Culemborg	1557
50	M	Damme	1558–1565
51	B	Damvillers	1560–1570
52	M	Delden	1557–1558
53	M	Delft	1556–1557
54	M	Den Haag	1560 (before)
55	M	Dendermonde	1565
56	M	Deventer	1557 (about)
57	M	Diksmuide	1550–1560
58	B	Dinant	1554–1566
59	M	Doesburg	1560
60	M	Doetinchem	1561 (about)
61	M	Dokkum	1562–1568
62	M	Dordrecht	1544–1558
63	M	Douai	1561–1561
64	M	Dunkerque	1548–1558
65	M	Echt	unknown
66	B	Echternach	1560

TABLE 3.229 The 226 cities: Alphabetically (*cont.*)

No.	Madrid or Brussel	City	Year or period
67	M	Edam	1560
68	B	Eindhoven	1557–1568
69	M	Elburg	1560 (about)
70	M	Enghien	1550
71	M	Enkhuizen	1558–1560
72	M	Enschede	1557–1558
73	M	Erkelenz	unknown
74	M	Franeker	unknown
75	M	Geertruidenberg	1561 (before)
76	M	Geervliet	1550–1572
77	M	Geldern	1560 (before)
78	B	Gembloux	1550–1560
79	M	Genemuiden	1557 (about)
80	M	Gent	1558–1564
81	M	Geraardsbergen	1551–1552
82	M	Gistel	1562–1566
83	M	Goedereede	1556–1557
84	M	Goes	1555–1563
85	M	Goor	1557 (about)
86	M	Gorinchem	1557 (after)
87	M	Gouda	1566 (before)
88	M	Gravelines	1555–1571
89	M	Groenlo	1561 (about)
90	M	Groningen	1561–1569
91	M	Grootebroek	1560 (about)
92	M	Haarlem	1560 (about)
93	M	Halle	1550–1554
94	M	Halluin	1567–1572
95	M	Hardenberg	1557 (about)
96	M	Harderwijk	1560 (about)
97	M	Harelbeke	1560–1560
98	M	Harlingen	1559–1562
99	M	Hasselt	1557 (about)
100	M	Hattem	1558 (about)
101	M	Heerenberg, 's-	1550 (before)
102	B	Helmond	1548
103	B	Herentals	1558
104	B	Hertogenbosch, 's-	1545
105	B	Herve	(unknown)
106	B	Herzogenrath	1564–1568
107	M	Hesdin	1559–1563
108	M	Heukelum	1566 (before)
109	M	Heusden	1555 (after)
110	M	Hindeloopen	1549–1570

TABLE 3.229 The 226 cities: Alphabetically (*cont.*)

No.	Madrid or Brussel	City	Year or period
111	M	Hondschoote	1558–1562
112	M	Hoorn	1558–1560
113	M	Hulst	1560 (before)
114	M	IJlst	unknown
115	M	IJsselstein	1557 (after)
116	M	Kampen	1557 (about)
117	M	Kortrijk	1560–1572
118	M	Kuinre	1557–1558
119	M	La Bassée	unknown
120	B	Landen	1560–1570
121	M	Landrecies	1545 (after)
122	M	Le Cateau-Cambrésis	1559
123	M	Le Quesnoy	1550–1569
124	M	Leerdam	1566 (before)
125	M	Leeuwarden	1554–1566
126	M	Leiden	1564 (before)
127	M	Lens	1554–1570
128	M	Le Roeulx	1550–1555
129	M	Lessines	1550–1570
130	B	Leuven	1543–1570
131	M	Leuze-en-Hainaut	1550–1560
132	B	Lier	1543–1570
133	M	Lille	1550–1565
134	M	Lillers	unknown
135	B	Limbourg	1550–1565
136	M	Lingen	1560
137	M	Lochem	1560 (about)
138	B	Luxembourg	1560–1560
139	B	Marche-en-Famenne	1550–1558
140	B	Mariembourg	1559–1565
141	M	Maubeuge	1550–1562
142	B	Mechelen	1550–1565
143	M	Medemblik	1560
144	B	Megen	unknown
145	M	Menen	1569–1573
146	M	Mesen	1552–1560
147	M	Middelburg (B)	1550–1558
148	M	Middelburg (Nl)	1550 (about)
149	M	Monnickendam	1560 (about)
150	M	Mons	1552–1555
151	M	Montfoort	1554 (about)
152	M	Montfort	1545 (about)
153	B	Montmédy	1560

TABLE 3.229 The 226 cities: Alphabetically (*cont.*)

No.	Madrid or Brussel	City	Year or period
154	M	Muiden	1560 (about)
155	M	Naarden	1560 (about)
156	M	Nieuwpoort (B)	1560–1570
157	M	Nieuwpoort (Nl)	1562 (or later)
158	M	Nijmegen	1557 (about)
159	M	Ninove	1555–1575
160	M	Oldenzaal	1557–1558
161	M	Oostburg	unknown
162	M	Oostende	1560–1562
163	M	Ootmarsum	1557–1558
164	M	Orchies	1561
165	M	Oudenaarde	1548–1565
166	M	Oudenburg	1562–1566
167	M	Oudewater	1575 (before)
168	M	Pernes	1546
169	M	Poperinge	1563
170	M	Purmerend	1560 (about)
171	M	Reimerswaal	1555 (about)
172	M	Rhenen	1568
173	M	Roermond	1554 (before)
174	M	Rotterdam	1557–1563
175	M	Saint-Ghislain	1563–1570
176	M	Saint-Omer	1564 (in or after)
177	M	Saint-Pol-sur-Ternoise	1546
178	M	Schiedam	1558 (after)
179	M	Schoonhoven	1562 (or later)
180	M	Sint-Anna-ter-Muiden	1560–1571
181	M	Sint-Maartensdijk	1556 (before)
182	M	Sloten	unknown
183	M	Sluis	1560–1571
184	M	Sneek	1560 (before)
185	M	Soignies	1550–1565
186	M	Stavoren	1549–1570
187	M	Steenwijk	1557 (about)
188	M	Straelen	1553 (after)
189	M	Thérouanne	1560–1569
190	B	Thionville	1550–1558
191	M	Tholen	1562 (before)
192	M	Tiel	1557–1558
193	B	Tienen	1560
194	M	Tournai	1562–1564
195	M	Utrecht	1564–1570
196	M	Valenciennes	1546–1570

TABLE 3.229 The 226 cities: Alphabetically (*cont.*)

No.	Madrid or Brussel	City	Year or period
197	B	Valkenburg	1563
198	M	Veere	1550–1551
199	M	Venlo	1570 (about)
200	M	Veurne	1567
201	M	Vianen	1563 (about)
202	M	Vieil-Hesdin	1559–1563
203	B	Vilvoorde	1561 (after)
204	M	Vlaardingen	1552–1574
205	M	Vlissingen	1548
206	M	Vollenhove	1557–1558
207	M	Wachtendonk	1572
208	M	Wageningen	1562
209	M	Warneton	1566 (before)
210	B	Wavre	1550–1560
211	B	Weert	1570
212	M	Weesp	1560 (about)
213	M	Wervik	1560
214	M	Westkapelle	1550–1551
215	M	Wijk bij Duurstede	1563 (about)
216	M	Woerden	1560 (about)
217	M	Workum	1562 (about)
218	M	Woudrichem	1557 (after)
219	B	Yvoix (Carignan)	1559
220	M	Zaltbommel	1557 (about)
221	M	Zevenbergen	1543–1559
222	B	Zichem	1550–1568
223	M	Zierikzee	1550 (before)
224	B	Zoutleeuw	1542–1558
225	M	Zutphen	1565 (about)
226	M	Zwolle	1557–1565

Table 3.229 lists all 226 cities in alphabetical order, including the measurement and analysis results. 'M' or 'B' refer to the library where it is kept: Madrid (189 maps) and Brussels (37 maps), respectively. Under 'B' also fall the three maps that are stored elsewhere: Limbourg, Valkenburg, and Weert. The year of inventarisation column lists the year or period in which the data for the map were collected for all cities. The data come from the most recent standard work on the work of Van Deventer: that of Rutte and Vannieuwenhuyze, *Stedenatlas Jacob van Deventer*. The exception is the example city Vlissingen, for which it could be shown that the map was inventoried in 1548. The measurements, counts, calculations, and analyses of the 226 individual city pages and will be presented in the tables that follow.

TABLE 3.230 The 226 cities: Dimensions

No.	City	Street sides: pixels	Scale stick in pixels (original maps)	Houses	Street sides	Houses per street side	Inhabitants (RF 55)
80	Gent	56,662	1359	9950	1286	7.7	54,724
38	Brugge	51,916	1357	9116	1125	8.1	50,140
39	Brussel	35,753	–	6278	767	8.2	34,530
194	Tournai	26,844	861	4714	558	8.4	25,926
130	Leuven	25,367	–	4454	456	9.8	24,499
142	Mechelen	24,783	1316	4352	535	8.1	23,935
195	Utrecht	22,616	1010	3971	517	7.7	21,843
14	Arras	21,938	897	3852	573	6.7	21,188
196	Valenciennes	20,517	897	3603	573	6.3	19,815
63	Douai	20,077	869	3526	426	8.3	19,390
92	Haarlem	18,845	1006	3309	430	7.7	18,201
133	Lille	18,621	912	3312	480	6.9	18,216
104	Hertogenbosch, 's-	18,081	–	3175	547	5.8	17,463
8	Amsterdam	18,009	1001	3162	548	5.8	17,393
176	Saint-Omer	17,216	884	3023	392	7.7	16,627
150	Mons	17,129	881	3008	304	9.9	16,543
127	Leiden	14,313	996	2513	346	7.3	13,823
90	Groningen	13,903	999	2441	309	7.9	13,428
53	Delft	12,839	1010	2255	234	9.6	12,400
7	Amersfoort	11,909	992	2091	263	8.0	11,502
117	Kortrijk	11,854	964	2082	269	7.7	11,449
193	Tienen	11,597	–	2036	256	8.0	11,200
158	Nijmegen	10,930	1003	1919	271	7.1	10,556
87	Gouda	9903	1001	1739	230	7.6	9564
62	Dordrecht	9812	1035	1723	284	6.1	9476
132	Lier	9730	–	1709	214	8.0	9397
54	Den Haag	9506	1013	1669	177	9.4	9181
24	Bergen op Zoom	9342	–	1640	264	6.2	9023
225	Zutphen	9111	1001	1600	205	7.8	8799
116	Kampen	9067	1004	1592	246	6.5	8757
103	Herentals	8799	–	1545	173	8.9	8498
56	Deventer	8563	997	1504	194	7.8	8270
126	Leeuwarden	8556	999	1502	248	6.1	8263
173	Roermond	8339	–	1464	161	9.1	8054
213	Wervik	7990	1027	1403	205	7	7717
223	Zierikzee	7962	1006	1398	151	9.3	7690
112	Hoorn	7811	1006	1372	212	6.5	7544
11	Armentières	7629	1027	1340	145	9.2	7368
226	Zwolle	7594	985	1334	238	5.6	7334
5	Alkmaar	7578	1006	1331	182	7.3	7319
148	Middelburg (Nl)	7328	1008	1287	124	10.4	7077
174	Rotterdam	7291	1016	1280	227	5.6	7042

TABLE 3.230 The 226 cities: Dimensions (*cont.*)

No.	City	Street sides: pixels	Scale stick in pixels (original maps)	Houses	Street sides	Houses per street side	Inhabitants (RF 55)
71	Enkhuizen	7165	1001	1258	217	5.8	6920
141	Maubeuge	6890	1031	1210	169	7.2	6654
13	Arnhem	6855	1009	1204	190	6.3	6621
1	Aalst	6812	1030	1196	133	9.0	6579
145	Menen	6800	1027	1194	113	10.6	6567
81	Geraardsbergen	6703	1023	1177	161	7.3	6474
138	Luxembourg	6362	–	1117	156	7.2	6144
156	Nieuwpoort (B)	6336	1027	1113	124	9.0	6119
70	Enghien	6142	1032	1079	118	9.1	5932
18	Axel	6124	1019	1075	151	7.1	5915
111	Hondschoote	6079	1026	1067	99	10.8	5871
25	Bergues	5993	1034	1052	133	7.9	5788
165	Oudenaarde	5960	979	1047	128	8.2	5756
162	Oostende	5933	1023	1042	136	7.7	5730
86	Gorinchem	5829	998	1024	152	6.7	5630
4	Aire-sur-la-Lys	5784	1035	1016	126	8.1	5586
55	Dendermonde	5767	1030	1013	126	8.0	5570
203	Vilvoorde	5677	–	997	168	5.9	5483
49	Culemborg	5626	991	988	112	8.8	5434
16	Ath	5581	1018	990	140	7.1	5445
183	Sluis	5560	–	976	161	6.1	5370
96	Harderwijk	5460	–	959	159	6.0	5273
224	Zoutleeuw	5364	–	942	147	6.4	5181
19	Bailleul	5246	1027	921	69	13.4	5067
123	Le Quesnoy	5174	1022	909	115	7.9	4997
184	Sneek	5163	1016	907	135	6.7	4986
220	Zaltbommel	5106	1012	897	114	7.9	4931
113	Hulst	5031	1019	883	120	7.4	4859
64	Dunkerque	5021	1034	882	111	7.9	4849
200	Veurne	4932	1026	866	120	7.2	4763
28	Binche	4911	1038	862	150	5.8	4743
59	Doesburg	4771	1000	838	72	11.6	4608
66	Echternach	4752	–	834	145	5.8	4589
26	Béthune	4567	1021	802	114	7.0	4411
211	Weert	4512	–	792	94	8.4	4358
57	Diksmuide	4317	1023	758	86	8.8	4169
179	Schoonhoven	4305	993	756	213	3.6	4158
58	Dinant	4287	–	753	163	4.6	4140
205	Vlissingen	4245	1025	745	97	7.7	4100
198	Veere	4229	725	743	112	6.6	4084
169	Poperinge	4198	1021	737	87	8.5	4054
160	Oldenzaal	4171	941	732	111	6.6	4028
91	Grootebroek	4151	1004	716	36	19.9	3939
199	Venlo	4084	–	717	96	7.5	3944

TABLE 3.230 The 226 cities: Dimensions (*cont.*)

No.	City	Street sides: pixels	Scale stick in pixels (original maps)	Houses	Street sides	Houses per street side	Inhabitants (RF 55)
17	Avesnes-sur-Helpe	4061	1015	713	119	6.0	3922
201	Vianen	4034	1009	708	79	9.0	3896
47	Comines	4010	1014	704	69	10.2	3873
155	Naarden	4006	–	703	89	7.9	3869
36	Brielle	3973	998	698	112	6.2	3837
61	Dokkum	3939	1004	692	144	4.8	3804
192	Tiel	3912	999	687	99	6.9	3778
178	Schiedam	3817	994	670	127	5.3	3686
190	Thionville	3737	–	656	133	4.9	3609
30	Bolsward	3682	–	647	94	6.9	3556
48	Conde-sur-l'Escaut	3536	1032	621	82	7.6	3415
129	Lessines	3508	1030	616	46	13.39	3388
167	Oudewater	3492	995	613	75	8.2	3373
215	Wijk bij Duurstede	3451	996	606	68	8.9	3333
74	Franeker	3331	1003	585	78	7.5	3217
149	Monnickendam	3323	989	584	95	6.1	3209
3	Aarschot	3317	–	582	75	7.8	3204
2	Aardenburg	3264	1030	573	84	7.0	3152
23	Beaumont	3252	1018	571	97	5.9	3141
122	Le Cateau-Cambrésis	3216	–	565	75	7.5	3106
20	Bapaume	3215	1023	565	67	8.4	3105
146	Mesen	3206	1027	563	80	7.0	3096
84	Goes	3188	995	560	59	9.5	3079
67	Edam	3185	1010	559	85	6.6	3076
10	Arlon	3156	–	554	122	4.5	3048
69	Elburg	3124	1012	549	116	4.7	3017
143	Medemblik	3084	1011	542	75	7.2	2979
159	Ninove	3066	1028	538	76	7.1	2961
216	Woerden	3043	–	534	84	6.4	2939
9	Appingedam	2973	935	522	77	6.8	2871
42	Cassel	2923	1021	513	69	7.4	2823
206	Vollenhove	2880	993	506	56	9.0	2782
98	Harlingen	2855	1003	501	81	6.2	2757
166	Oudenburg	2835	1021	498	55	9.1	2738
77	Geldern	2813	–	494	47	10.5	2717
139	Marche-en-Famenne	2806	–	493	94	5.2	2710
99	Hasselt	2759	998	484	65	7.5	2665
37	Brouwershaven	2758	999	484	51	9.5	2664
128	Lens	2726	1023	479	75	6.4	2633
93	Halle	2716	1029	477	82	5.8	2623
119	La Bassée	2699	1027	474	46	10.3	2607
21	Bastogne	2671	–	469	87	5.4	2580
172	Rhenen	2667	1002	468	76	6.2	2576
214	Westkapelle	2663	725	468	40	11.7	2572

TABLE 3.230 The 226 cities: Dimensions (*cont.*)

No.	City	Street sides: pixels	Scale stick in pixels (original maps)	Houses	Street sides	Houses per street side	Inhabitants (RF 55)
187	Steenwijk	2632	1004	462	60	7.7	2542
185	Soignies	2628	1034	461	75	6.1	2538
171	Reimerswaal	2617	1010	460	65	7.1	2527
222	Zichem	2598	–	456	64	7.1	2509
109	Heusden	2563	1006	450	47	9.6	2475
210	Wavre	2560	–	450	87	5.2	2472
33	Bouvignes-sur-Meuse	2505	–	440	77	5.7	2419
89	Groenlo	2483	1021	436	59	7.4	2398
124	Le Roeulx	2456	1038	431	77	5.6	2372
75	Geertruidenberg	2434	1007	427	49	8.7	2351
217	Workum	2401	996	422	43	9.8	2319
161	Oostburg	2389	1021	420	56	7.5	2307
88	Gravelines	2386	1010	419	55	7.6	2304
219	Yvoix (Carignan)	2386	–	419	84	5.0	2304
94	Halluin	2377	1027	417	36	11.6	2296
32	Bourbourg	2338	1010	411	51	8.1	2258
209	Warneton	2322	1015	408	61	6.7	2243
60	Doetinchem	2294	1007	403	47	8.6	2216
50	Damme	2257	1018	396	76	5.2	2180
151	Montfoort	2244	996	394	61	6.5	2167
27	Beverwijk	2215	1006	389	47	8.3	2139
163	Ootmarsum	2190	941	385	60	6.4	2115
131	Leuze-en-Hainaut	2188	1030	384	46	8.4	2113
208	Wageningen	2133	–	375	59	6.4	2060
68	Eindhoven	2080	–	365	49	7.5	2009
29	Bitburg	2072	–	364	75	4.9	2001
34	Braine-le-Comte	2069	1034	363	69	5.3	1998
82	Gistel	2058	1021	361	53	6.8	1988
73	Erkelenz	2048	1009	360	47	7.7	1978
191	Tholen	2045	1013	359	47	7.6	1975
204	Vlaardingen	2043	999	359	46	7.8	1973
85	Goor	2042	950	359	37	9.7	1972
212	Weesp	2003	1010	352	73	4.8	1934
105	Herve	1986	–	349	52	6.7	1918
137	Lochem	1972	1021	346	52	6.7	1905
121	Landrecies	1961	1015	344	75	4.6	1894
115	IJsselstein	1924	995	338	40	8.5	1858
40	Buren	1923	992	338	44	7.7	1857
65	Echt	1914	1008	336	46	7.3	1849
44	Chimay	1912	1018	336	93	3.6	1847
135	Limbourg	1906	–	335	66	5.9	3332
170	Purmerend	1897	997	333	61	5.5	1832
102	Helmond	1878	–	330	33	10.0	1814
177	Saint-Pol-sur-Ternoise	1848	1015	325	54	6.0	1785

TABLE 3.230 The 226 cities: Dimensions (*cont.*)

No.	City	Street sides: pixels	Scale stick in pixels (original maps)	Houses	Street sides	Houses per street side	Inhabitants (RF 55)
125	Leerdam	1839	1007	323	45	7.2	1776
78	Gembloux	1815	–	319	65	4.9	1753
107	Hesdin	1754	1026	308	62	5.0	1694
164	Orchies	1747	1021	307	46	6.7	1687
15	Asperen	1739	1007	305	39	7.8	1680
52	Delden	1736	1005	305	48	6.4	1677
12	Arnemuiden	1718	1008	302	27	11.2	1659
72	Enschede	1704	998	299	37	8.1	1646
22	Bavay	1666	1004	293	42	7.0	1609
97	Harelbeke	1598	1025	281	31	9.1	1543
51	Damvillers	1582	–	278	51	5.5	1528
106	Herzogenrath	1558	–	274	47	5.8	1505
79	Genemuiden	1544	993	271	33	8.2	1491
186	Stavoren	1540	1000	270	41	6.6	1487
101	Heerenberg, 's-	1503	1011	264	29	9.1	1452
152	Montfort	1476	1008	259	32	8.1	1426
221	Zevenbergen	1468	999	258	19	13.6	1418
175	Saint-Ghislain	1421	1029	250	40	6.2	1372
120	Landen	1396	–	245	49	5.0	1348
153	Montmédy	1383	–	243	56	4.3	1336
83	Goedereede	1373	1001	241	47	5.1	1326
207	Wachtendonk	1369	951	240	38	6.3	1322
188	Straelen	1357	1007	238	39	6.1	1311
6	Almelo	1336	950	235	34	6.9	1290
197	Valkenburg	1336	–	235	52	4.5	1290
118	Kuinre	1322	1004	232	27	8.6	1329
147	Middelburg (B)	1306	1021	229	35	6.6	1261
140	Mariembourg	1303	–	229	46	5.0	1258
100	Hattem	1275	1012	224	35	6.4	1231
144	Megen	1266	–	222	38	5.9	1223
182	Sloten	1232	989	216	50	4.3	1190
218	Woudrichem	1215	998	213	33	6.5	1173
46	Coevorden	1202	1018	211	31	6.8	1161
136	Lingen	1201	1007	211	33	6.4	1160
181	Sint-Maartensdijk	1142	1001	201	33	6.1	1103
157	Nieuwpoort (Nl)	1119	993	196	25	7.9	1081
43	Chièvres	1104	1030	194	26	7.5	1066
156	Nieuwpoort (B)	1085	993	191	24	7.9	1143
154	Muiden	1061	1010	186	33	5.7	1025
108	Heukelum	953	1007	167	21	8.0	920
114	IJlst	950	989	167	8	20.9	918
134	Lillers	861	1015	151	31	4.9	832
110	Hindeloopen	851	996	149	16	9.3	822

TABLE 3.230 The 226 cities: Dimensions (*cont.*)

No.	City	Street sides: pixels	Scale stick in pixels (original maps)	Houses	Street sides	Houses per street side	Inhabitants (RF 55)
45	Chiny	810	–	142	19	7.5	782
76	Geervliet	795	–	140	41	3.4	768
168	Pernes	689	1015	121	23	5.3	665
95	Hardenberg	623	1003	109	21	5.2	602
35	Bredevoort	622	1007	109	15	7.3	601
31	Bouchain	615	1005	108	30	3.6	594
180	Sint-Anna-ter-Muiden	328	–	58	2	28.8	317
41	Cambrai	–	1387	–	–	–	–
189	Thérouanne	–	1035	–	–	–	–
202	Vieil-Hesdin	–	–	–	–	–	–

In Table 3.230, the 226 cities are presented in order of size. It should be mentioned that the total number of pixels, houses, and inhabitants lead to exactly the same order. The number of pixels results from the measurements of the street sides. The houses and inhabitants are derived from these, using respectively the Passus coefficient (0.1756) and the reduction factor (5.5). The table shows that Ghent and Bruges stand out in terms of size in construction and inhabitation above the other cities. The third city, Brussels, is more than a third smaller than the two Flemish cities. Brussels, in turn, is significantly larger than the fourth city, Tournai. After that, the ranking goes down more gradually. The largest city from the Northern Netherlands, Utrecht, is the only one that is part of the top 10, which otherwise consists of six Belgian and four French cities (counted according to the current geographical situation). Places 11 to 20 include one Belgian city, seven Dutch cities, and two French cities. It is worth mentioning the current Dutch, but in the sixteenth century, Flemish city of Sint-Anna-ter-Muiden with the least number of houses (58), built street sides (2), and inhabitants (347).

TABLE 3.231 The 226 cities: Houses per street side

No.	City	Houses	No.	City	Houses
180	Sint-Anna-ter-Muiden	29.0	119	La Bassée	10.3
114	IJlst	20.9	47	Comines	10.2
91	Grootebroek	19.9	102	Helmond	10.0
221	Zevenbergen	13.6	150	Mons	9.9
129	Lessines	13.4	217	Workum	9.8
19	Bailleul	13.4	130	Leuven	9.8
214	Westkapelle	11.7	85	Goor	9.7
59	Doesburg	11.6	53	Delft	9.6
94	Halluin	11.6	109	Heusden	9.6
12	Arnemuiden	11.2	37	Brouwershaven	9.5
111	Hondschoote	10.8	84	Goes	9.5
145	Menen	10.6	54	Den Haag	9.4
77	Geldern	10.5	110	Hindeloopen	9.3
148	Middelburg (Nl)	10.4	223	Zierikzee	9.3

TABLE 3.231 The 226 cities: Houses per street side (*cont.*)

No.	City	Houses	No.	City	Houses
11	Armentières	9.2	155	Naarden	7.9
70	Enghien	9.1	220	Zaltbommel	7.9
101	Heerenberg, 's-	9.1	157	Nieuwpoort (Nl)	7.9
173	Roermond	9.1	15	Asperen	7.8
97	Harelbeke	9.1	204	Vlaardingen	7.8
166	Oudenburg	9.1	225	Zutphen	7.8
206	Vollenhove	9.0	3	Aarschot	7.8
156	Nieuwpoort (B)	9.0	56	Deventer	7.8
1	Aalst	9.0	80	Gent	7.7
201	Vianen	9.0	117	Kortrijk	7.7
103	Herentals	8.9	176	Saint-Omer	7.7
215	Wijk bij Duurstede	8.9	92	Haarlem	7.7
49	Culemborg	8.8	187	Steenwijk	7.7
57	Diksmuide	8.8	195	Utrecht	7.7
75	Geertruidenberg	8.7	205	Vlissingen	7.7
118	Kuinre	8.6	40	Buren	7.7
60	Doetinchem	8.6	162	Oostende	7.7
169	Poperinge	8.5	73	Erkelenz	7.7
115	IJsselstein	8.5	191	Tholen	7.6
194	Tournai	8.5	88	Gravelines	7.6
20	Bapaume	8.4	48	Conde-sur-l'Escaut	7.6
211	Weert	8.4	87	Gouda	7.6
131	Leuze-en-Hainaut	8.4	122	Le Cateau-Cambrésis	7.5
27	Beverwijk	8.3	74	Franeker	7.5
63	Douai	8.3	45	Chiny	7.5
79	Genemuiden	8.2	161	Oostburg	7.5
39	Brussel	8.2	199	Venlo	7.5
165	Oudenaarde	8.2	43	Chièvres	7.5
167	Oudewater	8.2	68	Eindhoven	7.5
142	Mechelen	8.1	99	Hasselt	7.5
38	Brugge	8.1	42	Cassel	7.4
152	Montfort	8.1	89	Groenlo	7.4
72	Enschede	8.1	113	Hulst	7.4
4	Aire-sur-la-Lys	8.1	5	Alkmaar	7.3
32	Bourbourg	8.1	65	Echt	7.3
55	Dendermonde	8.0	81	Geraardsbergen	7.3
132	Lier	8.0	35	Bredevoort	7.3
108	Heukelum	8.0	127	Leiden	7.3
7	Amersfoort	8.0	143	Medemblik	7.2
193	Tienen	8.0	200	Veurne	7.2
64	Dunkerque	7.9	125	Leerdam	7.2
25	Bergues	7.9	138	Luxembourg	7.2
90	Groningen	7.9	141	Maubeuge	7.2
123	Le Quesnoy	7.9	222	Zichem	7.1

TABLE 3.231 The 226 cities: Houses per street side (*cont.*)

No.	City	Houses	No.	City	Houses
18	Axel	7.1	36	Brielle	6.2
158	Nijmegen	7.1	24	Bergen op Zoom	6.2
159	Ninove	7.1	98	Harlingen	6.2
16	Ath	7.1	172	Rhenen	6.2
171	Reimerswaal	7.1	185	Soignies	6.2
146	Mesen	7.0	149	Monnickendam	6.1
26	Béthune	7.0	188	Straelen	6.1
2	Aardenburg	7.0	181	Sint-Maartensdijk	6.1
22	Bavay	7.0	62	Dordrecht	6.1
192	Tiel	6.9	126	Leeuwarden	6.1
133	Lille	6.9	183	Sluis	6.1
6	Almelo	6.9	96	Harderwijk	6.0
30	Bolsward	6.9	177	Saint-Pol-sur-Ternoise	6.0
213	Wervik	6.8	17	Avesnes-sur-Helpe	6.0
82	Gistel	6.8	203	Vilvoorde	5.9
46	Coevorden	6.8	23	Beaumont	5.9
9	Appingedam	6.8	144	Megen	5.9
86	Gorinchem	6.7	93	Halle	5.8
14	Arras	6.7	106	Herzogenrath	5.8
184	Sneek	6.7	71	Enkhuizen	5.8
105	Herve	6.7	104	Hertogenbosch, 's-	5.8
209	Warneton	6.7	8	Amsterdam	5.8
164	Orchies	6.7	28	Binche	5.8
137	Lochem	6.7	66	Echternach	5.8
198	Veere	6.6	33	Bouvignes-sur-Meuse	5.7
160	Oldenzaal	6.6	154	Muiden	5.7
186	Stavoren	6.6	174	Rotterdam	5.6
67	Edam	6.6	124	Le Roeulx	5.6
147	Middelburg (B)	6.6	226	Zwolle	5.6
112	Hoorn	6.5	170	Purmerend	5.5
116	Kampen	6.5	51	Damvillers	5.5
218	Woudrichem	6.5	21	Bastogne	5.4
151	Montfoort	6.5	178	Schiedam	5.3
163	Ootmarsum	6.4	34	Braine-le-Comte	5.3
224	Zoutleeuw	6.4	168	Pernes	5.3
100	Hattem	6.4	139	Marche-en-Famenne	5.2
136	Lingen	6.4	50	Damme	5.2
128	Lens	6.4	95	Hardenberg	5.2
216	Woerden	6.4	210	Wavre	5.2
52	Delden	6.4	135	Limbourg	5.1
208	Wageningen	6.4	83	Goedereede	5.1
13	Arnhem	6.3	120	Landen	5.0
207	Wachtendonk	6.3	219	Yvoix (Carignan)	5.0
196	Valenciennes	6.3	107	Hesdin	5.0
175	Saint-Ghislain	6.2	140	Mariembourg	5.0

TABLE 3.231 The 226 cities: Houses per street side (*cont.*)

No.	City	Houses	No.	City	Houses
190	Thionville	4.9	197	Valkenburg	4.5
78	Gembloux	4.9	153	Montmédy	4.3
134	Lillers	4.9	182	Sloten	4.3
29	Bitburg	4.9	44	Chimay	3.6
212	Weesp	4.8	31	Bouchain	3.6
61	Dokkum	4.8	179	Schoonhoven	3.6
69	Elburg	4.7	76	Geervliet	3.4
58	Dinant	4.6	41	Cambrai	–
121	Landrecies	4.6	189	Thérouanne	–
10	Arlon	4.5	202	Vieil-Hesdin	–

In Table 3.231, the 226 cities are ordered based on the number of houses per street side. In general, it can be said that a city with – proportionally – many long streets scores high here, and a city with many short streets scores low. The table shows at first glance that there is little difference between large and small cities. The smallest city from Table 3.230, Sint-Anna-ter-Muiden, leads the list with as many as 288 houses per street side. A look at the map of this city shows that there were only two street sides, with the church in the middle. Among the first 20 cities with the most houses per street side are also Mons, Leuven, and Delft, which were all among the top 20 in size. The same picture is seen at the bottom of the ranking. The small city of Geervliet counted 34 houses per street side, while the top 20 cities Amsterdam and 's-Hertogenbosch with 5.8 houses, were well below the average of 7.5.

TABLE 3.232 The 226 cities: Relations size and houses

	Number of cities	Size in pixels	Street-sides	Average size of cities	Average size of cities	Average number of pixels per street side	Average number of pixels per street side
Above the average of 7.5	100	703,496	14,605	7035	62%	48	58%
Below the average of 7.5	123	500,961	14,299	4073	38%	35	42%
Not counted	3	–	–	–	–	–	–
Total	**226**	**1,204,457**	**28,904**	–	**100%**	–	**100%**

There is of course an almost perfect correlation between the number of street sides and the size of the cities (correlation coefficient of +0.98). That makes sense, because large cities simply have more streets than small cities. This statistical relationship is less present when the number of houses per street side and the size of the cities (pixels, houses, and inhabitants) are used to calculate the correlation coefficient. That results in a correlation coefficient of +0.05. However, an alternative calculation, presented in Table 3.232, shows a stronger relationship. Here, the average of 7.5 houses per street side acts as a dividing line between the cities that are above and below it. It turns out that the number of street sides is almost the same, but that the average city size in the upper part is significantly higher than in the lower part: 7035 versus 4112. Moreover, these numbers are achieved with fewer cities (100 vs. 123). This relationship may have to do with the density of construction, but also with the average lengths of the streets. Table 3.232 shows that the latter is the case. There are no indications that the density of construction of the cities drawn by Van Deventer has any kind of relationship with the size of the cities.

TABLE 3.233 The 226 cities: Accuracies

No	City	Scale (1:)	Scale deviation in meters from Vlissingen en Leiden	Rotation (°)	Mean Position Error (± m)	Mean Position Error weighted (meters per house)	Recognizable reference points (Reliability 0–5)
39	Brussel	8940	8	15	16	0.0025	5
80	Gent	8040	−1	9	34	0.0034	5
194	Tournai	8760	6	12	17	0.0036	5
133	Lille	9070	9	11	12	0.0037	5
38	Brugge	7930	−2	10	35	0.0038	5
148	Middelburg (Nl)	8830	7	16	5	0.0039	5
130	Leuven	9190	10	9	18	0.0040	5
142	Mechelen	8430	3	13	18	0.0041	5
117	Kortrijk	7800	−3	13	9	0.0043	5
8	Amsterdam	8260	1	14	14	0.0044	5
195	Utrecht	8410	3	14	19	0.0048	5
176	Saint-Omer	9200	11	7	15	0.0050	5
226	Zwolle	7870	−3	9	7	0.0052	5
7	Amersfoort	8540	4	9	11	0.0053	5
193	Tienen	8510	4	13	12	0.0059	5
104	Hertogenbosch, ’s-	8670	5	14	19	0.0060	5
63	Douai	9070	9	14	22	0.0062	4
225	Zutphen	7490	−7	11	10	0.0063	5
224	Zoutleeuw	6900	−12	14	6	0.0064	5
158	Nijmegen	8490	3	12	13	0.0068	5
127	Leiden	8080	−1	8	18	0.0072	5
223	Zierikzee	9040	9	11	10	0.0072	5
5	Alkmaar	8260	1	16	10	0.0075	5
184	Sneek	7750	−4	10	7	0.0077	5
14	Arras	8920	8	8	30	0.0078	5
196	Valenciennes	9680	15	10	28	0.0078	3
92	Haarlem	8390	2	10	26	0.0079	4
87	Gouda	7890	−3	10	14	0.0081	5
53	Delft	8400	3	9	19	0.0084	5
54	Den Haag	8550	4	11	14	0.0084	5
62	Dordrecht	8630	5	11	15	0.0087	5
173	Roermond	8670	5	14	14	0.0096	5
90	Groningen	8230	1	11	24	0.0098	4
132	Lier	8890	7	19	17	0.0099	5
61	Dokkum	8370	2	12	7	0.0101	5
162	Oostende	8100	0	8	11	0.0106	5
116	Kampen	8050	−1	8	17	0.0107	5
156	Nieuwpoort (B)	7780	−4	9	12	0.0108	5
24	Bergen op Zoom	8740	6	20	18	0.0110	5
103	Herentals	8140	0	12	17	0.0110	5
203	Vilvoorde	8040	−1	10	11	0.0110	5
210	Wavre	7600	−5	6	5	0.0111	5

TABLE 3.233 The 226 cities: Accuracies (*cont.*)

No	City	Scale (1:)	Scale deviation in meters from Vlissingen en Leiden	Rotation (°)	Mean Position Error (± m)	Mean Position Error weighted (meters per house)	Recognizable reference points (Reliability 0–5)
59	Doesburg	8580	4	10	10	0.0119	5
56	Deventer	8480	3	7	18	0.0120	5
150	Mons	8590	4	11	37	0.0123	3
112	Hoorn	8660	5	10	17	0.0124	5
113	Hulst	8010	−1	8	11	0.0125	5
138	Luxembourg	7400	−7	11	14	0.0125	5
71	Enkhuizen	8500	4	8	16	0.0127	5
28	Binche	7310	−8	10	11	0.0128	5
204	Vlaardingen	8680	5	14	5	0.0139	5
1	Aalst	8090	−1	6	17	0.0142	4
10	Arlon	6840	−13	9	8	0.0144	5
220	Zaltbommel	8950	8	12	13	0.0145	5
64	Dunkerque	7800	−3	7	13	0.0147	3
13	Arnhem	8700	6	16	18	0.0150	5
160	Oldenzaal	8090	−1	12	11	0.0150	5
20	Bapaume	8310	2	12	9	0.0159	5
166	Oudenburg	8010	−1	5	8	0.0161	5
16	Ath	7480	−7	11	16	0.0162	5
198	Veere	8070	−1	14	12	0.0162	5
200	Veurne	7200	−9	11	14	0.0162	5
213	Wervik	7460	−7	5	23	0.0164	4
201	Vianen	7280	−9	9	12	0.0169	5
126	Leeuwarden	8740	6	13	26	0.0173	4
185	Soignies	7730	−4	6	8	0.0174	5
86	Gorinchem	7780	−4	9	18	0.0176	5
77	Geldern	8520	4	11	9	0.0182	5
190	Thionville	9090	9	15	12	0.0183	5
30	Bolsward	8220	1	12	12	0.0185	5
58	Dinant	7720	−4	10	14	0.0186	5
149	Monnickendam	8390	2	10	11	0.0188	5
93	Halle	7650	−5	10	9	0.0189	5
192	Tiel	8940	8	13	13	0.0189	5
25	Bergues	7860	−3	8	20	0.0190	5
18	Axel	8520	4	5	21	0.0195	0
211	Weert	8200	1	10	16	0.0202	5
66	Echternach	7300	−8	6	17	0.0204	5
111	Hondschoote	8240	1	7	22	0.0206	4
123	Le Quesnoy	7910	−2	12	19	0.0209	5
170	Purmerend	9190	10	11	7	0.0210	5
81	Geraardsbergen	7740	−4	8	25	0.0212	4
106	Herzogenrath	7660	−5	11	6	0.0219	5
165	Oudenaarde	7670	−5	10	23	0.0220	4

TABLE 3.233 The 226 cities: Accuracies (*cont.*)

No	City	Scale (1:)	Scale deviation in meters from Vlissingen en Leiden	Rotation (°)	Mean Position Error (± m)	Mean Position Error weighted (meters per house)	Recognizable reference points (Reliability 0–5)
17	Avesnes-sur-Helpe	7430	−7	11	16	0.0224	5
129	Lessines	8510	4	8	14	0.0227	5
205	Vlissingen	8030	−1	12	17	0.0228	5
75	Geertruidenberg	8630	5	8	10	0.0234	5
179	Schoonhoven	8360	2	12	18	0.0238	5
187	Steenwijk	7750	−4	12	11	0.0238	5
155	Naarden	8070	−1	10	17	0.0242	5
216	Woerden	7760	−4	11	13	0.0243	5
174	Rotterdam	8620	5	11	32	0.0250	3
197	Valkenburg	7670	−5	9	6	0.0255	5
36	Brielle	8570	4	16	18	0.0258	3
107	Hesdin	7760	−4	12	8	0.0260	4
19	Bailleul	7620	−5	6	24	0.0261	5
49	Culemborg	7850	−3	16	26	0.0263	4
219	Yvoix (Carignan)	7320	−8	18	11	0.0263	5
115	IJsselstein	8220	1	13	9	0.0266	5
4	Aire-sur-la-Lys	7300	−8	9	27	0.0266	3
33	Bouvignes-sur-Meuse	7190	−10	15	12	0.0273	5
102	Helmond	7840	−3	10	9	0.0273	5
183	Sluis	8180	0	12	27	0.0277	4
215	Wijk bij Duurstede	8310	2	13	17	0.0281	5
222	Zichem	8610	5	12	13	0.0285	5
131	Leuze-en-Hainaut	8200	1	11	11	0.0286	5
153	Montmédy	7300	−8	15	7	0.0288	5
57	Diksmuide	7170	−10	8	22	0.0290	4
69	Elburg	8050	−1	11	16	0.0291	4
207	Wachtendonk	8260	1	13	7	0.0292	5
199	Venlo	8700	6	13	21	0.0293	4
65	Echt	8680	5	13	10	0.0298	5
89	Groenlo	8340	2	12	13	0.0298	5
72	Enschede	8300	2	14	9	0.0301	5
48	Conde-sur-l'Escaut	7610	−5	7	19	0.0306	3
3	Aarschot	8410	3	15	18	0.0309	5
212	Weesp	8260	1	7	11	0.0313	5
47	Comines	7080	−11	11	22	0.0313	4
70	Enghien	7760	−4	9	34	0.0315	3
206	Vollenhove	8150	0	11	16	0.0316	5
137	Lochem	7890	−3	15	11	0.0318	5
172	Rhenen	8430	3	15	15	0.0321	5
182	Sloten	8380	2	4	7	0.0324	5
55	Dendermonde	7880	−3	10	33	0.0326	3
151	Montfoort	7580	−6	16	13	0.0330	5
163	Ootmarsum	7510	−6	12	13	0.0338	5

TABLE 3.233 The 226 cities: Accuracies (*cont.*)

No	City	Scale (1:)	Scale deviation in meters from Vlissingen en Leiden	Rotation (°)	Mean Position Error (± m)	Mean Position Error weighted (meters per house)	Recognizable reference points (Reliability 0–5)
98	Harlingen	8760	6	8	17	0.0339	5
74	Franeker	7970	−2	6	20	0.0342	4
214	Westkapelle	9100	10	12	16	0.0342	4
96	Harderwijk	7660	−5	9	33	0.0344	3
105	Herve	7370	−8	15	12	0.0344	5
40	Buren	7210	−9	11	12	0.0355	5
146	Mesen	7240	−9	8	20	0.0355	4
51	Damvillers	7450	−7	12	10	0.0360	5
139	Marche-en-Famenne	7050	−11	9	18	0.0365	5
169	Poperinge	8150	0	11	27	0.0366	4
99	Hasselt	8110	0	12	18	0.0372	5
109	Heusden	7710	−4	16	17	0.0378	5
188	Straelen	9050	9	11	9	0.0378	5
6	Almelo	8260	1	10	9	0.0383	5
23	Beaumont	8220	1	11	22	0.0385	4
27	Beverwijk	9020	9	12	15	0.0386	5
26	Béthune	8770	6	9	31	0.0387	1
11	Armentières	9410	13	4	53	0.0396	1
144	Megen	7300	−8	13	9	0.0405	5
88	Gravelines	8220	1	12	17	0.0406	5
122	Le Cateau-Cambrésis	8520	4	5	23	0.0407	4
73	Erkelenz	8370	2	11	15	0.0417	5
208	Wageningen	9480	13	20	16	0.0427	5
140	Mariembourg	7990	−2	12	10	0.0437	5
34	Braine-le-Comte	6870	−13	2	16	0.0441	2
85	Goor	7830	−3	13	16	0.0446	5
60	Doetinchem	8610	5	10	18	0.0447	5
178	Schiedam	9410	13	16	30	0.0448	4
83	Goedereede	7970	−2	13	11	0.0456	5
9	Appingedam	8150	0	10	24	0.0460	5
152	Montfort	8700	6	15	12	0.0463	5
125	Leerdam	8300	2	10	15	0.0464	5
84	Goes	9350	12	15	26	0.0464	4
67	Edam	8530	4	10	26	0.0465	3
68	Eindhoven	8710	6	7	17	0.0466	5
78	Gembloux	8170	0	6	15	0.0470	5
2	Aardenburg	7300	−8	2	27	0.0471	2
124	Le Roeulx	8110	0	9	21	0.0487	4
82	Gistel	8050	−1	9	18	0.0499	5
161	Oostburg	7660	−5	3	21	0.0500	4
119	La Bassée	8570	4	7	24	0.0506	4
164	Orchies	7410	−7	15	16	0.0521	5
42	Cassel	7870	−3	4	27	0.0526	4

TABLE 3.233 The 226 cities: Accuracies (*cont.*)

No	City	Scale (1:)	Scale deviation in meters from Vlissingen en Leiden	Rotation (°)	Mean Position Error (± m)	Mean Position Error weighted (meters per house)	Recognizable reference points (Reliability 0–5)
145	Menen	8680	5	7	63	0.0528	3
32	Bourbourg	7710	−4	8	22	0.0535	1
209	Warneton	7470	−7	8	22	0.0539	5
181	Sint-Maartensdijk	8130	0	10	11	0.0547	5
29	Bitburg	7260	−9	9	21	0.0577	4
217	Workum	8340	2	13	26	0.0616	4
44	Chimay	6800	−13	13	21	0.0625	4
186	Stavoren	8450	3	9	17	0.0630	5
191	Tholen	8450	3	8	23	0.0641	3
218	Woudrichem	7990	−2	14	14	0.0657	5
143	Medemblik	8610	5	11	36	0.0664	3
159	Ninove	8350	2	9	36	0.0669	3
147	Middelburg (B)	7240	−9	13	16	0.0699	5
37	Brouwershaven	9150	10	9	34	0.0702	5
100	Hattem	8620	5	8	16	0.0714	5
167	Oudewater	7690	−5	15	44	0.0718	1
121	Landrecies	8040	−1	23	25	0.0727	4
141	Maubeuge	7330	−8	13	88	0.0727	0
52	Delden	8100	0	7	23	0.0754	2
50	Damme	7730	−4	4	30	0.0758	2
21	Bastogne	7240	−9	10	36	0.0768	1
157	Nieuwpoort (Nl)	7510	−6	12	16	0.0816	5
22	Bavay	8270	1	16	24	0.0819	4
95	Hardenberg	8540	4	13	9	0.0826	5
128	Lens	8870	7	9	40	0.0835	1
136	Lingen	7960	−2	9	19	0.0900	5
154	Muiden	8780	6	4	17	0.0914	5
175	Saint-Ghislain	8130	0	9	24	0.0960	3
12	Arnemuiden	8610	5	10	29	0.0960	1
101	Heerenberg, 's-	7280	−9	19	26	0.0985	1
91	Grootebroek	7800	−3	10	73	0.1020	0
15	Asperen	8040	−1	12	32	0.1049	5
120	Landen	8710	6	9	26	0.1061	4
76	Geervliet	8540	4	15	15	0.1071	5
31	Bouchain	7110	−10	6	12	0.1111	4
110	Hindeloopen	7750	−4	8	17	0.1141	2
221	Zevenbergen	9070	9	11	30	0.1163	3
108	Heukelum	8720	6	8	20	0.1198	4
134	Lillers	9450	13	13	19	0.1258	3
46	Coevorden	9680	15	17	29	0.1374	2
35	Bredevoort	8810	7	5	15	0.1376	1
114	IJlst	8230	1	7	23	0.1377	4
168	Pernes	7120	−10	8	18	0.1488	5
79	Genemuiden	8270	1	12	42	0.1550	1

TABLE 3.233 The 226 cities: Accuracies (*cont.*)

No	City	Scale (1:)	Scale deviation in meters from Vlissingen en Leiden	Rotation (°)	Mean Position Error (± m)	Mean Position Error weighted (meters per house)	Recognizable reference points (Reliability 0–5)
45	Chiny	7080	−11	11	23	0.1620	4
43	Chièvres	7510	−6	9	32	0.1649	3
177	Saint-Pol-sur-Ternoise	8360	2	3	54	0.1662	0
94	Halluin	7460	−7	25	74	0.1775	1
135	Limbourg	7460	−7	25	74	0.2209	5
97	Harelbeke	9440	13	13	64	0.2278	3
118	Kuinre	6900	−12	3	62	0.2672	1
180	Sint-Anna-ter-Muiden	8180	0	12	27	0.4655	1
41	Cambrai	8160	0	12	50	–	1
189	Thérouanne	8170	0	14	23	–	4
171	Reimerswaal	–	–	–	–	–	–
202	Vieil-Hesdin	–	–	–	–	–	–

Table 3.233 serves as the basis for summary Table 3.227 and shows the deviations in the analyses of MapAnalyst and the addition made within the present study: the Reliability. In the table, the cities are arranged based on the Weighted Mean Position Error. The maps of Brussels and Ghent, also the third largest and the largest of all 226 cities, emerge as the most reliable from this exercise. Sint-Anna-ter-Muiden, the smallest city, emerges as the least reliable. A quick glance at these three cities reveals that the reliability scores given are completely in line: respectively, 5 and 1. This pattern is also evident in the correlation calculations where the column 'Recognizable Reference points (Reliability 0–5)' is consistently compared with the Weighted Mean Position Error. The image is, of course, the same as in Table 3.227: a strong correlation of 0.80 and 0.50 between the Reliability and the Mean Position Error and the Weighted Mean Position Error, and a weak one between the Reliability and the other three: respectively, 0.01, 0.01, and 0.15. The comparison between the Weighted Mean Position Error and the size of the cities yields a moderate correlation of 0.35 (not shown in Table 3.227), as evidenced in Table 3.233 where larger cities on average have a lower error score than smaller ones.

The cautious conclusion that could be drawn is twofold: (1) although all the maps excel in accuracy, there are indeed differences, which, however, should not be expressed in terms of 'good' or 'bad', but rather in terms of 'very good', 'good', and 'slightly less good', and (2) the quality differences are not tied to time periods, geographical location of the cities, or the place where the maps are stored. A good example is the map of Herentals, which is stored in Brussels and for which the inventory took place around 1548, scoring very well in MapAnalyst and receiving the maximum number of reliability points for the chosen reference points. A comparable city is Saint-Pol-sur-Ternoise, of which the Van Deventer map is stored in Madrid and for which the inventory took place around 1546. This town scores very low in MapAnalyst and receive zero reliability points for the chosen reference points. A second example is the combination of Haluin and Herzogenrath. These cities were only drawn in the late 1560s and show the same large differences in origin, storage, and quality as the combination of Herentals and Saint-Pol-sur-Ternoise.

Of course, many more analyses and comparisons can be carried out with the help of the tables in this section. The exercises carried out here already show that the research of the 226 maps as presented in Section 3.1 can be used for further research per city or group of cities, but also in its entirety can be useful for comparisons and total analyses. The material of Van Deventer is unique because of the great uniformity of research and production and the fact that they were all made within a relatively short period. The method presented and applied in the first three chapters of the present study makes grateful use of these advantages. The inevitable question that must now be asked is about the applicability of the method to other series of city maps.

CHAPTER 4

Applicability

4.1 Possibilities and Impossibilities

The Passus method was developed to gain more insight into the quality of the 226 city maps created by Jacob van Deventer and the possibilities of using these maps for historical demographic research. The question with which the previous chapter ended is addressed in this last chapter: can the Passus method also be used for other city maps and collections of city maps from the sixteenth, seventeenth, and eighteenth centuries? In other words, can it be detached from the Van Deventer case and the demarcations in time and place? To determine this, it is necessary to reduce the Passus method to its essence: mapping the pattern in the relationship between the lengths of the built-up street sides and the population size of the city. This enables an objective mutual comparison of city maps that were drawn by one single cartographer or one single team of cartographers within a delimited time period, using the same techniques and cartographic insights. It cannot be assumed that the coefficient of 0.1756 calculated for Van Deventer could also be applied to other collections in any period without modification. Secondly, it must be determined whether this pattern – stripped of the fixed number 0.1756 – is universal and therefore applicable to other collections of city maps in different time periods.

To this end, it is necessary to conduct an exploratory study with three variants, three types of city maps, three different periods and a variation in city sizes. A condition is that, for all cities to be studied, there is at least one alternative primary or secondary source known that gives an indication of the number of houses and/or the number of inhabitants in the period the map was made. A second requirement is that the three cities in each variant must differ in terms of (a) size and (b) geographical location. The last named requirement is applied on a slightly smaller scale for the third variant, which utilizes the work of the eighteenth-century cartographer family Hattinga, as their work covered a smaller geographical area compared to that of Braun and Hogenberg (the first variant) and Blaeu (the second variant). The following variants and their associated time periods and collections were chosen:

1. Variant I (sixteenth century), still applying the coefficient 0.1756: the city atlas by Georg Braun and Frans Hogenberg, *Civitates orbis terrarum* (1572–1617), a six-volume collection of maps, images, and descriptions of 475 European, North African, and West Asian cities. On these maps, both the streets and the houses are visible. The chosen cities are London, Zwolle, and Sloten with respectively 1559, 1572, and 1581 as their years of publication.[1]
2. Variant II (seventeenth century), with the application of a coefficient that can be applied to each of the individual maps: the city atlas by Joan Blaeu (*Novum ac Magnum Theatrum Urbium*, 1649), a two-part collection of maps and descriptions of 131 fortified cities in the Northern Netherlands (part 1: *Belgica Libera ac Foederata*) and 150 fortified cities in the Southern Netherlands (part 2: *Belgica Regiae*). On these maps, both the streets and the houses are visible. The chosen cities are Arnhem, Leiden, and Valenciennes, all with 1649 as their year of publication.[2]
3. Variant III (eighteenth century), with a coefficient calculated based on the data from one city where the houses can also be counted and which is representative for the entire collection: the maps from the third quarter of this century, drawn by one or more members of the Hattinga family of cartographers. Most maps do not indicate houses. One of the few cities where this is the case is Tholen, which therefore serves as the reference city. The chosen cities are Middelburg, Tholen, and IJzendijke with respectively 1740, 1744, and 1745 as their year of publication. For the map of Middelburg, the Hattingas used a previously drawn map from 1739.[3]

There are no comparable 16th-century collections of city maps when focusing solely on the graphical representation of urban areas. Van Deventer paid particular attention to the street patterns and the surfaces, while much less emphasis was placed on residential buildings, which are depicted as simple ^-shapes. The many available collections primarily use the so-called oblique

1 Braun and Hogenberg, 1572–1617, respectively folios 42–43 (Part I), 18 (Part IV), 35 (Part III).

2 Blaeu, 1649, respectively folios 31–35 (Part I), 119–132 (Part I), and 345–349 (Part II).

3 Zeeuws Archief, 157, Blad [29], Platte Grond van Tholen, gemeten door (W.T.) Hattinga anno 1744, gekaarteerd door D.W.C. Hattinga. Zeeuws Archief, 358, Blad [11], Plan der Stadt Ysendyke, D.W.C. Hattinga, 1746; Zeeuws Archief, 293, Blad [18], Platte grondt der stadt Middelburg, D.W.C. en/of A. Hattinga, 1740. Note: on the map itself, W.T. Hattinga is mentioned as the creator.

 | HTTPS://DOI.ORG/10.1163/9789004755031-005

parallel projection, in which houses and buildings play a significant role. However, all these city maps also depict street patterns, which can be compared to those in Van Deventer's maps.[4]

It is important, however, to take into account deviations in these patterns when analyzing measurement results, as they are influenced by the so-called 'bird-eyes view': the oblique parallel projection of houses and buildings. In this technique, all structures are drawn at the same angle on the map. The vertical lines are drawn parallel to each other and lack perspective. Often, significant buildings, such as churches, are depicted at a larger scale.In the seven-volume standard work *The History of Cartography* published in 2007, a comprehensive contribution was included about commercial cartography and atlas production in the Low Countries in the sixteenth and seventeenth centuries. The authors outlined the rise and decline of cartography and examined how the increasingly commercial objectives related to the quality of the productions. The approach of the early mapmakers was mainly aimed at representing reality as meticulously as possible. A good example was Jacob van Deventer, whose significant degree of accuracy was demonstrated in Chapter 2; his maps are not paragons of graphic beauty, but they are precise and – in the sixteenth century – functional. Some of Van Deventer's successors had a different motivation, namely to sell as many atlases, wall maps, and globes as possible. The authors of *The History of Cartography* therefore judged that the scientific value of these projects was subordinate to commercial objectives.

The first in this new tradition were the learned canon Georg Braun and the engraver Frans Hogenberg. In the sixteenth century, they were among the earliest commercial mapmakers in the Low Countries and set a model for the products of later seventeenth-century publishers of atlases, wall maps, and globes. Their examples were the cartographers in Leuven who had made a name with scientific maps, among whom were Gerardus Mercator, the already discussed Gemma Frisius and Jacob van Deventer. It is a general belief that the flourishing and quality of Dutch cartography in the seventeenth century would not have been possible without the Leuven influences.

According to *The History of Cartography*, the objections to the new presentation forms of maps also apply to Joan Blaeu, whose internationally acclaimed work *Atlas Maior* was indeed a magnificent publication, but substantively had little to offer. The authors had more appreciation for Joan's father, Willem Blaeu, in whose globes the scientific value prevailed over commercial interests. The article did not mention research investigating the accuracy of Blaeu's maps, in comparison with, for example, other maps, digital cartographic techniques, or alternative source material.

Finally, it has to be stated that the work of the Hattingas does not fall under the category of 'bird-eyes view'. Most city maps they drew were 'ground plans' in which the built-up blocks and thus the streets in between were made visible, but not the buildings themselves. On a few maps, houses and important buildings also appeared, but they were never depicted in 'bird-eyes view', as was the case with Braun and Hogenberg and Blaeu, but in the form of one-dimensional plotted parcels.

4.2 Braun and Hogenberg

The city atlas *Civitates orbis terrarum* by Frans Hogenberg and Georg Braun, who wrote the Latin texts, is a six-volume publication that appeared between 1572 and 1617. The first volume was published in Cologne by the authors themselves and in Antwerp by publisher and engraver Philips Galle, containing 138 city maps and illustrations. The following three volumes were published in 1575, 1581, and 1588.[5] After Hogenberg's death in 1588, his son Abraham took over the project. Under his leadership and still with the collaboration of Braun, the final two volumes were published. In total, the series contains 543 illustrations of 475 cities, mainly in Europe, with 27 depictions of non-European cities. Later, German and French editions were also released.

The first volume in 1572 followed the publication of *Theatrum Orbis Terrarum* in 1570 by traveler, merchant, and graphic artist Abraham Ortelius. Hogenberg had also contributed to this work, and in literature both works are considered complementary in mapping cities during the

4 Woodward, 2007, Volume II and III: Guillaume Guéroult (1507–1564), pp. 618, 789, 1334; Ferdinando Bertelli (1556–72), pp. 281, 273, 787, 788, 789, 833; Giulio Ballino (†1592), pp. 788, 789, 1334; Jacopo Filippo Foresti da Bergamo (1434–1520), pp. 616, 932; Hartman Schedel (1440–1514), pp. 83, 89, 92, 75, 68, 439, 392, 407, 612, 1187.

5 The six volumes: (I) *Civitates orbis terrarum* in 1572, (II) De *praecipuis, totius universi urbibus, liber secundus* in 1575, (III) *Urbium praecipuarum totius mundi, liber tertius* in 1581, (IV) *Urbium praecipuarum totius mundi, liber quartus* uit 1588, (V) *Urbium praecipuarum mundi theatrum quintum* in 1596 and (VI) *Theatri praecipuarum totius mundi urbium liber sextus* in 1617.

third quarter of the 16th century. It is even likely that their preparations overlapped.[6]

The maps and the illustrations in *Civitates* were largely based on the works of others. The main contributors were Joris Hoefnagel, who provided 63 city views, and Jacob van Deventer with 48 city maps. Other contributors included Simon van den Neuvel, Ortelius, Sebastian Münster, Pieter Breughel, Hieronymous Schol, Lucas van Valkenborch, Joost Murer, Diederich Bangs and Heinrich von Rantzau.[7] There are no indications that Hogenberg and/or Braun, like Jacob van Deventer and others, traveled to make observations in the cities they mapped.[8]

Another important source for the atlas was its readers. In the introduction to the second volume, there is a call for illustrations of cities not included in the first two volumes:

> Should anyone not find their home town or native city in either of these first two books, however, I would kindly invite them to draw it from life and send it to me. Then I will have it faithfully copied by the skilled Franz Hogenberg and include it in the 1st or 2nd book or keep it for the 3rd book.[9]

The strong sales of the first volume encouraged Braun and Hogenberg to continue the project. The many responses to their request for material ultimately led to additional volumes. It is likely that these also consisted of contributions from third parties, which may explain the lack of systematic organization in the series. The various continents, countries, and regions are randomly distributed across the six volumes, with Germany and the Low Countries being overrepresented. At least 50 cities appear in multiple volumes.[10] The quality of the maps varies greatly, as will become evident later in this section. There is still no systematic research analyzing the maps as has been done in this study with the Van Deventer maps.[11]

6 Koeman, 2007, p. 1334; Van der Krogt, 2008, pp. 1–2; Naidut, 2020, pp. 9–10.
7 Egmond, 2025; Naidut, 2020, p. 12.
8 Compare: Woodward, 2007, Volume III, part 1 and 2, pp. 405, 419–420, 613, 619, 645, 647, 680, 687, 689, 690, 807, 825, 1011–1012, 1017–1018, 1020, 1052, 1094, 1234–1235, 1237, 1275, 1303, 1305, 1334–1338, 1537, 1574, 1608, 1650–1651, 1657, 1790.
9 Naidut, 2020, p. 15.
10 Egmond, 2025; Van der Krogt, p. 8.
11 Van der Krogt, 2008, pp. 12–13.

For this initial exploratory study, three cities were chosen from the collection of 475 cities in the *Civitates orbis terrarum*: London, Sloten, and Zwolle. London is one of the largest cities in this city atlas. In the Van Deventer case, these were mainly the Flemish and Brabant cities like Bruges, Ghent, and Brussels. Compared to London, the town of Sloten in the Northern Netherlands is very small but distinguishes itself as the most accurately drawn small (< 500 houses) city in terms of both absolute and weighted scores in the Van Deventer research. It also makes the two maps of Sloten good for comparison with each other. Finally, Zwolle ranks high on the list of most reliable medium-sized maps by Van Deventer, with a thirteenth place. On all three maps, the built street sides can be measured in pixels, and the number of houses can be counted. The three cities were measured and analyzed using the same procedure:

1. Use the same digital version of the atlas for the three maps;
2. Check if the three maps have the same resolution;
3. If they differ: adjust the resolutions to the same value;
4. Determine five representative orientation points per city, spread over the entire map but within the city walls, ramparts, or canals;
5. Determine the scale and reliability indicators (Rotation, Weighted Mean Position Error, Reference Points) per city using MapAnalyst;
6. Compare the map scale per city with the average scale of the Van Deventer maps (1:8148) and determine the ratio between the two;
7. Multiply this ratio by the Van Deventer Passus coefficient (0.1756) to get the amended Passus coefficient for the Braun maps;
8. Count the number of houses on the map per city;
9. Measure all built street sides in pixels per city;
10. Determine the number of houses and thereby the number of inhabitants per city with the adjusted Passus coefficient, with several variants depending on the reduction factors;
11. Search for the number of houses and/or inhabitants in the middle of the eighteenth century in alternative sources for each city and compare the outcomes with the counts and measurements from steps 8 and 9.

The results are presented in Table 4.1.

TABLE 4.1 The measurements, counts, and analyses of the maps of Braun and Hogenberg of London, Sloten, and Zwolle in respectively 1559, 1572, and 1581

			Dimensions							
Map: Year of publi-cation	City	Passus coef-ficient (amen-ded)	Houses mea-sured	Houses counted	Deviation (measured / counted)	Reduction-factor	Inhabi-tants (mea-sured)	Inhabitants according to alter-native source	Year alter-native source	Deviation (measured / alternative source)
1559	London	0.1052	7357	6639	10%	8,0	58,859	60,000	approx. 1560	-2%
1572	Sloten	0.0437	155	164	-6%	5,5	850	750	approx. 1560	12%
1581	Zwolle	0.0280	1569	1430	9%	5,5	8631	7500	approx. 1560	13%
	Average Braun	0.0280			8%					9%
	Average vDeventer	0.1756								

		Accuracies					
Map: Year of publi-cation	City	Scale (1:)	Rotation (°)	Rotation (% of 180 °)	Mean Position Error absolute (± m)	Mean Position Error weighted (meters per house)	Recognizable reference points (Relia-bility 1–5)
1559	London	4880	9	5%	208	0,028	5
1572	Sloten	2030	4	2%	7	0,045	5
1581	Zwolle	1300	100	56%	79	0,050	5
	Average Braun	2737	38	21%	98	0,041	
	Average vDeventer	8148	11	6%		0,044	

Note: the negative numbers have been made positive for the calculation of the average.

London experienced spectacular growth in the sixteenth century. Whereas the city had only 20,000 inhabitants in 1530, this number had risen to 35,000 by 1560, and to 75,000 by 1605. When also taking into account the then-outlying areas and parishes – which would soon become part of the city itself – the population numbers in the benchmark years 1530 and 1605 were: 50,000 and 225,000.

The number for 1560 is calculated in Table 4.1 as 60,000.[12] The Passus method with a coefficient of 0.1756 (house per pixel), as shown in the table, gives for London including the outlying areas and parishes in 1559 a new Passus coefficient of 0.1052 and a number of houses of 7357. The count of the houses yields a number of 6639, 10 percent fewer. When using the average reduction factor from the Van Deventer study of 5.5 inhabitants per house, the population of London in the 1550s comes to 40,464 (not shown in the table), a quarter fewer than the estimated number of 50,000 in the 1530s. The cause of this

12 Pevsner, 1962, p. 52. The population of 60,000 in table 4.2.01 is calculated based on the hockey stick model: the growth showed a slower pace in the first decades than in the last decades. The estimate for 1560 in that model is approximately 60,000 inhabitants.

FIGURE 4.1 Braun and Hogenberg, Map of London (approx. 1559), measured

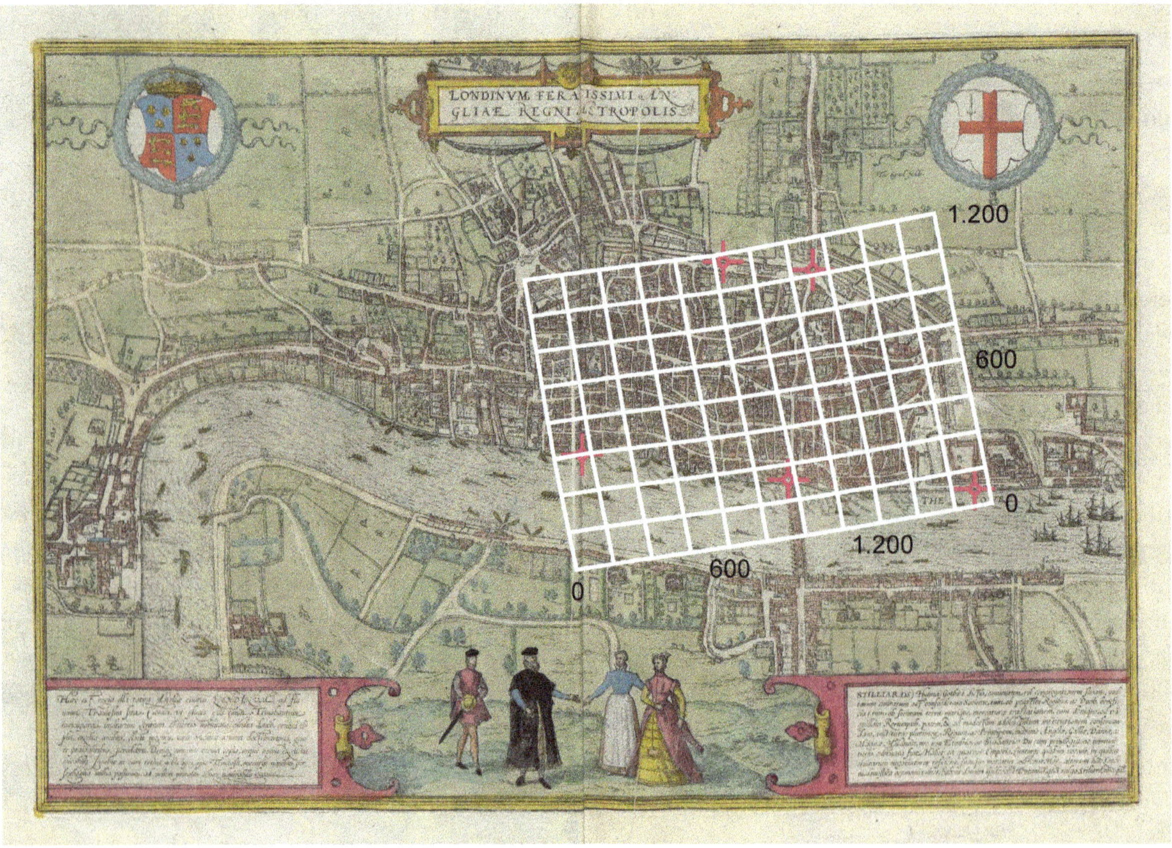

FIGURE 4.2 Braun and Hogenberg, Map of London (approx. 1559), MapAnalyst (5 reference points)

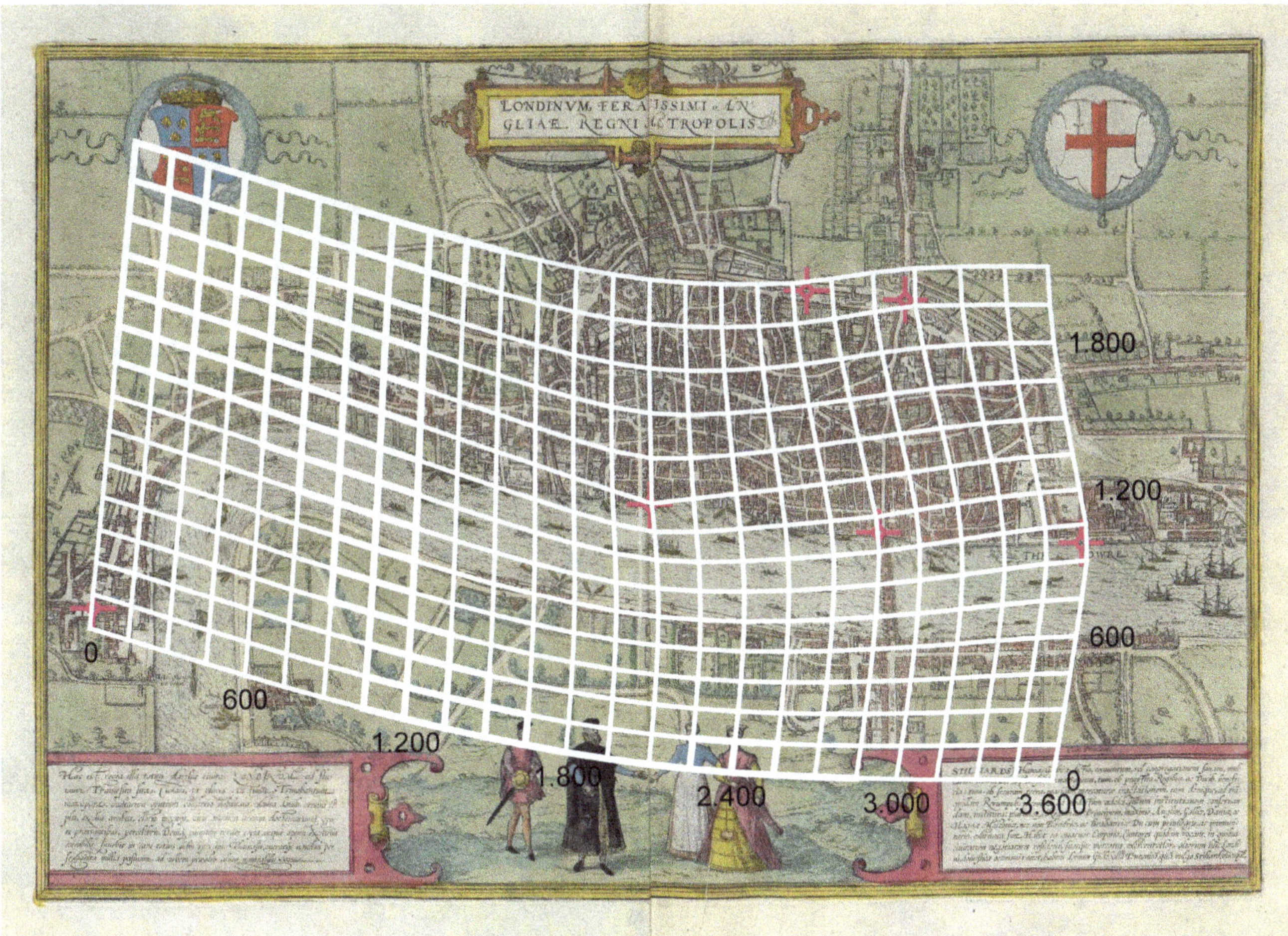

FIGURE 4.3 Braun and Hogenberg, Map of London (approx. 1559), MapAnalyst (6 reference points). The map with 5 reference points in the old city and surroundings shows much less deviation than the one with 6 points that also includes Westminster. This indicates that the open space between the built-up city parts was measured less precisely. This phenomenon was also observed in the Van Deventer maps.

difference should be sought in the reduction factor. It is plausible that this was significantly higher in the rapidly growing city where housing construction could not keep pace with population growth. At the same time, the city government tried to prevent construction outside the city walls. From the 1580s, this effort even led to special prohibition laws. In the seventeenth century, Graunt estimated the reduction factor for the situation in London in 1662 to be 8.[13] When this is applied to the Passus method, the population number in the 1550s comes to 58,859, which fits completely into the growth pattern of the city between 1530 and 1605. The quality of the map is higher with an average deviation of 3 centimeters per house than that of the Van Deventer maps – including comparable cities like Bruges and Ghent – where this is 4 centimeters. Additionally, the map of London also has a lower rotation (9 degrees) than the average of 11 degrees with Van Deventer's maps.

The measurement on the Braun map of the town of Sloten shows 155 houses. The count comes to 164, which is 6 percent less. The 155 houses housed – with a reduction factor of 5.5 – 850 people. The alternative secondary sources used here indicate between 500 and 1000 people, on average 12 percent less.[14] The map has a rotation of only 4 degrees and a Weighted Mean Position Error of 5 centimeters per house, thereby approaching the quality of Van Deventer's map of Sloten. Regarding the quality of the Braun map of Zwolle, there is something peculiar. The Weighted Mean Position Error is in line with that of Sloten at 5 centimeters, but the rotation amounts to as much as 100 degrees, which means the map is tilted more than a quarter turn relative to the north. There is only a 9 percent

13 Graunt, 1676, 1759, p. 31.

14 Ramaer, 1921, p. 35; Visser, 1985, p. 19. From: Lourens, 1996, p. 15.

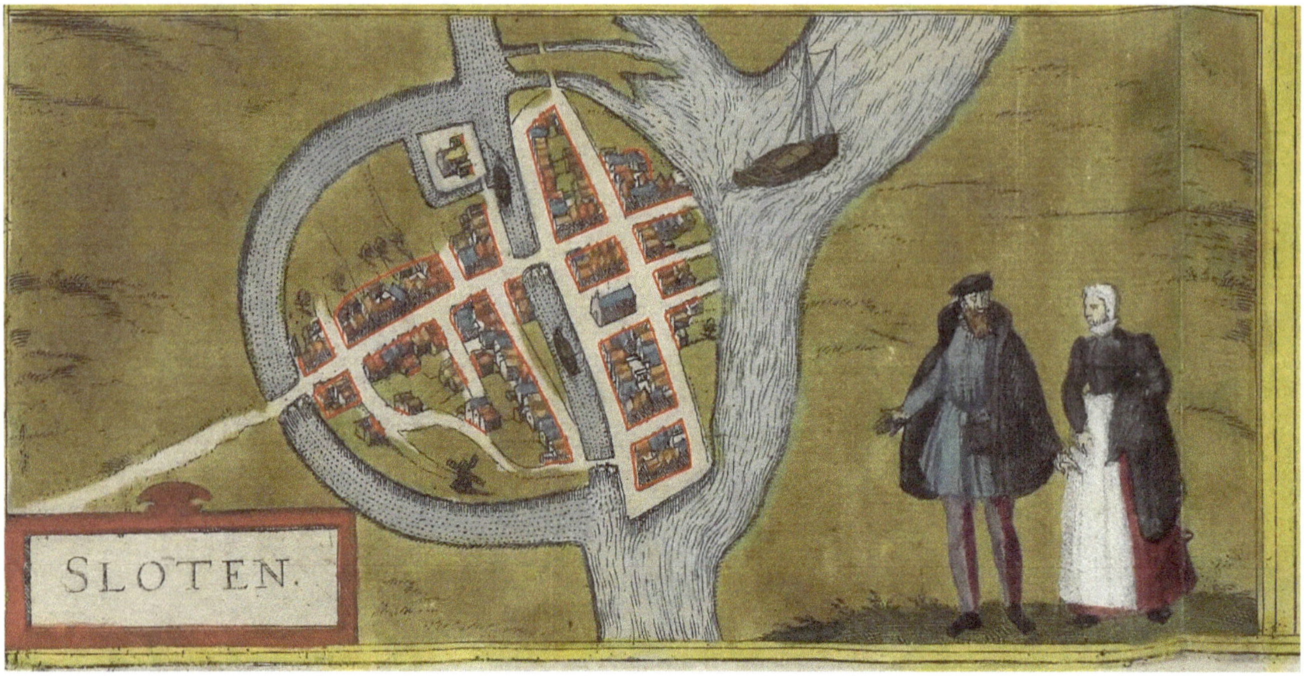

FIGURE 4.4 Braun and Hogenberg, Map of Sloten (approx. 1572), measured

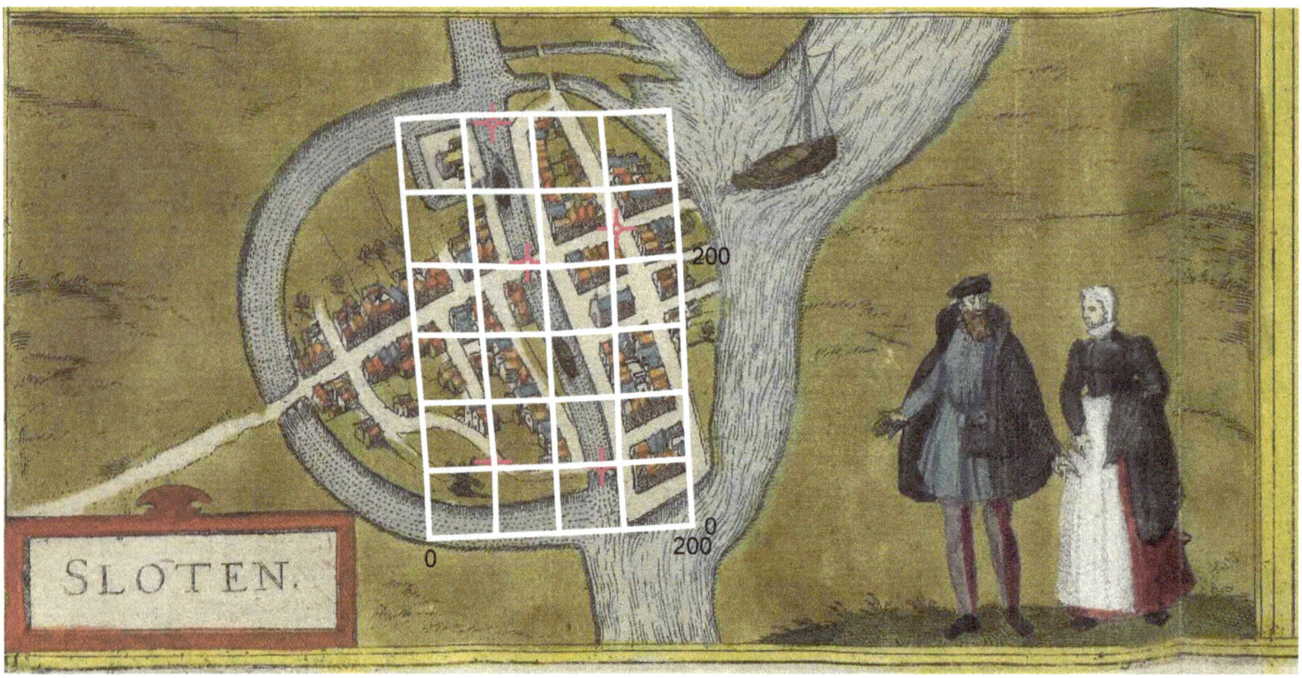

FIGURE 4.5 Braun and Hogenberg, Map of Sloten (approx. 1572), MapAnalyst (5 reference points)

difference between the number of measured and counted houses on the map. The difference with the alternative source is 13 percent.[15] It should be emphasized here that rotation and measurement do not necessarily correlate: a map can be rotated 180 degrees and yet have a very small deviation in measurements and thus maintain its high quality. The opposite is also true: a map with a small rotation can show very large deviations in measurements. In that case, the quality is low.

The conclusion for the three city maps of Braun and Hogenberg is as follows. The Weighted Mean Position is on average slightly better than that of the Van Deventer maps. The rotation of London and Sloten is lower than

15 Ramaer, 1921, p. 35; Visser, 1985, p. 20; Van der Woude, 1980, p. 136. From: Lourens, 1996, p. 83.

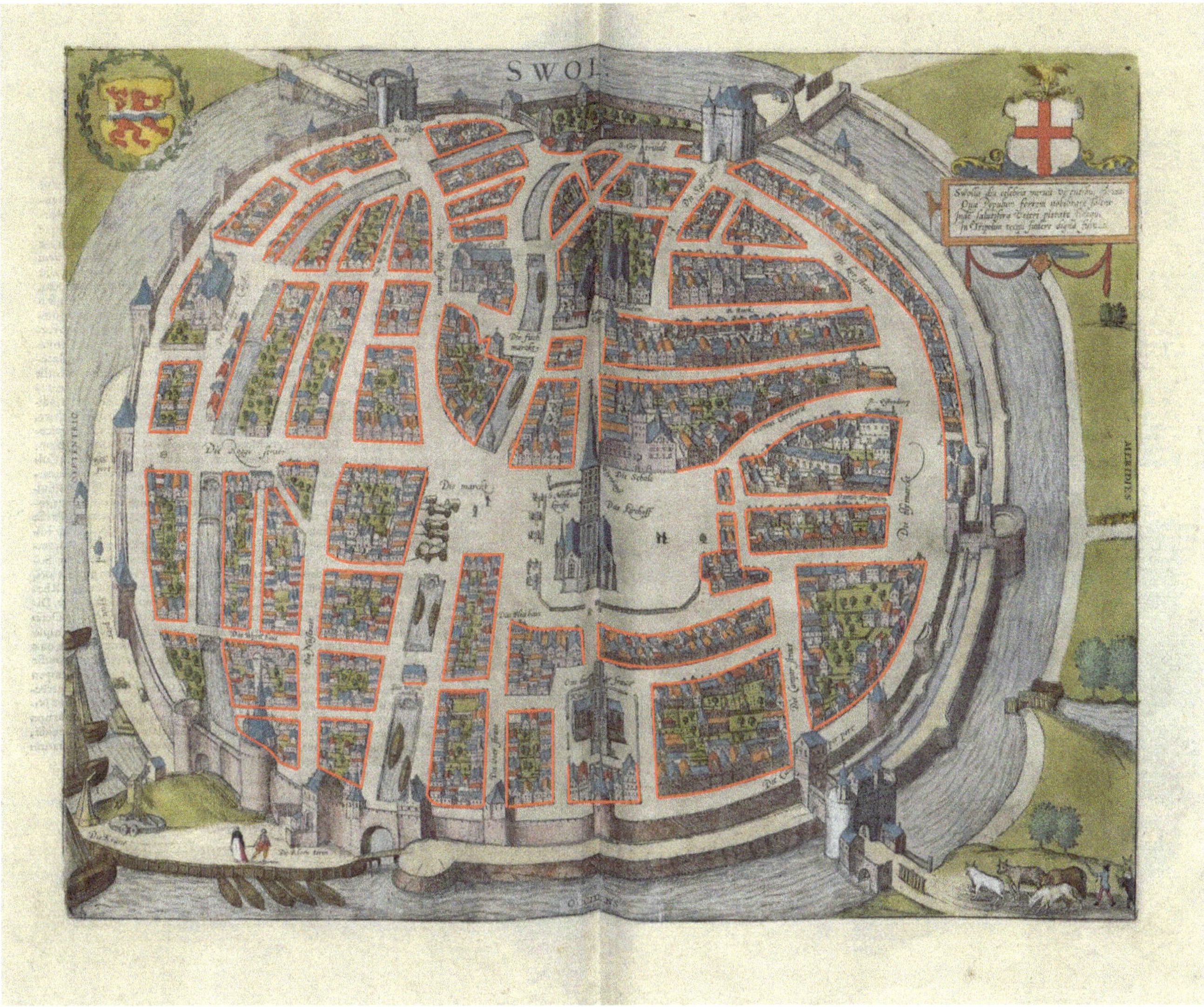

FIGURE 4.6 Braun and Hogenberg, Map of Zwolle (approx. 1572), measured

it is with Van Deventer. However, that average is brought down by the extreme deviation on the map of Zwolle of 100 degrees. It is important to mention that the rotation factor does not indicate the quality of the map but only reflects the choices made by the cartographers, as the north was not always precisely positioned at the top of the map. The numbers of measured and counted houses are close to each other with an average deviation of 8 percent, and a significant consistency in the percentages. The differences with the alternative sources (2, 12, and 13 percent) average out to 9 percent, with little consistency. However, the alternative sources may be of questionable consistency and reliability. All sources for the three chosen cities are secondary and based on assumptions, reasoning, and estimates. This is related to the time period in which the chosen Braun maps were drawn, that is to say, 1559–1580, and the fact that the available primary source material is very incomplete and inconsistent in all cases, both in terms of time and space. One of the underlying motivations for the present study is to seek source material that is complete and consistent in time and space.

This initial exploration has a positive result regarding the reliability of the three chosen maps and the balance between the lengths of built street sides and the number of houses drawn. However, a quick look at the other 472 maps and city views of Braun and Hogenberg shows a different picture in many places. Of a number of cities, only the city views are depicted and not the maps. Moreover, not all maps are suitable for both measuring the street sides and counting the houses. A good example of this is Vlissingen. The city plays an important role in Chapter 2 in measuring and analyzing the Van Deventer maps. The map of Vlissingen by Braun and Hogenberg mainly shows a fantasy version: the number of houses is a quarter of the number mentioned in the primary source. Additionally, the map is rotated 84 degrees relative to the

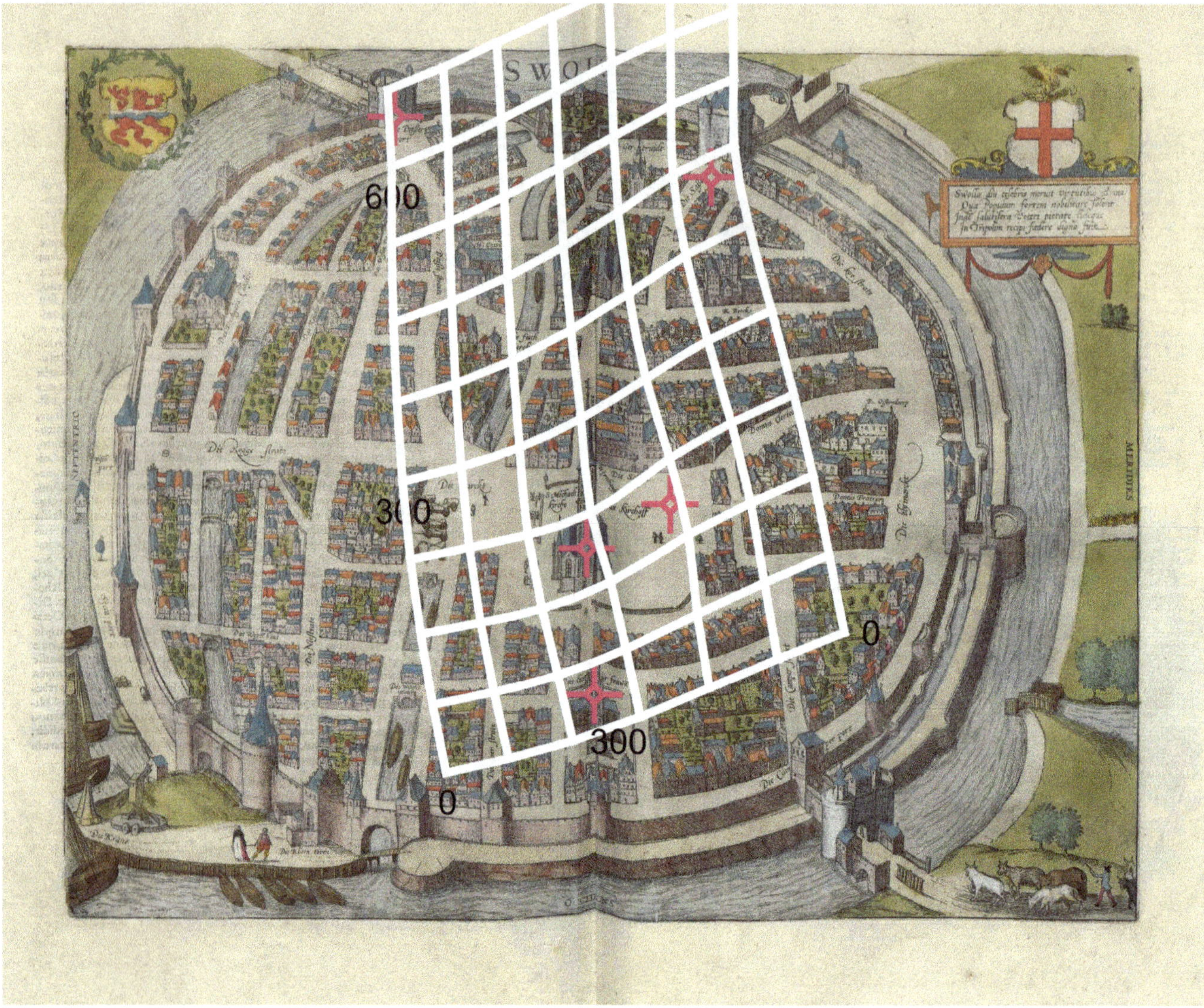

FIGURE 4.7 Braun and Hogenberg, Map of Zwolle (approx. 1572), MapAnalyst (5 reference points)

north, and the Weighted Mean Position Error is 11 centimeters per house, significantly higher than the three test cities London, Sloten, and Zwolle.[16] The advice for measurements and analyses in the atlas of Braun and Hogenberg is therefore to limit the final selection to the maps of which the deviation is not more than of 4–6 centimeters per house.

4.3 Blaeu

In 1618, the final volume of *Civitates orbis terrarum* by Braun and Hogenberg was published. During this period, a Dutch edition of the description of the Netherlands by the Antwerp-based Italian merchant Ludovico Guicciardini had also appeared.[17] The maps of Braun and Hogenberg, along with their sources, served as examples for this work. As such, most city atlases from the first half of the seventeenth century were based on older works and were therefore outdated, especially given the enormous growth that the Dutch Republic experienced during this time. This created an increasing demand for new cartographic works. It was Willem Blaeu, and especially his son Joan, who recognized this need and decided to create an atlas of the cities in the Low Countries.

In 1649, the first edition of Blaeu's new two-volume city atlas was published in Latin.[18] The Dutch-language version followed in 1652. Joan Blaeu had learned the craft of cartography from his father, who had established himself as a printer and publisher in Amsterdam around 1604.

16 Analysis in MapAnalyst dated January 31, 2024.

17 Guicciardini, 1612.

18 Blaeu, 1649, *Belgicae liberae* en *Belgicae regiae*.

Willem Blaeu focused on cartography, designing his maps and globes based on existing materials, travel reports, and conversations with sailors. After Willem Blaeu passed away in 1638, his sons Joan and Cornelis took over the business. Cornelis died shortly afterward, leaving Joan in sole control. The publishing prospered under his management, notably with the famous world atlas, the *Atlas Maior*, published in 1662.[19]

The city atlas consisted of two parts: the Southern and Northern Netherlands. The sources for Blaeu's *Toonneel der Steden* were diverse. Maps were engraved on copper plates, which publishers kept for future revisions in case of reprints. Sometimes publishers sold these plates, or they ended up in bankruptcy sales. The new owner could erase the previous publisher's name and replace it with their own information. While some of these copper plates – and therefore the maps – were altered over time, many remained unchanged. As a result, a map dated 1650 could have originally been created in 1600. It is certain that Joan Blaeu personally drew some of the maps.

However, a large number of maps were based on older versions, including those from the *Theatrum* by Marcus Zuerius Boxhorn from 1632.[20] This edition, however, only included the cities of Holland, meaning many maps from this province depicted the situation as it was in the 1630s. The section of the *Toonneel* covering the Southern Netherlands was based on a 1641 publication: *Flandria Illustrata*. This work was compiled by the Antwerp-born canon Antoon Sanderus, who had signed a contract for its publication in 1634 with the Amsterdam publisher Hendrik Hondius, son of the Flemish cartographer and publisher Jodocus Hondius. In 1641, the publishing rights were acquired by Joan Blaeu for 11,000 guilders. The second volume followed in 1644.[21]

From the two-part city atlas by Joan Blaeu from 1649, the cities of Arnhem, Leiden, and Valenciennes were chosen. The first two are from the part about the Northern Netherlands and the third from that about the Southern Netherlands. For all three cities, alternative sources are available with indications of the number of inhabitants in the middle of the seventeenth century. Here too, there is a diversity regarding the size of the cities: Leiden was among the largest cities that Blaeu mapped, Arnhem among the smaller ones, and Valenciennes was in between. The procedure followed largely corresponds to that of Braun and Hogenberg. The difference is that the three Blaeu maps stand entirely on their own. The Passus coefficient was generated on the maps themselves: on the north, east, south, and west sides, small representative built blocks were selected, the streets of which were measured and the houses counted. The procedure is as follows and applies to each of the three cities separately:

1. Use the same digital version of the atlas for the three maps;
2. Determine five representative orientation points, distributed over the entire map, but within the city walls, ramparts, or canals;
3. Determine with the help of MapAnalyst the scale and the reliability indicators (Rotation, Weighted Mean Position Error, Reference Points);
4. Select four representative built blocks on the north, east, south, and west side of the map where the houses and streets are clearly depicted;
5. Measure all built street sides in the four blocks in pixels;
6. Count the total number of houses in the four blocks;
7. Determine the specific Passus coefficient of the map by dividing the number of houses by the number of pixels in the four blocks;
8. Count the number of houses drawn on the map and compare these with the outcomes of the measurement;
9. Measure all built street sides in pixels;
10. Determine with the specific Passus coefficient the number of houses and thereby the number of inhabitants, with several variants, depending on the reduction factors;
11. Search in the alternative sources for the number of houses and/or inhabitants in the middle of the seventeenth century and compare the outcomes with the counts and measurements from steps 8 and 9.

The results are presented in Table 4.2.

Leiden was among the five largest cities of the Northern Netherlands in the seventeenth century and had undergone enormous growth, especially in the last quarter of the sixteenth century. The population had more than tripled between 1587 and 1622, while the built-up city area had only increased by about a third.[22] This means that the expansions largely took place within the existing buildings.

19 Blaeu, 1662.

20 Boxhorn, 1632.

21 Sanderus, 1641–1644. Van 't Hoff, 1966, p. 19.

22 Index Hoofdgeld 1622 stad Leiden, Stadsarchief Leiden, 0501A Inventaris, (1253) 1574–1816. (1897) Inv.nrs. 4021–4022 (Introduction to this archive: P.J.M. de Baar); Walda, 2020, pp. 25–46.

TABLE 4.2 The measurements and analyses of the Blaeu maps of Arnhem, Leiden, and Valenciennes as published in 1649

			Dimensions						
Map: Year of publication	City	Passus coefficient (amended)	Houses measured	Houses counted	Deviation (measured / counted)	Inhabitants (measured RF 5.5)	Inhabitants according to alternative source	Year alternative source	Deviation (measured / alternative source)
1649	Arnhem	0.0393	1539	1330	14%	8466	7150	1650	16%
1649	Leiden	0.0661	7433	7424	0%	40,884	44,743	1622	-9%
1649	Valenciennes	0.0566	3489	3597	-3%	19,188	19,000	1600	1%
	Average Bleau	0.0177			6%				8.6%
	Average vDeventer	0.1756			6%				

			Accuracies					
Map: Year of publication	City	Passus coefficient (amended)	Scale (1:)	Rotation (°)	Rotation (% of 180 °)	Mean Position Error absolute (± m)	Mean Position Error weighted (meters per house)	Recognizable reference points (Reliability 1–5)
1649	Arnhem	0.0393	1440	36	20%	10	0.006	5
1649	Leiden	0.0661	1140	31	17%	16	0.002	5
1649	Valenciennes	0.0566	1610	10	6%	46	0.013	3
	Average Bleau	0.0177	1397	26	14%	24	0.007	
	Average vDeventer	0.1756	8148	11	6%		0.044	

Note: the negative numbers have been made positive for the calculation of the average.

On Blaeu's map, this is clearly visible in the filled spaces in many blocks of houses where this was not or much less the case in 1587. The various structures are therefore much harder to recognize individually than the regular buildings that bordered the streets. The measurements and counts of the four representative blocks of houses in the north, east, south, and west of the city yielded a specific Passus coefficient of 0.0661. The measurement of the built street sides amounted to 112,481 pixels and, multiplied by 0.0661, results in 7433 houses. The count of the houses drawn on this map comes very close to this: 7424 houses, a negligible deviation from the number in the measurements. Multiplied by the reduction factor of 5.5 inhabitants per house, the population of Leiden in the 1640s amounts to 40,488.

For the verification of these numbers, a good alternative source is available, namely the census from 1622. This operation, organized in Holland, was intended to collect the so-called 'hoofdgeld' (poll tax), a tax meant to contribute to the costs of the war against Spain, resumed in 1621. For the 'hoofdgeld,' no complicated procedures were necessary because the amount was the same for everyone: one guilder. The only thing that was to be done was a census. This resulted in 44,743 'heads' in 1622.[23] Compared to the number of 40,484 inhabitants resulting from the Passus method on Blaeu's map in the 1640s, that was

23 Index Hoofdgeld 1622 stad Leiden, Stadsarchief Leiden, 0501A Inventaris, (1253) 1574–1816. (1897) Inv.nrs 4021–4022 (Introduction to this archive by P.J.M. de Baar).

FIGURE 4.8 Blaeu, Map of Leiden (1649), measured

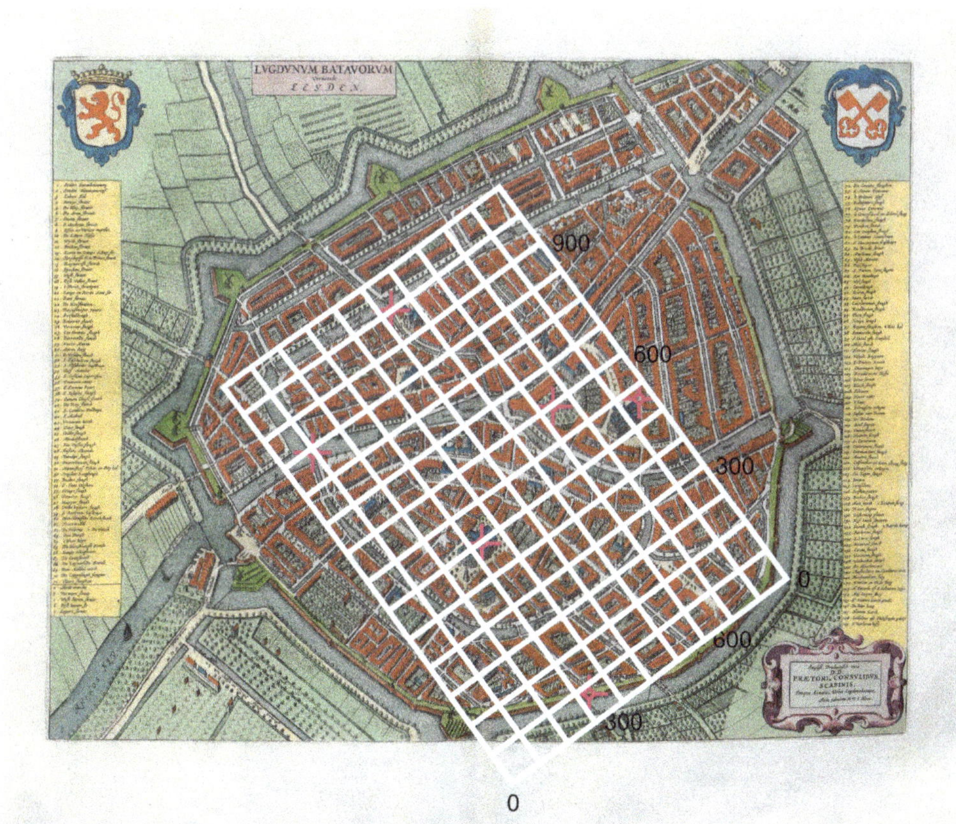

FIGURE 4.9 Blaeu, Map of Leiden (1649), MapAnalyst (5 reference points)

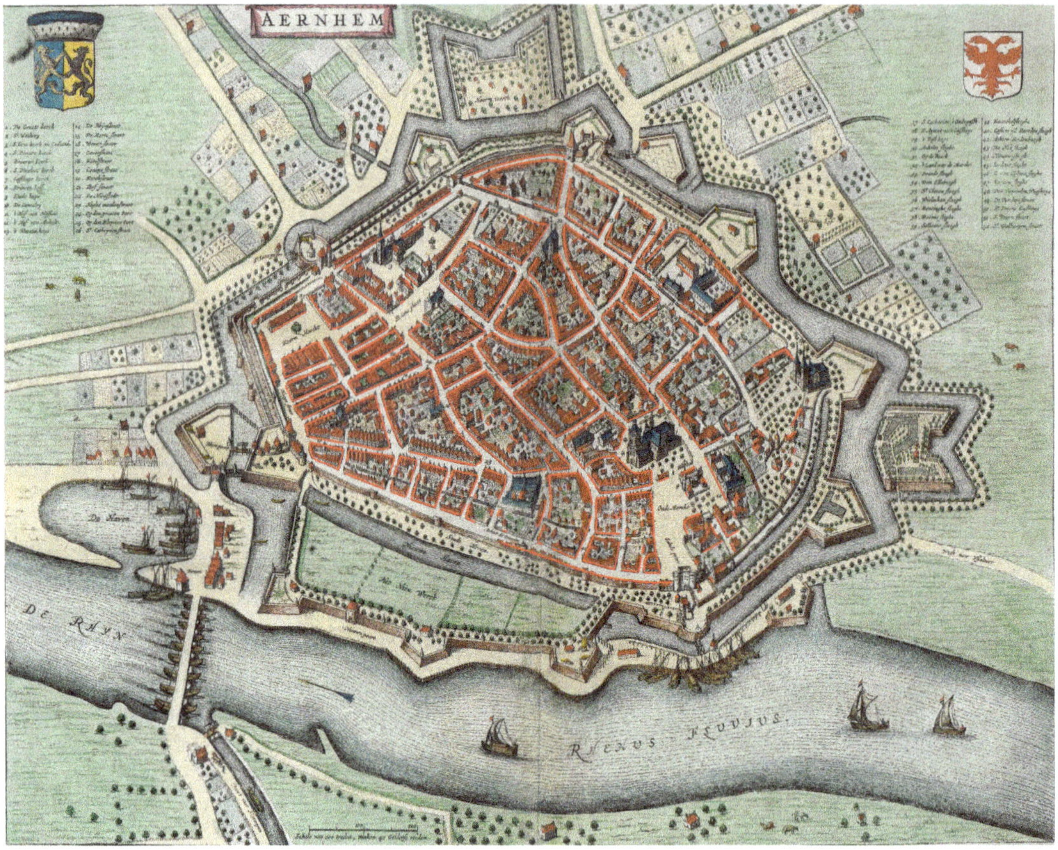

FIGURE 4.10 Blaeu Map of Arnhem (1649), measured

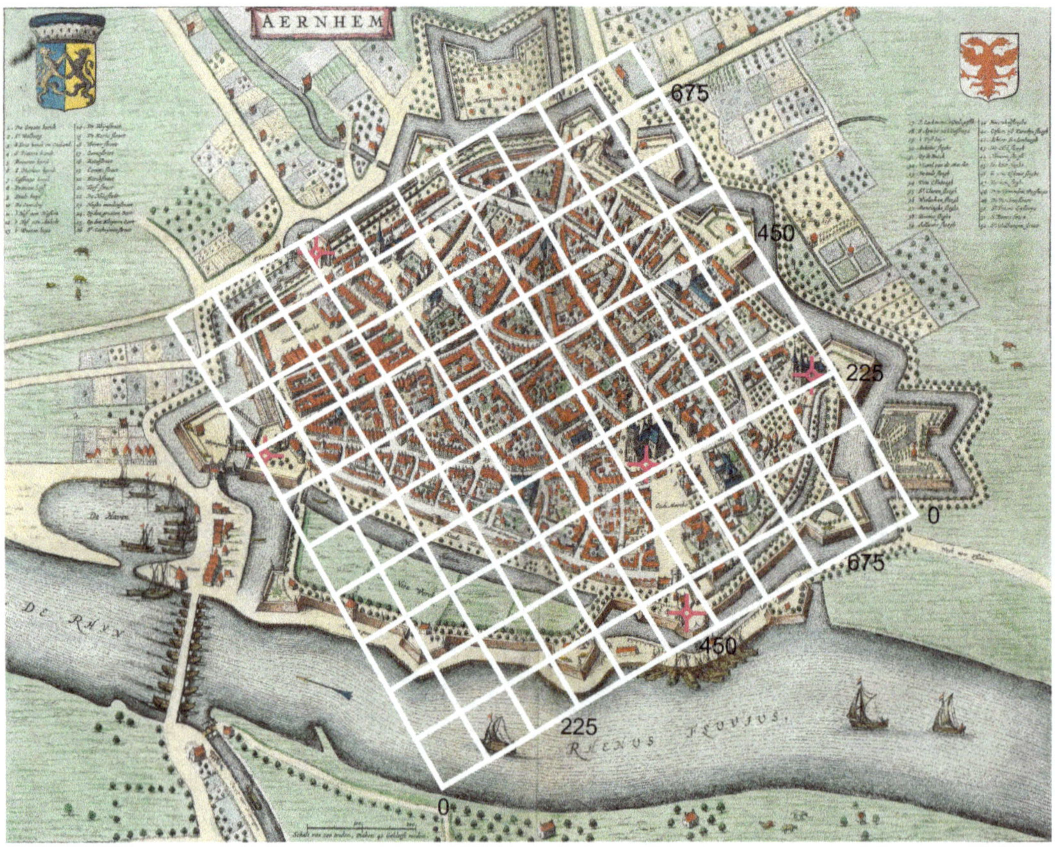

FIGURE 4.11 Blaeu, Map of Arnhem (1649), MapAnalyst (5 reference points)

FIGURE 4.12 Blaeu, Map of Valenciennes (1649), measured

FIGURE 4.13 Blaeu, Map of Valenciennes (1649), MapAnalyst (5 reference points)

9 percent less. It can be assumed that the numbers must have been closer to each other because – as mentioned earlier in this paragraph – the city could hardly expand its buildings, meaning the number of inhabitants per house must have been certainly greater than 5.5 during this period. With regard to quality, Van Deventer's map of Leiden is comparable to Blaeu's map, with a 6% deviation from the 1574 census.[24]

In the much smaller Arnhem, a house count took place in 1650 as part of a tax measure (the 'honderdste penning'). This tax on movable and immovable property was established in 1569 by the Duke of Alva. In 1650 a total of 1300 houses were counted.[25] Multiplied by the reduction factor of 5.5, this results in a population of 7150. In the measurements on Blaeu's map of Arnhem, where the four representative blocks of houses produced a Passus coefficient of 0.393, a house number of 1539 and a population of 8466 emerged. This was a difference of 16 percent compared to the alternative source. The final house count on Blaeu's map resulted in a total of 1330, which is close to the alternative source (1300), but a 14 percent difference compared to the number of houses emerging from the Passus method.

The city of Valenciennes in Northern France, part of the Habsburg Netherlands in the mid-seventeenth century and therefore included in Part I, Belgica Libera ac Foederata of Blaeu's city atlas, was somewhat smaller than Leiden but was still among the larger cities in the atlas. Around 1600, the city counted 3681 houses within its walls, housing around 21,000 people. A hundred years later, this number had dropped to just over 17,000.[26] The measurements and counts on the map of Valenciennes confirm this picture. In the 1640s, according to the Passus method, there were 3489 houses and according to the counts, 3597. Numbers that – as with Leiden – are close to each other. Multiplied by the reduction factor of 5.5, this results in 19,188 inhabitants, almost exactly between the 21,000 in 1600 and 17,000 in 1700 mentioned in the alternative sources. The Passus coefficient for Valenciennes, based on four representative blocks of houses, was determined at 0.0566.

24 Van Druenen, 2022, p. 143.

25 Roessingh, 1964, pp. 95–96. The 1300 houses are not visible in the table, but the number of inhabitants calculated on this basis is: 7150.

26 Junot, 2009, pp. 17–32.

4.4 The Hattingas

The third Passus method to be tested involves selecting a reference city from a collection of maps by the same cartographer, for which an alternative source is available and the number of houses on the same map can be counted. A special feature is that these maps, like those of Van Deventer, are equipped with a scale bar using the 'Blooise Roede' (approximately 3.6 meters) as the unit of length.[27] For most cities drawn by the Hattingas, only the street plan has been mapped. The cartographic oeuvre of Willem Tiberius Hattinga and his sons Anthony and David contains one map that meets these criteria, namely that of the city of Tholen in Zeeland, with the note that the houses are drawn in one dimension and not in oblique parallel projection.

The Hattingas' production, in terms of scope, is equivalent to that of Braun and Hogenberg and Blaeu. In the 1740s and 1750s, they produced four atlases with 649 maps, mainly of the border provinces of the Northern and Southern Netherlands, with a focus on the areas and cities in Zeeland.[28] Like Jacob van Deventer, they aimed to inventory, draw, and produce maps according to certain standards. In this collection too, the search was for maps of cities of various sizes for which alternative sources are available. The maps of IJzendijke, Tholen and Middelburg were chosen. The Hattingas carried out the measurements and mappings for part of their work themselves. Many maps were also copied from earlier material, such as the map of Middelburg used here, which was based on the layout by Abraham Zehender. The latter map, according to a note, is only partly by Hattinga: 'Recorded by Mr. Abraham Zehender anno 1739: after which this was copied and reduced by W.T. Hattinga in Middelburg 27 July 1740'.[29] The city map of Tholen was based on an older map for which David and his father Willem had carried out measurements a decade earlier.[30] The three cities were measured and analyzed using the same procedure:

1. Use the same source for the three maps (Zeeuws Archief in Middelburg) and choose the separate copies;[31]

27 The 'Blooise roede' was used on the Zeeland islands of Walcheren and Beveland Meertens Institute. https://mgw.meertens.knaw.nl/plaats/133 (accessed on February 24, 2024).

28 Nagtglas, F., *Levensberichten van Zeeuwen* (*Middelburg*, 1890, 4 delen), Deel II, p. 342–347; Vannieuwenhuyze, 2022, pp. 171–222.

29 Ibid., p. 215.

30 Ibid., pp. 184–185.

31 The maps exist in different versions, also published in atlases and chronicles (such as Tirion, I., *Current state of the United Netherlands*. Part IX and XX. Amsterdam, 1753.

2. Choose the reference map that depicts both the built street sides and the individual buildings (Tholen);
3. Check if all three maps have the same resolution;
4. If they differ, align the resolutions so they have the same value;
5. Check if the scale bars on the four maps have the same pixel dimensions among themselves;
6. If they differ, adjust the dimensions of the scale bars to the same value as that of the reference city, by adjusting the dimensions of the maps;
7. Determine five representative orientation points per city, distributed across the entire map, but within the city walls, -ramparts, or -canals;
8. Determine per city with the help of MapAnalyst the scale and the reliability indicators (Rotation, Weighted Mean Position Error, Reference Points);
9. Count the number of houses on the map of the reference city;
10. Measure all built street sides in pixels on the map of the reference city;
11. Determine the Passus coefficient for the Hattinga maps by dividing the number of houses by the number of pixels of the built street sides on the map of the reference city;
12. Measure per city the total number of pixels of the built street sides and calculate with the new Passus coefficient the number of houses and thereby the number of inhabitants, with several variants, depending on the reduction factors;
13. Search per city in the alternative sources for the number of houses and/or inhabitants in the middle of the eighteenth century and compare the outcomes with the counts and measurements from steps 9 and 10.

The results are presented in Table 4.3.

TABLE 4.3 The measurements and analyses of the Hattinga maps of Middelburg, Tholen, and IJzendijke as published in the 1740s and 1750s

		Dimensions							
Map: Year of publication	**City**	**Passus coefficient (amended)**	**Houses measured**	**Houses counted**	**Deviation (measured / counted)**	**Inhabitants (measured RF 5.5)**	**Inhabitants according to alternative source**	**Year alternative source**	**Deviation (measured / alternative source)**
1740	Middelburg	0.0337	4176	–	–	22,967	21,368	1739	7%
1744	Tholen	0.0305	363	392	-8%	1997	2156	1753	-8%
1745	IJzendijke	0.0124	91	–	–	501	460	1748	8%
	Average Hattinga	0.0260			8%				8%
	Average vDeventer	0.1756							

		Accuracies					
Map: Year of publication	**City**	**Scale (1:)**	**Rotation (°)**	**Rotation (% of 180 °)**	**Mean Position Error absolute (± m)**	**Mean Position Error weighted (meters per house)**	**Recognizable reference points (Reliability 1–5)**
1740	Middelburg	3050	66	37%	15	0.004	5
1744	Tholen	2970	17	9%	11	0.030	5
1745	IJzendijke	1210	68	38%	13	0.143	5
	Average Hattinga	2410	50	28%	13	0.059	
	Average vDeventer	8148	11	6%		0.044	

Note: the negative numbers have been made positive for the calculation of the average.

FIGURE 4.14 Hattinga, Map of Tholen (approx. 1744), measured

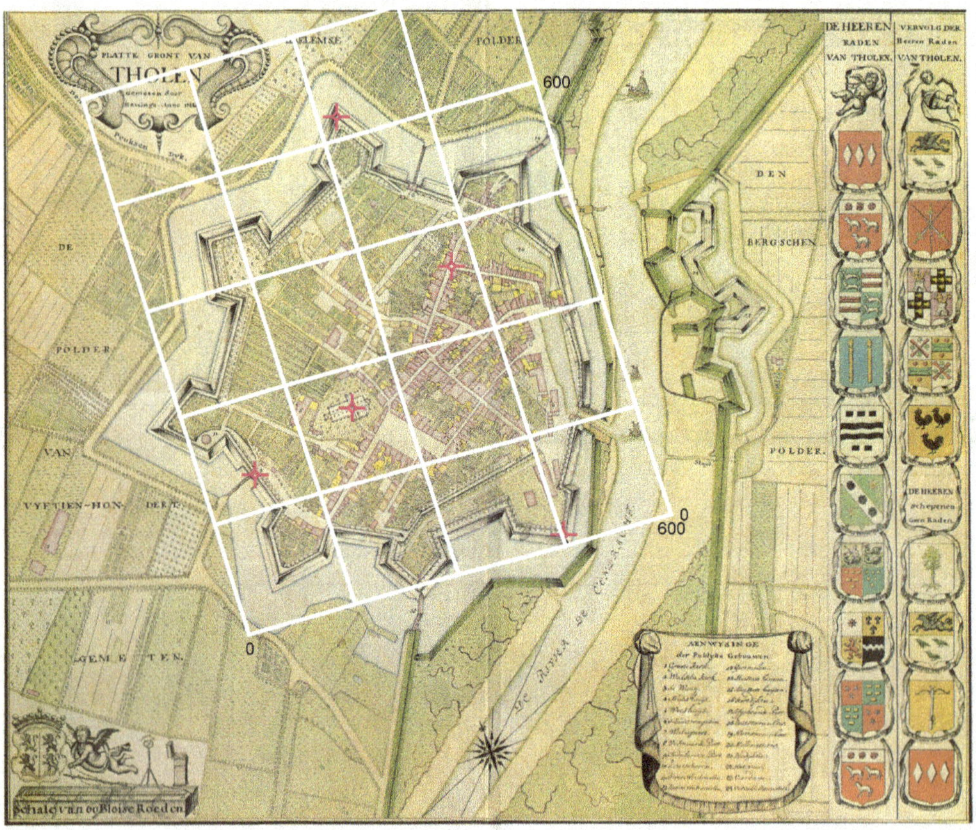

FIGURE 4.15 Hattinga, Map of Tholen (approx. 1744), MapAnalyst (5 reference points)

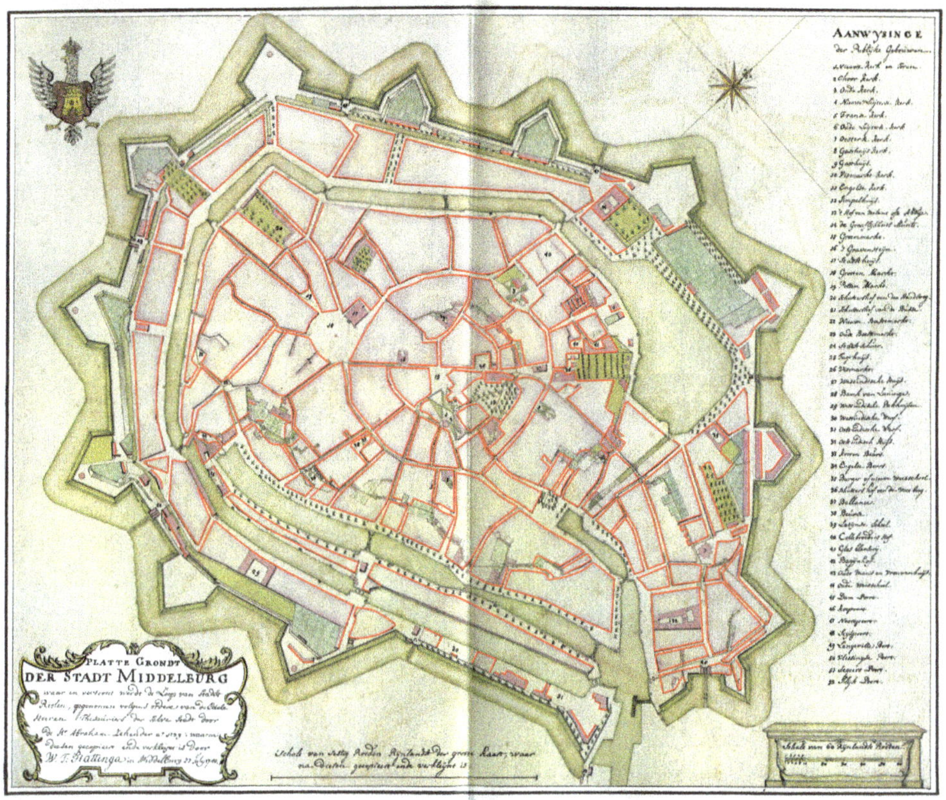

FIGURE 4.16 Hattinga, Map of Middelburg (approx. 1744), measured

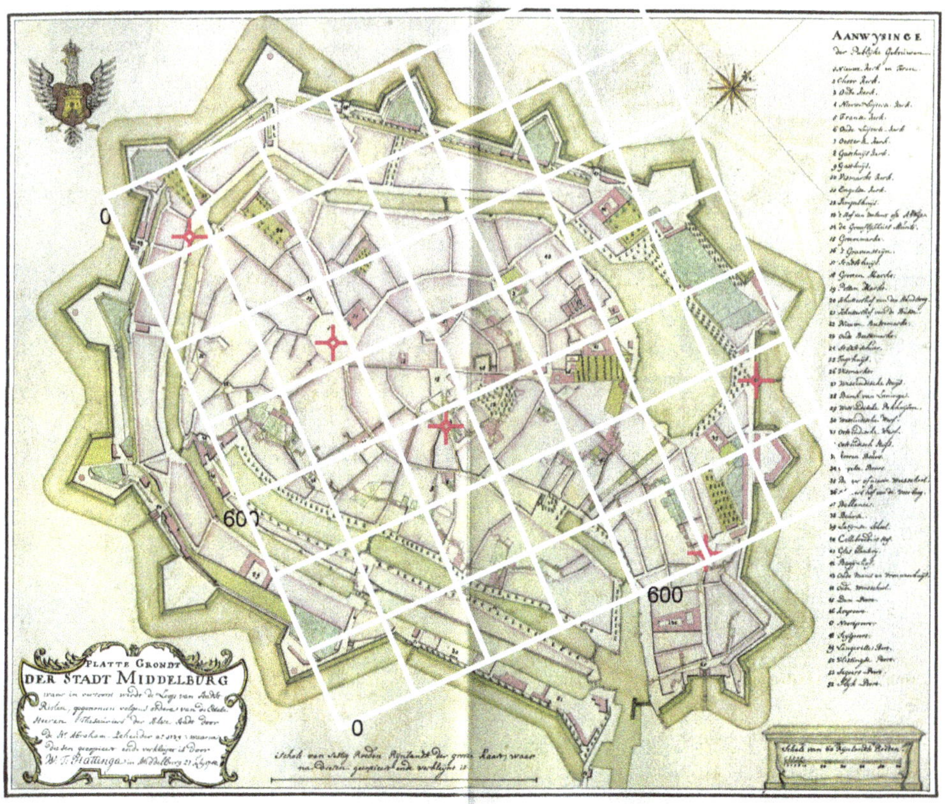

FIGURE 4.17 Hattinga, Map of Middelburg (approx. 1744), MapAnalyst (5 reference points)

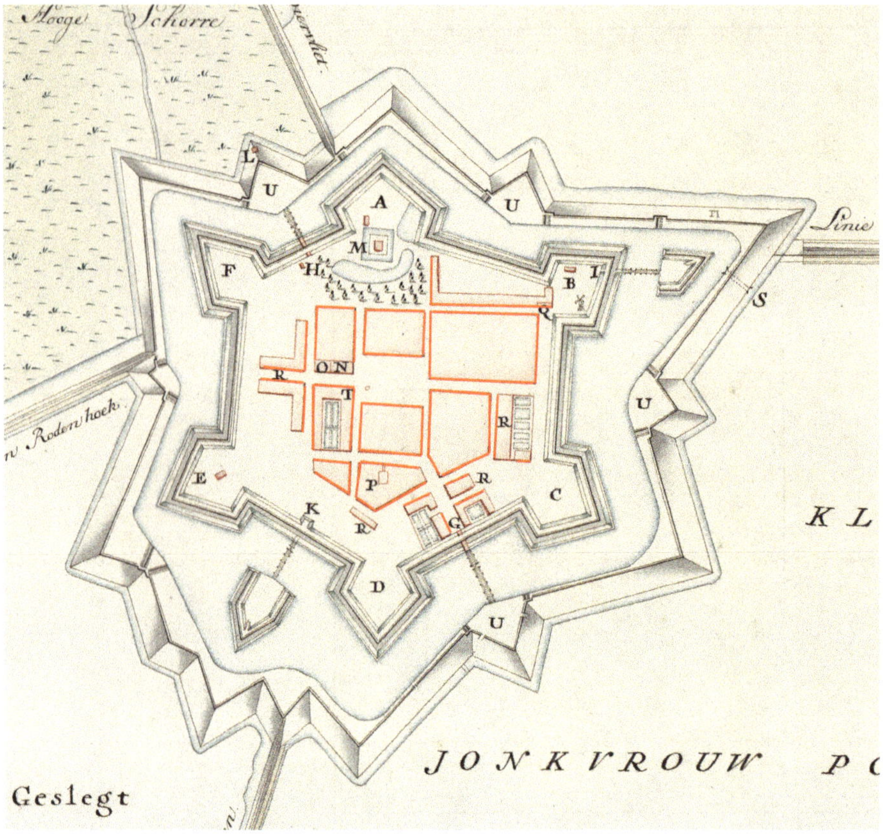

FIGURE 4.18 Hattinga, Map of IJzendijke (approx. 1744), measured

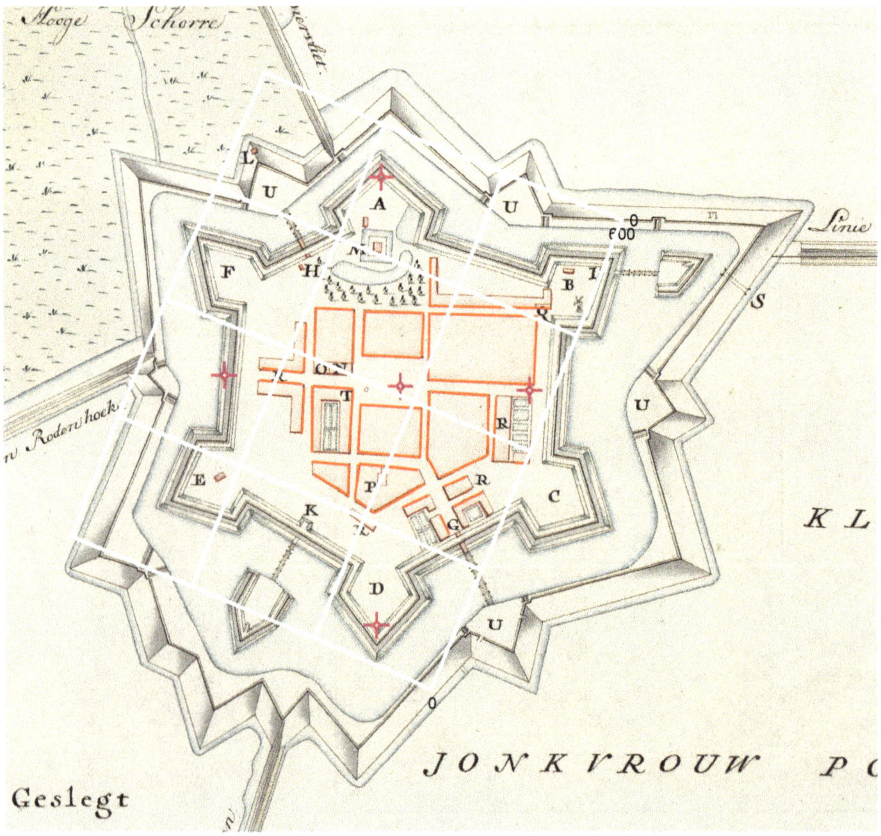

FIGURE 4.19 Hattinga, Map of IJzendijke (approx. 1744), MapAnalyst (5 reference points)

TABLE 4.4 Overview of the three variants

				Accuracies						
Variant	**Period map**	**Averages**	**Passus coefficient (amended)**	**Deviation (measured / counted)**	**Deviation (measured / alternative source)**	**Deviations (average)**	**Scale (1:)**	**Rotation (°)**	**Rotation (% of 180 °)**	**Mean Position Error weighted (meters per house)**
I	1572–1618	**Average Braun (3 cities)**	0.0280	8%	9%	9.0%	2737	38	21%	0.04
II	1630–1648	**Average Bleau (3 cities)**	0.0540	6%	9%	7.5%	1397	26	14%	0.01
III	1740–1755	**Average Hattinga (3 cities)**	0.0260	8%	8%	8.0%	2410	50	28%	0.06
		Average (9 cities)	0.036	7%	9%	8.2%	2181	38	21%	0.04
		Van Deventer	0.1756	6%			8148	11	6%	0.04

Note: the negative numbers have been made positive for the calculation of the average.

For the reference city of Tholen, both the streets were measured and the houses counted. The street sides have a total length of 11,886 pixels. This results in 363 houses. The alternative source for Tholen, the contemporary publication *Tegenwoordige staat der Vereenigde Nederlanden* by Izaak Tirion from 1753, gives also a total number of houses of 363, which comes from a 'Quohier' – a tax register – from 1738.[32] The number of counted houses on the map of Tholen is 392: 8 percent more than the Passus method and the alternative source indicates, a margin that is acceptable. The Passus coefficient for the other three Hattinga maps can be calculated with these results: 363 / 11,886 = 0.0305.

The application to the other two maps gives almost the same result. Middelburg, by far the largest city in Zeeland around 1750, counted 4176 houses according to the Passus method at that time. The alternative source, a house count from 1739, mentions 3885 houses and, multiplied by the reduction factor 5.5, a total of 21,368 inhabitants, an acceptable difference compared with the Passus method of 7 percent. This number fits into the demographic development that Middelburg underwent between 1550 and 1800: continuous growth, which was strongest in the first 100 years and then leveled off, but never – unlike many other cities in the Northern and Southern Netherlands – experienced a decline.[33]

The Flemish fortress town of IJzendijke, located in the area now called Zeeuws Vlaanderen, was administratively part of the Republic of the Seven United Netherlands in the eighteenth century. Unlike the provinces of Groningen, Friesland, Overijssel, Gelderland, Utrecht, Holland, and Zeeland, this region did not have a voice in the national government. IJzendijke was part of the Spanish-Dutch lines, the buffer zone between the Republic and the Spanish Netherlands. The results of the census held in the year 1748 for the town are available. In that year, a special tax was introduced to cover the costs of the French occupation as a result of the War of Austrian

32 Ibid., Part XX, p. 80; house valuation register of Middelburg, 1752 (Zeeuws Archief, 33.3 Verzameling Aanwinsten Rijksarchief in Zeeland 1955–1999, zestiende-twintigste eeuw. Inv.nr.: 1955.10).

33 The 3885 houses are not visible in the table. The number of inhabitants calculated on this basis is: 22,967; Van Druenen, 2022, p. 215.

TABLE 4.5 Three variants on the Passus method, applied to the map of London from 1559, of Zwolle from 1560, and Tholen from 1751

			Dimensions						
Map: Year of publication	City	Passus coefficient (amended)	Houses measured	Houses counted	Deviation (measured / counted)	Inhabitants (measured London RF 8; Zwolle RF 5,5)	Inhabitants according to alternative source	Year alternative source	Deviation (measured / alternative source)
1559	London, variant I	0.1052	7357	6639	10%	58,856	60,000	1560	-2%
	London, variant II	0.0962	6730	6639	1%	53,839	60,000	1560	-11%
1581	Zwolle, variant I	0.280	1569	1430	9%	8631	7500	1564	13%
	Zwolle, variant II	0.0251	1404	1430	–2%	7724	7500	1564	3%
1744	Tholen, variant III	0.0305	363	392	–8%	1997	2156	1753	-8%
	Tholen, variant II	0.340	404	392	3%	2224	1695	1753	24%
	Average vDeventer	0.1756							

			Accuracies				
Map: Year of publication	City	Scale (1:)	Rotation (°)	Rotation (% of 180 °)	Mean Position Error absolute (± m)	Mean Position Error weighted (meters per house)	Recognizable reference points (Reliability 1–5)
1559	London, variant I	4880	9	5%	208	0.028	5
	London, variant II	4880	9	5%	208	0.028	5
1581	Zwolle, variant I	1300	100	56%	79	0.050	5
	Zwolle, variant II	1300	100	56%	79	0.050	5
1744	Tholen, variant III	1210	68	38%	13	0.143	5
	Tholen, variant II	1210	68	38%	13	0.143	5
	Average vDeventer	8148	11	6%		0,044	

Succession. It was a personal tax, the amount of which depended on the wealth of a household. According to this census, 460 people lived in IJzendijke that year. The Passus method yielded 91 houses and, multiplied by the reduction factor of 5.5, 501 inhabitants, 8 percent more than the alternative source.[34]

4.5 Conclusions

The question now is whether these differences are caused by the quality of the cartographic work that was tested or by the variant of the Passus method that was used. The analyses by MapAnalyst show that Blaeu scores the highest by far on the reliability variables Rotation and Weighted Mean Position Error: 26 degrees (14 percent deviation from the maximum of 180 degrees) and 1 centimeter per house, respectively. Braun follows with 38 degrees (21 percent) and 4 centimeters, and the Hattingas with 50 degrees (28 percent) and 6 centimeters.

Overall, it *appears* that the integral application of the Passus coefficient 0.1756 on the maps of Braun and Hogenberg leads to good results. However, it should be noted that they based part of their maps on those of Van Deventer.[35] Hogenberg stayed in Cologne between 1572 and 1575 and must have surely been in contact with his colleague. It is possible that Van Deventer specifically went to Cologne to negotiate with Hogenberg about a license for the use of the maps. The copied maps of some Dutch cities (not all, as sometimes claimed) appear in part III (1581) and part IV (1588).[36] The test city Sloten (Part III) is one of them.[37] It is certain, however, that Van Deventer never drew a map of London.

In Table 4.5, for the cities of London, Zwolle, and Tholen, in addition to the original Variant, the second Variant was also tested. The results show that the three variants are close to each other, albeit with shifts in all cases. The number of measured and counted houses is almost identical in Variant II: only 1 percent difference for London, 2 percent for Zwolle, and 3 percent for Tholen. This indicates that in these three cities, the four chosen blocks are very representative of the entire city.

However, the difference with the number of houses and inhabitants from the alternative source has increased for London and especially Tholen: from 2 to 11 percent and from 8 to 24 percent, respectively. In Zwolle, it has decreased, from 13 to 3 percent. For the map of Zwolle, this could indicate that the work of Van Deventer had an influence; the data from his map of Zwolle around 1560 also deviated little (2 percent) from the alternative source. Variant I was also tested on the maps of Blaeu and the Hattingas, but this did not yield satisfactory results. Therefore, the conclusion for Variant I is that it *may* be applicable to sixteenth-century collections of city maps related to the Van Deventer maps.

Variants II and III have more possibilities in terms of application. They are independent of the specific measurements on the Van Deventer maps and assume a universal pattern in the relationship between the lengths of the built street sides and the number of houses. Variant II has the advantage that it can be applied to individual maps, whether or not they are part of a collection. The condition for this is that both the drawn houses and the street pattern are clearly visible.

Variant III, in terms of procedure, most closely resembles the Passus method as applied to the maps of Jacob van Deventer. It is suitable for collections of city maps where the houses are missing. The condition is that a copy of one of the maps is available on which the houses are drawn and/or there is an alternative source that displays the number of houses in a certain year. These will usually be fiscal sources with house counts. These alternative sources should always be used, for every city where they are available, to check the measurements and counts of the Passus method and to place the research results in the broader context of place and time.

Finally, it should be noted that Variant II and Variant III, when they both meet the aforementioned conditions, can be used in combination as each other's control mechanism, together with the alternative source or sources.

34 Van Cruyningen, 2000, pp. 385–386, 390.

35 Van 't Hoff, 1966, pp. 16–17.

36 Ibid., p. 17.

37 The maps of Van Deventer and Braun and Hogenberg of Sloten largely overlap. The Mean Position Error is 7 meters for both.

CHAPTER 5

Epilogue

5.1 Summary and Partial Conclusions

The field of historical demography has undergone tremendous development since World War II. Whereas for centuries, the focus was primarily on counting birth rates, marriages, deaths, and migrants, it has now evolved into a full-fledged discipline with numerous specializations. Moreover, there is intensive collaboration with other sciences. Strangely, in the midst of this development, the original focus – mapping population numbers – has been pushed into the background. This applies to both the research and the development of new methods and techniques. The first chapter of the present study demonstrates that this stagnation in the Dutch context has led to calculations in the existing literature being based on a very narrow but influential foundation laid in the 1960s. Furthermore, when it comes to calculating population figures for the period 1500–1800, and thus, for example migration and urbanization movements, these calculations are somewhat speculative. Nevertheless, they have become the standard and continue to serve as a source for new publications up to this day.

The judgment of 'speculative' stems, on the one hand, from the observation that figures from regions and cities where extensive research had been conducted in baptism, marriage, burial, and tax registers were extrapolated to areas and cities where such research had not been done. On the other hand, it also arises from the unreliability of this kind of source material. From Wageningen University – where historical demography had been practiced as a legitimate science since the early 1950s – early warnings were issued not to draw conclusions too hastily from these sources. They are rarely complete, scarcely comparable on a larger geographic scale or over a longer period, unreliable in terms of the administrative skills of the officials responsible for recording the data, and confusing when it comes to interpreting terms such as 'poorter' (burgher), 'haardstede' (hearth), 'raam' (window), 'communicants' (churchgoers), 'weerbare mannen' (able-bodied men), and 'muntsoorten' (coin types). Given that many researchers have relied on digitized files in recent decades, it is also necessary to take into account the quality of the input and software formats used.

From the above, it can be concluded that developing new methodologies that allow for the calculation of these numbers with a better and more differentiated foundation than before is essential. Only then can debates be held on topics such as de-urbanization, migration, or any other subject related to the development of population totals. In this regard, there should also be a systematic search for alternative sources, both within and outside the boundaries of the historical demographic field. This means that the pursuit of the most reliable quantitative foundation should be a key element in any research in this discipline, as well as in the resulting scientific debates. Various prominent historians, such as Henri Pirenne in 1903 and Rogier Mols and Bernhard Slicher van Bath in the 1950s, have previously advocated for the use of source collections that have a high degree of uniformity across both time and space. Mols explicitly pointed to the use of coherent collections of city maps, such as those produced in the sixteenth and seventeenth centuries in the Low Countries by figures like Van Deventer, Braun and Hogenberg, Janssonius, Blaeu, and De Wit. However, the only times this was actually done were long before Mols' call: In the seventeenth century, John Graunt counted the number of houses on the map of London and calculated the population based on that. In the 1920s, Johan Ramaer used Van Deventer's city maps to calculate the built-up areas of cities in the Northern Netherlands and derived the number of houses and population figures from them. After that, city maps were still used as supplementary sources by figures like Tertius Chandler, Paul Bairoch, and Daniel Pasciutti, but never again as a standalone, uniform source set.

In 2022, the study *Meten en Tellen* introduced a method for calculating urban population figures based on sixteenth- and seventeenth-century city maps: the Passus method. The key advantage of this method is that it allows cities under investigation to be compared with a high degree of consistency. At its core, the Passus method involves mapping the relationship between the lengths of built-up street sides, the population size of the city. This enables an objective comparison of city maps that are part of a collection or an atlas. *Meten en Tellen* tests the method on a limited number of cities in Zeeland, Holland, and Vlaanderen.

In the present study, the method was applied to all 226 available city maps created by Jacob van Deventer between 1542 and 1575. Most of these cities are located in

 | HTTPS://DOI.ORG/10.1163/9789004755031-006

present-day Netherlands and Belgium, some in Germany, France, and Luxembourg. These maps are distinguished by their remarkable accuracy and consistency – especially considering the period in which they were inventoried and drawn. Notable features include the consistent northward orientation of all maps and the use of a uniform scale, typically around 1:8000. Crucially for research in the present study is the premise that Van Deventer's maps can reveal more than just the topography of the cities depicted. Scholars from Ghent University, such as Bram Vannieuwenhuyze and Colin Dupont, have contributed significantly to this specialized form of historical cartographic research, particularly through the method of *digital thematic deconstruction*. This approach focuses on the sixteenth-century works of Van Deventer and contemporaries like Braun and Hogenberg, aiming to uncover thematic historical insights embedded in these detailed urban representations. Just as with the more traditional sources mentioned earlier, critical questions can also be raised about the degree of accuracy when using old city maps. For example, regarding the difference in dating between the field research and the drawing of the final result. Another question is whether the map in question was even intended to be accurate. Enlarging important buildings may also have been a primary goal, which almost always led to distortions in the proportions of the map. Since many maps were wholly or partially copies of earlier work, the researcher should also always attempt to trace this genealogy of cartographic material and examine how accurate those earlier drawings were.

In the last paragraph of this first chapter, building on the insights mentioned above, a number of case studies were developed across seven subdisciplines of historical research: military, political, judicial, socio-economic, ecclesiastical-cultural, demographic, and urban morphological history. Since historical demography is the primary 'end user' of the research outcomes in this study, this subdiscipline was compared with the other six. The conclusion was that the first six subdisciplines typically require the integration of other primary and secondary sources for meaningful interpretation. However, for historical-demographic applications, such supplementary material is not necessary, or at least far less so. One could even argue that it is preferable not to use any additional sources at all. After all, the objective is to determine the population figures of a city at a specific point in time in a way that is impartial, consistent, and objective.

The Passus method, as introduced in 2022 in *Meten en Tellen*, is refined in the second chapter of the present study based on insights gained from its application to all 226 city maps by Van Deventer and by various evaluations of the original procedure. However, this did not result in any fundamental changes to the method itself, nor to the results for the cities studied in *Meten en Tellen*. Instead, certain criteria – such as the delineation of the area to be measured – have been sharpened and applied more consistently.

For the Passus method, the second chapter introduces a detailed procedure that must be followed carefully when measuring street frontages and extrapolating from these measurements to estimate the number of houses and inhabitants. The procedure is based on the measurements taken from Van Deventer's map of the city of Vlissingen. For this city, an alternative house count is available, and, in addition to the Van Deventer map, a second map from the same period survives. This map is different: whereas Van Deventer's map allows only street frontages to be measured, it is here also possible to count individual houses and buildings. Using a representative portion of the city center – which today still retains its sixteenth-century street structure – a coefficient, the 'Passus coefficient', was calculated. This figure, 0.1756, was tested on six cities in Zeeland, Holland, and Vlaanderen: Ghent, Goes, Leiden, Middelburg, Veere, and Zierikzee. The results – compared with available alternative contemporary population counts – consistently fell within a margin of error of no more than 12 percent. Additional research based on the abundant source material from the city of Leiden further supports both the selected reduction factor of 4–7 and the Passus coefficient of 0.1756. When establishing the Passus coefficient for the map of Vlissingen, the software program MapAnalyst was used. This enabled researchers to determine and incorporate the deviations of the historical maps – both in relation to one another and to the modern situation in Google Maps and OpenStreetMaps – into the calculations that ultimately yielded the figure 0.1756.

MapAnalyst also plays a central role in the first part of Chapter 3, where all 226 city plans by Van Deventer are measured and analyzed. The software operates using at least five reference points per city, which must be visible both on the historical map and on the modern map of the city. These reference points may include significant buildings such as churches or castles, as well as street patterns, blocks of houses, squares, hills, harbors, or rivers. MapAnalyst calculates the scale, the rotation relative to the north, and the deviation in meters compared to the modern map. Because using this type of software carries the risk of selecting incorrect reference points on historical maps, a new variable 'Reliability' was introduced in this study. This variable, ranked on a scale from 1 to 5 per city, indicates the degree of reliability of the analysis relative to the other 225 cities in the dataset.

In the second part of Chapter 3, the measurement and analysis results of all 226 cities are compared with each other in five tables. Unlike the city-specific calculations of the number of houses and inhabitants – which require interpretation that takes into account variations in reduction factors (the number of houses per street side and the number of inhabitants per house) – the comparative analysis of the total street side lengths per city is notably free from assumptions regarding the values of such unknown variables. At most, one might question the relative quality of the maps, but those differences are now quantified through the quality analysis using MapAnalyst. In this context, the 'Weighted Mean Position Error' plays a key role: it reflects the weighted deviations from present-day city maps. In other words, the ratios between the total lengths of street sides provide a reliable representation of the relative size of the built-up areas of the 226 cities. However, when comparing numbers of houses and inhabitants, assumptions about the respective reduction factors are again necessary.

The first two tables include the totals and averages of all measurements, counts, calculations, and analyses. These figures show, among other things, that nearly 29,000 street sides were identified, on which – according to the Passus coefficient – over 210,000 houses once stood, housing approximately 1.2 million people. The average scale of the maps is 1:8147, with a standard deviation of only 7 percent, indicating a high degree of uniformity among the Van Deventer maps. This is further confirmed by the average deviation per house of less than 4 centimeters when comparing Van Deventer's maps with modern maps from OpenStreetMaps and Google Maps. An important insight emerges when comparing this deviation to the assessment of the quality of the reference points. These two data sets show a significant negative correlation of −0.7, suggesting that the precise determination of reference points on both the historical and modern maps is a critical success factor in this type of analysis. In cases where this is not always feasible – as with many of the cities analyzed in the present study – reporting the reliability factor is therefore a necessary requirement for valid interpretation.

The third table lists all 226 cities in alphabetical order, including the location of the maps and the year in which the data for the map were inventoried and the fourth table ranks the cities by size. The six largest cities are Ghent, Bruges, Brussels, Tournai, Leuven, and Mechelen – today all part of Belgium. In seventh place is Arras, now in France, and in eighth place Utrecht, the largest city in the present-day Netherlands. The fifth table ranks the cities based on the average number of houses per street side. Notably, the smallest city by size, Sint-Anna-ter-Muiden, scores the highest in this category. This is due to the fact that the city had only two street sides of considerable length. In larger cities, which often had many longer streets, the average is lowered by the presence of numerous small alleys. The sub-table here demonstrates that the number of street sides correlates perfectly with the overall size of the respective city.

The sixth table presents the results of the various quality analyses. Cities are ranked based on the 'Weighted Mean Position Error'. Here too, it becomes clear that the accuracy of the maps strongly correlates with the reliability of the selected reference points. The table also shows that, although all maps excel in accuracy, there are indeed differences which, however, cannot be expressed in terms of 'good' or 'bad', but rather in terms of 'very good', 'good', and 'slightly less good'. A second insight is that the differences in quality are not linked to time periods, the geographical location of the cities, or the place where the maps are stored.

The general conclusion of this research chapter is that the analyses and the mutual comparisons between the 226 cities, as presented in the five tables described above, can only be made on the basis of research methods like the Passus Method, which makes use of uniform source material produced under comparable conditions, using standardized protocols within a time span of at most a few decades.

In the fourth chapter, three sample sets of city maps – each consisting of three cities – are used to investigate the extent to which the Passus method can be applied to other series. From the six-volume atlas by Braun and Hogenberg, the selected cities are London, Zwolle, and Sloten. From the two-volume atlas by Blaeu, Arnhem, Leiden, and Valenciennes were chosen. From the four atlases compiled by the Hattinga family, Middelburg, Tholen, and IJzendijke were selected. For each of these three samples, a different variant of the method was applied.

On the sixteenth-century maps by Braun and Hogenberg (Variant I), the Passus procedure was applied in full – that is, including the Passus coefficient – but combined with directly counting the houses, as they are visible on each map. The same applies to the work of Blaeu from the seventeenth century (Variant II), for which a specific coefficient was calculated for each map. For the Hattinga maps from the eighteenth century (Variant III), a coefficient can be calculated based on the only city map that also shows individual houses: that of the town of Tholen in Zeeland.

In the final paragraph of this chapter, the three variants are compared side by side. The conclusion is that Variant I is reasonably applicable to the work of Braun and Hogenberg, particularly to the maps derived from Van Deventer. Applying the method to the atlases of Blaeu and Hattinga did not yield satisfactory results. Variant II is also applicable to the Braun and Hogenberg maps because a unique coefficient is calculated for each city. This makes Variant II also suitable for maps that are not part of a uniform collection, and is therefore the most versatile. The condition, however, is that the houses must be visible on the respective maps. Variant III can be applied to all three collections, and to other collections as long as at least one map within them includes visible and countable houses. A key finding is that all three variants can serve as cross-checks for one another.

5.2 Final Conclusion and Recommendations

The essence of the Passus method, as applied to the 226 city maps by Jacob van Deventer produced between 1542 and 1575, is to chart the pattern in the relationship between the lengths of built-up street sides and the population size of the city. This is achieved by determining reduction factors for each city, expressed in the number of dwellings and inhabitants. The method enables objective comparison between these and similar city maps that were drawn by a single cartographer or a single team of cartographers within a defined time period, using consistent techniques and cartographic principles. A key aspect of applying the method is the use of analytical software such as MapAnalyst, which allows for the determination of deviations in scale, orientation, and distance for each individual map in comparison to modern maps such as OpenStreetMaps and Google Maps. The quality of the research results can be further enhanced through the use of alternative maps for comparison and verification, through direct house counts on the maps themselves, and by drawing on additional sources such as historical population and housing censuses.

The case studies in Chapter 4, based on three other collections of city maps, show that the application of the Passus method in its original form – including the Passus coefficient of 0.1756 – is only moderately feasible for the maps of Braun and Hogenberg, with the caveat that satisfactory results are achieved only for maps derived from the work of Jacob van Deventer. For Braun and Hogenberg's other city maps, as well as for later collections such as those of Blaeu and the Hattinga family, this coefficient proves unusable. However, this does not apply to the core principle of the Passus method, as described in the previous paragraph: mapping the pattern in the relationship between the lengths of built-up street sides and the population size of the city. For all three case studies, this principle yields results that fall within an acceptable margin of error of 10 percent.

It is therefore recommended to apply the Passus method to the measurements, counts, and analyses of other collections of city maps from the sixteenth and seventeenth centuries. When combined with the major population censuses of the centuries that followed, it becomes possible to calculate, for each city, population developments over extended periods, developments that can, in turn, form the basis for interpretations concerning migration and urbanization trends. The analytical component of the Passus method, using software such as MapAnalyst, also makes it possible to map, on a large scale, the genealogies of city maps: identifying which maps were derived from the work of Jacob van Deventer and which, in turn, may have served as the basis for even later maps.

Literature and Sources

Literature

Abel, W., *Agricultural fluctuations in Europe. From the thirteenth to the twentieth centuries*. London, 1980.

Abrahamse, J.E., 'Cartograaf van de macht'. In: *Tijdschrift voor historische geografie* 6.2. 2021, pp. 196–200.

Alkhoven, P., 'Bestudering en analyse van oude stadsplattegronden met behulp van de computer. De kaart van Heusden uit het stedenboek van Joan Blaeu 1649'. In: *Caert-Thresoor*, 11, 4. 1992, pp. 87–93.

Baar, P.J.M. de, 'Jan van Hout en zijn Stratenboek en Grachtenboek'. In: *Ups en Downer. Bundel artikelen bij het afscheid van mr. W. Downer als gemeentearchivaris van Leiden*. 1985, pp. 5–18.

Bade, K.J., *Migration in European History*. New York, 2003.

Bairoch, P., *Cities and economic development from the dawn of history to the present*. Chicago, 1988.

Bairoch, P., *De Jéricho à Mexico. Villes et économie dans l'Histoire*. Parijs, 1985.

Blaeu, J., *Novum ac magnum theatrum urbium Belgicae liberae ac foederatae*. Amsterdam, 1649.

Blaeu, J., *Toonneel der steden van 's Konings Nederlanden*. Amsterdam, 1652.

Blaeu, J., *Toonneel der steden van de Vereenighde Nederlanden*. Amsterdam, 1652.

Blaeu, J., *Zelandiae pars transscaldina vulgo Beooster-Scheld*. Amsterdam, 1662.

Blakemore, M.J. and J.B. Harley, *Concepts in the History of Cartography. A Review and Perspective*. Toronto, 1980.

Blockmans, W.P., *Keizer Karel V 1500–1558, De utopie van het keizerschap*. Amsterdam, 2000.

Blockmans, W.P., *Metropolen aan de Noordzee. De geschiedenis van Nederland, 1100–1560*. Amsterdam, 2010.

Blok, D.P. et al. (ed.), *Algemene Geschiedenis der Nederlanden*. Haarlem, 1977–1983.

Blok, P.J. en P.C. Molhuysen, *Nieuw Nederlandsch biografisch woordenboek*. Leiden, 1939.

Blom, P., 'Demografie van een dode stad. Bronnen voor het onderzoek naar de bevolkingsontwikkeling van de stad Veere vanaf 1470'. In: *Nehalennia*, 96. 1993, pp. 2–10.

Blondé, J., I. Devos, J. Hanus, W. Ryckbosch, *Trend en toeval. Inleiding tot de kwantitatieve methoden voor historici*. Leuven, 2015.

Boxhorn, M.Z., *Chronijcke van Zeeland*. Middelburg, 1644.

Boxhorn, M.Z., *Theatrum sive Hollandiae Comitatues et Urbium nova descriptio*. Amsterdam, 1632.

Braudel, F., 'En Espagne au temps de Richelieu et d'Olivarès'. In: *Annales. Economies, sociétés, civilisations, 2, 3*. 1947, pp. 354–358.

Braudel, F., *La Méditerranée et le Monde Méditerranéen a l'époque de Philippe II*. Paris, 1949.

Braun, G. and F. Hogenberg, *Civitates orbis terrarum*. Part I, III, IV. Keulen, 1572.

Breure, L., W. Landman, S. Spruit, H. Voorbij, 'Modeling Portals for Cultural Landscapes'. In: *Museums and the Web, the international conference for culture and heritage on-line*, 5. 2008 (https://www.museumsandtheweb.com/mw2008/papers/breure/breure.html, accessed on April 2, 2026).

Briels, J.G.C.A., *De Zuidnederlandse immigratie 1572–1630*. Haarlem, 1978.

Briels, J.G.C.A., *Zuid-Nederlanders in de Republiek 1572–1630: Een demografische en cultuurhistorische studie*. Sint-Niklaas, 1985.

Brugmans, H., *Opkomst en bloei van Amsterdam*. Amsterdam, 1911.

Brusse, P.G., 'Bevolking en sociale verhoudingen'. In: P.G. Brusse, J. Dekker (ed.), *Geschiedenis van Zeeland*, deel III. Zwolle, 2013, pp. 117–168.

Brusse, P.G., 'Economie'. In: P.G. Brusse, J. Dekker (ed.), *Geschiedenis van Zeeland*, deel III. Zwolle, 2013, pp. 49–116.

Brusse, P.G., 'Het plotselinge verval van Middelburg na 1795'. In: *Archief. Mededelingen van het Koninklijk Zeeuwsch Genootschap der Wetenschappen, 2006*. Middelburg, 2006, pp. 55–69.

Brusse, P.G., *Gevallen stad, stedelijke netwerken en het platteland Zeeland 1750–1850*. Zwolle, 2011.

Burke, P., *The French Historical Revolution – The Annales School 1929–2014*. Cambridge, 2015.

Carton, C., *Oudenaarde op kaart. Een grondige analyse van de 16de-eeuwse kaart van Oudenaarde van Jacob van Deventer*. Gent, 2010.

Chandler, T. and G. Fox, *Four Thousand Years of Urban Growth*. Lewiston, 1974, 1987.

Cloet, M., Chr. Vandenbroeke, *Tien bijdragen tot de lokale en regionale demografie in Vlaanderen*. Brussel, 1989.

Commissie tot het ontwerpen van een plan van constitutie voor het volk van Nederland, *Volks-tellinge in de Nederlandsche republiek 1796*. Den Haag, 1796.

Cruyningen, P. van, 'Bevolking en sociale verhoudingen'. In: P.G. Brusse, J. Dekker (ed.), *Geschiedenis van Zeeland*, deel II. Zwolle, 2012, pp. 103–146.

Cruyningen, P. van, 'Nogmaals het inwonertal van Middelburg'. In: *Archief. Mededelingen van het Koninklijk Zeeuwsch Genootschap der Wetenschappen, 2006*. Middelburg, 2006, pp. 73–79.

Cruyningen, P.J. van, Behoudend maar buigzaam. Boeren in West-Zeeuws-Vlaanderen 1650–1850. Wageningen 2000.

D'haene, S., *Toponymie en stadscartografie als verklikkers voor de stadsontwikkeling van laatmiddeleeuws Kortrijk*. Gent, 2010.

 | HTTPS://DOI.ORG/10.1163/9789004755031-007

Dambruyne, J., *Mensen en centen. Het 16de-eeuwse Gent in demografisch en economisch perspectief.* Gent, 2001.

Darnton, R. 'Il faut savoir compter'. In: *French Historical Studies* 27(4). 2004, pp. 725–731.

Davids, K., *Wat kunnen we (nog) leren van de annales-school?* Amsterdam, 2011 (https://www.boomgeschiedenis.nl/media/14/dossier_alg.1_lessen_van_de_annales-school.pdf, accessed on April 2, 2026).

Deys, H.P., 'De stadsplattegronden van Jacob van Deventer. Resultaten van recent onderzoek te Madrid'. In: *Caert-Thresoor*, 8, 4. 1989, pp. 81–95.

Dommisse, P.K., 'De ambachtsheerlijkheid van Oud Vlissingen en de wording van Nieuw Vlissingen'. In: *Archief. Mededelingen van het Koninklijk Zeeuwsch Genootschap der Wetenschappen, 1910*. Middelburg, 1910, pp. 1–259.

Druenen, P.G. van, 'Dartelheid en Weelde. Beknopte historiografie betreffende de immigratie van Vlaamse vluchtelingen naar Holland en Zeeland in het laatste kwart van de zestiende eeuw'. In: *Archief. Mededelingen van het Koninklijk Zeeuwsch Genootschap der Wetenschappen.* Middelburg, 2024, pp. 49–90.

Druenen, P.G. van, 'Oprichting van het Zeeuwsch Genootschap der Wetenschappen in Vlissingen in 1769: incident of implicatie?' In: *Archief. Mededelingen van het Koninklijk Zeeuwsch Genootschap der Wetenschappen.* Middelburg, 2019, pp. 57–86.

Druenen, P.G. van, *Meten en tellen. Stedelijke inwonertallen in de Lage Landen, een alternatieve benadering* (Amsterdam, 2022).

Druenen, P.G. van, *Vissers, Kapers, Arbeiders. Vlissingen 700 jaar stadsrechten.* Vlissingen, 2015.

Dupâquier Jacques. 'J.A. Faber, H.K. Roessingh, B.H. Slicher van Bath, A.M. Van der Woude et H.J. Van Xanten. Évolution démographique et progrès économique aux Pays-Bas, vue perspective (Population changes and economic developments in the Netherlands: historical survey)'. In: *Annales de démographie historique.* 1965 (Etudes et Chronique), pp. 199–226.

Dupont, C., *Cartographie et pouvoir au XVI*^{e} *siècle : l'atlas de Jacques de Deventer.* Turnhout, 2019.

Eggen, J.L.M., *De invloed door Zuid-Nederland op Noord-Nederland uitgeoefend op het einde der XVIde en het begin der XVIIde eeuw.* Gent, 1908.

Egmond, M. van, 'Town atlas by Braun en Hogenberg. 16th century metropolises: Civitates orbis terrarum'. Utrecht University, Special Collections, (https://www.uu.nl/en/special-collections/, accessed on March 1, 2025).

Engelen, Th.L.M., 'M. Cloet, Chr. Vandenbroeke, ed., Tien bijdragen tot de lokale en regionale demografie in Vlaanderen'. In: *Bijdragen en Mededelingen betreffende de Geschiedenis der Nederlanden,* deel 107, 2. 1992, pp. 318–319.

Faber, J.A., H.K. Roessingh, B.H. Slicher van Bath, A.M. van der Woude, H.J. van Xanten, 'Population changes and economic developments in the Netherlands: a historical survey'. In: *A.A.G. Bijdragen* 12. Wageningen 1965, pp. 47–114.

Flinn, M.W., *The European demographic system, 1500–1820.* Baltimore, 1981.

Fokker, G., 'Iets over de bevolking van Middelburg vóór 1795'. In: *Archief. Mededelingen van het Koninklijk Zeeuwsch Genootschap der Wetenschappen,* 1875 deel III, tweede stuk, pp. 81–100.

Frijhoff, W., 'Migrations religieuses dans les Provinces-Unies avant le second Refuge'. In: *Revue du Nord.* Tome 80, no. 326–327. 1998, pp. 573–598.

Frijhoff, W., 'Van "Histoire de Léglise" naar "Histoire Religieuse". De invloed van de "Annales"-groep op de ontwikkeling van de kerkgeschiedenis in Frankrijk en de perspectieven daarvan voor Nederland', In: *Nederlands archief voor kerkgeschiedenis/Dutch Review of Church History,* nieuwe serie, 61, 2. 1981, pp. 113–153.

Gencúrová, G., 'Braudel's model of historical structures and interpretative significance to teach church history in the Slovak school system'. In: *IJRDO – Journal of Social Science and Humanities Research,* 1, 12. 2016, pp. 11–21.

Gerding, M.A.W., 'Vier eeuwen turfwinning. De verveningen in Groningen, Friesland, Drenthe en Overijssel tussen 1550 en 195'. In: *A.A.G. Bijdragen* 35. Wageningen, 1995, pp. 1–533.

Glass, D.V., 'Gregory King and the population of England and Wales at the end of the seventeenth century'. In: *The Eugenics Review,* 37, 4. 1946, pp. 170–183.

Gottschalk, M.K.E., *Stormvloeden en rivieroverstromingen in Nederland.* Assen, 1971–1977, 3 volumes.

Gouw, J. ter, *Amstelodamia,* volume II. Amsterdam, 1874.

Graunt, J., *Natural and Political Observations Made upon the Bills of Mortality.* London, 1662 (used edition: London, 1676, herdruk London, 1759).

Groen, P. (ed.), *De Tachtigjarige oorlog. Van opstand naar geregelde oorlog 1568–1648.* Amsterdam, 2013, pp. 79–86, 89–101.

Groenveld, S. and J. Vermaere, 'Zeeland en Holland in 1569. Een rapport voor de hertog van Alva'. In: *Nederlandse Historische Bronnen,* 2. 1980, pp. 103–141.

Grol, H.G. van, 'Iets uit de vonnissen van de Hooge Vierschaar te Vlissingen onder de regering van keizer Karel V'. In: *Vlissingsche Courant* 6 maart 1930 t/m 5 april 1930 (Vlissingen, 1930, I).

Guicciardini, L., *Beschrijvinghe van alle de Nederlanden.* Amsterdam, 1612 (originally 1567).

Haas, H.G. de, *A Review of How Migration Really Works: 22 Things You Need to Know About the Most Divisive Issue in Politics.* London, 2024.

Hélin, E., 'Demografische ontwikkeling van de Zuidelijke Nederlanden'. In: D.P. Blok, et al. (ed.), *Algemene Geschiedenis der Nederlanden*, deel 5. Haarlem, 1980, pp. 169–195.

Hoff, B. van 't, *Jacob van Deventer, keizerlijk-koninklijk geograaf.* Den Haag, 1953,

Hoff, B. van 't, *Toonneel der steden van de Vereenighde Nederlanden, met hare beschrijvingen (Holland en West-Vriesland/Utrecht), Inleiding op de facsimile-uitgave van een selectie van stadsplattegronden.* Amsterdam and Brussel, 1966.

Hofhuis, S.T., 'A Rose-Tinted View of Immigration'. In: *Quillette*, 19 november 2024.

Hooymans, M., *Toelichting bij DANS Dataset Volkstelling Leiden 1581* (deel 1). Leiden, 2018.

Hoven van Genderen, B. van den and R. Rommes, 'Rijk en talrijk. Beschouwingen over de omvang van de Utrechtse bevolking tussen circa 1300 en het begin van de zeventiende eeuw'. In: *Jaarboek Oud-Utrecht*. Utrecht, 1995, pp. 54–86.

Iong, J. de, *De koningin: Maria van Hongarije, landvoogdes der Nederlanden, 1505–1558*. Amsterdam, 1981.

Junot, Y., *Les Bourgeois de Valenciennes*. Villeneuve d'Ascq, 2009.

Koeman, C., G. Schilder, P. van der Krogt and M. van Egmond, 'Commercial Cartography and Map Production in the Low Countries, 1500–ca. 1672'. In: D. Woodward (ed.), *The History of Cartography, Cartography in the European Renaissance*, deel 3. Chicago, 2007, pp. 1296–1383.

Koeman, C., M. van Egmond, 'Surveying and Official Mapping in the Low Countries, 1500–ca. 1670'. In: D. Woodward (ed.), *The History of Cartography, Cartography in the European Renaissance*, deel 3. Chicago, 2007, pp. 1246–1295.

Kossmann, E.H., *De Lage Landen 1780–1980*. Amsterdam, 1986.

Koster, E.A., *Stadsmorfologie een proeve van vormgericht onderzoek ten behoeve van stedenbouwhistorisch onderzoek*. Groningen, 2001.

Kostov, S.G., *The City Shaped. Urban Patterns and Meanings Through History*. London, 1991.

Krogt, P. van der, 'Mapping the towns of Europe: The European towns in Braun en Hogenberg's Town Atlas, 1572–1617'. In: *Belgeo* [Online], 3–4 | 2008 (https://journals.openedition.org/belgeo/11877?lang=nl, accessed on April 2, 2026).

Krogt, P. van der, 'Welke steden karteerde Van Deventer voor Filips II in diens "landen van herwerts overe"?'. In: *Caert-Thresoor*, 25, 4. 2006, pp. 116–120.

Kuyper, J., *Gemeente-Atlas*. Leeuwarden, 1871.

Leeuwen, M.H.D. van & J.E. Oeppen, 'Reconstructing the demographic regime of Amsterdam 1681–1920'. In: *Economic and social history in the Netherlands*. Amsterdam, 1993, pp. 61–102.

Lisson, J., *Een papieren stad. Het ontstaan en de ontwikkeling van Zoutleeuw aan de hand van een morfologische analyse van Jacob van Deventers stadsplattegrond (10e–16e eeuw)* (unpublished). Leuven, 2012.

Lottum, J. van, *Across the Northsea. The impact of the Dutch Republic on international labour migration, c. 1550–1850*. Amsterdam, 2007.

Lourens, P. and J. Lucassen, *Inwonertallen van Nederlandse steden ca. 1300–1800*. Amsterdam 1997.

Lucassen, J., L. Lucassen, 'The mobility transition in Europe revisited, 1500–1900. Sources and methods'. *IISH – research paper*. Amsterdam, 2010.

Lucassen, J., *Immigranten in Holland 1600–1800 Een kwantitatieve benadering*. Amsterdam, 2002.

Lucassen, J., *Naar de kusten van de Noordzee. Trekarbeid in Europees perspectief.* Gouda, 1984.

Lucassen, L., H. van Houtum, *Voorbij Fort Europa, een nieuwe visie op migratie*. Amsterdam, 2016.

Lucassen, L., J. Lucassen, *Winnaars en verliezers. Een nuchtere balans van vijfhonderd jaar immigratie*. Amsterdam, 2011.

Maitland, W., *The history and survey of London, from its foundation to the present time*. London, 1756.

Martens, P., 'Nieuwe inzichten over de stadsplannen van Jacob van Deventer (ca. 1545–1575)'. In: *Stadsgeschiedenis* 17. 2022, 1, pp. 55–69.

Memon, J., 'Handwritten Optical Character Recognition (OCR): A Comprehensive Systematic Literature Review (SLR)'. In: *IEEE Access* July 2020. DOI: 10.1109/ACCESS.2020.3012542.

Modelski, G., *World Cities: -3000 to 2000*. Faros, 2000, 2003.

Moerman, M., *Diepgaande analyse van twee Ieperse kaarten. Het stadsplan van Thévelin-Destrée (1564) en de gravure over het beleg van Ieper van Guillaume du Tielt (1610)*. Gent, 2010.

Molders, K., 'De stippen van Jacob van Deventer'. In: *Caert Thresoor*, 39–2. 2020, pp. 19–27.

Mols, R.M., *Introduction à la démographie des villes d'Europe du XIVe au XVIIIe siecle*. Gemblouw, 1954–1956.

Monmonier, M., *How to Lie with Maps*. Chicago en London, 1996.

Nagtglas, F., *De Algemeene kerkenraad der Nederduitsch-Hervormde gemeente te Middelburg van 1574–1860*. Middelburg, 1860.

Năiduț, P., 'Maps, Fictions, and Facts: Representations of the City in Literature and Film'. In: *A Book of Cities: Mapping Urban Space in Braun and Hogenberg's Civitates Orbis Terrarum (1572–1617)*. Bucharest, 2020.

Nierop, L. van, 'De aanvang der Nederlandsche demografie'. In: *Economisch Historisch Jaarboek* 5. Den Haag, 1919, pp. 192–208.

Noordhoff Atlasproducties, *De Bosatlas van de geschiedenis van Nederland*. Groningen, 2011.

Nusteling, H.P.H., *Welvaart en werkgelegenheid in Amsterdam 1540–1860. Een relaas over demografie, economie en sociale politiek van een wereldstad*. Amsterdam, 1985.

Obdeijn, H., M. Schrover, *Komen en gaan. Immigratie en emigratie in Nederland vanaf 1550*. Amsterdam, 2008.

Paping, R., *General Dutch Population development 1400–1850: cities and countryside*. Paper presented at 1st ESHD conference, Alghero, Italy. Alghero, 2014.

Pasciuti, D. and C. Chase-Dunn, *Estimating the population sizes of cities. Institute for Research on World-Systems at University of California*. Riverside, 2002.

Perkins, Chr., 'Critical cartography'. In: A.J. Kent and P. Vujakovic (ed.), *The Routledge Handbook of Mapping and Cartography*. London and New York, 2018, pp. 80–90.

Pevsner, N., *The Cities of London and Westminster*. Vol. 1 (2nd ed.). Penguin Books. London, 1962.

Pirenne, H., *Démographie. Rapport Sur la Quatorzieme Question* (*Les Archives Au Point de Vue de la Démographie Historique*). Bruxelles, 1903.

Pleij, H., 'Volksfeest en toneel in de middeleeuwen I De zot draait de rollen om'. In: *De Revisor*. Jaargang 3. 1976, pp. 53–64 en Pleij, H. 'Volksfeest en toneel in de middeleeuwen II Entertainers en akteurs'. In: *De Revisor*. Jaargang 4. 1977, pp. 34–41.

Pleyte, W., *Leiden voor 300 jaren en thans; photolitographische afbeelding van een plattegrond van 1578 en chromolithographische afbeelding van het Chaertbouc van Straten binnen desen Stadt Leyden en Chaertbouc van de Stadts Wateren, gemeten bij Mr. Salomon Davidssoon van Dulmanhorst en Jan Pieterssoon Dou, voorafgegaan door eene geschiedkundige schets van den Platte-Grond en de voornaamste gebouwen der stad, met opgave der woonhuizen van eenige personen, die tijdens het beleg, de stichting der Akademie en thans op den voorgrond treden*. Leiden, 1874.

Posthumus, W., *Bronnen tot de geschiedenis van de Leidsche textielnijverheid 1333–1795*. Den Haag, 1912.

Prak, M., J. Luiten van Zanden, *Nederland en het poldermodel*. Amsterdam, 2013.

Prevenier, W. (intr.), *Demografische evoluties en gedragspatronen van de 9de tot de 20ste eeuw in de Nederlanden*. Gent, 1977.

Priester, P.R., *Geschiedenis van de Zeeuwse landbouw, circa 1600–1910*. Houten, 1998.

Puschmann, P., T. Riswick (ed.), *Building Bridges. Scholars, History and Historical Demography. A Festschrift in Honor of Professor Theo Engelen*. Arnhem, 2018.

Ramaer, J.C., 'De middelpunten van bewoning in Nederland voorheen en thans'. In: *Tijdschrift van het Koninklijk Nederlandsch Aardrijkskundig Genootschap*, XXXVIII–1, 2, 1921, pp. 1–38 and 174–215.

Reba, M., F. Reitsma, K.C. Seto, 'Spatializing 6,000 years of global urbanization from 3700 BC to AD 2000'. In: *Nature, Scientific Data*, 3, 160034 (2016).

Reygersberch van Cortgene, J.J., *Cronijcke van Zeelandt*. Middelburg, 1551.

Roessingh, H.K., 'The number of inhabitants in the Veluwe, 1526–1947'. In: *A.A.G. Bijdragen* 11. Wageningen 1964.

Rutte, R. and B. van Nieuwenhuyze, 'Medieval urban form in the Low Countries: state of research, comparative perspective and symbolic meaning'. In: Simms, A., H.B. Clarke (ed.), *Lords and Towns in Medieval Europe. The European Historic Towns Atlas Project*. 2015, pp. 375–398.

Rutte, R. and B. Vannieuwenhuyze, *Stedenatlas Jacob van Deventer. 226 Stadsplattegronden uit 1545–1575 – Schakels tussen verleden en heden*. Bussum, 2018.

Rutte, R. and J.E. Abrahamse, *Atlas van de verstedelijking in Nederland : 1000 jaar ruimtelijke ontwikkeling*. Bussum, 2014.

Rutte, R., 'Antoon van den Wijngaerde's drawings of cities in the Low Countries'. In: *Bulletin KNOB* 119. 2020, 1, pp. 1–24.

Rutte, R., 'Middeleeuwse nieuwe steden in Nederland. Aanzet tot een onderzoek naar oorsprong, verspreiding en betekenis'. In: *Bulletin KNOB*, 1996–6, pp. 187–202.

Rutte, R.J. and H. Engen, *Stadswording in de Nederlanden. Op zoek naar overzicht*. Hilversum, 2005.

Sanderus, *Flandria Illustrata*. Amsterdam, 1641–1644.

Santing, C., 'A Flemish-Centred Economic History of the Netherlands'. In: *BMGN – Low Countries Historical Review*, 127–2. 2012, pp. 75–80.

Schelven, A.A. van, *Omvang en invloed der Zuid-Nederlandse immigratie van het laatste kwart der zestiende eeuw*. Den Haag, 1919.

Schöffer, I., H. van der Wee, J.A. Bornewasser, *De Lage Landen van 1500 tot 1780*. Amsterdam, 1985.

Schoor, A. van der, *Stad in aanwas. Geschiedenis van Rotterdam tot 1813*. Zwolle, 1999.

Schrover, M., 'Dutch Migration History. Looking Back and Moving Forward'. In: *Het Tijdschrift voor Sociale en Economische Geschiedenis*. Jrg. 11–2. Utrecht, 2014, pp. 199–218.

Schuurman, A.J., *Historische demografie: bevolkings- en gezinsgeschiedenis*. Zutphen, 1991.

Slicher van Bath, B.H., *De Agrarische Geschiedenis van West-Europa (500–1850)*. Utrecht, 1960.

Slicher van Bath, B.H., *Een samenleving onder spanning: geschiedenis van het platteland in Overijssel*. Wageningen, 1957.

Struyck, N., *Inleiding tot de algemeene geographie, benevens eenige sterrekundige en andere verhandelingen*. Amsterdam, 1740.

Terlouw, K., 'A general perspective on the regional development of Europe from 1300 to 1850'. In: *Journal of Historical Geography*, 22, 2. 1996, pp. 129–146.

Terlouw, K., 'Five centuries of regional development in northwest Germany and the Neterlands'. In: *Journal of World-Systems Research*, 17(1). 2011, pp. 199–217.

Terlouw, K., 'Transnational regional development in the Netherlands and Northwest Germany, 1500–2000'. In: *Journal of Historical Geography* 35. 2009, pp. 26–43.

Tirion, *Tegenwoordige staat van de Nederlanden*. Part IX and XX. Amsterdam, 1753.

Tjaden, A.J., 'De reconquista mislukt. De opstandige gewesten 1579–1588'. In: D.P. Blok, et al. (ed.), *Algemene Geschiedenis der Nederlanden*, volume 6. Haarlem, 1977–1983, pp. 244–258.

Unger, W.S., J.J. Westendorp Boerma, 'De steden van Zeeland (derde stuk)'. In: *Archief. Vroegere en latere mededelingen voornamelijk in betrekking tot Zeeland. Middelburg*, 1956. pp. 1–27.

Unger, W.S., 'Hoeveel inwoners had Leiden tijdens het bezet van 1574?'. In: *Bijdragen voor Vaderlandsche Geschiedenis en Oudheidkunde*, 5de reeks, 2. 1915, pp. 86–92.

Unger, W.S., *Geschiedenis van Middelburg in omtrek*. Middelburg, 1966.

Vanhaute, E., Lambrecht, Th., Devos, I., Béaur, G., Fertig, G., Gadd, C.J., Karel, E., Limberger, M., Paping, R., Schofield, Ph., 'Making a living in rural societies in the North Sea area, 500–2000'. In: *Rural Economy and Society in Northwestern Europe, 500–2000*. Gent, 2012, pp. 323–347.

Vannieuwenhuyze, B., 'Op zoek naar Zeeuwse kaarten: isaak tirions kaarten van Zeeland, de Zeeuwse eilanden en de Zeeuwse steden'. In: *Archief 2022. Mededelingen van het Koninklijk Zeeuwsch Genootschap der Wetenschappen*. Middelburg, 2022, Deel I, pp. 171–222.

Vannieuwenhuyze, B., *Brussel, de ontwikkeling van een middeleeuwse stedelijke ruimte*. Gent, 2008.

Visser, J.C., 'De waarde van de stedenatlas van Jacob van Deventer voor de topografie van de laat-middeleeuwse stad'. In: H. Aubin, E. Ennen, H. Kellenbenz, T. Mayer, F. Metz, M. Miller, J. Schmithüsen (Eds.), *Beiträge zur Wirtschafts- und Stadtgeschichte. Festschrift für Hektor Ammann*. Wiesbaden, 1965, pp. 116–123.

Visser, J.C., 'Dichtheid van de bevolking in de laat-middeleeuwse stad'. In: *Tijdschrift voor Historische Geografie*, 3. 1985.

Visser, J.C., 'Inleiding'. In: *De stadsplattegronden van Jacob van Deventer*. Weesp, Alphen aan den Rijn, 1992–2001, pp. 10–21.

Vliet, A.P. van, *Vissers in oorlogstijd: De Zeeuwse visserij in de jaren 1568–1648*. Middelburg, 2003.

Vlist, E.T. van der, 'De bonboeken te Leiden'. In: *RGP, Broncommentaren 4. Bronnen betreffende de registratie van onroerend goed in Middeleeuwen en Ancien Régime*. Den Haag, 2001, pp. 82–107.

Vossius, I., *Variarum observationum liber*. London, 1685.

Vries, J. de, A.M. van der Woude, *Nederland 1500–1815. De eerste ronde van moderne economische groei*. Amsterdam, 1995.

Vries, J. de, A.M. van der Woude, *The First Modern Economy: Success, Failure, and Perseverance of the Dutch Economy, 1500–1815*. Cambridge 1997.

Vries, J. de, *European urbanization 1500–1800*. Cambridge, 1984 en London, 2013.

Walda, M., 'Bouwen in de krimpende stad. Renovatie, functieverandering en vergroening in Haarlem en Leiden rond 1800'. In: *Bulletin KNOB*, 119, 1. 2020, pp. 25–46.

Wee, H. van der, W. Prevenier (intr.), 'Demografische evoluties en gedragspatronen van de 9de tot de 20 ste eeuw in de Nederlanden'. In: *The Economic History Review*, 31, 4. 1978, pp. 697–698.

Werveke, H. van, 'Introduction à la démographie historique des villes d' Europe du XIVe au XVIIIe siècle'. In: *The Economic History Review*, 10, 1. 1957. pp. 154–155.

Winkelman, H.P., *Geschiedkundige plaatsbeschrijving van Vlissingen*. Vlissingen, 1873.

Wood, D. and J. Krygier, 'Critical Cartography', In: R. Kitchen, N.J. Thrift (ed.), *International Encyclopedia of Human Geography*. Amsterdam, London, Oxford. 2009, pp. 340–344.

Woodward, D. (ed.), *The History of Cartography, Volume three, Cartography in the European Renaissance*. Volume II and III. Chicago, 2007.

Woude, A.M. van der, 'De contractiefase van de seculaire trend in het Noorderkwartier nader beschouwd'. In: *BMGN – Low Countries Historical Review*, 103, 3. 1988, pp. 373–398.

Woude, A.M. van der, 'De demografische ontwikkeling van de Noordelijke Nederlanden 1500–1800'. In: D.P. Blok et al. (ed.), *Algemene Geschiedenis der Nederlanden*, deel 5. Haarlem, 1980, pp. 102–168.

Woude, A.M. van der, 'De weerbare mannen van 1747 in de dorpen van het Zuiderkwartier van Holland als demografisch gegeven'. In: *A.A.G. Bijdragen* 8. Wageningen, 1962, pp. 35–76.

Woude, A.M. van der, 'Les villes Neerlandaises', In: A. Lottin et al., *Études sur les villes en Europe occidentale (milieu du XVIIe siècle à la veille de la revolution française)*, deel 2. Parijs, 1983, pp. 309–385.

Woude, A.M. van der, 'Vijftig jaar agrarische geschiedschrijving in Nederland (1950–2000)'. In: A.M. van der Woude, *Leven met geschiedenis. Theorie, praktijk en toepassing van historische kennis*. Amsterdam, 2000, pp. 103–118.

Woude, A.M. van der, C. Vandenbroeke, F. van Poppel, 'De zuigelingen- en kindersterfte in België en Nederland in seculair perspectief'. In: *Tijdschrift voor Geschiedenis*, 94. 1981, pp. 461–491.

Woude, A.M. van der, De historische demografie in de ontwikkeling van de geschiedwetenschap. In: P.A.M. Geurts, F.A.M. Messing, *Theoretische en methodologische aspecten van de economische en sociale geschiedenis*. Den Haag, 1979, pp. 157–199.

Woude, A.M. van der, F.W.A. van Poppel, 'The role of migration in the demographic system of the Dutch Republic. In: A. Bideau, A. Perrenoud, K.A. Lynch, et G. Brunet (réd.), *Les systèmes Démographiques du Passé*. Lyon, 1996, pp. 182–201.

Woude, A.M. van der, *Het Noorderkwartier. Een regionaal onderzoek in de demografische en economische geschiedenis van westelijk Nederland van de late middeleeuwen tot het begin van de negentiende eeuw*. 3 delen. Wageningen, 1972.

Woude, A.M. van der, J. de Vries, A. Hayami, 'Introduction: the hierarchy, provisioning and demographic patterns of cities'. In: A.M. van der Woude, J. de Vries, A. Hayami, *Urbanization in History*. Oxford, 1990, pp. 1–19.

Xanten, H.J. van, A.M. van der Woude, 'Poll-tax and population in the bailiwick of Bois-le-Duc about 1700'. In: *A.A.G. Bijdragen* 13. Wageningen, 1965, pp. 41–256.

Sources

Biblioteca Nacional de España, Planos de ciudades de los Países Bajos. Parte II, parte III. Deventer, Jacob van (ca. 1505–1575).

CBS, Volkstellingen.nl. Census 1830 and 1839.

British Library. Plan of the town of Flushing, drawn by Robert Adams, 1588. Location: British Library, Shelfmark: Cotton MS Augustus I ii 105.

DANS Dataset Census Leiden 1574, Martin Hooymans, version 1.0, august 2018.

Erfgoed Leiden en omstreken (formerly: GA Leiden en Regionaal Archief Leiden)

Index Hoofdgeld 1622 stad Leiden, Stadsarchief Leiden, 0501A Inventaris, (1253) 1574–1816. (1897) Inv.nrs. 4021–4022.

Kohier betreffende de volkstelling in 1581, 1074.

SA II 5113, Grachtenboek. Kaarten van de grachten en singels met opgaven van de minimaal vereiste breedten, de breedten van de aanliggende percelen en de eigenaars daarvan. 1583–1589.

SA II inv. 5153, Straetbouc. Kaarten van alle straten, met vermelding van de straat- en perceelbreedten en van de eigenaars van de percelen. 1588–1597.

Koninklijke Bibliotheek België (KBR)

Atlas des villes des Pays-Bas : 73 places levées entre 1550 et 1565 sur les ordres de Charles Quint et de Philippe II / Deventer, Jacob van.

Museum Plantin-Moretus, Antwerpen.

Nationaal Archief, Collectie Hingman no. 3492, Kaart van Vlissingen door Cornelis Poel uit ca. 1570.

Regionaal Historisch Centrum Limburg, 18.BC Tekeningen-, prenten-, kaarten en fotocollectie RAL.

Royal Collection Trust, London, Leonardo Da Vinci, A map of Imola 1502, RCIN 912284.

Universiteitsbibliotheek Utrecht, online version of: J. Blaeu, *Toonneel der steden van de Vereenighde Nederlanden*. Amsterdam, 1652. 2 Volumes. OCLC: 974064627.

Vlissingen, Oud Archief (not available anymore), no. 1: Vertoogh 't wele Baillu, Burgemrs. enden Schepenen der stede Vlissynghe doende syn aen den Raedt ordin. van Syne Co Maj.

Websites

Bairoch, P., database (https://github.com/JakeRuss/bairoch-1988 pdf, accessed on April 2, 2026).

Erfgoed Leiden (www.erfgoedleiden.nl en www.leidseregioinkaart.nl/kaarten/1583 pdf, accessed on April 2, 2026).

European project Historic Town Atlases (www.uni-muenster.de/Staedtegeschichte/en/portal/staedteatlanten pdf, accessed on April 2, 2026).

Meertens Instituut. (https://mgw.meertens.knaw.nl/plaats/133, accessed on February 24, 2024).

Zeeuws Archief (ZA)

33.3 Verzameling Aanwinsten Rijksarchief in Zeeland 1955–1999, zestiende-twintigste eeuw. Inv.nr: 1955.10. Opgave van de wijkindeling van Middelburg, 1752.

Atlassen Hattinga, nr 157, Plattegrond van Tholen, gemeten door (W.T.) Hattinga anno 1744.

Index of Names

 | HTTPS://DOI.ORG/10.1163/9789004755031-008

Index of Places

 HTTPS://DOI.ORG/10.1163/9789004755031-009

Index of Subjects

 | HTTPS://DOI.ORG/10.1163/9789004755031-010

www.ingramcontent.com/pod-product-compliance
Lightning Source LLC
LaVergne TN
LVHW080800011026
849160LV00015B/1495